HANDBOOK OF
POLYCYCLIC AROMATIC HYDROCARBONS

HANDBOOK OF
POLYCYCLIC AROMATIC HYDROCARBONS

Edited by *ALF BJØRSETH*
Central Institute for Industrial Research
Oslo, Norway

MARCEL DEKKER, INC. New York and Basel

Library of Congress Cataloging in Publication Data

Main entry under title:

Handbook of polycyclic aromatic hydrocarbons.

Includes index.
1. Hydrocarbons--Handbooks, manuals, etc. 2. Polycyclic compounds--Handbooks, manuals, etc.
I. Bjørseth, Alf.
QD341.H9H317 1983 547'.61 83-1903
ISBN 0-8247-1845-3

MARCEL DEKKER, INC.
270 Madison Avenue, New York, New York 10016

Current printing (last digit):
10 9 8 7 6 5 4 3 2 1

PRINTED IN THE UNITED STATES OF AMERICA

Preface

During recent years, it has become increasingly evident that cancer in man is linked to environmental factors. In particular, attention has been focused on the importance of chemical carcinogens in the environment.

One of the largest classes of environmental carcinogens known today is the polycyclic aromatic hydrocarbons (PAH). These compounds have their main sources in fossil or synthetic fuels or from combustion or high-temperature reactions of organic materials. Since these sources are ubiquitous in an industrialized society, there are a large number of stationary as well as mobile sources of PAH. In less developed countries PAH from fires may also represent a pollution problem. Furthermore, PAH are subject to considerable aerial transport. This makes PAH one of the most widespread environmental contaminants known today. Since many PAH compounds exhibit carcinogenic or cocarcinogenic properties, the environmental concern for these compounds is well justified.

A prerequisite for the control of PAH in the environment is an understanding of the chemical and physical properties of PAH as well as the methodologies for their sampling, sample handling, and analysis. I have for a long time seen the need for a comprehensive and systematic treatise describing our present knowledge in these areas in such a way that it might be of value to active researchers as well as people involved in regulatory actions, monitoring work, etc. Furthermore, a handbook giving a state-of-the-art review and at the same time providing hints for the practical application of methods would be extremely valuable for those researchers beginning work in this area. It is also useful to show how modern-day analytical techniques are applied to various environmental problems.

This handbook consists of contributions from selected authors who are authorities in their subject areas and provides a description of the state of the art in these areas of research. It is my hope that it will encourage and motivate further research on PAH in order to provide a clearer understanding of sources and distributions of PAH in the environment.

I want to express my warm thanks to all the contributors to this handbook for their cooperation, their hard work, and their patience while their manuscripts were being prepared for the press. I also wish to express my indebtedness to Dr. Peter W. Jones for helpful discussions in the early planning stage and to my colleagues at the Central Institute for Industrial Research, Oslo, for their valuable advice and encouragement. Finally, I wish to thank my wife, Eva, and my daughter, Tone, for their understanding and patience when I occasionally gave higher priority to my professional life than to my family life.

Alf Bjørseth

Contents

Contributors

ALF BJØRSETH Central Institute for Industrial Research, Oslo, Norway

JOACHIM BORNEFF Institute of Hygiene, University of Mainz, Mainz, Federal Republic of Germany

JOHN E. CATON Oak Ridge National Laboratory, Oak Ridge, Tennessee

JOAN M. DAISEY Institute of Environmental Medicine, New York University Medical Center, New York, New York

BRUCE P. DUNN Department of Medical Genetics, University of British Columbia, Vancouver, British Columbia, Canada

THOMAS FAZIO Food and Drug Administration, Washington, D.C.

RICHARD B. GAMMAGE Oak Ridge National Laboratory, Oak Ridge, Tennessee

WAYNE H. GRIEST Oak Ridge National Laboratory, Oak Ridge, Tennessee

GERNOT GRIMMER Biochemisches Institut für Umweltcarcinogene, Ahrensburg, Federal Republic of Germany

JOHN W. HOWARD Food and Drug Administration, Washington, D.C.

JÜRGEN JACOB Biochemisches Institut für Umweltcarcinogene, Ahrensburg, Federal Republic of Germany

BJÖRN JOSEFSSON Chalmers University of Technology, University of Göteborg, Göteborg, Sweden

HELGA KUNTE Institute of Hygiene, University of Mainz, Mainz, Federal Republic of Germany

FRANK SEN-CHUN LEE* Ford Motor Company, Dearborn, Michigan

BJØRN SORTLAND OLUFSEN Central Institute for Industrial Research, Oslo, Norway

DENNIS SCHUETZLE Ford Motor Company, Dearborn, Michigan

E. L. WEHRY Department of Chemistry, University of Tennessee, Knoxville, Tennessee

*Current affiliation: Standard Oil Company of Indiana, Naperville, Illinois

CURT M. WHITE Pittsburgh Energy Technology Center, Bruceton, Pennsylvania

STEPHEN A. WISE National Bureau of Standards, Washington, D.C.

MAXIMILIAN ZANDER Rütgerswerke AG, Castrop-Rauxel, Federal Republic of Germany

Abbreviations and Acronyms

Anthr	Anthracene
ASTM	American Society for Testing and Materials
BaA	Benz[a]anthracene
BaP	Benzo[a]pyrene
BkF	Benzo[k]fluoranthene
EPA	Environmental Protection Agency
FID	Flame-ionization detector
GC	Gas chromatography
GC/MS	Gas chromatography/mass spectrometry
CGC/MS	Capillary gas chromatography/mass spectrometry
GLC	Gas-liquid chromatography
GSC	Gas-solid chromatography
HPLC	High-performance (or high-pressure) liquid chromatography
HMO	Hückel molecular orbital
HRMS	High resolution mass spectrometry
IR	Infrared
LC	Liquid chromatography
MOs	Molecular orbitals
Nap	Naphthalene
NMR	Nuclear magnetic resonance
PAH	Polycyclic aromatic hydrocarbons
Phen	Phenanthrene
ppb	Parts per billion
ppm	Parts per million
ppt	Parts per trillion
TLC	Thin-layer chromatography
Py	Pyrene
RE	Resonance energy
UV	Ultraviolet

HANDBOOK OF POLYCYCLIC AROMATIC HYDROCARBONS

1

Physical and Chemical Properties of Polycyclic Aromatic Hydrocarbons

MAXIMILIAN ZANDER / Rütgerswerke AG, Castrop-Rauxel, Federal Republic of Germany

---

I. Structural Principles of PAH

This chapter deals exclusively with PAH that are composed solely of six-membered or six- and five-membered rings and in which interlinked rings have at least two carbon atoms in common, i.e., bi- and polyaryls are not taken into consideration.

The entire group of PAH can be divided into kata-annellated and peri-condensed systems. In kata-annellated PAH, the tertiary carbon atoms are centers of two interlinked rings (e.g., tetracene [1]), whereas in peri-condensed PAH some of the tertiary C atoms are centers of three interlinked rings (e.g., pyrene [2]). Annellation can be linear (the six-membered rings are located on one straight line) or angular (the six-membered rings are located on different straight lines, with the angle between the lines always being 120°). Tetracene [1] is an example of a PAH with linear annellation, and triphenylene [3] that of a PAH with angular annellation.

Larger PAH may contain peri-condensed molecular regions in addition to linear and angular annellated regions (e.g., dibenzo[a,n]perylene [4]).

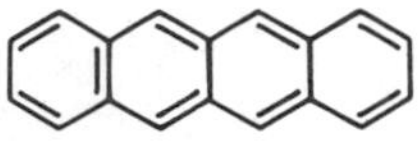

[1] 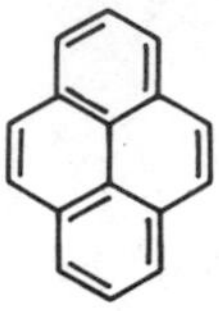 [2]

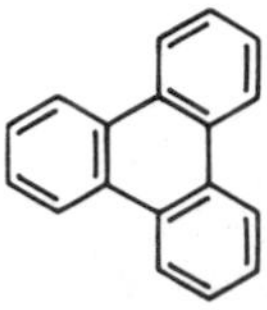

[3] 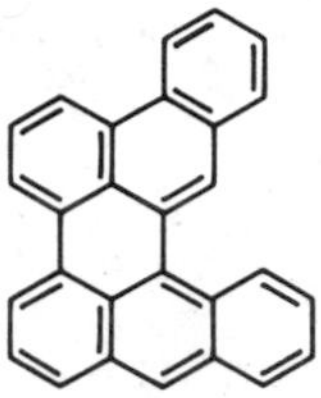 [4]

Table 1 shows the number of possible isomeric PAH which consist exclusively of six-membered rings, as a function of the number n of six-membered rings [1]. It will be seen that the sum total of possible PAH with two to eight rings comes to as much as 1896. There are 683,101 theoretical possibilities for the linking of 12 rings alone [2].

II. Fundamental Aspects Concerning the Correlation of Structure and Properties of PAH

In their ground state, PAH with an even number of carbon atoms are systems with "closed shells," i.e., all bonding π-orbitals are occupied by two electrons with opposite spin, and all antibonding orbitals are unoccupied.

The pairing theorem applies to alternant PAH (systems comprising six-membered rings only) [3]: the π-molecular orbitals (MOs) appear in pairs ψ_μ^+ and ψ_μ^- with energies $\alpha + E_\mu$ and $\alpha - E_\mu$, where α is the Coulomb integral (energy of a carbon-$2p_z$ orbital), i.e., bonding and antibonding π-MOs are in symmetry with one another. The MO diagram of an even-alternant PAH is shown in Fig. 1. The energies of the frontier orbitals, the HOMO (highest occupied molecular orbital) and the LUMO (lowest unoccupied molecular orbital), are particularly significant for the physical and chemical properties of PAH.

The properties of PAH depend on the *size* and *topology* of the systems. Here "size" is understood to mean the number of carbon centers, i.e., number of π electrons, while "topology" denotes the type of ring linkage (see Sec. I). Since the energies of the frontier orbitals determine the properties of PAH in decisive degree, the significance of the size of the systems is understandable: the number of occupied MOs increases in step with the growing total number of π electrons—it is always half the number of π electrons—and the energies of the frontier orbitals increase accordingly.

For PAH, correlations between topology and properties can be demonstrated in a variety of ways and frequently by very simple means. Figure 2 shows the characteristic graphs [1] of the isomeric, i.e., isoelectronic, kata-

Table 1. Number of PAH with n Six-Membered Rings

n	Kata-annellated PAH	Peri-condensed PAH	Σ
1	1	0	1
2	1	0	1
3	2	1	3
4	5	2	7
5	12	10	22
6	37	45	82
7	123	210	333
8	446	1002	1448

Source: Ref. 4.

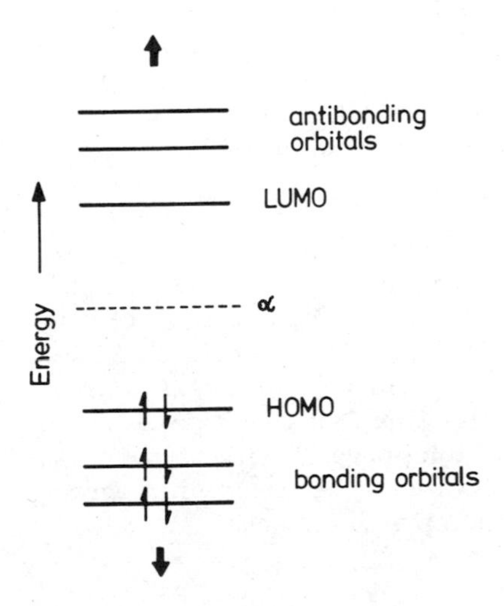

Figure 1. MO diagram of an even-alternant PAH.

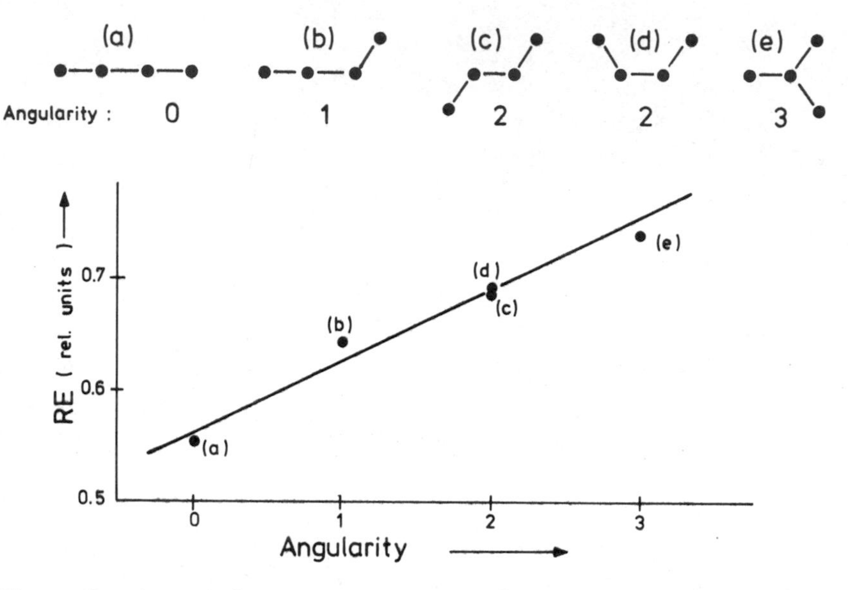

Figure 2. Correlation between resonance energy (RE) and angularity of isoelectronic PAH. (From Ref. 4.)

annellated PAH with four rings. The points mark rings; the lines, the edges of annellated rings. The characteristic topological property is the *angularity*, which is the sum of the 120° angles in the graphs. As an example of correlations between angularity and molecular properties, the resonance energy (RE) as the quantity characterizing the stability of PAH is plotted against the angularity in the figure. Note that the linear correlation is quite satisfactory [4].

The pronounced dependence of most PAH properties on the topology of the systems explains why all quantum-mechanical approximation methods, which are topological methods, reproduce PAH properties well. Methods using Kekulé structures are explicitly topological. These also include the methods developed using elements of graph theory [5-7]. The Hückel molecular orbital (HMO) method is also topological, since the HMO secular determinant reproduces the topology of the π-electron system [8]. The fact that the resonance theory (not strictly proven in physics [9]), which operates with structures, supplies results for PAH that coincide with HMO results is based on a correspondence conditioned by the topological character of both methods.

III. Clar's π-Sextet Model

A structural-chemical model introduced by Clar more than 20 years ago is eminently suitable for systematizing PAH topologies and representation of correlations between topology and properties of PAH [10]. Figure 3 gives formulas for pyrene (*a*) and benzo[ghi]perylene (*d*). The dots represent carbon-$2p_z$ electrons, whose interaction leads to the formation of π MOs. Clar's proposal is to divide the quantity of carbon-$2p_z$ electrons into sub-

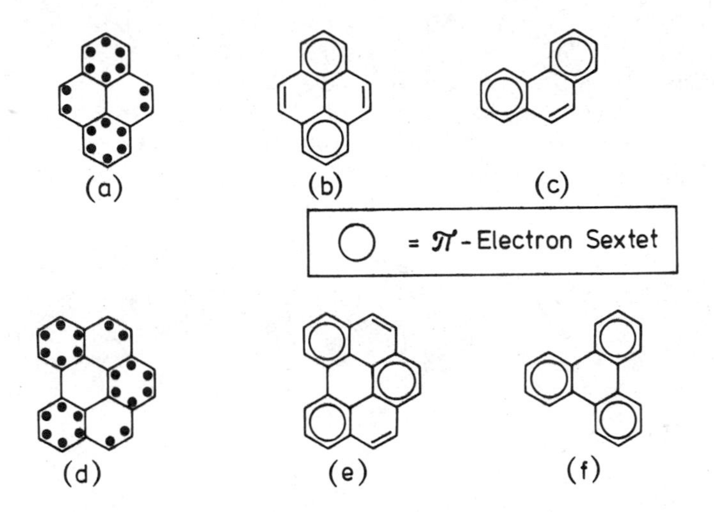

Figure 3. Clar's π-sextet model of PAH. (From Ref. 4.)

quantities in which π electrons are assigned to particular rings. This is accomplished in such a way that the maximum possible number of subquantities consisting of six π electrons, i.e., (benzoid) π-electron sextets, is obtained. Thus, Clar's formulas *b* and *e* are formed from the initial formulas *a* and *d*. Formulas *c* and *f* are Clar's models for two kata-annellated PAH.

The connection between Clar's model and the graph-theoretical methods employing Kekulé structures [5-7] is evident. Figure 4 shows the Kekulé structures of phenanthrene. Let each double bond be assigned only one ring. Then rings A and B in four of the five structures each contain three double bonds (the rings are "benzoid"), while ring C contains three double bonds in two structures only. Rings A and B are also "benzoid" in character in Clar's phenanthrene formula (six π electrons).

Numerous examples from the chemistry and physics of PAH [11] suggest that Clar's model is more than a useful formalism, i.e., the inherent π sextets postulated may possess physical reality. Only two examples will be discussed

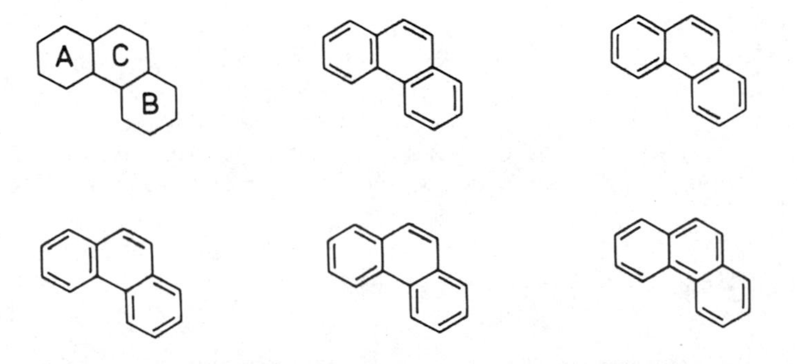

Figure 4. Kekulé structures of phenanthrene.

in detail here: (i) The PAH [5] is surprisingly basic for a hydrocarbon and together with hydrochloric acid forms a hydrochloride. Evidently the carbenium ion [6] formed by the addition of a proton possesses greater stability than the hydrocarbon. Moreover, precisely this result is obtained by writing the PAH and its carbenium ion with the maximum number of π sextets and applying Clar's postulate that the stability of the systems increases with the number of π sextets. (ii) There is a simple correlation between the positions

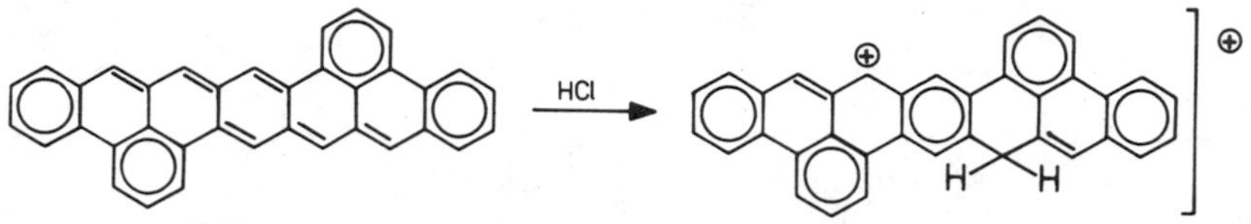

HCl

[5] [6]

at which a PAH adds maleic anhydride endocyclically on the lines of a Diels-Alder reaction and the π-sextet formulas. In cases where the addition can lead to several isomeric adducts, the adduct whose formula can be written with the largest number of inherent π-electron sextets is always the only one to be formed [12]. Figure 5 gives a number of examples. A necessary and usually also adequate condition for the endocyclic Diels-Alder reaction of PAH with maleic anhydride is that at least one π-electron sextet be "gained" by the addition, i.e., that the number of sextets in the product be at least one more than in the reactant. The application of this principle has also led to the discovery of the first example of an endocyclic Diels-Alder reaction with a PAH including dearomatization of the pyrene system: the PAH [7] reacts with maleic anhydride to form the Diels-Alder adduct [8] [13]. Recently, the rate constants of the Diels-Alder reaction with maleic anhydride of numerous kata-annellated PAH have been measured under identical conditions and correlated with various experimental and theoretical molecular parameters [14].

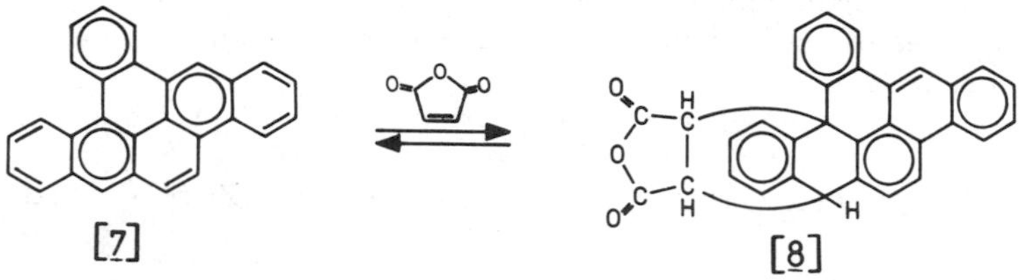

[7] [8]

Even though Clar's model permits a consistent description of the relationships between structure and properties of PAH (and has enabled many correct predictions to be made [15]), it cannot be utterly precluded that Clar's π-sextet formulas correspond to another parameter—in the model not explicitly expressed—which ultimately determines the structure-property relationships. Of course, the same applies to all graph-theoretical methods that are equivalent to Clar's model [5-7].

IV. Correlations among Properties, Size, and Topology of PAH

When considering the properties of PAH, it is useful to separate the effects of size and topology on those properties. Here, Clar's π-sextet model is eminently suitable: the properties of PAH of different sizes but with the same

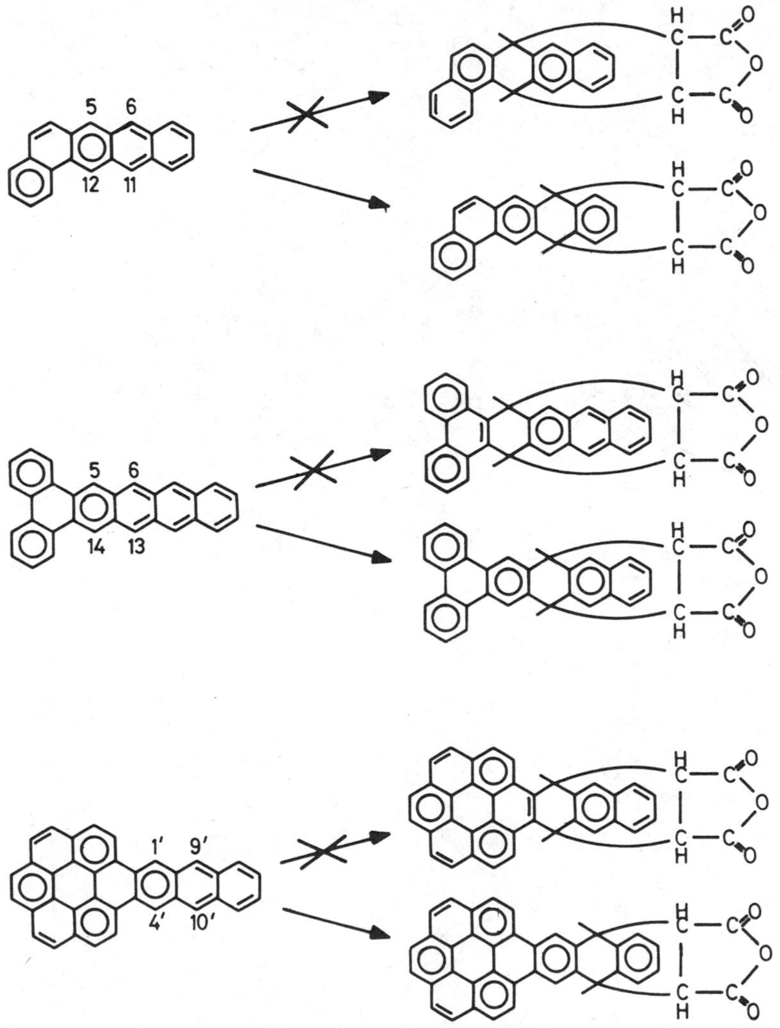

Figure 5. Regioselective maleic anhydride addition to PAH.

number of π sextets ("sextet isomers") are compared. The parameter $(n - 2)/n^2$ has proved useful for describing the size of PAH, where n is the total number of π electrons in the systems.

The size and topology of PAH extensively determine the energy of the HOMO. A particularly obvious example of HOMO-dependent properties is the energy of the para band in the ultraviolet (UV) absorption spectrum of alternant PAH. The para band corresponds to the transition of an electron from the HOMO to the LUMO and hence correlates directly with the energy of the HOMO in the case of alternant hydrocarbons. Figure 6 shows the energy of the para band (cm^{-1}) for various PAH topologies plotted against the parameter $(n - 2)/n^2$. The topologies of the annellation series R1 to R5 under discussion are shown on the right-hand side, with the formulas of the

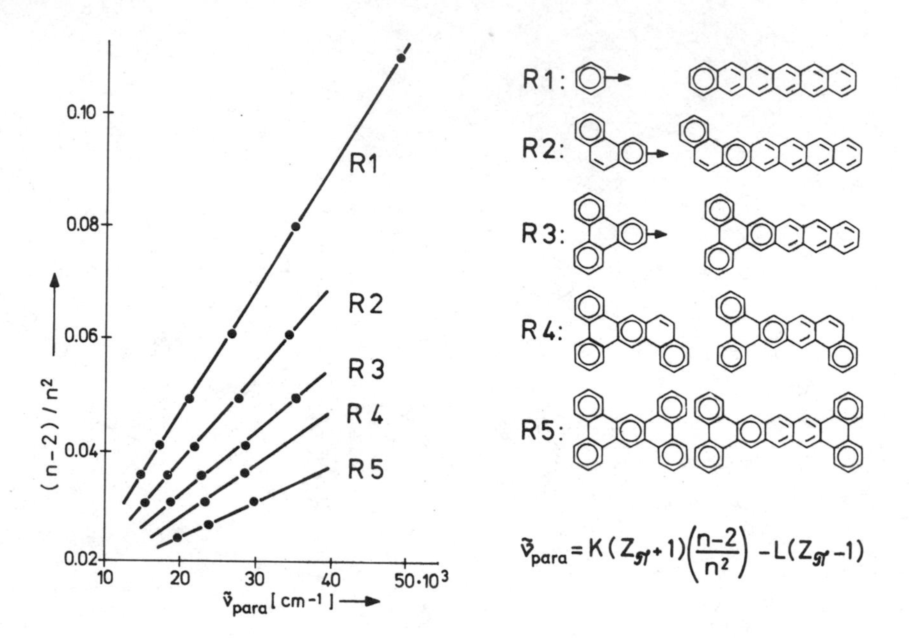

Figure 6. Size/topology dependence of the energy of the para band. (From Ref. 4.)

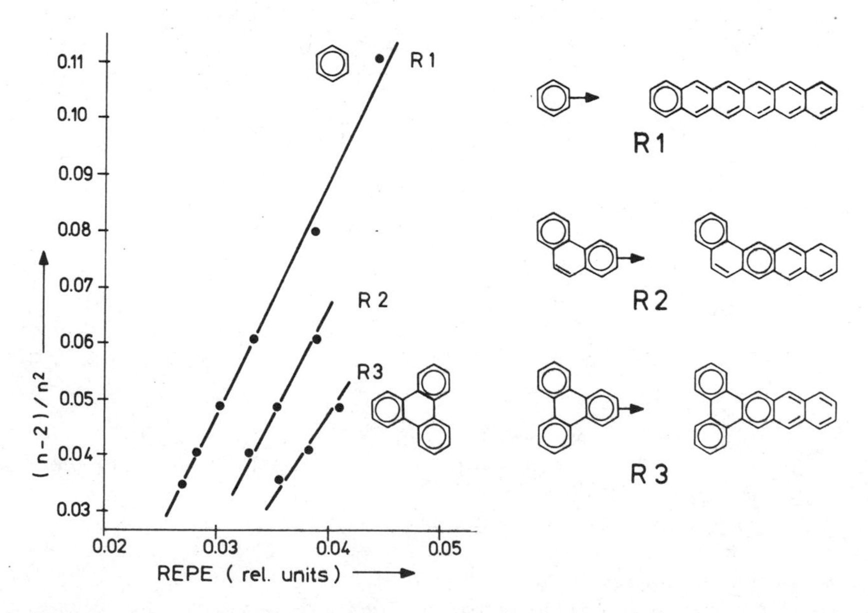

Figure 7. Size/topology dependence of the resonance energy per π electron (REPE). (From Ref. 4.)

first and last members of each annellation series being specified. With respect to the annellation series R1 to R3, the arrows indicate the annellation directions and each annellation step signifies an increment of four π electrons. In the case of annellation series R4 and R5, further rings are internally inserted into the system, and here one annellation step involves an increment of two π electrons. The number of Clar's π sextets remains constant in any one annellation series (sextet isomers). The correlations presented in Fig. 6 all satisfy the equation given therein, with Z_π being the number of Clar's π sextets and K and L having the same values for all PAH under discussion [16].

The resonance energy per π electron (REPE) is the suitable parameter for considerations referring to the stability of PAH. Figure 7 shows REPE (the RE are Dewar resonance energies [17]) for three PAH topologies plotted against $(n - 2)/n^2$. Again, linear correlations are obtained, and for isoelectronic PAH (roughly speaking) a constant amount of resonance energy is obtained with every newly acquired Clar's π sextet [4]. On the basis of Clar's model it may be expected that triphenylene has the same REPE as benzene, and that is indeed approximately the case. This can be viewed as a strong argument in support of the physical reality of Clar's π sextets.

Since of all aromatic hydrocarbons benzene possesses the greatest stability (has the highest REPE value), PAH of the triphenylene type, which can be imagined as structured exclusively of rings containing π sextets and linked by quasi-single bonds, should be the most stable of all PAH [10]. Precisely this is observed experimentally, and it is probably no coincidence that just these "all-benzoid" PAH have also been identified in interstellar matter [18]. Representatives of some of the all-benzoid PAH known to date are shown in Fig. 8.

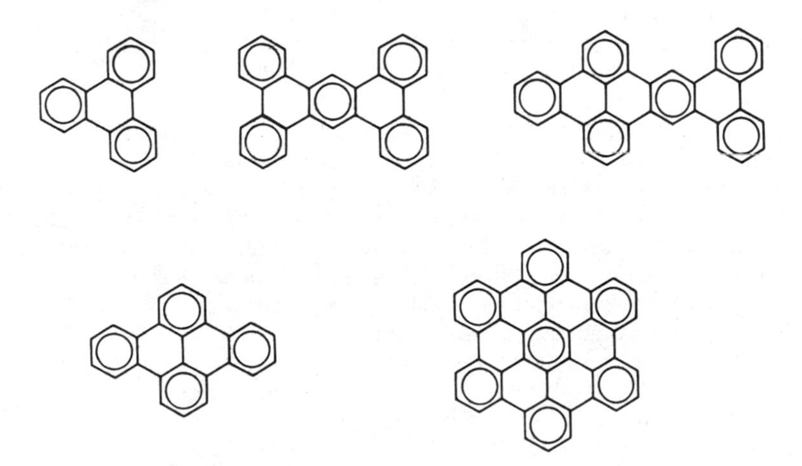

Figure 8. All-benzoid PAH. (From Ref. 4.)

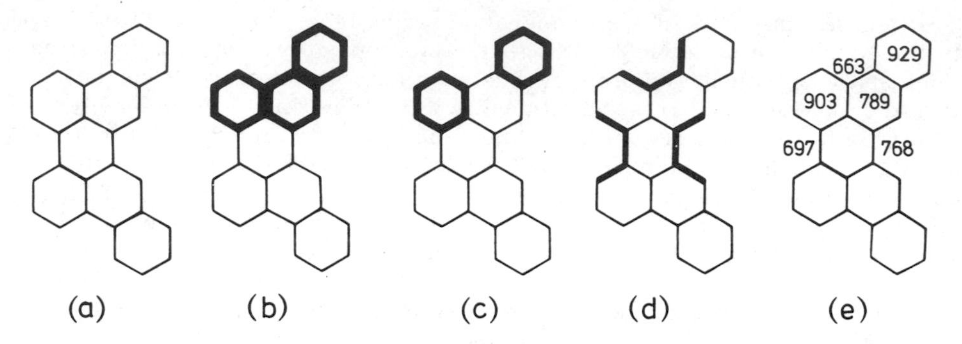

Figure 9. Polansky's pars-orbital method. (From Ref. 4.)

V. Polansky's Pars-Orbital Method

Polansky and Derflinger [19] have verified Clar's qualitative structural-chemical model by quantum mechanics. Their approach consists in formally dividing a molecule into partial structures L and defining a "character order" ρ whose value measures the analogy of the π-electron system of the partial structure L against that of a reference compound. In Fig. 9, *a* is the formula for dibenzo[b,n]perylene. In the formulas *b* to *d*, partial structures L are marked by thicker lines. Formula *e* gives the character orders of the benzoid partial structures and those of the exocisoid C_4 units. The character orders are defined such that their value is greater the more pronounced is the analogy of the π-electron system of the partial structure L with that of the reference compound—in the present case, benzene and butadiene, respectively. Polansky and Derflinger [19] identify high benzoid character orders in rings which in Clar's formulas contain π sextets, and low benzoid character orders in "quasi-empty" rings (e.g., the central ring in triphenylene; see Fig. 8).

For the linear combination (using the Hückel approximation), Polansky's mathematical formulation employs the HMO of benzene and other partial structures L in a PAH instead of the usual atomic orbitals as basic quantities. The character orders measure the contribution of the bonding orbitals of a partial structure L to the bonding orbitals of the whole molecule.

The pars-orbital concept has two important interrelated advantages over Clar's qualitative model: (1) There are several (degenerate) Clar's formulas with the same number of π sextets for numerous PAH. One example is shown in Fig. 10. Inferences drawn from Clar's formulas in respect of the properties of the PAH in such cases are unreliable. In Polansky characterograms of the type *e* in Fig. 9, each molecular zone is characterized by an index (character order). (2) This also permits a check to be made as to whether correlations exist between experimental data and the calculated indices, and consequently permits quantitative testing of the model.

For PAH, Polansky character orders have indeed been successfully correlated with NMR*-coupling constants [20], half-wave potentials [21], reaction

*Nuclear magnetic resonance.

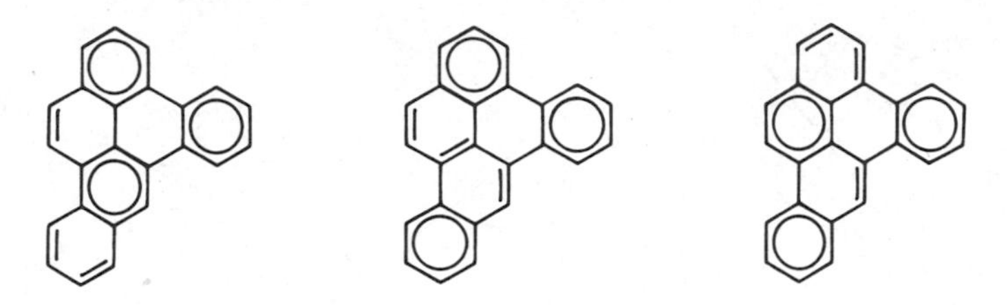

Figure 10. Degenerate π-sextet formulas.

rate constants [22,23], and triplet zero-field splitting parameters [24]. To illustrate this, the two last-mentioned correlations will be discussed in detail.

It has been known for some time [25] that PAH, which like benzo[ghi]perylene [9] contain a peripheral cisoid C_4 arrangement, react with maleic anhydride in the presence of a suitable dehydrogenating agent to form fully aromatic dicarboxylic acid anhydrides [11] ("benzogenic Diels-Alder reaction" [26]); in this process the rate-determining step is the formation of the Diels-Alder adduct [10] [22]. Figure 11 shows the logarithms of the relative reaction

- 4 H

[9] [10] [11]

rate constants of the benzogenic Diels-Alder reaction plotted against the butadienoid character orders of the reacting molecular zones. There is a linear Hammett-analogous relationship in which the character orders correspond to the Hammett σ values [22]. In principle [27], it is difficult to decide whether the character orders correspond to the relevant properties of the transition state or the reactant. Still, it follows from Fig. 11 that the character orders at least for comparable compounds—also permit quantitative predictions about their reactivity.

The magnetic dipole-dipole interaction between the unpaired electrons of a PAH in its lowest triplet state splits this into triplet sublevels, in the absence of an external magnetic field into zero-field sublevels. The triplet zero-field splitting is determined by two parameters D and E, which depend very heavily on the size and topology of the PAH [28]. Figure 12 shows the D parameter plotted against the arithmetic mean $\bar{p}_B$ of the benzoid character orders of PAH for the acene series (benzene, naphthalene, anthracene, tetracene, etc.) [24]. The observed linear correlation between $\bar{p}_B$ and D and also analogous experimental results have been used to obtain a detailed theoretical interpretation of triplet zero-field splitting [24].

The pars-orbital concept has so far not been applied to the calculation of atom localization energies, but it contributes directly toward understanding the interrelationships between localization energies and PAH topology.

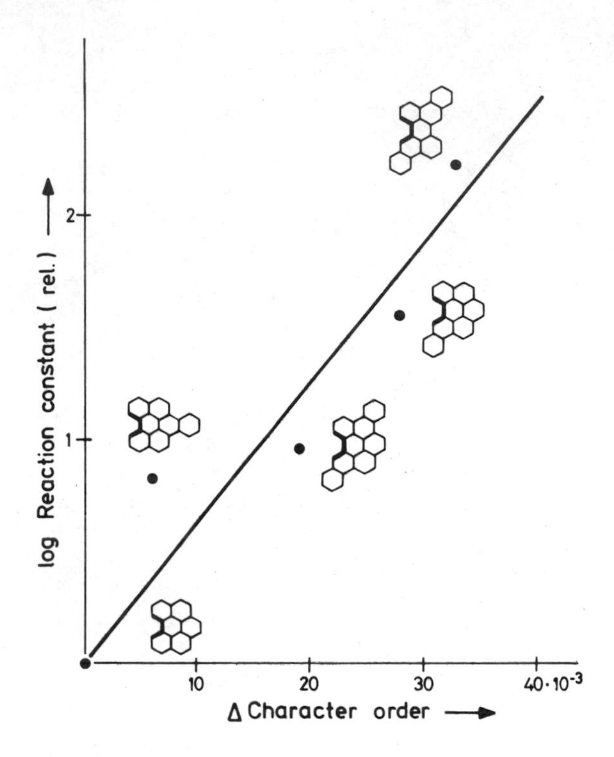

Figure 11. Correlation between reaction constant of the benzogenic Diels-Alder reaction and the character order of the reacting molecular zones (thicker lines in structures).

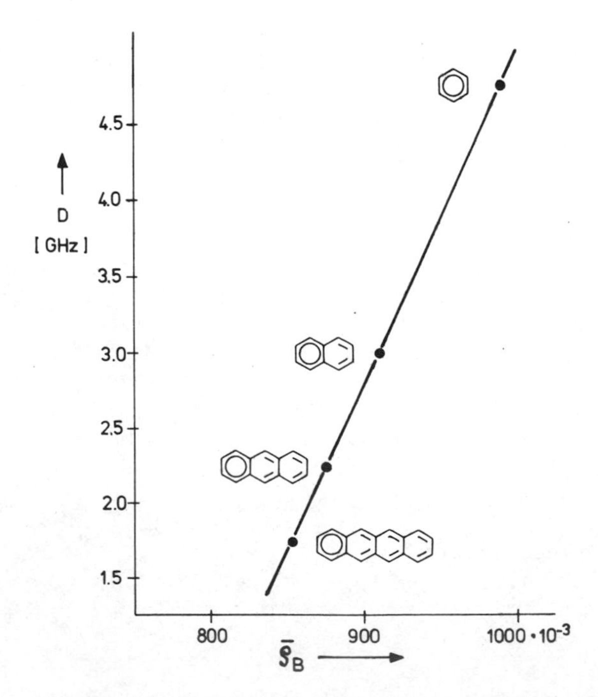

Figure 12. Correlation between zero-field splitting parameter D and $\bar{\rho}_B$. (From Ref. 24.)

VI. The Localization Energy Concept in PAH Chemistry

Many addition reactions are known in PAH chemistry (the Diels-Alder syntheses discussed earlier are examples), and in other reactions typical for PAH, e.g., electrophilic substitution, an addition step determines the rate of the reaction. For addition reactions of PAH, Wheland's concept [29] of localization energy has proved extremely useful. The localization energy L_u, given below in units of the resonance integral β, is the energy required to isolate a π electron at the center u of the PAH from the remaining π system. It is found that the smaller the value of L_u, the lower the activation energy and the greater the reaction rate constant of an addition at the center u. In the process, the attacking species can be a nucleophile, an electrophile, or a radical and the addition can be a one-center or pericyclic reaction. Of the various reactivity indices which are a measure of the localization energy [30], Dewar's reactivity number N_u [31] has two decisive advantages: the N_u values (1) correlate excellently with experimental data and (2) are very easy to calculate by perturbation theory [32] for systems with a very large number of centers. A drawback of Dewar's method of calculation lies in the fact that in its simple form it can only be applied to even-alternant PAH.

Figure 13 shows the localization energies N_u (in units of the resonance integral β) of all secondary carbon centers of benzene, naphthalene, anthracene, and phenanthrene. Of these hydrocarbons, benzene should according to the N_u values be the least reactive; with naphthalene, higher reactivity is expected for the 1-position than for the 2-position; anthracene should preferably react in the 9,10 positions; and preferred reactivity is also expected in the 9,10-positions for phenanthrene but should be markedly weaker than in the case of anthracene. All these inferences derived from the localization energies are fully consistent with experimental experience. For a given type of reaction, Hammett-analogous linear correlations exist between the logarithm of the relative reaction rate constants and the Dewar reactivity numbers N_u, which correspond to the Hammett σ values. The gradient of the straight lines is proportional to the resonance integral β, which assumes different values for different reactions and corresponds to the Hammett ρ values. As an example, Fig. 14 shows such a Hammett-analogous relationship with Dewar reactivity numbers N_u for the bromination of aromatic hydrocarbons. A β value of -65 kJ/mol is obtained as the gradient of the straight line [33].

The sum of Dewar's reactivity numbers ($N_u + N_t$) is used as the reactivity index for two-center reactions (e.g., Diels-Alder syntheses) at the centers u and t of a PAH. A Hammett-analogous relationship with Dewar's reactivity numbers (bislocalization energies $\times\ \beta^{-1}$) has been found, for example, for the benzogenic Diels-Alder reaction of PAH [34]. In special cases, the Dewar

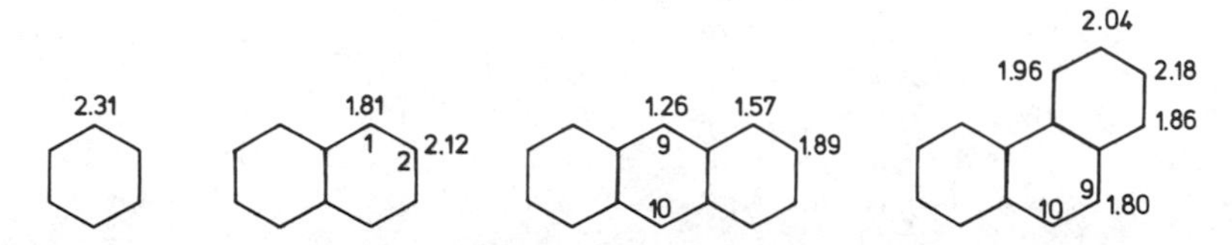

Figure 13. Reactivity numbers of PAH.

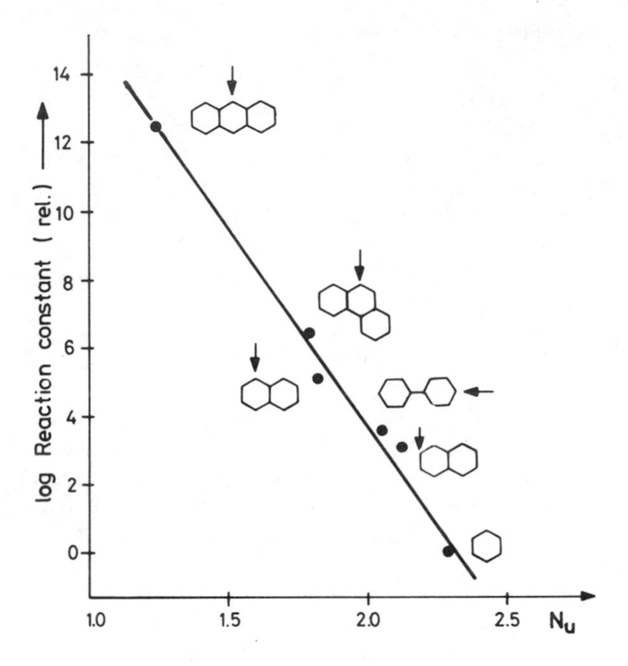

Figure 14. Correlation between reaction constant of the bromination of PAH and reactivity number (reactive positions marked by arrows). (From G. W. Klumpp, *Reaktivität in der organischen Chemie*, Vol. 1, p. 119, Thieme, Stuttgart, 1977.)

reactivity numbers are helpful in clarifying the mechanism of reactions with PAH [35].

The N_u values can be calculated for all centers of a PAH, and accordingly a N_u pattern is obtained for each PAH. Correlations between the N_u pattern and the topologies of PAH are not discernible to begin with but become evident when Clar's π-sextet model or, better still, Polansky's pars-orbital concept is taken into consideration.

VII. N_u Pattern and Topology of PAH

Figure 15 shows Clar's formulas for some PAH which undergo the benzogenic Diels-Alder reaction. The corresponding Dewar reactivity numbers N_u are given at the reactive centers (marked in the formulas). The following correlations are evident [34]: (1) In systems with different localization energies (in β^{-1}) at the two reacting centers, the greater localization energy is always found at the center belonging to a benzoid partial structure (marked by the solid circle) and the smaller localization energy always at the center belonging to an ethylenoid partial structure, in conformity with the fact that the localization energy of benzene is greater than that of ethylene. [The benzoid character orders calculated by the pars-orbital method constitute a more favorable basis for this approach than do Clar's formulas with the maximum

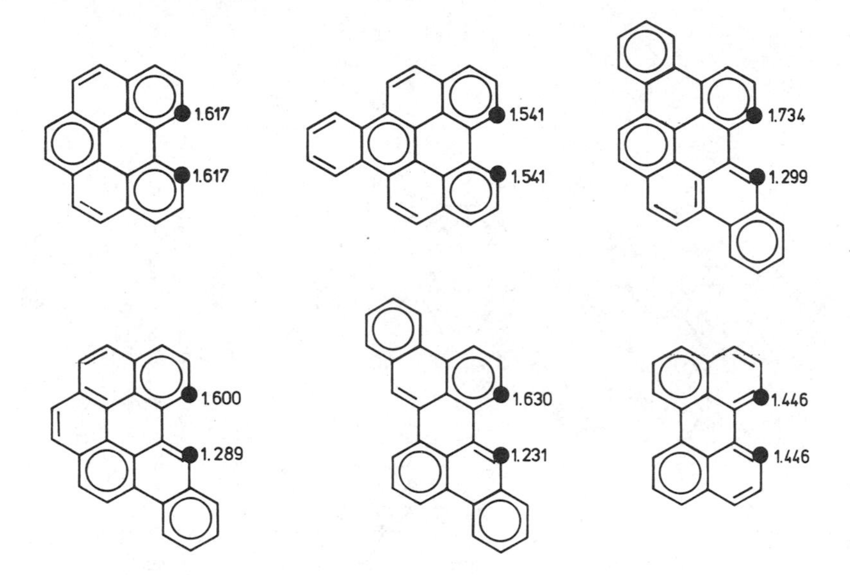

Figure 15. Correlation between π-sextet formulas and reactivity numbers. (From Ref. 34.)

number of inherent π sextets, since—as mentioned earlier—in many cases several degenerate Clar's formulas with the same number of π sextets can be used for one PAH (see Fig. 10).] (2) For centers of molecular zones, which in Clar's formulas can be represented both as benzoid and as non-benzoid, the localization energies are smaller than for centers of molecular zones, which in Clar's formulas can only be represented as benzoid; this corresponds to the increase of localization energies when going from naphthalene to benzene. (3) It seems plausible that the localization energies for centers of benzoid partial structures of PAH are smaller than in benzene (2.31 β) or in all-benzoid PAH such as triphenylene (2.12 β in the 2-position) and is consistent with the values of benzoid character orders [19] of the compared systems.

Evidently, plausible relationships do exist between the N_u pattern and Clar's formulas of PAH. Quantitative correlations are obtained with the benzoid character orders ρ_B [36]. Since ρ_B provides information on a partial structure consisting of six carbon centers in a PAH but the N_u values refer to individual carbon centers, a mean localization energy $\bar{N}_u$ for benzoid partial structures L_B must first be introduced which represents the arithmetic mean of the N_u of the secondary carbon centers ($\gtrdot$C–H) of an L_B. Figure 16 shows the ρ_B, the N_u of the secondary carbon centers, and the $\bar{N}_u$ of each L_B for 25 L_B in 15 PAH. For topologically comparable L_B, linear correlations between the relevant ρ_B and the N_u topologically comparable centers of this L_B are highly satisfactory, as the three examples in Fig. 17 show. Figure 18 shows the ρ_B from Fig. 16 plotted against the relevant $\bar{N}_u$. The linear correlation between ρ_B and $\bar{N}_u$ observed in the region of higher ρ_B has a correlation coefficient of 0.9676. The L_B not satisfying the relation all have

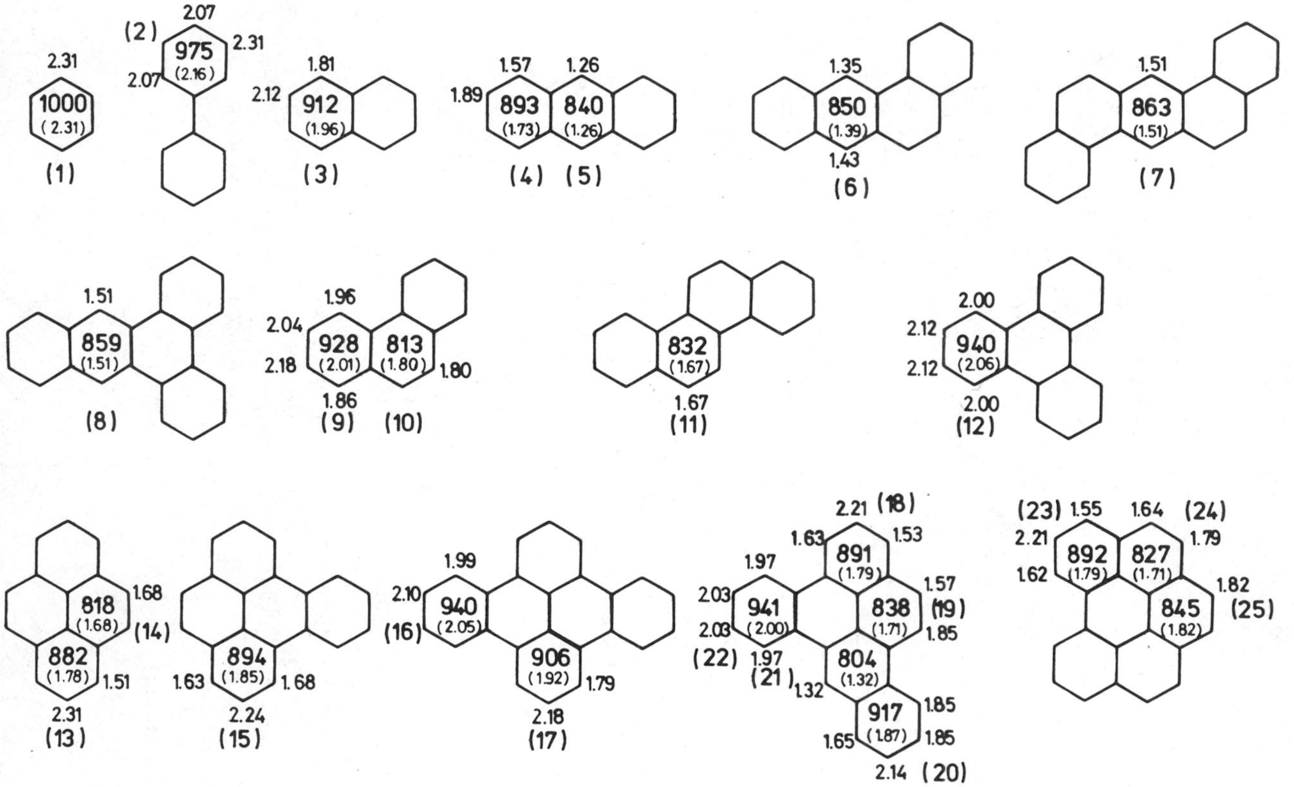

Figure 16. Benzoid character orders ρ_B (upper number in hexagon), mean localization energies $\bar{N}_u$ (lower number—in parens—in hexagon) and localization energies N_u for single centers (number outside the hexagon) for 25 benzoid partial structures (number—in parens—outside the hexagon) of 15 even-alternant PAH. (From Ref. 36.)

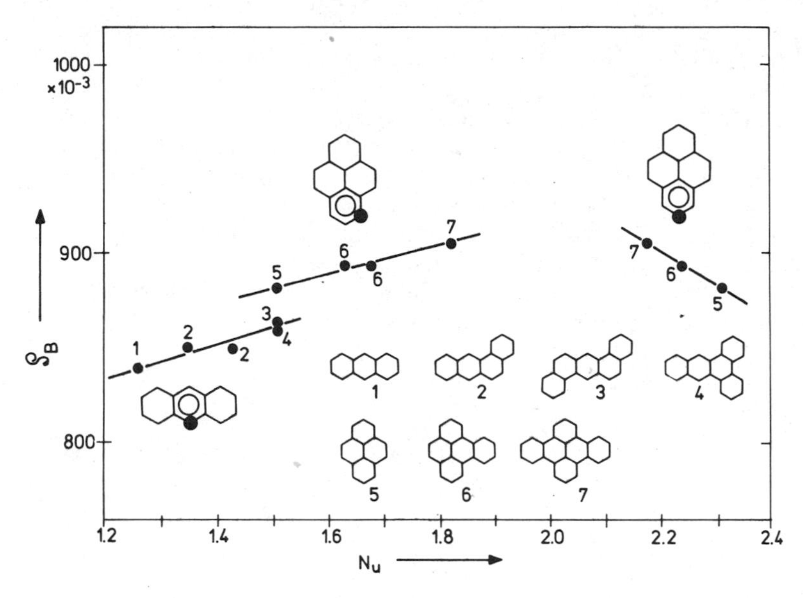

Figure 17. Correlation between ρ_B and the N_u topologically comparable centers. (From Ref. 36.)

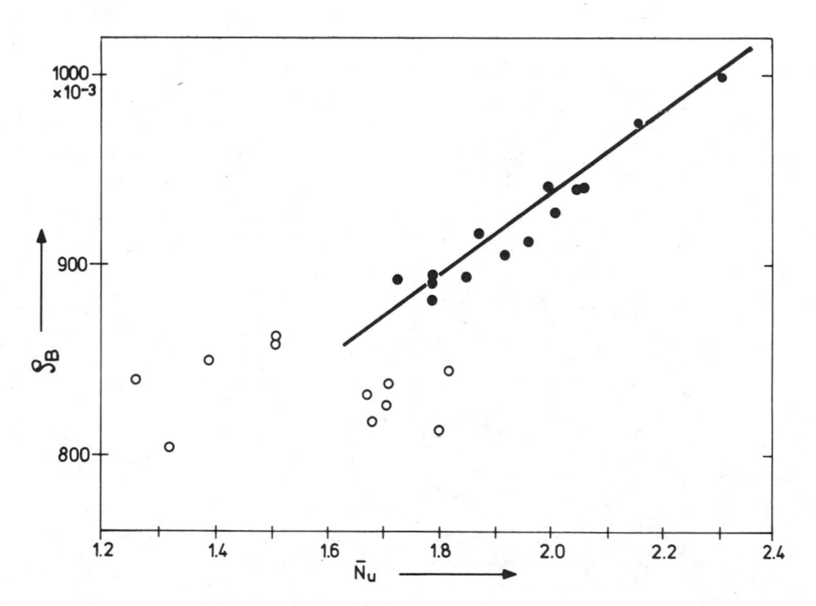

Figure 18. Correlation between ρ_B and $\bar{N}_u$. (From Ref. 36.)

less than three secondary carbon centers, so that the relevant $\bar{N}_u$ are the first to have no physical significance. The fact that these $\bar{N}_u$ are all located in the region of low ρ_B is plausible since the ρ_B have a tendency to decrease in step with the increasing number of tertiary carbon centers ($\gtrless$C−) in the L_B [19]. For topologically comparable L_B, the differences between the N_u of the individual centers become smaller as the ρ_B increases (see, for example, L_B structures 13, 15, and 17 in Fig. 16). The increasing equalization of the N_u represents an approximation to the bonding properties in benzene.

In summary, it can be said that there is extensive correspondence between the N_u pattern, the π-sextet formulas and the character orders of benzoid regions of PAH. The type of correspondence is also in keeping with the experience and intuition of the organic chemist and ultimately demonstrates once again the close connection between topology and properties of PAH.

VIII. PAH in Electronically Excited States

Discussions so far have mainly referred to PAH in the electronic ground state. Given the interaction with photons of suitable energy, a PAH passes from its electronic ground state into an electronically excited state—the simplest case being where an electron is transferred from the HOMO to the LUMO (para band; see Sec. IV).

The following are the main differences between the PAH in its ground state and in an electronically excited state: (1) The electronically excited PAH has an extremely short lifetime; after excitation, it returns very rapidly by nonradiative deactivation or light emission (fluorescence or phosphorescence) to the ground state or enters into photochemical reactions. (2) In the excited state, two orbitals are in each case occupied by only one electron. Since the electrons on the singly occupied orbitals have different orbital ring quantum numbers, they are not subject to any restriction as regards their spin quantum numbers (Pauli principle). This means that the electrons in the singly occupied orbitals can have either opposite spin (singlet excited state) or the same spin (triplet excited state). (3) The node properties, or LCAO (linear combination of atomic orbitals) pattern, of the electronic wave functions in the excited state are different from those in the ground state, wherein lies one of the fundamental causes for the different chemical behavior of PAH in the ground state and excited states.

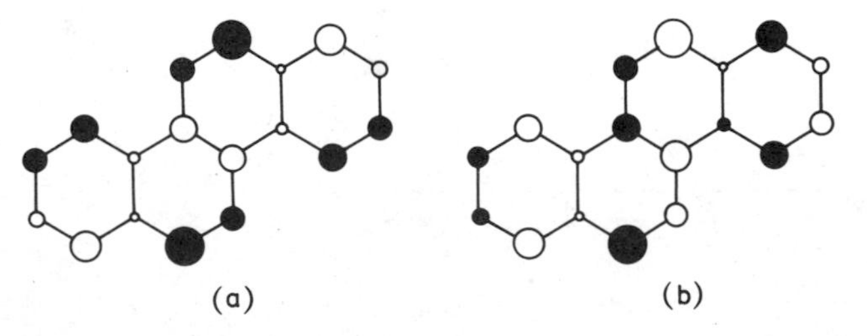

Figure 19. LCAO pattern of the highest occupied orbital of chrysene in its ground state (a) and in its S_1 - α state (b) (from Ref. 37).

As an example, Fig. 19 shows the LCAO pattern [37] of the highest occupied orbital of chrysene in its ground state (a) and the main configuration interaction (CI) contribution to the highest occupied orbital of chrysene in its $S_1 - \alpha$ state (b). Since chrysene is an alternant PAH, it is only the node properties that change on account of the pairing theorem, but this applies to virtually all interlinked centers of the molecule, from which it can be deduced that the chemical behavior of chrysene is vastly different in its ground and excited states [38].

Of the numerous photochemical reactions which PAH can enter, only a few will be discussed here in greater detail. Linear-annellated PAH such as anthracene [12] react in solution with molecular oxygen to form endoperoxides [13] [39]. The PAH is excited by absorption of light in its lowest singlet

$$[12] \xrightarrow{O_2} [13]$$

excited state. The next step is the nonradiative transition to the lowest triplet excited state (intersystem crossing [40]). This is followed by intermolecular energy transfer from the triplet state of the anthracene to the oxygen to form singlet oxygen:

$$^3M^* \text{ (anthracene)} + {}^3O_2 \rightarrow {}^1M \text{ (anthracene)} + {}^1O^*_2 \quad (1)$$

Singlet oxygen ($^1O^*_2$) then reacts with anthracene in the ground state (^1M) to form the endoperoxide [13]. In the case of peri-condensed PAH, such as benzo[a]pyrene (BaP), photooxidation frequently leads to quinones [41].

In the absence of oxygen and under photochemical conditions, PAH of the anthracene type yield dimers [14] [42]. The first reaction step is the formation of an excimer [40], which is followed by the formation of the photodimer:

$$^1M^* + {}^1M \rightarrow {}^1(MM)^* \rightarrow M_2 \quad \text{(see structure [14])} \quad (2)$$

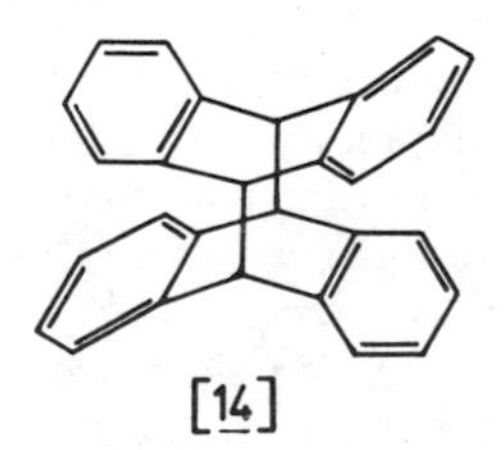

[14]

PAH of the anthracene type react photochemically [43] with dienophils, such as maleic anhydride (just as they do thermally; see Sec. III, Fig. 5), in the sense of a Diels-Alder synthesis.

Photochemical benzogenic Diels-Alder synthesis has only recently been discovered [44-46]. It will be discussed here in some detail, taking as an example the reaction of chrysene with maleic anhydride. According to the frontier orbital concept of Fukui [38], a thermal (4 + 2)-cycloaddition is not expected in chrysene: as will be seen from Fig. 19, the node properties of the peripheral C_4 units in the HOMO of the chrysene in its ground state do not match those of a 1,3-diene. And, indeed, neither has a thermal addition of maleic anhydride to chrysene been observed in experiments [45]. On the other hand, node properties analogous to those in a 1,3-diene have been calculated for the peripheral C_4 units of the electronically excited chrysene (Fig. 19). In agreement with this, formation of the compound [17] has been observed in the reaction of chrysene [15] with maleic anhydride under photochemical conditions and in the absence of air. The primary Diels-Alder adduct [16] has not been identified, probably because it stabilizes very rapidly to form [17] (by dehydrogenation and/or disproportionation). In the presence of oxygen, the main product formed is the fully aromatic [18] (by dehydrogenation from [17] [46].

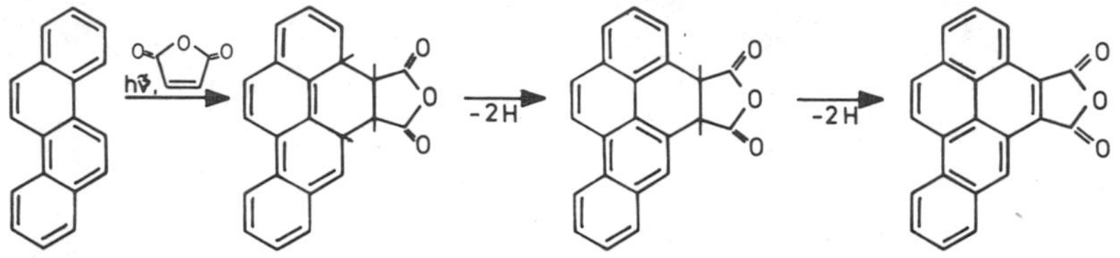

[15] [16] [17] [18]

The pars-orbital concept (see Sec. V) has also been applied to PAH in the electronically excited state [47]. Here, the character orders of the partial structures L are defined so as to describe the analogy between L in the excited molecule and the suitable reference compound (for example, butadiene for peripheral C_4 units in PAH) in the ground state. However, to date character orders for electronically excited molecules have only been calculated using the simple Hückel approximation. Accordingly, application of the concept remains for the time being limited to PAH, whose first absorption band is a para band (HOMO-LUMO transition; see Sec. IV). However, even by using the Hückel approximation the concept should be able to render valuable service in identifying the correlations between topology and properties of electronically excited PAH.

IX. Nonalternant PAH

In addition to six-membered rings, nonalternant PAH also contain five-membered rings (e.g., fluoranthene [19]). The main topologically conditioned difference between alternant and nonalternant PAH lies in the fact that the pairing theorem [3] does not apply to the latter, i.e., in nonalternant PAH the bonding and antibonding MOs are arranged asymmetrically about the energy reference point. The MO diagram of a nonalternant PAH is shown in Fig. 20. The different properties of these two classes of PAH result from the different HOMO/LUMO situation of alternant (see Fig. 1) and nonalternant PAH.

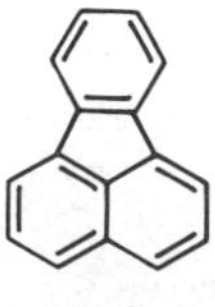

[19]

As the para band in the absorption spectrum of the hydrocarbons corresponds to the HOMO-LUMO transition (see Sec. IV), the energy of the para

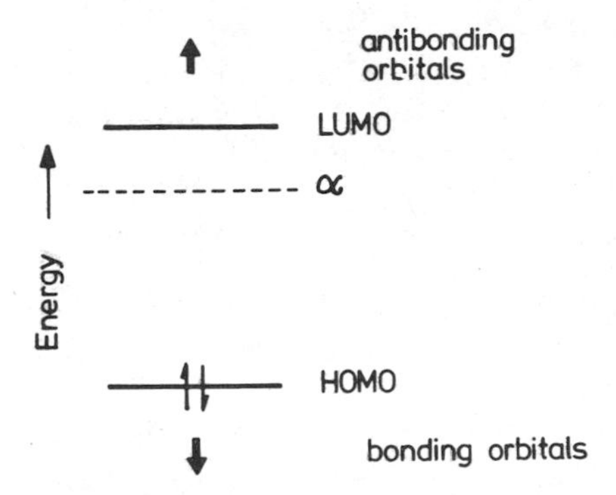

Figure 20. MO diagram of a nonalternant PAH.

band in systems A (alternant PAH) and N (nonalternant PAH) in Fig. 21 is identical (the same HOMO-LUMO difference). The ionization energy I_D must be raised to remove an electron from an orbital (to transfer it to the energy band of the "free" electrons), whereas the opposite is the case when a free electron enters an orbital, i.e., energy in keeping with the "electron affinity" E_A is gained. Koopman's theorem applies to the correlation between the energy ε_m of the orbital which releases an electron into the energy band of the free electrons, or takes it from the band, and the I_D and E_A in question [48]:

$$\varepsilon_m = -I_D = E_A \tag{3}$$

As is directly derivable from Fig. 21, the first ionization potential (the energy required to remove an electron from the double-occupied HOMO) is smaller for the alternant PAH (A) than for the nonalternant PAH (N).

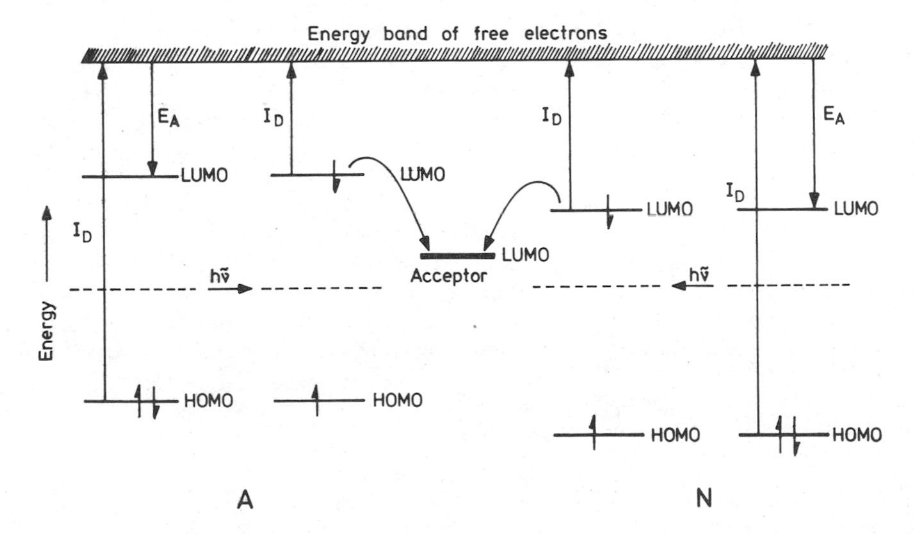

Figure 21. HOMO/LUMO diagram of an alternant (A) and a nonalternant (N) PAH before and after HOMO-LUMO excitation (for further details see text).

The polarographic reduction potentials of PAH measured in solution using the dropping mercury electrode [49] correlate with the electron affinity E_A [30]. In electrochemical reduction, the PAH (in most cases) changes primarily into a radical ion in that the unpaired electron occupies the LUMO. In accordance with the MO diagrams in Fig. 21, the reduction potentials of nonalternant PAH (N) are more positive by some 0.3-0.5 V than the alternant PAH (A) with the same energy of the para band in each case [49].

Following optical HOMO-LUMO excitation (Fig. 21), less energy is required to transfer the electron from the LUMO to the energy band of the free electrons in the case of alternant PAH (A) than in the case of nonalternant PAH (N): in the excited state, the alternant PAH is a "better" electron donor than the nonalternant PAH. Where an electron acceptor is present in the system, electron transfer can be accomplished from the singly occupied LUMO of the PAH to the LUMO of the acceptor (in its ground state) (Fig. 21). This process competes with the electron transition from PAH-LUMO to PAH-HOMO leading to fluorescence of the PAH, i.e., the electron acceptor acts as a fluorescence quencher [50]. The fluorescence-quenching effect is more pronounced in the alternant PAH (A), as it is a better electron donor than in the nonalternant PAH (N) [51].

Another substantial difference between alternant and nonalternant PAH (which also follows from the pairing theorem) is that alternant PAH exhibit a uniform charge distribution over all centers of the molecule whereas nonalternant PAH frequently possess a dipole moment.

X. Some Unusual Chemical Reactions of PAH

In the substitution reactions typical for PAH, the hydrocarbon acts as a nucleophile. (Since the bases of DNA are nucleophiles, reactions of nonderivated PAH with DNA do not take place. After opening of the epoxide ring in the "ultimate carcinogen," an electrophilic system is present which can then react with the bases of DNA.) Nonetheless, some reactions are known in which the PAH act as electrophiles. These include the base-catalyzed methylation with dimethyl sulfoxide (DMSO), where the methyl sulfinyl carbanion ($H_3C-SO-CH_2^-$) is the attacking species [52]. In this reaction, 9,10-dimethylanthracene is formed from anthracene in a yield of 96%.

Apart from the Diels-Alder reactions in which PAH function as dienes (see Sec. III and V), Diels-Alder reactions with PAH as dienophils are also known. In particular these are Diels-Alder reactions with an inverse electron demand, and hexachlorocyclopentadiene is the diene used most. A more recent example is the reaction of pyrene [2] with hexachlorocyclopentadiene to produce the adduct [20] [53]. Here, note that neither a double addition to [2] nor a single addition to the 9,10-double bond of phenanthrene occurs.

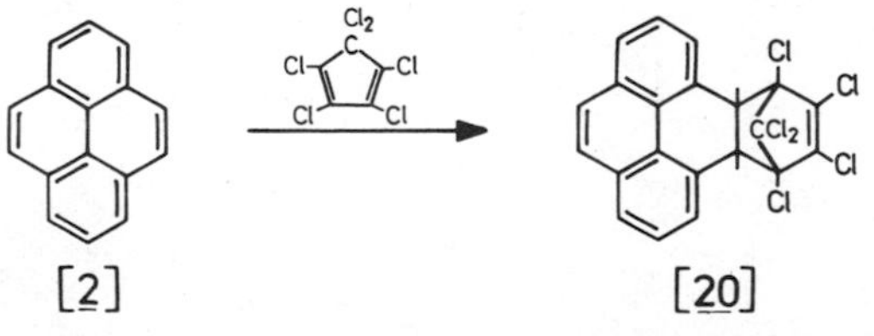

[2] [20]

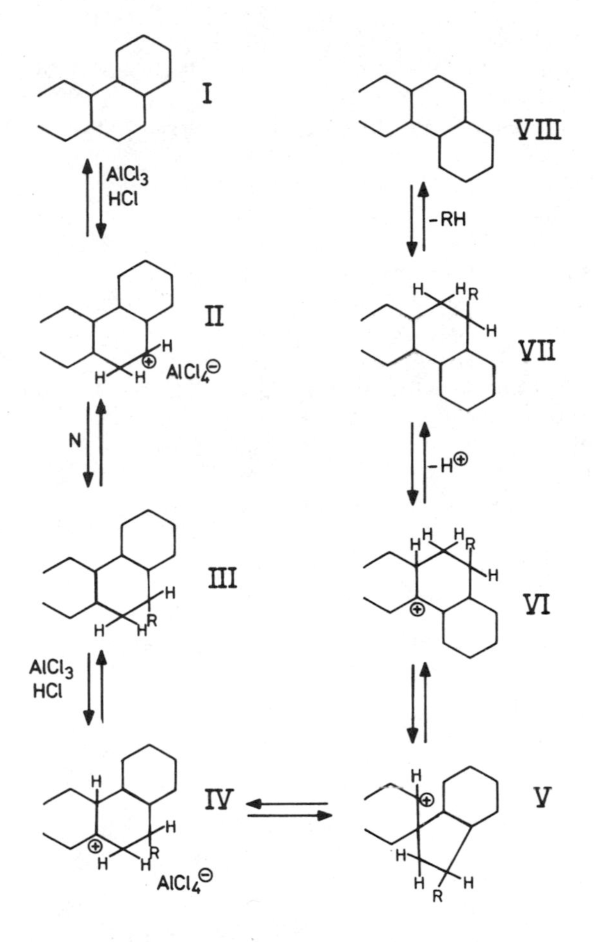

Figure 22. Mechanism of $AlCl_3$-catalyzed skeleton rearrangements of PAH.

$AlCl_3$-catalyzed skeleton rearrangements, e.g., from [21] to [22], frequently observed in PAH, are also rather remarkable [54]. Similar rearrangements are known for annellated carbazoles [55] and thiophenes [56]. The mechanism of such skeleton rearrangements shown in Fig. 22 is derived from detailed experimental investigations. The nucleophile N (which appears in structure III as the substituent R) may be benzene, the initial aromatic, or else the hydride ion [57].

$AlCl_3$

[21] → [22]

References

1. A. T. Balaban, in *Chemical Applications of Graph Theory* (A. T. Balaban, Ed.), Academic Press, New York, 1976, p. 63.
2. W. F. Lunnon, in *Graph Theory and Computing*, Academic Press, New York, 1972, p. 87.
3. C. A. Coulson and G. S. Rushbrooke, *Proc. Cambridge Phil. Soc. 36*: 193 (1940).
4. M. Zander, in *Polycyclische aromatische Kohlenwasserstoffe* (VDI-Bericht No. 358), VDI-Verlag, Düsseldorf, W. Germany, p. 11.
5. W. C. Herndon and M. L. Ellzey, *J. Amer. Chem. Soc. 96*:6631 (1974).
6. W. C. Herndon, *J. Chem. Educ. 51*:10 (1974).
7. I. Gutman and M. Randic, *Chem. Phys. 41*:265 (1979).
8. D. H. Rouvray, in *Chemical Applications of Graph Theory* (A. T. Balaban, Ed.), Academic Press, New York, 1979, p. 175.
9. M. J. S. Dewar, *The Molecular Orbital Theory of Organic Chemistry*, McGraw-Hill, New York, 1969, p. 241.
10. E. Clar and M. Zander, *J. Chem. Soc.*, p. 1861 (1958).
11. E. Clar, *The Aromatic Sextet*, Wiley, New York, 1972.
12. H. G. Franck and M. Zander, *Chem. Ber. 99*:1272 (1966).
13. G. P. Blümer, K. D. Gundermann, and M. Zander, *Chem. Ber. 109*: 1991 (1976).
14. D. Biermann and W. Schmidt, *J. Amer. Chem. Soc. 102*:3163, 3173 (1980).
15. E. Clar, C. T. Ironside, and M. Zander, *Tetrahedron 6*:358 (1959).
16. M. Zander, *Z. Naturforsch. 33a*:1398 (1978).
17. M. J. S. Dewar and H. C. deLlano, *J. Amer. Chem. Soc. 91*:789 (1969); B. A. Hess and L. J. Schaad, ibid. *93*:305 (1971).
18. B. Donn, *Astrophys. J. 152*:129 (1968).
19. O. E. Polansky and G. Derflinger, *Int. J. Quantum Chem. 1*:379 (1967).
20. H. Sofer and O. E. Polansky, *Mh. Chem. 102*:256 (1971).
21. H. Sofer, Thesis, University of Vienna, 1969.
22. M. Zander, *Ann. Chem. 723*:27 (1969).
23. H. Sofer, O. E. Polansky, and G. Derflinger, *Mh. Chem. 101*:1318 (1970).
24. Chr. Bräuchle, H. Kabza, and J. Voitländer, *Chem. Phys. 48*:369 (1980).
25. E. Clar, *Ber. Dtsch. Chem. Ges. 65*:846 (1932).
26. E. Clar and M. Zander, *J. Chem. Soc.*, p. 4616 (1957); M. Zander, *Angew. Chem. 72*:513 (1960).
27. K. Yates, *Hückel Molecular Orbital Theory*, Academic Press, New York, 1978, p. 218.
28. S. P. McGlynn, T. Azumi, and M. Kinoshita, *Molecular Spectroscopy of the Triplet State*, Prentice-Hall, Englewood Cliffs, N.J., 1969, p. 329.
29. G. W. Wheland, *J. Amer. Chem. Soc. 64*:900 (1942).
30. A. Streitwieser, *Molecular Orbital Theory for Organic Chemists*, Wiley, New York, 1961, p. 335.
31. M. J. S. Dewar, *J. Amer. Chem. Soc. 74*:3357 (1952).
32. M. J. S. Dewar and R. C. Dougherty, *The PMO Theory of Organic Chemistry*, Plenum Press, New York, 1975.
33. L. Altschuler and E. Berliner, *J. Amer. Chem. Soc. 88*:5837 (1966).
34. M. Zander, *Z. Naturforsch. 33a*:1395 (1978).

35. M. Zander, in *Analytical Techniques in Environmental Chemistry* (J. Albaiges, Ed.), Pergamon Press, Elmsford, N.Y., 1980, p. 83.
36. M. Zander, *Z. Naturforsch. 34a*:521 (1979).
37. V. Bachler, G. Olbrich, and O. E. Polansky, private communication, 1980.
38. K. Fukui, *Theory of Orientation and Stereoselection*, Springer-Verlag, New York, 1975.
39. C. Dufraise, *Bull. Soc. Chim. Fr. 6*:422 (1939).
40. J. B. Birks, *Photophysics of Aromatic Molecules*, Wiley (Interscience), New York, 1970.
41. *Particulate Polycyclic Organic Matter*, National Academy of Science, Washington, D.C., 1972, p. 68.
42. C. A. Coulson, L. E. Orgel, W. Taylor, and J. Weiss, *J. Chem. Soc.* p. 2961 (1955); A. S. Cherkasov and T. M. Vember, *Opt. Spectr. (USSR) (English Transl.) 6*:319 (1959).
43. G. Kaupp, *Ann. Chem.* 254 (1977).
44. H. Karpf, O. E. Polansky, and M. Zander, *Tetrahedron Lett.* p. 339 (1978).
45. H. Karpf, O. E. Polansky, and M. Zander, *Tetrahedron Lett.* p. 2069 (1978).
46. H. Karpf, O. E. Polansky, and M. Zander, *Mh. Chem. 112*:659 (1981).
47. O. E. Polansky, *Z. Naturforsch. 29a*:529 (1974).
48. T. Koopman, *Physica 1*:104 (1933).
49. I. Bergman, *Trans. Faraday Soc. 50*:829 (1954).
50. H. Leonhardt and A. Weller, *Z. Physik. Chem. [N.F.]* (Frankfurt) *29*: 277 (1961); D. Rehm and A. Weller, *Israel J. Chem. 8*:259 (1970).
51. U. Breymann, H. Dreeskamp, E. Koch, and M. Zander, *Chem. Phys. Lett. 59*:68 (1978).
52. G. A. Russel and S. A. Weiner, *J. Org. Chem. 31*:248 (1966).
53. G. P. Blümer and M. Zander, *Mh. Chem. 110*:1233 (1979).
54. M. Zander, *Naturwissenschaften 49*:300 (1962).
55. M. Zander and W. H. Franke, *Angew. Chem. Intern. Ed. 3*:755 (1964).
56. G. P. Blümer, K. D. Gundermann, and M. Zander, *Chem. Ber. 110*:269 (1977).
57. G. P. Blümer, K. D. Gundermann, and M. Zander, *Chem. Ber. 110*:2005 (1977); G. P. Blümer, Thesis, University of Clausthal, W. Germany, 1977.

2

Sampling, Extraction, and Analysis of Polycyclic Aromatic Hydrocarbons from Internal Combustion Engines

FRANK SEN-CHUN LEE* AND DENNIS SCHUETZLE / Ford Motor Company, Dearborn, Michigan

*Current affiliation: Standard Oil Company of Indiana, Naperville, Illinois

I. Introduction

Incomplete combustion of fuel in an internal combustion engine results in the formation of trace amounts of polycyclic aromatic hydrocarbons (PAH) along with other gaseous and particulate emissions. The concentration of exhaust PAH is a function of a number of complex factors including engine types, fuel and oil compositions, and engine operating conditions. Measurement of PAH from spark-ignition gasoline engines first appeared in the literature in the 1950s [1-3]. Hoffmann, Wynder, Begerman, and their co-workers reported a series of studies in the early 1960s on the measurement of PAH and their associated biological effects in gasoline engine exhaust [4-9]. Since these early studies, a number of investigators have reported on the analysis of PAH from a variety of engines under various operating conditions. To date, more than a hundred PAH compounds have been identified in vehicle exhaust.

In the past few years, exhaust measurements from diesel engines have received increasing attention because of the projected increasing usage of diesel engines in passenger cars. Concern over possible health effects associated with diesel particulates has prompted extensive studies on biological testing and analytical measurements. A number of new findings concerning exhaust PAH emissions have also been discovered from these investigations.

As recently as 1978, it was believed that the PAH themselves were primarily responsible for the biological activities observed for exhaust particulates in laboratory animal testing. In 1978, research by Huisingh et al. [10] and later by others using the Ames test [11-13] showed that most of the mutagenic activity in diesel as well as gasoline engine exhaust particulate extract is concentrated in chemical fractions other than those containing PAH.

The analysis of these non-PAH, biologically active fractions has since been under extensive investigation in a number of laboratories. These studies resulted in the identification of various oxygenated PAH derivatives in diesel particulate extract. Erickson et al. [14] reported the identification of alkyl-9-fluorenones. Chaigneau et al. [15] found in soot a compound with a molecular weight corresponding to nitronaphthalene and Rappaport et al. [16] identified cyclopenteno[c,d]pyrene dicarboxylic acid anhydride in a highly mutagenic fraction of a diesel particulate extract. Pitts et al. [17] and Lee et al. [18] reported the facile oxidation of PAH during air and exhaust sampling and analysis, as well as the finding of several oxygenated PAH compound classes. Schuetzle et al. [19] found that the most mutagenic fractions contain largely PAH derivatives, including compounds with hydroxy, ketone, quinone, carboxaldehyde, acid anhydride, dihydroxy, and nitro substituents on parent PAH. More recently, a number of highly mutagenic nitro-PAH species have been identified by Schuetzle [20], Gibson and Williams [21], and Tejada [22] in diesel particulate extracts.

The identification of oxygenated PAH is of crucial importance since some of these species, e.g., 1-nitropyrene, are known to be highly mutagenic and the biological activities of most of the others are still unknown. The ease of their formation from the oxidation of PAH also raises a question regarding the origin of these species, i.e., whether they are produced as "native" products during the engine combustion process or, instead, are formed as artifacts during sampling or analysis procedures. Determination of the extent to which these artifact formation problems may affect the results of subsequent chemical or biological studies also requires the development of analytical methods for oxygenated PAH measurements.

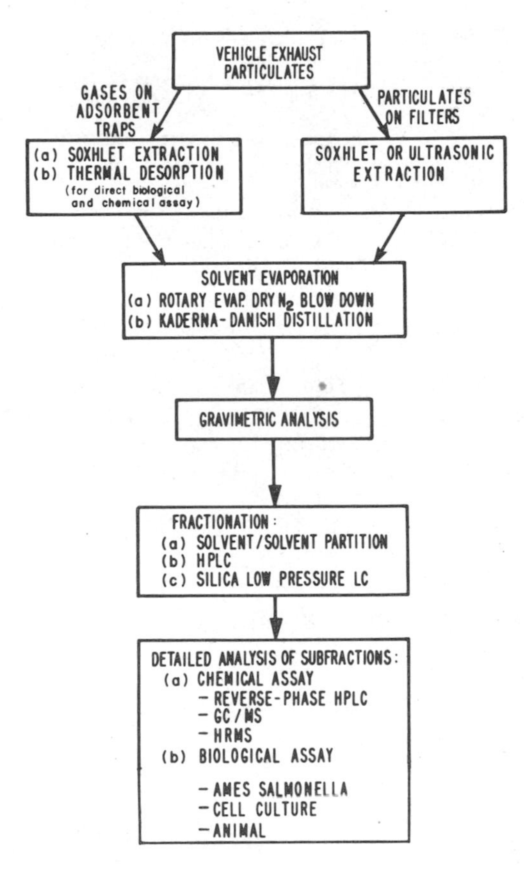

Figure 1. Scheme for the analysis of PAH and PAH derivatives in vehicle exhaust particulates.

Our emphasis in this chapter is placed on the review of current methods for the sampling and analysis of PAH and oxygenated PAH from vehicle exhaust emission. Quantitative comparison of PAH from different engine and fuel types is not included since such comparisons require a detailed discussion of the testing and engine operating conditions, which are outside the scope of this review. In Fig. 1 a schematic showing the typical analytical procedure is given, and in the following sections details of each analytical step are discussed individually.

II. Sampling

The emissions from vehicle exhaust include both gaseous and particulate phases. The exhaust temperature is very high, and a cooling step is therefore required before sample collection. Most of the PAH with five or more rings are expected to be particulate associated as the exhaust is cooled to near ambient temperatures. For the lighter PAH with two to four rings, however, concurrent sampling of both gases and particulates is required to ensure their efficient collection.

A. Exhaust Sampling Systems

Two types of approaches have been used for the sampling of exhaust PAH: the first involves the use of isokinetic sampling in an air dilution tube, whereas the second involves the sampling of raw exhaust after it has been passed through a cooling system. The vehicles are operated on a chassis dynameter with controlled speed and load, while regulated emissions such as hydrocarbons, carbon monoxide, carbon dioxide, and oxides of nitrogen are monitored [23-25].

Dilution tube sampling is the most widely used technique in the United States [23-27]. The technique is designed to simulate the dilution of vehicle exhaust to what would occur under real road conditions. Figure 2 gives a schematic for a typical dilution tube system used in this laboratory [28]. The system consists of an exhaust inlet tube, a dilution tunnel, and various downstream sampling probes and devices; it receives the exhaust as input and supplies a known and reproducible quantity of filtered ambient air for rapid dilution and cooling of the exhaust. The mixed stream flows at a measured velocity through the dilution tube before it is withdrawn from the tube isokinetically into the sampling devices.

A direct exhaust sampling method has also been used by Grimmer et al. [29] and others [9,30] for exhaust PAH collections. The exhaust is first

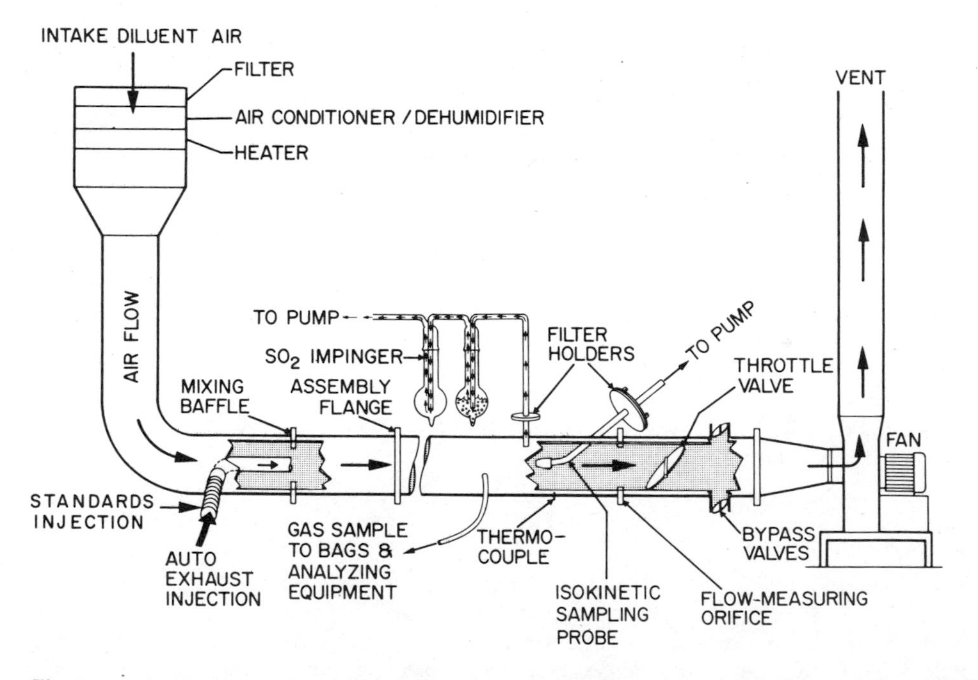

Figure 2. Diagram of a vehicle exhaust dilution tube. (Modified from Ref. 28.)

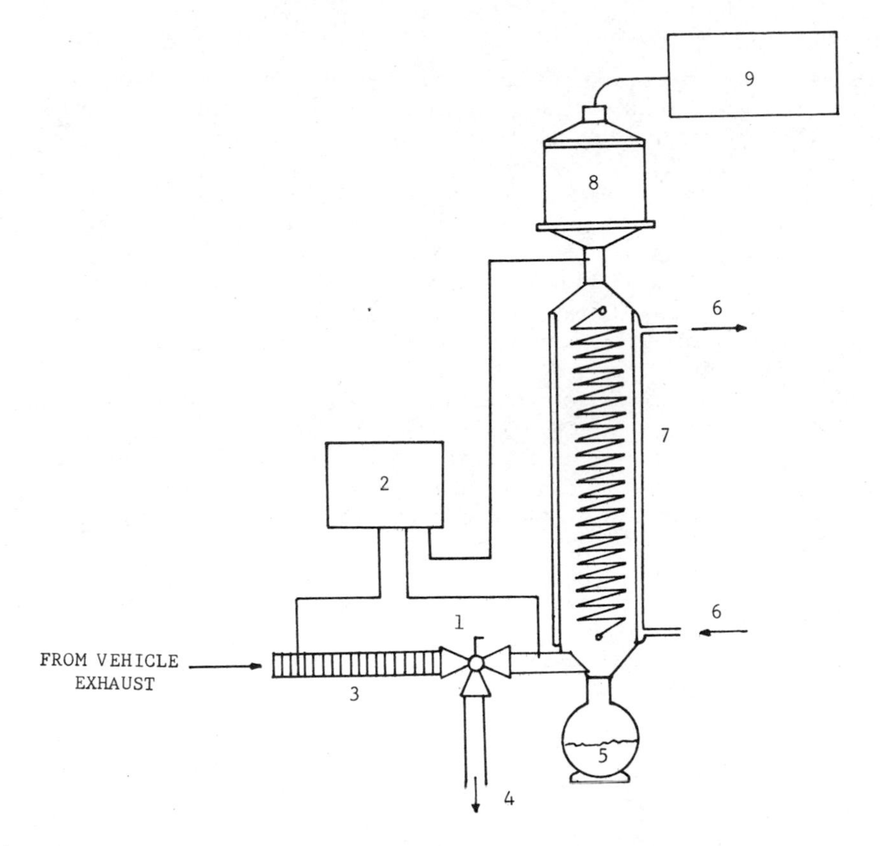

Figure 3. Condensation collection system for vehicle exhaust emissions: (1) valve, (2) thermocouple gauges, (3) connecting pipe, (4) waste pipe, (5) condensed water, (6) cooling water, (7) glass cooler, (8) filter, (9) CVS unit. (Modified from Ref. 30.)

cooled in a glass device with a large cooling surface (up to few cubic meters). The cooled exhaust is then allowed to pass through a filter for particulate collection. Both the condensed material in the cooler and the filter-collected particulates are analyzed for PAH, without any attempt to distinguish between the two phases. Figure 3 shows a collection system designed for the direct sampling of undiluted exhaust [30].

B. Particulate Sampling

Particulates emitted from vehicles may be collected using filters or electrostatic precipitators [31]. Filter collection is the most widely used method. The use of impingers and impactors is not suitable because of their poor efficiency for collecting submicrometer particles generated from internal combustion engines.

1. Filter Collection

Various types of filter media have been used. Teflon membrane filters are preferred because of their high collection efficiency and inertness. Selection of filter types will be discussed further in a later subsection.

The weight of total particulates should be measured in order to determine the concentrations of PAH. Prior to collection the filter is conditioned by allowing it to equilibrate about 24 hr in a desiccator filled with Drierite or in a constant-humidity room (~45% relative humidity) and weighed to the nearest 0.01 mg before use. If the relative humidity in the weighing area is high (>50%), it is desirable to equilibrate the filter for 24 hr in a desiccator containing a saturated solution of calcium nitrate to maintain a rough humidity control at approximately 50%.

After collection, the filters are placed in petri dishes and allowed to equilibrate in the dark under the same conditions as those under which the filter was initially weighed. For longer term storage the petri dishes are wrapped with aluminum foil and stored at temperatures below 0°C. The use of subambient temperature for prolonged storage is recommended in order to minimize losses due to evaporation and degradation.

Loss due to evaporation of the more volatile components of the particulates has been observed in an experiment whereby the weight of a diesel-particulate-laden filter stored at room temperature was monitored for a 17-day period beginning within hours after collection. A continual decrease in weight occurs throughout the monitoring period, leading to a weight loss of 4.8%

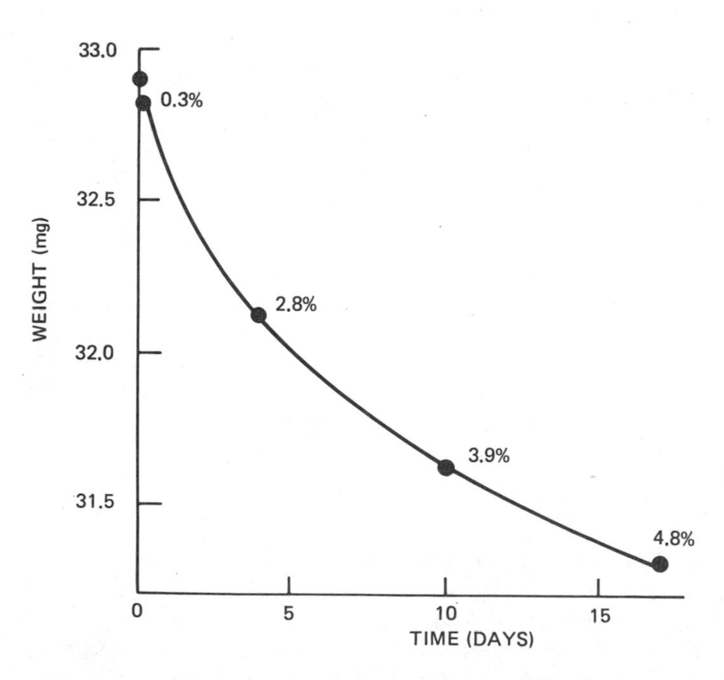

Figure 4. Loss of weight for a filter-collected diesel particulate sample as a function of time (percentage refers to percentage loss of the original weight).

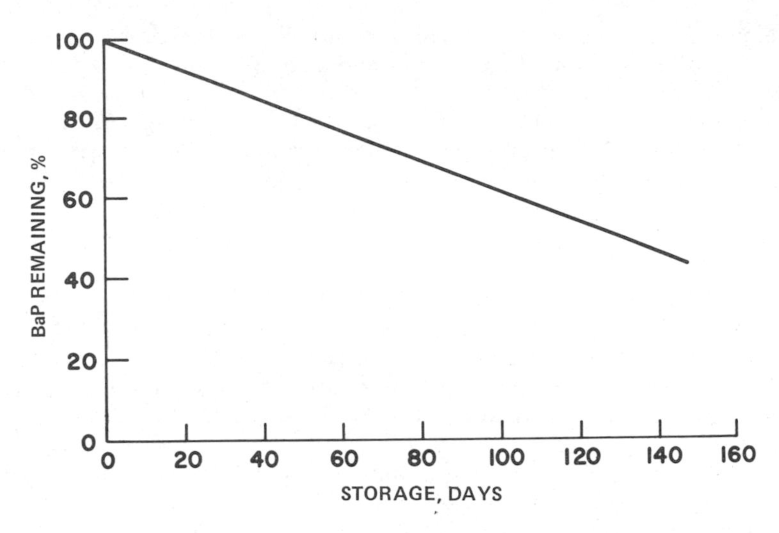

Figure 5. Effect of filter sample storage on BaP concentration. (From Ref. 32.)

of the original sample after 17 days (see Fig. 4). Such loss can probably occur for some PAH compounds, especially the lighter ones such as phenanthrene or anthracene.

Significant reaction loss of PAH may also occur. In a recent report by Swarin and Williams [32] it was found that a slow but consistent loss of BaP in diesel particulates during storage was observed (57% in 150 days). Their results are reproduced in Fig. 5. In a paper by Korfmacher et al. [33], it was reported that PAH containing benzylic carbon atoms such as 9,10-dimethylanthracene and 9,10-dihydroanthracene, fluorene, etc. undergo rapid degradation reactions (in the absence of light) when adsorbed on fly ash or charcoal. The observation of facile, non-photolytic conversion of PAH to corresponding oxygenated products from the results cited here and those discussed in later sections shows that long-term storage should be avoided whenever possible. When such storage is necessary, the sample should be stored at dry-ice temperatures in the dark in order to minimize PAH degradation.

2. PAH Degradation During Filter Collection

The possible degradation of PAH during filter sampling (sampling artifact) was first pointed out by Pitts et al. [17]. In their study, it was found that BaP, perylene, and other PAH undergo facile chemical transformation as they are exposed to gaseous pollutants, e.g., NO_X, O_3, or PAN, during ambient air sampling. In a similar study by Lee et al. [18,34] it was reported that conversion of BaP and benz[a]anthracene (BaA) occurred readily during filter sampling of diesel exhaust as well as ambient air. In Fig. 6 the effects of diesel exhaust gases on [^{14}C]BaP are illustrated. In this experiment, the [^{14}C]BaP was spiked on filters preloaded with diesel particulates. The spiked filter was then exposed to exhaust gases in the dilution tube for 15 min. A double-filter technique was employed in which an upstream blank filter was placed before the particulate-laden filter, thus allowing only gases passing through the spiked downstream filter.

Figure 6 shows that of all extracted radiocarbon activities, only 36% and 75 ± 5% were recovered as ^{14}C-BaP on quartz fiber and Fluoropore filters, respectively, after exposure to diesel exhaust. Much more loss of ^{14}C-BaP occurred in the former case due probably to the more reactive nature of the quartz surface in catalyzing oxidation reactions. More discussion on the filter effect is to be given in the next section.

The ease of PAH degradation during sampling is not too surprising in view of the reactive nature of some of the PAH species. It has been known for some time that PAH can be readily oxidized by O_3 [35]. In Pitts' work discussed earlier [17], it was shown that PAH can also be readily nitrated by flowing air spiked with ppm levels of NO_2 (accompanied by trace HNO_3) over the PAH preadsorbed on a glass fiber filter.

The vehicle exhaust is rich in various oxidant gases among which the more abundant ones include sulfur as well as nitrogen oxides. In addition to nitrogen oxides, sulfur oxides were also found to react with BaP adsorbed on the filter. Such an effect is illustrated in Fig. 7. In this experiment, a known amount of BaP (2 μg) was spiked directly on the surface of a 47 mm glass fiber filter. The BaP spiked filters were then each exposed to air alone (control run) or air doped with 2 ppm SO_2 or sulfuric acid aerosols for 15 min with a flow rate of 3 ft^3/min. The relative humidity was controlled at 45%.

It may be seen from Fig. 7 that the recovery of BaP remains the same with or without the presence of SO_2 in air. Extensive additional reaction loss was

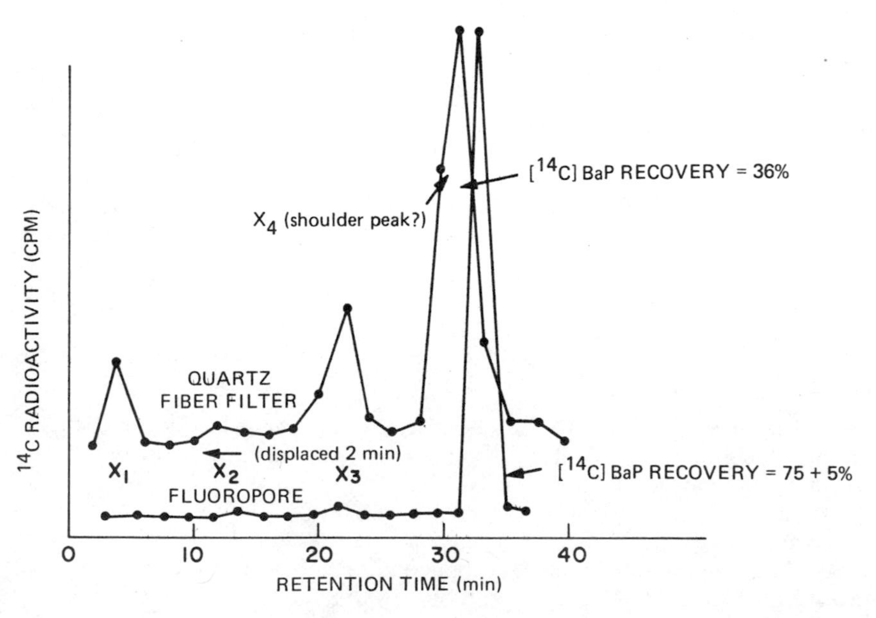

Figure 6. The effect of diesel exhaust gases on 0.5 μg of [^{14}C]BaP spiked on quartz fiber and Fluoropore filters preloaded with diesel particulates.

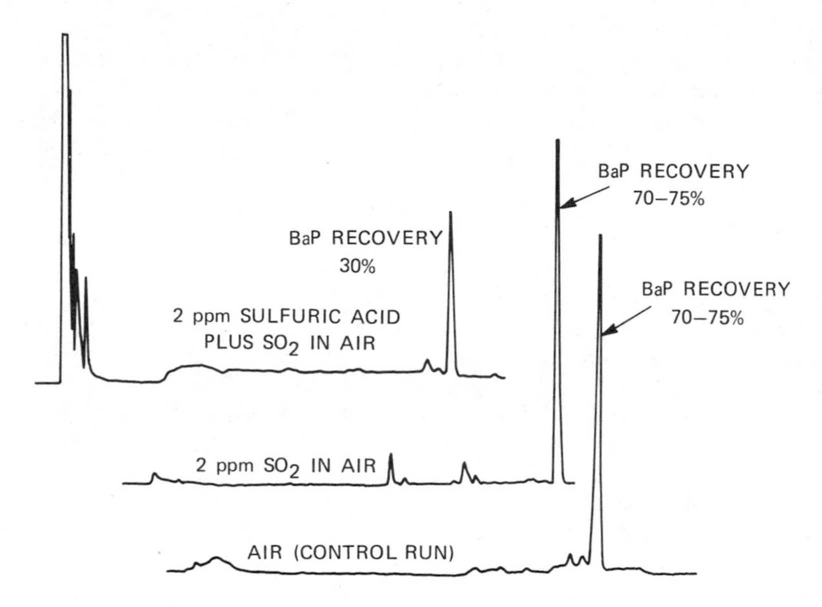

Figure 7. Reverse-phase HPLC chromatogram showing the recovery of 2 μg BaP on a 47-mm glass-fiber filter after being exposed to 2 ppm sulfuric acid aerosol or sulfur dioxide. Sample extracted with dichloromethane and analyzed on a 5 μm HI-EFF Micropart C_{18} column: linear gradient of 40% CH_3CN in water to 100 percent CH_3CN in 50 min.

observed, however, as the BaP was exposed to air carrying trace amounts of sulfuric acid aerosols (30% here vs. 70-75% recovery for the air exposure run). The degradation products, as shown in the figure, were not further analyzed.

It is evidenced from the discussions above that PAH react readily with nitrogen oxides and sulfur oxides during typical exhaust sampling procedures where a filter-collection technique was employed. However, in all of the aforementioned studies, the gas exposure experiments were carried out in model systems where PAH were spiked either directly on the filter surfaces, or on particulates already collected on the filters. While these experiments indicate strongly the probability of PAH degradation during sampling, whether and to what extent these reactions also occur for native PAH embedded in the diesel particulate matrix remained a subject of controversy.

The foregoing question was later addressed in a number of studies using a reexposure technique in which filter-collected particulates were reexposed to diesel exhaust gases. A double-filter arrangement similar to those described earlier was used. The PAH concentration on the particulates were then compared after exposure for a different time period.

In an extensive study of diesel exhaust PAH, Petersen et al. reported [36] that no discernible systematic decrease in PAH concentration was observed with increasing exposure time for each of the five PAH tested—pyrene, chrysene, BaA, BaP, and perylene. Their data for BaP, for instance, were measured as 2.9, 2.0, 1.7, and 2.3 ppm (wt/wt, BaP/particulates) after respectively, 0-, 5-, 10-, and 15-min exposure. The same type of experiments were also reported by Swarin and Williams [32]. In these experiments only BaP was measured, and the investigators reported that a constant BaP concentration was observed comparing unexposed and 10-min-exposed runs. The same type of experiment was repeated by Gibson and colleagues in their later work searching for NO_2-PAH [21]. Here, however, it was reported that BaP and pyrene indeed react with nitrogen oxides in the exhaust to form 6-nitro-BaP and 1-nitro-pyrene, respectively, and a decrease in PAH and an accompanying increase in NO_2-PAH were observed as the particulates were reexposed to exhaust gases in a time period of 86-92 min. In addition, an increase in Ames mutagenicity was also noted when reexposed diesel particulates were compared with unexposed ones. Similar results have been reported by Schuetzle [37].

The results obtained from these studies illustrate the difficulties involved in this type of experiment. The fact that hundreds of PAH are present in the particulates, each with varying reactivities toward the oxidant gases, makes it difficult to quantitatively assess the magnitude of such PAH oxidations merely from the analysis of minute concentrations of few PAH species. The extent of such PAH degradations most probably also depends on the nature of the particulate. The latter, being a complex function of the engine and operating conditions, would make sample-to-sample comparison still more difficult.

It would seem reasonable to assume that PAH adsorbed or encapsulated in particulates are more stable toward oxidation than pure PAH because here the PAH are less accessible to oxidant gases. In Korfmacher's study [33] referred to earlier, it was reported that such stabilization effects were observed for some PAH but rapid reactions were still observed for other PAH containing benzylic carbons such as fluorene, benzofluorenes, 9,10-dimethylanthracene, and 9,10-dihydroanthracene.

It should also be noted that the sampling artifact problem is a question not only of analytical recovery but also of whether significant biologically active conversion products, e.g., NO_2-PAH, are produced. Thus, care should be taken to minimize the occurrence of these reactions to the least possible extent. In addition to reduced sampling exposure time, the selection of inert filters is also recommended. The latter will be discussed next.

3. *Selection of Filter Media*

Two considerations should be given in selecting the filter media: the first is the efficiency of the filter in collecting the particles, and the second is the inertness of the filter media. Both factors have been evaluated in a number of studies as described below.

The types of filters commonly used for the sampling of airborne particulate matter are listed in Table 1, along with descriptions of their material, structure, and names of manufacturers. The measured collection efficiencies of these filters are also listed.

The absolute efficiency of a filter depends on a number of factors including the flow rate, the pore size distribution of the filter, and the sample loading [38]. In Table 1, the efficiencies measured for these filters by Walter and Reischl [39] for submicrometer aerosols and Truex [40] for diesel particulates are listed. For our purpose, the details about how they arrive at the effi-

Table 1. Selected Filter Types and Their Measured Efficiencies

Filter identification	Manufacturers[a]	Material & structure	Face velocity (liters/cm^2 min)	Efficiency[b] (%)	Ref.
Gelman A	a	Glass filter type A	1.61-4.09	>99.9	39
			0.41-1.87	94.1-99.2	40
Tissuquartz (2500 QAO)	b	Silica (quartz) fibers	—	—	—
T60A20	b	Micro glass fibers with Teflon binder	0.41-1.87	65.9-99.3	40
Fluoropore (1 μm)	c	PTFE membrane bonded to polyethylene net	0.27-7.91	>99.9	39
Zefluor (0.2-4 μm)	d	PTFE membrane supported by PTFE fiber	0.27-7.91	>99.9	39
Nuclepore (0.8 μm)	e	Polycarbonate	0.81-4.09	72-89	39

[a]Manufacturers: (a) Gelman Instrument Co., Ann Arbor, Mich.; (b) Pallflex, Putnam, Conn.; (c) Millipore, Bedford, Mass.; (d) Ghia Corp., Pleasanton, Cal.; (e) Nucleopore, Pleasanton, Cal.

[b]Ref. 39 measures efficiencies for collecting submicrometer aerosols; Ref. 40 measures efficiencies for collecting diesel exhaust particulates.

ciency calculations and measurements will not be discussed here. The face velocities (flow rate) used in their measurement are included in the table. These flow rates fall in the range of those typically used in exhaust particulate sampling.

Among all the tested filters, close to 100% efficiencies were found for Gelman A glass fiber, Tissuquartz, Fluoropore, and Ghia filters. The efficiencies for T60A20 and Nuclepore, however, are considerably lower. At no or light loadings, as high as 30-35% of the particulates may penetrate these filters.

The T60A20 filter has been used widely in a number of laboratories for particulate sampling. The filter is made from glass fiber impregnated with PTFE in order to eliminate the surface reactivity of the glass fibers. The filter was found to cause less PAH degradation problems compared to the glass fiber filters, as will be seen later in this section. Compared to the Teflon membrane filters, it also had the advantage of causing less pressure drop across the filter during sampling. Its inefficiency, however, makes it undesirable for the quantitative exhaust emission work.

The effect of filter media on the chemical integrity of the collected particulates has been reported recently [34]. It was observed that glass fiber and Tissuquartz filters could catalyze the oxidation of PAH. An example of such an effect was shown previously in Fig. 6.

The effect of filter media on BaP recovery reported in the aforementioned references is graphically illustrated in Fig. 8. The experiments were carried out under typical air-sampling conditions using the double-filter arrangement. The recoveries of [^{14}C]BaP spiked directly on the filter surfaces after sampling varying amounts of air are compared. The percentage recoveries plotted in the figure are calculated from the recovered [^{14}C]BaP relative to the total recoverable radiocarbon activity. The latter was generally in the range of 80-85% relative to the total ^{14}C-activity originally spiked.

It is seen that pure Teflon filter causes very little conversion of additional [^{14}C]BaP after the initial sharp loss (from 90 to 60%). Contrary to this, both glass fiber and Tissuquartz filters show severe [^{14}C]BaP conversions in the beginning, and an increasing amount of [^{14}C]BaP was decomposed as the volume of air passing through the filter increases. Recovery of [^{14}C]BaP with T60A20 is lower than with Fluoropore but considerably higher than with the other two filters.

The mechanisms of such PAH reactions on the filter surfaces are not understood. The surface properties of the filter media such as specific surface area, water and vapor affinity, and the polar nature of the surface are likely to play part of the role in filter reactions with PAH. The polar functionalities associated with glass filters also enhance their ability to adsorb or convert reactive gases during sampling. These adsorbed species may also react with PAH. The high adsorbility of glass filters compared to that of Teflon filters was demonstrated by Lee et al. [34].

The initial sharp loss of [^{14}C]BaP in all the filters, as shown in Fig. 8, is probably due to the presence of "reactive" sites on the "fresh" filter surfaces. In case of Fluoropore, these residue reactive sites were quickly removed and further air passage did not cause noticeable additional oxidation of BaP. Contrary to this, both glass fiber and Tissuquartz caused continuous [^{14}C]BaP oxidations until the complete depletion of [^{14}C]BaP. The T60A20 filter gave results somewhere between the two cases.

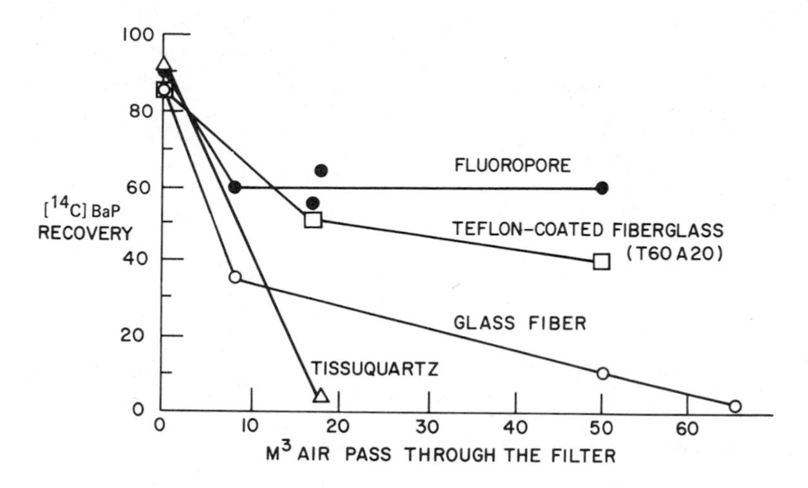

Figure 8. Percentage recovery of [^{14}C]BaP spiked on blank filters after sampling prefiltered urban air.

The foregoing discussion shows that Teflon membrane is superior to all the other filters tested in terms of efficiency as well as chemical inertness. The difficulty of high pressure drop of the filter can be overcome through the selection of large pore sizes. As was shown in Table 1, ⩾99% collection efficiency can still be achieved by these filters even with pore size in the range of 2-4 μm. The filter thus behaves somewhat like an impactor for the trapping of submicrometer particles. In this laboratory, Zefluor filters were used routinely for the collection of diesel particulates. This filter is made from the same Teflon membrane material as Fluoropore but is much easier to handle because of the use of hard Teflon fiber as the membrane backing material.

C. Gas Sampling

Polymeric absorbents have been used widely for the sampling of gas-phase PAH in air and stationary-source emissions. The commonly used polymer materials include Tenax, Chromosorb 102, and XAD-2, and their applications have been recently reviewed by Pellizzari et al. [41] and Strup et al. [42].

Very little work has been reported on the application of polymeric absorbents for the sampling of gaseous PAH in the exhaust. Spindt [43] reported the use of a Chromosorb 102 trap downstream of a filter for the sampling of diesel exhaust gases. No BaP was observed in the trap, but other PAH species were not determined.

We have used a high-volume XAD-2 trap (6 in. diameter × 2 in. long) for the collection of gas phase PAH emitted from a gasoline engine (Fig. 9). Before use, the XAD-2 polymer beads (20-50 mesh) were washed first with distilled water to remove inorganic impurities. They were then washed with methanol followed by dichloromethane to remove the organic contaminants [42]. To further prevent the adsorption of impurities from air, the polymer beads

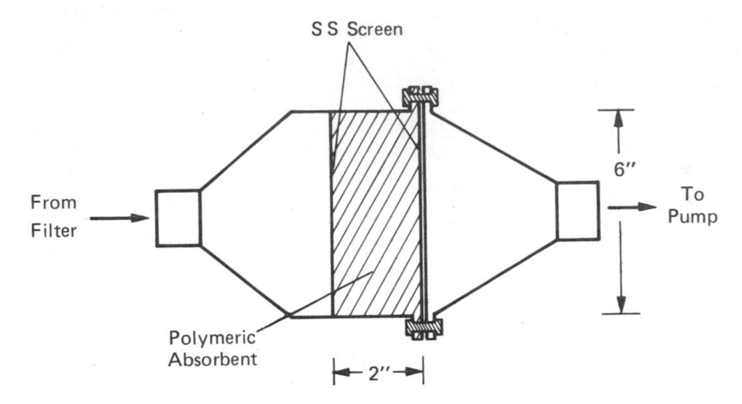

Figure 9. Adsorbent trap for the collection of gas-phase PAH from vehicle exhaust.

were stored in methanol solvent until use. It was found that a 2-3 in. deep XAD-2 bed was sufficient for the collection of three- and four-ring PAH with high efficiency even at a flow rate as high as 30 ft^3/min.

After sampling, the XAD-2 beads were placed in a large thimble made from stainless steel screens and extracted with dichloromethane in a Soxhlet apparatus. The analytical recovery of 14 spiked three- to seven-ring PAH averaged 82%, indicating a reasonable extraction efficiency for the procedure. Besides Soxhlet extraction, the trap can also be flushed, in a reverse gas-sampling direction, with dichloromethane. A solvent volume two to three times the volume of the polymer beads should be used to ensure the quantitative elution of the adsorbed organics.

Table 2 gives the distribution of PAH in the particulate and gas phase from a stratified-charge engine [44]. It is seen that substantial amounts of light three- and four-ring PAH indeed exist in the gas phase. No heavier PAH ($\geq$ five rings), however, were found in the polymer traps under the conditions studied.

III. Extraction of Particulates

The soluble organic fraction (SOF) of the particulates which contain PAH is generally separated from the carbonaceous or inorganic exhaust particulates by solvent extraction procedures. Extraction may be carried out using Soxhlet apparatus or by ultrasonic agitation. The qualitative and quantitative nature of the extracted organics is influenced by the types of solvent used and the temperature and the time used for extraction. The procedures and their quantitative comparisons will be discussed in the following subsections. Gaseous PAH collected on porous polymer traps are usually separated by solvent extraction procedures previously described in Sec. II.C.

Table 2. Distribution of PAH in the Particulate and Gas Phase from a Stratified-Charge Engine

PAH	Micrograms/sample	
	Filter	XAD trap
Phenanthrene	40	16
Anthracene	8	30
Fluoranthene	34	30
Pyrene	36	40
Chrysene + benz[a]anthracene	70	50
Benzo[e]pyrene	28	0.1
Benzo[a]pyrene	9	0.1
Benzo[ghi]perylene	31	0.2
Picene + dibenzoanthracenes	9	0.2
Anthanthrenes	13	0.2
Dibenzopyrene	3	0.2
Coronene	4	0.2

Source: Ref. 44.

A. Soxhlet

Soxhlet extraction is the most widely used technique in particulate extractions. The size of the apparatus should be adjusted according to the size of the filter used for particulate collection in order to minimize the solvent used. Generally, Soxhlet thimbles are not used in order to minimize possible contaminations. Instead, particulate-laden filters are folded with the particulates sheathed inside the wrapped filter and then placed directly inside the extraction chamber. This procedure can be used for glass fiber or Teflon filters for which the material is strong enough to remain unshredded during extraction. For more fragile filters such as Tissuquartz, it can be wrapped in another blank Teflon filter, e.g., Ghia Teflon filter (see Table 1), to avoid the shedding of particulate through the shredded filter into the extraction solution.

Normally, it was found that a 12-hr extraction time with a solvent cycling rate of 2-4 cycles/hr is able to extract on the average more than 85% of the PAH spiked on either diesel or gasoline engine particulates and more than 95% of the extractable material. To further enhance the efficiency of extraction, we have employed a modified Soxhlet extractor in this laboratory. The apparatus is equipped with an additional sealed glass jacket outside the extraction chamber whereby the heated vapor in the outside jacket would keep the extraction chamber at a temperature near the boiling point of the solvent. The apparatus should be wrapped in aluminum foil in order to avoid photodecomposition. It is known that PAH are susceptible to near-ultraviolet

(UV) light irradiation, and even exposure to fluorescent lamps used for laboratory illumination causes considerable decomposition [45].

As in any environmental trace analysis, care must be taken in avoiding the introduction of contaminants during extraction procedures. Robertson et al. [46] in a recent article described possible contaminations from filters and solvents in PAH analysis by HPLC. The selection and evaluation of solvent purity is of critical importance in exhaust PAH analysis because of the small sample size normally involved.

B. Ultrasonic Agitation

Ultrasonic agitation has been used for the solvent extraction of air particulates [47-49]. Using cyclohexane as an extraction solvent, Golden and Sawicki [49] reported that sonication was able to extract 15% more PAH from air particulates compared to the Soxhlet method.

We have applied the ultrasonic agitation method for the extraction of exhaust particulates. In our procedure, sonication and vacuum filtration for the separation of extract from particulates and shredded filter pieces are combined into one operation. This is achieved by carrying out the sonication step directly in a Millipore glass filter-funnel apparatus. The particulate-laden filter is first cut into strips and placed into the filter funnel. The sonication probe (Branson sonicator from Heat Systems Ultrasonics) is then placed slightly above the filter strips in the glass funnel. The vacuum used for filtration is then turned on, and the selected solvent is added to the filter funnel. The sonication probe is initially operated at relatively high frequencies. A few seconds after sonication has begun, the particulates attached on the filter will fall off and the extract will become opaque black because of the suspended particulates. At this point, the sonication probe frequency is lowered. Whenever part of the solvent is removed by filtration from the funnel chamber, additional solvent is added. This replenished solvent ensures the dilution of solvent extract, resulting in efficient partition of PAH and other organics. With each replenishing operation the probe frequency is increased for a few minutes to ensure mixing. The total time for the procedure ranges from 0.5 to 1.5 hr, depending on the filtration speed, which varies with the nature of the particulates and the viscosity of the solvent. The vacuum used for filtration can be adjusted to control the speed of filtration. During sonication, the temperature in the sonication vessel may rise rapidly. In our procedure, such temperature rise is offset by the cooling effect through vacuum filtering processes and the continuous adding of fresh solvent. The temperature rise is generally found to be less than 5°C.

In the filtration method described above, a 0.2- to 0.5 μm pore size Teflon membrane filter (Millipore or Ghia) was normally used. Ghia pure Teflon filter is favored since the polymeric backing materials used in Millipore filters dissolve partially in chlorinated solvents.

C. Extractable Mass

The mass of extracted material (SOF) after Soxhlet extraction may be determined by both a weight difference of the filter before and after extraction and by weighing the extract after evaporation of the solvents. Only the later method can be used for sonication extractions because it is difficult to weigh the shredded filter pieces.

The extraction solutions are evaporated to dryness at reduced pressures on a rotary evaporator. Chloroform is used to rinse and transfer the residue in the flask to preweighed (0.01 mg) 5.0-ml Wheaton minivials. Further concentration to dryness in the vials is accomplished by nitrogen-gas blowdown. The extracts in preweighed minivials are allowed to come to total dryness overnight in the constant humidity room. The sample vials are then weighed every 2 to 4 hr until constant weight (±0.05 mg) is established. This weight divided by the weight of total particulates for the given filter fraction yields the weight percentage of total extractable material for a given solvent.

The introduction of chemical contaminants may occur at any point in the analytical step. Common contaminants include silicon, phthalate esters, saturated hydrocarbons, and fatty acids. These contaminants do not usually represent more than 1 or 2% of the total extractable mass. However, the presence of certain contaminants (i.e., silicon and phthalate esters) greatly complicates the chromatographic and spectroscopic analysis for chemical constituents. Thus, for instance, it was our experience that contamination may occur not only in glassware but also during N_2 blowdown steps. Possible contamination due to cylinder gases used for solvent evaporation has recently been reported in detail by Nestrick and Lamparski [50]. Thus, it is recommended that only high-purity N_2 should be used. In addition, Teflon tubing should be used throughout the gas connecting and transfer lines.

After extraction and weighing, the SOF should be purged with an inert gas and stored in sealed vials at dry-ice temperatures in order to minimize sample degradation.

D. Extraction Efficiency

The extraction efficiency of airborne particulates depends on the physical and chemical nature of the particulates as well as the extraction procedure. The particulates emitted by leaded-gasoline engines consist of lead or other inorganics in addition to carbonaceous particles. The PAH and other extractable organics can be removed readily since most of the organics are coated on the surface of the particulates through condensation. Different from this, diesel particulates consist of mostly carbonaceous particles in the submicrometer range. The efficiency of extracting total SOF and PAH needs to be evaluated because of the possible existence of strong physical or chemical sorptions due to the large particulate surface areas available.

The extraction efficiencies for diesel particulates have been studied extensively in this laboratory and several others. In general, it has been found that although the total mass of SOF can be efficiently extracted, longer extraction time and proper selection of solvents are needed to ensure the quantitative removal of PAH.

The mass of total SOF extracted as a function of time was studied by Seizinger [51], and his data are presented in Table 3. It may be seen that 99% of the extractable material can be effectively extracted in 8 hr, and further prolonged extraction up to 24 hr resulted in the extraction of only ~1% additional material. In this laboratory, sequential extractions with various solvents have been routinely used to test the extractability of diesel particulates. Typical data are shown in the accompanying tabulation [52].

	Percentage of extractable mass	
	Sequential extraction with toluene/n-propanol, methylene chloride, and o-dichlorobenzene	Toluene/n-propanol (50:50)
Run 1	15.1	14.1
Run 2	42.2	43.4

Within the experimental errors, it is seen that exhaustive sequential extraction does not seem to extract significant amount of additional mass.

The extraction efficiency for PAH has been studied using [^{14}C]BaP spiked particulates [18]. In these experiments, a known amount of [^{14}C]BaP was spiked on the particulates collected on the filter. The spiked particulates were then allowed to equilibrate for 48 hr in the dark. The recovery of total radiocarbon activity after Soxhlet extraction with methylene chloride (~12 hr) followed by acetonitrile (~12 hr) are shown in Table 4. The recovery of [^{14}C]BaP from gasoline engine particulates is nearly quantitative, but small loss compared to the blank filter control runs was observed in ambient air and diesel particulates.

The recovery of native particulate-associated BaP versus Soxhlet extraction time was studied by Seizinger et al. [53]. Their data are shown in Table 5. Here it is seen that significant amounts of BaP were still detected after 4.5- to 6-hr extraction. It seems that some of the PAH are much more strongly adsorbed on the particulates than are others. Such sorption has been observed for polymeric carbons [54] and fly ash [55]. Thus, while the spiking experiment suggests that probably most of the surface-adsorbed PAH can be extracted effectively, a portion of the strongly adsorbed PAH may not be readily extractable and the selection of strong solvents and the use of longer (12-14 hr) extraction time are required.

We have studied and compared the relative extraction efficiencies of 11 commonly used solvent systems for the extraction of diesel particulates.

Table 3. The Relationship Between Soxhlet Extraction and Mass Percent of Solvent Extractable Material from a Diesel Particulate Sample[a]

Extraction time (hr)	Cumulative percentage extractable
1.0	77.9
2.0	93.0
3.0	97.0
8.0	99.0
24.0	100.0

[a]Soxhlet cycle time is once per hour.
Source: Data from Ref. 51.

Table 4. Total Radiocarbon Activity Recovery for [^{14}C]BaP Spiked on Particulate Matter after Sequential Soxhlet Extraction with Methylene Chloride Followed by Acetonitrile

Particulates	Total percentage recovery (No. of determination in parentheses)
Ambient air	80 ± 5 (4)
Diesel engine	85 ± 4 (3)
Gasoline engine (Proco)	88 ± 4 (6)
Blank filter[a]	92 ± 4 (6)

[a]Average recovery of [^{14}C]BaP spiked on various types of filters.

Table 6 gives the relative extraction efficiencies of these solvents employing a room temperature sonication method. In these [^{14}C]BaP recovery experiments, a known amount of [^{14}C]BaP was spiked onto the particulates collected on the filter. The spiked [^{14}C]BaP were then allowed to equilibrate overnight with the particulate-associate organics. After extraction, the total recovered activity was measured and percentage recovery was calculated relative to the original [^{14}C]BaP activity spiked.

It was reported that the total extractable material from ambient air particulates depends greatly on the dipole of the solvent used [56]. From Table 6 it may be seen that the solvent polarity and the presence of aromatic solvents are both important factors affecting the extraction efficiency of PAH and total extractable mass. The inclusion of an alcohol in the aromatic system, i.e., toluene/propanol and benzene/ethanol, increases the total mass extraction efficiency. However, it appears that this additional mass is due to the increased extraction for sulfates and other inorganics as well as polar organics. Halogenated solvents, although known to be excellent universal solvents, are not as efficient as the aromatic systems. Both chloroform and methylene chloride give 20% less SOF mass than do benzene and toluene. In addition,

Table 5. The Relationship Between Soxhlet Extraction Time and Mass of Benzo[a]pyrene Recovered from Two Diesel Particulate Samples

Extraction time (hr)	BaP recovered Sample 1 (ng)	BaP recovered Sample 2 (%)
1.5	389	10.7
3.0	108	6.9
4.5	71	5.6
6.0	72	5.4

Source: Data from Ref. 53.

Table 6. Efficiency of Solvents for the Extraction of Diesel Particulates

	Polarity index (P_i)[a]	Relative efficiency[b]		
		Total extractable mass	$[^{14}C]$BaP[c]	SO_4^{2-}
Toluene	2.3	1.0 1.0	1.0	1.0 N.D.[d]
Xylene	2.3	1.0	0.98	N.D.
Acetone/hexane (50:50)	2.7	0.8	0.36	0.3
Toluene/propanol (50:50)	3.3	1.1	0.95	7.3
Benzene/$MeCl_2$/MeOH (25:10:1)	3.2	1.1	0.94	N.D.
Benzene		—	0.92	N.D.
$MeCl_2$				
Diesel	3.4	0.8	0.87	1.5
Air particulates	—	—	0.73	N.D.
Benzene/EtOH (80:20)	3.4	1.1	0.84	4.2
$MeCl_2$/MeOH (97:3)	3.5	0.8	0.72	2.7
Cyclohexane	—	—	0.58	N.D.
Acetone	5.4	0.6	0.36	1.0

[a] $P = 0.01 \Sigma_i \%_i P_i$; $\%_i$ = % vol. of solvent i; P_i = polarity indices of pure solvents i (see Ref. 56).
[b] An average of three determinations for each solvents. Values are normalized to toluene = 1.0.
[c] Each sample spiked with 400 ng $[^{14}C]$BaP.
[d] N.D. = not determined.

chloroform and methylene chloride give lower recoveries for BaP. The results are in accordance with the published data from ambient aerosol studies in which it is generally observed that benzene is superior to other solvents in extracting total organics.

The extraction efficiency depends very much on the nature of the particulates. For instance, it was reported that a polar solvent such as methanol is very efficient in the extraction of PAH or other organics from ambient air particulates [57]. This is different from what was observed for diesel particulates. Here, it is seen that generally more polar solvent systems are actually less efficient in extracting $[^{14}C]$BaP. Presumably, the chemical nature of the adsorbed organics, i.e., a high percentage of nonpolar hydrocarbons in diesel particulates (see Sec. IV) and a high percentage of polar oxygenates in air particulates, plays an important role in determining the efficiency of the solvent used. The seemingly dependence of BaP extraction on solvent polarity is revealed from the data in Table 6. Thus, it seems that a systematic decrease in $[^{14}C]$BaP recovery with increasing polarity exists for aromatic/halogenated solvents. The two exceptions are acetone/hexane systems for which the efficiencies are lower for both [^{14}C]BaP and total extractable material.

Figure 10 compares the HPLC profiles of a diesel extract from two solvent systems. A reverse-phase column was used in which the compounds are eluted in the order of decreasing polarity. The chromatograms display considerable variations in the polar region (0-15 min elution), but close similarity in the moderate polar (γ_1, γ_2, and β: 15- to 23-min elution) and the PAH regions (α_1, α_2: 23- to 45-min elution).

Table 7 shows the effect of extraction solvents on the total fluorescence area of the PAH region (α_1, α_2), which should approximate the total PAH and other nonpolar organics extracted. Because of the approximate nature of the measurement, the total areas for these solvent extracts do not seem to follow exactly those shown in Table 6 for either total mass extractable or [^{14}C]BaP. However, it is obvious from the data that these solvents extract, within some 10-20%, about the same PAH and nonpolar organics. Similar results were observed for the γ_1, γ_2, and β fractions. Contrary to this, the polar materials extracted (δ) are quantitatively as well as qualitatively dependent very much on the solvent selected, as evidenced by the variations in the δ peaks shown in Fig. 10. The chemical nature of this polar fraction is not well known. It probably includes both highly oxygenated polar organics and heterocyclic species. We have indeed observed noticeable concentrations of sulfate and possibly also nitrate species in some of these extracts using ion chromatographic analysis. As seen from Table 6 the relative extraction

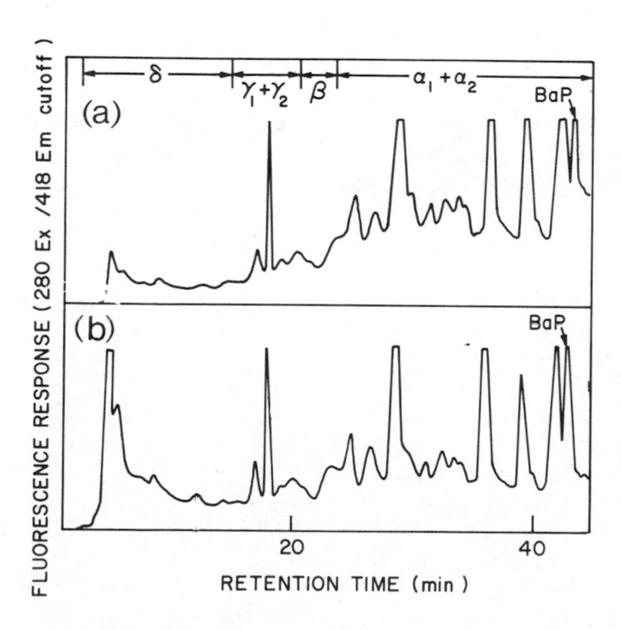

Figure 10. HPLC/fluorescence profile for a diesel particulate extracted with (a) xylene and (b) benzene/ethanol (80:20, v/v). Samples were separated on a reverse-phase Zorbax ODS column with 50% CH_3CN in water initially, followed by linear gradient to 100% CH_3CN in 50 min.

Table 7. The Effect of Extraction Solvents on the Recovery of Total PAH and Nonpolar Fluorescent Organics as Measured by Reverse-Phase HPLC

Solvent	Relative area: $\alpha_1 + \alpha_2$ fraction[a]
Toluene	1.0
Xylene	N.D.
Benzene/CH_2Cl_2/MeOH (25:10:1)	1.1
Toluene/Propanol (50:50)	1.1
CH_2Cl_2	1.0
Benzene/EtOH (80:20)	1.2
CH_2Cl_2/MeOH (97:3)	1.0
$CHCl_3$	1.2

[a]Total PAH is calculated from the summation of all the fluorescence peaks observed in the PAH region of the HPLC chromatogram (α_1, α_2, from 27 to 43 min elution) in the chromatograms shown in Fig. 10.

efficiencies for sulfate appear to follow the alcohol content or the polarity of the solvent system.

The foregoing extraction data are in qualitative agreement with those reported by other laboratories. Swarin and Williams [32] reported that methylene chloride extracts only 60-70% as much BaP and total organic from light-duty diesel particulates as a binary mixture of benzene/ethanol. Petersen et al. [37] compared three solvent systems—methylene chloride, toluene, and benzene/methanol (70:30, v/v)—for the extraction of a heavy-duty diesel exhaust particulates. The benzene/methanol system was found to extract the most material, but within the scatter of the data the quantities of BaP, chrysene, BeP, pyrene, and perylene extracted are about the same for all the three solvent systems. The slight discrepancies in PAH extraction efficiencies are not too surprising since the physical and chemical nature of the diesel particulates and their interaction with the associated PAH are probably dependent on the engine types and operating conditions.

It can be seen from the preceding discussion that for PAH recovery nonpolar aromatic solvents such as toluene are desirable. An aromatic in combination with an alcohol polar solvent is preferred for quantitative extraction of total organics. Of all the solvents tested, toluene/n-propanol (50:50; an azeotrope) is among the best in extracting both BaP and total organics. However, toluene with azeotropic mixtures of methanol or ethanol is more suitable because of its lower refluxing temperature during Soxhlet extraction. Its higher vapor pressure also facilitates removal during solvent evaporation steps. The extraction of sulfate and other inorganics in the later systems, however, should be considered because it may potentially lead to artifact formation and interferences during analysis. The separation of PAH and polar materials immediately after extraction before storage should be able to minimize this potential problem.

Table 8. Comparison of Relative Extraction Efficiencies for Sonication and Soxhlet Techniques[a]

	B.P. (°C)	Total extractables Son. (%)	Total extractables Sox. (%)	[^{14}C]BaP Son. (%)	[^{14}C]BaP Sox. (%)
Toluene	111	1.0	1.1	1.00	1.27
Toluene/propanol (50:50)	93	1.1	1.1	0.95	0.93 (1.03)
Benzene/ethanol (80:20)	68	1.1	1.2	0.84 (0.78)	0.99
Methylene chloride	41	0.8	1.1	0.87	0.62 (0.81)

[a]Numbers listed are efficiencies relative to toluene = 1.00. Numbers in parenthesis represent measurements obtained from a different set of samples.
Source: Ref. 8; copyright American Society for Testing Materials, Philadelphia, Pa. Adapted with permission.

The relative extraction efficiencies of several solvent systems obtained by Soxhlet and sonication techniques are compared in Table 8. It is seen that with some scattering of the data, Soxhlet extraction does not seem to provide noticeable improvement in efficiency as compared to the sonication method. Efficiencies in the sonication method would probably be further increased by the use of solvent at higher temperatures. The speed and less-involved nature of the latter method warrants its application in the extraction of exhaust particulates.

IV. Fractionation

The composition of the extract is extremely complex, consisting of hydrocarbons, oxygenated organics, PAH, N or S heterocyclics, and a variety of organic as well as inorganic compounds originating from fuel and oil additives. Depending on the information needed from the sample, a number of separation and fractionation schemes have been reported for subsequent class or individual compound analysis. Analysis for PAH is generally preceded by liquid chromatography (LC), thin-layer chromatography (TLC), or solvent-partitioning enrichment procedures. Direct analysis has so far been limited only to BaP, and in a few cases also BaA, utilizating high-performance liquid chromatography (HPLC) coupled with selective fluorescence detection.

In this section, the more general fractionation methods capable of isolating PAH and their oxygenated derivatives will be given. Generally, the separation schemes used for exhaust particulates characterizations are derived from the more extensively studied air particulates or cigarette smoke condensate work [58-61]. A typical scheme is shown in Fig. 11. The procedure involves the use of liquid chromatographic separation of the neutral SOF after the removal of acidic and basic fractions. The PAH are enriched in the aromatic fraction

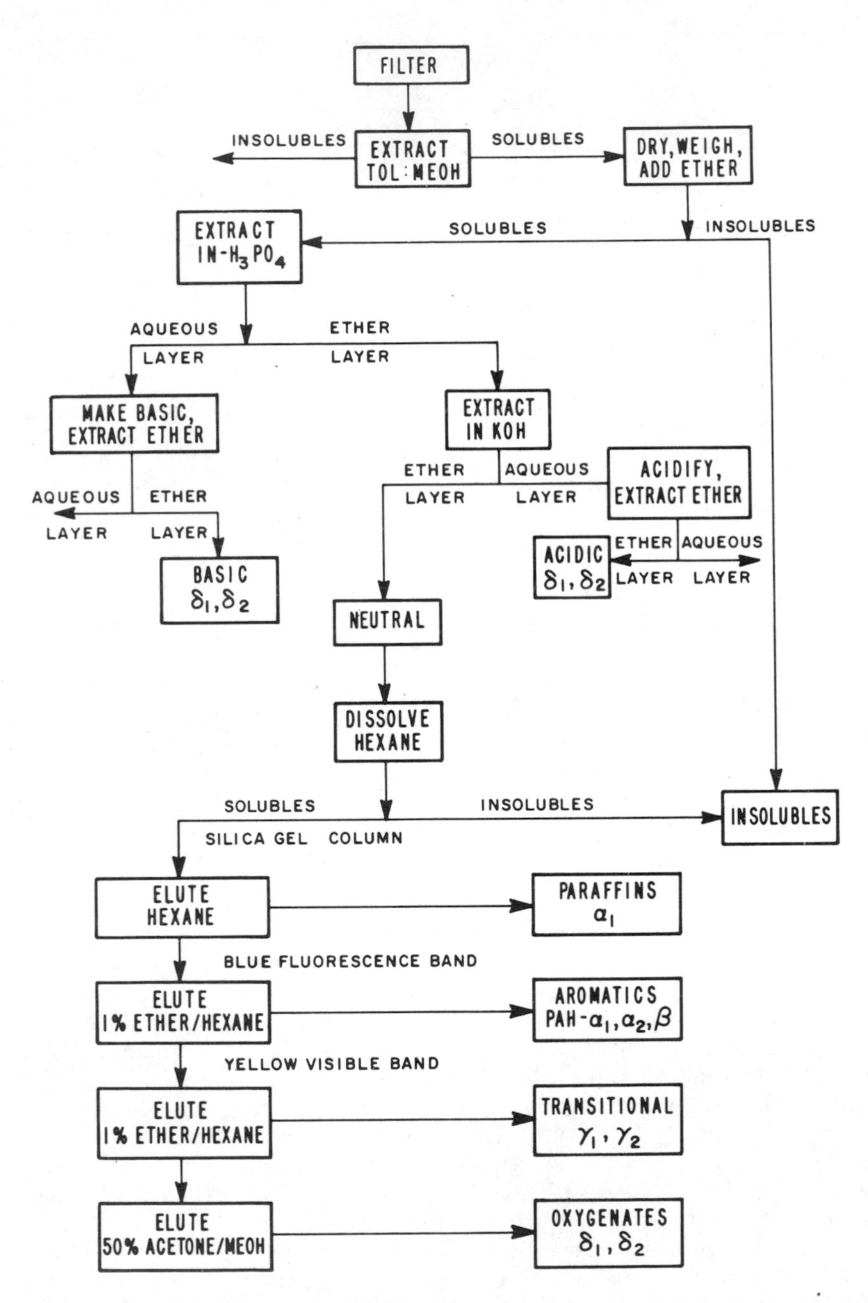

Figure 11. Fractionation scheme for diesel particulate extracts using liquid-liquid extraction and column liquid chromatography.

while most of the oxygenated PAH are concentrated in the transitional fraction. In addition to LC techniques, solvent/solvent partition, TLC and HPLC procedures have also been utilized for the enrichment of PAH in exhaust emission work. These fractionation methods will be described separately below.

A. Liquid-Liquid Partition

A number of liquid-liquid partition procedures have been reported for the isolation of PAH from exhaust particulate extract. This is accomplished by extracting with solvents having a high PAH partition ratio with respect to hydrocarbon solvents. The most widely used solvents include CH_3NO_2, N,N-dimethylformamide (DMF), and dimethyl sulfoxide (DMSO). The high solubility of these solvents toward PAH is probably due to their ability to interact with PAH through complexations.

Grimmer and Böhnke [62] and Grimmer [63] have used a DMF/H_2O/cyclohexane extraction method for their extensive work on exhaust PAH. In their scheme, the particulate extract is first partitioned between DMF/H_2O and cyclohexane phases (9:1:5, v/v). The PAH is enriched in the DMF/H_2O, and most of the aliphatic hydrocarbons are separated in the cyclohexane phase. The PAH in the above DMF/H_2O phase are then back-extracted into another cyclohexane phase in a second extraction with DMF/H_2O/cyclohexane ratio of 1:1:2 (v/v). The high water content in the DMF/H_2O phase during the second extraction step "drives" PAH into the cyclohexane phase while leaving the more polar materials behind. The method has been used by Grimmer for both gasoline and diesel engine exhaust and was also used extensively in our laboratory for the measurement PAH in a stratified gasoline-engine exhaust [44].

A similar procedure using CH_3NO_2 instead of DMF has been used for PAH isolations [6,64]. In this procedure, the PAH are first separated from polar compounds by extracting PAH in cyclohexane with equal volume of MEOH/H_2O (2:1, v/v). The PAH enriched in cyclohexane is then extracted into CH_3NO_2 in the second extraction (cyclohexane/CH_3NO_2 = 2:1, v/v) while leaving the aliphatic hydrocarbons in the cyclohexane phase.

It was found in both the aforementioned schemes that the final separated PAH fraction is still contaminated with considerable aliphatics. Thus, a further chromatographic separation is needed before gas chromatographic/mass spectrometric (GC/MS) analyses. In order to achieve this and also to separate PAH according to their ring numbers, the PAH enriched extract is further fractionated by LH-20 Sephadex column with 2-propanol elution. The procedure was found to provide clean PAH fractions suitable for GC/MS analysis, as will be discussed later.

In the diesel particulate work in this laboratory, we have also tested yet another extraction scheme developed by Natusch and Tomkins [65]. In this procedure, extraction was accomplished by the partition of PAH between DMSO and pentane. The dried extract is first dissolved in pentane. It was then extracted with equal volume of dimethyl sulfoxide. The PAH are extracted into the DMSO phase while most of the aliphatic hydrocarbons remain in pentane. An equal volume of water is then added into DMSO and reextracted with pentane. This allows the partitioning of PAH back into the pentane phase with polar materials remaining in DMSO/H_2O mixture. Using a [^{3}H]BaP radiotracer technique, we found that over 95% of the [^{3}H]BaP can be recovered in the final pentane fraction when a [^{3}H]BaP spiked diesel particulate extract was separated by the method.

When the three solvent extraction methods are compared, the DMSO procedure involves fewer handling steps and also uses less solvent. The method also provides the advantage that the final PAH solution in pentane is easier to further concentrate because of its low boiling point. As we will see later, some of the medium polar species are still present in the final pentane fraction, and like the DMSO and CH_3NO_2 methods described earlier, further cleanup by silica gel column is still needed for detailed PAH analysis.

B. Column Liquid Chromatography

The Rosen type of open-column LC separation has been the most widely used method for fractionation of complex environmental particulate samples according to functionalities [66]. In this method, a glass column of 10-20 cm length packed with 2-10% water deactivated silica gel or alumina was employed. The sample was then separated on the column into aliphatic, aromatic, PAH, and polar fractions through the use of solvents with increasing polarities (Figure 11).

The silica gel columns are prepared using Pyrex chromatographic columns with a length of 40 cm and inner diameter of 1 cm. Silica gel is activated overnight at 120°C, slurried with hexane and the column filled to a height of 10 cm. The diethyl ether is removed from the neutral fraction, dissolved in a minimum amount of hexane and placed at the head of the silica gel column. The column is eluted with hexane to separate the paraffin fraction (α_1) and is collected until the appearance of a blue fluorescence (using long wavelength UV light) in the column tip. The aromatic-PAH (α_1, α_2) fraction is eluted with 1% ether in hexane collected until a yellow band observable under visible light begins to appear. The transitional fraction (γ_1, γ_2) is eluted with another 25 ml of 1% ether in hexane and the oxygenated fraction (γ_2, δ_1, δ_2) eluted with 50% acetone/methanol.

The fraction designation in parentheses is given for the purpose of comparing this technique to the normal phase HPLC separation technique, as will be discussed in later sections.

The main disadvantages of open-column LC are its slow speed and lack of reproducibility. The separation is controlled manually and the column to column variation makes clean-cut separation of compound classes an art difficult to reproduce. Since normally a new column is packed for each sample, background contamination from the packing material is a serious problem for small sample sizes.

Low resolution glass or stainless steel columns adapted in HPLC instruments have recently received increasing popularity in the fractionation of environmental samples. The use of a silica gel column in a HPLC instrument for the fractionation of diesel particulate extract has been reported [67]. A typical chromatogram is shown in Fig. 12. In this procedure, 25-30 cm long, 4 mm i.d. stainless steel column was packed with 100-200 mesh Bio-Sil-A (Bio-Rad). The sample was injected when the column was eluted under hexane with 5% methylene chloride. After the complete elution of α_1 and α_2, the gradient starts and the solvent is programmed to 100% methylene chloride

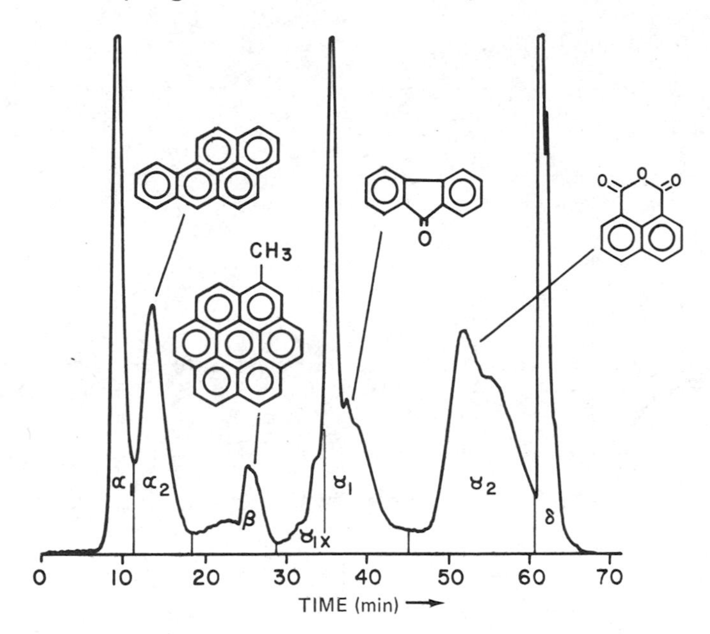

Figure 12. HPLC fractionation of a diesel particulate extract on a 3 mm × 30 cm Biosil A (20-44 μm) column with fluorescence detection at λ = 313 nm and λ ≥ 418 nm. Compounds shown in figure indicate their calibrated retention times from standard injections.

thus allowing the elution of other fractions (β, γ_{1x}, γ_1, and γ_2; for definition of γ_{1x}, see Sec. V.C.2).

The above procedure with different but small variations has been extensively used in many laboratories for the fractionation of diesel extracts. It has also been studied in a round-robin experiment organized by the Diesel Chemical Characterization Panel of Coordinating Research Council, Inc. [67]. In that procedure, an additional CH_3CN elution step was introduced after the elution of γ_2 peak (Fig. 12), thus allowing the elution of polar oxygenated fractions (δ). Data from this round-robin study is reproduced in Table 9. The results obtained from the 11 participating laboratories show that the procedure is impressively reproducible in terms of both retention times and fluorescent peak area measurements (see references cited in the table).

Typically, diesel particulates extracts contain very little basic material (<0.1%) and less than 1 to 2% of acidic materials (excluding phenolics) as separated by the aqueous acid/base extraction scheme given earlier in Fig. 11. Furthermore, it was also found that extract after silica gel separation showed no loss of Ames mutagenicity as compared to the original extract [19]. Thus, extract samples can be injected directly onto the silica gel column for separation without prior acid/base extractions. The latter materials can either be removed by the installation of a short precolumn or backflushed after the completion of the chromatographic elution.

Compared with open-column separation, the low-resolution LC procedure is superior in terms of both speed and reproducibility. Using the 3 mm

Table 9. Summary of Coordinating Research Council HPLC Round-robin Results

Sample[a]	Relative peak areas				
	$\alpha_1 + \alpha_2$	β	γ_1	γ_2	δ_1
NI-1 extract:					
Mean ($\bar{X}$)	1.00	0.21	0.77	0.70	0.66
SD (1σ)	—	0.06	0.14	0.11	0.18
% SD (%σ)	—	29	18	16	27
CU-1 extract:					
Mean ($\bar{X}$)	0.07	0.14	1.00	0.20	1.43
SD (1σ)	0.01	0.03	—	0.05	0.19
% SD (%σ)	14	21	—	25	13

Sample	Retention time (min)					
	α_1	α_2	β	γ_1	γ_2	δ_1
NI-1 extract:						
Mean ($\bar{X}$)	10.0	15.4	24.4	35.1	50.6	56.7
SD (1σ)	1.3	2.1	3.0	2.7	3.8	3.6
% SD (%σ)	13	14	12	8	8	6
CU-1 extract:						
Mean ($\bar{X}$)	9.3	14.7	26.0	34.3	49.3	56.5
SD (1σ)	1.8	2.7	2.7	4.5	5.1	0.5
% SD (%σ)	19	18	10	13	10	1

[a]SD = standard deviation.

diameter column, it has been found that up to 10 mg extract can be separated without noticeable sacrifice in resolution. Furthermore, the column can be readily upscaled in order to accommodate larger samples. A 8 mm × 25 cm Michel-Miller glass prep column, for instance, was used to separate adequate size diesel fractions for subsequent Ames biological testing and mass spectrometric analysis [18,19]. It was reported there that the peak elution profile for such a column is similar, with only slightly poorer baseline separation, compared to the one illustrated in Fig. 12. A diesel particulate extract was fractionated by such a prep-scale column, and a number of heavier PAH have been identified in the β fractions (as will be discussed when we consider spectrometric techniques in Sec. V.C).

C. HPLC

We have recently developed an HPLC procedure involving the use of a CH_3CN-deactivated microparticle silica column that is commercially available (RCSS-B; or Radial-Pak B from Waters Associates, Inc.) for the fractionation of particulate samples. The column was chosen because of its ability to perform adequate resolution in both analytical and semipreparative-scale separations needed in particulate analysis. We have applied the procedure for the separation of particulates collected from diesel engines and from the atmosphere near roadways.

The mobile phase solvents used for the procedure are hexane, dichloromethane (DCM) and acetonitrile. In some cases, chloroform was used instead of DCM. The two give similar elution profiles except the retention time of a given fraction or testing compound is always shorter in chloroform solvent systems (see Table 10). Because of the difficulties occasionally observed in the pumping of the more volatile hexane solvent, heptane was used to replace hexane occasionally. The difference observed for such solvent exchange is generally insignificant.

The detailed solvent programming profiles used for the fractionation varies slightly depending on the requirement of the analysis. The variation of the HPLC chromatogram caused by such slight variation in solvent programming is generally very small. For the chromatograms illustrated in Fig. 13 the procedure is as follows. The sample was injected when the column was flowed under hexane. It was then programmed from hexane to 100% methylene chloride in 20 min using gradient curve No. 9 preset in the Waters Model 440 solvent programmer. The solvent was then held isocratically at methylene chloride for 8 min before it is switched to acetonitrile. After isocratic acetonitrile flowing for 10 min, the column was finally switched to methanol flow for 10 min. The designations of fractions as α_1, α_2, γ_1, γ_2, and δ_1 are indicated in the figure. The methanol-eluted fraction (δ_2) was not shown.

The use of methanol and acetonitrile as solvents for the silica column causes noticeable column deactivation after two or three runs. The column, however, was then stabilized after this initial deactivation period and maintain a highly reproducible separation for a time period of 2 to 3 months under continuing use. The repeatibility of the retention times for a given peak is usually ±1%. In Fig. 14, the reproducibility of the peak areas from fluorescence measurement is shown. The samples shown were analyzed continuously under the same solvent programming procedures. It can be seen from the figure that the reproducibility is within 5% for α_1 and $\alpha_2 + \beta$ and within 7% for $\gamma_1 + \gamma_2$.

Figure 13 also shows that the effect of sample size on column resolution. As seen from the figure, up to 2-mg samples can be injected without noticeable loss of resolution of α_2, β, γ_1, γ_2, and δ peaks. The δ_1 peak was seen to overlap with a frontmost peak at larger sample sizes (1.3 and 2.0 mg runs). The latter was probably caused by the perturbation of the fluorescence detection as large quantities of aliphatic or small aromatic hydrocarbons were passing through the cell.

The ability of the column to separate PAH from the other compounds was tested by the injection of a series of PAH standards ranging from two to seven rings. The calibrated relative retention times for PAH standards and several other polar compounds are listed in Table 10. Most of the three or four rings PAH are eluted in the α_1 region. The four- to seven-ring PAH are concentrated in α_2 and β. All the calibrated PAH are in the α_1 to β region well separated from the medium or highly polar species in γ and δ fractions, respectively.

Table 10. Relative Retention Times of PAH on a Normal-Phase HPLC Column under Two Different Solvent-Eluting Conditions[a]

Compound	HPLC conditions I	HPLC conditions II
Naphthalene	0.35	0.80
Anthracene	0.54	0.81
9-Methyl anthracene	–	0.82
Pyrene	0.54	0.85
Fluoranthene	0.66	1.05
Xanthene	0.77	0.84
Benzo[a]pyrene	1.00 (12.3 min)	1.00 (3.26 min)
Benzo[e]pyrene	1.01	–
Perylene	1.02	1.01
Benzo[k]fluoranthene	1.02	–
Benzo[ghi]perylene	1.03	1.13
Chrysene	1.05	0.95
Coronene	1.11	–
1,2,3,4-Dibenzoanthracene	1.18	–
1,2-Benzocoronene	1.19	1.33
1,2,4,5-Dibenzopyrene	1.17	–
2-Nitrofluorene	1.38	–
9-Hydroxyfluorene	1.49	3.65
1-Nitropyrene	1.59	3.40
9-Fluorenone	1.71	3.82
Acridine	2.58	5.89
5,6-Benzoquinoline	2.60	5.91

[a]Detailed chromatographic conditions given in Fig. 13. Condition I uses a solvent elution sequence n-hexane/dichloromethane/acetonitrile. Condition II uses a solvent elution sequence: n-heptane/chloroform/acetonitrile.

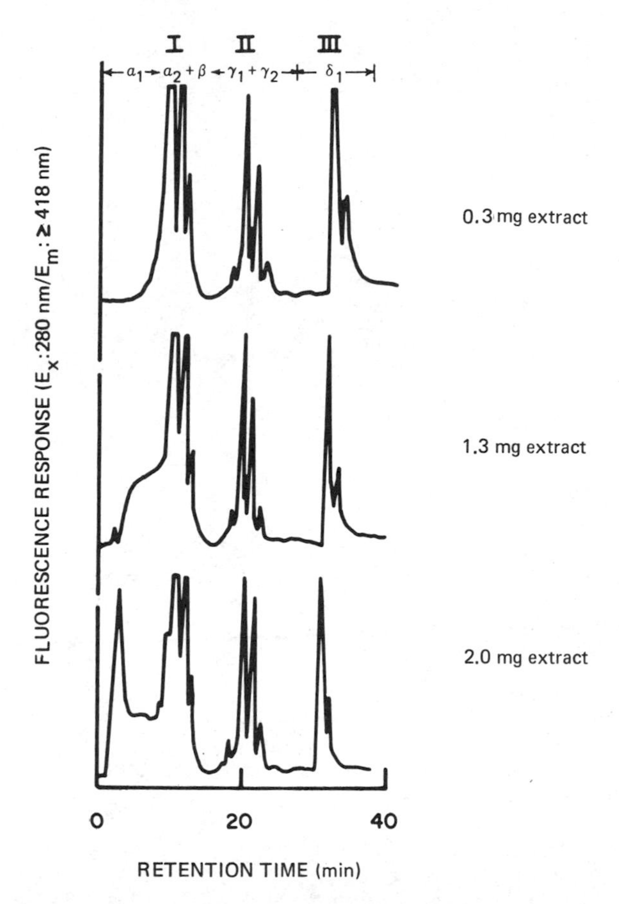

Figure 13. Diesel particulate extract fractionation by HPLC: the effect of sample loading on chromatographic separations. A Waters Radial Compression type B silica column with a solvent gradient system involving hexane/$MeCl_2$/CH_3CN was used. The sample was injected when the column was flowed under hexane. It was then programmed from hexane to 100% $MeCl_2$ in 20 min using gradient curve No. 9 on a Waters Model 440 solvent programmer; held at 100% $MeCl_2$ for 8 min before switching to CH_3CN and held at CH_3CN for 10 min. At the end of CH_3CN elution, the column was flowed with 20-30 ml CH_3OH for the elution of strongly held polar compounds.

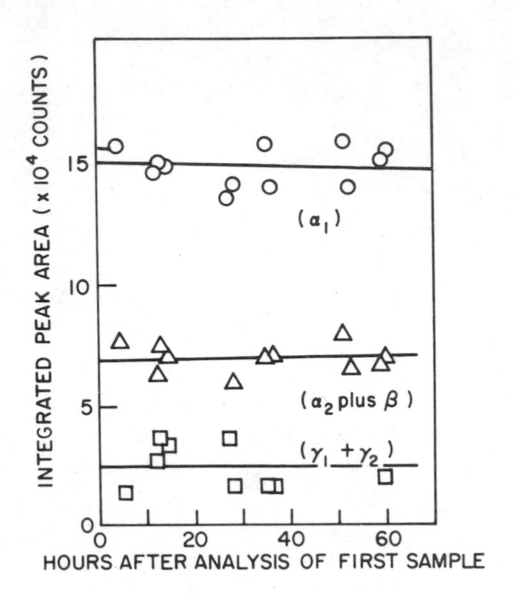

Figure 14. Reproducibility of HPLC/fluorescence profile for a diesel particulate extract (No. NI-1). For chromatographic conditions and peak designations, see Fig. 13. Peak areas integrated by a H/P Model 3380 printer/plotter.

Table 11. Weight Distribution of a Diesel Particulate Extract Fractionated by HPLC

Fractions	Major compound types	Weight[a] recovered (mg)	Weight recovered (%)
α_0[b]	Aliphatics: one- of two-ring aromatics	5.2	46
$\alpha_1 + \alpha_2$	PAH	0.31	2.8
β	Heavy PAH	0.14	1.3
$\gamma_1 + \gamma_2$	Oxy-PAH	0.81	7.3
δ_1	Heterocyclics	1.40	12.6
δ_2	Acids	3.45	31.1
Total		11.31	101.1

[a]A total of 11.19 mg Nissan diesel particulate extract was fractionated by three separate HPLC runs. Weight distribution is calculated from the combined total of the three runs.

[b]α_0 = nonfluorescent material eluted before α_1.

The utilization of this procedure for the fractionation of diesel particulate extracts has been discussed extensively in a recent report [19]. In this laboratory, the method was routinely used for the fractionation of exhaust samples for subsequent biological testings or chemical analysis. The weight distribution of a Nissan diesel particulates extract separated by the procedure is given in Table 11, and it can be seen that an excellent mass recovery from the HPLC procedure was obtained.

V. Analysis

Methods for the analysis of PAH in complex environmental samples have been extensively reported in the literature. Many of the state-of-the-art analytical techniques are also reviewed in other chapters in this book. In this section, our discussion will be limited only to those methods which have been applied in automobile exhaust work.

The complexity of the exhaust sample is compounded by the presence of both fuel components with a high proportion of alkyl-substituted PAH and polar-oxygenated species due to oxidation or incomplete combustion. In the earlier work by Hoffmann et al. and others, large samples (tens or hundreds of grams) collected over an extended time period were required in order to isolate sufficient quantities of material for identification and quantitation. Such practice is generally disfavored now because of the concern over possible sample degradations during collection, as already discussed in Sec. II.B.2. Thus, for instance, a typical sample size collected from a dilution tube on a 47 mm filter is on the order of tenths or hundredths of a gram. Such limited sample sizes dictate the need for sensitive analytical methods.

Table 12 summarizes the different analytical methods employed in exhaust PAH work reported in the literature. The summary is not meant to be complete but rather to provide an overview of the various techniques used in different laboratories. In the following sections, these methods will be discussed separately with emphasis placed on the more recent techniques such as HPLC, capillary GC, GC/MS, and high-resolution mass spectrometry (HRMS).

A. Thin-Layer Chromatography

Earlier work by Hoffmann and colleagues [4-6] employed paper chromatography for the isolation of individual PAH. With the use of acetylated paper chromatography in combination with UV spectrophotometric detection, they were able to quantitatively determine the concentrations of more than 20 three- to seven-ring PAH. The procedure was extremely time consuming, involving a number of solvent extraction steps followed by repeated column and paper chromatography. Besides the slow analysis speed, such lengthy sample preparation steps also tended to increase the possibility of PAH degradation. It was reported by Hoffman and Wynder that extensive PAH degradation was indeed observed during analysis, although the cause of such degradations was not discussed in detail.

The rapid development of the TLC techniques since the early 1960s has largely replaced the classical paper chromatographic methods in environmental analysis. Sawicki et al. [78,79] and Mackay and Latham [80] first reported

Table 12. Reported Analytical Methods for Vehicle Exhaust PAH Analysis

Sample type and sampling method	Analytical procedure and techniques	PAH analyzed	Ref.
Gasoline engine; total exhaust collected by condensation/filter	Nitromethane enrichment; SG and alumina col chrom; acetylated paper chrom; UV/vis identification (id.) of isolated compounds	About 20 four- to seven-ring PAH & few alkyl PAH; quantitative	4-7
Gasoline engine; fuels with different aromatic contents; total exhaust condensation and filtration for particulates	[^{14}C]BaP and [^{14}C]BaA int. std.; dil. HCl/NaOH wash; alumina col chrom; GC sep. on 3 m SE-30 S.S. col, 175-300°C at 4°C/min; separated GC components id. by UV/vis spec.	Pyrene, chrysene, BaA, BaP, BeP; quantitative	69-70
Gasoline engine; high and low aromatic fuels, w/o lead; European testing cycle, collected by condensation/filter	[^{3}H]BaP int std; SG and alumina col chrom; acetylated paper chrom; separated components id. by UV/vis spectroscopy	Three- to seven-ring PAH; quantitative	71
Gasoline engine; steady-state speed, leaded fuel; particulates collected by filters in dilution tube	SG col chrom; separated LC fractions analyzed by GC/MS/DS; GC on a 12-ft Dexsil 300 glass col from 150-310°C at 6°C/min	Two- to seven-ring PAH and methyl PAH and other organics; qualitative	72
Heavy-duty diesel; steady-state speed; air dil, prop. sampling; filter/chrom 102 trap	^{14}C-BaP int std; SG and alumina col chrom ; cellulose paper chrom; LC and acetylated cellulose TLC; isolated BaA and BaP fractions id. by UV spectroscopy	Quantitative for BaA and BaP	43
Gasoline engine; European testing cycle; total exhaust collected by condensation/filter	DMF/water/cyclohexane enrichment; SG col chrom; PAH fractionated by Sephadex LH-20; fractions analyzed by GC/MS; GC on a 20-m OV 101 glass col, 120-270°C at 1°C/min	Total of more than 80 three- to seven-ring PAH quantified	62,63, 73
Air near highways collected by low-vol sampler	Fractionated as above; fractions analyzed by GC^{2}/FID on a 25-m OV-17 col, 100-270°C at 4°C/min	15 three- to seven-ring PAH; quantitative	73

Light- to medium-duty diesel; steady-state speed; w/o exhaust treatment; filter collection of dil. exhaust particulates	Direct HPLC/fluorescence of extract; Dupont Zorbax ODS col, 70-99% methanol in water; ex 380 nm/em 454 nm; stop-flow fluorescence spec scan for BaA and BaP	Quantitative for BaA and BaP; qualitative for fluoranthene and BkF	75,76
Stratified-charge gasoline engine; steady-state speed, w/o oxid. cat.; filter/XAD-2 trap collection in dil. tube	DMF/water/cyclohexane enrichment; SG col chrom.; Sephadex LH-20 col chrom; fractions analyzed by HPLC/UV and GC/MS; HPLC on a Bondapak C_{18} column with 70-100% CH_3CN in water gradient; GC on a 3-m packed and 25-m capillary Dexsil 300 col	Total of 10 three- to seven-ring PAH quantified by HPLC; and GC/MS; 35 PAH qualitatively analyzed by GC/MS	44
Diesel and gasoline engine; with and without exhaust treatment; collected in dil. tube; U.S. testing cycle	Polar removal by hexane/dichloromethane/methanol partition, HPLC/fluorescence with ex 383 nm/em 430 nm; Dupont Zorbax ODS, 90% CH_3CN in water, isocratic	Quantitative for BaP	32
Light-duty diesel particulates collected in dil. tube	HPLC preseparation on Waters RCSS silica col; PAH fraction analyzed by HPLC/fluorescence and GC/MS; HPLC on P/E HC-ODS and Dupont Zorbax ODS col; GC on a 25 m GC^2 dexsil 300 col; HRMS and GC/MS data compared	Quantitative for 25 PAH and more than 40 PAH oxygenated derivatives	19 This work
Heavy-duty diesel; particulates collected in dil tube; U.S. testing cycle	Direct HPLC/fluorescence, Dupont Zorbax ODS col, 80% CH_3CN in water, isocratic, GC/MS confirmation	Quantitative for BaP few others identified	77
Diesel particulates, various speed and load, collected in dil. tube	Deuterated PAH and [^{14}C]BaP int std; dil. acid/base wash; SG col fractionation; GC/MS/DS; GC on a 20-m SE-52 GC^2, 150-250°C at 4°C/min	Quantitative for pyrene, BaA, chrysene, BaP and perylene	36
Gasoline and diesel engines; U.S. and European testing cycles, collection of total exhaust condensates	DMF/water/cyclohexane enrichment; Sephadex LH-20 col chrom; PAH fractions analyzed by TLC/fluorescence and GC/FID, TLC used acetylated cellulose plate; GC on OV 101, SE 30, SE 52 and Dexsil 300 capillary columns (25 m; 100-310°C at 40°C/min)	Quantitative for 12 three- to six-ring PAH, GC and TLC methods compared	30

the successful application of TLC in the analysis of air particulates. An acetylated cellulose plate was used for the isolation of BaP and other PAH for fluorescence detection. The method was later further improved by Pierce and Katz [81]. In their method, a two-dimensional TLC using a mixture of acetylated cellulose and aluminum oxide was employed for the separation of PAH in air particulate extracts. The separated PAH were then quantitatively determined by fluorescence spectrophotometry. Recently, the rapid advancement in high-resolution TLC has further established itself as a powerful tool in environmental analysis.

Despite the potential of the TLC method, only limited progress has been reported toward its application in exhaust PAH analysis. Spindt [43] first reported the application of TLC in diesel exhaust PAH analysis. In his method, separation of PAH was achieved through the use of multiple-acetylated cellulose plate TLC. The extract sample was first purified by two successive alumina column LC to remove the aliphatics and light aromatics. The PAH fraction was then further enriched on a cellulose plate TLC eluted with isoctane. The enriched PAH fraction was then separated on a RP-HPLC column (Bondapack C18/corasil), and the BaA and BaP fractions eluted from the column were collected separately. The two fractions were finally separated individually on an acetylated cellulose plate, and the isolated BaA and BaP compounds measured by UV spectrophotometry.

The above procedure developed by Spindt was quite complicated and did not seem to provide substantial improvement over the classical procedures used by Hoffman et al. Recently, a much-improved procedure was reported by Kraft and Lies for the analysis of PAH from both gasoline and diesel engine emissions [30]. In this method, a two-dimensional acetylated cellulose plate was used for the separation and in situ fluorescence spectrometry was used for the quantitative determination of 14 four- to six-ring PAH. The extract sample was first purified by a procedure similar to those used by Grimmer including solvent/solvent extractions and Sephadex gel chromatography, as described in Sec. IV.A. The enriched PAH fraction was then developed on the TLC plate first with toluene/n-hexane/n-pentane (5:90:5; v/v) and then with methanol/diethyl ether/water (6:4:1, v/v). Selective wavelength excitation and emission was used for the fluorescence detection of PAH with sensitivity ranging from 0.4 to 200 ng for each PAH.

Kraft and Lies [30] have also compared the TLC results with those obtained from capillary GC/FID* measurement (20-25 m OV 101, SE-52, and Dexsil 300 columns). Their results are reproduced here in Table 13. It is seen that excellent agreement between the two methods are obtained, especially for the heavier PAH such as BeP, BaP, perylene, and indeno[1,2,3-c,d]pyrene [8% SD (standard deviation)]. Such good agreement suggests the validity of both methods for the quantitative measurements of selected PAH species. The establishment of the validity of the capillary GC/FID procedure is particularly noteworthy since the method involves relatively inexpensive instrumentation and is thus well suited for routine analysis work.

---

*FID = flame-ionization detection.

Table 13. Comparison Between the Results of the Thin-Layer Chromatographic Procedure (TLC) and the Glass Capillary Gas Chromatographic Determination (GC^2) Using an Exhaust Gas Sample of a FTP Driving Cycle

Compound	GC^2 method (a) μg/test	TLC method (b) μg/test	$\frac{a - b}{\bar{x}} \cdot 100\%$
Fluoranthene	1671.8	1591.7	+ 4.91
Pyrene	1304.8	1587.5	-19.55
Benz[a]anthracene	136.2	111.5	+19.94
Chrysene	316.9[a]	231.8	—
Benzo[b]fluoranthene	169.3	143.2	+16.70
Benzo[k]fluoranthene	88.8	77.6	+13.46
Benzo[e]pyrene	78.0	81.3	- 4.14
Benzo[a]pyrene	46.6	45.0	+ 3.49
Perylene	10.6	9.8	+ 7.84
Indeno[1,2,3-cd]pyrene	56.4	59.8	- 5.85
Benzo[ghi]perylene	78.0	95.0	-19.65

[a]Chrysene plus triphenylene.
Source: Ref. 30.

B. HPLC

HPLC has been used extensively for the analysis of PAH in exhaust as well as other environmental samples. Its speed and relatively simple operation procedure makes it usually the method of choice for routine analytical work. The general techniques of utilizing HPLC in combination with fluorescence or UV/vis (visual light) spectrophotometric detection for airborne particulate analysis have been extensively reviewed in the literature [82-85].

The HPLC techniques have been applied with varying extent of success to three types of exhaust PAH analysis. These include: (1) BaP analysis, (2) multiple PAH analysis, and (3) exhaust fingerprinting. The fingerprinting method involves the use of the procedure described previously in Sec. IV.C. In the following two subsections, the first two types of analyses will be described.

1. BaP Analysis

The concentration of BaP has been used traditionally as the indicator of general PAH levels in exhaust as well as other environmental samples. Thus, rather than performing detailed PAH analysis, it would be desirable to develop a simple procedure aimed specifically toward BaP. The procedure should be reliable and with no or minimum sample pretreatment steps to permit the rapid processing of a large number of samples.

The use of HPLC coupled with fluorescence spectrophotometric detection for the measurement of BaP in diesel engine exhaust has been reported by Tejada [22], Seizinger [75,76], Swarin and Williams [32], and Perez [77]. The procedure was also used routinely in this and other laboratories for the analysis of BaP in vehicle exhaust as well as atmospheric particulates. The procedures used by these workers are very similar, with only minor variations in the fluorometer settings and column conditions. Typically, the particulate extract is either injected directly into HPLC or preextracted with methanol/water mixture for the removal of polar materials before injection. The BaP fluorescence was selectively monitored at 383- or 365-nm excitation (Ex) and 454- or 418-nm emission (Em).

The separation of BaP from other PAH can be readily achieved on several reverse-phase columns. The relative retention times of the three most commonly used columns are compared in Table 14. Columns a and b are seen to provide superior separations for the test PAH compounds than do column c and the one listed in Table 15. Of the former two columns, column b seems to give still better resolution for the separation of BaP, BeP, perylene, and BkF (benzo[k]fluoranthene). The observation is consistent with a recent report by Ogan et al. [87].

The BaP has excitation maxima at 295, 381, 362, and 280 nm and emission maxima at 404, 425, and 453 nm. At 381-nm excitation, the BaP fluorescence is selectively enhanced compared with other PAH. In addition, various methods such as the stop-flow technique for fluorescence spectrum scanning and standard addition have been used to further ensure that the BaP peak is free of interferences from quenching and absorption effects from coeluting species.

Seizinger [76] has used stop-flow and fluorescence scanning techniques for the identification of BaP in a diesel particulate extract. His results are reproduced here in Fig. 15. After stop flow, the trapped BaP peak is scanned for both excitation and emission spectra. The emission spectrum matches exactly with standard BaP. The excitation spectrum also matches with standard spectrum at wavelength >300 nm. The slight distortion at shorter (300-nm) wavelength could be due to interference peaks. Such interference was also frequently observed by us during routine diesel sample analysis. The chromatogram of a typical direct-injection diesel particulate extract is shown in Fig. 16. The fluorescence was monitored at both 365-nm excitation/≥418-nm emission and 280-nm excitation/≥418-nm emission using a tandem fluorescence detector setup. It can be seen that an interfering shoulder peak for BaP was observed at 280-nm but not at 365-nm excitation. The emission spectra of the peak (not given in the figure) at 365 nm again match with the standard.

The interfering peaks for BaP can be removed by either the selection of better chromatographic column separation or by the selection of appropriate excitation wavelengths, e.g., 365 nm or 380 nm. In Fig. 17, a diesel particulate extract sample was analyzed under different fluorescence monitoring conditions using a different column (column b of Table 14). The sample was analyzed twice using the tandem detector setup, thus giving responses at four different fluorescence conditions. Here, it is seen that the BaP peak was well separated from all the other interfering peaks and the fluorescence response ratios obtained at different wavelengths agree within ±10% with those observed for standard BaP.

Table 14. Comparison of Relative Retention Times of Selected PAH on Three Reverse-Phase HPLC Columns

Compound		Relative retention times (BaP = 1.00)				
	Columns:[a]	a	a	a	b	c
	HPLC conditions:[b]	1	2	3	4	5
	Reference:	This work	32	77	This work	
Phenanthrene		0.56	0.377	0.33	0.38	0.41
Fluorene		—	—	—	—	0.32
Anthracene		0.59	0.402	0.35	0.42	0.46
Fluoranthene		0.64	0.461	0.448	0.47	0.65
Pyrene		0.65	0.529	0.511	—	0.66
BaA		0.76	0.611	0.623	0.69	0.71
Chrysene		—	0.622	0.619	0.72	—
BkF		0.91	0.867	0.879	0.94	0.93
BeP		0.91	0.871	0.879	0.83	0.93
Perylene		0.92	0.882	0.879	0.86	0.97
BaP		1.00	1.00	1.00	1.00	1.00
Benzo[ghi]perylene		1.25	1.457	1.468	1.10	1.23
Dibenzo[a,l]pyrene		1.32	2.425	2.22	1.24	1.42
Anthanthrene		1.37	1.751	—	—	1.36
Coronene		1.71	—	—	1.65	1.60

[a]Column identification: (a) Dupont Zorbax; (b) Perkin/Elmer HC-ODS SIL X; (c) Whatman Partisil ODS-2.

[b]HPLC conditions:

(1) 50% CH_3CN in water linear gradient to 100% CH_3CN in 30 min, 1 ml/min BaP retention time = 29 min

(2) 90% CH_3CN in water, isocratic at 1 ml/min, BaP retention time = 18 min

(3) 80% CH_3CN in water, isocratic at 1 ml/min, BaP retention time unknown

(4) 50% CH_3CN in water linear gradient to 100% CH_3CN in 30 min, 1 ml/min, BaP retention time = 24 min

(5) 50% CH_3CN in water linear gradient to 100% CH_3CN in 30 min, BaP retention time = 28 min

Table 15. Summary of PAH Retention Times on a Reverse-Phase HPLC Column and Relative UV Absorption Responses at Different Wavelengths

Compound	Relative R.T.[a]	Relative UV responses at different wavelengths (nm)[b]					
		254	280	313	340	365	405
Naphthalene	1.00	1.00[c]	—	0.06	0	0	0
Biphenyl	1.13	4.35	—	—	0	0	0
Fluorene	1.17	4.43	2.10	0.01	0.01	0	0
Phenanthrene	1.22	11.1	3.5	0.2	0.05	0	0
Anthracene	1.26	29.8	0.2	0.66	0.88	0.6	0
Fluoranthene	1.34	4.06	6.04	1.25	1.41	1.38	0.28
1-Methyl phenanthrene	1.37	11.5	—	—	0.19	—	—
9-Methyl phenanthrene	1.37	—	—	—	—	—	—
Pyrene	1.38	1.99	1.71	2.12	3.79	0.06	0.01
Benzo[a]fluorene	1.43	6.40	—	—	0.29	—	0.1
Benzo[b]fluorene	1.44	—	—	—	—	—	—
Triphenylene	1.46	13.3	2.5	0.03	0	0	0
Chrysene	1.50	6±1	—	1±1	—	—	—
BaA	1.51	6.6	—	2.1	0.47	—	—
Perylene	1.66	3.35	0.3	0.13	0.01	0.42	2.44
BeP	1.66	—	—	1.88	—	—	0.28
BkF	1.68	—	—	—	0.24	—	0.39
BaP	1.70	5.34	5.26	0.53	0.79	3.94	0.30
Dibenzo[ac]anthracene	1.72	4.42	—	—	2.46	—	0
Dibenzo[ah]anthracene	1.75	—	—	—	—	—	—
Picene	1.76	12.5	18.1	4.3	1.54	—	—
Benzo[ghi]perylene	1.87	2.03	—	24.1	0.71	—	0.03
Anthanthrene	1.93	1.01	2.68	0.5	0.1	—	0.87
Dibenzo[al]pyrene	1.96	2.79	4.43	0.43	0.96	—	0
Coronene	2.10	0.45	2.08	2.28	1.67	1.67	⩽0.02
Benzo[a]coronene	2.46	0.50	—	—	—	—	—

[a] Retention times relative to naphthalene = 1.00; HPLC retention times calibrated on three Waters Bondapak C_{18} columns connected in series. A solvent system of 70% CH_3CN in water linear gradient to 100% CH_3CN in 40 min was used for the separation.

[b] UV responses relative to naphthalene = 1.00; "0" means below detection limit; a Waters Model 440 Dual UV detector was used for all the measurements.

[c] Detection limit for naphthalene = 5 ng.

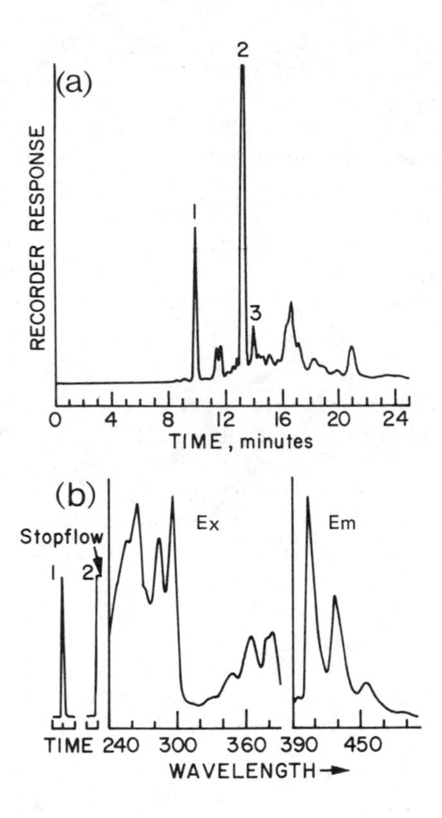

Figure 15. (a) LC-fluorescence chromatogram of a diesel exhaust sample. Chromatographic conditions: 70-99% methanol/water gradient set at 2 cc/min. Excitation and emission wavelength settings were 383 and 454 nm, respectively. Peak designation: (1) fluoranthene; (2) benzo[k]fluoranthene; and (3) benzo[a]pyrene. (b) Examples of chromatograms and fluorescence excitation (Ex) and emission (Em) scans of a trapped BaP sample. (From Ref. 76.)

Swarin and Williams [32] and Williams [86] have used both stop-flow/fluorescence scanning and standard addition for the verification of BaP measurement in diesel as well as gasoline engine exhaust. Using chromatographic conditions given in Table 14 and with fluorescence setting at 383-nm excitation/430-nm emission, their results demonstrate quite convincingly the validity of the procedure for the quantitative analysis of BaP.

2. PAH Analysis

The extension of the aforementioned HPLC fluorescence technique for BaP to the quantitative measurement of other PAH was attempted in this as well as other laboratories with limited success. The difficulty lies both in the limited HPLC column resolution (comparing to capillary GC, for instance) and the susceptible nature of fluorescence measurement. The extensive validation processes required so far restricts successful application of the technique to only selected PAH, e.g., anthracene, pyrene, BaA, BkF, and benzo-[ghi]perylene. Compared to the fluorescence technique, UV/vis spectrophotometric detection is on the average one to two orders of magnitude less sensitive. The latter, however, is generally also much less sensitive to the effects of sample matrices and instrumental conditions. This stability in day-to-day operations makes multicomponent analysis more reliable in routine analytical work. The applications of a HPLC/UV detector in airborne particulates have been reported by Golden and Sawicki [83], Krstalovic et al. [88], and Smillie et al. [89]. In the following discussion, we consider the application of both UV absorbance and fluorescence techniques in exhaust analysis.

a. HPLC/UV The use of an HPLC/UV detector for gasoline engine exhaust PAH analysis was reported by Lee et al. [44]. In this procedure, the particulate extract was first enriched for PAH by Grimmer and Böhnke's [62] procedure as described in Sec. IV.A. The enriched PAH fraction was then analyzed by HPLC using tandem UV detectors set at different wavelengths. The same sample was generally analyzed two to three times in order to obtain responses at three to five different wavelengths. The response ratios obtained for individual PAH were compared with standards for identification and quantitation. The method was further validated by comparing with the results obtained from GC/MS.

Despite the above PAH enrichment procedure, it was found that the sample was still too complex and only some 10-12 PAH can be adequately resolved through selective wavelength detection. The complication probably arose

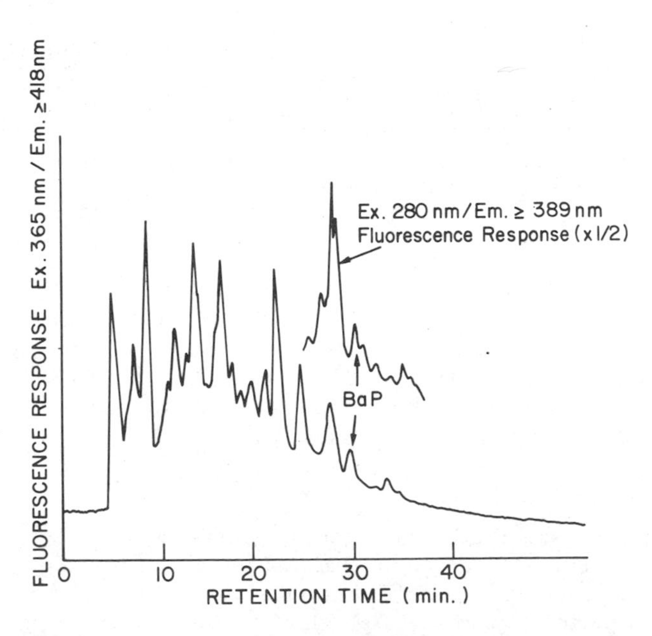

Figure 16. HPLC/fluorescence chromatogram of a diesel exhaust sample monitored at two different fluorescence settings using tandem detector set up. Chromatographic conditions listed in Table 14 (column a, condition 1).

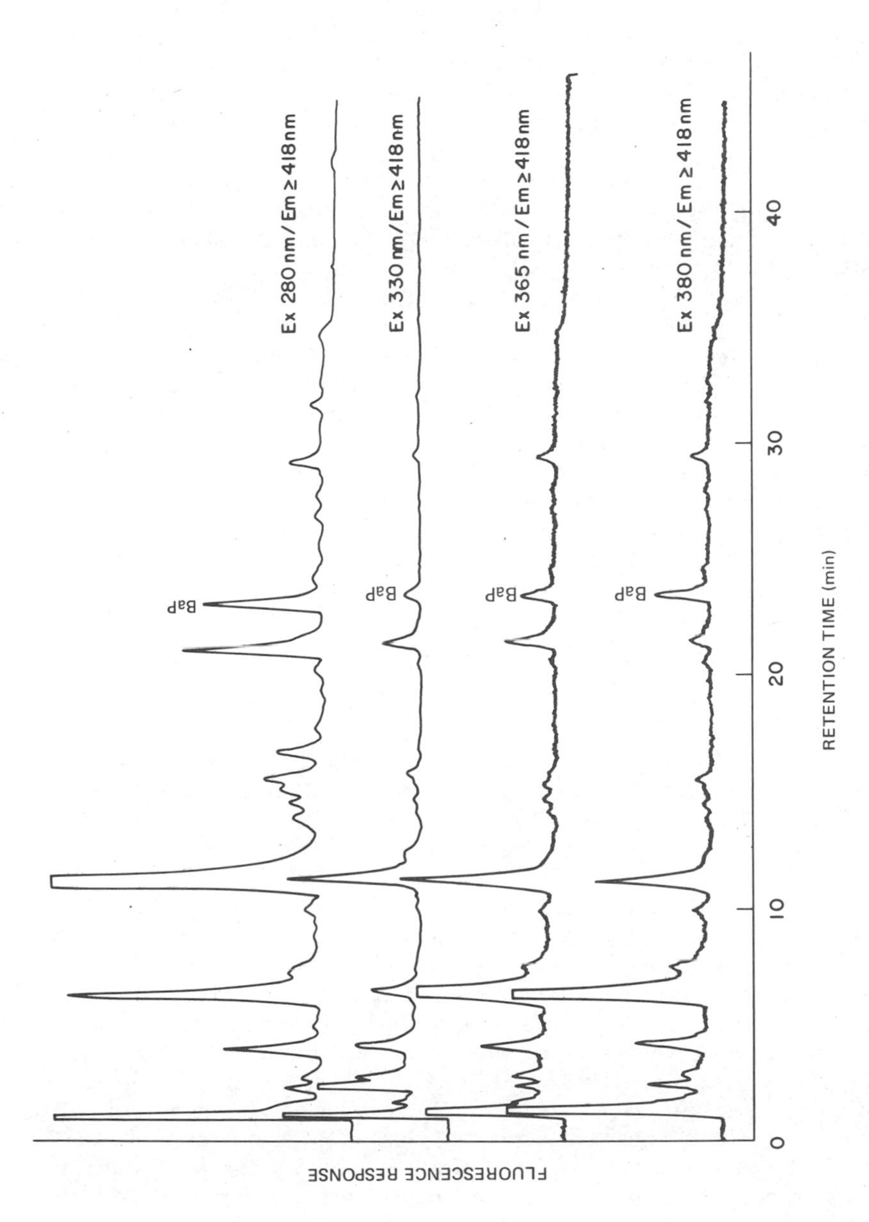

Figure 17. HPLC/fluorescence chromatogram of a diesel particulate extract monitored at different fluorescence conditions. Chromatographic conditions listed in Table 14 (column b, condition 4).

mostly from the alkyl-substituted PAH, which do not elute ideally from the reverse-phase columns according to ring numbers.

Table 15 summarizes the relative UV responses measured at different wavelengths and the retention times calibrated on a Waters C_{18} column. It is seen that methyl anthracenes and methyl phenanthrenes elute in the region of four-ring PAH such as pyrene. Similarly, it was also found later that methyl and ethyl pyrenes elute close to perylene and BaP (not given in the table). The exhaust sample generally contain abundant alkyl PAH, which makes complete separation of PAH species by current HPLC columns extremely difficult if not impossible.

The peak overlapping problems can be partially solved by the selection of different wavelengths. Thus, Table 15 shows that phenanthrene and anthracene can be quantitatively measured at 254/280 nm or 280/313 nm combinations, respectively. Pyrene and fluoranthene can be selectively analyzed at 280/340 nm. Chrysene and BaA are the most difficult to analyze, due partially to the severe overlapping of these with other unknown peaks which are possibly methyl derivatives of three- or four-ring PAH. This assumption is supported by the calibrated relative retention times of various methyl-BaA and methyl chrysenes on several different reverse-phase columns in an extensive study by Wise et al. [85].

The five- and more ring PAH could be adequately measured at longer wavelengths. Perylene and anthanthrene can be selectively detected and quantified by their strong absorptions at 405 nm and weak responses at 340 nm. Similarly, BaP, BeP, and BkF were well separated with the selection of 313/340 nm and 340/365 nm combinations. The coronene peak is well separated from the other PAH or interfering peaks at several wavelengths ranging from 280 to 365 nm. The peak was always observed in our analysis and served as a good marker for the calibration of the retention times of other PAH.

The foregoing discussion has shown that selective PAH can be quantitatively determined by HPLC in combination with UV detectors. Obviously, the relative response factors at different wavelengths as discussed here depend strongly on the type of detectors used. The values listed in Table 15 thus serve only as a general guide for wavelength selection. The recent availability of commercial scanning UV detector for HPLC may further facilitate the HPLC/UV technique in PAH analysis.

The HPLC/UV results have been quantitatively compared with those obtained from GC/MS, as illustrated in Table 16. In general, the agreement is within ±30% except for the perylene/BaP/BeP measurement in sample 1 and the phenanthrene/anthracene measurement in sample 2. The larger discrepancies in the latter two cases were probably due partly to the fact that the isomeric PAH peaks were not well separated from each other and from other interfering peaks under the GC/MS conditions. The packed column used for these comparisons was later replaced by capillary columns in our routine PAH analysis work; for more discussion regarding this work, see Sec. V.C.

The HPLC/UV method can probably be further improved by preseparating the PAH fraction according to ring numbers before final reverse-phase HPLC analysis. Wise et al. [90] have found that PAH and alkyl-substituted PAH can be fractionated according to their ring numbers on a normal-phase amino-bonded column. Such separation has been applied to the analysis of methyl-PAH air particulates and the authors were able to measure a number of PAH compounds by UV or fluorescence detection.

Table 16. Comparison of GC/MS and HPLC/UV for the Determination of PAH in Gasoline-Engine Exhaust

Compound	Sample 1[a] (μg/sample found)				Sample 2[a] (μg/sample found)			
	(a) HPLC		(b) GC/MS	$\frac{a-b}{\bar{x}}$ (100%)	(a) HPLC		(b) GC/MS	$\frac{a-b}{\bar{x}}$ (100%)
Phenanthrene	16	} 19	} 22	-14	9	} 11	} 22	-66
Anthracene	3				2			
Fluoranthane	25		37	-26	21		25	-17
Pyrene	29		29	0	25		33	-28
Chrysene	35	} ≤145	54	—	15	} ≤85	56	—
BaA	110				70			
Perylene	25	} 79	} 36	+57	8	} 44	} 38	+15
BaP	14				10			
BeP	40				26			
Benzo[ghi]perylene	12	} 18	} 20	-11	20	} 22	} 18	+20
Anthanthrene	6				2			
Coronene	6		6	0	4		4	0
Dibenzopyrenes	3		r	-29	0.3		0.4	-29

[a] Notes: [a]HPLC conditions given in Table 15, footnote b. [b]GC/MS analysis on a packed Dexsil 300 glass column, temperature programmed from 100°C to 300°C at 8°C/min.

Source: Data from Ref. 44.

b. HPLC/Fluorescence Very limited work has been reported on the application of the HPLC/fluorescence method for the quantitative determination of multiple PAH in vehicle exhaust. Such a procedure is currently being developed in our laboratory for the analysis of diesel particulates and vehicle exhaust particulates collected along roadways. The only other report we are aware of are those by Nielsen on gasoline-engine exhaust [91] and by Gibson et al. on diesel-engine exhaust [21]. The two procedures are described here.

We have used a tandem fluorescence detector setup for all our HPLC fluorescence measurements. The sample was first fractionated by the normal-phase silica column procedure described earlier (in Sec. IV.C). The two PAH fractions (see Fig. 13) α_1 and $\alpha_2 + \beta$ were collected separately along with other fractions γ_1, γ_2, and δ_1. The collected α_1 and $\alpha_2 + \beta$ fractions were each evaporated to dryness by N_2. They were then redissolved in CH_3CN for subsequent HPLC analysis. The polar fractions γ_1, γ_2, and δ_1 were further analyzed by GC/MS, for which detailed descriptions are given later in Sec. V.C.

The normal-phase HPLC procedure was found to greatly reduce the amount of interfering species in the sample. This is illustrated in Figs. 18 and 19, in which the chromatograms of an exhaust particulate extract sample without prefractionation steps are compared with those after HPLC fractionation and DMSO solvent/solvent extraction steps (see Sec. IV.B). By comparing Fig. 18 with Fig. 19a, we can see that nearly all the earlier-eluting polar species in the former chromatogram are effectively removed by the normal-phase

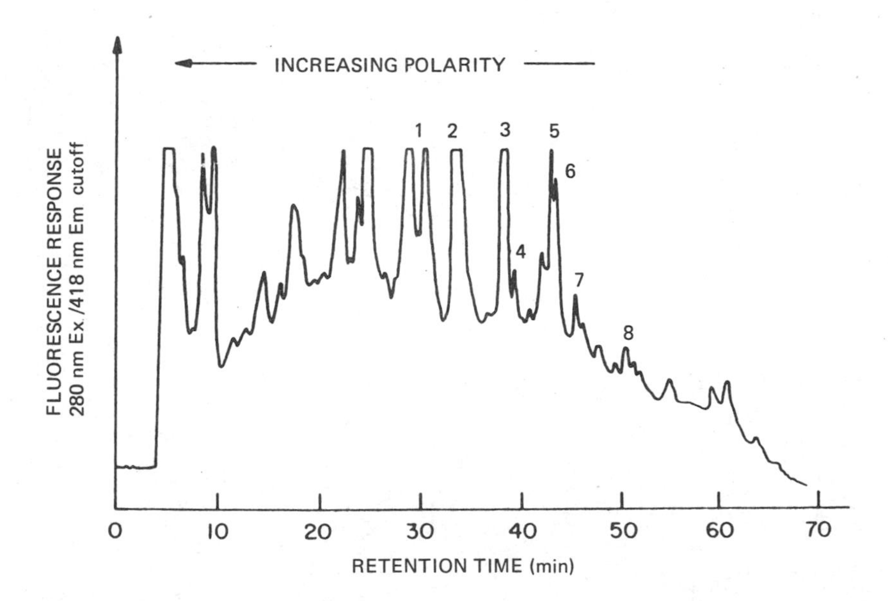

Figure 18. HPLC/fluorescence analysis of a diesel exhaust particulate extract. Sample was separated on a Zorbax ODS column with 50% CH_3CN in 40 min at 1 ml/min flow rate. Peak designation: (1) anthracene, (2) fluoranthene; (3) pyrene; (4) chrysene; (5) perylene; (6) BkF; (7) BaP; (8) benzo[ghi]perylene.

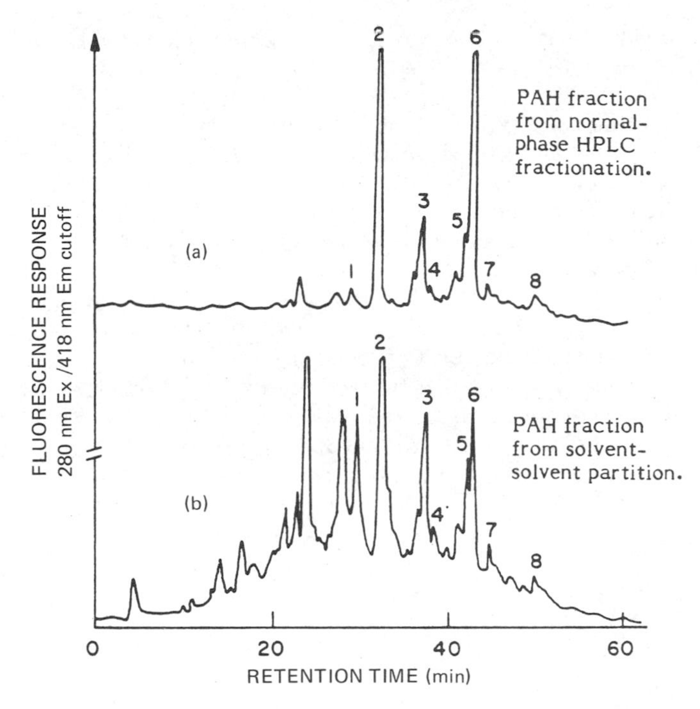

Figure 19. Reverse-phase HPLC/fluorescence analysis of the PAH fraction of an exhaust particulate extract: (a) PAH fractions ($\alpha_1 + \alpha_2 + \beta$) collected from normal phase HPLC fractionation; (b) PAH enrichment by DMSO solvent extraction (Sec. IV.B). Chromatographic conditions and peak designations same as in Fig. 18.

HPLC preseparation procedure, as illustrated in the latter figure. In addition, the "apparent" sizes of the anthracene, fluoranthene, pyrene, and perylene peaks in the raw sample were also greatly reduced after HPLC fractionation, presumably due to the elimination of coeluting interfering peaks. Similarly, the DMSO solvent extraction procedure was also found to be effective in enriching the PAH species, as illustrated in Fig. 19b.

A comparison of the two chromatograms given in Fig. 19 suggests that HPLC method is superior in its ability to isolate the PAH fraction for further analysis. Thus, it is seen that a hump due to residue impurities exists under the PAH peaks in the DMSO extract chromatogram (chromatogram b). This is not observed in the HPLC fractionated sample (chromatogram a), where a flat baseline was observed instead. The quantities of PAH recovered in the two procedures were in good agreement except for BkF (peak 6). The peak was probably contaminated by unknown peaks in the HPLC fractionated sample.

Figure 19a shows that, despite the normal-phase prefractionation step, most of the PAH of interest were still subject to interferences under the chromatographic conditions. The selection of 365-nm excitation (Ex) and

≥418-nm emission (Em) maxima removed most of these interference peaks, but the sensitivities for detection were also reduced for most of the PAH of interest.

A different chromatographic condition (column b in Table 14) was later used for the analysis of a similar sample, as shown in Fig. 20 (α_1 fraction) and Fig. 21 ($\alpha_2 + \beta$). Here, BeP, perylene, BkF, BaP, benzo[ghi]perylene, dibenzo[a,b]pyrene, and coronene peaks were well resolved. The BaA peak was partially resolved, but severe overlappings were still observed for light PAH such as anthracene, fluoranthene, and pyrene.

By comparing Fig. 20 (α_1) with Fig. 21 ($\alpha_2 + \beta$), we see that adequate fractionation of the light (three- or four-ring) and heavy PAH (four- or five-ring) was achieved by the normal-phase HPLC procedure. Such division is advantageous since most of the PAH of health interest can be concentrated in the $\alpha_2 + \beta$ fractions for further analyses.

As can be seen in Figs. 20 and 21, the simultaneous monitoring of fluorescence at 280/≥418 and 365/≥418 Ex/Em combinations can be used to check the

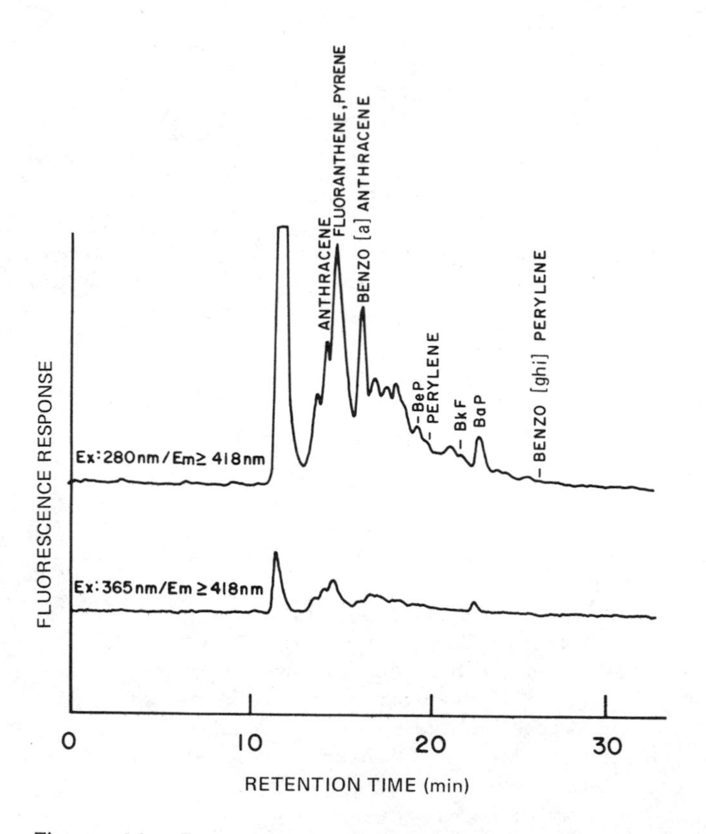

Figure 20. Reverse-phase HPLC/fluorescence chromatograms of the α_1 fraction of a vehicle exhaust particulate extract analyzed under two different fluorescence conditions using tandem detectors. Chromatographic conditions given in Table 14 (column b, condition 4).

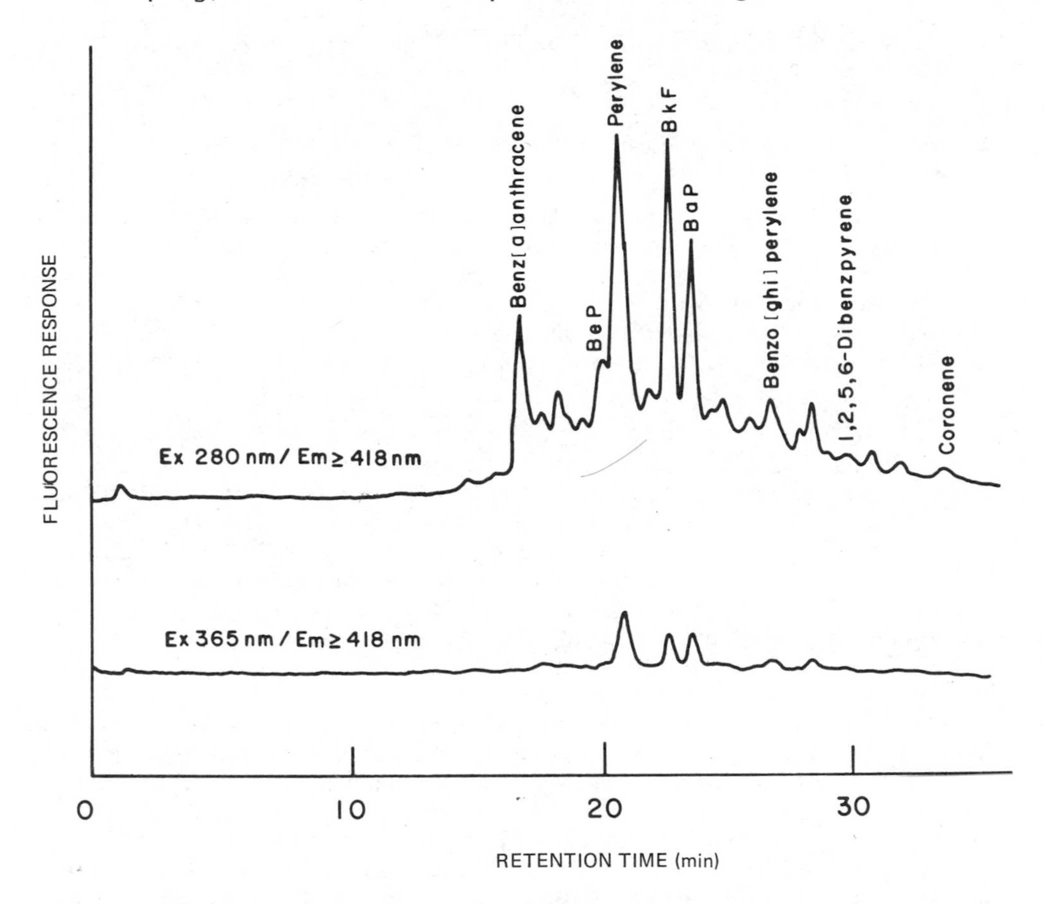

Figure 21. Reverse-phase HPLC/fluorescence chromatograms of the ($\alpha_2 + \beta$) fraction of a vehicle exhaust particulate extract analyzed under two different fluorescence conditions using tandem detectors. Chromatographic conditions given in Table 14 (column b, condition 4).

purity of the fluorescence peak measurement. Such combination of fluorescence settings is not implied to be superior to the other possible combinations but rather represent a compromise between sensitivity and selectivity in our particular situation where limited sample sizes were available.

Our results presented above are generally consistent with those reported by Nielsen [91], cited earlier in this subsection. In his work, selective detection for exhaust PAH were obtained by the combination of 340/425 nm and 363/435 nm. A stop-flow/scanning technique was also used for the identification of PAH. The sample was preseparated on a TLC silica gel plate developed by hexane, toluene/cyclohexane (1:1, v/v) sequence. The PAH fraction was eluted in the toluene/cyclohexane mixture. The isolated PAH fraction was then analyzed by HPLC on a reverse-phase ODS column. With this procedure, the author was able to measure anthracene, fluoranthene,

1-methylanthracene, pyrene, BaA, BaP, and benzo[ghi]perylene in a number of vehicle exhaust samples.

C. Gas Chromatography/Mass Spectrometry (GC/MS) and Other Mass Spectrometric Techniques

To date, only a limited number of studies have been reported on the application of GC/MS and other mass spectrometric techniques to the analysis of PAH and PAH derivatives in vehicle exhaust. At this time, little work has been reported on the use of GC techniques alone for these analysis [30]. A study was described in Sec. V.B.2 that compared the analysis of PAH using HPLC/UV and GC/MS techniques. In general, the agreement was good to within ±30%.

1. *PAH Analysis*

Lee et al. have used high-resolution glass capillary column gas chromatography/mass spectrometry (GCGC/MS) techniques for the identification and quantification of PAH in vehicle particulate extracts [44]. The advantages of this technique are specificity for unequivocal identification and versatility for the separation of large numbers of compounds. Normally, capillary columns are limited to 0.5- to 2.0-μl injections, making identification and quantitation difficult because of the dilute concentrations of PAH in exhaust samples. The use of a falling-needle solids injector allows from 2 to 20 μl to be injected into the capillary column.

Figure 22 shows the reconstructed ion chromatogram (RIC) for the GCGC/MS analysis of PAH in a particulate sample collected from a stratified-charge engine. The RIC is much like an FID trace in which the computer accumulates the total ion signal produced from each scan and plots this signal vs. the scan number. The mass chromatogram plots only the signal from a given ion vs. the scan number. The m/z 252 ion in the figure represents the parent ion of the PAH isomers such as BaP, BeP, and perylene. The mass spectra of these three PAH are indistinguishable by the computer, and so one must rely on accurate retention times for their identification.

The PAH components were quantitated by comparing the areas of the parent ion mass chromatograms to either internal or external standards, weighting each PAH with an appropriate response factor. The concentration of BaP, for example, in Fig. 22 is represented by the shaded area in the m/z 252 mass chromatogram.

Petersen et al. [36] have improved the aforementioned technique by operating the mass spectrometer in a selected ion monitoring mode (SIM) and coinjecting deuterated PAH with the sample. [The stable isotope used is the perdeutero derivatives of PAH, where each proton (^{1}H) on the PAH nucleus is replaced with a deuteron (^{2}H).] In this mode, the mass spectrometer concurrently monitors one or more ions characteristic of a specified compound during its expected elution time from the gas chromatographic column. The sensitivity of this technique is from two to three orders of magnitude more sensitive than the continuous scanning technique.

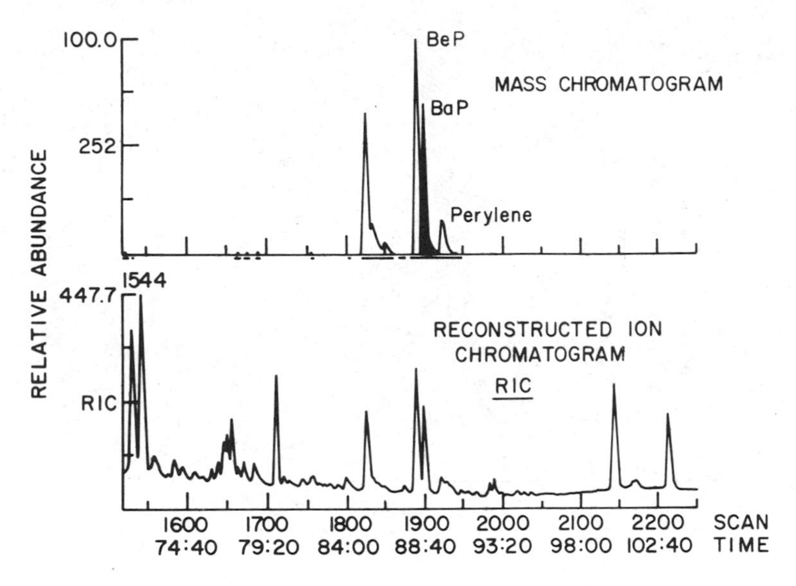

Figure 22. Reconstructed ion chromatogram (RIC) and a mass chromatogram from m/z 252 used to identify perylene, BeP, and BaP in the GC/MS analysis of a diesel particulate sample using a 25-m glass capillary Dexsil 300 column.

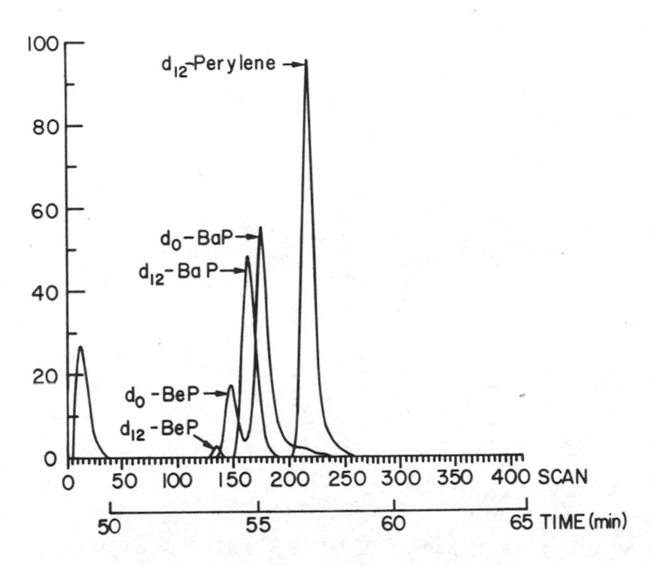

Figure 23. Computer reconstructed selected ion monitoring chromatogram of the native and deuterated benzopyrenes and perylene (SE-52 glass capillary column, 30 m × 0.25 mm. (From Ref. 37.)

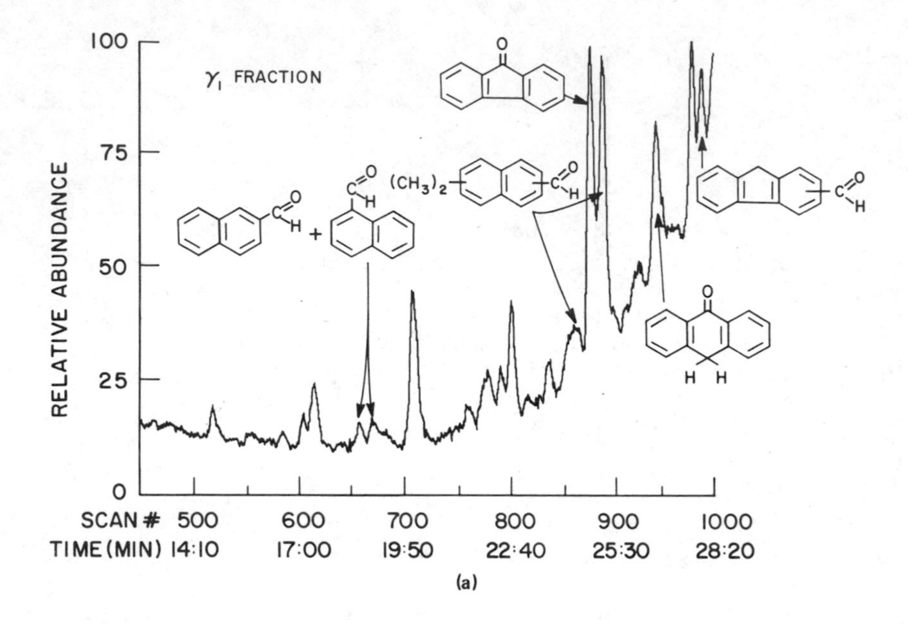

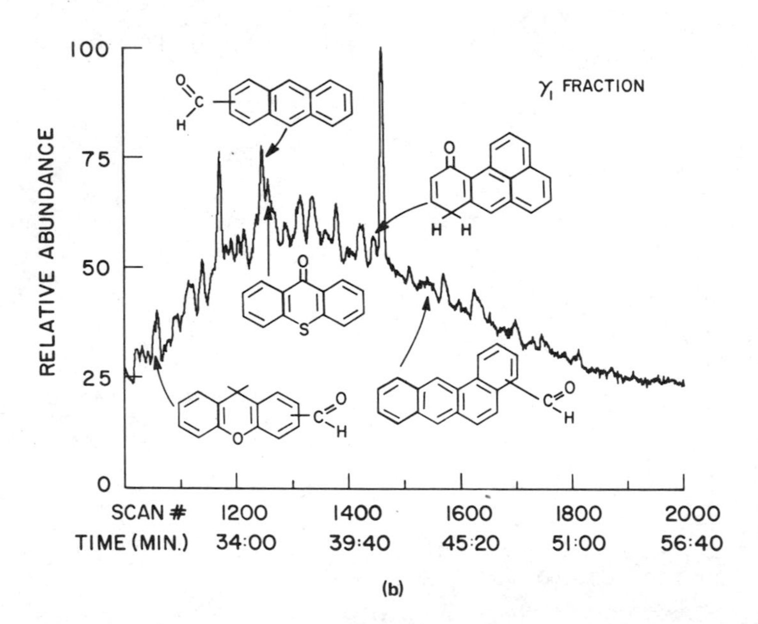

Figure 24. (a,b) Total ion chromatograms for the GC/MS analysis of the γ_1 HPLC fraction for a diesel particulate extract utilizing a 6-ft SP 2250 column. (From Ref. 19.)

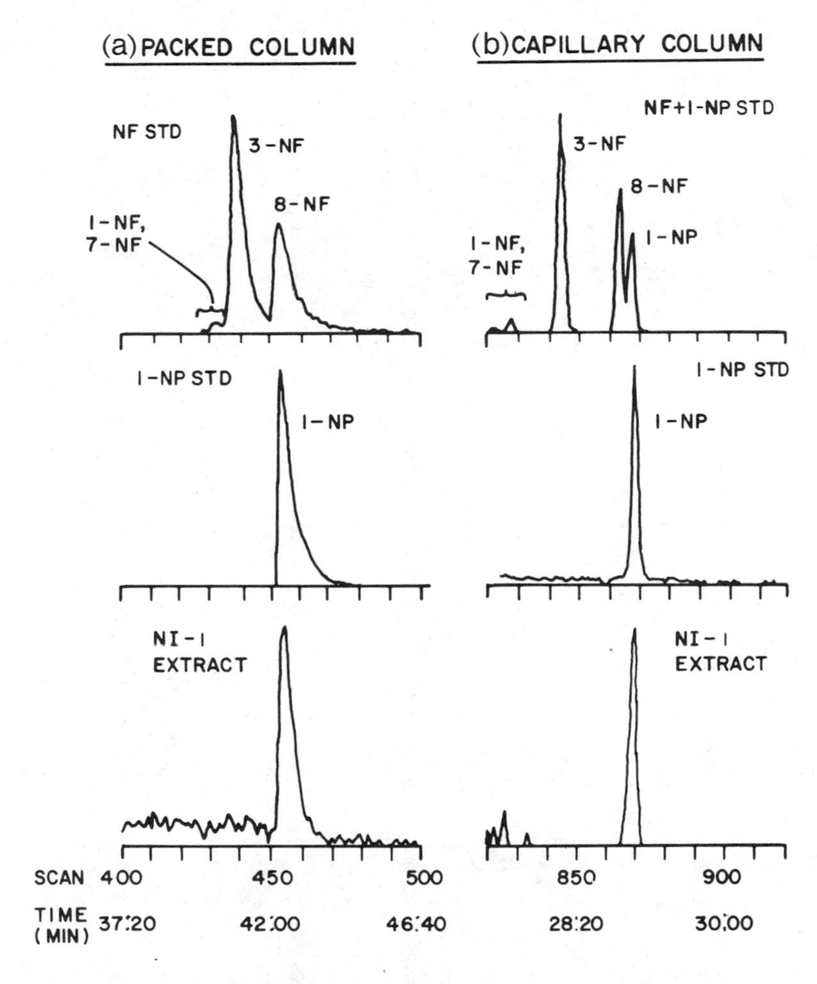

Figure 25. GC/MS Mass chromatograms of the m/z 247 ion generated from 3-mitrofluoranthene (3-NF), 8-nitrofluoranthene (8-NF), 1-nitrofluoranthene (1-NF), 7-nitrofluoranthene (7-NF), and 1-nitropyrene (1-NP) standards and a HPLC fractionated diesel particulate extract (NI-1): (a) 180-cm 1% SP 2250 packed column separation; (b) 30-m SE-54 capillary column separation. [From Ref. 20; reprinted with permission from D. Schuetzle et al., *Anal. Chem. 54*:265 (1982). Copyright 1981 American Chemical Society.]

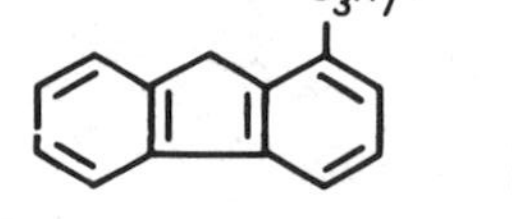
$C_{14}H_8O_2$ (m/z:208.052)
O
O
ANTHRACENE QUINONES
PHENANTHRENE QUINONES
C=O
H
FLUORENONE CARBOXALDEHYDES
CYCLOPENTENONE NAPHTHALENE CARBOXALDEHYDE
$C_{14}H_{12}N_2$ (m/z:208.100)
CH_3
CH_3
N=N
DIMETHYL BENZO [c] CINNOLINES
ETHYL BENZO [c] CINNOLINES
$C_{16}H_{16}$ (m/z:208.125)
C_3H_7
C_3H_7 - FLUORENES
HEXAHYDROPYRENES
$C_{10}H_8SO_3$ (m/z:208.019)
SO_3H
NAPHTHALENE SULFONIC ACIDS

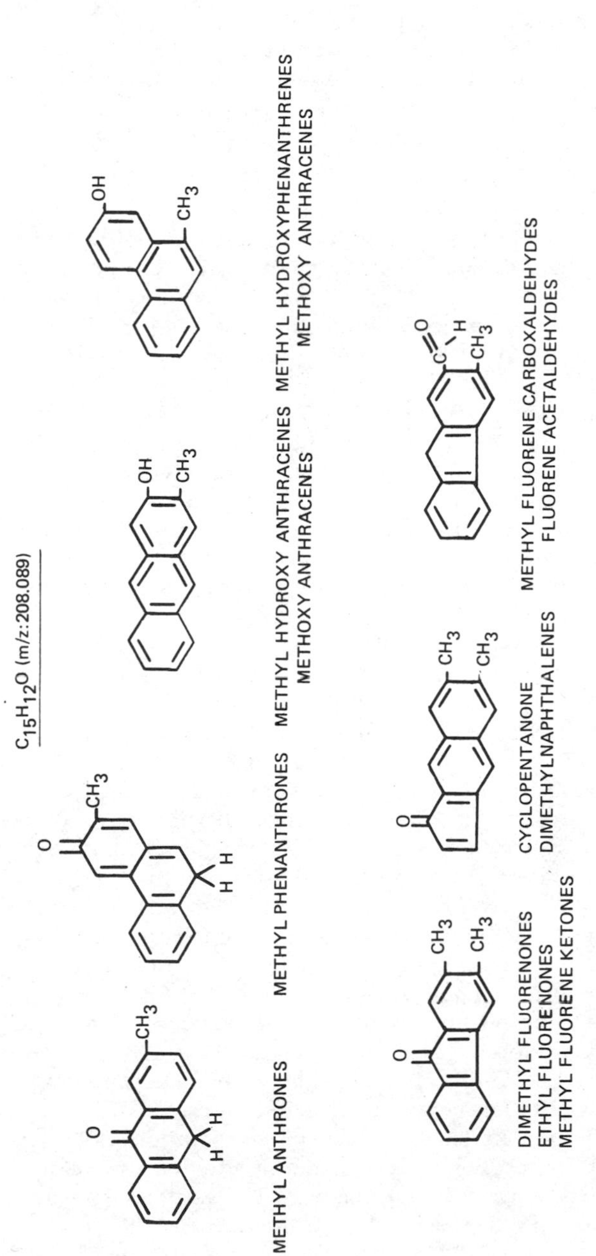

Figure 26. Possible structure of PAH and PAH derivatives with a molecular weight of 208.1 ± 0.1.

Since the gas chromatographic properties of the native PAH and deuterated PAH are similar, they coelute into the mass spectrometer. By monitoring these ions at their retention time, the native PAH can be unequivocally identified. Quantification is accomplished by comparison of the integrated ion current response of the native PAH to that of the deuterated PAH and relating to a standard curve. Figure 23 illustrates the excellent chromatographic resolution for deuterated and native benzopyrenes and perylene in an extract of diesel particulate.

2. *PAH Derivatives*

It has been found that the moderately polar HPLC fractions of diesel particulates account for most of the Ames *Salmonella* mutagenicity when compared to the nonpolar (PAH) and polar fractions [10,19]. These fractions mostly contain PAH derivatives which appear to be responsible for this observed mutagenicity. Therefore, analytical studies are beginning to focus on this important group of compounds.

Schuetzle et al. [19] reported the identification of more than a hundred PAH derivatives in moderately polar HPLC fractions using GC/MS and HRMS. The PAH derivatives identified included PAH substituted with hydroxy, ketone, quinone, carboxaldehyde, acid anhydride, acid, dihydroxy, and nitro groups. Other investigators have reported the identification of PAH derivatives in similar fractions using GC/MS including alkyl-9-fluoreneones [14], nitronaphthalene [15], and cyclopenteno[c,d]pyrene dicarboxylic acid anhydride [16]. PAH derivatives are eluted in the γ_1 (γ_{1x} and γ_{1y}), δ_1, δ_2 fractions which are separated using a normal-phase silica column procedure as described earlier in Sec. IV.C (see Fig. 12). Analysis of these fractions were undertaken using 6-ft packed SP2250 and glass capillary fused silica columns. Figure 24a,b gives the computer-reconstructed total ion current chromatogram for the packed column GC/MS analysis of HPLC fraction γ_1. Some representative PAH derivatives are shown in this chromatogram.

Analyses of these fractions are greatly enhanced through the use of fused silica glass capillary columns. Figure 25 gives the mass chromatograms for m/z 247 ($C_{16}H_9NO_2$) used for the identification of 1-nitropyrene in the γ_1 HPLC fraction of a diesel particulate extract utilizing both the packed and fused silica columns. The fused silica solumn was able to separate the 1-nitropyrene from the 8-nitrofluoranthene species, whereas these two species appear as one peak when using the packed column.

These fractions were found to be highly complex, each containing several hundred PAH derivatives. HRMS, in conjunction with GC/MS and direct-probe introduction, was required to determine the identity of specific compounds. HRMS was used to provide an accurate molecular formula to help substantiate GC/MS identifications. At least 12,000 mass resolution (10% valley) proved to be necessary to resolve mass fragments adequately. For instance, at mass 208.1 ± 0.1 there are a large number of possible structures of PAH and PAH derivatives, as shown in Fig. 26. High-accuracy mass chromatograms were generated for mass 208.1 ± 0.1, as shown in Fig. 27. The only ions present were at mass 208.089 and 208.052. Direct-probe HRMS confirmed that these were the only two ions present at mass 208.1 ± 0.1.

All of the species shown in Fig. 27 with the compositional formula $C_{15}H_{12}O$ and $C_{14}H_8O_2$ could be eluted in the γ_1 fraction, as based upon studies for the

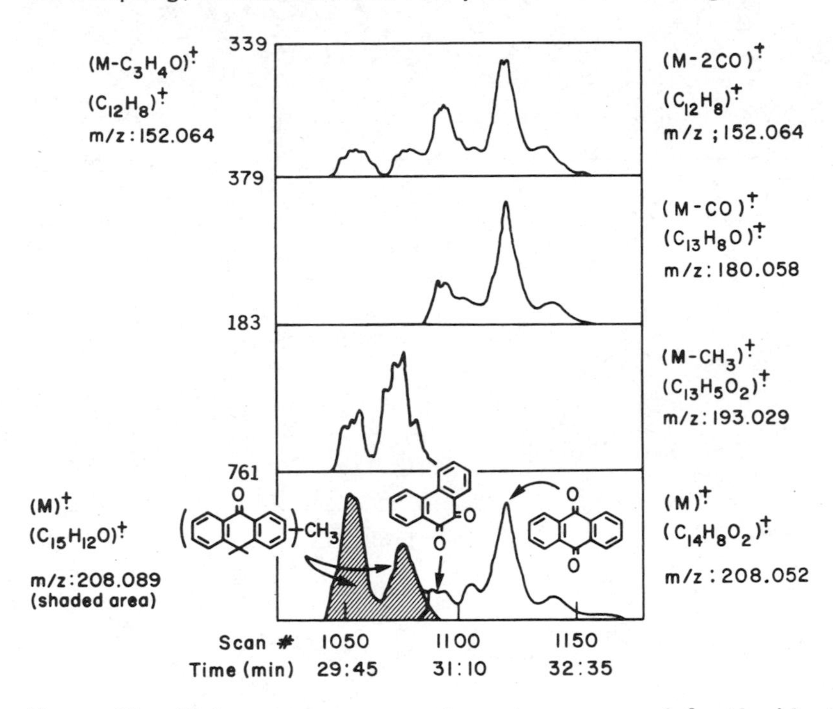

Figure 27. High-accuracy mass chromatograms used for the identification of 9,10-phenanthrene quinone, 9,10-anthracene quinone, and two methyl-PAH (anthrones and phenanthrones) in the γ_1 HPLC fraction of diesel particulates extract OL-1 utilizing a 6-ft SP2250 column. (From Ref. 19.)

elution of PAH derivatives through the HPLC column. The absence of ions with the compositional formula $C_{14}H_{12}N_2C_{16}H_{16}$ and $C_{10}H_8SO_3$ is confirmed by the HPLC elution of cinnolines and naphthalene sulfonic acids in the δ_1 fraction and the $C_{16}H_{16}$ hydrocarbon in the α_2 fraction. Identification of specific isomers requires matching of retention times and mass spectra with that of a synthesized standard. However, even with the data most identifications must still be considered tentative. Recent work from this laboratory has shown that in addition to the above, absolute identification of specific isomers for nitrated PAH derivatives requires a comparison of results from two or more GC columns and mass spectrometric ionization procedures [20].

The results of the work from this laboratory on the analysis of PAH derivatives are extensive, and some of these compounds are summarized in Tables 17 and 18.

D. Summary of PAH Vehicle Emission Analysis

Some results concerning the analysis of PAH emissions from gasoline and diesel engine exhaust are given in Tables 19 and 20, respectively. It is obvious that there is considerable variation in the level of PAH species in particulates generated from catalyzed and uncatalyzed gasoline engine exhaust.

Table 17. Derivatives of Some Polycyclic Compounds Identified in Diesel Particulate Extracts

Derivative	PAH[a]		
Derivative	Naphthalene(NA)	Methyl-NA(MNA)	Dimethyl-NA(DNA)
Carboxaldehyde	1-NA; 2-NA	1	1
Acetaldehyde	2	—	—
Dicarboxaldehyde	3	—	—
Dicarboxylic acid	1	—	—
Dicarboxylic acid anhydride	1	3	4
Nitro	1-NA; 2-NA	—	—
Dinitro	1,5-NA; 1,8-NA	—	—
Derivative	Fluorene(FL)	Methyl-FL(MFL)	Dimethyl-FL(DFL)
Hydroxy	9-FL	—	—
Ketone	9-FL	3	—
Quinone	4	3	3
Carboxaldehyde	3	4	—
Nitro	2-FL	2	—
Dinitro	2,7-FL	—	—
Derivative	Phenanthrene(PH) Anthracene(AN)	Methyl-PH(MPH) Methyl-AM(MAN)	Dimethyl-PH(DPH) Dimethyl-AN(DAN)
Ketone	2; 9-AN	2	—
Quinone	2; 9,10-AN; 9,10-PH	3	—
Carboxaldehyde	2	4	5
Nitro	3; 2-AN; 9-AN	5; 10,9-MAN; 9,1-MAN	6
Derivative	Pyrene(PY) Fluoranthene(FL)	Methyl-PY(MPY) Methyl-FL(MFL)	Dimethyl-PY(DPY) Dimethyl-FL(DFL)
Ketone	3	4	—
Quinone	2	2	—
Carboxaldehyde	1	3	—
Nitro	1-PY; 3-FL(trace)	6; 3,1-MPY; 6,1-MPY	—
Dinitro	1,3-PY; 1,6-PY; 1,8-PY		

[a]The first listing in the table designates the number of unidentified isomers present. Exact isomer designation is abbreviated as 1-NA (1-naphthalene carboxaldehyde) for the carboxaldehyde derivative.

Table 18. The Concentration of Various PAH Compounds and PAH-Derivatives in the Nonpolar and Moderately-Polar Fractions of a Diesel Particulate Extract[a]

Compound	Fraction[b] concentration (ppm/1000)
Nonpolar Fractions (α_1, α_2, β)[c]	
PAH	
Phenanthrenes and anthracenes	1.1
Methyl (phenanthrenes and anthracenes)	2.6
Dimethyl (phenanthrenes and anthracenes)	5.8
Pyrene	3.1
Fluoranthene	2.5
Methyl (pyrenes and fluoranthenes)	1.4
Chrysene	0.18
Cyclopenta[c,d]pyrene	0.03
Benzo[ghi]fluoranthene	0.24
Benzo[a]anthracene	0.95
Benzo[a]pyrene	0.07
Total	18.0
Other PAH, heterocyclics	62.0
Hydrocarbons and alkyl benzenes	920.0
Total	100.0
Moderately Polar Fractions (γ_1, γ_2)[d]	
PAH-ketones	
Fluorenones	43.2
Methyl fluorenones	4.8
Dimethyl fluorenones	1.8
Anthrones and phenanthrones	17.9
Methyl (anthrones and phenanthrones)	17.7
Dimethyl (anthrones and phenanthrones)	14.4
Fluoranthones and pyrones	13.4
Benzanthrones	2.1
Xanthones	3.6
Methylxanthones	1.8
Thioxanthones	17.0
Methyl thioxanthones	9.3
Total	147.0
PAH-carboxaldehydes	
Fluorene carboxaldehydes	17.6
Methyl fluorene carboxaldehydes	3.9
(Phenanthrene and anthracene) carboxaldehydes	28.2

(continued)

Table 18 (continued)

Compound	Fraction[b] concentration (ppm/1000)
Methyl (anthracene and phenanthrene carboxaldehydes)	17.7
Dimethyl (anthracene and phenanthrene carboxaldehydes)	4.8
(BaA, chrysene and triphenylene carboxyaldehydes)	4.4
Naphthalene dicarboxaldehydes	3.5
Dimethyl naphthalene carboxaldehydes	3.5
Trimethyl naphthalene carboxaldehydes	10.5
(Pyrene and fluoranthene) carboxaldehydes	17.3
Xanthene carboxaldehydes	6.6
Dibenzofuran carboxaldehydes	4.2
Total	122.2
PAH-acid anhydrides	
Naphthalene dicarboxylic acid anhydridies	31.2
Methyl naphthalene dicarboxylic acid anhydrides	11.3
Dimethyl naphthalene dicarboxylic acid anhydrides	5.0
(Anthracene and phenanthrene) dicarboxylic acid anhydrides	6.6
Total	54.1
Hydroxy-PAH	
Hydroxy fluorene	14.8
Methyl hydroxy fluorene	4.1
Dimethyl hydroxy fluorene	16.7
Hydroxy (anthracenes and phenanthrenes)	6.3
Hydroxy methyl (anthracenes and phenanthrenes)	9.5
Hydroxy dimethyl (anthracenes and phenanthrenes)	14.4
Hydroxy fluoreneone	22.4
Hydroxy xanthone	14.1
Hydroxy xanthene	10.8
Total	113.1
PAH-quinones	
Fluorene quinones	8.1
Methyl fluorene quinones	6.2
Dimethyl fluorene quinones	5.3
(Anthracene and phenanthrene) quinones	20.4
Methyl (anthracene and phenanthrene) quinones	22.4

(continued)

Table 18 (continued)

Compound	Fraction[b] concentration (ppm/1000)
(Fluoranthene and pyrene) quinones	2.1
Naphtho(1,8-c,d) pyrene-1,3-dione	6.8
Total	71.3
Nitro-PAH	
Nitrofluorenes	0.34
Nitroanthracenes	0.71
Nitrofluoranthrenes	<0.05
Nitropyrenes	1.5
Nitro methyl pyrenes	0.25
Total	2.9
Other Oxygenated PAH	83.4
PAH carry-over (from α_1, α_2, β)	66.0
Phthalates, hydrocarbon contaminants	340.0
Total	1000.0

[a]See Ref. 37 for experimental details concerning sample collection procedure.
[b]Fraction concentration determined from the summation of the three most abundant ions in the mass spectra.
[c]Divide fraction concentration by 1.8 to determine total extract concentration. Fraction cut defined from elution time zero to the elution time of methyl coronene.
[d]Divide fraction concentration by 10.9 to determine total extract concentration. Fraction cut defined from elution of 2-nitrofluorene to that of 7-benz[a]anthrone.

Limited information is available on the concentration of PAH in diesel exhaust particulates. Extensive studies have been undertaken [37] to develop suitable sampling and analysis techniques for PAH in these diesel exhaust particulates. It has been shown using tracer studies that certain PAH, such as pyrene and BaP, undergo substantial losses. The data in Table 20 represent the concentration of several PAH and PAH derivatives determined to date in a diesel particulate extract sample. Considerably more work will be required to determine the concentration of these species from a number of different engines operating under a variety of conditions with several fuels.

Table 19. PAH Emissions of Gasoline Engine Exhaust

Sample type:	Particulates/ condensates (ppm (wt/wt)	Particulates/ condensates μg liter^{-1} fuel burned		Particulates/ condensates μg liter^{-1} fuel burned		Particulates/ condensates μg liter^{-1} fuel burned	Exhaust tar ppm (wt/wt)	Particulates ppm (wt/wt)	
Compound		Low PAH fuel	High PAH fuel	Leaded fuel	Unleaded fuel			With oxid. catalyst	No oxid. catalyst
	Concentration								
Anthracene	44	—	—	18.2-268.1	25.9-392.5	588	—	1.4	21
Methyl anthracenes	—	—	—	—	—	98	—	—	⩽190
Phenanthrene	—	—	—	—	—	2643	—	7.4	70
Methyl phenanthrenes	—	—	—	—	—	1175	—	—	⩽190
Fluoranthene	972	—	—	17.8-183.8	24.0-235.5	1361	69.7	7.9	91
Chrysene	175	12.5	34.6	27.9- 50.9	29.8- 64.9	104	182.7	—	70
Methyl chrysenes	67	—	—	—	—	10	—	—	—
Pyrene	1770	73.2	182	28.2-192.7	38.6-570.5	2517	116.5	59	115
Methyl pyrenes	8	—	—	—	—	iden.	—	—	—
BaA	131	6.74	24.7	7.3- 29.7	8.6- 32.4	67	174	—	⩽280
BeP	422	10.34	33.0	4.6- 16.9	3.6- 38.9	48	322	17	119
BaP	73	2.30	8.45	1.1- 10.2	1.6- 16.6	66	228	9	42
Perylene	trace	—	—	—	—	11	—	20	59
Methyl BaPs	1	0.67	1.60	—	—	—	—	—	—
BkF	54	—	—	—	—	12	—	—	—
Benzo[ghi]perylene	—	16.1	39.9	—	—	224	—	—	63
Coronene	338	—	—	1.6- 41.8	1.4- 52.0	189	—	14	18
Reference	[30]	[68-70]		[71]		[74]	[92]	[44]	

Table 20. PAH Emissions of Diesel Engine Exhaust (ppm of Particulates) Collected in Dilution Tubes

Compound	Concentration (ppm in particulates)	
	Ref. 19[a]	Ref. 36
Phenanthrene and anthracene	428	–
Methyl (phenanthrenes and anthracenes)	18	–
Pyrene	217	19-1413
Fluoranthene	175	–
Dimethyl (phenanthrenes and anthracenes)	406	–
Methyl (pyrenes and fluoranthenes)	98	–
Trimethyl (phenanthrenes and anthracenes)	196	–
Cyclopenta[c,d]pyrene	2.1	–
Benzo[ghi]fluoranthene	17	–
Benz[a]anthracene (BaA)	66	3-55
Chrysene	98	16-307
Diphenylbenzene	58	–
Methyl (benz[a]anthracenes, chrysenes, triphenylenes)	223	–
Benzo[a]pyrene (BaP)	4.9	0.3-23
Benzo[e]pyrene (BeP)	4.0	0.5-47
Perylene	0.2	0.1-14
Coronene	32	–
Dibenzopyrenes	11	–

[a]Relative abundance only given in Ref. 19—absolute quantitative values were determined as a result of further work in this laboratory.

VI. Conclusions

An overview of the current status in exhaust PAH emission research is provided in this chapter. It is evident from our discussion that much advancement in analytical technique development has occurred in the area recently. HPLC is currently under extensive usage for the analysis of exhaust PAH. The method is simple, rapid, and in combination with fluorescence detection provides an extremely sensitive method for the analysis of selected PAH species. The method is, however, very limited in its application to detailed PAH analysis and to the analysis of oxygenated PAH species. The application of more powerful analytical tools such as GC/MS and HRMS in exhaust analysis has uncovered a large number of new PAH as well as oxygenated PAH species.

The major open issue remaining unanswered is the question of sampling degradation. Recent findings of a series of oxygenated PAH species suggest strongly the occurrence of such processes. Many of these oxy-PAH species are thermally and photolytically labile and present a challenging analytical task.

Acknowledgments

Thanks are due M. C. Paputa and T. J. Prater for their dedication to the analytical effort and above all to our wives, Sheng-Chen and Judith Mary, whose patience made such an endeavor possible.

References

1. P. Kotin, H. L. Falk, and M. Thomas, *A.M.A. Arch. Ind. Hyg. 9*:164 (1954).
2. M. L. Lyons, *Brit. J. Cancer 13*:126 (1959).
3. G. E. Moore and M. Katz, *J. Air Pollut. Control Assoc. 2*:221 (1960).
4. E. L. Wynder and E. C. Hammond, *Cancer 15*:79 (1962).
5. D. Hoffman and E. L. Wynder, *Cancer 15*: 93 (1962).
6. D. Hoffman and E. L. Wynder, *J. Air Pollut. Control Assoc. 13*: 322 (1963).
7. D. Hoffman, E. Theiss, and E. L. Wynder, *J. Air Pollut. Control Assoc. 15*:162 (1965).
8. C. R. Begeman and J. M. Colucci, Apparatus for determination of the contribution of the automobile to the benzene-soluble organic matter in air, *Nat. Cancer Inst. Monogr. 9*:17 (1962).
9. C. R. Begeman and J. M. Colucci, *Science 161*:271 (1968).
10. J. Huisingh, R. Bradow, R. Jungers, L. Claxton, R. Zweidinger, S. Tejada, J. Bumgarner, F. Duffield, M. Waters, V. F. Simmon, C. Hare, C. Rodriguez, and L. Snow, Application of bioassay to the characterization of heavy duty diesel particle emissions: Part I. Characterization of heavy duty diesel particle emissions; Part II. Application of a mutagenic bioassay to monitoring light duty diesel particle emissions. Presented at Symposium on Application of Short-Term Bioassays in the Fractionation and Analysis of Complex Environmental Mixtures, Williamsburg, Va. (1978).
11. Y.-Y. Wang, R. E. Talcott, R. F. Sawyer, S. M. Rappaport and E. Wei, Mutagens in Automobile Exhaust. Presented at Symposium on Application

of Short-Term Bioassays in the Fractionation and Analysis of Complex Environmental Mixtures, Williamsburg, Va. (1978).

12. J. Santodonats, D. Basu, and P. Howard, Health effects associated with diesel exhaust emissions: Literature review and evaluation. EPA-600-78-063 (1978).
13. Y. Wang, S. M. Rappaport, R. F. Sawyer, R. C. Talcott, and E. T. Wei. *Cancer Lett.* 5:39 (1978).
14. M. D. Erickson, D. L. Pellizzari, K. B. Tomer, and D. Dropkin, *J. Chromatogr. Sci.* *17*:450 (1979).
15. M. Chaigneau, L. Giry, and L. P. Ricard, *Chim. Anal.* *51*:487 (1979).
16. S. M. Rappaport, Y. Wang, and E. T. Wei, R. Sawyer, B. E. Watkins, and H. Rappaport, *Environ. Sci. Technol.* *14*:15 (1980).
17. J. N. Pitts, Jr., K. A. Van Cauwenberghe, D. Grosjean, J. P. Schmid, D. R. Fitz, W. L. Belser, G. B. Knudson, and P. M. Hynds, *Science* *202*:515 (1978).
18. F. S. C. Lee, T. M. Harvey, T. J. Prater, M. C. Paputa, and D. Schuetzle, Chemical analysis of diesel particulate matter and an evaluation of artifact formation, in *Proceedings of the ASTM Symposium on Sampling and Analysis of Toxic Organics in Source-Related Atmospheres*, ASTM STP 721, pp. 92-110 (1981).
19. D. Schuetzle, F. S.-C. Lee, T. J. Prater, and S. B. Tejada, *Int. J. Environ. Anal. Chem.* *9*:92 (1981).
20. D. Schuetzle, T. J. Prater, T. Riley, T. M. Harvey, and D. M. Hunt, *Anal. Chem.* *54*:265 (1982).
21. T. L. Gibson, A. I. Ricci, and R. L. Williams, Measurement of polynuclear aromatic hydrocarbons, their derivatives, and their reactivity in diesel automobile exhaust, in *Polynuclear Aromatic Hydrocarbons: Chemistry and Biological Effects*, A. Bjørseth and A. J. Dennis (Eds.), Battelle Press, Columbus, Ohio, 1980.
22. S. B. Tejada, R. B. Zweidinger, J. E. Sigsby, Jr., and R. L. Bradow, Identification and measurement of nitro derivatives of PAH in diesel particulate extract. Chemical Characterization of Diesel Exhaust Emissions Workshop, Coordination Research Council, Detroit, Mich. (March 2, 1981).
23. Particulate regulations for light duty diesel vehicles. *Federal Register* (February 1, 1979).
24. Gaseous emissions regulations for 1984 and later model year heavy duty engines. *Federal Register* (January 21, 1980).
25. E. Danielson, Draft recommended practice for measurement of engines under transient conditions. EPA Technical Report SDSB-79-18 (April 1979).
26. C. T. Hare, K. J. Springer, and R. L. Bradow, Fuel and additive effects on diesel particulate development and demonstration of methodology. SAE 760130 (1976).
27. C. T. Hare and R. L. Bradow, Characterization of heavy-duty diesel gaseous and particulate emissions, and effects of fuel composition. Paper No. 790490, Society of Automotive Engineers, Warrendale, Pa. (1979).
28. D. E. McKee, F. C. Ferris, and R. E. Goeboro, Unregulated emissions from a PROCO engine powered vehicle. Paper No. 780592, Society of Automotive Engineers, Warrendale, Pa. (1978).

29. G. Grimmer, A. Hildebrandt, and H. Böhnke, *Zbl. Bakteriol. Hyg., I. Abt. Orig. B158*:22 (1973).
30. J. Kraft and K. H. Lies, Polynuclear aromatic hydrocarbons in the exhaust of gasoline and diesel vehicles. Paper No. 810082, Society of Automotive Engineers Meeting, Detroit (March 1981).
31. T. L. Chan, P. S. T. Lee, and J. S. Siak, *Environ. Sci. Technol.* in press (1981).
32. S. J. Swarin and R. L. Williams, Liquid chromatographic determination of BaP in diesel exhaust particulates: Verification of the collection and analytical methods, in *Polynuclear Aromatic Hydrocarbons: Chemistry and Biological Effects*, A. Bjørseth and A. J. Dennis (Eds.), Battelle Press, Columbus, Ohio, 1980.
33. W. A. Korfmacher, D. F. S. Natusch, D. R. Taylor, E. L. Wehry, and G. Mamantov, Thermal and photochemical decomposition of particulate PAH, in *Polynuclear Aromatic Hydrocarbons*, P. W. Jones and P. Leber (Eds.), Ann Arbor Science Publs., Ann Arbor, Mich., 1979, pp. 165-170.
34. F. S. C. Lee, W. R. Pierson, and J. Ezike, The problem of PAH degradation during filter collection of airborne particulates: An evaluation of several commonly used filter media, in *Polynuclear Aromatic Hydrocarbons: Chemistry and Biological Effects*, A. Bjørseth and A. J. Dennis (Eds.), Battelle Press, Columbus, Ohio, 1980, pp. 543-563.
35. S. C. Barton, N. D. Johnson, and B. S. Das, Final Report to Air Resources Branch, Canadian Ministry of Pnvironment: A feasibility study of alternatives to high-volume sampling for labile constituents of atmospheric particulates. Ontario Research Foundation, Ontario, Canada, 1979.
36. B. A. Petersen, C. C. Chuang, G. W. Kinzer, and D. A. Trayser, Polynuclear aromatic hydrocarbon emissions from diesel engines. Coordinating Research Council Project CAPE-24-72 Progress Report (1980).
37. D. Schuetzle, Sampling of vehicle emissions for chemical analysis and biological testing, Karolinska Institute Symposium, Stockholm, Sweden, Feb. 8, 1982, Environmental Health Perspectives Journal. In press (1982).
38. B. Y. H. Liu and K. W. Lee, *Environ. Sci. Technol. 10*:345 (1976).
39. J. Walter and G. Reischl, *Atmos. Environ. 12*:2015 (1978).
40. T. Truex, Ford Motor Co., Dearborn, Mich., private communication, 1980.
41. E. D. Pellizzari, J. E. Buck, B. H. Carpenter, and E. B. H. Sawicki, *Environ. Sci. Technol. 9*:552 (1975).
42. P. E. Strup, R. D. Giammar, T. B. Stanford, and P. W. Jones, Improved measurement techniques for polycyclic aromatic hydrocarbons from combustion effluents, in *Carcinogenesis: A Comprehensive Survey*, R. I. Freudenthal and P. W. Jones (Eds.), Raven Press, New York, 1976.
43. R. S. Spindt, First and Second Annual Report: Polynuclear aromatic content of heavy duty diesel engine exhaust gases. U.S. EPA Contract No. 68-01-2116 and Coordinating Research Council, CRC-APRAC Project CAPE-24-72 Reports (1972-1974).
44. F. S. C. Lee, T. J. Prater, and F. Ferris, PAH emissions from a stratified-charge vehicle with and without oxidation catalyst, in *Polynuclear Aromatic Hydrocarbons*, P. W. Jones and P. Leber (Eds.), Ann Arbor Science Publs., Ann Arbor, Mich., 1979.

45. M. Kmastsme and T. Hirohata, Paper presented in Symposium on the Analysis of Carcinogenic Air Pollutants, Cincinnati, Ohio, 1961.
46. D. J. Robertson, R. H. Groth, D. G. Gardner, and E. G. Glastris, *J. Air Pollut. Control Assoc.* *29*:143 (1979).
47. G. Chatot, *Anal. Chem. Acta.* *53*:259 (1971).
48. G. Chatot, *J. Chromatogr.* *72*:202 (1972).
49. C. Golden and E. Sawicki, *Int. J. Environ. Anal. Chem.* *4*:9 (1975).
50. T. J. Nestrick and L. L. Lamparski, *Anal. Chem.* *53*:122 (1981).
51. D. Seizinger, U.S. Dept. of Energy, Bartlesville, Okla., private communication, 1980.
52. F. Ferris, Ford Motor Co., Dearborn, Mich., private communication, 1980.
53. D. Seizinger, U.S. Dept. of Energy, Bartlesville, Okla., private communication, 1980.
54. W. L. Fitc and D. H. Smith, *Environ. Sci. Technol.* *13*:341 (1979).
55. W. H. Griest, L. B. Yeatts, Jr., and J. E. Caton, *Anal. Chem.* *52*:199 (1980).
56. L. Rohrschreider, *Anal. Chem.* *45*:1241 (1973).
57. W. Cautreels and K. Van Cawenbergh, *Water, Air Soil Pollut.* *6*:103 (1976).
58. I. Schmaltz, C. J. Dooley, R. L. Stedman, and W. J. Chamberlain, *Phytochemistry* *6*:33 (1967).
59. A. P. Swain, J. E. Cooper, and R. L. Stedman, *Cancer Res.* *29*:579 (1969).
60. D. Hoffman and D. L. Wynder, Chemical analysis and carcinigenic bioassays of organic particulate pollutants, in *Air Pollution*, 2nd ed., Vol. 2, Academic Press, New York, 1968.
61. E. F. Funkenbusch, D. G. Leddy, and J. H. Johnson, Paper No. 790418, Society of Automotive Engineer Meeting, Detroit (1979).
62. G. Grimmer and H. Böhnke, *Environ. Pathol. Toxicol.* *1*:661 (1978).
63. G. Grimmer, Air pollution and cancer in man, *IARC Publ. No. 16*, U. Mohr, D. Schmähl, and L. Tomatis (Eds.), International Agency for Research on Cancer, Lyons, France, 1977.
64. M. Novotny, M. L. Lee, and K. D. Bartle, *J. Chromatogr. Sci.* *12*:606 (1974).
65. D. F. Natusch and B. A. Tomkins, *Anal. Chem.* *50*:1429 (1978).
66. A. A. Rosen and F. M. Middelton, *Anal. Chem.* *27*:790 (1955).
67. J. M. Perez, F. J. Hills, D. Schuetzle, and R. L. Williams, Informational report on the measurement and characterization of diesel exhaust emissions. Coordinating Research Council, Atlanta (Dec. 1980).
68. G. P. Gross, Gasoline composition and vehicle exhaust gas polynuclear aromatic content. U.S. Clearing-House Federal Science Technology Information, PB Report, Issue No. 200266 (1971), 124 pp.
69. R. A. Brown, T. D. Searl, W. H. King, Jr., W. A. Dietz, and J. M. Kelliher, Rapid methods of analysis for trace quantities of polynuclear aromatic hydrocarbons and phenols in automobile exhaust, gasoline and crankcase oil. Coordinating Research Council, CRC-APRAC Project CAPE-12-68 Final Report, Atlanta (1971).
70. G. P. Gross, Gasoline composition and vehicle exhaust gas polynuclear aromatic content. Final Report for Project CAPE-6-68, Coordinating Research Council, Atlanta (1972).

71. A. Candeli, V. Mastsandrea, G. Morozzi, and S. Toccaceti, *Atmos. Environ. 8*:693 (1974).
72. K. W. Boyer and H. A. Laitinen, *Environ. Sci. Technol. 9*:457 (1975).
73. G. Grimmer, H. Böhnke, and A. Glaser, *Zbl. Bakteriol. Hyg., I. Abt. Orig. B164*:218 (1977).
74. G. Grimmer, K. W. Naujack, and D. Schneider, Changes in PAH-profiles of different areas in a city during the year, in *Polynuclear Aromatic Hydrocarbons: Chemistry and Biological Effects*, A. Bjørseth and A. J. Dennis (Eds.), Battelle Press, Columbus, Ohio, 1980.
75. D. E. Seizinger, Analysis of carbonaceous diesel emissions, presented at Conference on Carbonaceous Particles in the Atmosphere, Berkeley, Cal. (March 28, 1978).
76. D. E. Seizinger, LC-fluorescence measurement of BaP in diesel exhaust, in *Trends in Fluorescence*, Vol. 1, No. 1, pp. 9-10, Perkin-Elmer Corp., 1978.
77. J. M. Perez, Paper presented at International Symposium on Health Effects of Diesel Engine Emission Cincinnati, Ohio (December 3, 1979).
78. E. Sawicki, T. W. Stanley, and W. C. Elbert, *Talanta 11*:1433 (1964).
79. E. Sawicki, T. W. Stanley, W. C. Elbert, and J. P. Pfaff, *Anal. Chem. 36*:479 (1964).
80. J. F. Mackay and D. R. Latham, *Anal. Chem. 44*:2132 (1972).
81. R. C. Pierce and M. Katz, *Anal. Chem. 47*:1743 (1975).
82. M. Dong, D. C. Locke, and E. Ferrand, *Anal. Chem. 48*:368 (1976).
83. C. Golden and E. Sawicki, *Anal. Letters 9*:957 (1976).
84. M. A. Fox and S. W. Staley, *Anal. Chem. 48*:992 (1976).
85. S. A. Wise, W. J. Bonnett, and W. E. May, Normal and reverse phase liquid chromatographic separations of PAH, in *Polynuclear Aromatic Hydrocarbons: Chemistry and Biological Effects*, A. Bjørseth and A. J. Dennis (Eds.), Battelle Press, Columbus, Ohio, 1980.
86. R. L. Williams, Chemical characterization of diesel particulates—Analysis of particulate extract: BaP. Information note to CRC-APRAC CAPI 1-64 Chemical Characterization Panel, Coordinating Research Council, Atlanta (1981).
87. K. Ogan, E. Katz, and W. Slavin, *Anal. Chem. 51*:1315 (1979).
88. A. M. Krstulovic, D. M. Rosie, and P. R. Brown, *Anal. Chem. 48*:1383 (1976).
89. R. D. Smillie, D. T. Wang, and O. Meresz, *J. Environ. Sci. Health A13*: 47 (1978).
90. S. A. Wise, S. N. Chesler, H. S. Hertz, L. R. Hilpert, and W. E. May, *Anal. Chem. 49*:2306 (1977).
91. T. Nielsen, *J. Chromatogr. 170*:147 (1979).
92. M. Zinbo, Ford Motor Co., Dearborn, Michigan, private communication.

3

Extraction of Polycyclic Aromatic Hydrocarbons for Quantitative Analysis

WAYNE H. GRIEST AND JOHN E. CATON / Oak Ridge National Laboratory, Oak Ridge, Tennessee

---

I. Introduction

Efficient extraction of polycyclic aromatic hydrocarbons (PAH) from the sample matrix is a crucial step in the preparation of a sample for PAH determination, yet it is one of the least appreciated steps. All too often, "semistandardized" extraction procedures are employed with little apparent regard for extraction recoveries of the PAH. Obviously, incomplete extraction without provisions for extraction recovery corrections will lead to a low bias in the analytical data. In the case of an exemplary highly sorptive matrix such as coal fly ash [1], extraction recoveries of some PAH can be as low as 10%, leading to an order-of-magnitude underestimation in PAH concentrations if the recoveries are not determined.

Numerous review papers [1-6] of the determination of PAH in various matrices have included short sections on PAH extraction, but few, if any, reviews appear to concern PAH extraction alone. The purpose of this chapter is to critically review the more widely used PAH extraction procedures reported over the last 10 to 15 years. This is not a comprehensive review in that it is impractical to cite each and every paper employing a PAH extraction of some type. Rather, the review is restricted in scope to the more prominent and important reports which would contribute to a critical comparison of PAH extraction procedures without being overly redundant. The attention allotted to each extraction procedure is roughly proportionate to its usage in the open literature.

II. Necessity for Extraction

Most analytical methods for quantifying PAH require a preliminary extraction step because few sample matrices can be analyzed directly without serious interferences. The few exceptions, for the most part involve analysis of the pure PAH themselves, or PAH in pure, analytically compatible matrices [7], or analyses by rapid scrrening procedures [8-10] which rely upon some unique molecular or spectral property for their specificity. However, the majority of the PAH analytical procedures are initiated with an extraction step to remove the PAH from the bulk of the sample matrix, which is, in a sense, a preliminary purification of PAH. This step places the PAH into a matrix more readily purified and analyzed; e.g., extraction can remove PAH from an aqueous matrix which would deactivate most adsorbents used for PAH isolation. With some sample matrices a simple extraction is the only sample preparation necessary before analysis, whereas others (e.g., coal tars) may require extensive fractionation to produce a suitable PAH isolate. The degree of sample handling and workup is a function of both the sample matrix and the analytical method employed for PAH measurement. In either case, the extraction procedure must quantitatively remove PAH from the sample matrix or the PAH extraction recovery must be known for the PAH analysis to be accurate and precise.

III. General Precautions in Extraction

The objective of the extraction procedure is to separate PAH from the bulk sample matrix in as high a yield as possible, with a minimum of coextraction of other classes of compounds or contamination or degradation of the extracted

PAH. To accomplish this goal, certain good laboratory practices must be observed. These are common to all the extraction procedures, which are reviewed later in this chapter.

Solvents, extraction thimbles, filter paper, and glass wool all represent potential major sources of contamination [4,11-15]. To minimize the introduction of contaminants and interferences, all filtration media should be preextracted and solvents must be at least freshly redistilled. Peroxide buildup in ethers used for extraction not only is a safety hazard but can also lead to oxidative degradation of the sample. Glassware must be scrupulously cleaned to avoid another source of contamination, and thought must be given as to minimizing contamination in the solvent- and extract-storage containers [4] and their cap closures [16].

It has long been recognized [17-19,55,101] that samples should be extracted as soon as possible after collection to minimize degradation, sublimation, and other routes for losses of PAH from the sample matrix. Losses of phenanthrene and fluoranthene from air particulate filter pads can exceed 50% after 3 weeks of storage in the dark (presumably at room temperature) and can approach quantitative loss after 1 year of storage [17]. Oven-drying of air particulate samples should be avoided [18] for the same reasons. In storage, PAH can be lost from aqueous samples by adsorption on glass container walls or by degradation [19-21] and can approach 77% [20] loss. In contrast, the PAH in solvent extracts appear stable [17,22].

To avoid photodecomposition of the PAH, sample handling and extraction (as well as the subsequent isolation and analysis) should be conducted in the absence of ultraviolet (UV) light [4,18]. Commercially available yellow filters and lights are useful in eliminating radiation below 450 nm.

Thermal degradation and sublimation also can cause losses of PAH during extraction. It has been recommended [4,22] that relatively volatile solvents be used to minimize extraction temperatures where reflux or distillation procedures are employed and to allow subsequent solvent extract concentration to proceed at lower temperatures. Keeping temperatures below 45°C reportedly [4] reduces or even eliminates PAH losses, but with relatively volatile diaromatics losses to 80% can be sustained in evaporative concentration [1,23], even under subambient temperatures.

IV. Measurement of Extraction Recoveries

The efficiency of the PAH extraction (and also isolation) procedure must be known to allow correction of the analytical measurements for PAH losses and to produce an accurate and precise PAH determination. Relatively few studies in the literature are concerned with the determination of the efficiency of a PAH extraction procedure. Of those which are, the method of spiking an exemplary sample matrix with cold, unlabeled PAH is employed most often. This procedure involves the addition of exactly known quantities of standard PAH (preferably synthetic in origin [4] to a sample matrix which either is free of PAH or contains known amounts of "native" PAH. This approach assumes that the added PAH are incorporated into the sample matrix and that they behave in the same manner as the "native" PAH. Achieving a uniform, homogeneous spiking of highly sorptive matrices is not a simple, straightforward problem, and inhomogeneities in spike distribution can influence [1,24] apparent recoveries.

The sample is then extracted, and the PAH in the extract are measured. Isolation or purification of the PAH in the extract may be necessary prior to measurement. These additional steps may introduce further losses of PAH. Comparison of the amounts of PAH originally present in the sample with those amounts found in the extract allows calculation of a percentage extraction recovery factor. The advantage of the cold PAH spiking procedure is that extraction recoveries can be measured in any laboratory equipped with instrumentation suitable for PAH measurements, which can include simple UV or fluorescence spectrophotometers or gas and liquid chromatographs. However, the main drawback to this procedure is that the extraction recovery of each individual sample cannot be monitored. The assumption must be made that the PAH extraction recoveries for similar samples are the same. This assumption is questionable because inadvertent errors or mistakes in sample handling and extraction and subtle differences among the matrices of separate samples can lead to unexpected differences in PAH extraction recoveries for apparently identical samples or aliquots of a given sample. Also, the PAH extraction recoveries can vary with the PAH content of the sample, generally decreasing with lower PAH levels [25]. The added PAH spikes must be similar in amount to those already present in the sample, thus requiring a prior estimate of the PAH concentrations in the sample.

In spite of these drawbacks, the cold PAH spike procedure has been employed to good advantage in determining the extraction (and isolation) recoveries of PAH (or more general organics measurements which would include PAH) from such varied matrices such as air particulates [26-31], water [32-37], cigarette smoke condensate [38,39], soot [40], and coal fly ash [41-42].

A less widely employed but more advantageous procedure employs PAH tracers labeled with ^{2}H, ^{3}H, or ^{14}C. The ^{14}C-label appears to be the most stable of this group, and exchange of ^{2}H- or ^{3}H-labels with ^{1}H has been observed when highly sorptive matrices such as coal fly ash [1,23,24,43] or sediment [23] are spiked. The ^{14}C-labeled PAH tracer thus is acknowledged [44] as the tracer of choice.

As with the cold PAH spikes, known amounts (^{2}H-labeled tracers) or activities (^{3}H- and ^{14}C-labeled tracers) of the PAH tracers are added to the sample and are measured in the extract to determine extraction recoveries. The ^{2}H-labeled tracers are usually added in amounts commensurate with the native PAH and are measured by mass spectrometric methods, most often in conjunction with a gas chromatographic separation. In contrast, only very small masses of high specific activity ^{3}H- or ^{14}C-radiolabeled PAH are added to the sample, and the native PAH act as carriers. Here, the amount of native PAHs in the sample does not have to be known, and the tracers are measured with very high sensitivity and specificity by liquid scintillation spectrometric techniques.

The isotopically radiolabeled tracer procedure has the great advantage that the PAH extraction recovery can be determined for each individual sample. However, as opposed to the cold spike procedure, only two (^{14}C or ^{3}H) or three (^{14}C, ^{3}H, and ^{2}H) tracers can be employed in a single sample because the liquid scintillation spectrometer cannot distinguish among ^{14}C- (or among ^{3}H-) labels in a mixture of different PAH tracers having the same radiolabel. Multiple PAH tracers having the same radiolabel can be used [45,126,137] if a fractionation procedure is employed to separate the tracers after extraction, but this approach is not often taken. The isotopically labeled tracer technique also has the disadvantage that the necessary expensive or specialized instrumentation (i.e., mass spectrometer or liquid scintillation spectrometer) and sample-handling procedures are not readily available to many laboratories. The ^{14}C-labeled PAH tracers have been employed in the PAH analyses of tobacco smoke [45,137], coal fly ash [46], water [23,47,48, 126], sediments [23,49], marine animals and plants [49,50], coal liquids [137], and automotive exhaust [51].

V. Solvent Extraction Methods

Solvent extraction methods are divided in this review into Soxhlet extraction, ultrasonic and high-intensity mechanical extraction, and solvent partitioning. High-intensity mechanical methods are included in the section (V.B) on ultrasound methods because both kinetic and ultrasonic energy is released by the homogenizer. Solvent partitioning methods which employ ultrasonic agitation to mix the solvent layers also are included in the ultrasonic section. Solvent extraction where shaking, stirring, or other less vigorous methods of agitation were employed are included in Sec. V.C.

A. Soxhlet Extraction

Soxhlet extraction has for many years been the standard method for preparing a PAH-containing solvent extract of solid matrices. The historical development of this procedure has been described in previous articles [2-6], and a detailed comprehensive review of the various solvents, extraction times, and applications is not attempted here.

In its role as a "standard reference" method, the Soxhlet procedure has been specified by various agencies and committees for the extraction of PAH from solid samples. The American Society for Testing and Materials (ASTM) recommends [52] a 5- to 6-hr extraction of air particulates with benzene. A similar procedure for air particulates has been proposed [53] by the U.S. Intersociety Committee on Recommended Methods, which also proposed the use of methylene chloride [54,55] for an unspecified time interval—"until all soluble matter has been removed" [55]—in a micro-Soxhlet, or cyclohexane for 6-8 hr in a micro-Soxhlet [56] or for 24 hr [57], presumably in a larger Soxhlet. The *Procedures Manual* for Level 1 environmental assessment of the U.S. Environmental Protection Agency (EPA) [58] specifies a 24-hr Soxhlet extraction of solid or particulate process stream effluents using methylene chloride, and the EPA manual for organics analysis of environmental samples by GC/MS [59] also suggests using methylene chloride for 25 cycles in a Soxhlet. None of these procedures specifies the Soxhlet solvent cycle time, which does affect the extraction efficiency. It is this plethora of "standard" procedures which do not agree upon uniform conditions, even for common matrices, that leads to the confusion and seemingly infinite number of variations in extraction procedures reported in the literature.

1. *Precautions*

One of the main precautions in the use of Soxhlet extraction is the purification of all solvents and also filtration media such as thimbles and glass wool which otherwise can contribute impurities [11-14] such as alkanes, alkylthiopenes, and phthalates to the solvent extract. Glass thimbles [13] reportedly offer less contamination than cellulose thimbles, but they can cause cross-contamination if not carefully cleaned between reuse. It has been recommended [12] that glass wool used to filter particles for air particulate extracts be treated with HCl and Soxhlet extracted with methylene chloride for 24 hr.

Thermal degradation of the sample extract in the Soxhlet pot [22,60] has been proposed as one route of PAH loss in Soxhlet extraction, suggesting the use of lower-boiling solvents such as methylene chloride, as opposed to benzene, to keep pot temperatures low. PAH degradation by light and exposure to air during Soxhlet extraction also have been claimed. Recoveries (overall, including extraction) of benzo[a]pyrene (BaP) from air particulates have been reported [25] to decrease and become less reproducible below 1 μg total BaP. Hence, for trace PAH level work and for sensitive or reactive samples, shielding of the apparatus from UV light and possibly flushing the apparatus with an inert, heavier-than-air gas such as argon should be considered.

For health and other safety reasons, Soxhlet extraction should be conducted in a fume hood. The incorporation of additional features such as a spray trap and catch basin have been recommended [61] to increase the safety of the procedure and also to prevent total loss of a sample in case of equipment malfunction or breakage.

2. *Comparative Studies with Other Extraction Procedures*

Most reviews of PAH analysis [2,3,6] include references to the inferior efficiency of Soxhlet vs. ultrasonic solvent extraction, which generally has focused upon air particulate samples. Ultrasonication [31,62,63] or high-intensity mechanical disruption [64] using cyclohexane, benzene, or other solvents for 12-30 min appears to be more effective than 6-8 hr with the Soxhlet for extracting PAH or benzene solubles which include PAH. Ultrasonic extraction recoveries of phenanthrene (Phen), anthracene (Anthr), and BaP of 95% or greater are reported [63], and the reproducibility also is far better than that attained by Soxhlet. Reports of the reproducibility of ultrasonic extraction of aromatics or mechanical extraction of benzene solubles reportedly have ranged from 1.3 to 4.3% relative standard deviation (SD), as opposed to 26% for the Soxhlet extraction [63,64]. Coal fly ash is another matrix for which ultrasonic extraction was found [46] more effective for extracting PAH than was Soxhlet extraction—although even with the former, recoveries are far from complete for the larger (more than four rings) PAH.

In spite of these studies reporting greater efficiency for ultrasonic vs. Soxhlet extraction, other studies (to be discussed next) have demonstrated acceptable PAH extraction recoveries using the Soxhlet. Additionally, it should be noted that a Soxhlet apparatus is far less expensive than a high-powered ultrasonicator, and multiple Soxhlet extractions can be set up and operated simultaneously.

3. *Comparative Studies with Different Solvents*

Several studies have compared the efficiency of different solvents for the Soxhlet extraction of PAH from air particles. Extractions were conducted for 16 hr, although recoveries were nearly quantitative after 4 hr. Benzene and carbon disulfide were found [27] superior to hexane or cyclohexane, and other investigators [26] have observed methanol also to be superior to cyclohexane for extraction of organics. Six-hour extractions of BaP-spiked air particulates have yielded quantitative recovery [28] using diethyl ether, ethyl acetate, acetone, and benzene/diethyl amine, and 95% recovery using ethanol, benzene, or methylene chloride. The last two solvents are preferred [46] for extraction of BaP from coal fly ash. Studies [30] of organic carbon extracted from air particulates indicated that 6-hr extractions with benzene can leave as much as 48% of the organic carbon (some possibly corresponding to large-ring PAH) unextracted. Binary (nonpolar/polar) solvent mixtures were found superior to benzene alone.

4. *Analytical Applications of Soxhlet Extraction*

As noted previously, Soxhlet extractions with benzene [52,53] methylene chloride [54,55], and cyclohexane [56,57] for periods ranging from 5 to 24 hr all are incorporated in standard procedures for the analysis of PAH in air particles. In addition to the solvent efficiency comparison studies [26-28] cited above, several investigators have included recovery measurements in their analyses. Several of these studies are cited in Table 1. The apparent differences among the results might be attributed in part to variations in unlisted factors which also affect recoveries, e.g., amount of PAH, percentage carbon in the particles, or Soxhlet cycling time. From these studies, it would appear that benzene is one of the preferred solvents; indeed, other investigators reporting quantitative analyses of air particles for PAH have employed 4-, 5-, 8-, and 24-hr extractions with benzene [73-76]. They also have used [77-79] a 4-hr extraction with benzene followed by 4 hr with methanol, especially when other classes of airborne organic compounds were measured. Cyclohexane is another solvent often employed for extraction of air particles. Extraction times ranging from 6 to 16 hr have been reported [80-84]. Some investigators [80,81] have remarked that cyclohexane (vs. benzene or methylene chloride) coextracts fewer compounds from air particles which would interfere with PAH isolation and analysis, but unfortunately few of these studies are backed up with data to indicate quantitative PAH extraction recovery.

Coal fly ash PAH are not particularly well extracted with the Soxhlet apparatus [41,46], notwithstanding the single published report [42] claiming complete extractions with benzene in 24 hr. However, the levels of PAH applied to this matrix were considerably higher than those of native PAH in fly ash.

PAH in river sediments are satisfactorily extracted [23] in 68 hr with acetone, a water-miscible solvent. Methylene chloride also is used [72,85] for 6- to 12-hr extractions, but acetone has the advantage over methylene chloride of not requiring a preliminary drying of the sediment before extraction, which can cause losses of PAH [72]. Biological media also are hydrophilic in nature, but PAH extraction procedures for corn and shellfish have been reported [86-88] in which the macerated tissue is directly extracted in a Soxhlet (sometimes with drying [86] with cyclohexane or acetone for 12-16 hr. Usually an alkaline digestion procedure is employed first (see Sec. VIII), but such steps were either omitted [86,87] or conducted after extraction [88].

Table 1. Soxhlet Extraction Recoveries of PAH from Various Sample Matrices

Sample matrix	PAH[a]	Level[b]	Percentage recovery[c]	Extraction conditions		Ref.
				Solvent[d]	Time (hr)	
Air particulates	BaP	4.57 μg	70-75[e]	CH	4	[25]
		1.96 μg	72-76[e]			
		0.37 μg	32-47[e]			
	3-5 ring	—	100	B or M	2-16	[27]
	BaP	1.12 mg/g	99-100	EA, A, EE, B/DEA	6	[28]
			95	B, MC, E	6	
			76-77	C, H, T	6	
			54	P	6	
	BaP	100 ng	94-95	B	16	[65]
	4-6 ring	—	95-100	B	8	[66]
	5-6 ring	2-11 μg	"Complete"	B	6	[67]
Polyvinyl chloride smoke particles	2-5 ring	20-100 μg	90-104[e,f]	CH	2	[68]
"Environmental"	2-7 ring	—	40-88[e,f]	M then B	24 (each)	[69]

Coke oven particles	2-6 ring	—	99-100	B	2-4	[70]
Auto exhaust particles	BaP	—	61-85[e]	MC	14-16	[71]
	BaA	—	80			
Coal fly ash	BaP	Radiotracer	18	B	64	[46]
	BaP	133 μg	60	B	32	[41]
	2-5 ring	90-250 μg	100	B	24	[42]
	4-6 ring	9.35 μg	32-91	MC	12	[119]
		31.7 μg	53-98			
River bottom sediment	Naphthalene	Radiotracer	100	A	68	[23]
	BaP	Radiotracer	100	A	68	
	2-7 ring	—	86-89[e,f]	MC	12	[72]

[a]PAH used for recovery determination.
[b]Amount or concentration of PAH in recovery study.
[c]Extraction recovery, unless denoted otherwise.
[d]CH = cyclohexane; B = benzene; M = methanol; EA - ethyl acetate; A = acetone; T = toluene; P = pentane; DEA = diethyl amine; EE = diethyl ether; E = ethanol; MC = methylene chloride.
[e]Overall analytical recovery.
[f]Recovery may not include extraction.

Soot and lampblack represents another difficult-to-extract matrix, and a variety of Soxhlet conditions have been employed, ranging from 8 to 18 hr with methylene chloride [89,90], 4 hr with cyclohexane [91], to 20 hr with benzene [92] or benzene/methanol (9:1) [93]. One of the papers [89] noted that the Soxhlet extraction had to be conducted under nitrogen, otherwise the extracted materials were not soluble in chloroform. This observation suggests that oxidation of the organics during Soxhlet extraction may be more of a problem than is currently appreciated.

Soxhlet extraction of PAH also has been applied to coal [94] using benzene or cyclohexane for 24 hr, solvent refined coal [95] using methylene chloride for 48 hr, and asbestos [96] using pentane for an unspecified period of time.

From these applications, and the myriad of others in the literature, the only uniformity in Soxhlet extraction conditions appears to be in the use of benzene, cyclohexane, or methylene chloride for periods of 8-24 hr. Other solvents (e.g., acetone) may be useful in special applications, but the main requirement is that PAH extraction recoveries be quantitative or measured. Unfortunately, very few of the published extraction procedures include such data.

5. *Unusual Soxhlet Procedures*

Poor PAH extraction recoveries with highly sorptive matrices have led to some unusual modifications of the Soxhlet procedure. To improve extraction recoveries of six- and seven-ring PAH from lampblack, a Soxhlet has been modified [97] with heating tape to allow the use of naphthalene (Nap) or methyl naphthalene as an extraction solvent. A single 6-hr Soxhlet run using Nap extracted almost twice the amount of benzo[ghi]perylene and coronene from lampblack than did two 20-hr extractions with benzene. Apparently, the higher extraction temperatures and presumably superior PAH solvent properties of Nap are responsible for the enhanced six- and seven-ring PAH recoveries. However the procedure has the disadvantage that smaller two-ring PAH are not recovered as well, probably from their greater volatilization losses at these higher temperatures. Also, the lampblack catalyzed the production of binaphthyls from the Nap solvent, and the introduction of such impurities is undesirable from both an analytical and a preparative standpoint.

On the other end of the working fluid scale, a low temperature, high-pressure Soxhlet extractor [98] employing carbon dioxide as the working fluid has been applied to several matrices, including fly ash. Considerable differences were noted in the gas chromatograms of the materials extracted by benzene and carbon dioxide, but no qualitative or quantitative data were reported. This is a promising new approach to PAH extraction, and it should be investigated further.

B. Ultrasonic and High-Intensity Mechanical Extraction

Ultrasonic agitation is a relatively new technique of achieving solvent extraction of PAH and involves the use of a power supply ranging from a few to several hundred watts, a converter and probe, for insertion into an extraction vessel, or a bath into which the extraction vessel is set. It differs from other solvent extraction methods in its use of high-intensity ultrasonic vibration (ca. 20 kHz) to produce solvent cavitation around the sample matrix particles, presumably leading to enhanced solvent contact and mixing with the sample. High-intensity mechanical extraction is included in this category, because the high-rpm rotors used for such extraction release both kinetic and ultrasonic energy to the extraction media [108].

For some time the Soxhlet apparatus has been used as a standard extraction method (see Sec. V.A), but now ultrasonic and high-intensity mechanical methods are replacing the Soxhlet. Reviews [2,4] have noted the greater speed and efficiency of the ultrasonic methods, and they are now included in newer standard procedures. Ultrasonic extraction is specified in a NIOSH procedure [99] for determining air particulate BaP and total air particulate aromatic hydrocarbons [100] and in the procedure proposed [101] by the U.S. Intersociety Committee on Recommended Methods for total air particulate hydrocarbons. It is specified for extraction of sediments in one EPA manual [59] and is listed as an option to the Soxhlet in extraction of air filters and particles rinsed from a cyclone, impactor, or probe of an air sampler in another EPA manual [102].

1. *Precautions*

The same precautions as noted for Soxhlet extractions generally also apply to ultrasonic extraction. It may be necessary to control the temperature of the extraction media because the ultrasonic energy released may raise the solvent temperature to unacceptable levels. For example, when ultrasonic extraction of air particles with tetrahydrofuran was conducted [105], the temperature was maintained at 15°C. A water-jacketed glass funnel has been designed [106] to conduct ultrasonic extraction and filtering at controlled temperatures.

Hearing protection may be needed for the operator when using high-powered units. This is a precaution not necessary with most other PAH extraction methods.

2. *Comparison with Other Extraction Methods*

As noted in the previous subsection, ultrasonic extraction procedures have been compared [31,62-64,100,101] with the Soxhlet for extraction of air particulate PAH, with 12- to 30-min ultrasonic extractions producing equivalent or better air particulate PAH or benzene solubles extraction recoveries than 6 to 8 hr of Soxhlet extraction. The ultrasonic procedure also has been found to be equally effective to the Soxhlet in PAH extraction of dust from the ventilation system of vehicular tunnels [103] and freshwater sediments [104] and to be superior to the Soxhlet in extraction of coal fly ash [1,46]. It has been compared with vacuum sublimation for extraction of air particulate PAH [105] and with "continuous solvent extraction" of aqueous PAH [19], and in both cases found to be equally effective but more rapid.

The main advantage of the ultrasonic procedure appears to be in its much more rapid extractions, which are a factor of 16 [31] to 30 or 40 [62,100,101] times more rapid than that of the Soxhlet. As noted previously it also appears to be far more reproducible: reports [62,100,101] claim a 1.33% relative SD for reproducibility in extraction of total air particulate aromatic hydrocarbons vs. the 26% for Soxhlet extraction. The conditions employed in these comparative studies are summarized in Table 2.

Table 2. Comparison of Ultrasonic and Soxhlet (Or Other) Extraction Conditions and PAH Recoveries

Sample matrix	Solvent[a]	Extraction time		Result[b]	Ref.
		Ultrasonic	Soxhlet		
Air particulates	CH	8 min × 2	6-8 hr	US extracted 14% more Repro: US, ± 1.33%, SOX, ± 26%	[62,100,101]
	Variety[c]	30 min	8 hr	US more efficient	[31]
	B	20 min	6 hr	US more efficient Repro: US, ± 4.3%	[64]
	THF	10 min	30 min[d]	Equal efficiency	[105]
Dust	B	5 min × 3	6 hr	Equal recovery (81%)	[103]
Sediment	CH	30 min × 2	18 hr	Equal recovery	[104]
Water	MC		—[e]	Equal recovery	[19]

[a]THF = tetrahydrofuran. See also the List of Abbreviations at the end of text.
[b]US = ultrasonic extraction; SOX = Soxhlet extraction.
[c]Solvents tested include benzene, cyclohexane, diethyl ether, methylene chloride, acetone, chloroform, and methanol.
[d]Vacuum sublimation.
[e]"Continuous solvent extraction."

3. *Analytical Applications of Ultrasonic Extraction*

Applications of ultrasonic and high-intensity mechanical solvent extractions of air particulate PAH abound in the literature. Some of the more prominent applications including studies of extraction efficiencies are listed in Table 3. As with the Soxhlet extraction, a variety of ultrasonic extraction conditions have been employed. Similar to Soxhlet extraction, cyclohexane [62,100,101, 107,108] and benzene [64,103,109] appear to be the most popular solvents, but acetonitrile [99], tetrahydrofuran [105], and methylene chloride [102, 110] are also used.

One group of investigators has reported [31] relative PAH extraction efficiencies of various solvents for air particulates. Their data suggest that benzene, chloroform, and diethyl ether are most efficient for extraction of BaP and that benzene or methylene chloride are most effective for extraction of "total aromatics" from air particulates. Another study [111] of ultrasonic extraction of BaP from pitch-impregnated paper suggests that benzene, tetrahydrofuran, cyclohexane, and methanol are all approximately equally effective, but no data are provided.

It would appear that most any solvent with good solubility characteristics for PAH can be employed for ultrasonic extraction of air particulate PAH if the other extraction parameters (e.g., time, temperature) are optimized. However, a word of caution is expressed for halogenated solvents. The use of chlorinated solvents may allow chlorination reactions during ultrasonic extraction. Investigators should consider this possibility when choosing extraction solvents.

Other groups [62,100,101] have included an in situ purification measure along with the extraction of air particulate PAH. By conducting the ultrasonic extraction with cyclohexane and controlled pore glass powder, the more polar compounds coextracted with PAH from air particles reportedly are adsorbed on the glass powder and their concentrations minimized, allowing direct high-performance liquid chromatographic measurement of the PAH in the supernatant liquid. However, the amount of the glass powder used must be optimized, and more recently it has been found necessary [106] to employ 35% methylene chloride in the cyclohexane to improve the extraction efficiencies of pentacyclic PAH. This latter observation suggests that earlier extraction recovery data [62,100,101] based on spiking experiments may not be realistic.

Extraction times ranging from 10 min [105] to 60 min [107] reportedly achieve essentially quantitative recovery of BaP. Generally, multiple batchwise extractions are conducted in favor of single, longer extractions. Ultrasonic units of 70 or 100 W power [99,100] or mechanical homogenizers capable of rotating disruptor blades at 2×10^3 rpm [60] have been employed for air particulate PAH extraction.

Ultrasonic extraction also has been applied to coal fly ash, but this particularly sorptive matrix poses some serious problems for PAH analysis. As is evident from Table 3, extraction recoveries of BaP can be as low as 25% [24], which suggests that BaP analyses of fly ash can have a very serious low bias if not properly corrected for extraction recoveries. It has been demonstrated [24] that the unextracted PAH remain with the fly ash and are not lost to the extraction vessel walls or other sites. Studies of the PAH extraction characteristics of fly ash show [1,24,46,119] that the rates and

Table 3. Recoveries of PAH from Various Matrices by Ultrasonic Extraction Procedures

Sample matrix	PAH[a]	Level	Percentage recovery	Extraction conditions: Solvent[b]	Extraction conditions: Time[c]	Ref.
Air particulates	Anthr	35 mg/pad	95	CH[d]	8 min × 2	[62,100,101]
	Phen	147 ng/pad	97.5			
	BaP	355 ng/pad	98.2			
	BaP	--	≥90	AN	5 min	[99]
	BaP	10 μg/pad	46.5	B	5 min × 1	[103]
			70.4		5 min × 2	
			83.1		5 min × 3	
			91.0		5 min × 4	
	BaP	200 ng/pad	99.8	THF	10 min	[105]
	BaP	10-30 μg/pad	96.6	CH	30 min × 1	[107]
			3.1		30 min × 2	
			96.8		60 min × 1	
			2.6		60 min × 2	

Coal fly ash	BaP	100 ng/g	25.2	B	30 sec × 2	[24]
River sediment	2-6 rings	0.5-8 μg/g	72-99[e]	CH	30 min × 2	[104]
Water	Py	0.3 μg/liter	84 68[f]	MC		[19]
	BghiP	0.1 μg/liter	82 53[f]			
	Fl	0.44 μg/liter	73	MC	5 min[g]	[117]
	Py	0.32 μg/liter	84			
	BaA	0.21 μg/liter	88			
	BkFl	0.15 μg/liter	64			
	BaP	0.13 μg/liter	70			
	Pery	0.16 μg/liter	69			
	InPy	0.07 μg/liter	80			
	BghiP	0.12 μg/liter	80			

[a]Anthr = anthracene; BaA = benz[a]anthracene; BaP = benzo[a]pyrene; BkFl = benzo[k]fluoranthene; BghiP = benzo[ghi]perylene; Fl = fluoranthene; InPy = indeno[1,2,3-cd]pyrene; Pery = perylene; Phen = phenanthrene; Py = pyrene.

[b]AN = acetonitrile. See previous tables for other abbreviations.

[d]Controlled pore glass included in extraction media.

[e]Overall analytical recovery including extraction, isolation, and concentration recovery.

[f]Extraction efficiency with 100 ppm of suspended solids present in water.

[g]After sonication MC layer left in contract with water layer overnight.

completeness of PAH extraction are inversely related to PAH ring size. The PAH-ring-size-dependence behavior, shown in Fig. 1, poses additional difficulties for accurate recovery corrections when several PAH are to be measured. The effects of solvent volume, polarity, PAH spike level, etc., on recoveries have been studied in some detail recently [119]. In addition to these problems, fly ash also can facilitate [1,23,24,43] the exchange with ^{1}H of ^{2}H or ^{3}H labels in PAH tracers, further complicating extraction recovery measurement.

PAHs in river sediment have been apparently successfully extracted, but considerably longer extraction times are required for these more organic hydrophilic sample matrices. A 20-g sample can be extracted ultrasonically [104] after freeze-drying, using cyclohexane and freshly activated copper powder and sonicating twice 30 min each. The copper removes elemental sulfur. Final analytical recoveries of two- to seven-ring PAH, estimated from spiking experiments, were in the range of 72-99%, with the five-ring and larger PAH yielding the lower recoveries. This was not felt to be an extraction problem because additional ultrasonic extractions did not increase recoveries and Soxhlet extractions gave the same results. An EPA manual for organics analysis [59] suggests a 20-sec ultrasonic homogenization using a mixed acetone/hexane solvent (1:1, presumably by volume) with 50 g of partially dried (air-dried in a pan) sediment. However, they did not feel that it is possible to spike a sediment with PAH in a manner simulating natural adsorption conditions. Direct ultrasonic extraction of 990 g of sediment with diethyl ether for approximately 2 hr (the longest ultrasonic extraction reviewed) [112] or ultrasonic homogenization of 25-100 g of sediment for 3-4 min using acetone and then benzene/methanol (ratio not specified) [113] also have been reported.

PAH in tars and particles collected on filters and cyclones from sampling of automotive exhaust [114,115] and forest fire smoke [116] also have been extracted by ultrasonic procedures. Cyclohexane is one of the preferred solvents [115] for the former because it is expected to extract fewer analytically interfering polar compounds. A methylene chloride extraction of the latter sample reportedly [116] was conducted at 50°C. This temperature measurement must be in error because it is above the boiling point of the solvent (at 1 atm); however, it does stimulate speculation that ultrasonic extraction efficiencies could be improved further by conducting the extraction under elevated pressures.

Finally, it should be noted that the application of ultrasonic procedures for PAH extraction is not solely limited to solids. Some applications reports [117] and a methods study [19] have appeared for solvent extraction of PAH from water, using an ultrasonic homogenizer to achieve intimate contact between the solvent (methylene chloride) and water. As shown in Table 3, extraction recoveries of four- to six-ring PAH at 0.1-0.4 μg/liter concentrations in water range from 64 to 88% with no obvious dependence of extraction efficiency upon PAH ring size [117]. The presence of particles at a 100-ppm concentration decreased extraction recoveries noticeably [19]. It was also recommended [19] that the water to be analyzed be directly taken into the extraction vessel and extracted as soon as possible to minimize PAH losses from adsorption on container walls or from degradation.

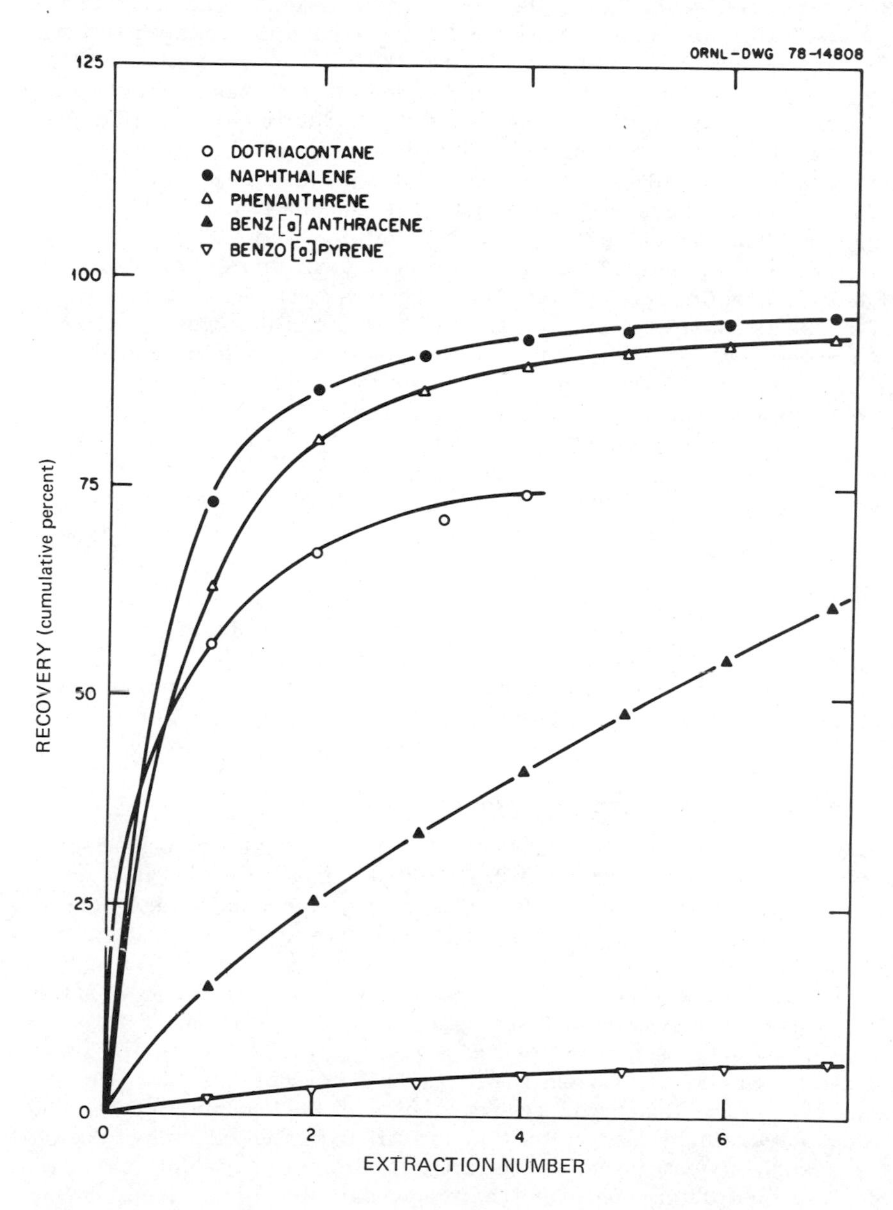

Figure 1. Cumulative extraction recoveries for multiple ultrasonic solvent extractions of coal fly ash. (From Refs. 1, 24, and 46.)

C. Solvent Partitioning and Shaking Extraction

Solvent partitioning, or liquid-liquid extraction, is the traditional procedure for PAH extraction from liquid sample matrices. It is one of the simplest yet most highly effective procedures for extracting PAH from liquids. All that is required is an immiscible extractant fluid with reasonable purity and PAH solubility properties. The extraction can be performed in most any glassware, with agitation by shaking, stirring, or other convenient means. Although PAH extraction by adsorbents (see Sec. VII) is rapidly gaining in popularity, solvent partitioning is still listed or specified in standard procedures, such as the EPA manual for Level 1 environmental assessment [58], the EPA manual for organics analysis using GC/MS [59], and the proposed ASTM method for determining PAH in water [122]. Solvent-solvent partitioning and liquid extraction have been included [2,3,5,6] in numerous reviews of the PAH analyses of specific sample matrices, usually from the standpoint of both extraction efficiency and PAH purification because the choice of solvents and conditions can greatly influence both factors. Solvent extraction and isolation of PAH also has been examined from a theoretical standpoint [123].

Included in this section also is the solvent extraction of PAH from solid and particulate matrices, where shaking or stirring was used in place of special means (e.g., Soxhlet or ultrasonic) of contacting the solvent with the sample.

1. Precautions

No special precautions beyond those noted previously are necessary for solvent partitioning or shaking. The user should consider possible solvent or sample toxicity from inhalation or skin contact, and also solvent flammability in conducting the extractions. Some solvents, such as dimethyl sulfoxide, can enhance transport of hazardous chemicals through the epidermis.

2. Comparison with Other Extraction Procedures

Direct solvent extraction is often used as a reference point for comparison with newer methods for extraction of PAH from aqueous matrices, in much the same manner as the Soxhlet has been used as a reference method for PAH extraction of solids. A few of these comparative studies are presented in Table 4 and are discussed below.

A continuous solvent extractor for aqueous PAH has been compared [19] with a batchwise ultrasonic solvent extraction of pure water and of water adulterated with particles. In both cases (one shown in Table 4), the ultrasonic procedure was found to yield somewhat higher recoveries than the continuous extractor. Other workers, comparing [121] various means of solvent and polyurethane foam extraction of PAH from fuel oil-spiked seawater, found the adsorbent foam to be more effective. However, absolute extraction recoveries were not determined, and there may have been significant differences among the PAH losses due to wall effects in the different volumes of water extracted and apparatus (different volume/wall surface area ratios) used for the various extraction procedures. In a more closely controlled comparison [124], solvent extraction has been found more effective than XAD-2 resin cartridges for extraction of PAH from simulated landfill leachates. However, the latter procedure included a direct 100-fold concentration step

Table 4. Comparison of PAH Recoveries by Solvent Extraction and Other Methods

Sample matrix	Matrix amount	PAH[a]	Level	Extraction conditions: Solvent[a]	Extraction conditions: Method	Percentage recovery	Ref.
Water	5 liters	Py	0.3 μg/l	MC	Ultrasonic, 300 ml, 5 min	84	[19]
	5 liters	Py	0.3 μg/l	MC	Continuous, 36 hrs	83	
	5 liters	BghiP	0.1 μg/l	MC	Ultrasonic, 300 ml, 5 min	82	
	5 liters	BghiP	0.1 μg/l	MC	Continuous, 36 hrs	55	
Seawater and fuel oil	12 liters[b]	Phen	—	C	Sep. funnel —	15[c]	[121]
	—	Phen	—	C	Poly. foam —	180[c]	
	—	Phen	—	C	Continuous —	150[c]	
	19 liters	Phen	—	C	Carboy —	140[c]	
	12 liters	Py	—	C	Sep. funnel —	83[c]	
	—	Py	—	C	Poly. foam —	330[c]	
	—	Py	—	C	Continuous —	240[c]	
	19 liters	Py	—	C	Carboy —	170[c]	
Paraffin waxes	100 g	BaP	Radiotracer	CH/DMF-H_2O (9:1)	Partition, 100 ml/200 ml	74-76	[118]
	100 g	BaP	Radiotracer	—	Ads. col. chrom.	79-93	
Aqueous leachate	0.5 liters	Nap	Radiotracer	CH or MC	Partition, 100 ml × 5	100	[124]
		BaP	Radiotracer	CH or MC	Partition, 100 ml × 5	100	
		Nap	Radiotracer	CH or MC	XAD-2 cartridge	100	
		BaP	Radiotracer	CH or MC	XAD-2 cartridge	82	

[a]C = chloroform, DMF = dimethyl formamide. See previous tables for other abbreviations.
[b]2 liters at a time
[c]Relative recovery (ng/liter) measured; not corrected for absolute recovery.

which was not conducted with the solvent extractions, and thus the final PAH recoveries after extraction and concentration may not differ significantly.

In the extraction of PAH from paraffin waxes, extraction with a silica gel adsorbent has been found [118] somewhat more effective than solvent partitioning the wax between cyclohexane and dimethyl formamide/water phases.

3. *Analytical Applications of Liquid Partitioning or Solvent Shaking*

For the extraction of PAH from water, direct partitioning with an immiscible organic solvent has been employed in most of the reported studies. Of the various solvents employed, methylene chloride appears to be the solvent of choice [19,58,59,120,122,125-129] because of its good solubility properties for PAH, low boiling point, high commercial purity, polarity, and relatively low toxicity. It is specified in standard or tentative procedures published by the EPA [58,59] and the ASTM [122]. In the application of this solvent to the PAH extraction of environmental and industrial waste water samples (see Table 5), a variety of conditions have been employed, ranging from two 250-ml extractions of 10 liters of water [58] to a novel three 0.3-ml extractions of 0.5 liter of water [125]. It also has been employed in continuous extractors [19] and compared with hexane [23] and methylene chloride/hexane (15:85) [125]. In both of the latter cases, methylene chloride was as good as or superior to the other solvent. Extraction at pH 7 reportedly [125] yields slightly higher PAH extraction recoveries, although some workers [e.g., 128,129] have acidified industrial wastewater and river water to pH 2 immediately after sampling to inhibit microbial degradation of PAH. In those studies, methylene chloride also was added to the water containers at the sampling site to begin the PAH extraction.

Cyclohexane is another popular solvent which has been found effective [23,130] for extraction of PAH from water in batchwise partitioning (Table 5). Continuous extractions with cyclohexane also have been reported [131]. Other solvents used in batchwise extractions include chloroform [121], benzene [113], and hexane [132].

Hydrophilic liquids or tars such as tobacco smoke condensate traditionally are diluted with an aqueous solvent such as methanol/water and are extracted with cyclohexane [e.g., 45,137] or hexane [e.g., 38]. Further solvent partitioning is conducted to purify the extracted PAH. Alternatively, the condensate can be dried and the oily residue extracted with methylene chloride [133]. Although very few, if any, PAH extraction recoveries are routinely reported for such procedures, the fairly high overall PAH recoveries [e.g., 134] after extraction and isolation would suggest high PAH extraction recoveries from the smoke condensate, as would be expected from their solvent partition coefficients [e.g., 193].

PAH in hydrophobic liquids, such as paraffin waxes [118,135], lubricating oils [115,136], and coal-derived liquids [137] can be extracted by first diluting with cyclohexane, heptane, or isooctane and then extracting with either dimethylsulfoxide or dimethylformamide/water. The water is used to make the dimethylformamide phase more selective for PAH. A reported [118] BaP extraction recovery of 74-76% from 100 g of paraffin wax by a partition with cyclohexane (100 ml) and dimethylformamide/water (9:1, 200 ml) followed by addition of 200 ml of water to the latter phase and back-extraction into 400 ml of cyclohexane undoubtedly could be improved by performing multiple extractions (at least three) at each step. Reduction in the initial volume of cyclohexane used also would improve recoveries [136]. In contrast, water is not added to dimethylsulfoxide until back-extraction, and recoveries of BaP can range 80-98% [137].

Solids and slurries also have been solvent-extracted using shaking or stirring. PAH in small samples (0.2 g) of marine sediments [132] have been rapidly extracted (45 sec) in 78-84% recoveries by two 1-ml hexane extractions using a vortex mixer. Particulate samples from air or automotive exhaust have been extracted by a variety of solvents, including methylene chloride [22] and benzene/methanol (no ratio given) [138]. Engine exhaust particulate sampling pads have been macerated for 8 hr on a Waring blender with the latter solvent; although the methanol reportedly [138] deactivates active sites on the glass fiber pads, it seems more likely that the binary solvent mixture is more effective in dissolving the auto exhaust particulate matrix than benzene alone. The efficiency of various solvents for recovering BaP [28] or total extractables [60] from air particles by shaking or stirring have been compared. As would be expected, the more polar solvents (such as acetone) extracted more BaP [28] at room temperature, but also more extraneous materials which could interfere in the analysis. In contrast, with Soxhlet extractions the less polar solvents (such as ether, benzene, or methylene chloride) were nearly as effective in extracting BaP but did not extract as much extraneous material.

4. *Novel Solvent Extraction Procedures*

Most PAH extractions of meat products are initiated by an alkaline digestion step. However, the possible effects of strong alkali on the more labile PAH is cause for concern, and less rigorous extraction procedures have been reported. In one case [139] the meat product was mixed with an equivalent volume of chloroform using kneading beaters and a food mixer. The sample was mixed with sodium sulfate and celite, and the chloroform was evaporated under vacuum. PAH were eluted by packing the sample into a column and passing propylene carbonate through. This solvent is claimed [139] to have the advantages of superlative PAH solubilization, and ready decomposition into propylene glycol for recovery of PAH. Recoveries of added BaP range from 95 to 100% but it is highly questionable that an externally applied PAH spike can accurately model the behavior of native PAHs in the meat.

An unusual procedure has been reported [140] for enrichment of PAH from air particulate pads. The pads are simply dissolved in hydrogen fluoride, diluted with water, and the PAH are extracted with cyclohexane. The dissolution procedure yielded very similar amounts of total extracts to Soxhlet extractions with cyclohexane, and recoveries of three PAH were 98% or better. Further, no ring or side chain rearrangements were observed for alkyl PAH spikes. However, the milligram level of PAH spikes applied to the pads is not realistic for most air particulate samples, and much lower levels of particle-associated PAH may behave quite differently. In spite of this shortcoming in the method evaluation, the procedure may be useful in samples where serious surface adsorption or occlusion of PAH hinders extraction recoveries.

A continuous, flow-blending procedure for rapidly extracting powdered solid samples with methylene chloride has been reported [141]. Although this procedure was not specifically applied to PAH extraction, the yields of

Table 5. PAH Extraction Recoveries from Various Matrices Using Liquid-Liquid Partitioning or Solvent Shaking

Sample matrix	Matrix amount	PAH[a]	Level	Percentage recovery	Extraction conditions			Ref.
					Solvent[a]	Volume	Time	
Water	1 1	Nap	Radiotracer	100	CH	100 ml × 3	–	[23]
		BaP		100				
	2.5 1	BaP	0.1 μg/liter	92.3	CH	125 ml	10 min	[130]
	2.5 1	BaP	0.1 μg/liter	96.3	CH	125 ml × 2	10 min	
Coal liquefaction wastewater	3 1	"PAH"	1 μg/liter	91	MC	250 ml 100 ml × 2	–	[120,126]
Estuarine water	4.5 1	Anthr	0.88 μg/liter	88	H	10 ml	1 hr	[132]
		Fl	0.88 μg/liter	89	H	10 ml	1 hr	
		BaA	0.88 μg/liter	84	H	10 ml	1 hr	
		BaP	0.88 μg/liter	59	H	10 ml	1 hr	
Water, pH 2	0.5 1	2-6 ring	40 μg/liter	66-96	MC/H (15/85)	0.3 ml × 3		[125]
	0.5 1	2-6 ring	40 μg/liter	86-106	MC	0.3 ml × 3		
Water, pH 7	0.5 1	2-6 ring	40 μg/liter	64-88	MC/H (15/85)	0.3 ml × 3		
	0.5 1	2-6 ring	40 μg/liter	99-101	MC	0.3 ml × 3		
Water, pH 10	0.5 1	2-6 ring	40 μg/liter	41-70	MC/H (15/85)	0.3 ml × 3		
	0.5 1	2-6 ring	40 μg/liter	63-101	MC	0.3 ml × 3		
Water, pH 6-8	3 1	2-6 ring	0.5-25 μg/liter	59-104	MC	125 ml, 75 ml × 2	3 min	[59]
Paraffin wax	100 g	BaP	Radiotracer	74-76	CH/DMF-H_2O	100 ml/200 ml	–	[118]
	100 g	4-6 ring	0.1 μg/liter	62-96	i-O/DMSO	100 ml/50 ml × 3	3 min	[135]

Air particles	1 g	BaP	1.12 mg/g	68	CH	50 ml × 6	30 min	[28]
	1 g	BaP	1.12 mg/g	80	B			
	1 g	BaP	1.12 mg/g	75	C			
	1 g	BaP	1.12 mg/g	72	i-O			
	1 g	BaP	1.12 mg/g	100	A			
	1 g	BaP	1.12 mg/g	95	M			
	1 pad	Phen	1.1 mg	100	HF, then			[140]
		BaA	1.0 mg	98	H_2O and CH			
		BaP	1.05 mg	100				
Marine sediment	0.2 g	Anthr	5 μg/g	78	H	1 ml × 2	45 sec	[132]
		Fl	5 μg/g	80	H	1 ml × 2		
		BaA	5 μg/g	80	H	1 ml × 2		
		BaP	5 μg/g	84	H	1 ml × 2		
Aqueous leachate	0.5 liter	Nap	Radiotracer	100	CH or MC	100 ml × 5	—	[130]
		BaP	Radiotracer	100	CH or MC	100 ml × 5	—	
Meat products	100-200 g	BaP	0.05-0.1 μg/g	95-100	C then PC	—		[139]

[a]See previous tables for abbreviations. Others: i-O = isooctane; PC = propylene carbonate; DMSO = dimethyl sulfoxide.

solvent-extractable material obtained in 10 min of extraction of limestone, shale, marl, or coal was equivalent to and three times more reproducible than that obtained by 20 hr of Soxhlet extraction with the same solvent. PAH spiked onto fly ash have been eluted in a somewhat similar fashion by packing the fly ash into a cell and pumping solvent through [119].

D. Recommendations for Solvent Extraction

Various methods of solvent extraction are available for recovery of PAH from a variety of sample matrix types. The analyst must choose the one most suitable for his particular application. The following guidelines are suggested.

For PAH extraction of solids, it is recommended that ultrasonic extraction be the first choice, based on its highly efficient, rapid recovery of PAH. In those cases where the cost of the ultrasonication unit is a limiting factor, Soxhlet extraction should be considered. Extraction by shaking with a solvent is a last choice only. The following recommendations are made for these applications:

1. For air particulate samples, benzene or benzene/methanol are among the preferred solvents. Cyclohexane may coextract less interfering material, but PAH recoveries may not be as complete. Chlorinated solvents such as methylene chloride may provide highly efficient extractions, but with ultrasonication the possibility of chlorination reactions must be considered.
2. For coal fly ash, carbon black, and other highly sorptive sample matrices, the strongest solvents should be used.
3. For hydrophilic materials, such as sediments, water-miscible solvents such as acetone should be employed with undried samples to avoid PAH losses by volatilization. Separate aliquots of sediment may be dried to determine dry/wet weight ratios.
4. With extended ultrasonic extractions, provisions for cooling the sample may be necessary to minimize volatilization losses of the lighter PAH.
5. If at all possible, extraction recoveries should be estimated with radiolabeled tracers or cold PAH spikes.

For hydrophilic liquid samples, direct solvent partitioning by shaking is the preferred method, although ultrasonic agitation can be used to good advantage in attaining good solvent contact with the sample. Water-immiscible solvents with good PAH solubility characteristics such as methylene chloride are preferred. Likewise, hydrophobic liquids are best diluted with a hydrocarbon solvent and extracted with dimethylsulfoxide [193], dimethylformamide, or nitromethane. As with solid samples, extraction recoveries should be measured if at all possible.

VI. Thermal Methods of Extraction

Thermal methods of PAH extraction, including vacuum and flowing-carrier sublimation and solvent distillation generally receive only limited attention in the few reviews (e.g., Refs. 5 and 6) noting their existence. Distillative methods for separating PAH-bearing fractions from a bulk sample have been in use for a considerable time (e.g., Ref. 5), as have sublimation methods for extracting air particulate PAH. The use of vacuum sublimation for PAH extraction has been mistakenly attributed in some reviews to cases where it actually was used only for solvent removal from a PAH extract.

In spite of their less widespread use, thermal methods of PAH extraction offer certain unique advantages over more conventional methods of extraction.

A. Sublimation Methods of Extraction

Extraction of PAH from the sample matrix by sublimation has been achieved both under vacuum [105,144-151] and under a nitrogen or hydrogen carrier [142,152-154]. In both cases the sublimed PAH are collected on glass surfaces or in solution for further purification or analysis. Vacuum sublimation is a standard procedure which has been used [105] as a reference point for evaluating an ultrasonic extraction procedure for extracting PAH in air particles. Although the extraction recoveries were not reported, the overall recovery of BaP and benzo[k]fluoranthene (BkFl) both were approximately 97% (Table 6) when the extraction of a section of an air particulate filter was conducted at 305°C for 30 min and 0.001-0.005 mmHg. However, the ultrasonic extraction procedure required one-third the time and produced essentially equivalent analytical results. The entire filter pad also has been subjected [146] to vacuum sublimation of 300°C, and 0.03-0.05 mmHg for 50 min. The overall recovery (after isolation) of spiked BaP averaged 102%, which would require a quantitative extraction recovery. The advantages of the total extraction and analytical method were listed as low equipment costs (ca. $3500 in 1973), speed (six samples/2.5 hr), sensitivity (to less than 1 ng of BaP), and minimal operator training and sample handling.

Extraction of PAH as large as coronene from TLC plates has been achieved [145] in reasonably high recoveries, using lower temperatures and pressures. However in applications [148,149] to air particles from various sources, it is possible that PAH are sublimed more easily from a TLC plate layer than from air particles, and the actual recoveries in applications may be lower. In most of these procedures, the sublimed PAH were separated by TLC and measured by fluorescence spectrophotometry.

An early report [150] of the use of vacuum sublimation for extraction of particulate PAH employed considerably milder conditions, 100°-175°C, at 1 mmHg (Table 6). Recoveries of Pyrene (Py) from air particulate filter pads were nearly quantitative, although the microgram levels spiked were considerably higher than those in other studies. It is interesting to note that the report claimed to be able to obtain essentially pure Py and coronene by vacuum sublimation from a carbon black sample which previously had received prolonged benzene and chloroform extraction. Rather than marveling at the extractive power of the sublimation procedure, one questions the efficiency of the unspecified solvent extraction procedure.

Sublimation of PAH in a heated nitrogen carrier also has been applied for the purification of PAH [152] and for the direct extraction of BaP and BkF from air filter pad disks [142]. At 230°C and a nitrogen flow of 10 cm^3/min, recovery of BaP and BkF was reportedly complete in 40 min, although the level of the spike and the numerical recovery were not listed. The fluorescence spectra of PAH obtained from extracts prepared in this manner were cleaner than those obtained from Soxhlet extraction of the filters, and no further isolation was required before analysis. This advantage may compen-

Table 6. PAH Extraction Recoveries from Various Sample Matrices by Thermal Methods

Sample matrix	Extraction method[a]	PAH	Level	Percentage recovery	Extraction conditions	Ref.
Air particles	VS	BaP BkFl		97[b] 97[b]	305°C, 0.001-0.005 mmHg, 30 min	[105]
	VS	BaP	22 ng	100-103	300°C, 0.03-0.05 mmHg, 50 min	[146]
	VS	Py	1.24 μg	96-99	100°C, 1 mmHg, 5 min	[150]
Dust	VS	Py	2.5 μg	93-94	175°C, 1 mmHg, 20 min	[150]
TLC plate	VS	Py		>90	100-165°C, 0.1 mmHg, 5 min	[145]
		BaP		>90	165°C, 0.1 mmHg, 15 min	
		BghiP		>90	200°C, 0.1 mmHg, 15 min	
		Coro		90	220°C, 0.1 mmHg, 60 min	
Crude nap	CS	Nap	30 g	74	100°C, 300 liters/hr of N_2, 80 min	[152]
Air filter pad	CS	Py Chry BkFl BaP	102 ng 77 ng 7.4 ng 64 ng	87 93 90 83	300°C, 50 cm^3/min of N_2, overnight	[153]

Water	SD	Nap	4 mg/l	92	24 hr	[156]
		Phen	4 mg/l	94		
		Chry	4 mg/l	78		
		BaP	4 mg/l	45		
	SD	MeNap	8 ng/kg	50	20 hr	[143]
		F	8 ng/kg	42		
		Anthr	8 ng/kg	82		
	SD	BaA	Radiotracer	100	24 hr	[124]
		BaP	Radiotracer	40		
Metal processing waste		BaA	Radiotracer	60		
Soybean process cake		BaA	Radiotracer	<1		

[a]VS = vacuum sublimation; CS = sublimation in carrier gas stream; SD = steam distillation.
[b]Overall recovery, including extraction.

sate for the much longer time required for PAH extraction than by ultrasonic methods. It would be expected that far less potentially interfering material (only volatile compounds) would be recovered by sublimation vs. direct solvent extraction.

For GC analysis of PAH extracted in a heated carrier gas, the sublimed PAH can be collected in a cartridge loaded with a conventional coated GC packing material. This approach has been taken [153] for the GC analysis of air particulate PAH. Roughly ground (20 mesh) air filter pads were thermally desorbed overnight at 300°C and 50 cm^3/min of nitrogen into a 6.5 mm × 120 mm cartridge (held at room temperature) of either 5% Dexsil 300 or 10% OV-1 on 80/100 Chromosorb G HP. Analysis was initiated by direct insertion of the cartridge into a heated GC inlet. The recovery of PAH spiked onto unused filters ranged from 83 to 93%, as noted in Table 6. However, the influence of air particles on the actual recoveries was not determined. It is likely that the sublimation of PAH occluded in other organics and spread out over carbonaceous surfaces would be considerably different from that achieved with PAH spread over glass fiber surfaces. Further, the overall method suffers from requiring a specific detector (gas phase fluorescence). However, as regards PAH extraction, the method offers the advantage of allowing multiple simultaneous extractions and high sensitivity.

A more direct PAH extraction approach [154] is to directly sublime the PAH onto the GC column. In a semiquantitative analytical procedure, the PAH in strips of air particulate filter pads were placed in glass tubes which were inserted into a GC inlet and connected with the GC column. Desorption of PAH was conducted at 350°C for 5 min to "remove all volatiles" from the air filter strip. Desorption efficiencies were not reported. It is interesting to note that the glass capillary column GC profiles of the desorbed organics were very similar to those obtained from 24-hr Soxhlet extractions with methylene chloride but were achieved in a fraction of the time required for the latter.

Although most analytical applications of the sublimation procedures have been to air particulate PAH, it seems feasible that PAH in other sample matrices (e.g., coal liquids) may be advantageously extracted. Indeed, recently the application of sublimation to the extraction of PAH from coal fly ash has been reported [155].

B. Distillation Methods of Extraction

Several extractive distillation methods have been used to recover PAH. Most of these involve stream distillation [37,120,124,143,156], but direct distillation of crude, wide-boiling range mixtures also have been used to obtain fractions suitable for solvent extraction of the PAH. In a sense, then, such distillations are a preextraction or concentration step.

Focusing, however, on the steam distillation methods, most of the analytical applications have been for the recovery of PAH from aqueous matrices. The methods produce fairly "clean" extracts [34,104,124] which lack much of the potentially analytically interfering material which is picked up in direct solvent extracts. However, the PAH ultimately are extracted into an organic solvent using one commercial apparatus [124] which provides for separate distillation pots but cocondensation of aqueous and organic vapors.

As shown in Table 6, the PAH extraction recoveries are quite good but fall off for the larger PAH ring systems which have lower vapor pressures. A comparison [120,143] of steam distillation with stripping onto an adsorbent with an air carrier showed superior PAH recoveries for steam distillation, particularly with less volatile PAH (anthracene vs. fluorene or methylnaphthalene). However, direct solvent extraction of water produces [37] better recoveries of large PAH ring systems such as benzo[ghi]perylene.

Steam distillation also has been applied in scoping studies [124] to the extraction of PAH from solid wastes (Table 6). Recoveries of benz[a]anthracene (BaA) were low from highly sorptive matrices, as might be expected, but were considerably better for less sorptive matrices. Perhaps conducting the steam distillation at elevated pressure and temperature or switching to other immiscible solvent pairs would improve PAH extraction recoveries from highly sorptive solid materials.

C. Recommendations for Thermal Extraction

Thermal methods of PAH extraction are not yet as well developed as the solvent extraction procedures and require somewhat longer periods for extraction. Newcomers to the field may find Soxhlet or ultrasonic extraction procedures more readily adoptable and perhaps should utilize them as first choices. However, the methods involving direct sublimation of PAH from an air filter pad onto a high-resolution GC column appear to have much potential in a rapid screening procedure and possibly even in a quantitative analytical procedure.

VII. Extraction onto Solid Phases

The extraction of PAH onto solid phases will be divided into four general approaches: (1) the extraction onto carbon or charcoal; (2) the extraction onto macroreticular resins such as the XAD-series, Tenax, and polyurethane; (3) the extraction onto such sorbents as silica, alumina, or Florisil by conventional liquid chromatographic techniques; and (4) the extraction by bonded-silica solid phases employing high-performance liquid chromatographic (HPLC) techniques.

A. Carbon

Activated carbon has several properties which may make it a sorbent of choice in some applications. Carbon particles with very high specific surface areas can be prepared. Such high specific surface areas optimize the volume of sorbent required relative to the total amount of organic material to be adsorbed. In addition, carbon shows very high thermal stability and it is available in a very high degree of purity. Carbon has been used to extract organic compounds from both air and water samples [34,36,174-177]. One novel design [34] employs a filter containing approximately 1.5 mg of active carbon assembled in a cylindrical disk 2.5 mm in diameter and 1 mm thick in a closed-loop stripping procedure. This carbon had a rather large particle size (0.05-0.1 mm), but by repeatedly stripping a liter of water through the filter for 2 hr many classes of organic compounds were adsorbed. Subsequently the organic compounds were removed from the charcoal by carbon disulfide. However, the recovery of aromatic compounds by this procedure appears to be low.

Other investigators [36] have used spherical carbon molecular sieve material that had an extremely high surface area (1200 m^2/g) to extract organic compounds from water. This carbon was packed into a small column (1.2 mm i.d. and 2.5 cm long). Water samples were extracted by making a single pass through the column. The columns were subsequently eluted with combinations of acetone and carbon disulfide. Here, again, recoveries of many classes of organic compounds were very good; however, the recovery of aromatics such as acenaphthene, acenaphthylene, and methylnaphthalene appears to be lower than that obtained for a macroreticular resin employed in a similar fashion.

Carbon filters also have been used extensively to extract organic compounds from air for subsequent analysis. A report in 1977 [176] describes a typical application and reviews the work of others.

In summary, although activated carbon appears to be an excellent solid phase extractant for many classes of organic compounds both in air and water, it would not seem to be the extractant of choice for the PAH as a class.

B. Macroreticular Resins

These resins have been used extensively for the extraction of PAH [21,35, 36,47,78,158-165] from either air or water. Perhaps the most widely used of these resins is Tenax GC, which is a porous polymer of 2,6-diphenyl paraphenylene oxide. This polymer has proven quite efficient in trapping organic volatiles, and the trapped sample can be removed by either heat desorption (often directly into a gas chromatograph) or by solvent extraction [78]. The disadvantages of Tenax are its solubility in some nonhydrocarbon solvents and its tendency to release background contamination upon desorption [157]. It also is decomposed by nitrogen oxides. Because of this tendency toward background contamination, it is important to "preextract" the Tenax with pure solvents before use.

With a column of Tenax 10 cm long and 0.4 cm i.d., almost quantitative recovery of PAH can be obtained [160] from water when the aqueous sample is forced through the column at a pressure of 5 $lb/in.^2$. Another apparatus for the extraction of liquid samples has been described [161]. Tenax can also be used to extract large volumes of water. Up to 20 liters of water can be extracted [35] with 1.5 g of Tenax in a column 7.5 cm high by allowing the water to flow over the column at the rate of 3 liters/hr. Recoveries of PAH by this method were 85% or better. One important precaution in using Tenax to extract PAH from water is the adjustment of the pH to a value ranging from 6.7 to 7.3. Recoveries will decrease as the pH of the water sample increasingly varies from this neutral range.

By far the most extensive use of Tenax as a solid extractant is in the extraction of organic compounds from gas phase samples. Here volumes of the gaseous sample are pumped through a cartridge filled with Tenax. After sample collection, direct thermal desorption onto a gas chromatography column fashions a very convenient analytical procedure. A typical procedure for Tenax extraction of air may be found in Ref. 158. Using Tenax extraction of air followed by an exhaustive desorption of the Tenax with methanol, one group [78] identified approximately 20 PAH in an air sample typical of a city residential area.

The XAD series of resins originally marketed by Rohm & Haas are undoubtedly the most widely used solid phases for the extraction of organic compounds from aqueous media. These XAD resins are generally available as four different materials, XAD-2, -4, -7, and -8. In general, the resins can be described as insoluble spheres of high surface area porous polymers. However, there are differences which should be indicated. XAD-2 and XAD-4 are polystyrene and are essentially nonpolar. The main difference between XAD-2 and XAD-4 is a much higher surface area for XAD-4. XAD-7 and XAD-8 are acrylic ester polymers that are of intermediate polarity. For all these XADs, the basis for extraction is adsorption on the surface; no ion exchange or pore exclusion mechanisms are involved [21].

Perhaps the major disadvantage of XAD resins is that the resins supplied in bulk generally require extensive cleanup to remove impurities [21]; however, purified resin and purified resin in disposable columns are now available commercially. The important point to note here is that resin which has not been purified will not deliver the successful results described in the literature.

A typical procedure for the extraction of organic compounds from water using XAD resins [21,36,47,164,165] involves passing a water sample over a column (approximately 3 cm long containing 5 ml of XAD) at a flow rate of 2 to 4 bed volumes per minute (10 to 20 ml/min). The extract is then eluted from the column with solvents such as acetone, methylene chloride, ether, and/or methanol. Two precautions should be mentioned: the column should not be allowed to go dry, and the first eluting solvent should be miscible with water.

The pH of the water sample is very important when extracting with XAD resins. The nonpolar resins tend to be most efficient in extracting neutral compounds. Thus PAH should be most efficiently extracted from water at pH 7 by the polystyrene resins, and under such conditions better than 90% recovery is possible for some compounds [157]. An interesting method of enhancing extracting efficiency is to mix a nonpolar resin and a more polar resin. It has been found [21] that a mixture of XAD-4 and XAD-8 is most efficient. Such a mixture should tend to reduce the effect of pH.

Polyurethane foams also have been used to extract PAH from water [20, 32,33,166]. When the foams are used in the form of flexible plugs, large volumes of water can be rapidly extracted [20,32,166], with PAH recoveries generally exceeding 80 percent. An interesting modification of this approach is extraction onto in situ polymerized polyurethane columns [33].

We can summarize our conclusions concerning preferred methods of PAH extraction by macroreticular resins as follows:

1. Tenax is the sorbent of choice for gas phase PAH samples where desorption is to be used for the analytical method. XADs can be used where solvent extraction of the resin is used to recover the collated PAH.
2. XAD-4 is the sorbent of choice for PAH in neutral aqueous media.

C. Silica Gel, Florisil, and Alumina

The previous discussions of solid extractants have previewed solid matrices that were employed to extract PAH from either the gas phase or aqueous media. In contrast, silica gel, Florisil, and alumina are generally employed to selectively remove PAH from organic media. Here the extraction is generally part of a scheme in which a complex mixture of organic compounds dissolved in a nonaqueous solvent is placed on the sorbent. The PAH are generally adsorbed on the solid extractant. The adsorbed PAH are then selectively removed by liquid partition chromatography.

Because the recovery of PAH from these extractants generally involves a liquid-liquid partitioning, the use of a radiochemical tracer can greatly facilitate the extraction by locating the fraction of interest and providing data to estimate recovery [170]. The recovery for such a solid extraction should generally exceed 90%. However, such a solid extraction is generally part of a much larger separation scheme [170], so the recovery for one particular step is incorporated into the overall analytical procedure. Thus little comparative recovery data are available.

These solid extractants have been used to isolate PAH fractions from coal solids [172], shale oil [170,173], heavy liquids derived from coal [170,172], finished fuels [170], cigarette smoke particulate matter [39], and petroleum substitutes [170].

The extraction onto alumina, Florisil, or silica gel has been employed quite extensively in analytical schemes involving the isolation of PAH fractions from complex organic mixtures. However, for analytical-scale samples the isolation methods based on HPLC (discussed in the next subsection) offer improved procedures.

D. Bonded Silica Solid Phases

This extraction method involves the use of solid phases generally employed in HPLC. Typical solid extractants are aminosilane [112], octadecylsilane, C_{18} [171], or fully porous (nonbonded) silica [177].

For complex organic mixtures in nonaqueous media the use of chemically bonded aminosilane as a solid extractant appears to be an excellent way to isolate a PAH sample for further detailed analysis. In this technique the complex organic mixture is placed on a semipreparative HPLC column (e.g., 25 cm long and 1 cm i.d.). The column is eluted with a nonpolar solvent such as pentane, hexane, or heptane in order to remove aliphatic hydrocarbons. PAH and all polar compounds are sorbed on the solid phase. The PAH can be removed by slightly increasing the polarity of the mobile phase. The more polar liquid phase is generally achieved by adding methylene chloride (5% or less by volume) to the aliphatic hydrocarbon solvent initially employed as the mobile phase. Such a PAH fraction can generally be quantitatively recovered and does not require additional cleanup before final analysis. Requiring no additional cleanup is a significant advantage because most of the solid extraction techniques discussed above generally require additional cleanup steps before the PAH fraction is ready for the final analytical step (see Table 7).

Isolation of PAH on an aminosilane bonded phase is closely related to several other solid phase extraction methods involving an oxypropionitrile bonded phase or simply porous silica [177]. However, all these solid phases perform in a similar manner with their affinity for organic compounds increasing with the increasing polarity of the compound. Recoveries are quantitative and no additional sample cleanup is needed. Another variation on this tech-

Table 7. Summary of Solid Phase Extractants for PAH

Solid phase	Major uses	Major advantages	Major disadvantages	Recommendation
1. Carbon	Air, water	High surface area; does not retain water	Poor recovery	Best for compounds other than PAH
2. Macroreticular resin:				
XAD	Water, air	Good flow characteristics; mechanically rugged	Requires cleanup and careful storage	Use for water samples
Tenax	Air, water	Can be thermally desorbed	Soluble in some solvents; cleanup difficult	Use for air samples
Polyurethane	Water	High recovery		
3. Sorbents:				
Silica gel Alumina Florisil	Complex organic mixtures	Low cost; can be used on large scale	Time consuming; separation requires tracer	Use for large-scale samples such as fractionation for biotesting
4. HPLC materials:				
Porous silica	Complex organic mixtures	Fast		Use to prepare analytical samples from oils, tars, etc.
Aminosilane	Complex organic mixtures	Fast, quantitative recovery		
Octadecylsilane	Water	Fast, less sample handling		Use for higher ring number PAH in water

nique involves coupled-column chromatography [178], where the PAH are extracted onto a C_{18} solid phase while many of the remaining organic compounds are eliminated in the mobile liquid phase. Subsequently the PAH in the solid phase are further purified by elution onto a size exclusion solid phase from which the PAH fraction or a specific portion of that fraction is selectively removed. This procedure is probably more complicated than necessary for most samples.

C_{18} solid phase may be used to extract PAH from aqueous media [167-169, 171]. Here large volumes of water can be pumped across the C_{18} solid phase. Recoveries generally increase with the number of rings in the PAH, although recoveries of naphthalene often are as low as 50%. However, recoveries of PAH with four or more rings generally exceed 80%. Fortunately, this method of extracting PAH from water has one major advantage: the C_{18} solid phase can be configured in the form of a precolumn that can be plumbed directly into an analytical HPLC system, thus eliminating the need for additional sample handling and concentration.

E. Recommendations for Use of Solid Phase Extractants

The solid phase extractants are summarized in Table 7, where the major uses, advantages, and disadvantages are listed. Typical PAH extraction recoveries are shown in Table 8 for applications of the various solid phase extractants. The recommendations listed in Table 7 are only suggestions, and it is quite likely that another method might work as well or even better when considered with respect to unique samples and specific laboratory experience. Nevertheless, the following guidelines are suggested for use of solid extractants:

1. For air samples, use Tenax.
2. For comprehensive water samples, use XAD-4 with pH of water adjusted to 6.7-7.3. For PAH of four rings and greater, one might use C_{18}.
3. For complex organic mixtures, use aminosilane for preparation of analytical samples. For large-scale fractionation of organic mixtures, use the more classical sorbents such as alumina, silica, and Florisil.

VIII. Digestion Methods of Extraction

Digestion methods for PAH extraction are meant here to refer to methods for extracting PAH from materials such as foodstuffs and biological tissue samples. The major problems with such samples is not the extraction per se but the digestion of the sample material in order to put it into a form that lends itself to one of the previously discussed PAH extraction techniques. In general, the approach to such sample digestion involves the digestion of the sample matrix in an alcoholic solution of a strong base followed by a liquid-liquid extraction of the digested material [179-192]. Two problems are inherent in such a procedure. First of all, it has yet to be definitely established whether or not the harsh digestive treatment has altered the chemical form of the PAH initially present in the sample. A second problem arises when a tracer is added to facilitate the calculation of recoveries, because it is quite likely that the tracer will not be incorporated into the sample material in the same manner as the native PAH which are present in the sample.

Table 8. Recoveries of PAH Extracted onto Solid Phase Extractants

Sample matrix	Amount	Solid phase extractant	PAH[a]	Level	Percentage recovery	Ref.
Water	1 liter	Activated carbon	Nap	0.5 μg/liter	53	[34]
			2MeNap	0.5 μg/liter	43	
Water	50 ml	XAD-4	Nap	2-10 μg/liter	83	[36]
			Nap	100 μg/liter	85	
			BP	2-10 μg/liter	87	
			BP	100 μg/liter	85	
Tap water	4 liters	Polyurethane	BaP	≤25 μg/liter	84-87	[32]
	10 liters	Foam (1 plug)	BaP	0.05 μg/liter	73	
	20 liters		BaP	0.05 μg/liter	67	
	40 liters		BaP	0.05 μg/liter	49	
	20 liters	(4 plugs)	BaP	0.05 μg/liter	85	
Lake water	4 liters	(2 plugs)	BaP	0.1 μg/liter	81	
Water	30 liters	Tenax	Anthr	0.13 μg/liter	96.6	[35]
	30 liters		Pery	0.13 μg/liter	92.0	
	30 liters		Inpy	0.13 μg/liter	86.0	
	25 liters		Py	0.08 μg/liter	98.5	
	25 liters		Fl	0.08 μg/liter	96.5	
	25 liters		BaP	0.08 μg/liter	96.6	
Lake water	100 ml	Tenax	Phen	500 μg/liter	98.7	[160]
Cigarette smoke condensate	2 g	XAD-2	BaP	0.96 μg/liter	85.1[b]	[163]

(continued)

Table 8 (continued)

Sample matrix	Amount	Solid phase extractant	PAH[a]	Level	Percentage recovery	Ref.
Sea water	100 liters	XAD-2	Phen	100 μg/liter	61.8	[162]
				10 μg/kg	42.5[c]	
				10 μg/kg	(102.)[d]	
				1 μg/kg	57	
Sea water	2 liters	C_{18}-Pellicular	Nap	28 μg/kg	19	[168]
			Phen	1.5 μg/kg	92	
			Py	2 μg/kg	78	
			BaP	1.5 μg/kg	58	
			DBA	3 μg/kg	14	
	600 ml	Headspace onto Tenax	Nap	3.5 μg/kg	38	
			n-PrNap	3.3 μg/kg	51	
			Phen	3.3 μg/kg	19	
Water	6-18 liters	C_{18}-Pellicular (various)	2-4 rings	22-1000 ng/liter	42-110	[167]

[a] 2MeNap = 2 methylnaphthalene; BP = biphenyl; N-PrNap = n-propylnaphthalene; DBA = dibenzanthracene.
[b] Recovery in isolated PAH fractions, including extraction recovery.
[c] Recovery from nonparticulate phase of preextracted seawater.
[d] Total recovery (dissolved and particulate phase) of preextracted seawater.

Table 9. Summary of Several Digestion Procedures

Sample	Digestion solution	Digestion conditions	Extraction solvent	Ref.
Meat, fish	2 N KOH in methanol/water (9:1)	Boil 2-4 hr	CH	[179]
Animal tissue	4 N NaOH	95°C for 2 hr	H	[180]
Mussels	4 N NaOH	90°C for 2 hr	E	[181]
Liquid smoke	KOH/ethanol/water	60°C for 30 min	CH	[182]
Nonfat dry milk	NaOH in water, acetone, and ethanol		i-O	[184]
Food	2 N NaOH in ethanol and water	Reflux 2 hr	H	[185]
Tissue	4 N NaOH	90°C for 2 hr	EE	[186]
Marine tissue	0.9 N KOH in ethanol	Digest for 2 hr	i-O	[187]
Smoked food	Ethanol, KOH	Extract with ethanol; saponify with KOH in Soxhlet	i-O	[188]
Cooked sausage	2 N KOH in methanol/water (9:1)	Boil 2-4 hr	CH	[190]
Charcoal-broiled food	2 N KOH in methanol/water	Boil 2-4 hr	CH	[191]
Fish	Alcoholic KOH	Digest	CH	[192]

Table 10. Recoveries of PAH by Alkaline Digestion Methods

Sample matrix	Amount	PAH	Level	Percentage recovery[a]	Ref.
Meat	200 g	5-7 rings	5.9-22 ng/g	98-102	[179]
Sunflower oil	200 g	3-6 rings	7.6-14 ng/g	94-105	
Dry milk	100 g	4-6 rings	2-20 ng/g	73-90	[184]
Root vegetables	100 g	4-6 rings	2-20 ng/g	71-92	
Chicken	100 g	4-6 rings	2-20 ng/g	70-91	
Frankfurters	100 g	4-6 rings	2-20 ng/g	71-83	
Shellfish	100 g	4-6 rings	2-20 ng/g	66-91	
Tea	10 g	BaP	2 ng/g	76	[185]
Chicken	50 g	BaP	0.4 ng/g	85	
Clam	50 g	BaP	0.4 ng/g	70	
Mackerel	50 g	BaP	0.4 ng/g	70	
Frankfurters	500 g	4-6 rings	2 ng/g	70-83	[188]
Cheese	500 g	4-6 rings	2 ng/g	75-80	
Fish	500 g	4-6 rings	2 ng/g	70-88	
Oyster	10 g	2-3 rings	10 μg/g	70-110	[186]
			2 μg/g	77-103	
			0.4 μg/g	53-93	
			0.1 μg/g	40-80	

[a]Final recovery; including extraction.

The method described in Ref. 179 is typical of digestion/extraction procedures for PAH. In this procedure 200 g of a foodstuff such as meat or fish is ground in a food grinder and digested in boiling 2 M potassium hydroxide (in 9:1 by volume methanol/water) for 2-4 hr until the sample is completely hydrolyzed. Following dissolution the PAH are partitioned into cyclohexane, and this cyclohexane extract is subjected to typical sample cleanup procedures for PAH in organic solutions. Table 8 summarizes the conditions for a variety of digested samples.

Recommendations

It is difficult to determine whether or not the differences in the methods listed in Table 9 are significant. All report good recoveries, as shown in Table 10. Thus only the following guidelines are suggested:

1. Use the lowest concentration of hydroxide that yields complete sample hydrolysis in a reasonable time period.
2. Use the lowest digestion temperature consistent with a reasonable digestion time.
3. For safety purposes an extraction solvent other than ethyl ether is recommended.

Appendix: Analytical Applications of the Various PAH Extraction Procedures[a]

	Extraction method									
	Solvent extraction			Thermal		Solid phase extractants				
Sample matrix type	Soxhlet	Ultrasonic	Solvent partition	Sublimation	Distillation	Carbons	Resins	Inorganic adsorbents	Bonded phases	Alkaline digestion
Air particles, dust	15,25-28, 30,31,52-57,62-68, 70,73-85, 103,158, 177	31,62-64, 99-103, 105-110	—	105,142, 144,146, 148,150, 151,153, 154	—	—	—	—	—	—
Volatile organics	—	—	—	—	—	176	29,78,158, 159	—	—	—
Automobile exhaust	71	114	58,138	145,149	—	—	—	—	—	—
Cigarette smoke condensate	—	—	38,45,133, 134,137, 163	—	—	—	163	39	—	—
Lube oil, paraffin	—	115	115,118, 135-137	—	—	—	—	118	—	—
Coal and shale liquids	95,98	—	170	—	—	—	—	170,172, 173	—	—
Soot, carbon black	89-93,97	—	—	150	—	—	—	—	—	—
Coal, fly ash	41,42,46, 93,94,119	1,24,43, 46	—	155	—	—	—	—	—	—
Marine and fresh-water sediment	23,72,85, 104,181	104,112, 113	132	—	—	—	168	—	—	49,180
Freshwater and seawater		117	19,34,37, 47,48,59, 113,120-122, 124-132, 143,180	—	37,120, 124,143, 156	21,34, 36,143, 175,177	20,21,32-37,47,48, 120,121,124, 160,162,164, 166,168,169	171	167,169, 171,178	—
Plant and animal tissue, foodstuffs	86-88	139	—	—	—	—	—	—	—	49, 179-191

List of Abbreviations

Category	Symbol	Meaning
Instrumental	GC/MS	Gas chromatography/mass spectrometry
	HPLC	High-performance liquid chromatography
	TLC	Thin-layer chromatography
	UV	Ultraviolet
Chemical	A	Acetone
	AN	Acetonitrile
	Anthr	Anthracene
	B	Benzene
	BaA	Benz[a]anthracene
	BaP	Benzo[a]pyrene
	BghiP	Benzo[ghi]perylene
	BkFl	Benzo[k]fluoranthene
	BP	Biphenyl
	^{14}C	Carbon-14 (radiolabel)
	C	Chloroform
	CH	Cyclohexane
	Chry	Chrysene
	Coro	Coronene
	DBA	Dibenzanthracene
	DEA	Diethylamine
	DMF	Dimethylformamide
	DMSO	Dimethylsulfoxide
	E	Ethanol
	EA	Ethyl acetate
	EE	Diethyl ether
	F	Fluorene
	Fl	Fluoranthene
	H	Hexane
	^{1}H	Hydrogen
	^{2}H	Deuterium
	^{3}H	Tritium (radiolabel)
	Inpy	Indeno[1,2,3-cd]pyrene
	M	Methanol
	MC	Methylene chloride
	2MeNap	2-Methylnaphthalene
	Nap	Naphthalene
	i-O	Isooctane
	P	Pentane
	PAH	Polycyclic aromatic hydrocarbon
	PC	Propylene carbonate
	Pery	Perylene
	Phen	Phenanthrene
	nPrNap	n-Propyl naphthalene
	Py	Pyrene
	T	Toluene
	THF	Tetrahydrofuran
Extraction	CS	Sublimation in carrier gas
	SD	Steam distillation
	Sox	Soxhlet
	US	Ultrasonic
	VS	Vacuum sublimation
Miscellaneous	i.d.	Inner diameter
	M	Molar (concentration)
	N	Normal (concentration)

Acknowledgments

The authors wish to thank G. M. Caton for valuable assistance in conducting the literature search. This review was sponsored jointly by the Electric Power Research Institute and the U.S. Department of Energy, Office of Health and Environmental Research, under contract W-7405-eng-26 with the Union Carbide Corporation.

REFERENCES

1. W. H. Griest, J. E. Caton, M. R. Guerin, L. B. Yeatts, Jr., and C. E. Higgins, Extraction and recovery of polycyclic aromatic hydrocarbons from highly sorptive matrices such as fly ash, in *Polynuclear Aromatic Hydrocarbons: Chemistry and Biological Effects*, A. Bjørseth and A. J. Dennis (Eds.), Battelle Press, Columbus, Ohio, 1980.
2. E. J. Baum, Occurrence and surveillance of polycyclic aromatic hydrocarbons, in *Polycyclic Hydrocarbons and Cancer*, H. V. Gelboin and P. O. P. T'so (Eds.), Vol. 1, Academic Press, New York, 1978, p. 45.
3. K. Kay, Liquid chromatography analysis in air pollution, in *Chromatographic Analysis of the Environment*, R. L. Grob (Ed.), Dekker, New York, 1975, p. 130.
4. D. Hoffmann and E. L. Wynder, Organic particulate pollutants: Chemical analysis and bioassays for carcinogenicity, in *Air Pollution*, A. C. Stern (Ed.), Vol. 2, Academic Press, New York, 1977, p. 364.
5. E. Sawicki, The separation and analysis of polynuclear aromatic hydrocarbons present in the human environment. *Chem. Anal. 53*:24 (1964); and other papers in the same volume.
6. A. A. Herod and R. G. James, A review of methods for the estimation of polynuclear aromatic hydrocarbons with particular reference to coke oven emissions. *J. Inst. Fuel 51*:164 (1978).
7. H. S. Hertz, L. R. Hilpert, W. E. May, S. A. Wise, S. N. Chesler, J. M. Brown, F. Guenther, and E. Zimmerman, Analysis of potentially deleterious constituents of coal and oil shale. *Preprints of the Symposium on Analytical Chemistry of Tar Sands and Oil Shale*, American Chemical Society, Washington, D.C., 1977, p. 808.
8. J. S. Fruchter, J. C. Laul, M. R. Petersen, P. W. Ryan, and M. E. Turner, High precision trace element and organic constituent analysis of oil shale and solvent refined coal materials, in *Analytical Chemistry of Liquid Fuel Sources, Tar Sands, Oil Shale, Coal, and Petroleum*, P. C. Uden, S. Siggia, and H. B. Jensen (Eds.), Advances in Chemistry Series No. 170, American Chemical Society, Washington, D.C., 1978, p. 255.
9. B. S. Causey, G. F. Kirkbright, and C. G. Dehima, Detection and determination of polynuclear aromatic hydrocarbons by luminescence spectrometry utilizing the Shpol'skii Effect at 77/7. Part II. An Evaluation of Excitation Sources, Sample Cells and Detection Systems. *Analyst 101*:367 (1976).

10. T. Vo-Dinh, R. B. Gammage, and A. R. Hawthorne, Analysis of organic pollutants by synchronous luminescence spectrometry, in P. W. Jones and P. Leber (Eds.), *Polynuclear Aromatic Hydrocarbons*, Ann Arbor Science Publ., Ann Arbor, Mich., 1979, p. 111.
11. D. J. Robertson, R. H. Groth, D. G. Gardner, and E. G. Glastris, Interferences from filters and solvents in PNA analysis by high performance liquid chromatography. *J. Air Pollut. Control Assoc. 29*:143 (1979).
12. P. Schwartz, Glass wool as a potential source of artifacts in chromatography. *J. Chromatogr. 152*:514 (1978).
13. H. G. Nowicki, C. A. Kieda, R. F. Devine, V. Current, and T. H. Schaefers, Identification of organic compounds solvent extracted from paper and glass Soxhlet thimbles. *Anal. Lett. 12*(A7):769 (1979).
14. R. E. Clement, F. W. Karasek, W. D. Bowers, and M. L. Parsons, Correction for artifacts in the analysis of atmospheric aerosols. *J. Chromatogr. 190*:136 (1980).
15. W. Cautreels and K. Van Cauwenberghe, Determination of organic compounds in airborne particulate matter by gas chromatography-mass spectrometry. *Atmos. Environ. 10*:447 (1976).
16. D. W. Denney, F. W. Karasek, and W. D. Bowers, Detection and identification of contaminants from foil-lined screw-cap sample vials. *J. Chromatogr. 151*:75 (1978).
17. B. T. Commins, Interim report on the study of techniques for determination of polycyclic aromatic hydrocarbons in air. *Nat. Cancer Inst. Monogr. 9*:225 (1962).
18. G. E. Moore, J. L. Monkman, and M. Klatz, Some problems concerning analysis of polycyclic hydrocarbons in particulate pollution samples. *Nat. Cancer Inst. Monogr. 9*:153 (1962).
19. M. A. Acheson, R. M. Harrison, R. Perry, and R. A. Wellings, Factors affecting the extraction and analysis of polynuclear aromatic hydrocarbons in water. *Water Res. 10*:207 (1976).
20. D. K. Basu and J. Saxena, Monitoring of polynuclear aromatic hydrocarbons in water. II. Extraction and recovery of six representative compounds with polyurethane plugs. *Environ. Sci. Technol. 12*:791 (1978).
21. P. Van Rossum and R. G. Webb, Isolation of organic water pollutants by XAD-resins and carbon. *J. Chromatogr. 150*:381 (1978).
22. E. Sawicki, T. W. Stanley, W. C. Elbert, J. Meeker, and S. McPherson, Comparison of methods for the determination of benzo[a]pyrene in particulates from urban and other atmospheres. *Atmos. Environ. 1*:131 (1967).
23. W. H. Griest, Multicomponent polycyclic aromatic hydrocarbon analysis of inland water and sediment, in *Hydrocarbons and Halogenated Hydrocarbons in the Aquatic Environment*, B. K. Afghan and D. Mackay (Eds.), Plenum Press, New York, 1980, p. 173.
24. W. H. Griest, L. B. Yeatts, Jr., and J. E. Caton, Recovery of polycyclic aromatic hydrocarbons sorbed on fly ash for quantitative determination. *Anal. Chem. 52*:199 (1980).
25. F. DeWiest, D. Rondia, and H. D. Fiorentina, Estimation of atmospheric benzo[a]pyrene using florimetry coupled with liquid scintillation spectrometry. *J. Chromatogr. 104*:399 (1975).

26. H. H. Hill, Jr., K. W. Chan, and F. W. Karasek, Extraction of organic compounds from airborne particulate matter for gas chromatographic analysis. *J. Chromatogr. 131*:245 (1977).
27. W. Cautreels and K. Van Cauwenberghe, Extraction of organic compounds from airborne particulate matter. *Water Air Soil Pollut. 6*:103 (1976).
28. T. W. Stanley, J. E. Meeker, and M. J. Morgan, Extraction of organics from airborne particulates: Effects of various solvents and conditions on the recovery of benzo[a]pyrene, benz[c]acridine, and 7H-benz[de]anthracene-7-one. *Environ. Sci. Technol. 1*:927 (1967).
29. A. M. Krstulovic, D. M. Rosie, and P. R. Brown, Distribution of some atmospheric polynuclear aromatic hydrocarbons. *Amer. Lab.*, p. 11 (July 1977).
30. D. Grosjean, Solvent extraction and organic carbon determination in atmospheric particulate matter: The organic extraction-organic carbon (OE-OCA) technique. *Anal. Chem. 47*:797 (1975).
31. G. Chatot, M. Castegnaro, J. L. Roche, and R. Fontanges, Etude comparée des ultra-sons et du Soxhlet dans l'extraction des hydrocarbures polycycliques atmosphériques. *Anal. Chim. Acta 53*:259 (1971).
32. J. Saxena, J. Kozuchowski, and D. K. Basu, Monitoring of polynuclear aromatic hydrocarbons in water. I. Extraction and recovery of benzo[a]pyrene with porous polyurethane foam, *Environ. Sci. Technol. 11*:682 (1977).
33. J. Navratil, R. E. Sievers, and H. F. Walton, Open-pore polyurethane columns for collection and preconcentration of polynuclear aromatic hydrocarbons from water. *Anal. Chem. 49*:2260 (1977).
34. V. Leoni, G. Puccetti, and A. Grella, Preliminary results on the use of Tenax for the extraction of pesticides and polynuclear aromatic hydrocarbons from surface and drinking waters for analytical purposes. *J. Chromatogr. 106*:119 (1975).
35. K. Grob, K. Grob, Jr., and G. Grob, Organic substances in potable water and in its precursor. III. The closed-loop stripping procedure compared with rapid liquid extraction. *J. Chromatogr. 106*:299 (1975).
36. A. Tateda and J. W. Fritz, Mini-column procedure for concentrating organic contaminants from water. *J. Chromatogr. 152*:329 (1978).
37. R. D. Smillie and D. T. Wang, A comparison of extraction techniques for PAH analysis of industrial effluents and natural waters, in *Polynuclear Aromatic Hydrocarbons: Chemistry and Biological Effects*, A. Bjørseth and A. J. Dennis (Eds.), Battelle Press, Columbus, Ohio, 1980.
38. H. J. Davis, L. A. Lee, and T. R. Davidson, Fluorimetric determination of benzo[a]pyrene in cigarette smoke condensate. *Anal. Chem. 38*:1753 (1966).
39. N. M. Sinclair and B. E. Frost, Rapid method for the determination of benzo[a]pyrene in the particulate phase of cigarette smoke by high performance liquid chromatography with fluorimetric detection. *Analyst 103*:1199 (1978).
40. K. Wettig, A. Ja. Chesina, A. B. Linnik, L. W. Kriwoschejewa, and W. H. Dotre, Problems of adsorption and desorption of 3,4-benzopyrene in-vitro and in-vivo. *Staub-Reinheit Luft 29*:25 (1969).
41. J. Jager, Behavior of Polycyclic aromatic hydrocarbons adsorbed on solid carriers. Part I. Extraction of polycyclic aromatic hydrocarbons from solid carriers. *Ceskoslov. Hyg. 14*:135 (1969).

42. A. H. Miguel, W. A. Korfmacher, E. L. Wehry, G. Mamantov, and D. P. S. Natusch, Apparatus for vapor-phase adsorption of polycyclic organic matter onto particulate surfaces. *Environ. Sci. Technol. 13*: 1229 (1979).
43. T. W. Sonnichsen, M. W. McElroy, and A. Bjørseth, Use of PAH tracers during sampling of coal-fired boilers, in *Polynuclear Aromatic Hydrocarbons: Chemistry and Biological Effects*, A. Bjørseth and A. J. Dennis (Eds.), Battelle Press, Columbus, Ohio, 1980.
44. National Research Council, *Particulate Polycyclic Organic Matter*, National Academy of Sciences, Washington, D.C., 1972, p. 617.
45. D. Hoffman, G. Rathkamp, K. D. Brunneman, and E. L. Wynder, *Sci. Total Environ. 2*:152 (1973).
46. W. H. Griest, M. R. Guerin, L. B. Yeatts, Jr., R. R. Reagan, T. K. Rao, J. L. Epler, M. P. Maskarinec, M. V. Buchanan, and B. A. Tomkins, Identification and quantification of polynuclear organic matter on particulates from a coal-powered plant. EA-1092, Electric Power Research Institute, Palo Alto, Cal., 1979.
47. W. H. Griest, M. P. Maskarinec, S. E. Herbes, and G. R. Southworth, in *Analysis of Waters Associated with Alternative Fuel Production*, L. P. Jackson and C. C. Wright (Eds.), ASTM-STP-720, American Society for Testing and Materials, Philadelphia, 1980, p. 168.
48. B. Bickford, J. Bursey, L. Michael, E. Pellizzari, R. Porch, D. Rosenthal, L. Sheldon, C. Sparacino, K. Tomer, R. Wiseman, S. Yung, J. Gebhart, L. Rando, D. Perry, and J. Ryan, Master scheme for the analysis of organic compounds in water. Part III. Experimental development and results. Preliminary draft report. U.S. Environmental Protection Agency, Athens, Ga., 1980.
49. B. P. Dunn, in *Polynuclear Aromatic Hydrocarbons: Chemistry and Biological Effects*, A. Bjørseth and A. J. Dennis (Eds.), Battelle Press, Columbus, Ohio, 1980, p. 367.
50. B. P. Dunn, Benzo[a]pyrene in the marine environments: Analytical techniques and results. In *Hydrocarbons and Halogenated Hydrocarbons in the Aquatic Environment*, B. K. Afghan and D. Mackay (Eds.), Plenum Press, New York, 1980, p. 109.
51. A. Candeli, G. Morozzi, A. Paolacci, and L. Zoccolilo, Analysis using thin layer and gas-liquid chromatography of polycyclic aromatic hydrocarbons in the exhaust products from a European car running on fuels containing a range of concentrations of these hydrocarbons, *Atmos. Environ. 9*:843 (1975).
52. ASTM D 2682-71, Standard test method for polynuclear aromatic hydrocarbons in air particulate matter. American Society for Testing and Materials, Philadelphia, 1971.
53. E. Sawicki, R. C. Corey, A. E. Dooley, J. B. Gisclard, J. L. Monkman, R. E. Neligan, and L. A. Ripperton, Tentative method of analysis for polynuclear aromatic hydrocarbon content of atmospheric particulate matter, *Health Lab. Sci.* 7:31-44 (1970).
54. E. Sawicki, R. C. Corey, A. E. Dooley, J. B. Gisclard, J. L. Monkman, R. E. Neligan, and L. A. Ripperton, Tentative method of microanalysis for benzo(a)pyrene in airborne particulates and source effluents, *Health Lab. Sci.* 7:56-59 (1970).

55. E. Sawicki, R. C. Corey, A. E. Dooley, J. B. Gisclard, J. L. Monkman, R. E. Neligan, and L. A. Ripperton, Tentative method of spectrophotometric analysis for benzo(a)pyrene in atmospheric particulate matter, *Health Lab. Sci.* 7:68-71 (1970).
56. E. Sawicki, R. C. Corey, A. E. Dooley, J. B. Gisclard, J. L. Monkman, R. E. Neligan, and L. A. Ripperton, Tentative method of chromatographic analysis for benzo(a)pyrene and benzo(k)fluoranthene in atmospheric particulate matter, *Health Lab. Sci.* 7:60-67 (1970).
57. E. Sawicki, R. C. Corey, A. E. Dooley, J. B. Gisclard, J. L. Monkman, R. E. Neligan, and L. A. Ripperton, Tentative method of routine analysis for polynuclear aromatic hydrocarbon content of atmospheric particulate matter, *Health Lab. Sci.* 7:45-55 (1970).
58. D. E. Lentzen, D. E. Wagoner, E. D. Estes, and W. F. Gutknecht, *IERL-RTP Procedures Manual: Level 1 Environmental Assessment*, 2nd ed., EPA-600/7-78-201, U.S. Environmental Protection Agency, Washington, D.C., October 1978.
59. W. L. Budde and J. W. Eichelberger, *An EPA Manual for Organics Analysis Using Gas Chromatography-Mass Spectrometry*, EPA-600/8-79-006, U.S. Environmental Protection Agency, Cincinnati, Ohio, March 1979.
60. J. L. Monkman, G. E. Moore, and M. Katz, Analysis of polycyclic hydrocarbons in particulate pollutants. *J. Amer. Ind. Hyg. 23*:487 (1962).
61. K. D. Cowan and E. J. Eisenbrau, Soxhlet extraction as a safety feature in the synthesis of polynuclear aromatic hydrocarbons. *Chem. Ind.* p. 46 (1975).
62. C. Golden and E. Sawicki, Ultrasonic extraction of total particulate aromatic hydrocarbons (TpAH) from airborne particles at room temperature. *Intern. J. Environ. Anal. Chem. 4*:9 (1975).
63. J. O. Jackson and J. A. Cupps, Field evaluation and comparison of sampling matrices for polynuclear aromatic hydrocarbons in occupational atmospheres, in *Carcinogenesis*, Vol. 3: *Polynuclear Aromatic Hydrocarbons*, P. W. Jones and R. I. Freudenthal (Eds.), Raven Press, New York, 1978, p. 183.
64. J. L. Bove and V. P. Kukreja, A new mechanical disruption technique for organic enrichment of hi-vol samples. *Environ. Lett. 10*:89 (1975).
65. M. A. Fox and S. W. Staley, Determination of polycyclic aromatic hydrocarbons in atmospheric particulate matter by high pressure liquid chromatography coupled with fluorescence techniques. *Anal. Chem. 48*: 992 (1976).
66. R. C. Pierce and M. Katz, Dependency of Polynuclear aromatic hydrocarbon content on size distribution of atmospheric aerosols. *Environ. Sci. Technol. 9*:347 (1975).
67. R. C. Pierce and M. Katz, Determination of atmospheric isomeric polycyclic arenes by thin-layer chromatography and fluorescence spectrophotometry. *Anal. Chem. 47*:1743 (1975).
68. J. C. Liao and R. F. Browner, Determination of polynuclear aromatic hydrocarbons in poly(vinyl chloride) smoke particulates by high-pressure liquid chromatography and gas chromatography-mass spectrometry. *Anal. Chem. 50*:1683 (1978).
69. W. Giger and M. Blumer, Polycyclic aromatic hydrocarbons in the environment: Isolation and characterization by chromatography, visible, ultraviolet, and mass spectrometry. *Anal. Chem. 46*:1663 (1974).

70. G. Broddin, L. Van Vaeck, and K. Van Cauwenberghe, On the size distribution of polycyclic aromatic hydrocarbon containing particles from coke oven emission source. *Atmos. Environ. 11*:1061 (1977).
71. F. S.-C. Lee, T. J. Prater, and F. Ferris, PAH emissions from a stratified-charge vehicle with and without oxidation catalyst: Sampling and analysis evaluation, in *Polynuclear Aromatic Hydrocarbons*, P. W. Jones and P. Leber (Eds.), Ann Arbor Science Publs., Ann Arbor, Mich., 1979, p. 83.
72. E. D. John and G. Nickless, Gas chromatographic method for the analysis of major polynuclear aromatics in particulate matter. *J. Chromatogr. 138*:399 (1977).
73. W. Cautreels and K. Van Cauwenberghe, Determination of organic compounds in airborne particulate matter by gas chromatography-mass spectrometry. *Atmos. Environ. 10*:447 (1976).
74. R. B. King, A. C. Antoine, J. S. Fordyce, H. E. Neustadter, and H. F. Leibecki, Compounds in airborne particulates: Salts and hydrocarbons. *J. Air Pollut. Control Assoc. 27*:867 (1977).
75. L. Van Vaeck and K. Van Cauwenberghe, Cascade impactor measurements of the size distribution of the major classes of organic pollutants in atmospheric particulate matter. *Atmos. Environ. 12*:2229 (1978).
76. R. J. Gordon and R. J. Bryan, Patterns in airborne polynuclear hydrocarbon concentrations at four Los Angeles sites. *Environ. Sci. Technol. 7*:1050 (1973).
77. W. Cautreels and K. Van Cauwenberghe, Fast quantitative analysis of organic compounds in airborne particulate matter by gas chromatography with selective mass spectrometer detection. *J. Chromatogr. 131*:253 (1977).
78. W. Cautreels and K. Van Cauwenberghe, Experiments on the distribution of organic pollutants between airborne particulate matter and the corresponding gas phase. *Atmos. Environ. 12*:1133 (1978).
79. W. Cautreels, K. Van Cauwenberghe, and L. A. Gazman, Comparison between the organic fraction of suspended matter at a background and an urban station. *Sci. Total Environ. 8*:79 (1977).
80. M. Dong, D. C. Locke, and E. Ferrand, High pressure liquid chromatographic method for routine analysis of major parent polycyclic aromatic hydrocarbons in suspended particulate matter. *Anal. Chem. 48*:368 (1976).
81. T. D. Searl, F. J. Cassidy, W. H. King, and R. A. Brown, An analytical method for polynuclear aromatic compounds in coke oven effluents by combined use of gas chromatography and ultraviolet absorption spectrometry. *Anal. Chem. 42*:954 (1970).
82. J. M. Daisey and M. A. Leyko, Thin-layer gas chromatographic method for the determination of polycyclic aromatic and aliphatic hydrocarbons in airborne particulate matter. *Anal. Chem. 51*:24 (1979).
83. R. C. Lao, R. S. Thomas, H. Oja, and L. Dubois, Application of a gas chromatograph-mass spectrometer-data processor combination to the analysis of the polycyclic aromatic hydrocarbon content of airborne pollutants. *Anal. Chem. 45*:908 (1973).
84. A. Bjørseth, Analysis of polycyclic aromatic hydrocarbons in particulate matter by glass capillary gas chromatography. *Anal. Chim. Acta 94*:21 (1977).

85. W. Giger and C. Schaffner, Determination of polycyclic aromatic hydrocarbons in the environment by glass capillary gas chromatography. *Anal. Chem. 50*:243 (1978).
86. F. J. Onuska, A. W. Wolkoff, M. E. Comba, R. H. Larose, M. Novotny, and M. L. Lee, Gas chromatographic analysis of polynuclear aromatic hydrocarbons in shellfish on short, wall-coated glass capillary columns. *Anal. Lett. 9*:451 (1976).
87. R. C. Lao, R. S. Thomas, and J. L. Monkman, The analysis of environmental samples by mass spectrometry. *Intern. J. Environ. Anal. Chem. 1*:187 (1972).
88. E. Winkler, A. Buchele, and O. Muller, Method for the determination of polycyclic aromatic hydrocarbons in maize by capillary-column gas-liquid chromatography. *J. Chromatogr. 138*:151 (1977).
89. B. B. Chakraborty and R. Long, Gas chromatographic analysis of polycyclic aromatic hydrocarbons in soot samples. *Environ. Sci. Technol. 10*:828 (1967).
90. M. L. Lee and R. A. Hites, Characterization of sulfur-containing polycyclic aromatic compounds in carbon blacks. *Anal. Chem. 48*:1890 (1976).
91. H. Tausch and G. Stehlik, Determination of polycyclic aromatic hydrocarbons in soot by gas chromatography-mass spectroscopy, *Chromatographia 10*:350 (1977).
92. L. Wallcave, D. L. Nagel, J. W. Smith, and R. D. Waniska, Two pyrene derivatives of widespread environmental distribution: Cyclopenta[cd]-pyrene and acepyrene. *Environ. Sci. Technol. 9*:143 (1975).
93. W. F. Fitch and D. H. Smith, Analysis of adsorption properties and adsorbed species on commercial polymeric carbons. *Environ Sci. Technol. 13*:341 (1979).
94. C. W. Woo, A. P. D'Silva, V. A. Fassel, and G. J. Oestrich, Polynuclear aromatic hydrocarbons in coal-identification by their x-ray excited optical luminescence. *Environ. Sci. Technol. 12*:173 (1978).
95. R. V. Schultz, J. W. Jorgenson, M. P. Maskarinec, M. Novotny, and L. J. Todd, Characterization of polynuclear aromatic and aliphatic hydrocarbon fractions of solvent-refined coal by glass capillary gas chromatography-mass spectrometry. *Fuel 58*:783 (1979).
96. J. Hilborn, R. S. Thomas, and R. C. Lao, The organic content of international reference samples of asbestos. *Sci. Total Environ. 3*:129 (1974).
97. W. L. Fitch, E. T. Everhart, and D. H. Smith, Characterization of carbon black adsorbates and artifacts formed during extraction. *Anal. Chem. 50*:2122 (1978).
98. L. Sucre, W. Jennings, G. L. Fisher, O. G. Raabe, and J. Olechno, Polynuclear aromatic hydrocarbons associated with coal combustion, in *Trace Organic Analysis: A New Frontier in Analytical Chemistry*, NBS Special Publ. No. 519, National Bureau of Standards, Washington, D.C., 1979, p. 109.
99. Measurements Research Branch, Benzo[a]pyrene in particulates: Analytical method, in *NIOSH Manual of Analytical Methods*, D. G. Taylor (Manual Coordinator), Vol. 1, U.S. Dept. of Health, Education, and Welfare, Cincinnati, Ohio, April 1977, p. 251-1.

100. Physical and Chemical Analysis Branch, Total particulate aromatic hydrocarbons (TPAH) in air, in *NIOSH Manual of Analytical Methods*, D. G. Taylor (Manual Coordinator), Vol. 1, U.S. Dept. of Health, Education, and Welfare, Cincinnati, Ohio, April 1977, p. 206-1.
101. E. Sawicki, T. Belsky, R. A. Friedman, D. L. Hyde, J. L. Monkman, R. A. Rasmussen, L. A. Ripperton, and L. D. White, Total particulate aromatic hydrocarbons (TPAH) in air. Ultrasonic extraction method, *Health Lab. Sci. 12*:407, 1975.
102. P. W. Jones, A. P. Graffeo, R. Detrick, P. A. Clake, and R. J. Jakobsen, *Technical Manual for Analysis of Organic Materials in Process Streams*, EPA-600/2-76-072, U.S. Environmental Protection Agency, Research Triangle Park, N.C., March 1976.
103. E. P. Lankmayr and K. Muller, Polycyclic aromatic hydrocarbons in the environment: High-performance liquid chromatography using chemically modified columns. *J. Chromatogr. 170*:139 (1979).
104. Y. L. Tan, Rapid simple sample preparation technique for analyzing polynuclear aromatic hydrocarbons in sediments by gas chromatography-mass spectrometry. *J. Chromatogr. 176*:319 (1979).
105. H. Matsushita, K. Arashidani, and T. Handa, Simple microanalysis of BaP in air suspended particulates using ultrasonic extraction. *Bunseki Kagaku 25*:263 (1976).
106. C. Golden and E. Sawicki, Determination of benzo[a]pyrene and other polynuclear aromatic hydrocarbons in airborne particulate material by ultrasonic extraction and reverse phase high pressure liquid chromatography. *Anal. Lett. A11*:1051 (1978).
107. B. Seifert and I. Steinback, Thin-layer chromatographic routine determination of benzo[a]pyrene in suspended particulate matter with in situ evaluation. *Z. Anal. Chem. 287*:264 (1977).
108. J. Mulik, M. Cooke, M. F. Guyer, G. M. Semeniuk, and E. Sawicki, A gas liquid chromatographic fluorescent procedure for the analysis of benzo[a]pyrene in 24-hour atmospheric particulate samples. *Anal. Lett. 8*:511 (1975).
109. A. H. Miguel and S. K. Friedlander, Distribution of benzo[a]pyrene and coronene with respect to particle size in Pasadena aerosols in the submicron range. *Atmos. Environ. 12*:2407 (1978).
110. G. Chatot, R. Dangy-Caye, and R. Fontanges, Study of atmospheric polycyclic hydrocarbons. II. Application of the combination of gas phase chromatography and thin layer chromatography to determination of polynuclear arenes. *J. Chromatogr. 72*:202 (1972).
111. P. Burchill, A. A. Herod, and R. G. James, A comparison of some chromatographic methods for estimation of polynuclear aromatic hydrocarbons in pollutants, in *Carcinogenesis*, Vol. 3: *Polynuclear Aromatic Hydrocarbons*, P. W. Jones and R. I. Freudenthal (Eds.), Raven Press, New York, 1978, p. 35.
112. S. A. Wise, S. N. Chesler, H. S. Hertz, L. R. Hilpert, and W. E. May, Chemically-bonded aminosilane stationary phase for the high-performance liquid chromatographic separation of polynuclear aromatic compounds. *Anal. Chem. 49*:2306 (1977).
113. M. T. Strosher and G. W. Hodgson, Polycyclic aromatic hydrocarbons in lake waters and associated sediments: Analytical determination by gas chromatography-mass spectrometry, in *Water Quality Parameters*,

ASTM-STP-573, American Society for Testing and Materials, Philadelphia, p. 259, 1975.
114. P. S. Pederson, J. Ingwersen, T. Nielsen, and E. Larsen, Effects of fuel, lubricant, and engine operating parameters on the emission of polycyclic aromatic hydrocarbons. *Environ. Sci. Technol. 14*:71 (1980).
115. T. Nielsen, Determination of polycyclic aromatic hydrocarbons in automobile exhaust by means of high-performance liquid chromatography with fluorescence detection. *J. Chromatogr. 170*:147 (1979).
116. C. K. McMahon and S. N. Tsoukalas, Polynuclear aromatic hydrocarbons in forest fire smoke, in *Carcinogenesis*, Vol. 3: *Polynuclear Aromatic Hydrocarbons*, P. W. Jones and R. I. Freudenthal (Eds.), Raven Press, New York, 1978, p. 61.
117. R. M. Harrison, R. Perry, R. A. Wellings, Effect of water chlorination upon levels of some polynuclear aromatic hydrocarbons in water. *Environ. Sci. Technol. 10*:1151 (1976).
118. S. Monarca, Polycyclic aromatic hydrocarbons in petroleum products for medicinal and cosmetic uses: Analytical procedure. *Sci. Total Environ. 14*:233 (1980).
119. P. A. Soltys, The extraction behavior of PAH from coal fly ash. M.S. Thesis, Colorado State University, Fort Collins, Colo., Summer 1980.
120. F. C. McElroy, T. D. Searl, and R. A. Brown, in *Trace Organic Analysis: A New Frontier in Analytical Chemistry*, NBS Special Publ. No. 519, National Bureau of Standards, Washington, D.C., 1979.
121. B. W. DeLappe, R. W. Risebrough, A. M. Springer, T. T. Schmidt, J. C. Shropshire, E. F. Letterman, and J. R. Payne, The sampling and measurement of hydrocarbons in natural waters, in *Hydrocarbons and Halogenated Hydrocarbons in the Aquatic Environment*, B. K. Afghan and D. Mackay (Eds.), Plenum Press, New York, 1980, p. 29.
122. American Society for Testing and Materials Subcommittee D19.06, Method of test for polynuclear aromatic hydrocarbons in water. Preliminary draft, May 29, 1979.
123. W. K. Robbins, Solvent extraction of polynuclear aromatic hydrocarbons, in *Polynuclear Aromatic Hydrocarbons: Chemistry and Biological Effects*, A. Bjørseth and A. J. Dennis (Eds.), Battelle Press, Columbus, Ohio, 1980, p. 841.
124. J. L. Epler, F. W. Larimer, T. K. Rao, E. M. Burnett, W. H. Griest, M. R. Guerin, M. P. Maskarinec, D. A. Brown, N. T. Edwards, C. W. Gehrs, R. E. Milleman, B. R. Parkhurst, B. M. Ross-Todd, D. S. Shriner, and H. W. Wilson, Jr., Toxicity of leachates. EPA-600/2-80-057, U.S. Environmental Protection Agency, Municipal Environmental Research Library, Cincinnati, Ohio, 1980.
125. J. E. Wilkinson, P. E. Strup, and P. W. Jones, Quantitative analyses of selected PAH in aqueous effluent by high-performance liquid chromatography, in *Polynuclear Aromatic Hydrocarbons*, P. W. Jones and P. Leber (Eds.), Ann Arbor Science Publs., Ann Arbor, Mich., 1979, p. 217.
126. W. K. Robbins, T. D. Searl, D. H. Wasserstrom, and G. T. Boyer, ASTM D-19, ASTM-STP, American Society for Testing and Materials, Philadelphia, 1980, p. 149.

127. K. Ogan, E. Katz, W. Slavin, Determination of polycyclic aromatic hydrocarbons in aqueous samples by reversed-phase liquid chromatography, *Anal. Chem. 51*:1315, 1979.
128. G. A. Jungclaus, L. M. Games, and R. A. Hites, Identification of trace organic compounds in tire manufacturing plant wastewaters, *Anal. Chem. 48*:1894, 1976.
129. L. S. Sheldon and R. A. Hites, Organic compounds in the Delaware River. *Environ. Sci. Technol. 12*:1188 (1978).
130. S. Monarca, B. S. Causey, and G. F. Kirkbright, A rapid method for quantitative determination of benzo[a]pyrene in water by low temperature spectrofluorimetry. *Water Res. 13*:503 (1979).
131. D. Quaghebeur and E. DeWulf, Polynuclear aromatic hydrocarbons in the main Belgian aquifers. *Sci. Total Environ. 10*:231 (1978).
132. W. S. Gardner, R. F. Lee, K. R. Tenore, and L. W. Smith, Degradation of selected polycyclic aromatic hydrocarbons in coastal sediments: Importance of microbes and polychaete worms. *Water Air Soil Pollut. 11*: 339 (1979).
133. M. L. Lee, M. Novotny, and K. D. Bartle, Gas chromatography/mass spectrometric and nuclear magnetic resonance studies of carcinogenic polynuclear aromatic hydrocarbons in tobacco and marijuana smoke condensate. *Anal. Chem. 48*:405 (1976).
134. I. Schmeltz, J. Tosk, and D. Hoffmann, Formation and determination of naphthalenes in cigarette smoke. *Anal. Chem. 48*:645 (1976).
135. J. W. Howard, E. O. Haenni, The extraction and determination of polynuclear hydrocarbons in paraffin waxes. *J. Assoc. Offic. Anal. Chem. 46*:933 (1963).
136. G. Grimmer and H. Böhnke, Enrichment and gas chromatographic profile-analysis of polycyclic aromatic hydrocarbon lubricating oils, *Chromatographia 9*:30 (1976).
137. H. Kubota, W. H. Griest, and M. R. Guerin, Determination of carcinogens in tobacco smoke and coal-derived samples: Trace polynuclear aromatic hydrocarbons, in *Trace Substances in Environmental Health-IX.*, D. D. Hemphill (Ed.), University of Missouri, Columbia, Mo., 1975, p. 281.
138. R. E. Jentoft and T. H. Gouw, Analysis of polynuclear aromatic hydrocarbons in automobile exhaust by supercritical fluid chromatography. *Anal. Chem. 48*:2195 (1976).
139. K. Potthast and G. Eigner, A new method for the rapid isolation of polycyclic aromatic hydrocarbons from smoked meat products. *J. Chromatogr. 103*:173 (1975).
140. V. P. Kukreja and J. L. Bove, An enrichment method for polycyclic aromatic hydrocarbons (PAHs) collected on glass fiber filters using hydrofluoric acid. *J. Environ. Sci. Health A11*:517 (1976).
141. M. Radke, H. G. Sittardt, and D. H. Welte, Removal of soluble organic matter from rock samples with a flow-through extraction cell. *Anal. Chem. 50*:663 (1978).
142. J. L. Monkman, L. Dubois, and C. J. Baker, the rapid measurement of polycyclic hydrocarbons in air by microsublimation. *Pure Appl. Chem. 24*:731 (1970).
143. K. Grob, Organic substances in potable water and in its precursor. Part I. Methods for their determination by gas-liquid chromatography. *J. Chromatogr. 84*:255 (1973).

144. R. B. Gammage, T. Vo-Dinh, A. R. Hawthorne, J. H. Thorngate, and W. W. Parkinson, A new generation of monitors for polynuclear aromatic hydrocarbons from synthetic fuel production, in *Carcinogenesis*, Vol. 3: *Polynuclear Aromatic Hydrocarbons*, P. W. Jones and R. I. Freudenthal (Eds.), Raven Press, New York, 1978, p. 155.
145. A. Colmsjo and U. Stenberg, Vacuum sublimation of polynuclear aromatic hydrocarbons separated by thin-layer chromatography for detection with Shpol'skii low-temperature fluorescence. *J. Chromatogr. 169*:205 (1979).
146. M. J. Schultz, R. M. Orheim, and H. H. Bovee, Simplified method for the determination of benzo[a]pyrene in ambient air. *J. Amer. Ind. Hyg. Assoc. 34*:404 (1973).
147. P. T. Perdue and E. T. Arakawa, A stable organic phosphor for the vacuum ultraviolet. *Nucl. Instr. Methods 128*:201 (1975).
148. U. Stenberg, T. Alsberg, L. Blomberg, and T. Wannman, Gas chromatographic separation of high-molecular polynuclear aromatic hydrocarbons in samples from different sources, using temperature-stable glass capillary columns, in *Polynuclear Aromatic Hydrocarbons*, P. W. Jones and P. Leber (Eds.), Ann Arbor Science Publs., Ann Arbor, Mich., 1979, p. 313.
149. A. Colmsjo and U. Stenberg, The identification of polynuclear aromatic hydrocarbon mixtures in high-performance liquid chromatography fractions utilizing the Shpol'skii effect, in *Polynuclear Aromatic Hydrocarbons*, P. W. Jones and P. Leber (Eds.), Ann Arbor Science Publs., Ann Arbor, Mich., 1979, p. 121.
150. W. L. Ball, G. E. Moore, J. L. Monkman, and M. Katz, An evaluation of microsublimation separation of atmospheric polycyclics. *J. Amer. Ind. Hyg. Assoc. 23*:222 (1962).
151. J. F. Thomas, E. N. Sanborn, M. Mukai, and B. D. Tebbens, A fractional sublimation technique for separating atmospheric pollutants. *Anal. Chem. 30*:1954 (1958).
152. V. S. Shved, G. I. Bydrin, L. N. Pungina, and N. F. Stepanov, Naphthalene sublimation in a nitrogen current. *Koks Khim. 8*:35 (1977); Engl. transl.: *Coke Chem. 8*:45 (1977).
153. H. P. Burchfield, E. E. Green, R. J. Wheeler, and S. M. Billedean, Recent advances in the gas and liquid chromatography of fluorescent compounds. I. A direct gas-phase isolation and injection system for the analysis of polynuclear arenes in air particles by gas-liquid chromatography. *J. Chromatogr. 99*:697 (1974).
154. E. Wauters, P. Sandra, and M. Verzele, Qualitative and semi-quantitative analysis of the non-polar organic fraction of air particulate matter, *J. Chromatogr. 170*:125 (1979).
155. D. R. Kalkwarf and S. R. Garcia, Sublimation of polynuclear aromatic hydrocarbons from coal fly ash. Presented at the 34th American Chemical Society Northwest Regional Meeting, Richland, Wash. (June 13-15, 1980).
156. M. P. Maskarinec, unpublished data, 1979.
157. P. E. Strup, J. E. Wilkinson, and P. W. Jones, Trace analysis of polycyclic aromatic hydrocarbons in aqueous systems using XAD-2 resin and capillary column gas chromatography-mass spectrometry analysis, in *Carcinogenesis* Vol. 3: *Polynuclear Aromatic Hydrocarbons*, P. W. Jones and R. I. Freudenthal (Eds.), Raven Press, New York, 1978, p. 131.

158. W. W. Bunn, E. R. Deane, D. W. Klein, and R. D. Kleopfer, Sampling and characterization of air for organic compounds. *Water Air Soil Pollut.* *4*:367 (1975).
159. B. I. Brookes, S. M. Jickells, and R. S. Nicolson, Atmospheric sampling for public health investigations using adsorption tubes with quantitative GC or GC/MS analysis. *J. Assoc. Publ. Analysts* *16*:101 (1978).
160. M. P. Shiaris, T. W. Sherill, and G. S. Sayler, Tenax-GC extraction technique for residual polychlorinated biphenyl and polyaromatic hydrocarbon analysis in biodegradation assays. *Appl. Environ. Microbiol.* *39*:165 (1980).
161. B. Olufsen, Adsorption and continuous extraction of resin without drying in a modified Soxhlet apparatus. *Anal. Chim. Acta* *113*:393 (1980).
162. C. Osterrohi, Development of a method for the extraction and determination of non-polar, dissolved organic substances in sea water. *J. Chromatogr.* *101*:289 (1974).
163. J. L. Robinson, M. A. Marshall, M. E. Draganjac, and L. C. Noggle, Determination of benzo[a]pyrene in cigarette smoke condensate by liquid chromatography on amberlite XAD-2. *Anal. Chim. Acta* *115*:229 (1980).
164. A. K. Burnham, G. V. Calder, J. S. Fritz, G. A. Junk, H. J. Svek, and R. Vick, Trace organics in water: Their isolation and identification. *Water Technol./Quality* p. 722 (November 1973).
165. A. K. Burnham, G. V. Calder, J. S. Fritz, G. A. Junk, H. J. Svek, and R. Willis, Identification and estimation of neutral organic contaminants in potable water. *Anal. Chem.* *44*:139 (1972).
166. D. K. Basu and J. Saxena, Polynuclear aromatic hydrocarbons in selected U.S. drinking waters and their raw water sources. *Environ. Sci. Technol.* *12*:795 (1978).
167. A. R. Oyler, D. L. Bodenner, K. J. Welch, R. J. Liukkonen, R. M. Carlson, H. L. Kopperman, and R. Caple, Determination of aqueous chlorination reaction products of polynuclear aromatic hydrocarbons by reversed phase high performance liquid chromatography-gas chromatography. *Anal. Chem.* *50*:837 (1978).
168. W. E. May, S. N. Chesler, S. P. Cram, B. H. Gump, H. S. Hertz, D. P. Enagonio, and S. M. Dyszel, Chromatographic analysis of hydrocarbons in marine sediments and seawater. *J. Chromatogr. Sci.* *13*:535 (1975).
169. C. G. Creed, LC simplifies isolating organics from water. *Research/Development* *27*(9):40 (1976).
170. B. A. Tomkins, H. Kubota, W. H. Griest, J. E. Caton, B. R. Clark, and M. R. Guerin, Determination of benzo[a]pyrene in petroleum substitutes. *Anal. Chem.* *52*:1331 (1980).
171. F. Eisenbeiss, H. Heim, R. Joester, and G. Naundorf, The separation by LC and determination of polycyclic aromatic hydrocarbons in water using an integrated enrichment step. *Chromatogr. Newslett.* *6*(1):8 (1978).
172. J. E. Schiller and D. R. Mathiason, Separation method for coal-derived solids and heavy liquids. *Anal. Chem.* *49*:1225 (1977).
173. R. J. Hurtubise and G. T. Skar, Determination of benzo[a]pyrene in shale oil by solid-surface fluorescence. *Anal. Chim. Acta* *97*:13 (1978).
174. H. Braus, F. M. Middleton, and G. Walton, Organic chemical compounds in raw and filtered surface waters. *Anal. Chem.* *23*, 1160 (1951).

175. C. W. Louw, J. F. Richards, and P. K. Faure, The determination of volatile organic compounds in city air by gas chromatography combined with standard addition, selective subtraction, infrared spectrometry and mass spectrometry. *Atmos. Environ. 11*:703 (1977).
176. G. Lunde, J. Gether, N. Gjøs, and M. S. Lande, Organic micropollutants in precipitation in Norway. *Atmos. Environ. 11*:1007 (1977).
177. W. C. Eisenberg, Fractionation of organic material extracted from suspended air particulate matter using high pressure liquid chromatography. *J. Chromatogr. Sci. 16*:145 (1978).
178. K. Ogan and E. Katz, Coupled-column chromatography used for sample preparation for chromatographic analysis of oil and coal samples. Presented at the 2nd Symposium on Environmental Analytical Chemistry, Brigham Young University, Provo, Utah (June 18-20, 1980).
179. G. Grimmer and H. Bohnke, Polycyclic aromatic hydrocarbon profile analysis of high-protein foods, oils, and fats by gas chromatography. *J. Assoc. Offic. Anal. Chemists 58*:725 (1975).
180. J. W. Anderson, J. W. Blaylock, and J. Barnwell-Clark, Fate of polycyclic aromatic hydrocarbons in controlled ecosystem enclosures. *Environ. Sci. Technol. 12*:832 (1978).
181. M. Heit, C. S. Klusek, and K. M. Miller, Trace element, radionuclide and polynuclear aromatic hydrocarbon concentrations in Unionidae mussels from northern Lake George. *Environ. Sci. Technol. 14*:465 (1980).
182. A. Radecki, H. Lamparczyk, J. Grybowski, and J. Halkiewicz, Separation of polycyclic aromatic hydrocarbons and determination of benzo[a]pyrene in liquid smoke preparations. *J. Chromatogr. Sci. 150*:527 (1978).
183. W. H. Swallow, Survey of polycyclic aromatic hydrocarbons in selected foods and food additives available in New Zealand. *New Zealand J. Sci. 19*:407 (1976).
184. F. L. Joe, E. L. Roseboro, and T. Fazio, Survey of some market basket commodities for polynuclear aromatic hydrocarbon content. *J. Assoc. Offic. Anal. Chemists 62*:615 (1979).
185. Y. Saito, H. Sekita, M. Takeda, M. Uchiyama, Determination of benzo[a]pyrene in foods, *J. Assoc. Offic. Anal. Chemists 61*:129 (1978).
186. J. S. Warner, Determination of aliphatic and aromatic hydrocarbons in marine organisms. *Anal. Chem. 48*:578 (1976).
187. R. J. Pancirov and R. A. Brown, Polynuclear aromatic hydrocarbons in marine tissues. *J. Environ. Pollut. 11*:989 (1977).
188. J. W. Howard, R. T. Teague, R. H. White, and B. E. Fry, Extraction and estimation of polycyclic aromatic hydrocarbons in smoked foods. I. General Method. *J. Assoc. Offic. Anal. Chemists 49*:595 (1966).
189. G. C. Lawler, W. Loong, and J. L. Laseter, Accumulation of aromatic hydrocarbons in tissues of petroleum-exposed mallard ducks (*Anas platyrhynchos*). *Environ. Sci. Technol. 12*:51 (1978).
190. K. Fretheim, Carcinogenic polycyclic aromatic hydrocarbons in Norwegian smoked meat sausages. *J. Agr. Food Chem. 24*:976 (1976).
191. T. Panalaks, Determination and identification of polycyclic aromatic hydrocarbons in smoked and charcoal-broiled food products by high pressure liquid chromatography. *J. Environ. Sci. Health B11*(4):299 (1976).

192. J. J. Black, Measurement of polynuclear aromatic hydrocarbon pollution in aquatic systems. Presented at the 2nd Symposium on Environmental Analytical Chemistry, Brigham Young University, Provo, Utah (June 18-20, 1980).
193. E. O. Haenni, J. W. Howard, and F. L. Joe, Jr., Dimethyl sulfoxide: A superior analytical extraction solvent for polynuclear hydrocarbons and for some highly chlorinated hydrocarbons. *J. Assoc. Offic. Anal. Chemists 45*:67 (1962).

4

Profile Analysis of Polycyclic Aromatic Hydrocarbons in Air

GERNOT GRIMMER / Biochemisches Institut für Umweltcarcinogene, Ahrensburg, Federal Republic of Germany

I. Type and Number of PAH

Suspended atmospheric particulate matter contains a large number of different polycyclic aromatic hydrocarbons (PAH). Hence PAH exist in urban air as a complex mixture clearly demonstrated by the glass capillary gas chromatogram of PAH shown in Fig. 1.

Lao et al. [1] have characterized by their mass spectra about 130 PAH in an airborne particulate sample collected on glass fiber filters of which 56 PAH were confirmed by reference substances. Likewise Lee et al. [2] characterized 120 PAH in urban air pollution by mass spectrometric (MS) investigations, complemented by nuclear magnetic resonance (NMR). The relative composition of the PAH mixture, i.e., the PAH profile, of an air sample is influenced by various "nearby-emission," and even by distant but larger emitters. The term *profile* is used in presenting the relative composition of the mixture of PAH. The profile usually is recorded by a mass-dependent detector; e.g., a glass capillary gas chromatogram recorded by a flame-ionization detector (FID) represents a PAH profile. In order to compare the PAH profiles of different sources, it is helpful to standardize the concentration of different PAH by comparing with the concentration of benzo[e]pyrene (BeP) in the sample. In this case the BeP concentration is defined as 1.0 and all other

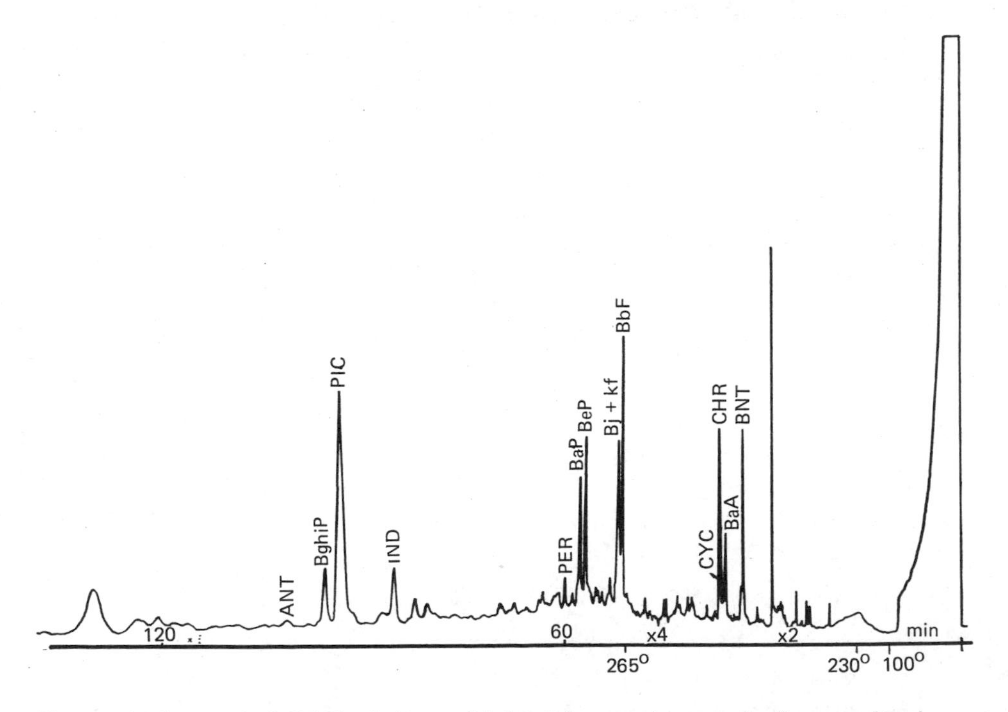

Figure 1 Separated PAH mixture obtained in an air sample from a city in Germany (Essen), November 1979: a coal heating area. Glass capillary (0.27 mm × 25 m) coated with Silicone OV 17, splitless injection at 100°C column temperature; injection port: 250°C, carrier gas: He. (See the List of Abbreviations at end of chapter.)

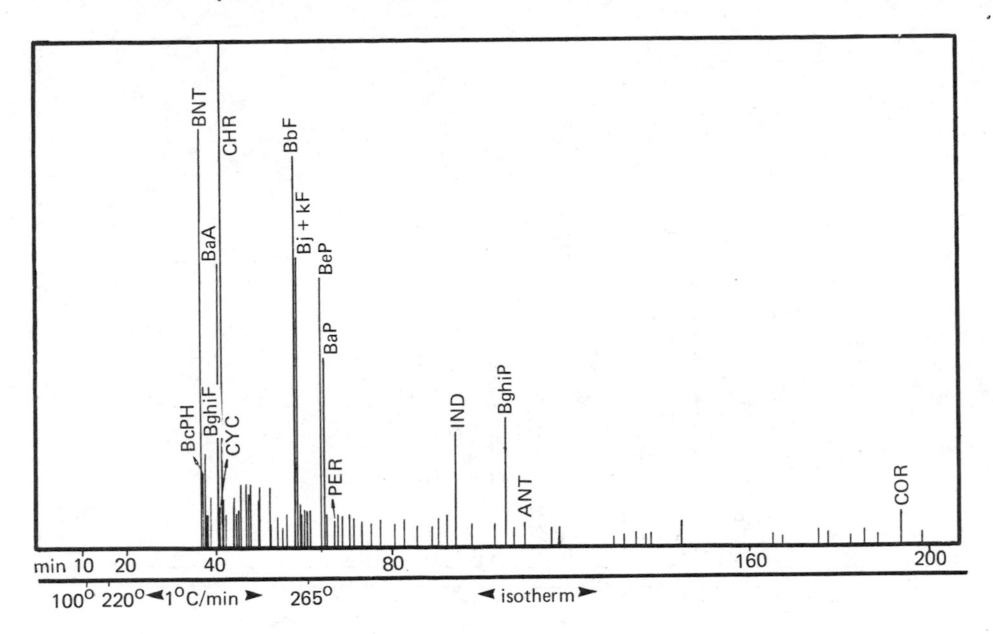

Figure 2. Standardized PAH profile of the glass capillary gas chromatogram of Fig. 1 (PAH profile of urban air).

PAH are ratioed to BeP, arranged according to the retention time (also relative to BeP). The relative concentration of the PAH, exprcssed by the different areas of the FID signals, are recorded one-dimensionally as lines. The profile of the glass capillary gas chromatogram shown in Fig. 1 is given in Fig. 2.

II. Sources of Emission in a City

Figures 3 to 6 show the characteristic PAH profiles of emissions from different sources, such as coal heating [hand-stoked residential coal furnaces (Fig. 3)], automobile traffic [passenger car, spark-ignited (Fig. 4); Diesel engine (Fig. 5)], and oil heating [low-pressure, air-atomized (Fig. 6)].

Figure 3 shows a standardized PAH profile of the emission from a hand-stoked residential stove, fired with briquets of hard coal under standardized conditions (a bench). A typical compound of this emission is benzo[b]naphtho[2,1-d]thiophene (BNT), which is present in high concentration owing to the sulfur content of the coal.

Figure 4 gives the standardized PAH profile of the emission from a passenger car, driven on a chassis dynamometer, during a test performed according to ECE regulation 15 [3]. In this case the test simulates driving conditions in city traffic. The PAH with the highest concentration in this range is cyclopenta[cd]pyrene, which is produced more than 10-fold relative to BeP. Furthermore, benzo[ghi]perylene, benzo[c]phenanthrene, and coronene are present in higher concentrations than BeP.

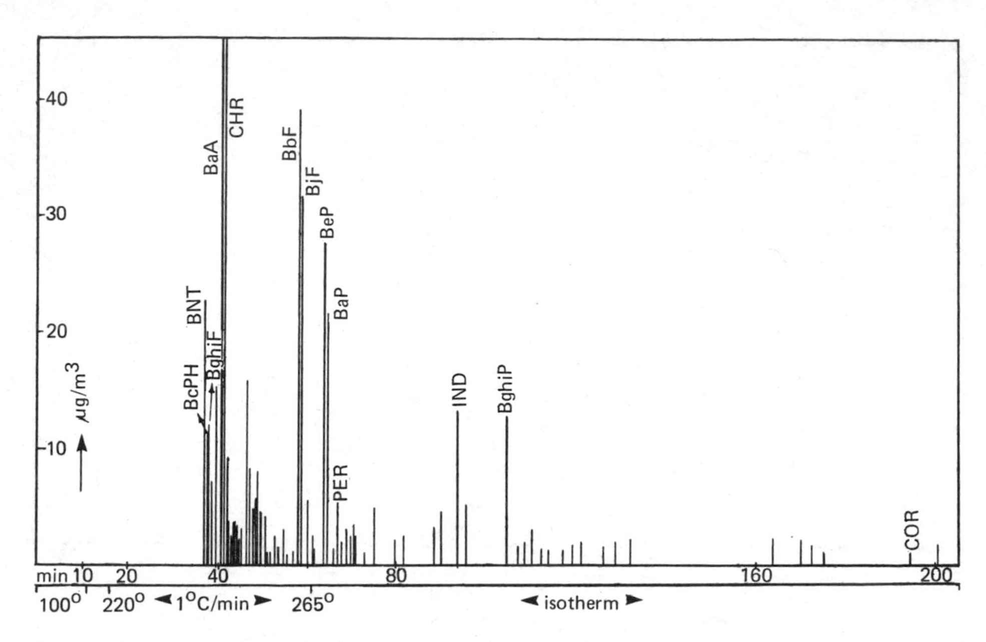

Figure 3. Flue gas emission by burning of hard coal briquets, standardized PAH profile as in Fig. 2.

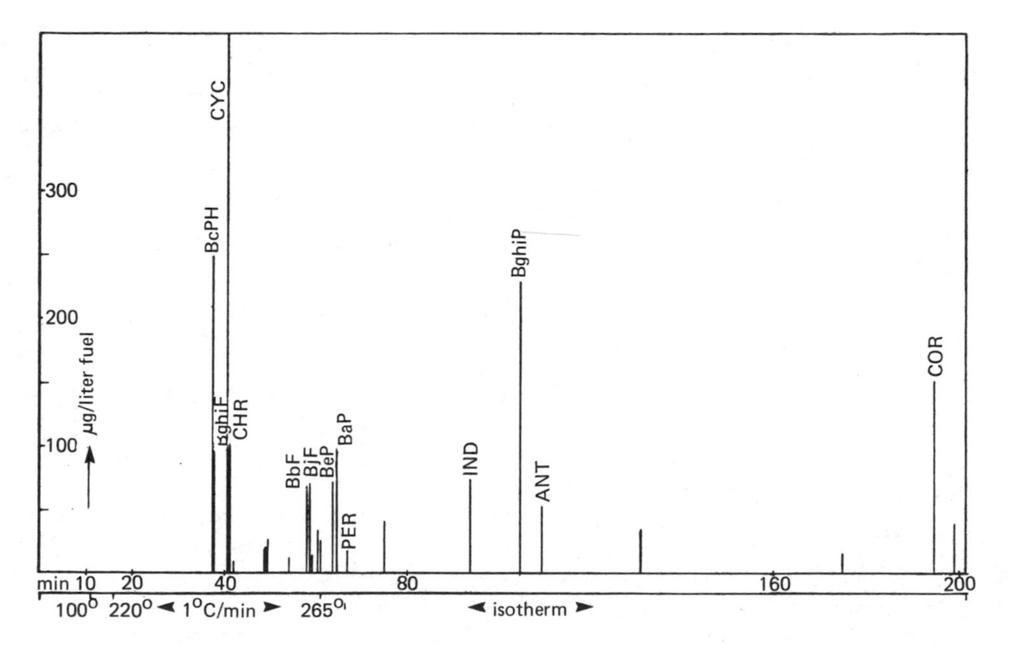

Figure 4. Automobile exhaust emission by a gasoline engine (spark plug ignited); standardized PAH profile as in Fig. 2.

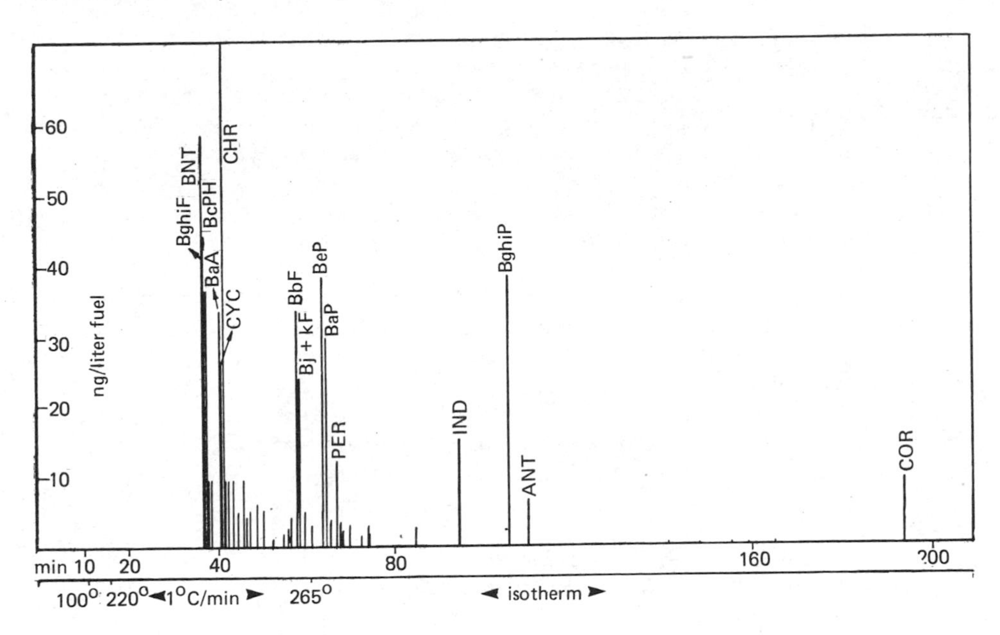

Figure 5. Automobile exhaust emission by a diesel engine (passenger car); standardized PAH profile as in Fig. 2.

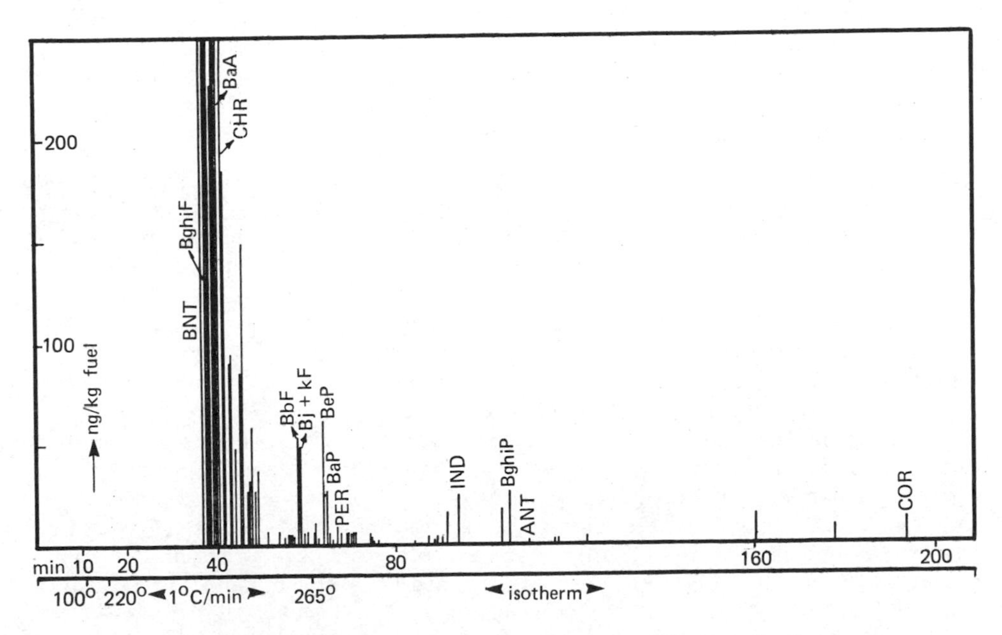

Figure 6. Flue gas emission by burning of light oil in an oil-fired central heating system; standardized PAH profile as in Fig. 2.

Figure 5 shows the PAH profile of a passenger car with a diesel engine that has been tested under the same conditions as the above (ECE regulation 15 on a bench) [3]. Unlike the exhaust gas of a gasoline-powered engine, the exhaust gas of a diesel engine contains benzo[b]naphtha[2,1-d]thiophene. The content of this compound in the exhaust correlates with the content of BNT in the fuel.

Figure 6 represents a PAH profile of the emissions from an oil burner (central heating, up to 60 kW heat output) under standardized conditions (performance, 34 kW; consumption, about 2.7 kg fuel/hr). The PAH profile contains a large number of compounds in the boiling point range between chrysene and benzo[b]fluoranthene, mostly alkyl derivatives of PAH in concentrations comparable with BeP.

The aforementioned figures show distinct differences in the composition of the PAH mixtures emitted by these sources. The registration of different PAH profiles may therefore be helpful in enabling investigators to recognize the main sources of urban air pollution in the future.

III. Comparison of the PAH Mass Profiles of Coal and Oil Heating, Gasoline- and Diesel-Fuel-Powered Engines, and Power Plants

From bench tests under standardized conditions [or conditions of DIN (Deutscher Industrie Norm) and ISO (International Standardization Organization), respectively] mass emissions are measured for benzo[a]pyrene (BaP) produced by 1 kg of fuel in each case, and the results are shown in Table 1.

From the ratio of the other PAH to BaP, demonstrated by the PAH profiles, it is possible to compile the mass emissions for other PAH as well.

IV. Carcinogenic Impact of Air Pollution

Today the only available method for estimating the carcinogenicity of suspended particulate matter in the atmosphere is the chronic toxicity testing of the potency of the sample to develop tumors in animals. In general, the carcinogenic potency of a sample exceeds the effect provoked by the sample's BaP content. Therefore, to produce the same number of tumors in a group of animals, it is necessary to apply a multiple amount of BaP compared with that present in the sample. In other words, BaP accounts only for a part of the total carcinogenicity of an environmental sample of the emissions mentioned before, of air particulate matter collected in a city.

The animal tests of a sample or of an extract from the sample are performed mainly by skin painting, or by subcutaneous or intratracheal application. If doses are administered in suitable amounts, these tests give a significant hint of tumorigenic effects. As animal experiments show, it is not sufficient to determine the amount of BaP only; for example, comparison of the carcinogenic effect of automobile exhaust condensate by skin painting of mice with the BaP content of the condensate solution reveals that the BaP portion represents 9% of the effect, i.e., to obtain the same effect with pure BaP, 11 times more pure BaP than is contained in the condensate must be applied [4-6]. The same comparison holds true for the particle phase of cigarette smoke. In this case, the BAP content of the condensate represents only

Table 1. Emission of Benzo[a]pyrene Produced by 1 kg Fuel

Source of BaP	BaP emission (mg/kg)	Consumption[a] per year (tons)
Hard coal briquets (domestic heating)	5-380.0	1,128,000
Oil heating (light) (domestic heating)	0.0001	42,600,000
Passenger cars	0.10	20,400,000
Cars with diesel engines	0.03	7,650,000
Power plants	0.005	26,800,000

[a]In the Federal Republic of Germany.

about 1% of the effect of skin [7,8]. The portion of the total carcinogenic effect of BaP in various particulate emissions is shown in Table 2.

Thus, it is not enough to define the BaP content of a sample only. It is important to record a total profile of all carcinogens. These are, e.g., in automobile exhaust gas condensate, other PAH which in all represent 90% of the tumorigenic effect of the condensate.

The aim of PAH determination is to estimate the carcinogenic impact obtained from chemical-analytical data of a sample taken from the environment. It assumes:

a. that the carcinogenic properties of PAH most commonly found in the environment are known
b. that the effects of individual PAH are additive but at the same time do not influence one another
c. that the effects can be neither enhanced nor lessened by foreign substances

Table 2. Portion of Carcinogenic Effect of BaP Related to Total Carcinogenicity

Source of BaP	Carcinogenicity of the BaP portion (%)
Automobile exhaust condensate (gasoline engine)	9.6
Automobile exhaust condensate (diesel engine)	16.7
Domestic hard coal heating[a]	~6
Domestic brown coal heating (briquets)[a]	~9
Lubricating oil of cars (used)	18
Sewage sludge (extract)	22.9
Cigarette smoke condensate	~1

[a]Preliminary results, as yet unfinished.

According to these pre-conditions, a chemical-analytical inventory of the carcinogens permits the assumption that in a certain animal model the total carcinogenic effect corresponds e.g., to ten times the benzo[a]pyrene content of the sample since benzo[a]pyrene accounts for one tenth of the carcinogenicity of the sample. The decisive advantage of the chemical method in this context is its short-term duration.

V. Benzo[a]pyrene Concentration in Air

For methodological reasons, only BaP was determined in most investigations. As mentioned before, in general BaP only indicates the presence of other PAH and does not permit assessment of the carcinogenic potency of an air sample. It is still an open question whether the mixture of PAH present in the air of different cities is similar or not. Only by the assumption of similarity of the PAH profiles it is possible to take the BaP concentration as a relative yardstick of the carcinogenic burden of different areas. It is an open question too whether other carcinogenic compounds such as aza-arenes, aromatic amines, and nitrosamines play an important role in the carcinogenic effects of air pollution in humans.

Unfortunately most published data on PAH concentrations in the atmosphere are restricted to BaP, and even these are incomplete. Sawicki [9] compiled a comprehensive survey of measured values of BaP immissions (Table 3).

Table 4 shows measured values of BaP immissions in the Federal Republic of Germany [10]. BaP concentrations in the atmosphere of many German cities are much higher than in most cities of North America. BaP concentrations in particularly polluted atmospheres and several emissions in various countries are presented in Table 5 [9].

A comparison of the BaP concentrations reported by various authors can be carried out only with great reservations because a large number of different sampling procedures and analytical methods have been applied. This means that the results may have to be modified considerably, as mentioned earlier. The site at which airborne particulate matter is collected must also be taken into account. A single sampling station in a big city cannot provide representative data for the whole city. Therefore, a comparison of data from two cities is actually merely a comparison of observations made at two sites in those cities. Thus, the data compiled in the tables are not comparable if we do not take the different sampling methods into account. Nevertheless, the present data on BaP concentrations demonstrate that the concentration of BaP in most cases is much higher in winter than in summer. Figure 7 demonstrates this for a single location in the city of Hamburg [11]. During the summer of 1962 the average concentration was about 10 ng BaP/m^3, whereas in December 1961 the same place was burdened by an average of 336 ng BaP/m^3. This pattern holds for all areas that largely use coal for domestic heating.

VI. Collection of PAH-Containing Air Samples

As is the case for each chemical determination of air pollutants, PAH analysis requires a controlled sampling. During collection of PAH from automobile exhaust, cigarette smoke, or air, various kinds of oxidative, photochemical

Table 3. Benzo[a]pyrene Concentration in the Atmosphere of Cities (ng/m^3)

Country–city	Year	Concentration[a]	
		Winter	Summer
Australia:			
Sydney	1962-1963	8	0.8
Belgium:			
Liège	1958-1962	110	15
Ontario, Canada	1961-1962		
Sarnia		3.5	1.6
Windsor		15.0	7.8
Chatham		5.0	2.3
London		3.2	1.7
Kitchener		2.7	1.2
Brantford		5.2	2.2
Hamilton		9.4	5.7
St. Catharines		9.1	3.8
Toronto		5.4	6.4
Oshawa		2.6	1.7
Peterborough		10.0	1.8
Belleville		2.0	1.7
Kingston		11.0	4.0
Brockville		1.5	1.6
Cornwall		20.0	18.5
Ottawa		2.6	0.6
Orillia		14.0	1.3
North Bay		4.9	2.9
Sudbury		11.0	1.2
Sault St. Marie		3.9	4.0
Port Arthur			1.3
Czechoslovakia:			
Prague		122 (high)	19 (low)
100 meters away from pitch battery at Orlova, Lazy coke kilns	1964	1800-3000	
Denmark:			
Copenhagen	1956	17	5
England:			
Bilston		27	
Bristol		13	
Burnley		27	
Cannock		19	
Hull		18	
Leicester		29	
Sheffield		42	
Salford	Feb. 1953	210	

(continued)

Table 3 (continued)

Country—city	Year	Concentration[a] Winter	Summer
[England]			
Salford		110	
Burnley		32	
Darwen		35	
Gateshead		62	
Lancaster		20	
Merseyside (St. George's Dock)		31	
Ripon		15	
Salford		108	
Warrington		31	
York		24	
Northern England and Wales		11-108	
Finland:			
Helsinki	1962-1963	5	2
France:			
Paris	1958	300-500	—
Lyon	March 1972	1.3	
Hungary:			
Budapest	1968	1000[b]	32
	1971-1972	27[c]	
Iceland:			
Reykjavik	1955	3	
India:			
Bombay	1973		
Near gas plant (coal as fuel)		170-860	
Street (traffic density: 60 vehicles/min)		15-36	
Near street and kiln for firing pottery		17-230	
Residential suburbs		0.8-3.9	
Iran:			
Teheran	1971	6	0.6
Ireland:			
Belfast	1961-1962	51	9
Dublin	1961-1962	23	3
Italy:			
Bologna		212 (high)	6 (low)
Genoa		37	1
Milan	1958-1960	610 (high)	3 (low)
Rome	1963-1966 (winter)	20-147	

(continued)

Country–city	Year	Concentration[a]	
		Winter	Summer
Japan:			
Muroran		110-160	
Osaka	1965	50	
	1970	15	
	1971	11	
Sapporo	Feb. 1961	200	
Tokyo	Feb. 1964	15	
Netherlands:			
Amsterdam	1968	22	
	1969	18	2
	1970	5	2
	1971	8	
Delft	1968	20	3
	1969	18	1
	1970	12	3
	1971	6	3
Rotterdam	1968	15	3
	1969	23	1
	1970	19	3
	1971	23	2
Vlaardingen	1968	35	3
	1969	32	4
	1970	13	5
	1971	16	9
The Hague	1968	23	
	1969	12	4
	1970		4
	1971	13	
Norway:			
Oslo	1956	15	1
	1962-1963	14	0.5
Poland (average for 10 large cities):	1966-1967	130	30
Gdansk	1966-1967	84.64	
Katowice		76.75	
Krakow		63.63	
Lodz		45.45	
Opole		54.47	
Poznan		48.49	
Szczecin		44.62	
Warszawa		29.29	
Wroclaw		57.64	
Zabrze		130.100	

(continued)

Table 3 (continued)

Country—city	Year	Concentration[a] Winter	Summer
South Africa:			
Durban	June 1964	5, 14, 28	
Johannesburg	May 1964	49	
	May 1964	1100[d]	
Pretoria	1963-1964	10	22
Spain:			
Madrid	1969-1970	120	0
	1969	9	0
Sweden:			
Stockholm	1960	10	1
	1967	27 (high)	2 (low)
Switzerland			
Basle	1963-1964	8-80	
United States:			
100 large urban communities	1958-1959	6.6	
	1962	5	
32 large urban stations	1966-1967	3.3	
	1968	2.7	
	1969	2.9	
	1970	2	
Los Angeles	June 1971-June 1972	1.1, 0.5, 3.5, 0.03	
USSR:			
Leningrad	1965	15	
Kiev	1965	9	
Tashkent	1965	110	
100 m away from a pitch boiling plant of a cardboard factory		129	
Near coke furnaces		570	
500 m to the south of the furnaces		120	

[a]Average values. Where one value is reported it is an annual average value, unless otherwise stated. In some cases high and low values of the year are reported. (BaP values taken in Germany are reported in Table 4.)
[b]Taken during a period of heavy smog.
[c]Winter values.
[d]Near a road-tarring operation.
Source: Sawicki [9].

Table 4. Benzo[a]pyrene Concentration in the Atmosphere: Germany[a]

Measuring station	Time of sampling Year	Month	Duration	BaP (ng/m^3)	References
Bonn	1965	Feb.	Per station:	133	[11]
Düsseldorf		Feb.	20 days	125	
Bochum		Feb.		144	
Bonn	1965	July	Per station:	4	
Düsseldorf		July	20 days	5	
Bochum		July		19	
Ruhr district	1966	Feb. or March	Per station: 1 day	133	
Ruhr district 12 stations	1967	Feb. or March	Per station: 1 day	103	
Essen 3 stations	1967	Jan.	4 × 1 day per station	333	
Ruhr district 13 stations	1968	Feb.-Dec.	Per station: 28 days all year round	110	[33]
Berlin (East) 2 stations	1970	July (st. 1)	Average of 6 hr for 11 days	18	[34]
		Sept. (st. 2)	for 9 days	18	
Duisburg	1969	Per year:	24 hr per day	23	[35 and additional data]
Düsseldorf	until	April-	and station,	6	
Krahm	1973	Sept.	analysis of filters per month	1	
Duisburg	1969	Per year:	As above	121	
Düsseldorf	until	Oct.-		50	
Krahm	1974	March		7	
Duisburg	1969	All monthly values combined		72	
Düsseldorf	until	(average of 5 years)		28	
Krahm	1974			4	
Cologne	1970 until 1973	Nov.-Feb.	5 sampling periods, total of 31 days	60	[36]
Gelsenkirchen	1970	Jan.-	24 hr per day	91	[37]
Mannheim (Rhine shore)	until 1973	Dec.		11	
Westerland (Sylt)			Analysis of collected filters of 1 month	3.4	

(continued)

Table 4 (continued)

Measuring station	Time of sampling Year	Month	Duration	BaP (ng/m^3)	References
Waldhof (Lüneburg Heath)				2.3	
Deuselbach (Hunsruck)	(Given are average values of 4 years; monthly average values are reported in publication).			1.5	
Brotjacklriegel (Bavarian Forest)				0.7	
Schauinsland (Black Forest)				0.4	
Duisburg	1973	Dec.	On 6 days	93	[38]
Düsseldorf	1975	Dec.	On 13 days	40	
Duisburg 15 stations	1974-1975	Dec.-March	Once or twice every 1-2 hr per station	5-737 $\bar{x}$ = 190	[39]
Düsseldorf 3 stations	1975	Jan.	For 28 days per station	19 44 16	[40]
Düsseldorf 3 stations	1975	Feb.	For 28 days per station	3 31 11	
Karlsruhe			3 × weekly per station		[41]
nuclear research	1974/75	Nov.-March		1.8	
center park	1975	May-June		0.1	
subway	1975	May-June		2.9	
park	1975/76	Oct.-March		4.8	
subway	1975/76	Oct.-March		9.5	
Duisburg (center)	1977/78	Dec.-March	3 × per 4 weeks	5	[42]

[a]In addition to these immission data, information is available on particularly heavy pollutions of the atmosphere notably at the workplace and other potentially hazardous emissions [9].
Note: "Immission" applies to pollutants in ambient air, while "emission" applies to pollutants emanating from a source.

Table 5. Benzo[a]pyrene Concentrations in Particularly Polluted Atmospheres and Emissions (ng/m^3)

Area	Concentration (ng/m^3)
Coal-fired residential furnaces	2200-1,500,000
Coal-fired power plants	30-930
Coal-fired unit (intermediate size)	49-7900
Coke oven, above gas works retorts	216,000
Coke-oven battery:	
on battery locations	172-15,900
off battery locations	21-1200
battery roof	6700
larry car	6300
pusher	960
pump house	260
brick shed	380
cortez van	150
Garage air (Cincinnati downtown)	33
Roof-tarring operations	90, 870, 14,000
Gas works:	
retort houses	3000 (average)
above the retorts	220,000
Gas-fired heat generation units	20-350
Incinerators:	
municipal[a]	17, 19, 2700
commercial	11,000, 52,000
Oil-fired heat generation units	20-1900
Open burning	2800, 4200, 173,000
Retort houses:	
maximal results	2,300,000
above horizontal retorts	220,000
Sidewalk tarring operations	52,110, 78,000
Silicon carbide (carborundum) plant:	
air in crusher shop	300-900
air from coke ovens	400-730
100 m from coke ovens	200-410
500 m from coke ovens	72-180
100 m from plant	28-56
Smoky atmosphere:	
beer hall in Prague	28-144[b]
arena	0.7-22

(continued)

Table 5 (continued)

Area	Concentration (ng/m^3)
Tar paper plant:[c]	
mass boiling shop	1100-1500
plant territory air	230-290
100 m from plant	125-135
500 m from plant	38-61
Tunnel:	
Blackwell	350
Sumner	690
Wall-tarring operations	520×10^3, 640×10^3 1600×10^3, 6000×10^3 [d]

[a]In a study of the combustion of municipal refuse in a continuous feed incinerator it was found that some 9, 3450, and 3.5 mg of BaP and BeP were emitted per day in the stack gases, the residues, and the water effluents, respectively.

[b]Dependent on the number of people smoking and the ventilation. At the same time urban air in Prague contained 2.8-4.6 ng BaP/m^3.

[c]The coal tar, coal tar pitch, and boiling mass utilized in the plant contained 0.35-1.0, 0.4-2.0, and 0.3-1.4% BaP, respectively.

[d]Within 1 hour workers could inhale 3 mg of BaP, the amount of BaP equivalent to smoking approximately 300,000 cigarettes or breathing polluted air containing 10 ng BaP/m^3 for approximately 70 years.

Source: Sawicki [9].

and metal-catalyzed decompositions can be expected. Furthermore, the varying vapor pressure of the PAH being collected on the filter (glass fiber, impregnated or nonimpregnated) may cause losses of PAH. This should be tested by a backup control filter. Such effects as the chemical destruction or penetration of the PAH through the particle filter and reevaporization of the already collected PAH are readily recognized during collection. Provided a constant mass stream exists, an accurate collecting process results in a linear relation between the time of collection and the loading of the filter with PAH, independent of the molecular size of the PAH or their sensitivity to oxidation. If a completely accurate collection of PAH can be guaranteed, the PAH profile will be identical after the first hour and the second; only the amount—the mass of each component—should have doubled.

The reevaporation of already collected PAH from the filter is relevant in case of PAH with a boiling point below 400°C, such as fluoranthene, pyrene, benzfluorenes, and their methylderivatives. This is shown in the Fig. 8. During long-term sampling for 4 to 12 weeks without changing the filters, the loss of PAH, especially of PAH with three and four rings, becomes more evident. For example, in the case of fluoranthene, after 1 week about 25% are evaporated from the first filter onto the second and third filters, and after 12 weeks more than 75% is lost from the first nonimpregnated filter [12].

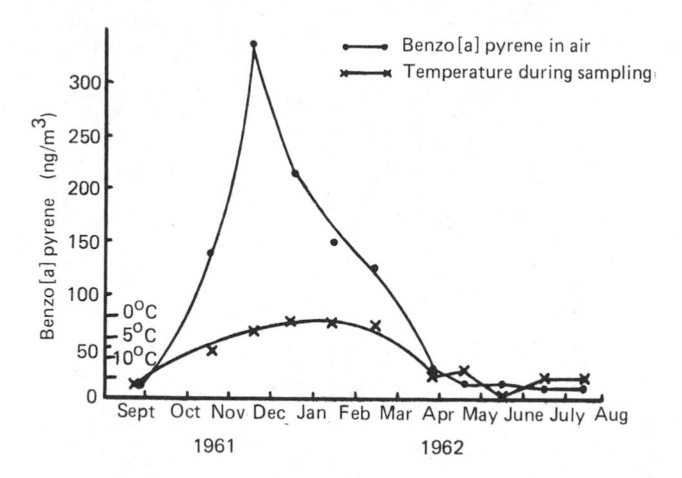

Figure 7. Average concentration of benzo[a]pyrene each month from September 1961 through August 1962 in Hamburg, Germany; collecting station on Jungius Street.

Besides the reevaporation in the case of PAH with 2, 3 and 4 rings, destruction of sensitive PAH such as benz[a]anthracene, benzo[a]pyrene, cyclopenta[cd]pyrene, etc., by light and/or oxidating agents (SO_3) or by nitration (NO_2) is to be expected [13,14]. This effect of destruction at the surface of the particle filter is demonstrated by the ratio of BaP to BeP. After 1 week of collecting, the ratio was about 1:1; after a 4-week collection period using the same filter, the ratio was 1.0:0.5. This means that about 50% of the BaP collected on the filter was destroyed by oxidation [the boiling point of BeP (492.9°C) and BaP (495.5°C) cannot explain this difference in the collection]. The same can be observed for the ratio of benz[a]anthracene (BaA) to chrysene. After 4 weeks, about one-half of the BaA was destroyed [12].

VII. Investigation of a "Representative" PAH Concentration in a City

To estimate the carcinogenic burden on humans by measuring the concentration of carcinogenic PAH in an area such as a large city, it is necessary to keep in mind the following facts:

1. A single sampling station in a large city cannot provide representative data for the whole city.
2. Because of the different sensitivity of the PAH against photochemical and chemical decomposition and because of the different vapor pressures of the PAH at the environmental temperature, it is not possible to collect the atmospheric particulate matter for a long period.

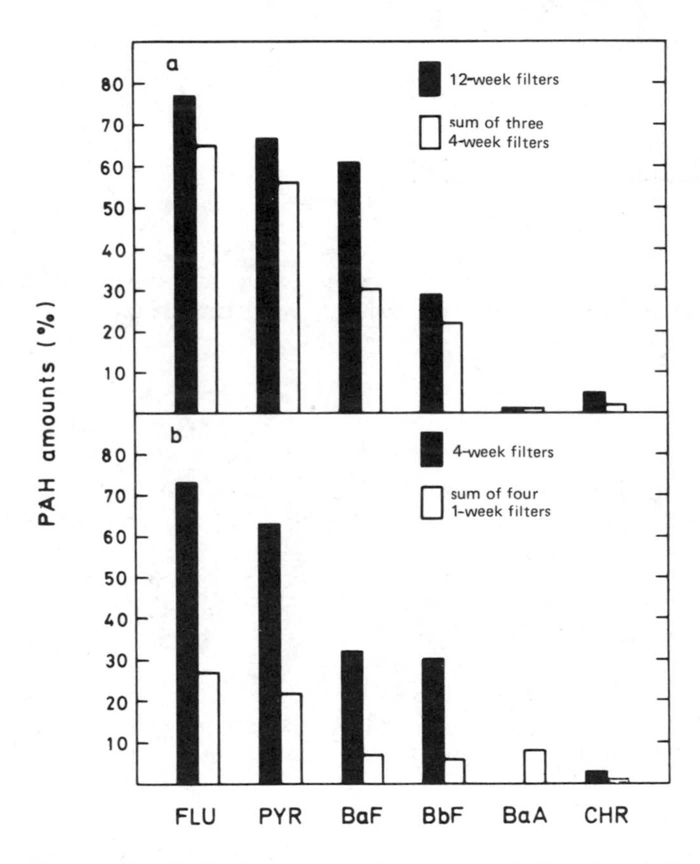

Figure 8. PAH amounts on impregnated No. 2 and No. 3 filters as a percentage of the total PAH (No. 1, No. 2, and No. 3 filter = 100%); nonimpregnated No. 1 filters measured during varying collection periods.

The chemical decomposition of sensitive PAH, such as BaP, depends on the concentration of oxidants in the air which react with the PAH collected on the filter. On the other hand, photochemical decomposition on the surface of the filter must not be disregarded. Therefore, it is not possible to suggest a period in which no loss of PAH can be guaranteed. To control the extent of chemical decomposition of BaP, it is helpful to determine the BeP concentration too, since in most emissions of fossil fuels the ratio of BaP to BeP is higher than 1.0.

VIII. Local and Temporary Variations of the PAH Concentration in a City

To estimate the environmental burden of a local area, a city, or a country it is of course not sufficient to know the total amount of PAH emitted. It is also necessary to know the local PAH concentration in the residential area. To measure the real PAH burden on the inhabitants of an area, it is necessary to compare local concentrations of several selected PAH and to record the temporary variation in PAH profiles of different areas in a city during the year. Such an extensive investigation in a city of 700,000 inhabitants in the Federal Republic of Germany, recording 250 PAH profiles by glass-capillary gas chromatography, has been reported [15]. Four areas inside and one area outside the city have been selected. The selected areas have the following characteristics:

I. An area with hand-stoked residential coal heating
II. An area with oil heating preferentially
III. Stations in a tunnel with automobile traffic
IV. An area with coke ovens

Each area was surrounded by four sampling stations. The distances between the various areas range from 3.5 to 8.0 km, the sampling time was 1 hr, and the sampling volume 10 m^3.

A. Variations in Concentration at a Selected Collecting Station in the City During the Day and the Year

The variations in concentration of several PAH compounds during a day are shown in Fig. 9. The area being studied is residential, mainly with domestic coal heating. The concentration of PAH is plotted on the left side of the figure, and the groups of lines represent the various PAH. Each of the seven lines represent one sampling period, as explained in the upper-left corner of Fig. 9. Not only in the case of BaP is there a big difference in the concentration between the first 2 hr in the morning and the following hours. Early in the morning the BaP concentration is more than five times higher than in the afternoon. Between 7:45 A.M. and 10:00 A.M. for most PAH the concentration is some three to four times higher than from 10:00 A.M. to 4:00 P.M.

These findings are also confirmed by long-term measurements. Figure 10 shows the variation for BaP concentration, measured each week for 12 months in the same area. The height of the lines represents the concentration of BaP. The highest BaP concentration, 300 ng BaP/m^3, was recorded in the 22nd week of 1979, the last week in May. The lowest concentration was detected 2 weeks later, in June.

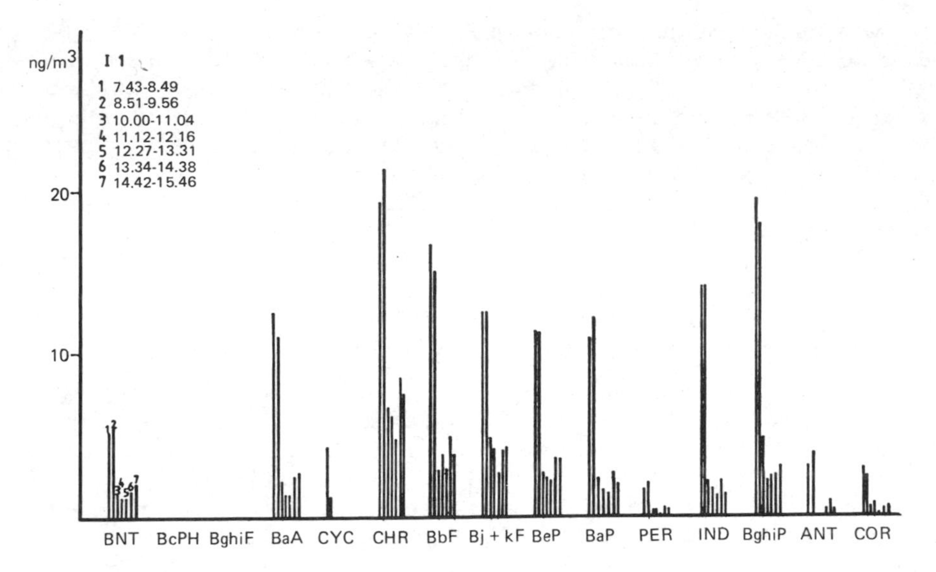

Figure 9. Concentration of different PAH (see list of abbreviations) in a residential area, predominantly with domestic coal heating. The seven lines represent the collecting time during the day, explained on the figure.

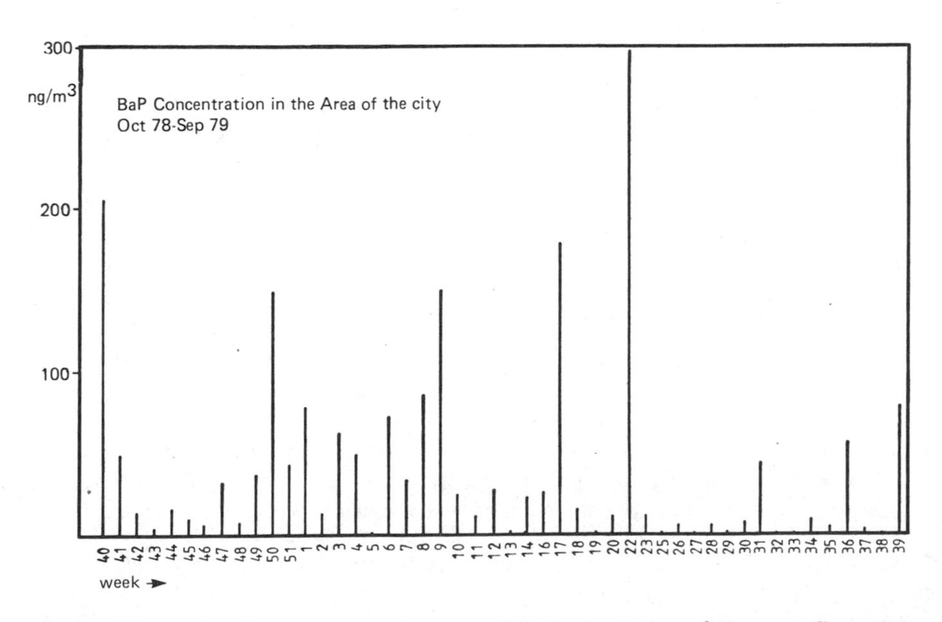

Figure 10. Benzo[a]pyrene concentration in an area of Essen, Germany, from October 1978 through September 1979, collected each week (collecting time: 1 hr).

The preliminary results show that the concentration of PAH in a selected location in the city varies considerably, even in a residential area. Nevertheless it is possible to calculate the total amount and the average concentration for this location.

B. Local Variations in Different Areas of the City

The second question is the differences among various localities of the city. Figure 11 shows the average concentration of BaP in the four areas burdened by typical pollution sources. In each area about 50 samples were collected during the period of October 1978 and September 1979. In area I, burdened mainly by domestic coal heating, the average value for 50 weeks was 15.4 ng BaP/m^3, and the range was 0.3-72.5 ng BaP/m^3. The other areas, II,

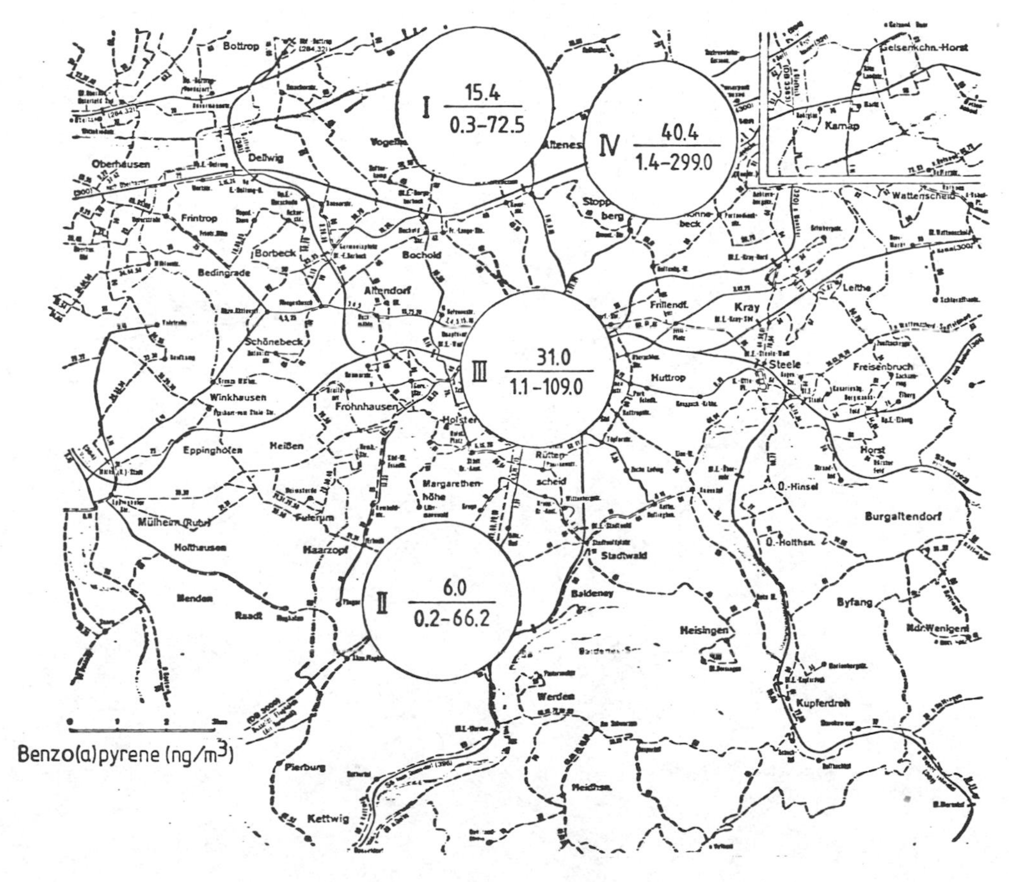

Figure 11. Map of an industrial city in which four areas polluted by typical emittents are labeled showing the distribution pattern of benzo[a]pyrene (ng/m^3) in the city. Average of 50 weeks (collecting time: 1 hr, different days and different times of day). Below the line the range (ng BaP/m^3).

III, and IV, shows averages of 6, 31, and 40.4 ng BaP/m^3, respectively. The most polluted area, 40 ng BaP/m^3, surrounds the coke plant. At a distance of about 3 km from this area the annual concentration drops to about one-third of this value.

Completely different to this pattern is the distribution of cyclopenta[cd] pyrene (CYC), a PAH especially produced by automobiles driven by gasoline engines (Fig. 12). As expected, the tunnel with automobile traffic shows the highest annual concentration. The average concentration is 88 ng CYC/m^3, from 0.1 to 440 ng/m^3. Surprisingly at a distance of about 4 km from the tunnel, the concentration drops to 1.6 ng CYC/m^3. This is about 2% of the value in the tunnel. This study leads to the following conclusions:

1. The distributions of different PAH in the city do not necessarily correlate with each other.
2. The different emittants produce different PAH profiles.

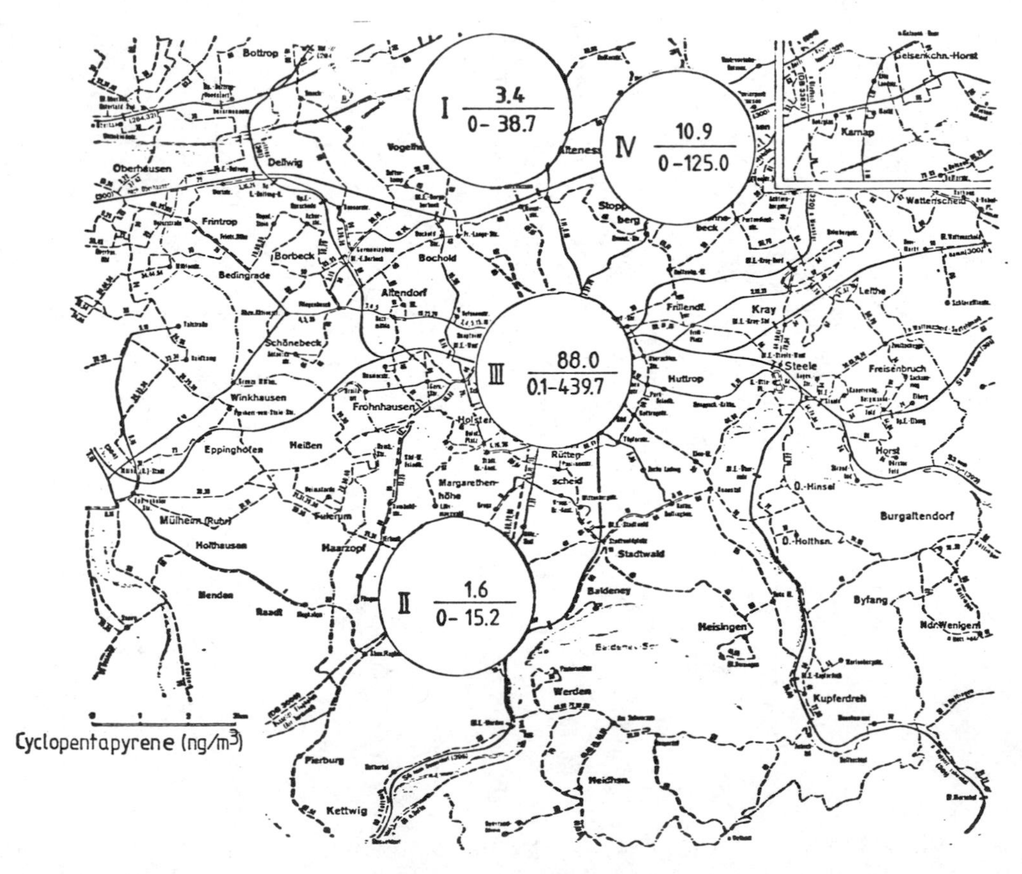

Figure 12. The distribution-pattern of cyclopenta[cd]pyrene (ng/m^3) in the same city as shown in Fig. 11. Average of 50 weeks, below the line the range.

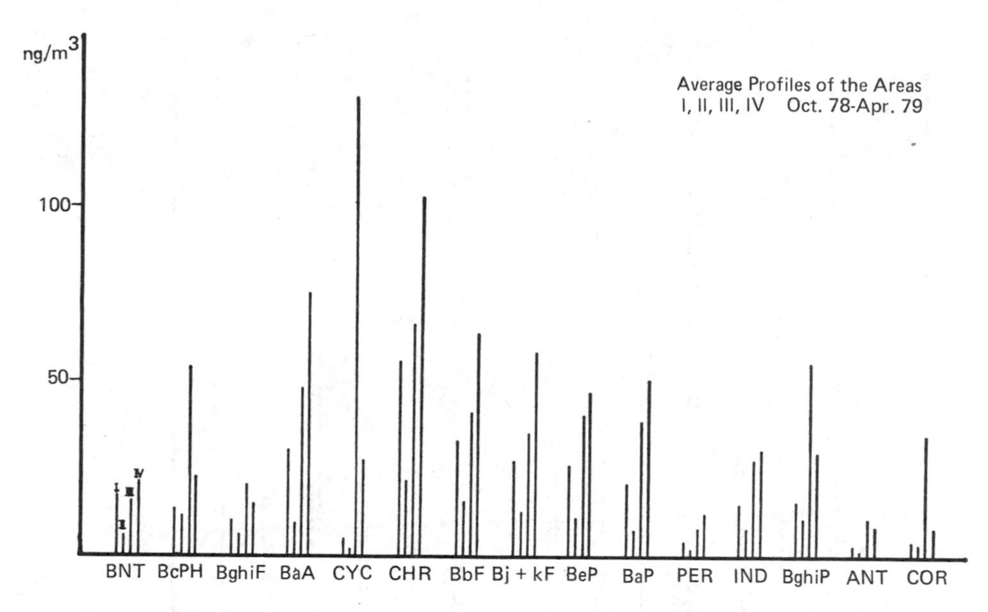

Figure 13. The average-profile of the four areas. Average of 30 weeks from October 1978 to April 1979 (collecting time: 1 hr in the week, on different days and different times of day).

Figure 13 compares the concentrations in the four areas: I, coal heating; II, oil heating; III, automobile traffic; and IV, around the coke plant. The ratio of the concentrations within the four areas is similar in the case of CYC, BghiP, or COR. These PAH are predominant in area III. In the case of CHR, BbF, BjF+BkF, BeP, and BaP, the ratios of the concentrations in the four areas are very similar. In this, area IV shows the highest concentration.

Figure 13 indicates that there are two main sources of emission in the city: (1) pyrolysis and combustion of coal and briquets, and (2) emission of automobile exhaust, especially in the tunnel. Figure 14 confirms this tentative conclusion: the PAH profile of the tunnel with automobile traffic is compared with the PAH emission of a passenger car [16-20] driven on a chassis dynamometer during a cycle which simulates city traffic [3]. In this case CYC is the PAH with the highest concentration. This is in variance with coal-heating emissions, which produce only small amounts of CYC. Furthermore, Fig. 14 demonstrates that the profile inside the tunnel and that of the bench test are very similar. A difference can be observed in the content of BNT; this compound is absent in the case of automobile exhaust from gasoline engines, because the gasoline contains no BNT or other sulfur-containing compounds. The source of BNT is not quite clear. Presumably, most of it originates from coal combustion. On the other hand, BNT could originate to a small extent from diesel engines, which emit thiophenes. It is likely that the use of the different PAH profiles, as illustrated in this study, may be helpful in enabling investigators to recognize the main sources of air pollution.

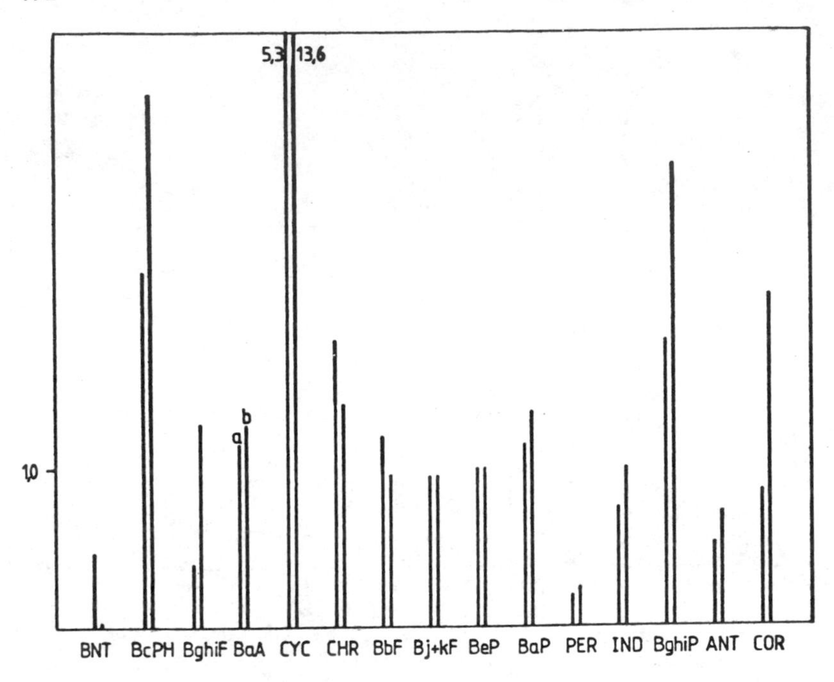

Figure 14. Comparison of the PAH profiles of (a) collecting station in a tunnel with automobile traffic and (b) PAH emission of a passenger car on a chassis dynamometer simulating city traffic (ECE regulation No. 15).

IX. Discussion of Methods Used to Determine PAH and Inherent Problems

A. Extraction

A critical study on the methods of extraction of samples of airborne particulate matter was performed by Stanley et al. [21]. For extracting filters loaded with airborne particulate matter by means of cyclohexane—a weak eluting solvent—Seifert and Steinbach proposed that ultrasonic extraction be used [22], but this method is not recommended in all cases (see Table 6). Furthermore, a sublimation under reduced pressure is used [23-25]. In the case of six-ring PAH such as benzo[ghi]perylene, dibenzofluoranthenes, or dibenzopyrenes, the recovery under these conditions (200°C, 10^{-5} bar) from coal soot or diesel-exhaust condensate is not satisfactory [26]. If necessary, the completeness of extraction with these methods may be checked by an additional extraction with boiling toluene or xylene—in particular if the sample contains soot particles.

The completeness of PAH extraction of particulate organic matter, emitted from gasoline or diesel engines as well as from coal combustion depends on the adsorption-desorption equilibrium between the insoluble residue and the solvent. In most cases acetone is a suitable solvent, but in the case of soot or carbon black the extraction of PAH is not complete with this solvent.

Table 6. Amount of the First Extraction (Percentage of Both Extractions) of Air Pollutants (Predominantly Diesel Engine Emission, Samples of 20 mg) by Various Solvents (100 ml, 1 hr refluxed), Followed by a Second Extraction with Toluene (100 ml, 1 hr Refluxed)[a]

	Methanol	Acetone	Cyclohexane	Xylol	Toluene
Benzo[b]chrysene[b] (300 ng)	3.1	24.6	81.2	92.8	96.4
Benzo[b]naphtho[2,1-d]thiophene	65.1	70.3	53.7	78.2	91.2
Benzo[c]phenanthrene	65.2	74.2	53.9	89.3	93.3
Benz[a]anthracene + chrysene	57.6	70.3	70.2	82.5	89.5
Benzofluoranthenes (b+k+j)	20.3	44.2	46.0	81.3	92.2
Benzo[e]pyrene	20.6	43.7	46.9	81.1	88.2
Benzo[a]pyrene	19.5	42.1	44.2	80.5	90.1
Indeno[1,2,3-cd]pyrene	0.5	8.0	21.3	72.1	91.0
Benzo[ghi]perylene	0.5	8.7	17.7	61.6	89.2

[a] Both extracts = 100%.
[b] Percent recovery of the added internal standard benzo[b]chrysene (300 ng).

Aromatic hydrocarbons such as toluene or xylene are the most effective extracting agents for this type of sample, presumably due to their structural similarity to PAH. As shown in Table 6, the amount of PAH extractable from diesel engine exhaust condensate varies depending on the solvent, e.g. from 3.1% to 96.4% recovery of benzo[b]chrysene (BbC). The latter compound was added as reference standard and is not originally present in diesel engine soot. BbC thus characterizes the extraction of PAH from the soot surface as equilibrium between adsorbing material and the solution. A further characteristic finding of this experiment is that the adsorption equilibrium depends largely on the molecular size; for example, chrysene and benz[a]anthracene are extracted with the first solvent to a considerably higher degree than benzo[ghi]perylene.

Under the same conditions extraction with cyclohexane at room temperature by ultrasound (1 hr) yields 65.3% of the added 300 ng BbC. About 100 ng BbC adsorbed to 20 mg of dust could be desorbed by boiling toluene only. (Further discussion of extraction of PAH is given in Chaps. 2 and 3.)

B. Enrichment

In most cases chromatographic methods are used to enrich the PAH by cleanup of the extract of the filter. Column packing materials such as silica gel and aluminium oxide permit separation of PAH from highly polar compounds of the matrix using nonpolar solvents such as cyclohexane, isooctane, or benzene, or mixtures of these solvents. The activity of inorganic adsorbents such as silica gel, alumina, or Florisil [27] depends on the water content. Thus reproducible retention volume of alumina with a certain particle size can be achieved only at an exactly defined water content. However, these column packings can be utilized without difficulty for a rough separation of PAH mixture and accompanying substances.

Thin-layer plates of pure inorganic material should not be used for preliminary separation since occasional decomposition of PAH on silica gel plates has been reported [22].

In the last few years hydrophobic Sephadex has predominantly been used as an organic adsorbent in column chromatography because its retention volumes are not dependent on the water content and even at high PAH concentrations the elution volumes do not change. It is particularly suitable for isolating PAH because nonaromatic, nonpolar compounds such as paraffins and triglycerides are eluted with alcohol directly behind the solvent front whereas aromatic compounds are considerably retarded according to the number of rings. This effect can be seen in Table 7 [28]. The utilization of high-performance liquid chromatography (HPLC) has been described for the separation of PAH mixtures isolated from samples of airborne particulate matter [29-31] and automobile exhaust gas [32]. So far, satisfactory separation of the complex mixture of airborne particulate matter has not been achieved due to the low number of separation stages [about 20,000 HETP (height equivalent to theoretical plates)]. According to comparative investigations carried out by Novotny et al. [29], the HPLC will "hardly be a final solution" for the separation of PAH mixtures obtained from environmental samples.

Table 7. Elution Volumes of PAH Adsorbed to Sephadex LH 20 (10 g, Isopropyl Alcohol)

PAH	Volume (ml)
Aliphatic hydrocarbons	20-35
Phenanthrene/anthracene	38-50
Pyrene/fluoranthene	48-65
Benz[a]anthracene/chrysene	60-78
Benzofluoranthenes (b+j+k)	70-90
Benzo[a]pyrene	73-93
Benzo[e]pyrene	78-98
Indeno[1,2,3-cd]pyrene/perylene	81-105
Anthanthrene/benzo[ghi]perylene	89-118
Benzo[b]chrysene	84-115
Coronene	105-140

C. Methods Used for Separating Complex PAH Mixtures

1. *Gas Chromatography with Glass Capillaries*

Gas-chromatographic separation using glass capillaries is presently the most efficient method for separating complex PAH mixtures. In routine operation, columns exhibiting 70,000 or more theoretical plates are routinely achieved. This number of theoretical plates indicate that about 200 baseline separated peaks fit in between the retention time of fluoranthene and that of coronene.

In case of "splitless" sample injection, 5-10 ng BaP is usually needed to record an amplitude of the FID signal extending over the whole recording width. (The detection limit is about 0.1 ng, defined as three times the noise level.)

Besides its high sensitivity, the FID recording offers the same advantages as a mass detector. Both identified and unidentified PAH are recorded in proportion to their quantity and, without needing correcting factors, give a direct presentation of the quantitative composition of the PAH mixture isolated from a sample (PAH profile). Further discussion of gas chromatographic analysis of PAH is given in Chap. 6.

2. *Thin-Layer Chromatography*

In the case of most sample types, one-dimensional thin-layer chromatography (TLC) is suitable for the identification of individual components only. Since the size of the substance spot does not depend upon the diameter of the application point alone but also upon the ratio to the solvent front (R_F value), only some 10-20 substances can be arranged between the starting point and the solvent front even at complete separation of the spots.

A possible quantitation by measuring ultraviolet (UV) light or fluorescence requires information on characteristic properties of the substance. The concentration of an unidentified PAH, therefore, cannot be determined.

Two-dimensional TLC considerably improves the separation of PAH but does not achieve the separation capacity of high-capacity columns used in gas chromatography. TLC analysis of PAH is described in Chap. 9.

3. *High-Performance Liquid Chromatography*

The presently used HPLC columns reach a maximum separation number of 20,000. This corresponds to the separation power accomplished with packed high-performance gas chromatographic columns of length 7-10 m. In practice this means that the predominating main components (some unsubstituted base components) can be separated completely from each other, whereas mixtures of (say) isomeric methyl derivatives of a parent component cannot be separated. Using UV detectors may also be of limited value. Since there is but a slight shift in the wavelengths of their UV adsorption maxima or minima, respectively, measurements are often without informative value as regards mixtures containing isomeric methyl derivatives. Further discussion of HPLC analysis is given in Chap. 5.

X. A Method for Glass-Capillary Gas Chromatographic Profile Analysis of PAH in the Air [43]

A. Scope and Field Application

This method specifies a procedure used in the author's laboratory for the determination of PAH containing more than three rings in airborne particulate matter and different exhaust gas condensates such as those emitted by gasoline and diesel engines or coal combustion in hand-stoked residential heating units or power plants. The method has limits of detection ranging from 0.05 to 0.2 ng PAH per sample, e.g., 0.05 ng for chrysene, 0.1 ng for benzo[a]pyrene, and 0.2 ng for coronene.

B. Definition

A glass-capillary gas chromatogram affords a graphic record of the quantitative composition of the PAH mixture in a sample when a flame ionization detector is used. Such a recording with a mass-dependent detector is referred to as the profile of the sample.

C. Principle

The PAH, collected on the filter, are extracted with toluene. Before heating, 100 ng benzo[a]chrysene (or benzo[b]chrysene) is added to the solution as an internal standard. For cleanup, the cyclohexane solution of this extract is filtered on a silica gel column. For further enrichment of the PAH, chromatography using 10 g Sephadex LH 20 with isopropanol is used. The fraction ranging from 45 to 180 ml contains only PAH with more than three rings (fluoranthene to coronene) or from 55 to 180 ml (benzo[b]naphtha[2,1-d]thiophene/chrysene to coronene).

To separate the mixture of PAH, glass-capillary gas chromatography is used. The capillary (e.g., 0.3 mm × 25 m) is coated with polydimethylsiloxane or polyphenylmethylsiloxane (see Chap. 6).

D. Reagents

All reagents should be of high-quality, analytical grade and distilled in glass. To check the purity of all analytical materials used (solvents, silica gel, Sephadex LH 20), a blank test should be carried out following the procedure described in Sec. X.G (below). Reference compounds are available from several commercial sources. Some of these compounds are also available from the European Communities, Community Bureau of Reference BCR, rue de la loi 200, B-1049 Brussels, Belgium, or from the author's laboratory.

E. Apparatus

For concentration of PAH before gas chromatography, commercial tubes (Prolabo 96725) or tubes specially made (Fig. 15) may be used. The glass columns used for silica gel (10 g) chromatography were 200 mm × 10 mm (i.d.) and for Sephadex LH 20 (10 g) were 100 mm × 30 mm (i.d.).

F. Sampling

The collection system consists of a low air sampler, equipped with a flowmeter which collects 10 m^3/hr (Sartorius or Ströhlein GmbH & Co., D-4000 Düsseldorf, Federal Republic of Germany: Porticon or equivalent collecting system). The air sampler is connected with a container for the filter. The glass fiber filter with a collecting area of about 500 cm^2 is not impregnated. The impregnation is not necessary if low flow rates are used and the collecting time is short (e.g., 1 hr). The container for the filter and the filter is shown in Fig. 16.

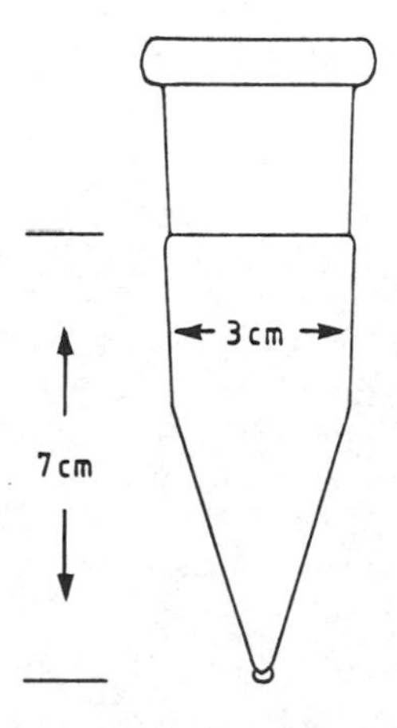

Figure 15. Tube for concentration of the PAH mixture before the injection in the GC injection port.

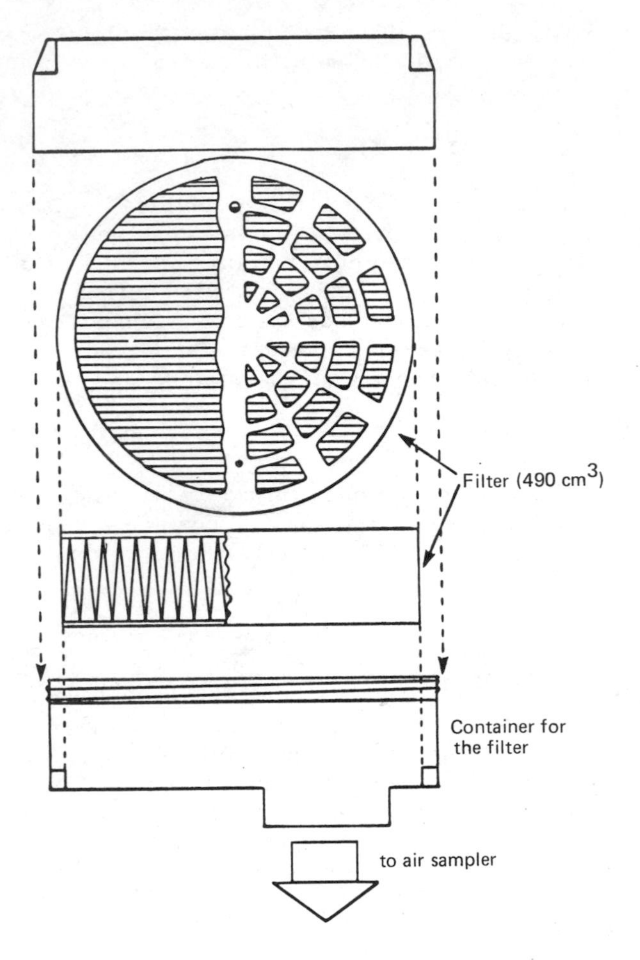

Figure 16. Collecting arrangement for air-suspended matter. Container for the filter, glass fiber filter (490 cm^2), Drägerwerke AG, D-2400 Lübeck, Germany—Type: Feinstaubfilter 909 ST.

To control the degree of the separation of the PAH on the glass fiber filter it is necessary to arrange a silica gel filter (1.0-1.2 mm particle size, diameter 100 mm, layer height 30 mm) between the air sampler and the glass fiber filter. This control filter can be omitted if the degree of separation on the glass fiber filter is known.

Samples should be stored in such a way that deterioration and change in composition are prevented. But even in the dark, it may happen that benzo[a]pyrene and other unstable PAH are destroyed by oxidation or nitration in a few days.

G. Procedure

Note: PAH are degraded by UV light, and exposure of samples, extracts, or standard solutions to sunlight or other strong light sources should be avoided.

A procedure for PAH analysis is shown schematically below:

filter, extracted with toluene + internal standard
↓
evaporation of toluene
↓
solved in cyclohexane, filtration on silica gel column
↓
evaporation of cyclohexane
↓
solved in isopropanol, chromatography on Sephadex LH 20
↓
evaporation of isopropanol
↓
solved in toluene, glass-capillary gas chromatography

The foregoing procedure is described in detail elsewhere [43].

H. Repeatability

The toluene extract of a filter was divided in five parts. Each part was added to the internal standard and analyzed according to the procedure just described. For sets of five analyses, the coefficient of variations for 11 PAH ranged from 2.3% (for benzo[e]pyrene) to 9.4% (for chrysene).

List of Abbreviations

ANT	Anthanthrene
BaA	Benz[a]anthracene
BaP	Benzo[a]pyrene
BbF	Benzo[b]fluoranthene
BcPH	Benzo[c]phenanthrene
BeP	Benzo[e]pyrene
BghiF	Benzo[ghi]fluoranthene
BghiP	Benzo[ghi]perylene
BjF	Benzo[j]fluoranthene
BkF	Benzo[k]fluoranthene
BNT	Benzo[b]naphtho[2,1-d]thiophene
CHR	Chrysene
COR	Coronene
CYC	Cyclopenta[cd]pyrene
IND	Indeno[1,2,3-cd]pyrene
PER	Perylene

References

1. R. C. Lao, R. S. Thomas, H. Oja, and L. Dubois, *Anal. Chem. 45*:908-915 (1973).
2. M. L. Lee, M. Novotny, and K. D. Bartle, *Anal. Chem. 48*:1566-1572 (1976).
3. EG-Richtlinie 74/290/EWG, Amtsblatt der Europäischen Gemeinschaft No. L159 (1974).
4. G. Grimmer and H. Böhnke, *J. Environ. Pathol. Toxicol. 1*:661-667 (1978).
5. H. Brune, M. Habs, and D. Schmähl, *J. Environ. Pathol. Toxicol. 1*: 737-746 (1978).
6. J. Misfeld and J. Timm, *J. Environ. Pathol. Toxicol. 1*:747-772 (1978).
7. W. Dontenwill, H.-J. Chevalier, H. P. Harke, and H.-J. Klimisch, *Z. Krebsforsch. 85*:155-167 (1976).
8. J. Misfeld and K. H. Weber, *Planta Med. 22*:282-292 (1972).
9. E. Sawicki, in *Environmental Pollution and Carcinogenic Risks*, C. Rosenfeld and W. Davis (Eds.), IARC Sci. Publ. No. 13, Lyon, France, 1976, pp. 297-354.
10. F. Pott, in *Luftqualitätskriterien für ausgewählte polycyclische Kohlenwasserstoffe*, Ber. 1/79, Umweltbundesamt, Schmidt Verlag, Berlin, 1979, p. 91-93.
11. H. O. Hettche and G. Grimmer, *Schriftenreihe Landesanstalt Immissions- u. Bodennutzungsschutz Landes Nordrhein-Westfalen 12*: 92-108 (1968).
12. J. König, W. Funcke, E. Balfanz, B. Grosch, and F. Pott, *Atmos. Environ. 14*:609-613 (1980).
13. M. M. Hughes, D. F. S. Natusch, M. R. Schure, and D. R. Taylor, in *Polynuclear Aromatic Hydrocarbons: Chemistry and Biological Effects*, A. Bjørseth and A. J. Dennis (Eds.), Battelle Press, Columbus, Ohio, 1980.
14. F. S. Lee, T. M. Harvey, D. Schuetzle, T. J. Prater, F. C. Ferris, and W. R. Pierson, in *Polynuclear Aromatic Hydrocarbons: Chemistry and Biological Effect*, A. Bjørseth and A. J. Dennis (Eds.), Battelle Press, Columbus, Ohio, 1980.
15. G. Grimmer, K.-W. Naujack, and D. Schneider, *Intern. J. Environ. Anal. Chem. 10*:265-276 (1981).
16. G. Grimmer, A. Hildebrandt, and H. Böhnke, *Zbl. Bakteriol. Hyg., I. Abt. Orig. B158*:22-34 (1973).
17. G. Grimmer, A. Hildebrandt, and H. Böhnke, *Zbl. Bakteriol. Hyg., I. Abt. Orig. B158*:35-49 (1973).
18. G. Grimmer and A. Hildebrandt, *Zbl. Bakteriol. Hyg., I. Abt. Orig. B161*:104-124 (1975).
19. G. Grimmer, H. Böhnke, and A. Glaser, *Zbl. Bakteriol. Hyg., I. Abt. Orig. B164*:218-234 (1977).
20. G. Grimmer, A. Hildebrandt, and H. Böhnke, in *Environmental Carcinogens: Selected Methods of Analysis*, Vol. 3: *Analysis of PAH in Environmental Samples*, IARC Sci. Publ. No. 29, Lyon, France, 1979.
21. T. W. Stanley, J. E. Meeker, and M. J. Morgan, *Environ. Sci. Technol. 1*:927-931 (1967).
22. B. Seifert and I. Steinbach, *Z. Anal. Chem. 287*:264-270 (1977).
23. H. Matsushita and Y. Esumi, *Bunseke Kagaku 21*:722-729 (1972).

24. J. L. Monkman, L. Dubois, and C. J. Baker, *Pure Appl. Chem. 24*:731-738 (1970).
25. R. Tomingas and A. Brockhaus, *Staub-Reinhaltung Luft 33*:481-482 (1973).
26. G. Grimmer, unpublished results.
27. J. W. Howard, E. W. Turicchi, R. H. White, and T. Fazio, *J. Assoc. Offic. Anal. Chemists 49*:1236-1244 (1966).
28. G. Grimmer and H. Böhnke, *Z. Anal. Chem. 261*:310-314 (1972).
29. M. Novotny, M. L. Lee, and K. D. Bartle, *J. Chromatogr. Sci. 12*:606-612 (1974).
30. M. Dong and D. C. Locke, *Anal. Chem. 48*:368-371 (1976).
31. M. A. Fox and S. W. Staley, *Anal. Chem. 48*:992-998 (1976).
32. J. A. Schmidt, R. A. Henry, R. C. Williams, and J. F. Dieckman, *J. Chromatogr. Sci. 9*:645 (1971).
33. K.-H Friedrichs, J. Stuke, A. Brockhaus, and H. Steiger, *Staub-Reinhaltung Luft 31*:323-326 (1971).
34. W. Prietsch, K. Wettig, and H. Kahl, *Z. Ges. Hyg. 17*:573-575 (1971).
35. A. Brockhaus, H. Weisz, K. H. Friedrichs, and U. Krämer, *Schriftenreihe Ver. Wasser-, Boden- u. Lufthyg. 42*:183-196 (1974).
36. M. Deimel, Kohlenmonoxid-, Blei-, Stickoxid- und Benzo[a]pyrenbelastung in Kölner Strassen, in *Immissionssituation durch den Kraftverkehr in der Bundesrepublik Deutschland*, F. Meinck (Ed.), Fischer, Stuttgart, 1974, pp. 149-163.
37. A. Brockhaus, G. Rönicke, and H. Weisz, *Benzpyrenpegel des Luftstaubs in der Bundesrepublik Deutschland*. Deutsche Forschungsgemeinschaft, Bonn-Bad Godesberg, Kommission zur Erforschung der Luftverunreinigung, No. XIII, p. 26.
38. R. Tomingas and A. Brockhaus, *Staub-Reinhaltung Luft 34*:87-89 (1974).
39. W. Schneider, L. Matter, and E. Jerrmann, *Umwelthyg. 9*:273-276 (1975).
40. R. Tomingas, unpublished results (1978).
41. G. Heinrich and H. Gusten, *Staub-Reinhaltung Luft 38*:94-100 (1978).
42. J. König, unpublished results (1978).
43. G. Grimmer, K.-W. Naujack, and D. Schneider, *Fresenius Z. Anal. Chem. 311*:475-484 (1982).

5

High-Performance Liquid Chromatography for the Determination of Polycyclic Aromatic Hydrocarbons

STEPHEN A. WISE / National Bureau of Standards, Washington, D.C.

I. Introduction

Since its birth in the early 1970s, high-performance liquid chromatography (HPLC) has been used extensively for the separation of polycyclic aromatic hydrocarbons (PAH). At present, HPLC does not approach the high separation efficiency of capillary column gas chromatography (GC). However, HPLC does offer several advantages for the determination of PAH. First, HPLC offers a variety of stationary phases capable of providing unique selectivity for the separation of PAH isomers that are often difficult to separate by GC. In HPLC, selectivity is achieved due to interactions of the solute with both the stationary phase and the mobile phase rather than only the stationary phase as in GC. Ultraviolet (UV) absorption and fluorescence spectroscopy provide extremely sensitive and selective detection for PAH in HPLC. Finally, HPLC provides an extremely useful fractionation technique for the isolation of PAH for subsequent analysis by other chromatographic and spectroscopic techniques. Because of these characteristics, HPLC has been employed extensively for the determination of PAH in water, sediments, marine biota, air particulates, cigarette smoke, fuels, etc.

In this chapter the modes of liquid chromatography and detection systems applicable to PAH measurement will be described. The application of HPLC to the analysis of complex mixtures will also be described. Finally, an extensive review of the literature concerning the determination of PAH utilizing HPLC techniques is presented.

A basic discussion of HPLC in general, i.e., instrumentation, column preparation, separation theory, etc., is not intended except as it relates to the unique problems of PAH separations. Recent books by Snyder and Kirkland [1] and others [2-5] provide excellent discussions of the technique of modern liquid chromatography. Several other authors [6-10] have briefly reviewed the use of HPLC in PAH analysis.

II. Liquid Chromatographic Modes for the Separation of PAH

Several modes of liquid chromatography (LC) can be used for the separation of PAH: (1) adsorption chromatography on "classical adsorbents" such as silica and alumina; (2) reverse-phase LC on chemically bonded nonpolar phases such as C_{18}; (3) normal-phase LC on chemically bonded polar phases such as CN, NH_2, and NO_2; and (4) steric exclusion or gel permeation chromatography. The application of each of these modes of LC for the separation of PAH will be described in the following sections with emphasis on the chemically bonded C_{18} and polar phases. In addition, the developments in capillary liquid chromatography will be described briefly.

In all of the above LC modes except steric exclusion, the general elution order for the parent PAH is very similar, i.e., retention increases as the number of aromatic carbon atoms increases. The retention characteristics of over 100 PAH on several different LC columns are summarized in Table 1 for comparison. These retention data are reported as retention indices as first described by Popl et al. [11,12], and later employed by others [13-15]. Popl and co-workers [11,12] described this retention index system for studying the elution behavior of PAH on alumina and silica. In this system, the elution volume of the solute is measured simultaneously with the elution volumes of standards (benzene, naphthalene, phenanthrene, benz[a]anthracene, and benzo[b]chrysene) representing one- to five-condensed-ring PAH. The retention index, I, was calculated using the following equation:

$$\log I_x = \log I_n + \frac{\log R_x - \log R_n}{\log R_{n+1} - \log R_n} \quad (1)$$

where x represents the solute, n and n + 1 represents the lower and higher standards, and the R values are the corresponding corrected retention volumes. The standards were assigned the following values (log I): benzene (1),

Table 1. HPLC Retention Characteristics for Polycyclic Aromatic Hydrocarbons (log I)

Compound	Normal-phase[a]			Reverse-phase C_{18}	
	Silica [12]	Alumina [11]	NH_2 column [14,33]	Polymeric[b] [14,33,40]	Monomeric[c] [14,33,40]
Naphthalene	2.00	—	2.00	2.00	2.00
Acenaphthene	2.30	—	2.10	2.64	2.51
Acenaphthylene	2.59	—	2.59	2.26	—
Fluorene:	2.25	2.79	2.55	2.70	2.75
1-Methylfluorene	3.46	3.08	2.64	3.15[d]	3.33
2-Methylfluorene	—	—	2.59	3.24[d]	3.42
4-Methylfluorene	—	—	—	3.10[d]	3.34
Anthracene:	2.95	3.00	2.94	3.20	3.14
1-Methylanthracene	—	—	—	3.41[d]	3.55
2-Methylanthracene	2.97	—	3.01	3.71[d]	3.69
9-Methylanthracene	3.02	3.14	3.02	3.41[d]	3.53
9,10-Dimethylanthracene	—	—	3.08	3.63	3.90
Phenanthrene:	3.00	3.00	3.00	3.00	3.00
1-Methylphenanthrene	3.26	3.18	3.02	3.39[d]	3.51
2-Methylphenanthrene	3.26	3.23	3.00	3.71[d]	3.71
3-Methylphenanthrene	3.24	3.21	3.12	3.29[d]	3.46
4-Methylphenanthrene	—	—	—	3.26[d]	3.40
9-Methylphenanthrene	3.18	3.18	3.02	3.34[d]	3.51
9-Ethylphenanthrene	3.17	3.05	2.97	3.55[d]	3.87
9-n-Propylphenanthrene	2.99	3.10	2.94	3.83[d]	4.26
9-Isopropylphenanthrene	3.03	2.96	2.91	3.66[d]	4.09
3,6-Dimethylphenanthrene	3.42	3.23	3.11	3.57	3.91
1,8-Dimethylphenanthrene	—	—	3.11	3.79	3.99
2,7-Dimethylphenanthrene	—	—	—	4.01[d]	4.23
9-Methyl-10-ethylphenanthrene	—	—	3.11	3.71	4.15
9,10-Dimethyl-3-ethylphenanthrene	—	—	3.14	4.02[d]	4.64

(continued)

Table 1 (continued)

Compound	Normal-phase[a] Silica [12]	Normal-phase[a] Alumina [11]	Normal-phase[a] NH_2 column [14,33]	Reverse-phase C_{18} Polymeric[b] [14,33,40]	Reverse-phase C_{18} Monomeric[c] [14,33,40]
Benzo[a]fluorene:	3.95	—	3.51	3.79[d]	3.76
9-Methylbenzo[a]fluorene	—	—	3.38	3.84	3.87
Benzo[b]fluorene	4.14	—	3.54	3.82[d]	3.78
Benzo[c]fluorene	—	—	—	3.49[d]	3.64
4,5-Methylenephenanthrene	3.06	3.23	3.10	3.16	3.35
Pyrene:	3.06	3.31	3.37	3.58[d]	3.65
1-Methylpyrene	3.23	—	3.46	3.98[d]	4.15
2-Methylpyrene	—	—	—	4.05[d]	4.19
2,7-Dimethylpyrene	—	—	3.47	4.42[d]	4.77
1-Ethylpyrene	—	—	3.41	4.07[d]	4.46
1-n-Butylpyrene	—	—	—	4.39[d]	5.33
Fluoranthene	3.42	3.41	3.51	3.37	3.43
Benzo[c]phenanthrene:	—	—	3.64	3.63	3.92
2-Methylbenzo[c]phenanthrene	—	—	—	3.83	4.23
3-Methylbenzo[c]phenanthrene	—	—	—	4.04	4.39
4-Methylbenzo[c]phenanthrene	—	—	—	4.01	4.38
5-Methylbenzo[c]phenanthrene	—	—	—	3.97	4.36
6-Methylbenzo[c]phenanthrene	—	—	—	3.94	4.38
Benz[a]anthracene:	4.00	4.00	4.00	4.00	4.00
1-Methylbenz[a]anthracene	—	—	3.90	4.14	4.38
2-Methylbenz[a]anthracene	—	—	—	4.09	4.40
3-Methylbenz[a]anthracene	—	—	—	4.39	4.53
4-Methylbenz[a]anthracene	—	—	—	4.33[d]	4.46
5-Methylbenz[a]anthracene	—	—	4.04	4.28	4.48

6-Methylbenz[a]anthracene	—	—	4.03	4.10	4.39
7-Methylbenz[a]anthracene	—	—	—	4.14	4.35
8-Methylbenz[a]anthracene	—	—	4.03	4.19	4.39
9-Methylbenz[a]anthracene	—	—	4.08	4.39	4.53
10-Methylbenz[a]anthracene	—	—	—	4.24	4.47
11-Methylbenz[a]anthracene	—	—	3.91	4.13	4.41
12-Methylbenz[a]anthracene	—	—	—	4.10	4.35
7,12-Dimethylbenz[a]anthracene	—	—	3.90	4.19[d]	4.83
Chrysene:	4.00	4.06	4.01	4.10	3.99
1-Methylchrysene	—	—	4.07	4.43	4.46
2-Methylchrysene	—	—	4.08	4.52	4.52
3-Methylchrysene	—	—	4.12	4.29	4.42
4-Methylchrysene	4.10	—	3.95	4.18	4.35
5-Methylchrysene	—	—	3.94	4.14	4.35
6-Methylchrysene	—	—	4.10	4.14	4.36
Triphenylene	4.04	3.87	4.07	3.73[d]	3.83
Naphthacene	3.95	—	3.95	4.51[d]	—
Cyclopenta[cd]pyrene	—	—	—	3.97[d]	4.03
Benzo[ghi]fluoranthene	—	—	3.84	3.95[d]	4.07
3-Methylcholanthrene	4.75	—	4.31	4.78	5.23
Benzo[b]fluoranthene	—	—	4.48	4.29	4.46
Benzo[j]fluoranthene	—	—	4.56	4.24	4.37
Benzo[k]fluoranthene	4.42	—	4.45	4.42	4.52
Benzo[a]pyrene:	4.11	—	4.38	4.53	4.68
6-Methylbenzo[a]pyrene	—	—	4.55	4.67	5.03
Benzo[e]pyrene	4.18	—	4.46	4.28	4.48

(continued)

Table 1 (continued)

Compound	Normal-phase[a]			Reverse-phase C_{18}	
	Silica [12]	Alumina [11]	NH_2 column [14,33]	Polymeric[b] [14,33,40]	Monomeric[c] [14,33,40]
Perylene	4.20	—	4.61	4.33	4.50
Benzo[ghi]perylene	4.21	—	4.83	4.73	5.16
Anthanthrene	4.34	—	4.80	4.93	5.38
Indeno[1,2,3-cd]pyrene	4.45	—	4.90	4.83	5.13
Dibenz[a,c]anthracene	4.93	—	4.93	4.40[d]	4.73
Dibenz[a,j]anthracene	—	—	—	4.51[d]	4.82
Dibenz[a,h]anthracene	—	—	4.94	4.72[d]	4.85
Dibenzo[a,e]fluoranthene	—	—	—	4.93[d]	—
Benzo[b]chrysene	5.00	—	5.00	5.00	5.00
Picene	5.07	—	5.03	5.10	5.31
Dibenzo[def,p]chrysene	5.07	—	>5	4.92	>5
Naphtho[1,2,3,4-def]chrysene	5.04	—	>5	>5	>5
Benzo[rst]pentaphene	4.10	—	>5	>5	>5
Biphenyl	2.92	—	2.16	2.37	2.38
o-Terphenyl	4.46	—	2.50	2.97[d]	3.31
m-Terphenyl	4.50	—	3.12	3.16[d]	3.56
p-Terphenyl	4.57	—	3.28	3.74[d]	3.75
m-Quaterphenyl	5.98	—	4.06	3.81[d]	4.49

p-Quaterphenyl	—	—	4.50	5.04[d]	5.48
m-Quinquephenyl	—	—	5.00	4.25[d]	5.29
1,1'-Binaphthyl	—	—	2.99	3.35[d]	3.92
1,2'-Binaphthyl	—	—	3.29	3.56[d]	4.13
2,2'-Binaphthyl	—	—	4.01	4.05[d]	4.23
9,9'-Biphenanthryl	—	—	4.64	4.19[d]	5.30
9,9'-Bianthryl	—	—	4.30	3.88[d]	4.44
9,9-Bifluorenyl	4.02	—	—	3.52[d]	4.04
9-Phenylanthracene	4.02	—	2.97	3.55[d]	4.21
9,10-Diphenylanthracene	5.13	—	3.07	4.08[d]	5.41
1-Phenylphenanthrene	—	—	3.49	3.65[d]	4.23
9-Phenylphenanthrene	—	—	3.40	3.58[d]	4.20
9-Methyl-10-phenylphenanthrene	—	—	3.27	3.81[d]	4.56

[a] μBondapak NH_2 column, n-hexane as mobile phase; silica and alumina/n-pentane.
[b] Vydac 201TP column, 85% acetonitrile in water as mobile phase except for PAH with log I < 3.00.
[c] Zorbax ODS column, 80% acetonitrile in water as mobile phase.
[d] Log I determined on a Vydac 201TP column from different lot.

naphthalene (2), phenanthrene (3), benz[a]anthracene (4), and benzo[b]-chrysene (5). Thus, a PAH with a log I value of 3.50 elutes between phenanthrene and benz[a]anthracene. Lee et al. [16] recently described a similar system for GC using PAH as the standards rather than n-alkanes as in the traditional Kovats retention index system.

A wide variety of different column packing materials from different manufacturers are available for each of these types of LC. The physical characteristics and availability of these materials have been summarized by Majors [17] and Snyder and Kirkland [1].

A. Liquid-Solid Chromatography

Liquid-solid (or adsorption) chromatography on the classical adsorbents, i.e., silica and alumina, is the oldest of the various LC modes. These adsorbents have traditionally been employed for the isolation and separation of PAH using open-column chromatography. In addition, PAH have often served as model solutes for study of the mechanisms of adsorption chromatography. Early studies using large particle adsorbents to study the adsorption energies of PAH are typified by the work of Klemm et al. [18] and Snyder [19,20]. An excellent summary of the early work in adsorption chromatography is the book by Snyder [21].

In the adsorption process, the size and shape of the PAH influences the retention by determining the number of sites capable of interacting with the solute. The influence of the molecular structure of PAH on their retention on silica and alumina has been extensively studied by Popl et al. [11,12,22]. In adsorption chromatography the retention of PAH generally increases with increasing number of aromatic rings or the number of aromatic carbon atoms as illustrated in Fig. 1. The retention data of Popl et al. [11,12] for numerous PAH on silica and alumina are summarized in Table 1. The retention of PAH on silica and alumina is somewhat different due to the nature of the adsorbents. The active sites on silica are hydroxyl groups which interact with the electron donor solute to form hydrogen bridges, whereas surface fields provide the active sites on alumina.

Snyder [21] proposed the following empirical relationship for the adsorption energies, S°, of PAH (determined from the retention volume);

$$S° = 0.31 nC_a \qquad (2)$$

where nC_a is the number of aromatic carbon atoms in the molecule. Popl [11] found this relationship suitable for planar compounds of one to three aromatic rings; however, deviations toward higher adsorption energies were observed as the number of rings increased. Deviations toward lower energies were found for nonplanar molecules such as terphenyls and benzo[c]phenanthrene. Snyder [21] describes the preferential adsorption of linear or near linear PAH on alumina as a consequence of "weak localization." Because of the crystalline structure of alumina, the adsorptive sites are arranged in a regular and linear fashion; thus the PAH tend to line up along the surface. This phenomenon does not occur on silica, which explains some of the differences in retention on the two adsorbents.

Alkyl-substituted PAH exhibit somewhat different behavior on silica and alumina [11,22]. In the alumina-pentane system, the addition of an alkyl

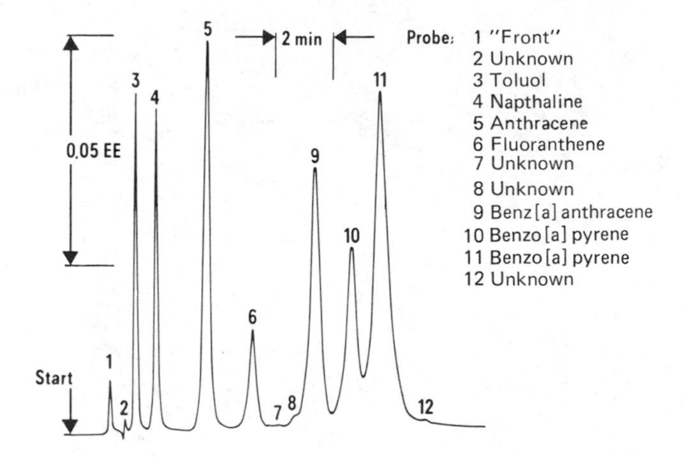

Figure 1. Normal-phase separation of a mixture of PAH on silica. Column: Spherosil XOA 400, 4-8 μm, 300 mm × 3 mm i.d. Mobile phase: isooctane at 1.4 ml/min. UV detection at 250 nm. (From Ref. 133.)

group generally increases the retention in a linear manner with increasing chain length of the n-alkyl group up to C_8-C_{18}. Methyl groups are an exception with greater retention than ethyl and propyl groups. Branching of an alkyl group decreases adsorption energy presumably due to steric hinderance. On silica gel, methyl substitution results in an increase in retention, but increasing the alkyl chain tends to decrease the retention. Branching of the alkyl chain was found to have virtually no effect on the retention. The position of alkyl substitution was observed to have a greater effect on adsorption on alumina than on silica [22]. Popl explained the differing behavior of alkyl PAH on these two adsorbents as a result of the different surface interactions.

Traditionally, adsorption chromatography has been employed extensively for sample cleanup and isolation of PAH fractions. Recently, however, chemically modified silicas have been used to achieve more selective and efficient separations. Silica and alumina are relatively inexpensive when compared to these other packing materials and therefore will continue to find applications in PAH analyses.

B. Reverse-Phase Liquid Chromatography

Reverse-phase packing materials generally consist of hydrocarbons with chain lengths of C_2, C_8, or C_{18} chemically bonded to silica particles. The C_{18} (octadecyl) supports are by far the most popular for the separation of PAH. In reverse-phase LC the mobile phase, which is more polar than the stationary phase, generally consists of mixtures of water with an organic solvent.

In 1971, Schmit et al. [23] first described the use of a chemically bonded octadecyl silicone for the separation of PAH. This separation, shown in Fig. 2, illustrates the excellent selectivity of reverse-phase LC for PAH isomers, i.e., phenanthrene/anthracene, fluoranthene/pyrene, and

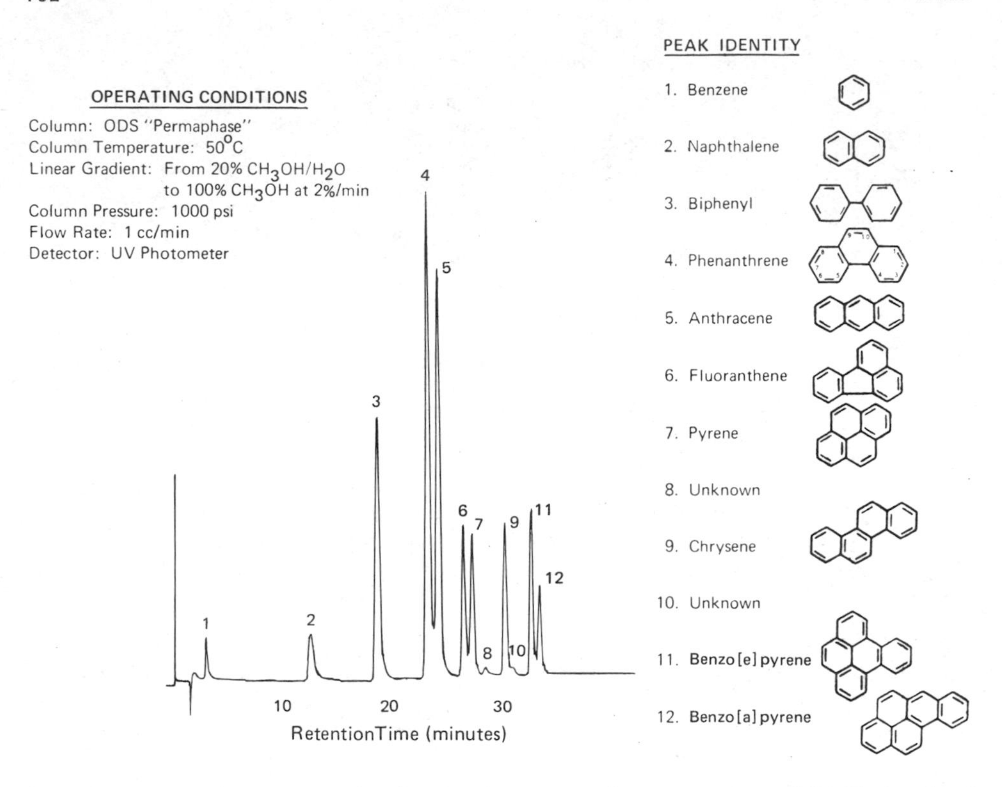

Figure 2. Reverse-phase liquid chromatographic separation of PAH standards. Column: ODS Permaphase. Mobile phase: 20% CH_3OH/H_2O to 100% CH_3OH at 2%/min at 1 ml/min. Column temperature: 50°C. (From Ref. 23, by permission of Preston Publications, Inc.)

benzo[e]pyrene/benzo[a]pyrene. The chromatographic support employed for this separation was a controlled surface-porosity support of 37-44 μm. This type of support, known as a "pellicular" or "superficially porous" support, consists of spherical siliceous particles with a porous surface of controlled thickness and pore size, thereby providing favorable mass transfer properties to achieve moderately high column efficiencies at moderate inlet pressures. However, due to the sample size limitations of these packings and later the availability of higher efficiency microparticulate supports, the pellicular packings currently have only limited use in modern LC.

In the early 1970s improvements in column efficiencies were achieved using totally porous microparticulate supports of 5-10 μm. In 1975 Wheals and coworkers [24] compared pellicular and microparticulate C_{18} packings and reported approximately an order of magnitude improvement in plate heights for the microparticulate over the pellicular packings. These microparticulate packings now typically provide columns of 60,000 plates per meter or more.

Reverse-phase LC provides unique selectivity for the separation of PAH isomers that are often difficult to separate by other modes of LC and often even capillary column GC. In addition, the compatibility of reverse-phase

LC with gradient elution techniques and the rapid equilibration of these columns to new mobile phase compositions make reverse-phase LC a convenient separation technique. As a result, reverse-phase LC is by far the most popular LC mode for analytical separations of PAH. The separation of a number of PAH on a 10-μm microparticulate column is shown in Fig. 3.

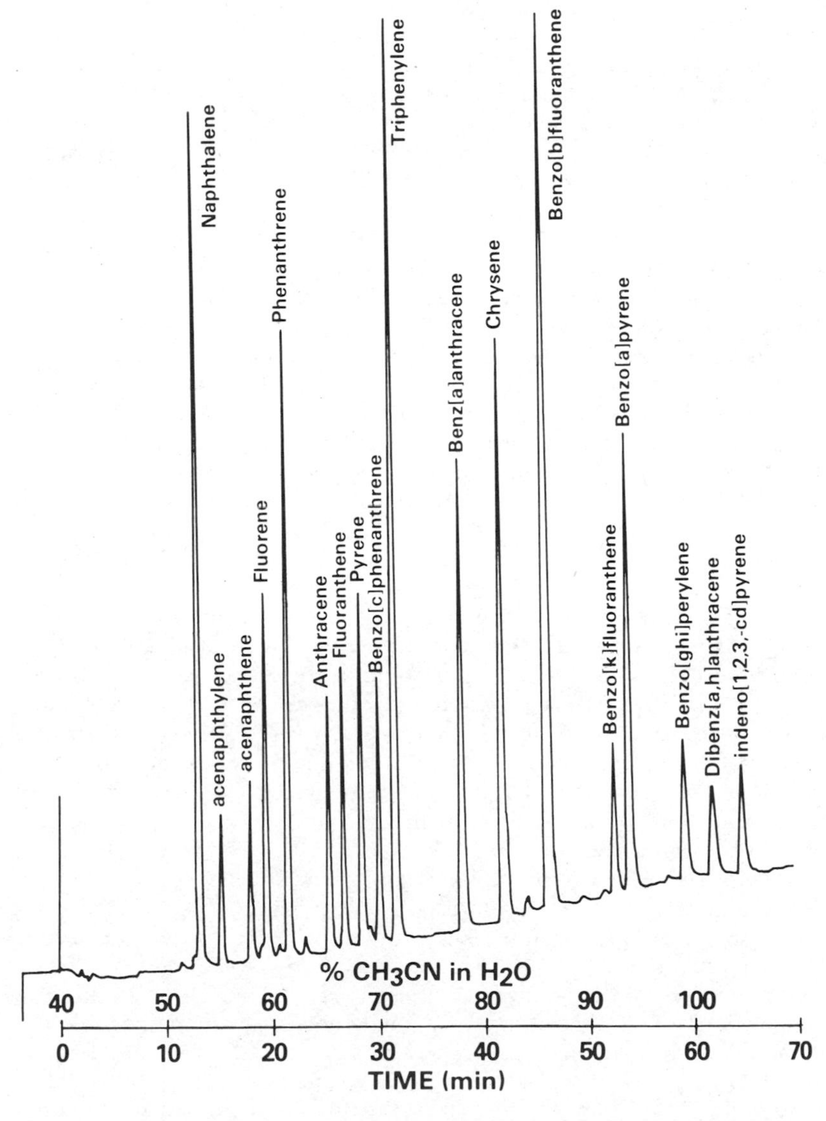

Figure 3. Reverse-phase HPLC separation of priority pollutant PAH. Column: Vydac 201TP. Mobile phase: linear gradient from 40 to 100% CH_3CN in H_2O at 1%/min. at 1 ml/min. UV detection at 254 nm. (From Ref. 14.)

1. Column Selectivity Differences

The majority of the HPLC separations of PAH reported in the literature have employed reverse-phase C_{18} columns. Several of these papers have included retention data for a number of PAH on specific C_{18} columns [13-15,25,26]. A comparison of these retention data and recent reports by Wise et al. [14], Ogan and Katz [27], and Colmsjö and MacDonald [28] indicate that C_{18} columns from different manufacturers provide not only different separation efficiencies but different retention characteristics and selectivities for PAH.

The selectivity differences for a number of PAH [14 of which are on the U.S. Environmental Protection Agency (EPA) Priority Pollutant list] on seven columns from different manufacturers are summarized in Table 2. These retention data are presented as the log of the retention index I, as described earlier in Eq. (1). For nearly complete resolution of two components on these C_{18} columns ($R_S = 1.0$), the log I values must differ by approximately 0.07 units depending on the efficiency of the particular column. The data in Table 2 indicate that, of the columns evaluated, only the Vydac 201TP column was successful in partially resolving all of the 16 PAH. Ogan et al. [29] recently reviewed the literature regarding the separations of three groups of PAH isomers that have traditionally been difficult to separate, i.e., (1) benz[a]anthracene and chrysene; (2) benzo[k]fluoranthene, benzo[b]-fluoranthene, benzo[a]pyrene, and benzo[e]pyrene; and (3) benzo[ghi]-perylene and indeno[1,2,3-cd]pyrene. They described an LC method to separate all of these isomers using an HC-ODS column [29]. The HC-ODS columns contain "selected" lots of material obtained from the manufacturer of Vydac 201TP. Ogan and Katz [27] later described the selectivity differences for seven difficult-to-separate PAH on eight different C_{18} columns (four of which were not included in the study by Wise et al. [14]). The liquid chromatograms for five of these columns obtained using identical mobile phase compositions are shown in Fig. 4. The k' values of each of the PAH were plotted as a function of the k' value for benzo[a]pyrene on each column (Fig. 5). By plotting the data in this manner, Ogan and Katz empirically ordered the data as a function of some pertinent characteristic(s) of these octadecyl silicas. Thus, if benzo[a]pyrene exhibits increased retention on one column relative to another column, the other PAH should also exhibit an increase. This general trend is evident for all the columns except the Vydac 201TP and the HC-ODS. Ogan and Katz [27] indicated that if changes in the retention of the PAH on the different columns were exactly proportional to the changes in retention of benzo[a]pyrene, i.e., if the retention of each of these compounds were linear functions of the same characteristic(s) of the bonded-phase material, then the data points for a given compound would be on a straight line passing through the origin of Fig. 5. Such lines, indicated by the dashed lines in Fig. 5, represent constant selectivity factors relative to benzo[a]pyrene. The actual data points do not fall on these lines, indicating differences in the selectivity for these compounds on these columns. Benz[a]anthracene and chrysene were completely separated only on these columns. However, as shown in Table 2, the Partisil 5-ODS and MicroPak CH-10 are also capable of nearly complete resolution of these isomers. The selectivity of benzo[ghi]perylene relative to indeno[1,2,3-cd]pyrene varies from column to column.

Table 2. Column Selectivity for Reverse-Phase Separation of PAH

	Partisil[b] 5-ODS	MicroPak[c] MCH-10	Nucleosil[c] 10 C_{18}	Zorbax[b] ODS	LiChrosorb[b] RP-18	MicroPak[b] CH-10	Vydac[a] 201TP
Naphthalene	2.00	2.00	2.00	2.00	2.00	2.00	2.00
Fluorene	2.76	2.70	2.71	2.74	2.73	2.70	2.73
Phenanthrene	3.00	3.00	3.00	3.00	3.00	3.00	3.00
Anthracene	3.11	3.12	3.11	3.14	3.14	3.19	3.24
Fluoranthene	3.45	3.43	3.42	3.44	3.42	3.44	3.38
Pyrene	3.61	3.60	3.59	3.66	3.62	3.69	3.56
Chrysene	3.96	3.98	3.98	3.99	4.00	4.04	4.10
Benz[a]anthracene	4.00	4.00	4.00	4.00	4.00	4.00	4.00
Benzo[b]fluoranthene	4.47	4.47	4.47	4.45	4.40	4.41	4.30
Benzo[k]fluoranthene	4.53	4.48	4.52	4.52	4.48	4.49	4.45
Benzo[e]pyrene	4.48	4.50	4.50	4.48	4.40	4.43	4.25
Benzo[a]pyrene	4.64	4.65	4.64	4.68	4.63	4.66	4.52
Dibenz[a,h]anthracene	4.92	4.89	4.90	4.85	4.78	4.74	4.69
Benzo[b]chrysene	5.00	5.00	5.00	5.00	5.00	5.00	5.00
Indeno[1,2,3-cd]pyrene	5.13	5.14	5.09	5.13	5.03	5.04	4.83
Benzo[ghi]perylene	5.14	5.18	5.17	5.16	5.05	5.05	4.71

[a]Mobile phase: 90% acetonitrile in water.
[b]Mobile phase: 80% acetonitrile in water.
[c]Mobile phase: 70% acetonitrile in water.
Source: Ref. 14.

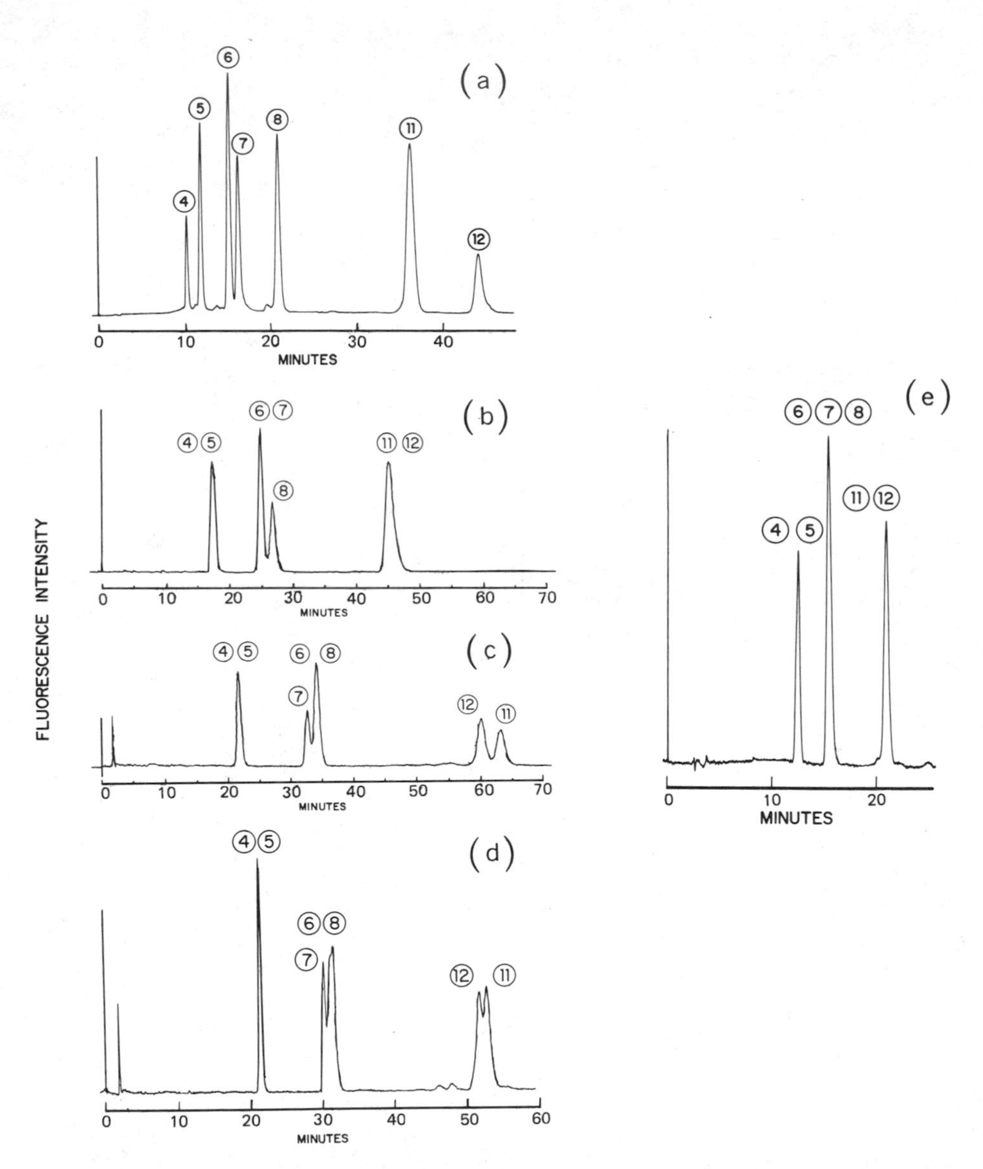

Figure 4. Comparison of liquid chromatographic separation of seven PAH on five different commercial C_{18} columns. Columns: (a) HC-ODS; (b) LiChrosorb RP-18; (c) Partisil ODS-2; (d) Zorbax ODS; (e) μBondapak C_{18}. Detection: fluorescence at λ_{ex} = 305 nm and λ_{em} = 430 nm. Mobile phase: CH_3CN/H_2O (80:20). PAH standards: (4) benz[a]anthracene; (5) chrysene; (6) benzo[e]pyrene; (7) benzo[b]fluoranthene; (8) benzo[k]fluoranthene; (11) benzo[ghi]perylene; and (12) indeno[1,2,3-cd]pyrene. (From Ref. 27.)

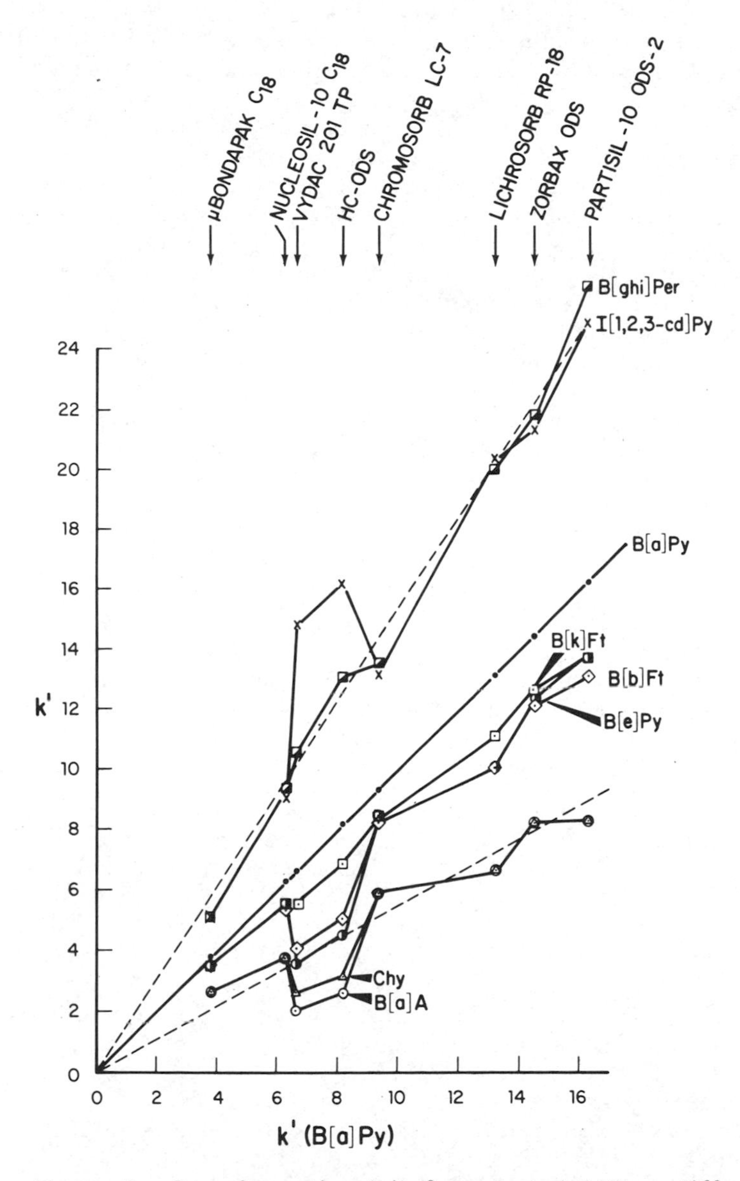

Figure 5. Capacity ratios (k') for selected PAH on different commercial C_{18} columns as a function of k' of benzo[a]pyrene. Conditions: same as for Fig. 4. (From Ref. 27.)

Selectivity factors for several PAH pairs from the data in Table 2 are summarized in Table 3. The differences in chromatographic retention and selectivity, described in the previously mentioned studies [14,27,28], are a result of the utilization of different silica materials as supports and a variety of reagents and procedures to produce the C_{18} bonded phase. Several recent reports have reviewed the state of the art of chemically bonded alkyl reverse-phase packings [30-32].

Some of the physical and chromatographic characteristics of seven C_{18} materials are compared in Table 4 [33]. A number of workers have examined the parameters which affect the chromatographic characteristics of alkyl bonded materials, i.e., carbon content, length of the alkyl chain, surface area, and unreacted silanol groups [32,34-36].

In general, the capacity ratio, k' increases with increasing carbon content as shown in Table 4. However, as pointed out by others [27], carbon content alone is often misleading due to differences in the surface area of the silica, resulting in different surface coverages. Ogan and Katz [27] utilized carbon content and specific surface area of the underivatized silica to calculate the surface coverage of different C_{18} materials but observed little correlation between k' data for benzo[a]pyrene and the surface coverage of the C_{18} material; they suggested that the use of the surface area of the chemically modified silica would provide a more accurate value for surface coverage. Table 4 compares surface coverages calculated from the specific surface area of the bonded C_{18} materials and the measured carbon contents rather than the literature values. It is interesting to note the significant reductions in surface area (20-70%) of the silica support after bonding of the C_{18} layer, presumably due to blocking of the silica pores by the hydrocarbon layer. As shown in Table 4, the capacity ratios for benzo[a]pyrene and benzo[b]-chrysene generally increase with increasing surface coverage.

2. *Comparison of Monomeric vs. Polymeric C_{18} Stationary Phases*

An important characteristic of the C_{18} bonded materials is the nature of the organic layer attached to the silica surface, i.e., monomeric or polymeric, which depends on the silane reagent and the reaction conditions employed for the reactions [30]. The majority of the commercially available C_{18} materials are monomeric in nature according to their manufacturers. Several researchers have determined that the maximum C_{18} monomeric surface coverage for silica is 3.4 $\mu mol/m^2$ [34-36]. Based on this value, the LiChrosorb RP-18, MicroPak CH-10, and Vydac 201TP materials definitely have polymeric layers. The Zorbax ODS material, described by the manufacturer as monomeric, appears to have the maximum monomeric coverage possible.

The monomeric C_{18} phases are generally produced by exhaustive reaction with monofunctional silanes, resulting in a high degree of reproducibility in surface coverage [30,31]. Reactions with di- or trifunctional silanes result in a polymeric layer which is often more difficult to reproduce. The non-reproducibility of surface coverage for different lots of a polymeric C_{18} packing from one manufacturer is illustrated in Table 5. Two monomeric C_{18} columns from different lots were found to have identical retention indices for the PAH listed in Table 2 [33].

Table 3. Selectivity Factors (α) for PAH on Different C_{18} Columns[a]

Compounds	Partisil 5-ODS	MicroPak MCH-10	Nucleosil 10 C_{18}	Zorbax ODS	LiChrosorb RP-18	MicroPak CH-10	Vydac 201TP
Anthracene/phenanthrene	1.07	1.10	1.08	1.11	1.12	1.22	1.42
Pyrene/fluoranthene	1.11	1.14	1.12	1.18	1.19	1.27	1.29
Benz[a]anthracene/pyrene	1.38	1.37	1.33	1.31	1.38	1.32	1.81
Chrysene/benz[a]anthracene	0.96	0.99	0.98	0.99	1.00	1.06	1.27
Benzo[e]pyrene/ benz[a]anthracene	1.35	1.53	1.49	1.53	1.55	1.72	1.77
Benzo[a]pyrene/ benzo[e]pyrene	1.11	1.14	1.12	1.18	1.29	1.34	1.89
Benzo[k]fluoranthene/ benzo[b]fluoranthene	1.03	1.01	1.04	1.05	1.10	1.12	1.40
Benzo[ghi]perylene/ dibenz[a,h]anthracene	1.17	1.27	1.24	1.32	1.32	1.40	1.05
Indeno[1,2,3-cd]pyrene/ benzo[b]chrysene	1.10	1.10	1.11	1.10	1.02	1.04	0.66

[a]Mobile phase conditions same as Table 2.
Source: Ref. 14.

Table 4. Reverse-Phase C_{18} Column Characteristics

Column	Surface area (m^2/g) Silica	Surface area (m^2/g) Bonded silica	Percent carbon	Surface conc. ($\mu mol/m^2$)	Normal-phase k' [a] nitrobenzene	Reverse-phase k' [b] BbC	Reverse-phase k' [b] BaP
Partisil 5-ODS	400	235	10.0	1.8	0.3	7.2	4.0
MicroPak MCH-10	400	259	12.2	2.0	5.6	8.1	—
Nucleosil 10 C_{18}	300	220	12.4	2.4	3.8	7.5	5.7[c]
Zorbax ODS	275-300	166	13.1	3.3	4.4	14	12.5[c]
LiChrosorb RP-18	300	149	19.6	5.5	1.2	14	13.2[c]
MicroPak CH-10	400	120	19.0	7.3	1.9	27	13.2
Vydac 201TP	100	57.7	8.2	6.6	3.0	32	—

[a] Mobile phase: n-hexane.
[b] Mobile phase: 80% acetonitrile in water.
[c] Data from Ref. 38.
Source: Ref. 33.

Table 5. Surface Characteristics of Different Lots of Polymeric C_{18} Material

Column	Surface area[c] (m^2/g)	Percent carbon	Surface coverage ($\mu mol/m^2$)
Vydac 201TP (Altex)[a]	73.6	7.7 ± 0.9	4.8 ± 0.6
Vydac 201TP (1382)[a]	64.6	7.5 ± 0.3	5.4 ± 0.2
Vydac 201TP (1390)[a]	57.7	8.2	6.6
Vydac 201TP (1402)[a]	81.2	8.3	4.7
HC-ODS Lot 25[b]	49.5	7.8	7.3
HC-ODS Lot 36[b]	50.8	9.0	8.2
HC-ODS Lot 38[b]	73.8	7.7	4.8

[a]Data from Ref. 33.
[b]Data from Ref. 37.
[c]Specific surface area determined by a modified, single-point B.E.T. method [S. Branauer, P. H. Emmett, and E. Teller, *J. Amer. Chem. Soc. 60*:309 (1938)] using nitrogen as the adsorbate.

Differences in selectivity are also evident for various lots of polymeric C_{18} material, as reported by Ogan et al. [27] and shown in Fig. 6. A wide range of α values is evident in Fig. 6; the range increases with increased retention, e.g., 11% SD (standard deviation) for anthracene and 23% for indeno[1,2,3-cd]pyrene. These differences in selectivities on different C_{18} columns appear to be related to the surface coverage of the C_{18} layer. The values for surface coverage on several lots of Vydac 201TP/HC-ODS material vary from 4.7 to 8.2 $\mu mol/m^2$ [33,37]. Katz and Ogan [38] compared the selectivity ratios α for several PAH on several different lots of HC-ODS material and on several other C_{18} materials; their work showed that HC-ODS materials with low surface coverage (4.8 $\mu mol/m^2$) exhibited selectivity ratios for several PAH pairs which were similar to all the other C_{18} columns (even the polymeric LiChrosorb RP-18 and Partisil ODS-2) whereas the HC-ODS materials with high surface coverage (8.2 $\mu mol/m^2$) exhibited the most suitable selectivity for their separations. As a result of Katz and Ogan's studies of selectivity, they selected lots of HC-ODS material that have certain selectivity characteristics for columns designed especially for PAH separations [39].

Monomeric and polymeric C_{18} phases exhibit somewhat different selectivities for several of the PAH. The retention characteristics of more than 100 PAH on both monomeric (Zorbax ODS) and polymeric (Vydac 201TP) C_{18} materials are summarized in Table 1. Several interesting trends are observed in Tables 2 and 3 for the monomeric vs. polymeric materials. (In both tables the different columns are arranged in order of increasing surface coverage except for the Vydac 201TP which exhibits unique behavior.) The separation of chrysene and benz[a]anthracene varies with the nature of the C_{18} layer. For the monomeric materials chrysene elutes prior to benz[a]anthracene with the greatest separation for the material with the lowest surface coverage. As the surface coverage approaches polymeric, these two compounds elute closer

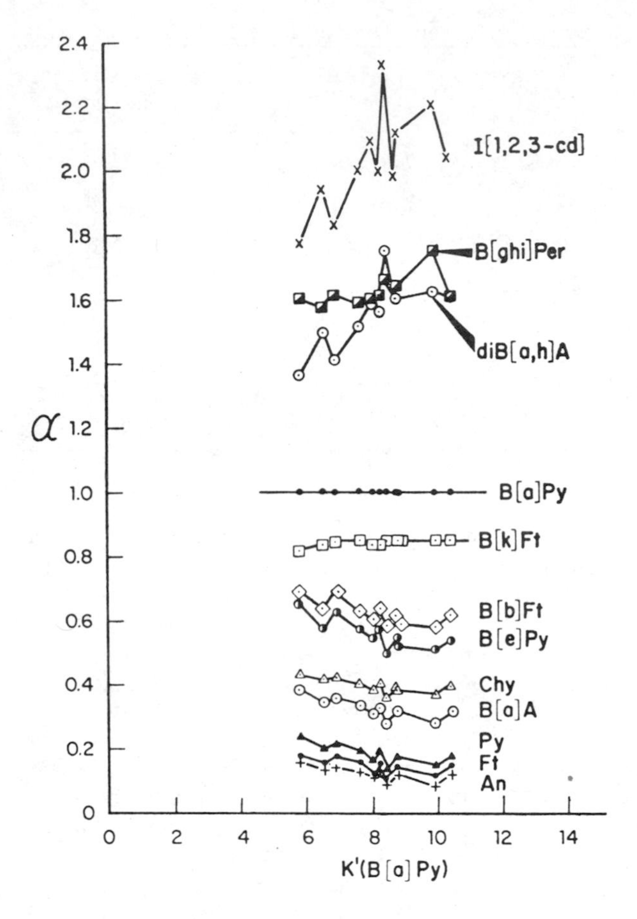

Figure 6. Selectivity factors (α), relative to benzo[a]pyrene, plotted as a function of k' of benzo[a]pyrene for different manufacturing lots of C_{18} silica used in HC-ODS columns. (From Ref. 27.)

together until they reverse elution order on the polymeric materials. The unique behavior of the Vydac 201TP material with respect to the elution order of benzo[b]chrysene, indeno[1,2,3-cd]pyrene, and benzo[ghi]perylene does not appear to be related solely to the polymeric nature of the C_{18} layer in view of the fact that the other two polymeric materials have elution orders similar to the monomeric materials. The polyphenyl arenes exhibit significantly different retention characteristics on monomeric vs. polymeric C_{18} materials. The polyphenyl arenes are generally retained longer, relative to the fused ring retention index standards, on the monomeric than on the polymeric materials; e.g., m-quaterphenyl, 1,2'-binaphthyl, 2,2'-binaphthyl, 9,9'-bianthryl, 9-phenylanthracene, 1-phenylanthracene, and 9-phenylphenanthrene all elute prior to benz[a]anthracene on the polymeric material but after it on the monomeric material.

The unusual behavior of the polyphenyl arene 9,10-diphenylanthracene on different C_{18} columns is illustrated in Fig. 7 for the various C_{18} columns compared by Katz and Ogan [38] and Wise et al. [14]. On the monomeric columns 9,10-diphenylanthracene elutes after benzo[a]pyrene and dibenz[a,h]anthracene, whereas on the polymeric materials it elutes prior to both benzo[a]pyrene and dibenz[a,h]anthracene. A significant change in the elution behavior of 9-phenylanthracene (a phenyl arene) on monomeric and polymeric materials is also observed. Smaller changes are exhibited for anthracene, 2-methylanthracene and dibenz[a,c]anthracene.

Elution order reversal, presumably due to nonplanarity, is observed for triphenylene and benzo[c]phenanthrene on monomeric and polymeric C_{18} materials [40] (see Table 1). The retention data for PAH on silica (see Table 1) indicate that polyphenyl arenes exhibit greater retention than corresponding fused ring PAH with equivalent number of rings. The increased retention of polyphenyl arenes and nonplanar PAH may indicate a greater contribution of adsorption on the C_{18} layer for the monomeric packings [40].

3. *Reverse-Phase Retention Mechanism*

The mechanism of retention on these chemically bonded C_{18} phases has not been established and is the topic of much discussion and research. In a recent review of chemically bonded phases, Colin and Guiochon [32] (see references therein) discussed the proposed mechanisms of retention in reverse-phase LC: (1) partition between two liquids (mobile phase and bonded phase), (2) adsorption on the nonpolar hydrocarbon bonded phase, and (3) partition of the solute between the mobile phase and a "mixed" stationary phase formed by adsorption of the organic modifier on the stationary phase. A detailed discussion of these mechanisms is beyond the scope of this chapter; however, several studies relating to the retention of PAH in particular will be described.

Sleight [41] studied structure-retention relationships of PAH on a C_{18} column and derived the following simple expression relating the retention to the number of carbons for unsubstituted PAH:

$$\log k' = a + bC_N \tag{3}$$

where k' = capacity ratio, C_N = number of carbon atoms, and a and b are constants. Blumer and Zander [25] measured k' values of a number of unsubstituted PAH and biaryls and correlated log k' with the number of carbon atoms. However, as indicated by the data in Table 1, most isomeric PAH have significantly different retention characteristics in reverse-phase HPLC to achieve separation (e.g., the kata-condensed four- and five-ring isomers; see Chap. 1).

Sleight [41] suggested that the PAH retention was dependent to some degree on their solubility in the polar mobile phase. Thus, alkyl substitution results in a significant increase in retention due to the decreased solubility in the polar mobile phase. Sleight [41] reported retention data for alkyl-substituted PAH which illustrate the increase in retention with increasing number of aliphatic carbon atoms in the alkyl group. The retention data in Table 1 for 9-methyl-, 9-ethyl-, and n-propylphenanthrene and 1-methyl-, 1-ethyl-, and 1-n-butylpyrene also illustrate the increased retention due to increased number of aliphatic carbon atoms.

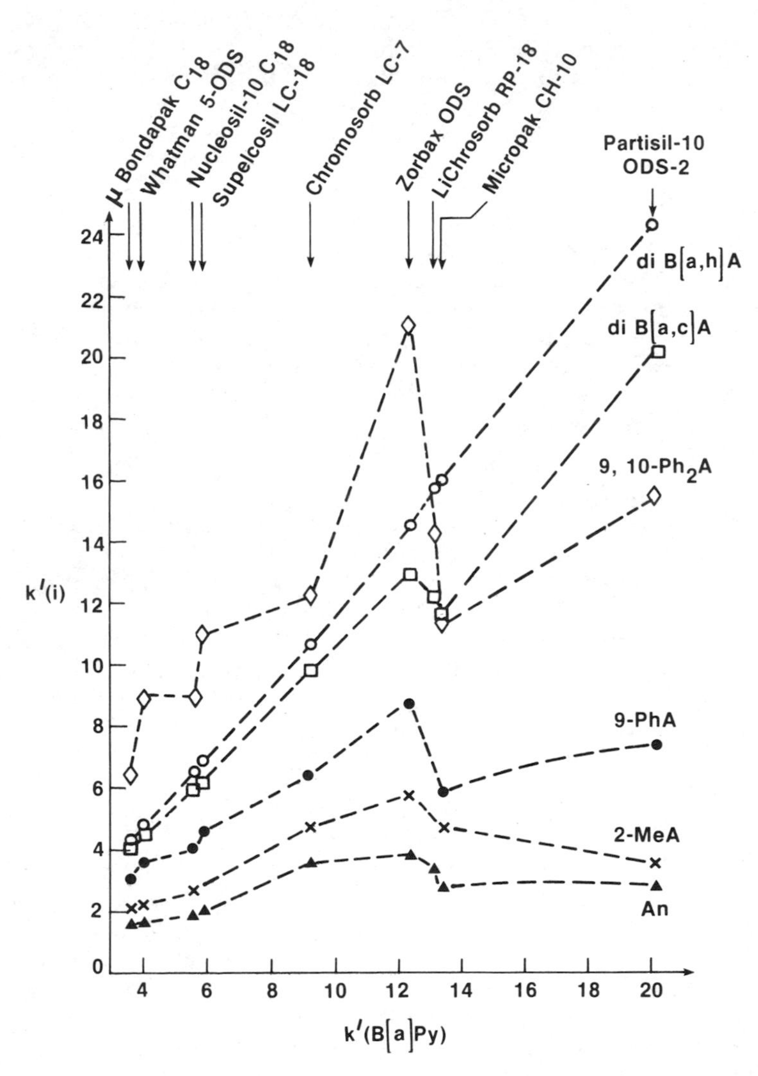

Figure 7. The k' values of several PAH on several C_{18} columns plotted as a function of k' of benzo[a]pyrene on the different columns. The index i refers to the PAH identified on the right: diB[a,h]A = dibenz[a,h]anthracene; diB[a,c]A = dibenz[a,c]anthracene; 9,10-Ph_2A = 9,10-diphenylanthracene; 9-PhA = 9-phenylanthracene; 2-MeA = 2-methylanthracene; An = anthracene. (Data from Refs. 33 and 38.)

Locke [42] also suggested that the basis of reverse-phase selectivity for PAH was the relative solubilities of the PAH in the mobile phase. Locke formulated the following postulates: selectivity is governed by the solubilities of similar solutes in the mobile phase; group selectivity (i.e., for a group of PAH with similar number of aromatic rings) is determined by both the stationary and mobile phases; and selectivity within a group is governed by the mobile phase. Since the solubilities of PAH are similar in organic solvents such as methanol and acetonitrile, Locke suggested that the selectivity differences are based on the aqueous solubility of the individual PAH [42]. Solubility data alone, however, do not adequately account for the reverse-phase chromatographic selectivity of many PAH [40]. A retention mechanism which completely describes the selectivity of PAH solutes in reverse-phase LC undoubtedly contains contributions from several factors.

Recently, Wise et al. [40] described a relationship between the shape of PAH solutes, particularly the length-to-breadth ratio, and the reverse-phase LC retention on C_{18} bonded phases. In this study the length-to-breadth ratios for more than 40 unsubstituted and 40 methyl-substituted PAH were compared with the LC retention characteristics. In nearly all cases, this ratio was successful in predicting the elution order of isomeric PAH, i.e., the retention increases with increasing length-to-breadth ratio. As an example, the length-to-breadth ratios and retention data for the kata-condensed four- and five-ring isomers and several methyl-substituted PAH are summarized in Table 6. The length-to-breadth ratios were found to be particularly useful in predicting the unique selectivity of reverse-phase LC for methyl-substituted PAH. The linear correlation for LC retention data [40] vs. length-to-breadth ratios for 23 methyl-substituted isomeric benzo[c]-phenanthrenes, benz[a]anthracenes, and chrysenes is illustrated in Fig. 8. This relationship suggests the importance of shape in predicting the unique selectivity on C_{18} columns of PAH and particularly methyl-substituted PAH. In GC a similar relationship exists between the shape and the retention on liquid crystal stationary phases [43], suggesting the possibility of a liquid crystal effect or "ordering" of the C_{18} phase in LC. The retention characteristics on the polymeric phases exhibit greater selectivity and a greater correlation with the shape data than on the monomeric phases [40]. This observation indicates that the polymeric materials may be more "ordered" than the monomeric materials.

4. *Effect of the Mobile Phase Composition*

Differences in retention and selectivity in reverse-phase LC can also be achieved by variation of the mobile phase composition. In reverse-phase systems the mobile phase is generally a mixture of water with various organic solvents such as methanol or acetonitrile. Retention is decreased by increasing the proportion of organic solvent in the mobile phase. Gradient elution techniques—increasing the organic composition of the mobile phase with time—are extremely useful for the analysis of PAH mixtures with a wide range of k' values.

Sleight [41] reported variation in the retention of alkyl-substituted and unsubstituted PAH with different mobile phase compositions. More recently, Katz and Ogan [44] have studied the effect of the mobile phase composition

Table 6. Comparison of Length-to-Breadth (L/B) Ratios and LC Retention Indices for PAH

Unsubstituted PAH	L/B	LC retention (log I)
Triphenylene	1.12	3.70
Benzo[c]phenanthrene	1.22	3.64
Benz[a]anthracene	1.58	4.00
Chrysene	1.72	4.10
Naphthacene	1.89	4.51
Dibenz[a,c]anthracene	1.24	4.40
Dibenz[a,j]anthracene	1.47	4.51
Dibenz[a,h]anthracene	1.79	4.72
Benzo[b]chrysene	1.84	5.00
Picene	1.99	5.10

Methyl-substituted PAH	L/B	LC retention (log I)
Phenanthrene:		
2-Methyl	1.58	3.71
1-Methyl	1.45	3.39
3-Methyl	1.37	3.29
9-Methyl	1.25	3.34
4-Methyl	1.25	3.26
Anthracene:		
2-Methyl	1.74	3.71
1-Methyl	1.41	3.41
9-Methyl	1.38	3.41
Pyrene:		
2-Methyl	1.37	4.05
1-Methyl	1.27	3.98
4-Methyl	1.10	3.96

Source: Ref. 40.

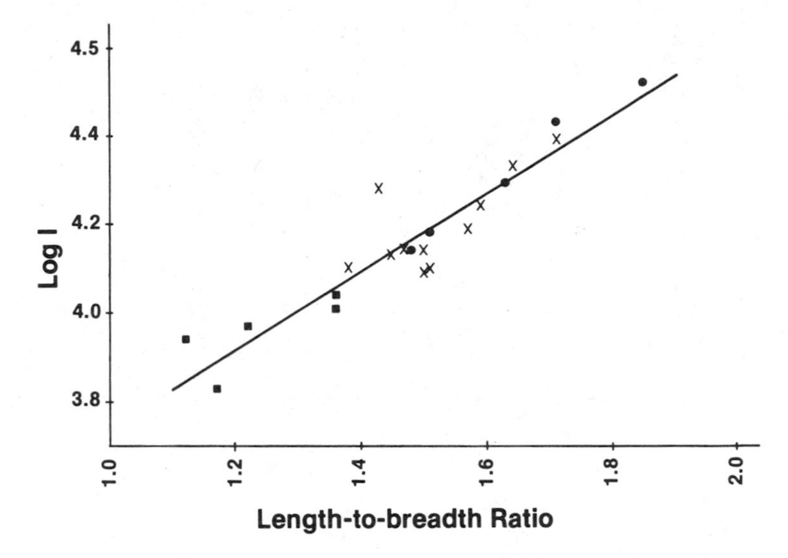

Figure 8. Linear correlation of reverse-phase LC retention (Log I) vs. length-to-breadth ratio for methylbenzo[c]phenanthrenes (■), methylbenz[a]-anthracenes (×), and methylchrysenes (●). (From Ref. 40.)

on the selectivity factors for PAH on different C_{18} columns. Plots of the selectivity factors for several PAH as a function of acetronitrile content of the mobile phase are shown in Fig. 9 for LiChrosorb RP-18 (polymeric), Zorbax ODS (monomeric), and HC-ODS (polymeric). As shown in the figure, selected PAH can usually be separated on these different columns by appropriate selection of the mobile phase composition. These data also suggest the reason for many contradicting reports in the literature of different elution orders of PAH on similar columns.

Because of the dominance of solute-mobile phase interactions in reverse-phase LC, the use of different organic modifiers has a dramatic effect on capacity factors for PAH. The k' values for several PAH were increased by a factor of 5 by replacement of acetonitrile with methanol, and appreciable retention of PAH could be attained only with a mobile phase composition of 40% or less tetrahydrofuran in water [44]. The effect of different organic modifiers in the mobile phase on the selectivity factors is illustrated in Fig. 10.

Roumeliotis et al. [45] recently reported an optimization study on the mobile phase composition for the separation of several PAH on several reverse-phase columns. They evaluated various mixtures of water with methanol, n-propanol, and dioxane. They concluded that a mobile phase composition of 65:35 (v/v) methanol/water was a reasonable compromise in relation to resolution and analysis time and that the other mixtures offered no significant improvement in selectivity. It should be noted, however, that at this mobile phase composition many PAH of interest would have extremely large k' values, especially on such polymeric materials as the Vydac 201TP/HC-ODS.

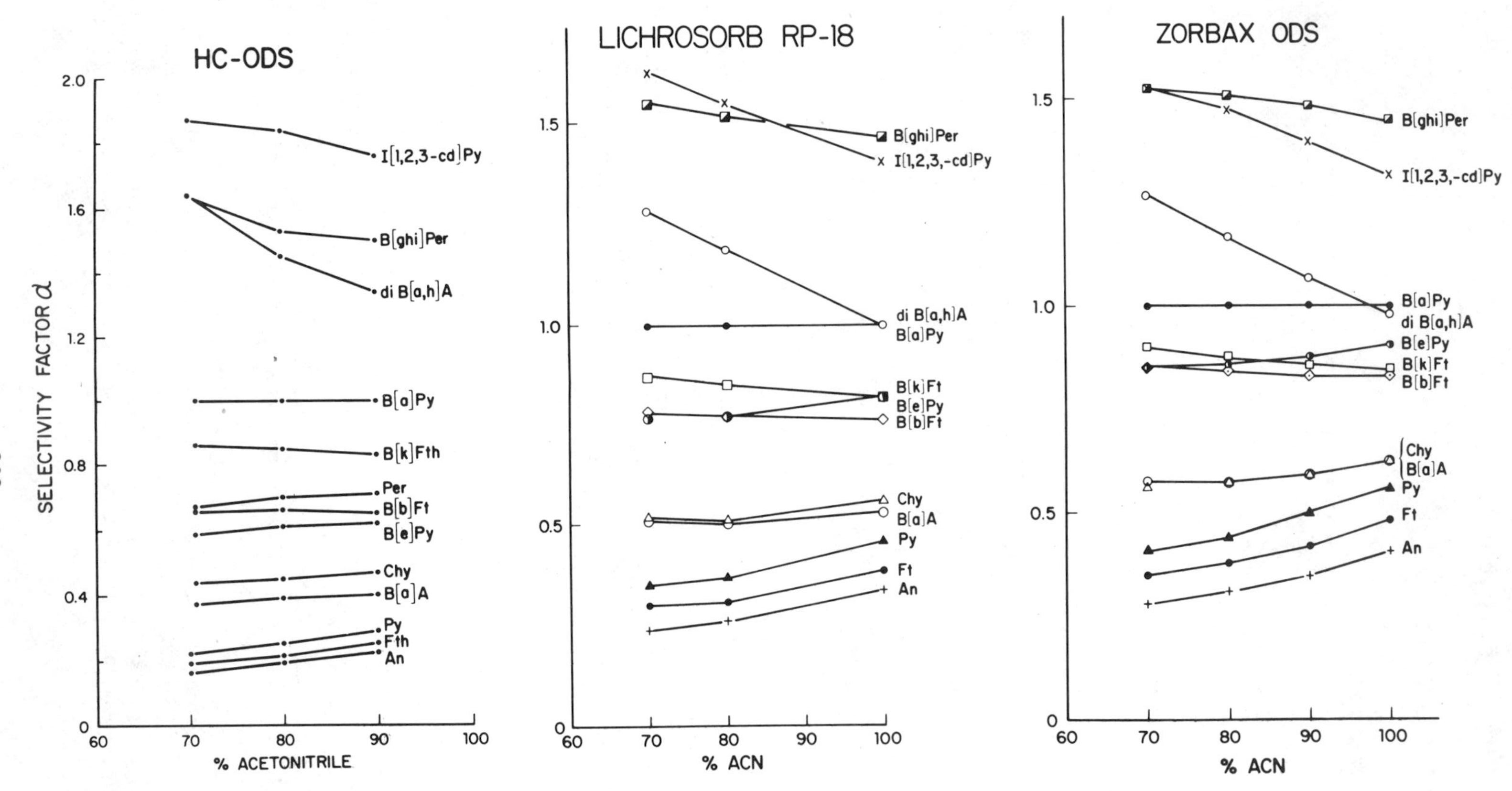

Figure 9. Selectivity factor α, relative to benzo[a]pyrene, as a function of acetonitrile content of the mobile phase (isocratic elution) for three C_{18} columns: HC-ODS, LiChrosorb RP-18, and Zorbax ODS. (From Ref. 44.)

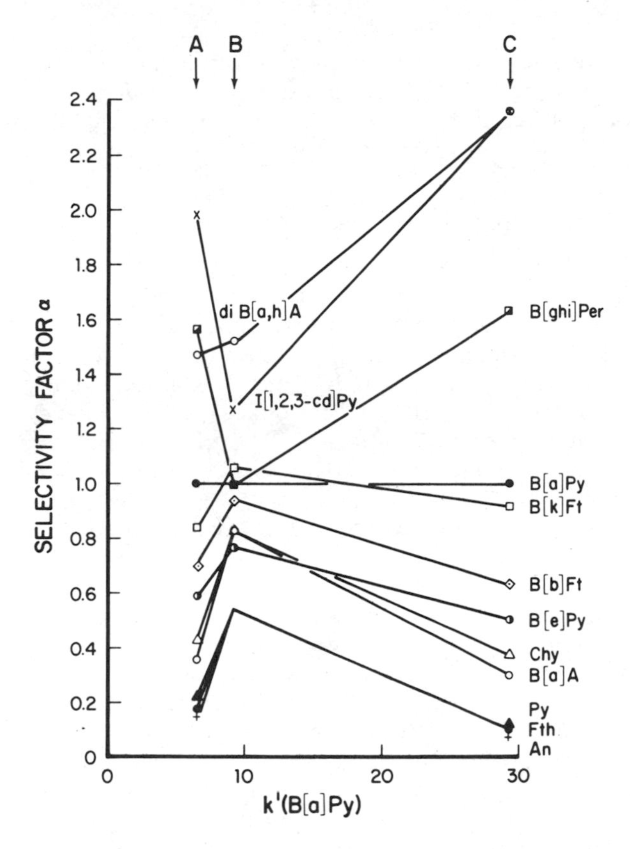

Figure 10. Selectivity factors for individual PAH relative to benzo[a]pyrene as a function of k' for benzo[a]pyrene with three different mobile phase compositions: (A) 80% CH_3CN in H_2O; (B) 35% tetrahydrofuran in H_2O; and (C) 80% CH_3OH in H_2O. Column: HC-ODS. (From Ref. 44.)

5. *Effect of Temperature on Selectivity*

In the early work of Schmit et al. [23] with chemically bonded C_{18} phases, these investigators described the effect of column temperature on solute retention and column efficiency. An increase in temperature resulted in decreased retention and increased column efficiency. Recently, Chmieloweic and Sawatzky [46] demonstrated the use of temperature as a separation parameter for the liquid chromatographic separation of PAH. The retention characteristics of a number of PAH on a C_{18} phase at various column temperatures were reported. Changes in the elution sequence with temperature changes were observed. They suggested that temperature changes might be substituted for changes in the mobile phase composition. Chmielowiec and Sawatzky [46] found that, with an increase in temperature, retention of the more compact PAH (e.g., benz[a]anthracene) decreased at a greater rate than that of the less compact compounds (e.g., polyphenyl arenes). These differences resulted in reversals of the elution sequence with temperature changes. Snyder [47] suggested that these differences also correlated with the differences in molecular structure and with the ordered nature of the stationary phase. Snyder described the "irregularity" of the polyphenyl arenes as a departure from flat, straight molecules vs. bulky three-dimensional or spherical shape.

C. Normal-Phase Liquid Chromatography on Polar Chemically Bonded Stationary Phases

Polar chemically bonded stationary phases, employed in conjunction with nonpolar mobile phases, are useful for the separation of PAH. Several polar phases are available containing such functional groups as amine (NH_2), diamine [$R(NH_2)_2$], nitrile (CN), diol [$R(OH)_2$], ether (ROR), and nitrophenyl (NO_2) bonded to the silica particles. When a nonpolar mobile phase (e.g., hexane, heptane, or isooctane) is used with these polar phases, PAH separations are achieved similar to those obtained on the classical adsorbents such as alumina and silica, i.e., saturated hydrocarbons elute prior to olefinic and aromatic hydrocarbons and the retention of PAH increases with the number of condensed aromatic rings.

Extending the work of Hupe and Schrenker [48] in liquid-liquid LC, Novotny et al. [49] employed a bonded oxydipropionitrile phase for the fractionation of complex PAH mixtures prior to GC/MS analysis. Novotny [50] also reported the separation of several PAH standards on a polar aminosilane column with a hexane mobile phase. Recently, the application of the polar aminosilane phase for normal-phase LC separations of PAH has been described by Wise et al. [13]. The liquid chromatogram in Fig. 11 demonstrates the normal-phase separation of several PAH on an aminosilane column. The retention characteristics of over 90 PAH on this column, as reported by Wise et al. [14], are summarized in Table 1.

On the aminosilane column the retention of PAH increases with increasing number of condensed aromatic rings (or number of aromatic carbon atoms). Other polar bonded phases appear to have similar retention characteristics, e.g., nitrophenyl [25,51], nitrile, diol, and diamine [15]. The presence of an alkyl group, particularly a methyl group, has only a slight effect on the retention on an amine phase (see Fig. 11). In contrast, with classical adsorbents the addition of an alkyl group generally increases the retention significantly (e.g., see Table 1; compare the retention of the alkylphenanthrenes with phenanthrene on silica and alumina with retention on the amine column). Normal-phase LC on this phase provides a more distinct class separation, i.e., all PAH with three rings, including alkyl-substituted members, have similar retention. This characteristic is often advantageous in prefractionation techniques for the analysis of complex PAH mixtures, as will be described later. The use of normal-phase LC on polar bonded-phases also eliminates one of the major difficulties encountered with classical adsorbents, i.e., nonreproducible retention due to small changes in the moisture content of the eluent. Finally, the use of volatile nonpolar mobile phases facilitates the concentration of collected fractions by evaporation.

Blumer and Zander [25] studied the retention of a number of PAH, biaryls, and aza-aromatics on silica modified with nitrophenyl groups. A linear relationship between log k' and carbon number was observed for these three

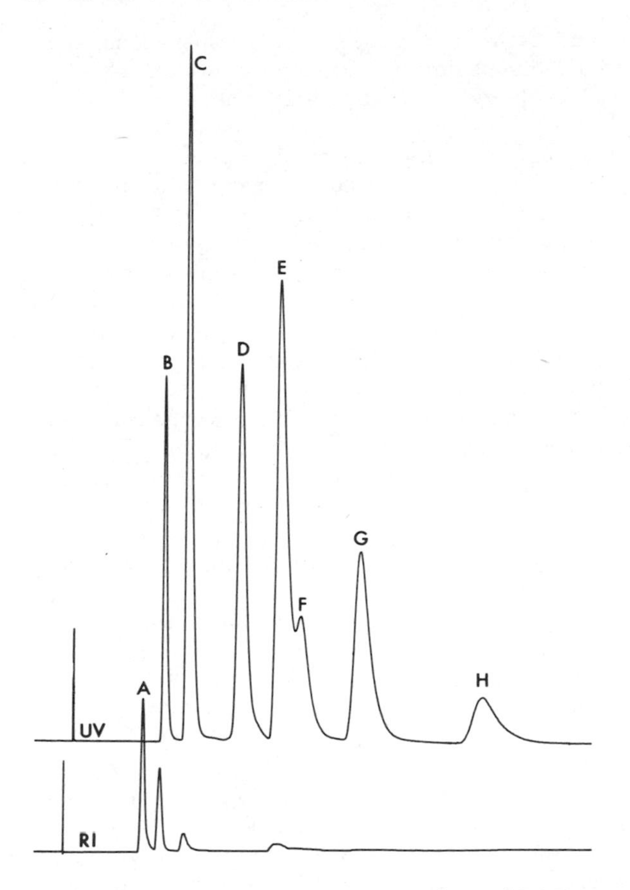

Figure 11. Normal-phase separation of PAH on a bonded amine stationary phase. Column: Bondapak NH_2. Mobile phase: cyclohexane at 1 ml/min. Standards: (A) 5-methyltetradecane, 7-methylhexadecane, and 2-methyloctadecane; (B) benzene, mesitylene, and m-xylene; (C) naphthalene, 2-methylnaphthalene, 2,3-dimethylnaphthalene, and 2,3,6-trimethylnaphthalene; (D) anthracene and 2-methylanthracene; (E) pyrene; (F) fluoranthene; (G) benz[a]anthracene; (H) benzo[a]pyrene. Detection: upper chromatogram, UV at 254 nm; lower chromatogram, refractive index detection. (From Ref. 33.)

groups of compounds. In addition, a class separation of the PAH and aza-aromatics could be achieved on this material due to the more polar nature of the N-containing compounds. Blumer et al. [52] separated more than 100 components of coal tar using this polar stationary phase.

Lankmayr and Müller [51] compared nitrophenyl, amine, and C_{18} columns for the separation of 17 PAH commonly found in dust samples; they reported that the nitrophenyl phase provided the best separation for those compounds evaluated. Several differences in selectivity were noted for the amine and nitrophenyl columns, e.g., elution order reversal for pyrene/fluoranthene, benz[a]anthracene/chrysene, and benzo[a]pyrene/benzo[k]fluoranthene. An extensive characterization and discussion of the retention of PAH and alkyl-substituted PAH on the nitrophenyl phase has yet to appear in the literature.

Recently, Chmielowiec and George [15] investigated the performance of several polar bonded phases (i.e., amine, nitrile, diol, ether, diamine, and quaternary ammonium) for normal-phase separations of 10 PAH. These authors suggested that the diamine column was superior to the other polar bonded phases for PAH separations and reported retention indices for about 50 PAH on this phase. In some instances the diamine column provides a slight increase in the selectivity for some isomeric PAH and for alkyl-substituted vs. parent PAH. However, equivalent differences in selectivity have been observed when different amine columns were compared [33] just as with the C_{18} columns indicating that the differences may be due only to the increased number of polar groups available for interaction with the solute.

Several others [53-58] have recently described new polar bonded phases for the determination of PAH. Mourey et al. [53] described the use of a chemically bonded pyrrolidone stationary phase for both reverse- and normal-phase separation of PAH. Retention data for about 30 PAH indicate that the separation is based on the number of aromatic rings and the type of ring condensation. Various nitrofluorenimine type stationary phases were evaluated by Lochmüller et al. [54] for the separation of 1-, 2-, and 3-methylcholanthrenes and 7- and 12-methylbenz[a]anthracenes. Selectivity differences were observed depending on the number and position of the nitro groups on the fluorenimine moiety. Hunt et al. [55,56] prepared a phthalimidopropylsilane stationary phase for use in both the normal- and reverse-phase separation of PAH. In the reverse-phase mode, the phthalomidopropylsilane phase exhibits different selectivity compared to C_{18} phases; in particular, benzo[a]pyrene elutes prior to benzo[e]pyrene. In the normal-phase mode, separation was achieved of the six PAH of interest to the World Health Organization (WHO), i.e., fluoranthene, benzo[k]fluoranthene, benzo[b]fluoranthene, benzo[a]pyrene, indeno[1,2,3-cd]pyrene, and benzo[ghi]perylene. Recent reports indicate that the nitrophenyl [51] and certain C_{18} columns [14] are also capable of resolving all of these PAH. Nondek and colleagues [57] studied the chromatographic properties of a chemically bonded 3-(2,4-dinitroanilino)propyl (DNAP) phase for the retention of PAH and suggested that charge transfer interactions of the solute with the DNAP groups were responsible for the retention. Martin et al. [58] suggested that a charge transfer interaction was responsible for enhanced retention of PAH on a picramidopropyl modified silica when compared to an aminopropyl phase.

D. Size Exclusion and Gel Chromatography

Size exclusion chromatography (SEC), also referred to as gel filtration or gel permeation chromatography (GPC), has found only limited use for analytical separation of PAH. In this chromatographic mode, separation is achieved on the basis of molecular size. Large molecules cannot penetrate the small pores of the packing material and, therefore, elute prior to smaller molecules which can penetrate the pores. For small molecules such as PAH, large numbers of plates are required to achieve separations of solutes with small molecular weight differences. Graffeo [59] reported nearly complete resolution of perylene, chrysene, phenanthrene, and benzene by connecting five 30-cm columns of a small pore size packing.

Edstrom and Petro [60] correlated the elution behavior of PAH in GPC with the molecular size. In general, the PAH were found to behave as expected with the longer, linear molecules eluting prior to the smaller, more compact molecules.

Due to the limited separation capabilities for small molecules, SEC is employed primarily as a prefractionation or group isolation technique for PAH analyses. Ogan and Katz [61] described the use of SEC for the isolation of various compound classes, including PAH, from coal liquids and shale oil samples prior to reverse-phase LC analyses. With continued advances and improvements in packing materials, SEC will find increased use in similar applications.

A number of miscellaneous gel materials have been employed for PAH separations, particularly for isolation techniques. These materials are generally large particles (40-100 μm) and often provide different elution orders depending on the mobile phase employed. Snook et al. [62] recently reviewed the literature concerning gel chromatography, on such materials as Sephadex LH 20 and Bio-Beads, for the isolation of PAH in complex mixtures.

Klimisch [63,64] and Klimisch and Ambrosius [65] demonstrated the application of cellulose acetate columns for the separation of benzo[a]pyrene from benzo[e]pyrene and the other PAH of the "benzpyrene" fraction. The separation of PAH on a cross-linked polyvinylpyrrolidone using polar eluents was described by Goldstein [66]. The order of elution was based on the number of aromatic rings. Popl et al. [67] investigated the adsorption effect of PAH in GPC on styrene-divinylbenzene gel and related the elution order to molecular size. Later, Popl and co-workers [68] studied the retention of more than 80 aromatic compounds, summarized as retention indices, on macroporous polystyrene gel using water/methanol/diethyl either as the eluent. Again, the retention was observed to increase with increasing number of aromatic carbons. The nonequivalence of different alkyl positions, which had a considerable effect on the elution of alkyl aromatics on silica and alumina, was not observed on the polystyrene gel.

E. Capillary Liquid Chromatography

The complexity of naturally occurring PAH mixtures requires either preisolation techniques or higher efficiency columns. Recently, interest has grown in the development of capillary columns for LC to obtain efficiencies comparable to capillary GC. Currently, three approaches in capillary columns are being pursued by various groups: packed microbore columns [69-71], packed microcapillary columns [72,73], and open-tubular microcapillary columns [74].

Scott [71] recently reviewed the progress of microbore columns (1 mm i.d.) for LC and illustrated the separation of aromatic compounds extracted from coal on a 1 mm i.d. × 2 m column. Novotny and co-workers [72,73] advocate the use of packed glass microcapillary columns prepared by drawing glass tubes of larger diameter, which have been prepacked with small-diameter particles, to the desired diameter. The column packing can then be modified to obtain chemically bonded stationary phases. Hirata and Novotny [73] reported some preliminary results for the separation of PAH in coal tar using a chemically bonded C_{18} capillary column (70 μm i.d. × 55 m). The separation required more than 20 hr for completion. Ishii et al. [74] reviewed the development of open-tubular microcapillary columns (30-60 μm i.d. and lengths of 3-10 m). The inner wall of the column is prepared with a thin layer of solid adsorbent, liquid phase, or chemically bonded layer.

The use of capillary columns in LC requires the modification of conventional instrumentation (e.g., injectors and detectors) to allow for the small flow rates (several microliters per minute), injection volumes, and quantities of material. In addition to providing a greater number of theoretical plates than conventional HPLC, capillary LC employs low flow rates which consume very little solvent and offer a convenient alternative for interfacing with mass spectrometry and infrared spectroscopy.

III. Detection of PAH in HPLC

A major advantage of HPLC for the determination of PAH is the availability of extremely sensitive and selective detectors. Ultraviolet (UV) absorption and fluorescence detectors are ideally suited for the detection of PAH. UV and fluorescence detectors are generally used in series; the UV detector is more universal for PAH measurement, whereas the fluorescence detector provides high sensitivity and specificity.

A. UV Absorption Detectors

Most HPLC systems employ a fixed wavelength UV detector at 254 nm. At this wavelength most PAH exhibit some absorption, thereby providing nearly universal detection for PAH. However, it is usually possible to improve the sensitivity and selectivity by monitoring at another wavelength. Fixed wavelength UV detectors are filter instruments which are limited to the wavelengths of the Hg emission spectrum (i.e., 254, 340, 365, and 436 nm) or 280 nm for a phosphor that is excited at 254 nm.

Multiwavelength or variable wavelength UV detectors utilize a monochromator for wavelength selection. These detectors permit selection of the wavelength to maximize sensitivity and minimize interferences from other compounds. The selectivity achieved for detection of PAH at different wavelengths is illustrated in Fig. 12. At 240 and 254 nm each of the 10 PAH in Fig. 12 respond. At 290 nm benzo[a]pyrene (peak 9) exhibits nearly maximum absorbance with very little interference from perylene (peak 8). Coronene (peak 10) also shows maximum response at 290 nm with low absorbance at the other wavelengths. Fluoranthene (peak 5) and pyrene (peak 6) can each be monitored selectively at the appropriate wavelength, i.e., 340 nm for fluoranthene and 360 nm for pyrene. These examples illustrate the selectivity possible using different wavelengths. Absorption wavelengths for a number of common PAH are listed in Table 7, indicating the possible selectivity that can be achieved. In general, the UV absorption maxima are shifted to longer wavelengths as the number of condensed aromatic rings increases. This is the reason for the low response at 254 nm for PAH having

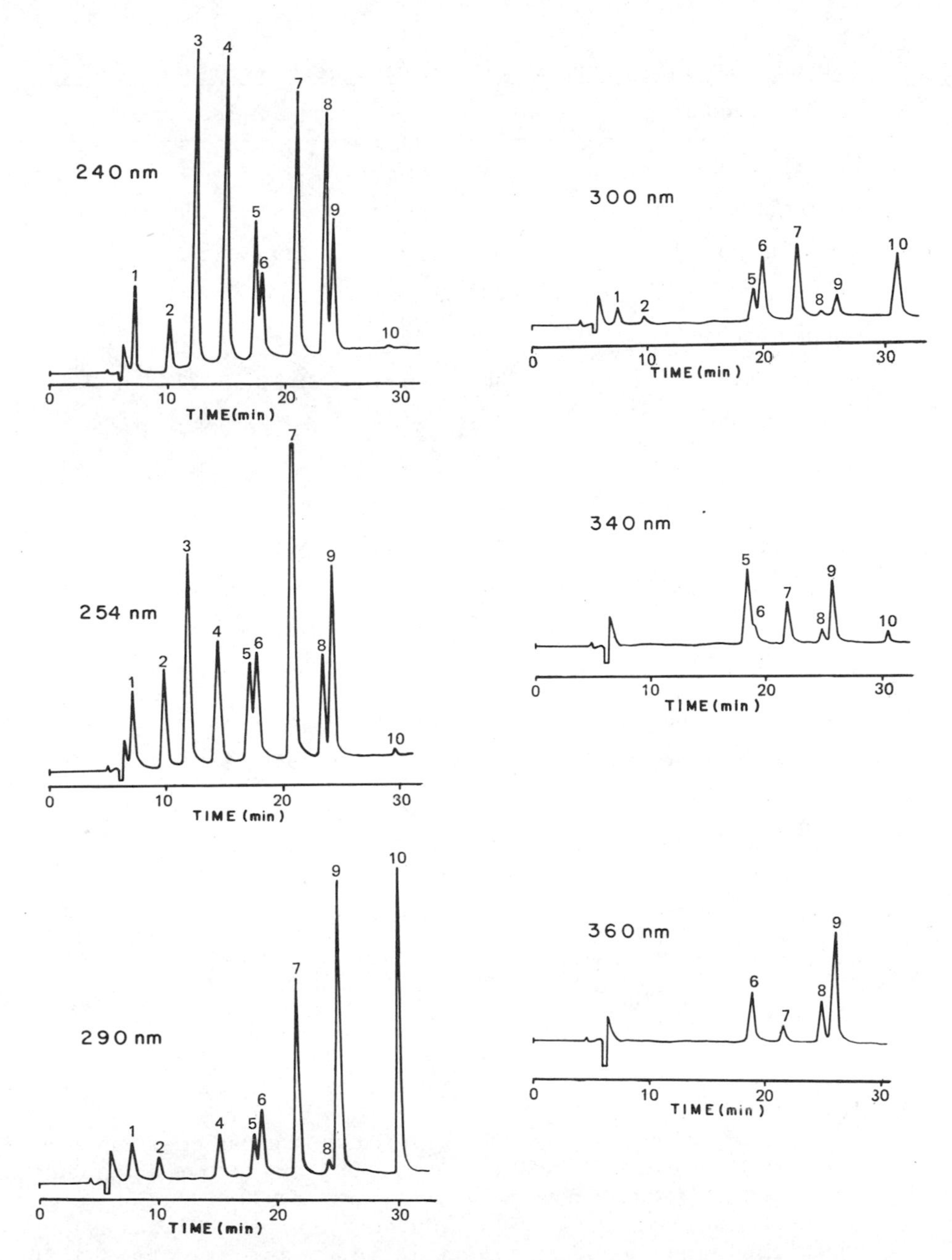

Figure 12. Detection of PAH at various wavelengths. Order of elution: (1) benzene (37.9 μg); (2) naphthalene (1.55 μg); (3) biphenyl (1.50 μg); (4) phenanthrene (0.65 μg); (5) fluoranthene (1.00 μg); (6) pyrene (0.65 μg); (7) chrysene (0.65 μg); (8) perylene (1.00 μg); (9) benzo[a]pyrene (0.35 μg); and (10) coronene (0.91 μg). Column: Partisil 10-ODS. Mobile phase: linear gradient from 60:40 CH_3OH/H_2O to 100% CH_3OH in 34 min. (Reprinted with permission from A. M. Krstulovic, D. M. Rosie, and P. B. Brown, Selective monitoring of polynuclear aromatic hydrocarbons by high pressure liquid chromatography with a variable wavelength detector. *Anal. Chem. 48*: 1383. Copyright 1976, American Chemical Society.)

Table 7. UV Absorption Wavelengths for Several PAH

Compound	Wavelengths (nm)[b]					
Naphthalene[a]	220	275	267	285	—	—
Acenaphthene[a]	227	290	279	300	305	—
Acenaphthylene[a]	324	308	339	265	—	—
Fluorene[a]	261	221	300	289	—	—
Phenanthrene[a]	252	247	258	223	293	275
Anthracene[a]	254	248	223	357	375	339
Fluoranthene[c]	288	277	282	358	342	322
Pyrene[c]	241	273	336	320	263	252
Benz[a]anthracene[a]	287	277	267	256	223	300
Chrysene[a]	267	258	220	320	307	295
Benzo[e]pyrene[c]	290	278	332	268	317	303
Benzo[a]pyrene[c]	296	265	284	255	284	365
Perylene[c]	253	247	436	410	388	—
Benzo[b]fluoranthene[c]	256	302	292	277	350	367
Benzo[k]fluoranthene[c]	307 380	245 361	296	283	268	402
Benzo[ghi]perylene[c]	300	289	381	362	277	346
Dibenz[a,h]anthracene[a]	300	290	323	336	351	373

[a]Data from Ref. 75.
[b]In decreasing order of intensity from left to right.
[c]Data from Ref. 82.

six or more rings. Absorption spectra for many PAH have been reported by Friedel and Orchin [75].

A variable-wavelength detector with scanning capabilities can also be used to record UV spectra of the chromatographic peaks for possible identification. Thoms and Zander [76] have reported the use of UV absorption spectra for the identification of PAH in HPLC effluents. However, UV absorption spectra have limited utility for identification of PAH. A more useful method for the identification of PAH is absorbance ratioing at several wavelengths. The peak heights or areas are recorded at two different wavelengths. These ratios, when obtained under standard conditions, are unique for each compound and can be used to identify the chromatographic peaks and to determine their purity. Krstulovic et al. [77,78] used absorbance ratios for identification of PAH, and Sorrell and Reding [79] employed this technique for PAH identification and determination of peak purity. A number of workers have reported the use of multiwavelength UV detection in the determination of PAH. Krstulovic et al. [77,78], Smillie et al. [80], and Fechner and Siefert [81,82] have employed these detectors in the analysis of air-particulate PAH. Fechner and Siefert [81,82] used fixed wavelength UV detectors, at 340 and 365 nm, in combination with fluorescence detection (λ_{ex} = 436 nm; $\lambda_{em} \geq$ 450 nm) for the quantitation of PAH in dust samples. This procedure allowed the quantitation of nine PAH from three chromatograms obtained from a single run. Boden [83] used selective UV monitoring at 383 nm for the determination of benzo[a]pyrene in coal tar pitch volatiles.

The detection limits of UV detectors are dependent on the extinction coefficient of the PAH at a particular wavelength. Christensen and May [84] compared a fixed-wavelength detector with several variable-wavelength detectors for PAH determinations. They reported detection limits (amount injected) for phenanthrene (25 pg), pyrene (85 pg), chrysene (46 pg), and benzo[a]pyrene (31 pg) for a fixed-wavelength detector at 254 nm. Variable-wavelength detectors set at 254 nm generally have detection limits an order of magnitude less than a fixed-wavelength detector at 254 nm. Thus, to improve the detection limits for a variable-wavelength over a fixed-wavelength detector, the PAH must exhibit an increase of a factor of 10 in the extinction coefficient at the optimum wavelength over 254 nm. However, selectivity at a different wavelength may be more useful than maximized sensitivity.

B. Fluorescence Detectors

Fluorescence detection is ideally suited for the determination of PAH separated by HPLC. Fluorescence provides sensitivity, selectivity, and the possible identification of individual PAH. Spectroscopists have long recognized that fluorescence provides greater sensitivity and selectivity than UV absorption for compounds such as PAH. Researchers first modified various existing fluorometers for use in HPLC. Pellizzari and Sparacino [85] and Slavin et al. [86] modified commercial spectrofluorometers for use to monitor chromatographic effluents of PAH and to obtain fluorescence emission spectra for peak identification. Wheals et al. [24,87] also used modified spectrofluorometers for HPLC detection.

At present, a number of commercial HPLC fluorescence detectors are available. There are two types of fluorescence detectors—filter fluorometers and spectrofluorometers. Filter fluorometers employ filters to select broad bands (10-40 nm) of radiation, thereby providing very high sensitivity. However, they lack the specificity that a spectrofluorometer can provide, and thus the number of interfering compounds will increase. Spectrofluorometers utilize monochromators to allow scanning for the recording of excitation and emission spectra. Monochromators provide the capability of adjusting both the excitation and emission wavelengths. Most have adjustable slit widths—narrow for high resolution, accurate scanning and wide for greater sensitivity. The sensitivity of fluorescence detection for several PAH is illustrated in Table 8 for both a filter fluorometer and a spectrofluorometer. In most cases the filter fluorometers are more sensitive than the spectrofluorometers. Christensen and May [84] compared the sensitivity of several filter fluorometers and a spectrofluorometer for PAH determinations.

Table 8. Comparison of Fluorescence Detection Limits (pg)

Compound	Filter fluorometer[a] ($\lambda_{ex}/\lambda_{em}$)	Spectrofluorometer[b] ($\lambda_{ex}/\lambda_{em}$)
Fluoranthene	0.5 (280/>389)	3 (360/460)
Benz[a]anthracene	0.6 (280/>389)	—
Chrysene	2.3 (250/>370)	—
Benzo[e]pyrene	5.1 (280/>389)	—
Benzo[a]pyrene	1.1 (280/>389)	0.8 (365/407)
Perylene	0.6 (280/>389)	1.0 (380/440)
Benzo[k]fluoranthene	0.4 (280/389)	—
Dibenz[a,h]anthracene	2.3 (280/>389)	—
Benzo[ghi]perylene	3.0 (280/>389)	6 (365/410)

[a]Ref. 91.
[b]Ref. 120.

Recently, Ogan et al. [88] compared the use of a cutoff filter and a monochromator in the fluorescence detection of PAH in environmental samples. The cutoff filter provided an improvement in sensitivity of three to five times that of the monochromator; however, these gains were often negated by the spectral interferences from other compounds in the samples. Thus, they recommended using a monochromator to reduce spectral interferences from overlapping peaks.

The extreme sensitivity of fluorescence detection permits the injection of very dilute solutions of PAH, often reducing the need for concentration techniques. Fluorescence detection has been employed following a trace enrichment procedure which will be described later.

The specificity of fluorescence allows for the selective determination of PAH in the presence of nonfluorescing interferences and for the determination of individual PAH in complex mixtures of PAH. Because only a limited number of compounds fluoresce, PAH can be analyzed in the presence of many other classes of compounds with little or no sample cleanup. The major advantage of fluorescence detection is the selectivity achieved for individual PAH and the possible identification of specific PAH in complex PAH mixtures from air particulates, diesel exhaust, cigarette smoke condensate, petroleum products, coal tar pitch, etc.

The fluorescence spectral characteristics of a number of PAH are listed in Table 9 [89,90]. By selection of the appropriate excitation and emission wavelengths, a high degree of specificity can be achieved. This spectral selectivity often permits the determination of individual PAH in a mixture even when LC resolution of the components is not achieved.

As described earlier, several PAH generally present in natural samples cannot be resolved by normal-phase or reverse-phase HPLC. For example, chrysene and benz[a]anthracene are unresolved on many C_{18} columns. However, chrysene can be measured in the presence of benz[a]anthracene by excitation at 260 nm and emission at 360 nm. Benz[a]anthracene can be selectively determined by excitation at 285 nm and emission at 405 nm with minimal response from chrysene.

The six isomeric PAH of molecular weight 252, which are generally found in all complex PAH mixtures, are not completely resolved from each other on most C_{18} columns. The specificity of fluorescence detection in HPLC, however, allows for the determination of all six isomers as summarized in Table 10. Fluorescence selectivity can also be employed for pairs of PAH which are often only partially resolved, such as benzo[ghi]perylene/indeno[1,2,3-cd]pyrene, or for close-eluting isomers such as phenanthrene/anthracene, and fluoranthene/pyrene.

Das and Thomas [91] described the selectivity achieved using a filter fluorometer. Even with the limitations of filters to select the excitation and emission wavelengths rather than monochromators, they were able to selectively determine 5 pg of benz[a]anthracene in the presence of 200 pg chrysene (λ_{ex} = 280 nm; λ_{em} > 389 nm).

The selectivity of fluorescence detection is illustrated by the chromatograms of a coal tar extract shown in Fig. 13. In this figure, the UV detector is compared with fluorescence selective for perylene (λ_{ex} = 405 nm; λ_{em} = 440 nm) and with more universal fluorescence conditions (λ_{ex} = 300 nm; λ_{em} = 400 nm).

Fluorescence detection has been used extensively for the selective determination of benzo[a]pyrene in complex mixtures such as diesel exhaust [92], cigarette smoke [93], shale oil [94], and petroleum oils [95,96]. The selective monitoring of benzo[a]pyrene in a reprocessed oil fraction is shown in Fig. 14.

The spectral characteristics of several methyl-substituted PAH are also listed in Table 9. The addition of a methyl group to a PAH generally shifts the emission maxima to slightly longer wavelengths. Thus, it is generally possible to monitor for a specific PAH and its methyl-substituted homologs at a particular wavelength. The selective monitoring for chrysene and alkyl-substituted chrysenes, isolated from a sediment sample, is illustrated in Fig. 15. Peak *a* was identified as chrysene by retention time and fluorescence emission spectrum. The other peaks, *b*-*g*, were tentatively identified as methyl- and dimethyl-substituted chrysenes by the fluorescence spectra.

The spectra in Fig. 15 were obtained using a low-volume switching valve to trap the LC peak in the flow cell of the spectrofluorometer. This permits the use of the UV detector as a universal PAH monitor and the use of the spectrofluorometer either to selectively monitor at a particular wavelength or to obtain fluorescence spectra for identification of the chromatographic peaks without stopping the flow and disturbing the UV trace.

When using fluorescence detection in HPLC, the mobile phase should be deoxygenated to avoid fluorescence quenching of the PAH. The importance

Table 9. Fluorescence Spectral Characteristics of PAH[a]

Compound	Fluorescence excitation spectra λ (nm)								Fluorescence emission spectra λ (nm)					
Naphthalene[b]	269	278	288						324	338	350			
Fluorene[b]	268	275(s)	293	303					303	310				
Phenanthrene:	249	273	280	292					345	353	362	382		
1-Methylphenanthrene	248	253	278	286	298				350		367	388	408	
3,6-Dimethylphenanthrene	247(s)	253	277	286	297				353		370	391	413(s)	
Anthracene	250	308	320	337	355	374			376	298	423	448		
Fluoranthene	260	275	280	284	307(s)	332	340	356	463	(425-475)				
Pyrene:	261	274	290(s)	305	318	333			370	377	382	387	390	
1-Methylpyrene	233(s)	242	254(s)	264	275	312	325	340	374			395	415	
11H-Benzo[a]fluorene[b]	255	265	296	306	318				347	365	382(s)			
11H-Benzo[b]fluorene[b]	270	288	306	319	326	342			342	351	359	369		
Triphenylene	245	254	269	281					344	351	358	367	377	387(s)
Benz[a]anthracene:	255	265	275	285	324	337	355		384	406	432	460		
7,12-Dimethylbenz[a]anthracene	263	275	284	295	327(s)	343(s)	358		406	425	450(s)			
Chrysene:	255	263	282	303	318				360	378	401	424		
1-Methylchrysene	260	268	287	298	310	324			362	380	403	427		
4-Methylchrysene		272		302	316	328			368	387	408	432		
Naphthacene	262(s)	272	292	410	439				470	504	543			
Benzo[e]pyrene	266	276	286	302	315	328			374	385	394	406	417	
Benzo[j]fluoranthene	290	312	305	328	362	381			507	(490-520)				
Benzo[b]fluoranthene	255	274	289	296	348				433	(415-460)				

Perylene	251	263(s)	363	383	406				435	463	498		
Benzo[k]fluoranthene	245	266	294	304	357	376			407	430	457(s)		
Benzo[a]pyrene	265	284	297	348	363	383			404	427	453		
Dibenz[a,c]anthracene[b]	269(s)	279	289						377	388	398	409(s)	
Dibenz[a,h]anthracene	286	295	317	332	347				392	402(s)	414	439	470(s)
Dibenzo[c,g]phenanthrene[b]	239	272	312	332					405	424	446(s)		
Benzo[b]chrysene	275	284	302						394	417	445		
Picene[b]	287	304	328						377	398	421	449	
Benzo[ghi]perylene	272(s)	284	295	326	343	357	378		404 455(s)	413 460(s)	417	427	441(s)
Indeno[1,2,3-cd]pyrene	239 380	247 404	273	288(s)	298	311	356	372	472	496	(465-515)		
Anthanthrene[b]	260	296	308	384	401	407	422	430	432	459	494		
Dibenzo[g,p]chrysene[b]	280	292	303	340	353				395	409			
Dibenzo[b,def]chrysene[b]	272	310	312	399	422	428			451	480	518		
Naphtho[1,2,3,4-def]chrysene[b]	276	293	305	330	342	358	376		397	408	420	446	
Benzo[rst]pentaphene[b]	247 395	274	285	297	316	332	355	373	434	450	462	480	494

[a]Data from Ref. 89. Note: The most intense peak is underlined with solid line, peaks which have intensity greater than 70% of the most intense peak are underlined with broken line. Shoulders are indicated by (s). Broad peaks are followed by wavelength range in which the intensity is greater than approximately 80% of the maximum. Spectra were obtained in 80:20 acetonitrile/water mixture and are uncorrected.
[b]Data from Ref. 90; spectra were obtained in cyclohexane.

Table 10. Fluorescence Spectral Conditions for Selectivity of Chromatographically Unresolved PAH

Unresolved or partially resolved PAH	Selective for:	Excitation wavelength (nm)	Emission wavelength (nm)
Perylene/	Perylene	405	435
benzo[b]fluoranthene	BbF	290	450
Benzo[e]pyrene/	BbF	290	450
benzo[b]fluoranthene	BeP	285 or 330	375 or 385
Benzo[j]fluoranthene/	BeP	285 or 330	275 or 385
benzo[e]pyrene	BjF	310	505

of excluding dissolved oxygen from the mobile phase has been reported by several workers [25,97,98]. Nielsen [26] reported that pyrene was particularly susceptible to oxygen quenching, with a fivefold increase in response after deoxygenating the mobile phase by bubbling argon through it for 1 hr, deaerating ultrasonically, and storing under argon. Similar enhancement has been reported for benzo[a]pyrene after deaerating the mobile phase [95].

Selective fluorescence quenching of certain PAH in the presence of nitromethane has been investigated as a selective HPLC detection system [99,100]. Sawicki et al. [101] and later Dreeskamp et al. [102] reported that in the presence of nitromethane the fluorescence emission of six-membered ring PAH was quenched to a much greater degree than of those PAH containing a fluoranthenic structure. Blumer and Zander [99] briefly described the application of this selective quenching as a detection system for HPLC. The selectivity achieved by the addition of 5% nitromethane to the mobile phase was illustrated for the analysis of a technical mixture of fluoranthene and pyrene.

Recently, Konash et al. [100] further investigated the use of selective fluorescence quenching as a detection system to determine benzofluoranthene isomers in the presence of benzo[e]pyrene, perylene, and benzo[a]pyrene. Selective fluorescence quenching is illustrated in Fig. 16 for the analysis of a coal tar sample. The addition of 0.5% nitromethane significantly quenches all the nonfluoranthenic PAH (except for a small percentage of anthracene), resulting in a selective chromatogram of only fluoranthenic PAH. This technique was found to be particularly useful when only an inexpensive filter fluorometer was utilized to achieve selectivity for fluoranthenic PAH as a group.

A recent development in spectroscopic detection of PAH in chromatographic effluents is the use of multichannel rapid scanning spectrometers. These detectors permit the recording of fluorescence spectra "on the fly," thereby eliminating stop-flow or valving to trap the chromatographic peak in the flow cell. Jadamec et al. [103] described the use of such a system for the characterization of petroleum fractions in the determination of the source of oil spills. More recently, Shelly and co-workers [104] reported a two-dimensional "video fluorimeter" as an HPLC detector. They recorded the fluorescence intensity as a function of emitting and exciting wavelengths every 25 sec to produce an emission-excitation matrix for PAH standards as they eluted from the LC.

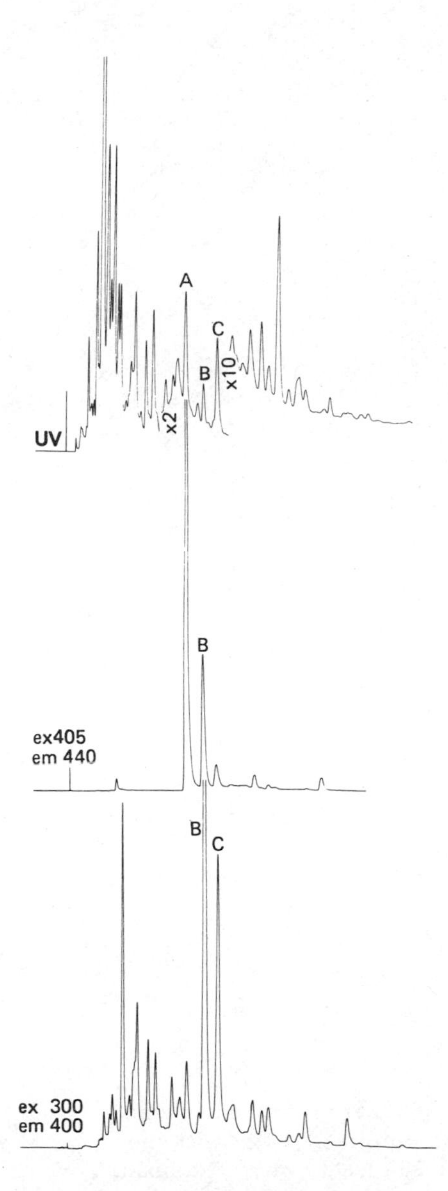

Figure 13. Separation of PAH from coal tar extract with (A) UV detection at 254 nm; (B) fluorescence at λ_{ex} = 405 and λ_{em} = 440; and (C) fluorescence at λ_{ex} = 300 and λ_{em} = 400. Column: Vydac 201TP 5 μ. Mobile phase: linear gradient from 80 to 100% CH_3CN in H_2O at 1%/min at 1 ml/min. (From Ref. 33.)

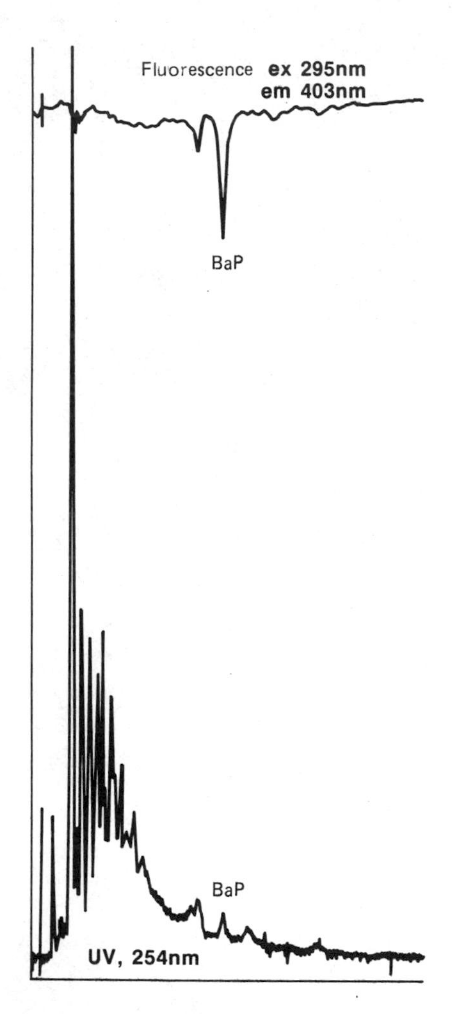

Figure 14. Selective determination of benzo[a]pyrene in reprocessed oil pentacyclic fraction with fluorescence detection at λ_{ex} = 295 nm and λ_{em} = 405 nm (upper) and UV adsorption detection at 254 nm (lower). Column: Zorbax ODS. Mobile phase: 80% CH_3CN in H_2O at 2 ml/min. The pentacyclic fraction was obtained by normal-phase LC separation on μBondapak NH_2 [94,96]. (From J. M. Brown and W. E. May, unpublished data, 1979.)

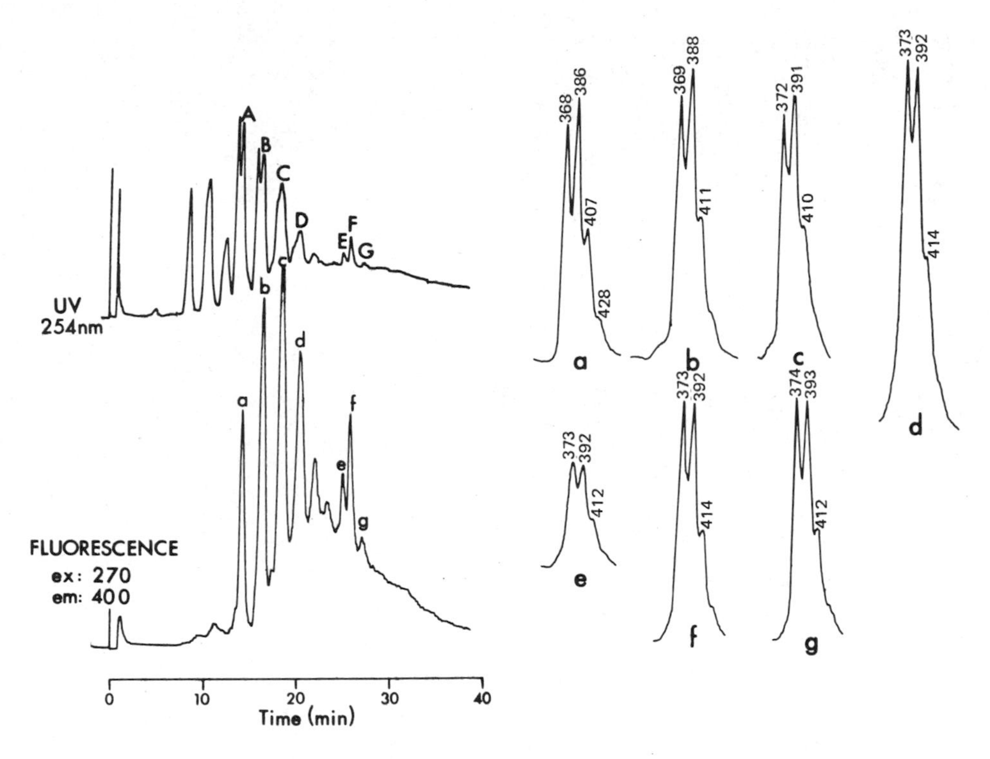

Figure 15. Reverse-phase analysis of tetracyclic fraction from sediment extract. Column: μBondapak C_{18}. Mobile phase: linear gradient of 50-100% CH_3CN in H_2O at 30 min at 2 ml/min. Upper chromatogram: UV absorption detection at 254 nm. Lower chromatogram: fluorescence at λ_{ex} = 270 nm and λ_{em} = 400 nm. Spectra *a* thru *g*: fluorescence spectra at λ_{ex} = 270 nm (numbers indicate wavelengths in nanometers). (From Ref. 13.)

Such systems, when combined with computers and floppy disks for more data storage, will allow for more rapid acquisition of an enormous quantity of data during one chromatographic run. These data can then be evaluated to identify and quantify the PAH in the complex chromatograms.

Time-resolved fluorescence to enhance the selectivity for PAH in HPLC was recently reported by Richardson et al. [105]. Using a pulsed laser as an excitation source, only PAH with long fluorescence lifetimes (e.g., fluoranthene) respond if the delay between excitation and detection is sufficiently long. Detection limits of 1-10 pg were reported.

C. Liquid Chromatography/Mass Spectrometry

The application of mass spectrometry for the analysis of PAH mixtures is described in Chap. 7. The potential of the combination of LC/MS for the separation and identification of organic compounds has generated considerable research in the past few years. The inherent problems of introducing a liquid

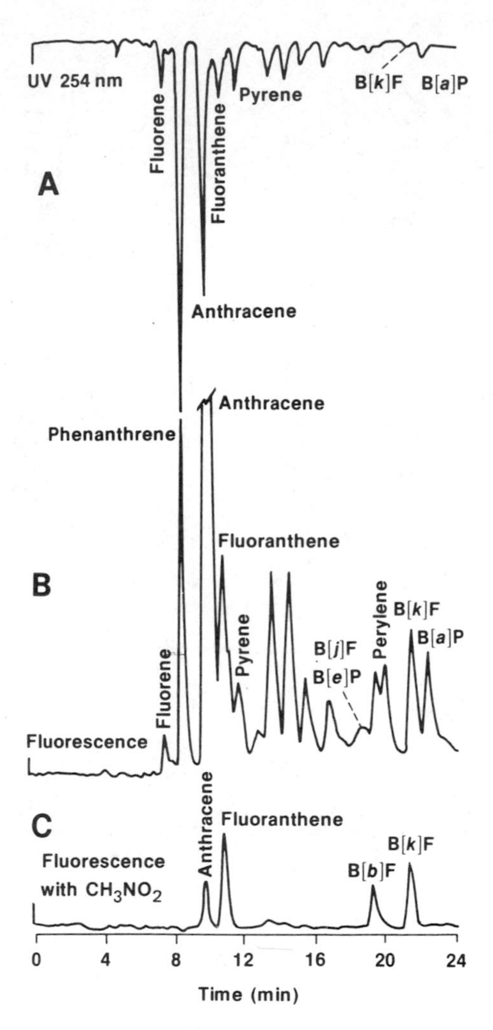

Figure 16. Comparison of (A) UV detection at 254 nm, (B) fluorescence detection, and (C) fluorescence detection with 0.5% nitromethane in the mobile phase for PAH in coal tar extract. Column: Vydac 201TP. Mobile phase: linear gradient from 50% CH_3CN in H_2O to 100% CH_3CN at 2%/min at 2 ml/min. (From Ref. 100.)

stream (~1 ml/min) into the high vacuum system of the mass spectrometer have led to several approaches. Arpino and Guiochon [106] recently reviewed the various LC/MS interfaces and described the construction, operating principles, and performance of each approach. Another review of the LC/MS techniques by McFadden [107] emphasized the current applications of this technique.

At present two LC/MS approaches are commercially available, i.e., direct liquid injection and depositing the effluent onto a moving belt from which the mobile phase is evaporated prior to introduction into the mass spectrometer. Both of these approaches transfer only a small portion of the LC solute into the MS. Christensen et al. [108] recently described a new LC/MS interface that combines several of the advantages of both the direct liquid injection and moving belt techniques. The LC effluent is concentrated by evaporation as it flows down an electrically heated wire. The concentrated effluent then flows through a small needle valve and is sprayed into the MS ion source. The application of this approach to the analysis of PAH standards resulted in a 20-fold enrichment.

Dark et al. [109], using the moving belt method, demonstrated the LC fractionation of coal liquid samples with structural characterization by LC/MS. Using LC retention data on two different columns and the molecular weight data, these investigators identified a number of aromatic compounds (molecular weight up to ~250) in the coal liquid. In the future, LC/MS will find greater application in the determination of higher-molecular-weight PAH which are not amenable to GC analysis. Other applications of LC/MS involving PAH have been reported [110].

IV. Applications of HPLC for the Determination of PAH

A. Sequential or Multidimensional HPLC Techniques

As described in the previous sections, various LC modes are useful for the separation of PAH. However, even the most selective and efficient HPLC columns available are unsuccessful in resolving all of the PAH in complex mixtures such as from air particulates, coal tar, cigarette smoke condensate, petroleum, or alternate fuels. The determination of individual PAH in such mixtures generally requires the use of several chromatographic modes (e.g., TLC, GC, and various LC modes) for sample preparation and for determination of the compounds of interest. LC is extremely useful for both prefactionation/isolation techniques and for analytical determinations.

Traditionally in PAH determinations sample preparation has consisted of separation on silica or alumina either by column chromatography, TLC, or more recently HPLC. After isolation of the PAH fraction, gas chromatographic or spectroscopic techniques have often been used for the final analytical step. This section will emphasize the sequential combination of different LC modes for the complete determination of individual PAH in complex mixtures. As described in Sec. II, different mechanisms of separation can be achieved using the various LC modes. For example, liquid-solid adsorption chromatography provides a separation based on the number of aromatic carbons or molecular weight, whereas reverse-phase LC is capable of separating many isomeric PAH.

In their early work with a chemically bonded C_{18} phase, Schmit et al. [23] recognized the potential of reverse-phase HPLC separation of PAH in auto exhaust condensate after an initial separation step involving GPC. Sleight [41] suggested the possible application of silica/alumina to obtain an aromatic class separation, followed by reverse-phase C_{18} analysis to separate the alkyl-substituted PAH.

More recently, Wise et al. [13,14] have advocated the use of normal-phase chromatography on a polar chemically bonded amine phase for PAH separation based on the number of condensed aromatic rings, followed by separation of the various PAH isomers and/or alkyl substituted PAH by reverse-phase chromatography on a C_{18} phase. The application of this multidimensional HPLC technique is illustrated in Figs. 17 and 18 for the determination of PAH in an air particulate extract. The normal-phase separation on the amine phase (Fig. 17) provides a separation based on the number of aromatic carbons, as indicated by the elution times of representative PAH. The fractions labeled in Fig. 17 were collected, concentrated, and analyzed by reverse-phase HPLC. The reverse-phase chromatograms for fractions 4 and 5 are shown in Fig. 18. Fraction 4 contains the isomeric four-condensed-ring PAH (benz[a]anthracene, chrysene, and triphenylene) and their alkyl derivatives. Fraction 5 contains the five-ring isomers benzo[e]pyrene, benzo[a]pyrene, perylene, benzo[b]-fluoranthene, and benzo[k]fluoranthene. In the reverse-phase analysis the methyl-substituted chrysenes and benz[a]anthracenes would coelute with the five-ring PAH if they were not previously isolated by normal-phase chromatography (see retention data in Table 1). The use of the bonded amine phase for prefractionation offers several advantages over classical adsorbents: (1) a more distinct class separation based on number of aromatic rings, i.e., alkyl-substituted PAH elute closer to parent PAH (see Table 1); (2) rapid equilibration of the chemically bonded phase; and (3) elimination of non-reproducible retention due to small changes in the moisture content of the eluent. In addition to the amine phase, other recently available chemically bonded polar phases (e.g., nitrophenyl, nitrile, diol, or diamine) could be employed for similar fractionation.

This sequential combination of normal-phase and reverse-phase HPLC has been employed for the quantitation of benzo[a]pyrene in some virgin, waste,

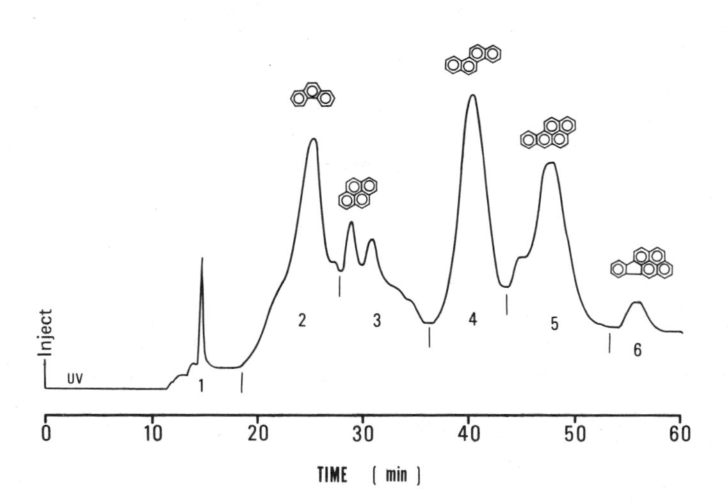

Figure 17. Normal-phase liquid chromatogram of PAH extract from urban-air-particulate extracts. Column: semipreparative μBondapak NH_2. Mobile phase: n-hexane at 1 ml/min. UV detection at 254 nm. Numbers indicate fractions collected for subsequent analysis, and the PAH structures indicate elution times of representative compounds. (From Ref. 14.)

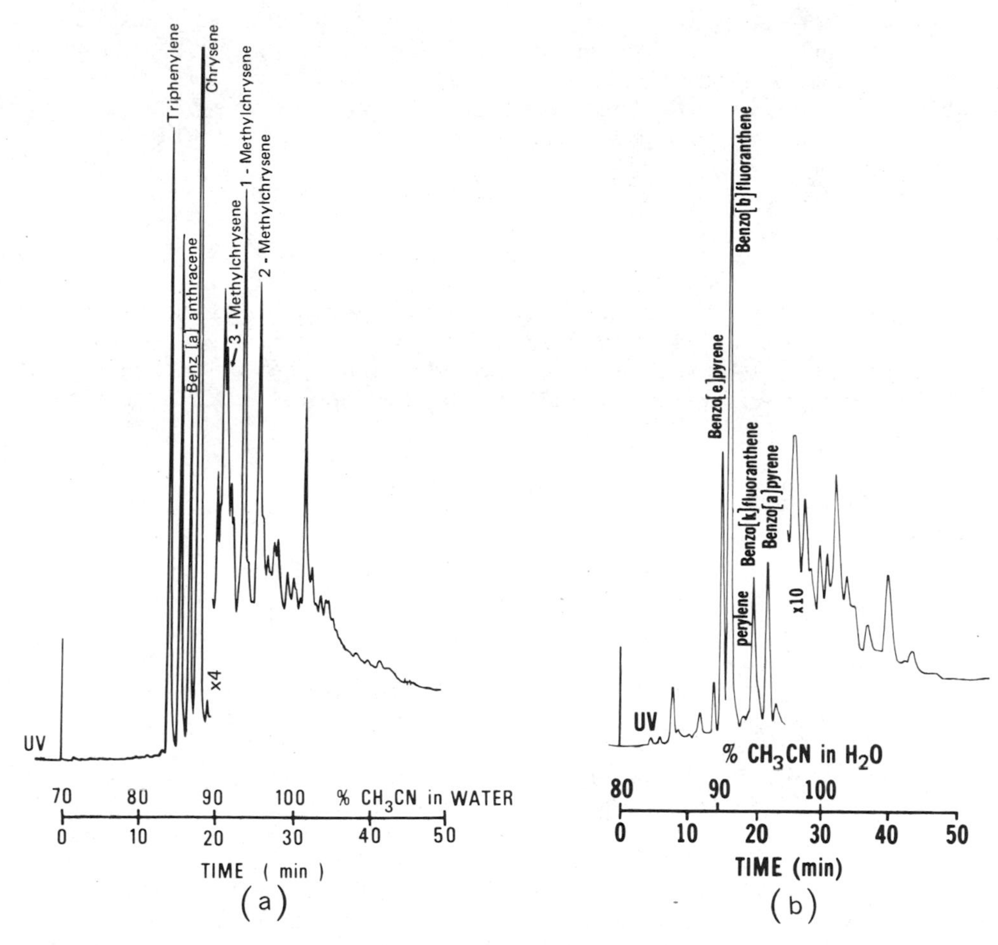

Figure 18. Reverse-phase LC analysis of Fractions 4 and 5 from air-particulate extracts (see Fig. 17). Column: Vydac 201TP. Mobile phase: linear gradient of CH_3CN in H_2O at 1 ml/min. (From Ref. 14.)

and recycled oils [96] and the quantitation of several PAH in shale oil [94]. In the shale oil sample the amounts of pyrene, fluoranthene, benzo[e]pyrene, and benzo[a]pyrene determined using this technique were comparable to the values obtained using either classical acid/base extraction isolation followed by GC/MS or direct sample injection using GC/MS with single-ion monitoring for selectivity. Dark et al. [109] and Dark and McFadden [111] have also utilized similar sequential techniques with silica or the amine phase followed by reverse-phase separation and mass spectrometric detection for the characterization of coal liquids. Even after LC prefractionation, these complex mixtures require the use of selective detectors such as LC/MS or fluorescence for individual compound identification and quantitation.

Other examples of combining different LC modes for PAH analyses include: silica gel/alumina gel chromatography for petroleum samples [112], alumina/reverse-phase HPLC for water samples [113], alumina- or silica/reverse-phase HPLC for air particulate extracts [114,115], and size exclusion chromatography/reverse-phase HPLC for alternate fuels [61]. TLC on silica or alumina, which many have used as an inexpensive alternative to HPLC is often employed for isolation of PAH fractions prior to HPLC analyses [116, 117].

Gel filtration chromatography has been used extensively for isolation of PAH fractions from cigarette smoke condensate (CSC) prior to GC analysis. Recently, Snook et al. [62] incorporated reverse-phase HPLC into their analytical scheme for CSC. In their scheme GC peaks from the analysis of gel chromatographic fractions were collected and analyzed by reverse-phase HPLC to separate isomeric PAH and alkyl PAH which were difficult to separate by GC.

B. Water Analysis

The low levels of PAH normally found in water samples necessitates the use of a sample-enrichment step and sensitive detection methods. LC satisfies both these criteria. Traditional enrichment procedures involve concentration of an organic solvent extract by evaporation. However, PAH enrichment directly on a reverse-phase LC column has been described [118-120].

In 1975, May et al. [118] first described the use of a "coupled column" procedure for the enrichment and separation of trace-level PAH in water. In this procedure a large water sample (1-2 liters) is passed through a short LC column containing 37- to 50 μm pellicular C_{18} material to concentrate the nonpolar compounds. After this enrichment step, the short column is connected at the head of an analytical column containing 10-μm totally porous microparticulate C_{18} material. A mobile phase gradient of 30-100% methanol was used to desorb the PAH from the precolumn (enrichment column) and to separate them on the analytical column. At the initial mobile phase composition of 30% methanol, the PAH are rapidly desorbed from the pellicular packing (~1% carbon loading) but are retained as a narrow band at the inlet of the microparticulate column, thereby resulting in negligible peak broadening during the analytical separation. Recently, Eisenbeiss and co-workers [119] and Ogan et al. [120] described similar approaches for the concentration and determination of PAH in aqueous samples. In these procedures an LC injection valve was employed to place the enrichment column on line with the analytical column after the enrichment step. A detection limit of 4 ng/liter of benzo[a]-pyrene in water was reported by Eisenbeiss et al. [119].

Some of the advantages of a two-column trace-enrichment approach for the analysis of water samples are the following:

1. There is minimal sample handling and no concentration of organic solvents, thereby minimizing sample contamination.
2. Each column can be optimized for its respective purpose, i.e., sample concentration or separation of PAH.
3. Transport of large volumes of water samples can be avoided by on-site sample enrichment.

4. The large-particle-size pellicular enrichment column allows high flow rates.
5. The bulk of the water and nonadsorbent compounds bypass the analytical column, thus extending its lifetime.

Adsorption of trace-level PAH onto glass and metal surfaces is a major problem in these techniques. Ogan et al. [120] found 20% methanol added to the samples resulted in significant improvement in the recoveries, whereas Eisenbeiss et al. [119] used similar percentages of isopropanol to provide optimum recoveries.

More conventional analytical procedures, involving extractions with cyclohexane or methylene chloride and TLC or column chromatographic sample cleanup, have been reported for determination of PAH in aqueous samples using HPLC for the analytical determination [113,117,121-123]. The proposed EPA method for the determination of the 16 PAH included on the EPA Priority Pollutant list employs HPLC [124]. Separation of these 16 PAH is accomplished using a reverse-phase C_{18} column (see Fig. 3). However, as shown in Table 2, not all C_{18} columns provide a suitable separation for all 16 of these compounds.

C. Air Particulate, Automobile Exhaust, and Sediment Analysis

In recent years reverse-phase HPLC has become a popular technique for the analysis of the relatively simple mixtures of unsubstituted PAH from air particulates [14,52,77,78,80-82,97,114-116]. Dong et al. [116], Fechner and Seifert [81,82], and Eisenberg [114] all used reverse-phase C_{18} columns for separations after sample cleanup on a silica column or TLC. Krstulovic et al. [77,78] and Smillie et al. [80] reported HPLC analyses of air particulate extracts with no sample cleanup but employed selective multiwavelength UV and/or fluorescence detection. Golden and Sawicki [115] used sequential analysis on alumina followed by reverse-phase HPLC on a C_{18} column to determine benzo[a]pyrene in air particulates. Lankmayr and Müller [51] found normal-phase HPLC on a nitrophenyl modified silica to be suitable. Similar HPLC methods using reverse-phase systems have been reported for the determination of PAH in sediments [13,123] and automobile exhaust [23,25,46,92, 125].

D. Petroleum and Alternate Fuels Analysis

The determination of PAH in complex petroleum matrices has always offered a challenge to the analyst. Analytical methods generally combine at least two LC modes and a selective detection system. Brown et al. [96] employed the combination of normal- and reverse-phase columns with selective fluorescence for the measurement of benzo[a]pyrene in recycled oils. Similar procedures were used for quantitation of four PAH in a shale oil sample [94]. Normal-phase isolation of PAH on silica or an amine phase, followed by reverse-phase separations with selective MS detection, was used for the characterization of a coal liquid by Dark et al. [108]. Mourey et al. [53] used a chemically bonded pyrrolidone phase in both normal- and reverse-phase modes to obtain profiles of shale oil.

Recently, an HPLC method for the direct determination of benzo[a]pyrene in petroleum oils was described by Durand and Petroff [95]. A liquid-liquid chromatographic column was prepared by modification of the microsilica support with a 45% ratio of dimethyl sulfoxide/silica. Dimethyl sulfoxide is a highly selective solvent for PAH. In this HPLC system, alkyl-substituted PAH had significantly less retention than the parent compounds. Benzo[a]-pyrene was quantitated directly with the aid of fluorescence detection.

E. Determination of High-Molecular-Weight PAH

Many of the PAH mixtures described in this section could be analyzed using capillary column GC. However, volatility limits the application of GC to the determination of high-molecular-weight PAH (molecular weight > 300). Presently, HPLC offers the most advantageous technique for the separation of these compounds.

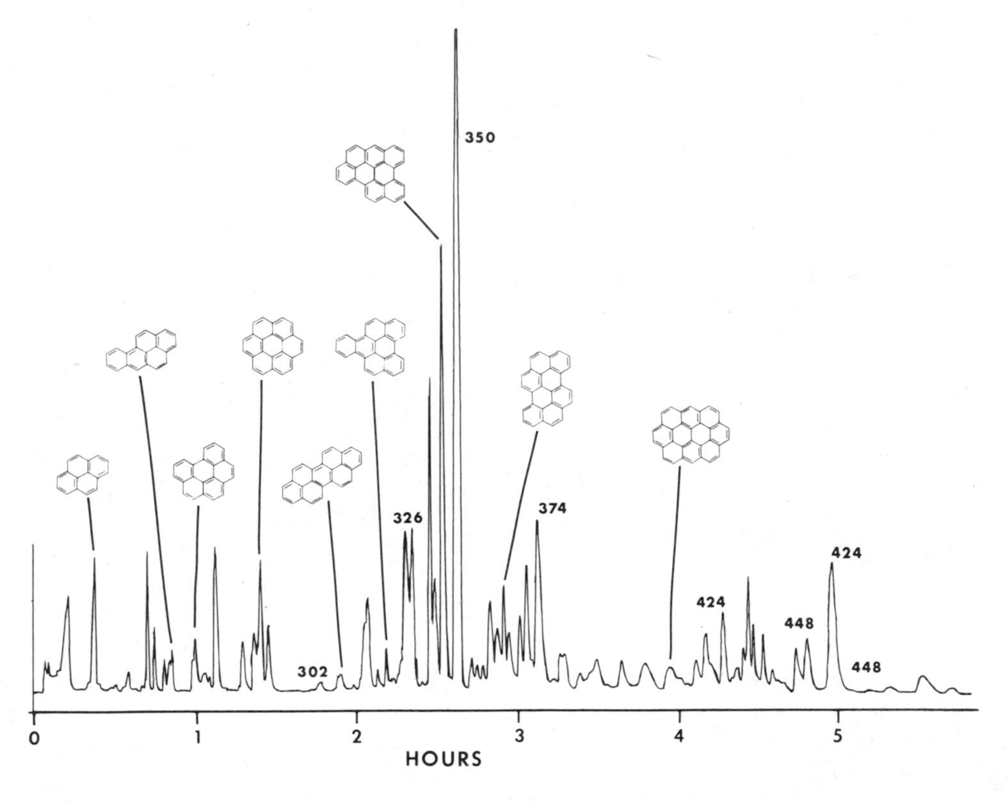

Figure 19. Liquid chromatogram of high-molecular-weight PAH in carbon black. Column: Vydac 201TP. Mobile phase: 50% CH_3CN in H_2O (15 min isocratic) to 100% CH_3CN in 70 min, then to 100% ethyl acetate in 130 min, and finally to 100% CH_2Cl_2 in 60 min. Detection: Fluorescence at λ_{ex} = 335 nm and λ_{em} = 480 nm. (Reprinted with permission from P. A. Peaden, M. L. Lee, Y. Hirata, and M. Novotny, High-performance liquid chromatographic separation of high-molecular weight polycyclic aromatic compounds in carbon black. *Anal. Chem.* *52*:2268. Copyright 1980, American Chemical Society.)

Table 11. Applications of HPLC for the Determination of PAH

Publication year [ref.]	Author(s)	Application	Column	Detection	Comments
1971 [23]	Schmit et al.	Standards and auto exhaust GPC fractions	ODS Permaphase	254 nm	GPC fractions analyzed by reverse phase
1972 [131]	Martinu and Janak	Standard mixtures	Porapak T	254 nm	
1972 [132]	Ives and Giuffrida	Standards, foods	Durapak OPN, MN cellulose	UV	
1972 [48]	Hupe and Schrenker	Standards	Merckosorb Si 60 coated with 50% w/w Fractonitril	UV 254 nm	
1973 [85]	Pellizzari and Sparacino	Standards	Corasil II	Fluorescence	
1973 [41]	Sleight	Standards	Zipax Permaphase ODS		Relative retention of 21 PAH and alkyl PAH
1973 [133]	Strubert	Standards	Spherosil XOA 400	250 nm	
1973 [134]	Karger et al.	Standards	Corasil impregnated with 2,4,7-trinitrofluorenone		
1973 [87]	Vaughan et al.	Used engine oil	Corasil II Corasil C_{18}		Retention for 13 PAH
1974 [63,64]	Klimisch	Coronene and standards	30% cellulose acetate and Polyamide-6	304 nm UV	

(continued)

Table 11 (continued)

Publication year [ref.]	Author(s)	Application	Column	Detection	Comments
1974 [11]	Popl et al.		Alumina		Retention data ~100 PAH
1974 [112]	Popl et al.	Tars and petroleum	Alumina,	Fluorescence: $\lambda_{ex} = 377$ $\chi_{em} = 403$	Selective detection for benzo[a]-pyrene
1974 [22]	Popl et al.	Alkyl naphthalenes, phenanthrenes, and anthracenes	Alumina-silica		
1974 [135]	Grant et al.	Standards	Zipax coated with Bentone 34		
1975 [118]	May et al.	Water and sediment	μBondapak C_{18} pellicular pre-column	UV at 254 nm	Trace enrichment
1975 [24]	Wheals et al.	Used engine oil	Silica-C_{18}, 7-μm	Fluorescence	Retention for 18 PAH, compares pellicular and microparticulate packings
1975 [136]	Haeberer et al.	Cigarette smoke condensate	Permaphase ODS	UV scanning	
1976 [65]	Klimisch and Ambrosius	Benzopyrene isomers	Cross-linked cellulose acetate	Fluorescence	Retention data for 9 PAH, extreme selectivity for BaP and BeP

1976 [76]	Thoms and Zander	Synthetic mixtures BaP, BbF, and DiBa,hA	LiChrosorb SI 100, 10 μm Nucleosil 5-C_{18}, 5 μm	UV 296 nm	Scanning UV for identification of high-molecular-weight PAH
1976 [116]	Dong et al.	Air-particulate extracts; quantitation of 18 PAH	Zorbax ODS	UV	Retention for 20 PAH
1976 [83]	Boden	BaP in coal tar pitch volatiles	Porasil/ LiChrosorb/ μBondapak C_{18}	Multiwavelength UV, 383 nm for BaP	Retention data on 3 columns
1976 [115]	Golden and Sawicki	Air-particulate filters	Alumina Spherisorb ODS	UV at 254 nm	Pentacyclic fraction from alumina collected, then analyzed on Spherisorb ODS
1976 [12]	Popl et al.	113 standards	Silica		Retention indices for 113 PAH
1976 [66]	Goldstein	Coal liquids	Polyvinylpyrrolidone, 63-90 μm	UV at 254 nm	Retention of 27 PAH
1976 [67]	Popl et al.	Standards	Styrene-divinylbenzene		Retention data for 30 PAH
1976 [97]	Fox and Staley	Air-particulate extracts	Zorbax ODS	Fluorescence, filter fluorometer	Quantitation of 11 PAH
1976 [137]	British Carbonization Research Assoc.	Coke oven pitch, air-particulate extracts	Partisil Bentone 34	UV at 385 nm	UV absorption for identification

(continued)

Table 11 (continued)

Publication year [ref.]	Author(s)	Application	Column	Detection	Comments
1976 [138]	Guerrero et al.	Shellfish	μBondapak C_{18}	UV at 254 nm	
1976 [139]	Williams and Slavin	Air-particulate extracts	ODS-SIL-X-1	Fluorescence	
1977 [108]	Dark et al.	Coal liquids–anthracene oil	μPorasil μBondapak C_{18}	UV at 254 and 313 nm	Fractions from silica, then analyzed on C_{18} with MS identification
1977 [77,78]	Krstulovic et al.	Air-particulate extracts	Partisil 10-ODS	Multiwavelength UV	Absorbance ratios
1977 [8]	Thoms and Zander				Review article
1977 [55]	Hunt et al.	Standards, mussel extract	3-Phthalimidopropyltrichlorosilane	Fluorescence filter	
1977 [6]	Hunt et al.				Review article
1977 [42]	Blumer and Zander	Standards	Nucleosil NO_2 Nucleosil C_{18}		Retention for 38 PAH
1977 [140]	Popl et al.	Standards	Polystyrene gel		Retention indices for 80 PAH
1977 [127]	Grant and Meiris	Tar and petroleum products	Partisil 5	Multiwavelength UV scanning	Retention data for 22 PAH
1977 [86]	Slavin et al.	Standards	ODS-SIL-X-I	Fluorescence	General use of fluorescence

1977 [122]	Hagenmaier et al.	Water extracts	LiChrosorb RP-8, RP-18	Filter fluorometer	
1977 [141]	Johnson et al.	Standards	MicroPak-CH-10	Fluorescence	General use of fluorescence
1977 [81,82]	Fechner and Seifert	Dust extracts: method for 13 PAH	μBondapak C_{18}, LiChrosorb RP-18, 5 μm	Multiwavelength UV and fluorescence	Single run provides 3 chromatograms
1977 [13]	Wise et al.	Sediment and petroleum	μBondapak NH μBondapak C_{18}	UV at 254 nm, fluorescence	Retention data for over 40 PAH on both columns
1978 [92]	Seizinger	Diesel exhaust	C_{18}	Fluorescence Ex: 383 nm Em: 454 nm	Determination of BaP
1978 [127]	Blumer and Zander	Coal tar, high-molecular-weight PAH	Nucleosil NO_2, 5 μm; Nucleosil C_{18}	UV	
1978 [53]	Blumer et al.	Coal tar	5 μm; Nucleosil NO_2	UV	Separated 100 components
1978 [114]	Eisenberg	Air-particulate extracts	Porasil, Bondapak C_{18}	UV at 254 nm	Fractionation on silica, then C to identify BeP, BkF, and BaP
1978 [93]	Sinclair and Frost	Cigarette smoke condensate	Partisil, 5 μm	Filter fluorometer	Determination of BaP
1978 [55,56]	Hunt et al.	6 PAH on WHO list	Bonded phthalimodopropyltrichlorosilane		

(continued)

Table 11 (continued)

Publication year [ref.]	Author(s)	Application	Column	Detection	Comments
1978 [84]	Christensen and May	Oysters, standards	μBondapak C_{18}	Multiwavelength UV and fluorescence	Comparison of detectors for PAH
1978 [91]	Das and Thomas		Zorbax ODS	Filter fluorometer	
1978 [80]	Smillie et al.	Air-particulate extracts	μBondapak C_{18}, Partisil 10-ODS MicroPak CH and Vydac 201TP	Fluorescence: λ_{ex} = 365 nm; λ_{em} = 415 nm Variable wavelength UV	Compared four columns for performance
1978 [142]	Kasiske et al.	Water	Nucleosil C_{18}, 5 μm	Filter fluorometer	
1978 [120]	Ogan et al.	Water--trace enrichment	HC-ODS	Spectrofluorometer: λ_{ex} = 383 nm; λ_{em} = 454 nm	
1978 [7]	Herod and James	Coke oven emissions			Review article
1978 [81,82]	Fechner and Seifert	Air-particulate extracts	LiChrosorb 5 RP-18	Multiwavelength UV and fluorescence	Quantitation of 9 PAH

1979 [46]	Roumeliotis et al.	Diesel exhaust	Silica, then C_{18}	UV at 254 nm	10 different columns compared: C_{18}, C_8, NH_2, and NO_2
1979 [143]	Hanus et al.	Standards and oysters	μBondapak C_{18}	Fluorescence	Clean-up on Styragel 100Å
1979 [79]	Sorrell and Reding	Water extracts	μBondapak C_{18}	Multiwavelength UV and spectrofluorometer	Absorbance ratios for identification
1979 [144]	Euston and Baker	Water--trace enrichment	LiChrosorb RP-18	UV scanning for identification	
1979 [128]	Felscher and Stein	Coal tar	LiChrosorb RP-18	UV at 300 nm	
1979 [9]	Bartle et al.	Coal-derived products			Review article
1979 [25]	Nielsen	Auto exhaust	Zorbax ODS	Filter fluorometer	TLC prefractionation
1979 [62]	Snook et al.	Cigarette smoke condensate	Zorbax ODS	UV at 254 nm	Analysis of GC peaks for alkyl PAH
1979 [113]	Wilkinson et al.	Water	HC-ODS	Fluorescence	
1979 [88]	Ogan et al.	Effluent samples	HC-ODS	Fluorescence, filter and monochrometer	Comparison of fluorescence selectivity

(continued)

Table 11 (continued)

Publication year [ref.]	Author(s)	Application	Column	Detection	Comments
1979 [104]	Shelly et al.	Standards	Spherisorb ODS, 10 μm	Rapid scanning fluorometer	Obtain excitation-emission matrices
1979 [47]	Chmielowiec and Sawatzky	Standards	Chromegabond C_{18}	UV at 254 nm fluorescence	Studied effect of temperature
1979 [52]	Lankmayr and Müller	Dust particles	LiChrosorb RP-18, NH_2 and Nucleosil 5 NO_2	UV at 384 nm for pentacyclic PAH	Comparison of different columns
1979 [54]	Lochmüller et al.	Methylcholanthrenes and methyl-BaA	LiChrosorb SI-60 with nitroaromatic bonded group		
1979 [117]	Preston and Macaluso	Refinery water	Spherisorb ODS	Variable UV and fixed 254 nm	Use peak ratios at different wavelengths
1979 [57]	Nondek et al.	Standards	3-(2,4-Dinitroanilino)-propyl group	UV at 254 nm	Retention data for 25 PAH
1979 [122]	Hagenmaier and Jaeger	Water, sludge, and sediment			
1979 [125]	Takami et al.	Marine sediments	LiChrosorb RP-18	UV and fluorescence	TLC cleanup
1979 [123]	Black et al.	Sediments	μBondapak C_{18}	UV at 254 nm	Silica cleanup

1980 [14]	Wise et al.	Air-particulate extracts	μBondapak NH_2, 8 different C_{18} columns		Comparison of 8 different C_{18} columns
1980 [15]	Chmielowiec and George	Standards	Diamine, C_{18}, and other polar-bonded phases		Comparison of polar bonded phases, retention data for 50 PAH on diamine and C_{18} columns
1980 [145,146]	Dunn	Marine samples	HC-ODS	UV and fluorescence	
1980 [121]	Katz and Ogan	Refinery waste water	HC-ODS	UV and fluorescence	
1980 [45]	Katz and Ogan	Standards	HC-ODS		Studied effect of mobile phase
1980 [95]	Durand and Petroff	Petroleum oils	Silica-DMSO	Filter fluorometer	Determination of BaP
1980 [58]	Matlin et al.	Standards	Picramidopropyl silica		
1980 [105]	Richardson et al.	Standards and condensed steam distillate from coal gasification	Partisil ODS	Time-resolved fluorescence detection	
1980 [28]	Colmsjö and MacDonald	Standards	LiChrosorb RP-18	UV	Comparison of 3 different C_{18} columns

(continued)

Table 11 (continued)

Publication year [ref.]	Author(s)	Application	Column	Detection	Comments
1980 [147]	Greenberg et al.	Air-particulate filters	HC-ODS	UV (280 and 365 nm) and fluorescence λ_{ex} = 360 nm; λ_{em} > 440 nm)	
1980 [148]	Maya and Danis	Standards	Bis(octadecyl phosphate) Zirconium(IV)	UV	Novel stationary phase with heavy carbon coverage
1980 [53]	Mourey et al.	Shale oil & standards	Bonded pyrrolidone and C_{18}	UV at 254 nm	Retention data for about 40 PAH
1980 [149]	Siefert and Lahmann	Dust samples	LiChrosorb RP-18	Multiwavelength UV	Preseparation on silica
1980 [150,151]	Müller and Rohbock	Particulate matter	LiChrosorb RP-8	UV at 254 nm	
1980 [152]	Katoh et al.	Coal liquid	μBondapak NH_2 LS-410 ODS	Synchronous fluorescence	Fractionation on amine column followed by reverse-phase
1980 [153]	Atwood and Goldstein	Quality control of C_{18} columns	Vydac TP material	UV at 254 nm	Study of lot-to-lot variations of selectivity of C_{18} material
1980 [154]	Liphard	Coal liquids	LiChrosorb NH_2	UV at 265 nm	Retention data for over 40 PAH and hydro-PAH

1980 [155]	Colmsjö and Ostman	Petroleum pyrolysis product		UV at 289 nm	Fractionation by reverse-phase LC with identification of PAH using Shpol'skii fluorescence spectroscopy
1980 [156]	Dunn and Armour	Marine samples	HC-ODS	Filter fluorometer and multi-wavelength UV	
1980 [157]	Belinsky	Coal tar pitch	Vydac TP	UV at 254 nm	Optimization of separation with ternary solvent system
1980 [158]	Smith and Strickler	Extracts of coal liquid process waters	Zorbax ODS	UV at 254 nm fluorescence filter	Quarternary solvent system
1980 [159]	Hellman	Water and soil samples	HC-ODS	Fluorescence	Selective determination of perylene at 430 nm excitation and 466 nm emission
1980 [160]	Van Heddeghem et al.	Fat products	LiChrosorb RP-18	Fluorescence	
1980 [161]	Szepesy and Czencz	Standards and water	μBondapak C_{18}	UV at 254 nm	
1981 [162]	Amos	Standards	30 different columns compared	UV	23 reverse-phase and 7 normal-phase columns compared for 15 PAH

(continued)

Table 11 (continued)

Publication year [ref.]	Author(s)	Application	Column	Detection	Comments
1981 [163]	Grant and Meiris	High temperature pitch	Partisil 5	Spectrofluorometer, spectra used for identification	Comparison of 5 C_{18} columns, multidimensional analysis using LH-20/silica/TLC
1981 [164]	Obana et al.	Marine samples: sediment, oysters, and seaweed	LiChrosorb RP-18	Fluorescence	Silica cleanup
1981 [165]	Chmielowiec et al.	Oil	Diamine $[R(NH_2)_2]$ C_{18}	UV/Fluorescence	
1981 [166]	Deymann and Holstein	Coal-derived recycled oil	Silica, alumina, amine, C_{18} and tetrachlorophthalic acid (TCI)		Retention data for C_1-C_6, C_{16}-substituted pyrene on 5 columns, discussion of group-type separations
1981 [167]	Fogarty et al.	Standards	Ultrasphere ODS	Video fluorometry	
1981 [168]	Miller et al.	Process oil	Size exclusion/HC-ODS	Fluorescence	
1981 [169]	Crosby et al.	Food, water, mussels, and smoke	LiChrosorb ODS and phthalimidopropylsilane	Fluorometer	

1981 [170]	Sepaniak and Yeung	Coal liquid	C_{18}	UV, visible fluorescence, and two photon-excited fluorescence	
1981 [171]	Lee et al.	Oysters	Spherisorb ODS, 5/10 μm	Fluorometer	Identified 9 PAH in samples
1981 [172]	Daisey et al.	Arctic aerosol	Radial-Pak C_{18}	UV at 254/313 nm and fluorometer	Quantitation of 11 PAH
1981 [173]	Readman et al.	Sediment	Hypersil ODS	Scanning UV/UV at 254 nm	
1981 [174]	Berthou et al.	Oysters	LiChrisorb NH_2 5 μm	UV at 254 nm	Normal-phase LC to prefractionate samples prior to GC
1981 [175]	Holstein	Coal liquids	Tetrachlorophathalimidopropyl silica	UV at 280 nm	Retention data for 40 PAH and 20 nitrogen aromatics
1981 [176]	Chmielowiec	Standards	Mercury phenylacetate column	UV at 254 nm	Retention data for over 60 PAH and 30 S-, N-, and O-containing aromatics
1981 [177]	May et al.	Shale oil	μBondapak NH Vydac TP/Zorbax ODS	UV at 254 nm Spectrofluorometer	Discussion of methods for quantitation of PAH in shale oil

(continued)

Table 11 (continued)

Publication year [ref.]	Author(s)	Application	Column	Detection	Comments
1981 [178]	Katz and Ogan	Petroleum and coal liquids	Size exclusion (SEC)/HC-ODS	Fluorescence	Fractionation on SEC column followed by separation of PAH on C_{18}
1981 [179]	Choudhury	Air-particulate samples	HC-ODS/Zorbax ODS	Scanning UV	Identification of PAH by UV spectra
1981 [180]	Choudhury and Bush	Air-particulate samples	Zorbax ODS	Scanning UV	Identification of PAH by UV spectra
1981 [181]	Marsh and McNair	Water	Vydac TP	UV/Fluorescence	Concentration of PAH from water samples
1981 [182]	Seizinger	Standards		Video fluorometry	
1981 [183]	Sirota and Uthe	Marine shellfish	Vydac TP	Multiwavelength UV and fluorescence	Quantitation of PAH in shellfish
1981 [184]	Black et al.	Sediment and fish tissue		UV at 254 nm and fluorescence	
1981 [185]	Cole et al.	16 priority pollutant PAH spiked in water and refinery effluent	HC-ODS PAH column	UV and fluorescence	Evaluation of EPA Method 610 for PAH

1981 [186]	Gibson et al.	Diesel-exhaust particulate material		Spectro-fluorometer	Determination of PAH, nitro-PAH and amino-PAH
1981 [100]	Konash et al.	Coal tar and shale oil	Vydac TP Zorbax ODS	Spectro-fluorometer and fluor-ometer	Use of nitromethane to quench non-fluoranthenic PAH
1981 [40]	Wise et al.	Standards	Vydac TP Zorbax ODS	UV at 254 nm	Correlation of re-tention and shape for over 100 PAH including numer-ous methyl sub-stituted PAH
1981 [187]	Isaaq et al.	Standards	Vydac TP	UV at 254 nm	Separation of methyl-BaP iso-mers on liquid crystal GC and reverse-phase LC
1981 [10]	Bartle et al.	Review			Review of analytical methods for PAH (includes LC)
1981 [188]	Hirata et al.	Carbon black extract	Capillary-bonded alumina C_{18}	Fluorescence	Comparison of capillary LC and GC for high-molecular-weight PAH
1981 [189]	Joe et al.	Beer	LiChrosorb RP-18	UV at 287 nm	
1981 [190]	Tomkins et al.	Natural and synthetic crudes	Partisil PAC Zorbax ODS	Fluorometer	Quantitation of BaP in oils

(continued)

Table 11 (continued)

Publication year [ref.]	Author(s)	Application	Column	Detection	Comments
1981 [191]	Colin et al.	White oils	MicroPak MCH-10	Spectro-fluorometer	Identification of peaks by Shpol'skii fluorescence
1981 [192]	Voigtman	Standards	None	Laser-excited fluorescence and photo-acoustic	Comparison of detection limits in flow cells for PAH
1981 [193]	Hershberger et al.	Shale oil, standards	Spherisorb ODS	Video fluorometry	Quantitation of BaP in shale oil
1981 [194]	Takeuchi and Ishii	Standards	Packed capillary ODS SC-01 5 μm (10 cm × 0.25 mm i.d.)	UV at 254 nm	Separation of standards
1981 [195]	Wolkoff and Creed	Water	Sep-Pak C_{18}	UV at 254 nm Filter fluorometer	
1981 [196]	DiCesare et al.	Standards Solvent refined coal	C_{18} packing, 3 μm	UV at 254 nm	
1982 [197	Sonnefeld et al.	Coal liquid and shale oil	μBondapak NH Vydac TP, 5 μm	UV at 254 nm and fluorescence	On-line multidimensional quantitation of PAH in coal liquid and shale oil
1982 [198]	Wise et al.	Air-particulate material	Vydac TP, 5 μm	UV at 254 nm and fluorescence	Quantitation of PAH in air material

Coal tar contains a relatively high proportion of high-molecular-weight PAH and aza-aromatics (only about 30% of coal tar is amenable to analysis by GC). Several groups of workers (e.g., Refs. 126 and 127) have reported the use of silica and alumina for the analysis of coal tar extracts. Grant and Meiris [127] described a normal-phase LC method on silica for the quantitation of a number of PAH using UV absorption detection at four different wavelengths. Boden [83] measured benzo[a]pyrene in coal tar volatiles by reverse-phase LC with selective UV detection after prefractionation on silica. Recently, Blumer et al. [52,28] reported the separation of approximately 100 constituents of coal tar pitch using a nitrophenyl phase. In this application approximately 70% of the tar could be eluted compared to only 30% by GC. On-line UV spectroscopy provided identification of many of the PAH [76]. The nitrophenyl phase provides a group separation of the PAH and aza-aromatics [42].

Separations of high-molecular-weight PAH in both normal- and reverse-phase systems require stronger mobile phase compositions than are generally employed in order to achieve sufficient solubility of the sample. Blumer and Zander [128] achieved normal-phase separations of PAH of molecular weight 400-600 on the nitrophenyl phase using 60-80% chloroform in hexane. In a reverse-phase system using a C_8 column, Felscher and Stein [129] used a mobile-phase gradient from 40 to 100% acetonitrile/dimethylformamide (75:25) in water to separate the coal tar constituents. Numerous PAH up to molecular weight 448 were separated and identified in a carbon black extract using a C_{18} column with a gradient from 50% water/acetonitrile to 100% ethylacetate and finally to 100% methylene chloride [130] (see Fig. 19).

F. Review of HPLC Applications for Determination of PAH

Table 11 summarizes the applications of HPLC for the determination of PAH. As indicated by the publication year of each application, the use of this technique has grown steadily since 1971 with a predominance of reverse-phase applications in recent years. Additional references for 1980-1981 were included in Table 11 after this chapter was completed. Thus, these references are not discussed in the text, but are included to provide more recent applications [131-198].

References

1. L. R. Snyder and J. J. Kirkland, *Introduction to Modern Liquid Chromatography*, 2nd ed., Wiley (Interscience), New York, 1979.
2. E. L. Johnson and R. Stevenson, *Basic Liquid Chromatography*, 2nd ed., Varian Associates, Palo Alto, Calif., 1978.
3. N. A. Parris, *Instrumental Liquid Chromatography: A Practical Manual*, Elsevier, New York, 1976.
4. R. J. Hamilton and P. A. Sewell, *Introduction to High Performance Liquid Chromatography*, Chapman & Hall, London, 1978.
5. A. Pryde and M. T. Gilbert, *Applications of High Performance Liquid Chromatography*, Wiley (Halsted Press), New York, 1978.
6. D. C. Hunt, P. J. Wild, and N. T. Crosby, *Rappt. Proces-Verbaux Reunions Conseil Perm. Intern. Exploration Mer. 171*:41 (1977).

7. A. A. Herod and R. G. James, *J. Inst. Fuel 408*:164 (1978).
8. R. Thoms and M. Zander, *Erdöl Kohle Erdgas Petrochem. 30*:403 (1977).
9. K. D. Bartle, G. Collin, J. W. Stadelhofer, and M. Zander, *J. Chem. Technol. Biotechnol. 29*:531 (1979).
10. K. D. Bartle, M. L. Lee, and S. A. Wise, *Chem. Soc. Rev. 10*:113 (1981).
11. M. Popl, V. Dolanský, and J. Mostecky, *J. Chromatogr. 91*:649 (1974).
12. M. Popl, V. Dolanský, and J. Mostecky, *J. Chromatogr. 117*:117 (1976).
13. S. A. Wise, S. N. Chesler, H. S. Hertz, L. R. Hilpert, and W. E. May, *Anal. Chem. 49*:2306 (1977).
14. S. A. Wise, W. J. Bonnett, and W. E. May, in *Polynuclear Aromatic Hydrocarbons: Chemistry and Biological Effects*, A. Bjørseth and A. J. Dennis (Eds.), Battelle Press, Columbus, Ohio, 1980, p. 791.
15. J. Chmielowiec and A. E. George, *Anal. Chem. 52*:1154 (1980).
16. M. L. Lee, D. L. Vassilaros, C. M. White, and M. Novotny, *Anal. Chem. 51*:768 (1979).
17. R. Majors, *J. Chromatogr. Sci. 15*:333 (1977).
18. L. H. Klemm, D. Reed, L. A. Miller, and B. T. Ho, *J. Org. Chem. 24*: 1468 (1959).
19. L. R. Snyder, *J. Phys. Chem. 67*:234 (1963).
20. L. R. Snyder, *J. Phys. Chem. 67*:240 (1963).
21. L. R. Snyder, *Principles of Adsorption Chromatography*, Dekker, New York, 1968.
22. M. Popl, V. Dolanský, and J. Mostecky, *Coll. Czech. Chem. Commun. 39*: 1836 (1974).
23. J. A. Schmit, R. A. Henry, R. C. Williams, and J. F. Dieckman, *J. Chromatogr. Sci. 9*:645 (1971).
24. B. B. Wheals, C. G. Vaughan, and M. J. Whitehouse, *J. Chromatogr. 106*:109 (1975).
25. G.-P. Blumer and M. Zander, *Z. Anal. Chem. 288*:277 (1977).
26. T. Nielsen, *J. Chromatogr. 170*:147 (1979).
27. K. Ogan and E. Katz, *J. Chromatogr. 188*:115 (1980).
28. A. L. Colmsjö and J. C. MacDonald, *Chromatographia 13*:350 (1980).
29. K. Ogan, E. Katz, and W. Slavin, *Anal. Chem. 51*:1315 (1979).
30. N. H. C. Cooke and K. Olsen, *Amer. Lab. 11(8)*:45 (1979).
31. J. J. DeStefano, A. P. Goldberg, J. P. Larmann, and N. A. Parris, *Ind. Res. Dev. 22(4)*:99 (1980).
32. H. Colin and G. Guiochon, *J. Chromatogr. 141*:289 (1977).
33. S. A. Wise, National Bureau of Standards, unpublished data (1979).
34. K. K. Unger, N. Becker, and P. Roumeliotis, *J. Chromatogr. 125*:115 (1976).
35. H. Hemetsberger, M. Kellermann, and H. Ricken, *Chromatographia 10*: 726 (1977).
36. H. Hemetsberger, P. Behrensmeyer, J. Henning, and H. Ricken, *Chromatographia 12*:71 (1979).
37. K. Ogan and E. Katz, Chromatogr. Rept. No. 72, Perkin-Elmer Corp., Norwalk, Conn. (1979).
38. E. Katz and K. Ogan, *J. Liquid Chromatogr. 3*:1151 (1980).
39. K. Ogan, personal communication (1979).
40. S. A. Wise, W. J. Bonnett, F. R. Guenther, and W. E. May, *J. Chromatogr. Sci. 19*:457 (1981).

41. R. B. Sleight, *J. Chromatogr.* *83*:31 (1973).
42. D. C. Locke, *J. Chromatogr. Sci.* *12*:433 (1974).
43. A. Radecki, H. Lamparczyk, and R. Kaliszan, *Chromatographia* *12*:595 (1979).
44. E. Katz and K. Ogan, *Chromatogr. Newslett.* *8*:20 (1980).
45. P. Roumeliotis, K. K. Unger, G. Tesarek, and E. Mühlberg, *Z. Anal. Chem.* *298*:241 (1979).
46. J. Chmielowiec and H. Sawatzky, *J. Chromatogr. Sci.* *17*:245 (1979).
47. L. R. Snyder, *J. Chromatogr.* *179*:167 (1979).
48. K.-P. Hupe and H. Schrenker, *Chromatographia* 5:44 (1972).
49. M. Novotny, M. L. Lee, and K. D. Bartle, *J. Chromatogr. Sci.* *12*:606 (1974).
50. M. Novotny, in *Bonded Stationary Phases in Chromatography*, E. Gruska (Ed.), Ann Arbor Science Publs., Ann Arbor, Mich., 1974, p. 199.
51. E. P. Lankmayr and K. Müller, *J. Chromatogr.* *170*:139 (1979).
52. G.-P. Blumer, R. Thoms, and M. Zander, *Erdöl Kohle Erdgas. Petrochem.* *31*:197 (1978).
53. T. H. Mourey, S. Siggia, P. C. Uden, and R. J. Crowley, *Anal. Chem.* *52*:885 (1980).
54. C. H. Lochmüller, R. R. Rydall, and C. W. Amoss, *J. Chromatogr.* *178*:298 (1979).
55. D. C. Hunt, P. J. Wild, and N. T. Crosby, *J. Chromatogr.* *130*:320 (1977).
56. D. C. Hunt, P. J. Wild, and N. T. Crosby, *J. Chromatogr.* *130*:643 (1978).
57. L. Nondek, M. Minárik, and J. Málek, *J. Chromatogr.* *178*:427 (1979).
58. S. A. Martin, W. J. Lough, and D. G. Bryan, *J. High-Resolution C Chromatogr.-Chromatogr. Commun.* *3*:33 (1980).
59. A. P. Graffeo, Assoc. Offic. Anal. Chem. Meeting, Washington, D.C., Oct. 19, 1977; see Ref. 1, p. 500.
60. T. Edstrom and B. A. Petro, *J. Polymer Sci. C.* *21*:171 (1968).
61. K. Ogan and E. Katz, *Anal. Chem.* *54*:169 (1982).
62. M. E. Snook, R. F. Sevenson, H. C. Higman, R. F. Arrendale, and O. T. Chortyk, in *Polynuclear Aromatic Hydrocarbons*, P. W. Jones and P. Leber (Eds.), Ann Arbor Science Publs., Ann Arbor, Mich., 1979, p. 231.
63. H.-J. Klimisch, *J. Chromatogr.* *83*:11 (1973).
64. H.-J. Klimisch, *Anal. Chem.* *45*:1960 (1973).
65. H.-J. Klimisch and D. Ambrosius, *J. Chromatogr.* *120*:299 (1976).
66. G. Goldstein, *J. Chromatogr.* *129*:61 (1976).
67. M. Popl, J. Fähnrich, and M. Stejskal, *J. Chromatogr. Sci.* *14*:537 (1976).
68. M. Popl, V. Dolanský, and J. Coupek, *J. Chromatogr.* *130*:195 (1977).
69. R. P. W. Scott and P. Kucera, *J. Chromatogr.* *125*:251 (1976).
70. R. P. W. Scott and P. Kucera, *J. Chromatogr.* *169*:51 (1979).
71. R. P. W. Scott, *J. Chromatogr. Sci.* *18*:49 (1980).
72. Y. Hirata, M. Novotny, T. Tsuda, and D. Ishii, *Anal. Chem.* *51*:1807 (1979).
73. Y. Hirata and M. Novotny, *J. Chromatogr.* *186*:521 (1980).
74. D. Ishii, T. Tsuda, K. Hibi, T. Takeuchi, and T. Nakanishi, *J. High-Resolution Chromatogr.-Chromatogr. Commun.* *2*:372 (1979).

75. R. A. Friedel and M. Orchin, *Ultraviolet Spectra of Aromatic Compounds*, Wiley, New York, 1951.
76. R. Thoms and M. Zander, *Z. Anal. Chem. 282*:443 (1976).
77. A. M. Krstulovic, D. M. Rosie, and P. R. Brown, *Anal. Chem. 48*:1383 (1976).
78. A. M. Krstulovic, D. M. Rosie, and P. R. Brown, *Amer. Lab. 9(7)*:11 (1977).
79. R. K. Sorrell and R. Reding, *J. Chromatogr. 185*:655 (1979).
80. R. D. Smillie, D. T. Wang, and O. Meresz, *J. Environ. Sci. Health A13*: 47 (1978).
81. D. Fechner and B. Seifert, *Z. Anal. Chem. 292*:199 (1978).
82. D. Fechner and B. Seifert, in *Polynuclear Aromatic Hydrocarbons*, P. W. Jones and P. Leber (Eds.), Ann Arbor Science Publs., Ann Arbor, Mich., 1979, p. 191.
83. H. Boden, *J. Chromatogr. Sci. 14*:391 (1976).
84. R. G. Christensen and W. E. May, *J. Liquid Chromatogr. 1*:385 (1978).
85. E. D. Pellizzari and C. M. Sparacino, *Anal. Chem. 45*:378 (1973).
86. W. Slavin, A. T. Rhys Williams, and R. F. Adams, *J. Chromatogr. 134*: 121 (1977).
87. B. B. Wheals, C. G. Vaughan, and M. J. Whitehouse, *J. Chromatogr. 78*:203 (1973).
88. K. Ogan, E. Katz, and T. J. Porro, *J. Chromatogr. Sci. 17*:597 (1979).
89. R. A. Velapoldi and S. A. Wise, National Bureau of Standards, unpublished data (1979).
90. J. F. McKay and D. R. Latham, *Anal. Chem. 44*:2132 (1972).
91. B. S. Das and G. H. Thomas, *Anal. Chem. 50*:967 (1978).
92. D. E. Seizinger, *Trends Fluorescence 1*:9 (1978).
93. N. M. Sinclair and B. E. Frost, *Analyst 103*:1199 (1978).
94. H. S. Hertz, J. M. Brown, S. N. Chesler, F. R. Guenther, L. R. Hilpert, W. E. May, R. M. Parris, and S. A. Wise, *Anal. Chem. 50*: 1650 (1980).
95. J. P. Durand and N. Petroff, *J. Chromatogr. 190*:85 (1980).
96. J. M. Brown, S. A. Wise, and W. E. May, *J. Environ. Sci. Health A15(6)*:613 (1980).
97. M. A. Fox and S. W. Staley, *Anal. Chem. 48*:992 (1976).
98. T. Bradley, *Chromatogr. Rev. 4(3)*:8 (1978).
99. G.-P. Blumer and M. Zander, *Z. Anal. Chem. 296*:409 (1979).
100. P. L. Konash, S. A. Wise, and W. E. May, *J. Liquid Chromatogr. 4*: 1339 (1981).
101. E. Sawicki, T. W. Stanley, and W. C. Elbert, *Talanta 11*:1433 (1964).
102. H. Dreeskamp, E. Koch, and M. Zander, *Z. Naturforsch. 30a*:1311 (1975).
103. J. R. Jadamec, W. A. Saner, and Y. Talmi, *Anal. Chem. 49*:1316 (1977).
104. D. C. Shelly, W. A. Ilger, M. P. Fogarty, and I. M. Warner, *Altex Chromatogram 3*(1):4 (1979).
105. J. H. Richardson, K. M. Larson, G. R. Haugen, D. C. Johnson, and J. E. Clarkson, *Anal. Chim. Acta 116*:407 (1980).
106. P. J. Arpino and G. Guiochon, *Anal. Chem. 51*:682A (1979).
107. W. H. McFadden, *J. Chromatogr. Sci. 18*:97 (1980).

108. R. G. Christensen, H. S. Hertz, S. Meiselman, and E. White V, *Anal. Chem. 53*(2):171 (1981).
109. W. A. Dark, W. H. McFadden, and D. L. Bradford, *J. Chromatogr. Sci. 15*:454 (1977).
110. W. H. McFadden, *J. Chromatogr. Sci. 17*:2 (1979).
111. W. A. Dark and W. H. McFadden, *J. Chromatogr. Sci. 16*:289 (1978).
112. M. Popl, M. Stejskal, and J. Mostecky, *Anal. Chem. 46*:1581 (1974).
113. J. E. Wilkinson, P. E. Strup, and P. W. Jones, in *Polynuclear Aromatic Hydrocarbons*, P. W. Jones and P. Leber (Eds.), Ann Arbor Science Publs., Ann Arbor, Mich., 1979, p. 217.
114. W. C. Eisenberg, *J. Chromatogr. 16*:145 (1978).
115. C. Golden and E. Sawicki, *Anal. Lett. 9*:957 (1976).
116. M. Dong, D. C. Locke, and E. Ferrand, *Anal. Chem. 48*:368 (1976).
117. H. G. Preston and A. Macaluso, in *Measurement of Organic Pollutants in Water and Wastewater*, ASTM Publ. No. STP 686, C. E. Van Hall (Ed.), American Society for Testing and Materials, Philadelphia, 1979, p. 152.
118. W. E. May, S. N. Chesler, S. P. Cram, B. H. Gump, H. S. Hertz, D. P. Enagonio, and S. M. Dyszel, *J. Chromatogr. Sci. 13*:535 (1975).
119. F. Eisenbeiss, H. Hein, R. Joester, and G. Naundorf, *Chem.-Tech. 6*: 227 (1977); Engl. transl. in *Chromatogr. Newslett. 6*(1):8 (1978).
120. K. Ogan, E. Katz, and W. Slavin, *J. Chromatogr. Sci. 16*:517 (1978).
121. E. Katz and K. Ogan, *Chromatogr. Newslett. 8*:18 (1980).
122. H. Hagenmaier, R. Feierabend, and W. Jäger, *Z. Wasser-u. Abwasser-Forsch. 10*:99 (1977).
123. J. J. Black, P. P. Dymerski, and W. F. Zapisek, *Bull. Environ. Contam. Toxicol. 22*:278 (1979).
124. Method 610: Polycyclic Aromatic Hydrocarbons, *Federal Register, 44*(233): 69514 (Dec. 3, 1979).
125. S. J. Swarin and R. L. Williams, in *Polynuclear Aromatic Hydrocarbons: Chemistry and Biological Effects*, A. Bjørseth and A. J. Dennis (Eds.), Battelle Press, Columbus, Ohio, 1980, p. 771.
126. M. Popl, V. Dolansky, and J. Mostecky, *J. Chromatogr. 59*:329 (1971).
127. D. W. Grant and R. B. Meiris, *J. Chromatogr. 142*:339 (1977).
128. G.-P. Blumer and M. Zander, *Compendium 78/79*, Suppl. to *Erdöl Kohle Erdgas Petrochem.* p. 1472 (1978).
129. D. Felscher and J. Stein, *Z. Chem. 19*:303 (1979).
130. P. Peaden, M. L. Lee, Y. Hirata, and M. Novotny, *Anal. Chem. 52*: 2268 (1980).
131. V. Martinu and J. Janak, *J. Chromatogr. 65*:477 (1972).
132. N. F. Ives and L. Giuffrida, *J. Assoc. Offic. Anal. Chem. 55*:757 (1972).
133. W. Strubert, *Chromatographia 6*:205 (1973).
134. B. L. Karger, M. Martin, J. Loheac, and G. Guiochon, *Anal. Chem. 45*: 496 (1973).
135. D. W. Grant, R. B. Meiris, and M. G. Hollis, *J. Chromatogr. 99*:721 (1974).
136. A. F. Haeberer, M. E. Snook, O. T. Chortyk, *Anal. Chim. Acta 80*: 303 (1975).
137. Carbonization Research Rept. No. 28, The British Carbonization Research Assoc., Chesterfield, Derbyshire, England (1976), 19 pp.

138. H. Guerrero, E. R. Biehl, and C. T. Kenner, *J. Assoc. Offic. Anal. Chem.* *59*: 989 (1976).
139. A. T. R. Williams and W. Slavin, *Chromatogr. Newslett.* *4*: 28 (1976).
140. M. Popl, V. Dolansky, and J. Coupek, *J. Chromatogr.* *130*: 195 (1977).
141. E. Johnson, A. Abu-Shumays, and S. R. Abbot, *J. Chromatogr.* *134*: 107 (1977).
142. D. Kasiske, K. D. Klinkmuller, and M. Sonneborn, *J. Chromatogr.* *149*: 703 (1978).
143. J. P. Hanus, H. Guerrero, E. R. Biehl, and C. T. Kenner, *J. Assoc. Offic. Anal. Chem.* *62*: 29 (1979).
144. C. B. Euston and D. R. Baker, *Amer. Lab.* *11*(3): 91 (1979).
145. B. Dunn, *Chromatogr. Newslett.* *8*: 10 (1980).
146. B. Dunn, in *Polynuclear Aromatic Hydrocarbons: Chemistry and Biological Effects*, A. Bjørseth and A. J. Dennis (Eds.), Battelle Press, Columbus, Ohio, 1980, p. 367.
147. A. Greenberg, R. Yokoyama, P. Giorgio, and F. Cannova, in *Polynuclear Aromatic Hydrocarbons: Chemistry and Biological Effects*, A. Bjørseth and A. J. Dennis (Eds.), Battelle Press, Columbus, Ohio, 1980, p. 193.
148. L. Maya and P. O. Danis, *J. Chromatogr.* *190*: 145 (1980).
149. B. Seifert and E. Lahmann, *VDI-Ber.* *358*: 127 (1980).
150. J. Muller and E. Rohbock, *Talanta* *27*: 673 (1980).
151. J. Muller, *VDI-Ber.* *358*: 133 (1980).
152. T. Katoh, S. Yokoyama and Y. Sanada, *Fuel* *59*: 845 (1980).
153. J. G. Atwood and J. Goldstein, *J. Chromatogr. Sci.* *18*: 650 (1980).
154. K. G. Liphard, *Chromatographia* *13*: 603 (1980).
155. A. L. Colmsjö and C. E. Östman, *Anal. Chem.* *52*: 2093 (1980).
156. B. P. Dunn and R. J. Armour, *Anal. Chem.* *52*: 2027 (1980).
157. B. R. Belinky, in *Analytical Techniques in Occupational Health Chemistry*, American Chemical Society Symposium Ser., 1980, p. 149.
158. T. R. Smith and V. A. Strickler, *J. High Resolution Chromatogr.-Chromatogr. Commun.* *3*: 634 (1980).
159. H. Hellmann, *Z. Anal. Chem.* *302*: 115 (1980).
160. A. Van Heddeghem, A. Huyghebaert, and H. DeMoor, *Z. Lebensm. Untersuch. Forsch.* *171*: 9 (1980).
161. L. Szepesy and M. Czencz, *Periodica Polytech.* *24*: 123 (1980).
162. R. Amos, *J. Chromatogr.* *204*: 469 (1981).
163. D. W. Grant and R. B. Meiris, *J. Chromatogr.* *203*: 293 (1981).
164. H. Obana, S. Hori, and T. Kashimoto, *Bull. Environ. Contam. Toxicol.* *26*: 613 (1981).
165. J. Chmielowiec, J. E. Beshai, and A. E. George, *Fuel* *59*: 838 (1981).
166. H. Deymann and W. Holstein, *Erdöl Kohle Erdgas. Petrochem.* *34*: 353 (1981).
167. M. P. Fogarty, D. C. Shelly, and I. M. Warner, *J. High-Resolution Chromatogr.-Chromatogr. Commun.* *4*: 561 (1981).
168. R. L. Miller, K. Ogan, and A. F. Poile, *Amer. Lab.* *13*(7): 52 (1981).
169. N. T. Crosby, D. C. Hunt, L. A. Philp, and I. Patel, *Analyst* *106*: 135 (1981).
170. M. J. Sepaniak and E. S. Yeung, *J. Chromatogr.* *211*: 95 (1981).
171. R. F. Lee, D. Lehsau, and M. Madden, in *Proc. 1981 Oil Spill Conf. (Prevention, Behavior, Control, Cleanup)*, Atlanta, Mar. 2-5, 1981, p. 341.

172. J. M. Daisey, R. J. McCaffrey, and R. A. Gallagher, *Atmos. Environ. 15*:1353 (1981).
173. J. W. Readman, L. Brown, and M. M. Rhead, *Analyst 106*:122 (1981).
174. F. Berthou, Y. Gourmelun, Y. Dreano, and M. P. Friocourt, *J. Chromatogr. 203*:279 (1981).
175. W. Holstein, *Chromatographia 14*:468 (1981).
176. J. Chmielowiec, *J. Chromatogr. Sci. 19*:296 (1981).
177. W. E. May, J. Brown-Thomas, L. R. Hilpert, and S. A. Wise, in *Chemical Analysis and Biological Fate: Polynuclear Aromatic Hydrocarbons*, M. Cooke and A. J. Dennis (Eds.), Battelle Press, Columbus, Ohio, 1981, p. 1.
178. E. Katz and K. Ogan, in *Chemical Analysis and Biological Fate: Polynuclear Aromatic Hydrocarbons*, M. Cooke and A. J. Dennis (Eds.), Battelle Press, Columbus, Ohio, 1981, p. 169.
179. D. R. Choudhury, in *Chemical Analysis and Biological Fate: Polynuclear Aromatic Hydrocarbons*, M. Cooke and A. J. Dennis (Eds.), Battelle Press, Columbus, Ohio, 1981, p. 265.
180. D. R. Choudhury and B. Bush, *Anal. Chem. 53*:1351 (1981).
181. D. G. Marsh and H. McNair, in *Chemical Analysis and Biological Fate: Polynuclear Aromatic Hydrocarbons*, M. Cooke and A. J. Dennis (Eds.), Battelle Press, Columbus, Ohio, 1981, p. 297.
182. D. E. Seizinger, in *Chemical Analysis and Biological Fate: Polynuclear Aromatic Hydrocarbons*, M. Cooke and A. J. Dennis (Eds.), Battelle Press, Columbus, Ohio, 1981, p. 307.
183. G. R. Sirota and J. F. Uthe, in *Chemical Analysis and Biological Fate: Polynuclear Aromatic Hydrocarbons*, M. Cooke and A. J. Dennis (Eds.), Battelle Press, Columbus, Ohio, 1981, p. 329.
184. J. J. Black, T. E. Hart, Jr., and E. Evans, in *Chemical Analysis and Biological Fate: Polynuclear Aromatic Hydrocarbons*, M. Cooke and A. J. Dennis (Eds.), Battelle Press, Columbus, Ohio, 1981, p. 343.
185. T. Cole, R. Riggin, and J. Glaser, in *Chemical Analysis and Biological Fate: Polynuclear Aromatic Hydrocarbons*, M. Cooke and A. J. Dennis (Eds.), Battelle Press, Columbus, Ohio, 1981, p. 439.
186. T. L. Gibson, A. I. Ricci, and R. L. Williams, in *Chemical Analysis and Biological Fate: Polynuclear Aromatic Hydrocarbons*, M. Cooke and A. J. Dennis (Eds.), Battelle Press, Columbus, Ohio, 1981, p. 707.
187. H. Issaq, G. M. Janini, B. Poehland, R. Shipe, and G. M. Muschik, *Chromatographia 14*:655 (1981).
188. Y. Hirata, M. Novotny, P. A. Peaden, and M. L. Lee, *Anal. Chim. Acta 127*:55 (1981).
189. F. L. Joe, Jr., E. L. Roseboro, and T. Fazio, *J. Assoc. Offic. Anal. Chem. 64*:641 (1981).
190. B. A. Tomkins, R. R. Reagan, J. E. Caton, and W. H. Griest, *Anal. Chem. 53*:1213 (1981).
191. J. M. Colin, G. Vion, M. Lamotte, and J. Joussot-Dubien, *J. Chromatogr. 204*:135 (1981).
192. E. Voigtman, A. Jurgensen, and J. D. Winefordner, *Anal. Chem. 53*:1921 (1981).
193. L. W. Hershberger, J. B. Callis, and G. D. Christian, *Anal. Chem. 53*:971 (1981).

194. T. Takeuchi and D. Ishii, *J. Chromatogr. 213*:25 (1981).
195. A. W. Wolkoff and C. Creed, *J. Liquid Chromatogr. 4*:1459 (1981).
196. J. L. DiCesare, M. W. Dong, and L. S. Ettre, *Chromatographia 14*:257 (1981).
197. W. J. Sonnefeld, W. H. Zoller, W. E. May, and S. A. Wise, *Anal. Chem. 54*:723 (1982).
198. S. A. Wise, S. L. Bowie, S. N. Chesler, W. F. Cuthrell, W. E. May, and R. E. Rebbert, in *Proc. 6th Intern. Symposium on Polynuclear Aromatic Hydrocarbons*, Columbus, Ohio, Oct. 27-29, 1981 (in press).

6

Analysis of Polycyclic Aromatic Hydrocarbons by Gas Chromatography

BJØRN SORTLAND OLUFSEN AND ALF BJØRSETH / Central Institute for Industrial Research, Oslo, Norway

I. Introduction

The PAH fraction of many environmental samples is usually of a high compositional complexity; several hundred compounds spread over a wide volatility and concentration range. Over the past 40 years several methods have been developed for the analysis of PAH, as single components (typically benzo[a]-pyrene), as a selection of compounds (e.g., the six chosen by the World Health Organization as a standard for PAH in drinking water [1] or the 14 recommended to the U.S. Environmental Protection Agency as major PAH components which should be measured in emissions from all energy-related process installations [2]), or as "total" PAH (e.g., spot tests which screen samples for the presence of PAH [3]). In order to completely characterize the whole range of PAH as well as detecting minor compounds with potential adverse biological effects, the analytical method must possess high separation efficiency and good sensitivity. Gas chromatography is a technique that has the potential of meeting these requirements. In this chapter, techniques and applications of gas chromatography in PAH analysis will be discussed.

A. Packed Columns

Gas chromatography (GC) has been used in the separation of PAH since the late 1950s and early 1960s [4-8], and most of the early work was done on packed columns. The attention in the late 1960s was, with a few scattered exceptions [9-12], still centered on packed columns. Work with packed columns in gas-liquid chromatography (GLC) includes the determination of PAH in airborne particulates [13-16], soot from an aluminum electrolysis furnace [17], airborne particulates from the work atmosphere of a Søderberg carbonpaste plant [18], unleaded fuel and automobile exhaust [19], crude oil [20], bitumen [21], high-protein foods, oils, and fats [22], marine organisms [20], lake and river sediments [23,24], carbon black [25], creosote [26], acetylene and ethylene flames [27], water [28,29], and several other environmental samples [30]. Severson and co-workers [31] have demonstrated good separations of PAH using packed columns. By coupled gas chromatography mass spectrometry (GC/MS) they characterized the PAH fraction of tobacco smoke condensate [32]. An example of this work is shown in Fig. 1. Similar work has been carried out by Grimmer [33] on cigarette smoke condensate. A few applications of gas-solid chromatography (GSC) have also been reported. Vidal-Madjar and Guiochon [34] achieved separation of PAH using a copper phthalocyanine adsorbent, Zane [35] applied graphitized carbon black. Fryčka [36] utilized the steric effects of Bentone 34 and graphitized carbon black to separate naphthalene homologs.

In the continuous effort for higher resolution and sensitivity and with the difficulties encountered in developing wall-coated open-tubular (WCOT) glass capillary columns many analysts sought the solution in support-coated open-tubular (SCOT) columns. In their much-referred-to work of 1973, Lao et al. [13] reported comparison between conventional packed columns and SCOT columns which showed better separation on the SCOT column in the region from fluoranthene to methylchrysene, while higher molecular weight compounds were late in eluting, if eluted at all. Furthermore, excellent separations of PAH in pitch covering a mass spectrum range from methylnaphthalene to anthanthrene were made on a commercially available 39 m × 0.5 mm i.d. SCOT glass column by Burchill et al. [37].

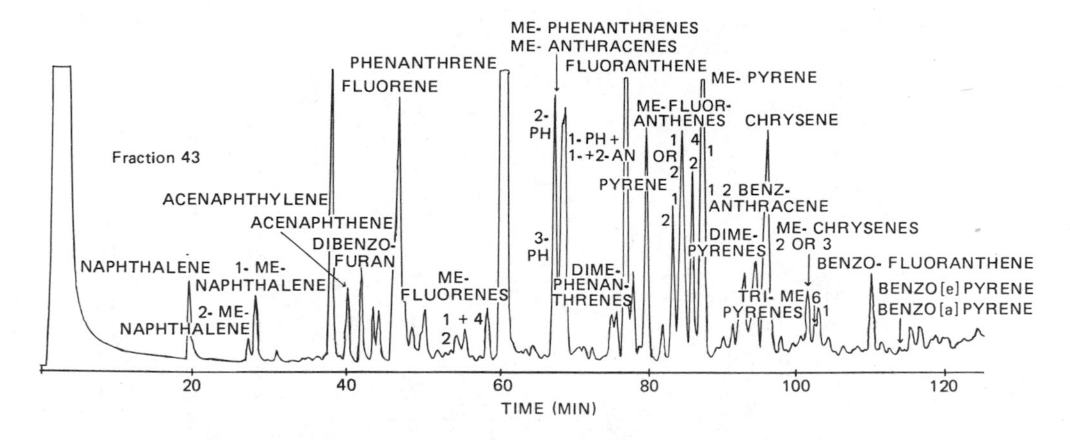

Figure 1. Gas chromatogram of GF fraction 43 derived from cigarette smoke condensate on a packed column of 5% Dexsil 300 programmed from 90° to 325°C at 2°C/min. [From R. F. Severson, M. E. Snook, H. C. Higman, O. T. Chortyk, and F. J. Akin, in *Carcinogenesis*, Vol. 1: *Polynuclear Aromatic Hydrocarbons: Chemistry, Metabolism, and Carcinogenesis*, R. I. Freudenthal and P. W. Jones (eds.), New York, Raven Press, 1976.]

B. Capillary Columns

When Golay [38] in 1958 published the very first capillary chromatograms, a mixture of C_6 separated on a Tygon tubing, he opened up a new area of chromatography. Desty and coworkers later encountered difficulties with adsorption on the wall of metal capillaries, and they invented the glass capillary drawing machine [39]. Thus the first glass capillary column for GC was developed. Today, more than 20 years later, the tool has proved its ability in numerous applications and the columns are in continuous use in research as well as routine analyses throughout the world.

Grimmer et al. [40] analyzed automobile exhaust gas on a 100 m × 0.5 mm i.d. capillary column. Early GC work on "polynuclear arenes" was performed on 60 m × 0.5 mm i.d. SE-30 capillary columns by Wilmshurst [8]. A 1.0 mm i.d. capillary was also used by Wilmshurst for fractions collection for spectrophotometry [8]. Gouw and coworkers [41] succeeded in separating anthracene from phenanthrene and benzo[a]pyrene from benzo[e]pyrene on a temperature programmed 10 m OV-101 capillary column. Lao et al. [13], Grimmer et al. [40], Wilmshurst [8], and Gouw et al. [41] did, however, use stainless steel columns, and the question is open as to whether catalytic effects would affect PAH at higher temperatures. To our knowledge, no comprehensive study has been reported on this matter. Adsorption effects of PAH on steel capillaries were discussed by, e.g., Doran and McTaggart [42]. Yet a compromise was offered by Grimmer et al. [40]. Packed glass columns of 20 m length were developed which exhibited resolution comparable to glass capillary columns

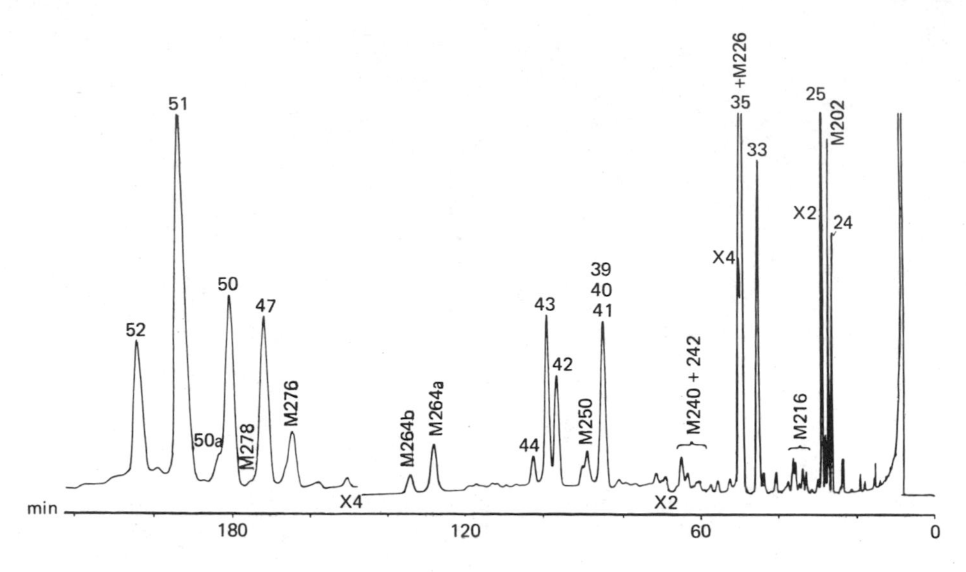

Figure 2. PAH-profile fractogram from automobile exhaust gas: 5% OV-101 on GasChrom Q 125-150 μm, glass column 20 m × 2mm i.d. 270°, 8.5 bar. (24) fluoranthene, (25) pyrene, (33) benzo[ghi]fluoranthene, (35) chrysene + (M226) cyclopenteno[cd]pyrene, (39-41) benzo[b/j/k]fluoranthenes, (42) benzo[e]pyrene, (43) benzo[a]pyrene, (44) perylene, (M 264a) 8,9-methylenebenzo[e]pyrene, (M 264b) 10,11-methylenebenzo[a]pyrene, (47) indeno[1,2,3-cd]pyrene, (50) internal standard = benzo[b]chrysene, (51) benzo[ghi]perylene, (52) anthanthrene. (From Ref. 242.)

of the same length. The drawbacks were, however, long retention times and higher column temperatures, as shown in Fig. 2. Such columns also have a considerable back pressure.

C. Packed Columns vs. Capillary Columns

When comparing packed columns and capillary columns in GC of complex mixtures most applications will come out in favor of the latter. Capillary GC also has apparent advantages when factors such as analysis time, sensitivity, and resolution are considered. We will make no attempt to prove this point but will instead refer to two specific papers which deal with how to initiate use of capillary GC [43] and a systematic approach to practical capillary GC [44].

Once the choice of capillary GC has been made, the importance of the practical limitations of the method should be recognized. A thorough report on this subject has been given by Schomburg [45]. There are four major areas where capillary columns have been particularly useful, as pointed out by Novotny [46]:

1. Resolution of isomers
2. Analysis of complex mixtures
3. Chromatographic "fingerprinting"
4. Fast separations

All these uses have been proven effective for the analyses of PAH. In particular, this is so for the first two mentioned because of the several hundred compounds, many of which are isomeric molecules, present in a PAH fraction. Giger and Schaffner [47] stated that "if extensive characterization of PAH mixtures is based on analyses of individual constituents rather than unresolved mixtures, we can improve our ability to describe the sources and the fates of these chemicals in the environment. . . ." The use of high-resolution data to characterize the samples has been discussed by many authors. Blokzijl and Guicherit [48] presented ratios between certain PAH in outdoor air which were dependent on the source (automobile traffic and domestic fuel); Bjørseth [49] used the parent PAH profile to illustrate the relative distribution of PAH compounds in working environments; and the total GLC profile has provided data on the differences between tobacco and marijuana smoke condensates [50]. Chapter 4 of this handbook gives several examples of PAH profiles in air samples. A nitrogen-selective detector was used by Lee et al. [51] to fingerprint used engine oils. The advantages of fast separations in analyses of PAH were demonstrated by Wright and Lee [52] on short capillary columns (2-4 m) with a high temperature programming rate. The fraction between napththalene and coronene was analyzed in 15-20 min, as shown in Fig. 3.

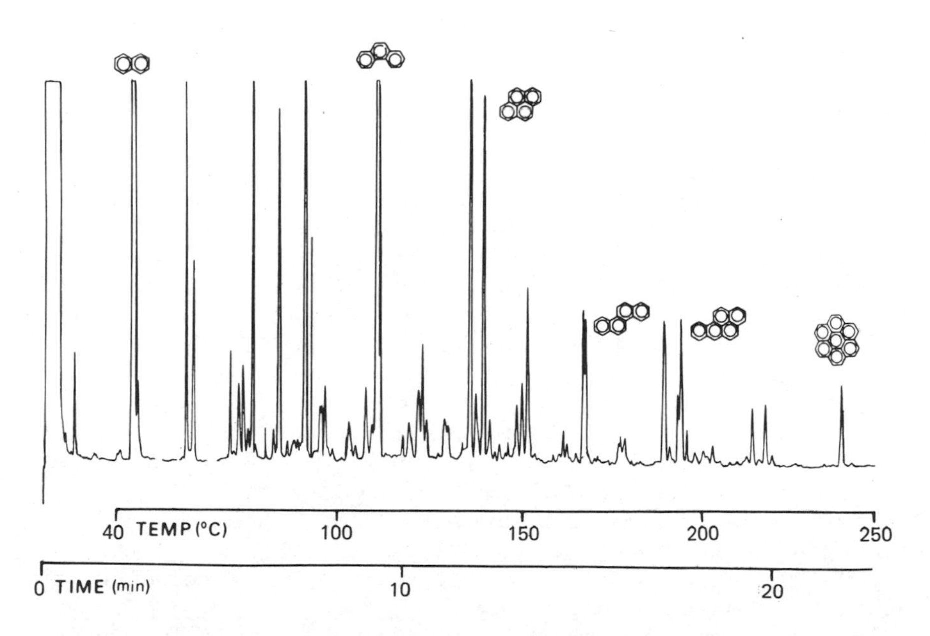

Figure 3. Capillary column gas chromatogram of coal tar: 4m × 0.30mm i.d. glass column coated with 0.25 μm SE-52. Oven held at 40°C for 2 min and then programmed to 250°C at 10°C/min. Hydrogen carrier gas at a linear velocity of 100 cm/sec. (From Ref. 52.)

II. Choice of Stationary Phase

The choice of stationary phase has been the dispute between chromatographers over the years. A large number of phases and mixtures thereof have been utilized to achieve the "ultimate" separation. For packed columns a list of applications includes Apiezon L [19,53,54], Apiezon W [55], Dexsil 300 [13-16,18,30,31,56-63], Dexsil 400 [30,64], Dexsil 410 [30], FFAP [63], OS-138 [65], OV-1 [30,37,66], OV-7 [15,17,67], OV-11 [68], OV-17 [20,29,37,61], OV-22 [69], OV-25 [37,55], OV-101 [22,33,70,71], PMPE [37], PS-176 [70], QF-1 [8], SE-30 [8,54,72-74], SE-52 [8,27,59,71,75,76], SP-2100 [56], and SP-2250 [77].

For capillary columns the choice of stationary phase seems to be less critical than for packed columns because the total efficiency of the column is much higher. Table 1 summarizes applications for PAH analyses on different stationary phases on glass capillary columns.

For glass capillary columns gum phases are generally superior to liquids. The advantages of gum phases, as pointed out by Grob [78], are the following:

1. Increased film homogeneity ultimately leads to higher separation efficiency.
2. The broadest practical temperature range.
3. Extended film thickness range.

If our wish is to separate "as many PAH as possible" in one single run, the best choice is a gum phase. SE-52 gives a better resolution than SE-30 or OV-1, and SE-54 is even superior to SE-52 with slightly better separation of the benzofluoranthenes and complete separation of 3-methylphenanthrene, 2-methylphenanthrene, 2-methylanthracene, 4,5-methylenephenanthrene (45MP), 4-methylphenanthrene (+ 9-methylphenanthrene), and 1-methylphenanthrene. OV-73, a relatively new polysiloxane gum with 5.5% phenyl [79] has been shown to give separations comparable to SE-54 (see Ref. 80 and chromatograms in this chapter) in addition to higher thermal stability and low bleeding. The separation of 45MP from methylphenanthrenes and methylanthracenes should be noted. According to Blumer [81] compounds with "highly strained molecules in which a single CH_2 group forms a bridge between two aromatic carbon atoms, as in 45MP, form with different relative abundance in various pyrolysates. . . ." The quantitation of such compounds might then "reflect the character of the source materials, the process of formation, the temperatures at which the compounds form and the reaction time. . . ."

Sandra and co-workers [82] recently introduced a new gum phase, RSL-110, which might prove selective for PAH. Gum phases such as OV-215 and SP-2125X have to our knowledge not been used to separate PAH.

The recent introduction of immobilized/crosslinked phases has made possible the cleaning of capillary columns by solvent washing. This is particularly useful for columns which have been contaminated in the inlet section due to repeated splitless and on-column injections.

For laboratories producing their own columns the aspect of film thickness of the stationary phase should be taken into consideration. Grob and Grob [83] have shown separation of the PAH fraction of a lake sediment sample on a column coated with 0.018 μm SE-52 (Fig. 4). Reduced elution temperatures and enhanced separations are obtained compared with commonly used film thicknesses (ca. 0.2 μm). However, there is not much to be gained below

Table 1. Glass Capillary Column Systems Reported Used in the GC Separation of PAH[a]

Stationary phase	Column length (m)	Internal diameter (mm)	Surface treatment	Film thickness (μm)	Injection technique	Carrier gas	Temperature program	Detection	Sample	Year	Origin of column	Reference
BBBT	20	0.25	sil		fn	H_2	Isothermal 220	FID	Standard mixture	80	lm	70
BBBT	17.5	0.2	sil		s	H_2	Isothermal 220	FID	Standard mixture	80	lm	99
BBBT/SE-30 mixture	20	0.25	sil		fn	H_2	Isoth. 220/265	FID	Standard mixture	80	lm	70
BBBT/SE-52 mixture	17.5	0.2	sil		s	H_2	Isothermal 220	FID	Standard mixture	80	lm	99
BMBT	50	0.25		0.05	s	N_2	200-2-240	FID	Standard mixture	79	c	106
Carbowax 20M	50	0.20	e/cw		sl	H_2	60-5-300	FID	Tar	78	lm	89
CP-Sil 5	25	0.26					130-1.6-250	FID	Diesel exhaust from automobiles	79		198
CP-Sil 5	25	0.25		0.1	s	N_2	165-8-330	FID	Standard mixture	79	c	106
CP-Sil 5	25	0.22	fs	0.12	sl	He	80-4-?	FID	Standard mixture	80	c	199
Dexsil 300	30				s	He	60-4-270	FID	Particulates and vapor phase of a coal gasifier	79		200
Dexsil 300	30				sl	He	60-2-270	MS	Particulates and vapor phase of a coal gasifier	79		200
Dexsil 300	25	0.25			fn	He	90-2-280	MS	Automobile exhaust	79		130
Dexsil 300	50	0.25			s	N_2	160-4-270	FID	Airborne particulates; coal tar; industrial process sludge	79	c	62

(continued)

Table 1 (continued)

Stationary phase	Column length (m)	Internal diameter (mm)	Surface treatment	Film thickness (μm)	Injection technique	Carrier gas	Temperature program	Detection	Sample	Year	Origin of column	Reference
Dexsil 400	50	0.27	e	0.086		He	50-4-320	FID/ND	Coal tar	79	lm	85
Dexsil 400	30	0.2	sil			He	100-3-320	FID	Multialkylated PAH	79		201
Methylpolysiloxane, polymerized	20	0.22	cl		sl	H_2	70-7-300	FID	Cyclone dust from an aluminium plant; tar combustion products	79	lm	92
Methylsilicone gum, chemically bonded	20	0.20	e		sl	H_2	70-7-300	FID	Cyclone dust; wood & peat combustion products	79	lm	91
Methylsilicone gum, chemically bonded	20	0.2	e		sl	H_2	70-7-300	FID/MS	Engine exhaust; wood & peat combustion products; airborne particulates	79	lm	202
OV-1	22	0.28			s			MS	Standard mixture	75	lm	203
OV-1	50	0.34			sl	H_2	100-3-250	FID	Long-range transported aerosols	76	c	204
OV-1	50	0.34			sl	H_2	100-3-250	FID	Long-range transported aerosols	77	c	205
OV-1	50	0.34			sl	H_2	100-3-250	FID	Particulate matter from an aluminium reduction plant	77	c	144
OV-1	50	0.38			sl	H_2	100-3-250	FID	Working atmosphere of an aluminium reduction plant	78	c	206

OV-1	15	0.3	sil	0.25	fn	H_2	Isothermal 200	FID	Standard mixture	80	lm	82
OV-3	12	0.24	w		pc/sl	He	60-2-230	FID	Standard mixture	76	lm	207
OV-7	30	0.27	e			H_2	?-?-280		Coal tar	79	lm	119
OV-17							?-?-265	FID	Airborne particulates; automobile exhaust	80		208
OV-17	25	0.27			sl	He	230-1-265	FID	Urban particulates, coal heating area	80		209
OV-17, Chemically bonded	20	0.22	e		sl	H_2	70-7-310	FID	Tar; wood combustion products	78	lm	91
OV-25	22						100-1-150	FID	Naphthalenic fraction of crude oils	78		64
OV-61	30	0.27	e		s	H_2	Isothermal 250	MS	Coal tar	79	lm	119
OV-73	50	0.32	psil	0.1	sl	H_2	120-3-300	FID	Workplace atmosphere	80	lm	80
OV-101	50				sl	H_2	amb-2.5-240	FID	Lake sediment	74	c	210
OV-101	40				sl	H_2	amb-2.5-240	MS	Lake sediment	74	c	210
OV-101	5.5	0.32			sl	H_2	200-4-260	FID	Coronene, dibenzo[a,e]-pyrene, truxene, rubrene (standard mixture)	74	lm	211
OV-101	50					N_2	Isothermal 300	FID	Tar; standard mixture	74	lm	183
OV-101	50	0.25			s	He	100-2(6)-265	FID	Standard mixture; automobile exhaust condensate	74	lm	40
OV-101	45	0.35			hd/ct	He	0-2-220	MS	Coal-derived fluids	76	lm	212
OV-101	30	0.35			sl	H_2	60-2.5-240	FID/ECD	Rainwater	76	c	152
OV-101	20	0.25			sl	H_2	100-3-260	FID	Standard mixture	77	c	144

(continued)

Table 1 (continued)

Stationary phase	Column length (m)	Internal diameter (mm)	Surface treatment	Film thickness (μm)	Injection technique	Carrier gas	Temperature program	Detection	Sample	Year	Origin of column	Reference
OV-101	25	0.26			sl	He	70-3-240	MS	Snow	78		213
OV-101	30	0.25			s	He	Isothermal 230	FID	Standard mixture	78		71
Poly S-179	90	0.27	e		s	H_2	200-1-400	FID	Coal tar	78	lm	84
Poly S-179	92	0.27	e		s	H_2	200-1-390	MS	Coal tar	79	lm	119
RSL-110	15	0.3	sil	0.25	fn	H_2	Isothermal 200	FID	Standard mixture	80	lm	82
SE-30	35	0.35			s	N_2	Isothermal 200	FID	Atmospheric dust	64		9
SE-30	30				s	N_2	Isothermal 200	FID	Standard mixture	65		10
SE-30	25	0.32				He	160-2-320	MS	Standard mixture; coal tar	75	c	214
SE-30	26	0.30			s	N_2	40-6-280	FID	Air & dust from iron foundries	78		215
SE-30	30	0.4-0.6		0.5-0.6	hd	H_2	160-2-225	FID	Air particulates	79	lm	131
SE-30	20	0.25	sil		fn	H_2	Isothermal 220	FID	Standard mixture	80	lm	70
SE-30	30	0.32			oc	H_2	100-6-270	FID	Standard mixture	80	lm	216
SE-52	50				s	N_2	100(150)-2.5-300	FID/ECD	Atmospheric dust; standard mixture	65		10
SE-52	65	0.30			s	N_2	Isothermal 200	FID	Standard mixture	67		11

SE-52	65	0.30			s	N_2	100-1.8-300	FID/ECD	Standard mixture; cigarette smoke condensate	67		11
SE-52	35	0.25			pc/ct	He	150-2-260	FID	Airborne particulate matter; standard mixture	74	c	217
SE-52	70	0.29			pc/ct	He	100-2-260	FID	Airborne particulate matter; standard mixture	74	c	217
SE-52	22	0.26		0.29	pc/sl	He	100-2-260	FID/ND	Engine oils	75	lm	51
SE-52	11	0.26			pc/ct	He	70-2-240	MS	Tobacco & marijuana smoke condensates	76		50
SE-52	11	0.26			pc/ct		70-2-240	MS	Airborne particulates	76	lm	218
SE-52	19	0.26					70-2-250	MS	Carbon black	76		158
SE-52	12	0.26	e/sil	0.27	pc/sl		60-4-240	FID	Shellfish	76	lm	219
SE-52					sl			MS	Airborne particulates	76		220
SE-52	13	0.3	$BaCO_3$	0.018	sl	H_2	80-3-230	FID	Lake sediment	77	lm	83
SE-52	19	0.26					70-2-250	MS	Airborne particulates; combustion products	77		221
SE-52	20	0.3	$BaCO_3$	0.1	sl	H_2	60-2.5-250	FID	Lake & river sediments; airborne particulates; street dust	78	lm	47
SE-52	25	0.22					?-?-250	FID/MS	River water	78		222
SE-52	20	0.28	$BaCO_3$		sl	He	50-6-240		Ocean sediments	78	lm	223
SE-52	19	0.26					70-2-250	FID	River & ocean sediments; soil; combustion products of kerosene	78	lm	224
SE-52	10	0.29		0.36			90-2-250	FID	Airborne particulate matter	78	lm	225

(continued)

Table 1 (continued)

Stationary phase	Column length (m)	Internal diameter (mm)	Surface treatment	Film thickness (μm)	Injection technique	Carrier gas	Temperature program	Detection	Sample	Year	Origin of column	Reference
SE-52	10	0.29		0.36		He	100-2-250	MS	Airborne particulate matter	78	lm	225
SE-52	17	0.5		0.3	fn	He	70-4-320	FID	Standard mixture	79	lm	167
SE-52	12	0.30	e/sil	0.34		He	50-2-250	FID	Standard mixture	79	lm	169
SE-52	12	0.28	e/sil	0.17		He	50-2-250	FID	Standard mixture	79	lm	169
SE-52	12	0.29	e/sil	0.34		He	50-2-250	FID	Standard mixture	79	lm	169
SE-52	30	0.25			s	N_2	160-4-270	FID	Airborne particulates; industrial process sludge; coal tar	79	lm	62
SE-52	12	0.30					50-2-250	MS	Coal liquefaction product	79		226
SE-52	20	0.25			pc/ct	He	100-2-250	MS	Solvent-refined coal & its recycle oil	79	lm	227
SE-52	15		psil		sl	H_2	100-3-330	FID	Airborne particulates; fly ash; secondary alumina; bottom ash from power plant	80	lm	228
SE-52	2-6	0.30	sil	0.25	sl	H_2	40-(4-15)-250	FID	Coal tar	80	lm	52
SE-52	15	0.27				H_2	110-2-350	FID	Carbon black	80	lm	125

SE-52	15,20,30	0.30	0.10		He	80-2-250	FID	Coal tar spiked with coronene	80	lm	125
SE-52	15	0.20, 0.25, 0.27	0.25		He	80-2-250	FID	Coal tar spiked with coronene	80	lm	125
SE-52	15	0.29	0.25, 0.50, 0.70		He	80-2-250	FID	Coal tar spiked with coronene	80	lm	125
SE-52	17	0.25			He		UV	Aromatics in crude coal tar	80		161
SE-52	12	0.28				50(90;120)-2(4)-230	FID/FPD	Coal gasification tar	80		159
SE-52	30	0.28				50-2-250	FID/MS	Coal gasification tar	80		159
SE-52	20-30	0.28				40(125)-2-250	FID/MS	Sulfur heterocycle fraction of coal gasification tar	80		159
SE-52	20	0.25 sil		fn	H_2	Isothermal 220	FID	Standard mixture	80	lm	70
SE-52	17.5	0.2 sil		s	H_2	Isothermal 220	FID	Standard mixture	80	lm	99
SE-52	20	0.3 $BaCO_3$	0.15	sl	H_2/He	70-2-250	FID/MS	Recent lake sediments	80	lm	229
SE-52	20	0.3 $BaCO_3$	0.15	sl	H_2/He	70-2-250	FID/MS	Recent lake sediments	80	lm	230
SE-54	50	0.36		sl	H_2	115-3-250	FID	Particulate matter from a coke plant	77	c	231
SE-54	50	0.38		sl	H_2	180-2.5-250	FID	Maize	77	c	232
SE-54	45	0.27 e		s	H_2	80-2-290	FID	Coal tar	78	lm	84
SE-54	20			s	He	30-6-260	MS	Air & dust from iron foundries	78		215

(continued)

Table 1 (continued)

Stationary phase	Column length (m)	Internal diameter (mm)	Surface treatment	Film thickness (μm)	Injection technique	Carrier gas	Temperature program	Detection	Sample	Year	Origin of column	Reference
SE-54	20	0.25					150-3-275	MS	Industrial effluent	78		233
SE-54	50	0.34-0.36			sl	H_2	100(115)-3-250	FID	Working atmospheres; sediment; mussels	78	c	234
SE-54	50	0.38			sl	H_2	100-3-250	FID	Working atmosphere of a coke plant	78	c	235
SE-54	48	0.27			s	H_2	69-3-290	MS	Coal tar	79	lm	119
SE-54	50	0.35			sl	H_2	120-3-250	FID	Fjord sediment; mussels	79	c	179
SE-54	50	0.38			sl	H_2	115-3-250	FID	Airborne particulates	79	c	180
SE-54	50	0.35			sl	He	120-3-250	FID/ECD	Airborne particulates	79	c	153
SE-54	50	0.34			sl	H_2	115-3-260	FID	Particulates from working atmospheres	79	c	236
SE-54	50	0.34			sl	He	115-3-260	MS	Particulates from working atmospheres	79	c	236
SE-54	50				sl	H_2	120-3-250	FID/MS	Particulate matter from an aluminium plant	79	c	49
SE-54	30						80-5-250	FID	Coal tar pitch; crude oil	79		237
SE-54	50	0.35			sl	H_2	120-3-260	FID	Tap water; industrial waste water	80	c	238

SE-54	50	0.34		sl	H_2	120-3-260	FID	Algae; invertebrates	80	c	239
SF-96	30	0.3				40-6-200	MS	Coal tar contaminated potable water	80		77
SP-2100	16	0.20	e/cw	sl	H_2	70-5-300	FID	Tar	78	lm	89
SP-2100	10	0.25			He	100-2-275	FID	Standard mixture	78	c	240
SP-2100	10	0.25				100-2-265	MS	Molten coal tar exudate; coke	78	c	240
SP-2100	22	0.20	fs/cw	sl		80-7-280	FID	Standard mixture	79	c	93
SP-2100	25	0.2	fs/cw	sl	H_2		FID	Sediment	80	c	241
XE-60	35			s	N_2	Isothermal 200	FID	Standard mixture	65		10

[a]Surface treatment: cl = chlorination; cw = Carbowax 20M deactivation; e = etching; fs = fused silica; psil = persilylation; sil = silylation; w = whisker growth.
Injection technique: ct = cold trapping; fn = falling needle (solid); hd = heat desorption in the injector; oc = on-column; pc = precolumn; s = split; sl = splitless.
Detection: ECD = electron capture detector; FID = flame ionization detector; FPD = flame photometric detector; MS = mass spectrometer; ND = nitrogen detector; UV = ultraviolet detector.
Origin of column: c = commerically available; lm = laboratory made.

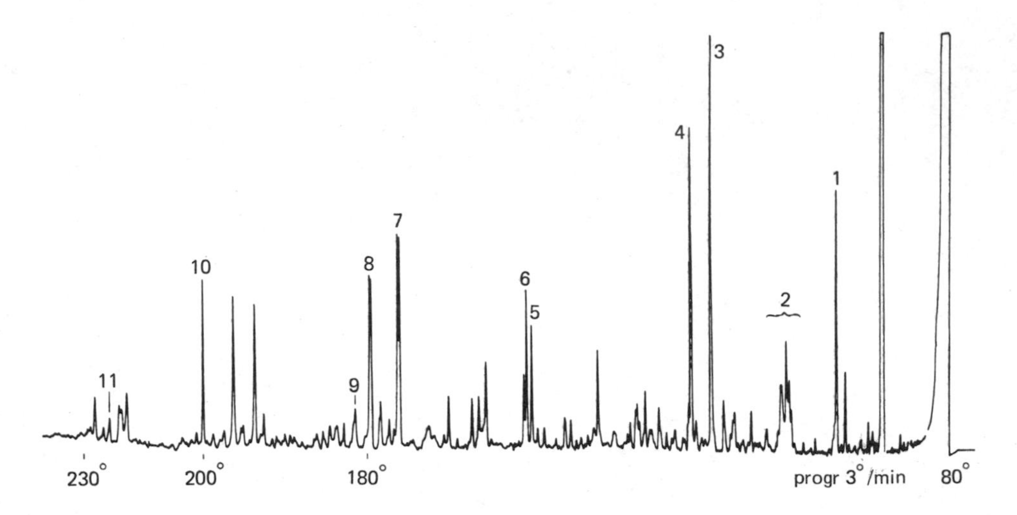

Figure 4. PAH fraction of extract from lake sediment separated on 13m × 0.3mm i.d., 0.018 μm SE-52 column. Carrier H_2, 0.7 at, producing 6.9 ml/min at 25°C, 3.0 ml/min at 230°C. Attenuation × 2, full-scale peaks corresponding to ca. 1 ng, which is maximum load yielding symmetrical peaks: (1) phenanthrene, (2) methylphenanthrenes and methylanthracenes, (3) fluoranthene, (4) pyrene, (5) benz[a]anthracene, (6) chrysene, (7) benzofluoranthenes, (8) benzopyrenes, (9) perylene, (10) benzo[ghi]perylene, (11) coronene. (From Ref. 83.)

ca. 0.03 μm, which is recommended for high boiling compounds. These extremely thin films require proper treatment of the glass surface to avoid adsorption phenomena causing peak tailing or compound breakdown. During the last few years several methods for surface treatment have been reported to yield inert columns with low adsorption and high-temperature operating range (up to 400°C). Schomburg and co-workers [84,85] applied gaseous etching with HCl and/or HF, or heating of the coated column. Grob et al. [86-88] demonstrated persilylation (i.e., complete silylation) for highly inert columns, and Blomberg et al. [89-92] prepared thermostable phenyl silicone-coated columns with in situ synthesis of gum phases. An example of the latter is visualized in Fig. 5. Noteworthy is the separation of triphenylene from chrysene. The introduction of fused silica columns [93] made available a glass surface with extremely low content of metal ions. The potential of these columns has not been fully exploited to date.

III. Selective Stationary Phases

Although most PAH analyses by GC have been carried out on silicone phases, several attempts to develop selective stationary phases for special applications have proved useful. Sato et al. [94] dispersed lithium nitrate in SE-30 and thus had a selective stationary phase for 15 aromatic hydrocarbons. Sauerland

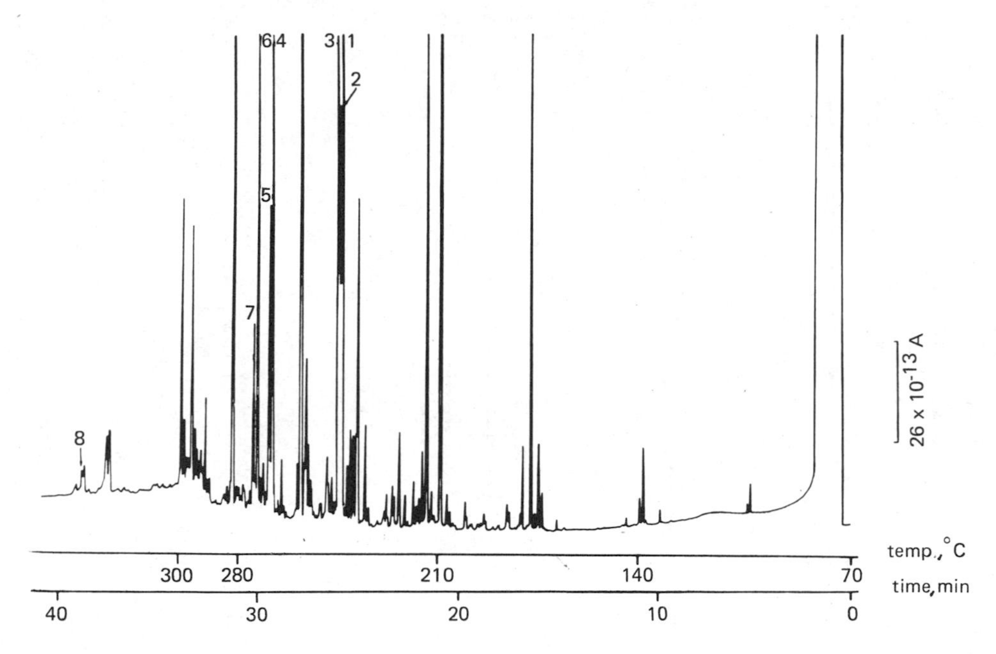

Figure 5. Gas chromatogram (FID) of a sample containing PAH derived from cyclone dust collected at an aluminum plant. Pyrex glass capillary column 20 m × 0.22 mm i.d. treated with hydrochloric acid/tetrachlorosilane and coated with methylpolysiloxane polymerized under ammonia. Carrier gas velocity (H_2) at 70°C, 70 cm/sec. Inlet splitter opened 1 min after injection: (1) benz[a]anthracene, (2) chrysene, (3) triphenylene, (4) benzo[b]fluoranthene, (5) benzo[k]fluoranthene, (6) benzo[e]pyrene, (7) benzo[a]pyrene, (8) coronene. (From Ref. 92.)

and Zander [95] employed poly-m-phenoxylene in coal tar analysis. Janini et al. [96-98] and others [70,99] used nematic liquid crystals for the selective separation of PAH [70,99]. In variance to conventional stationary phases which (to simplify matters) "distil" the sample and elute the solutes according to their boiling points, the liquid crystals also "consider" the geometry of the solute molecule. Of two isomers the one with the larger length-to-breadth ratio should have the longer retention on a liquid crystal column. Two review articles on the subject have been given by Zielinski [100,101]. Of the several applications of liquid crystals, the majority have been made with standards on packed columns [102-107]. Coal tar pitch was, however, separated on a SP-301 column [108]. Other applications (PAH in cigarette smoke condensate [109] and liquid smoke preparations [110]) revealed columns of low separation efficiency. This seems to be one severe drawback with liquid crystal phases, as well as their narrow temperature range (which restricts temperature programming) and the high bleeding rate.

Laub and co-workers [111-113] have formulated and established predictive techniques for quantitative selection of all conditions required to achieve a given separation by gas chromatography. The method, called the plenary optimization strategy, has been used to mix phases for enhanced GC separa-

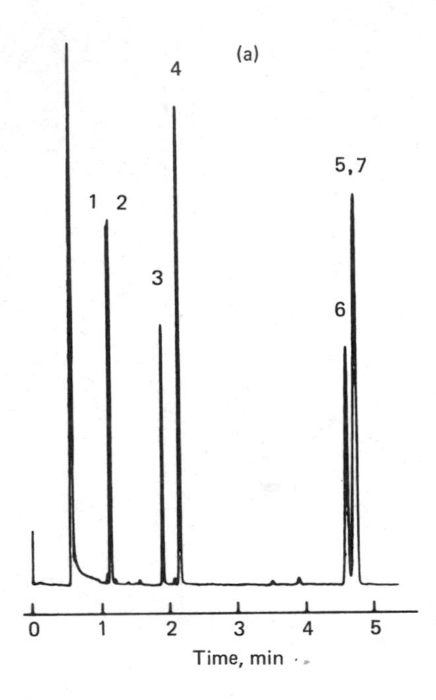

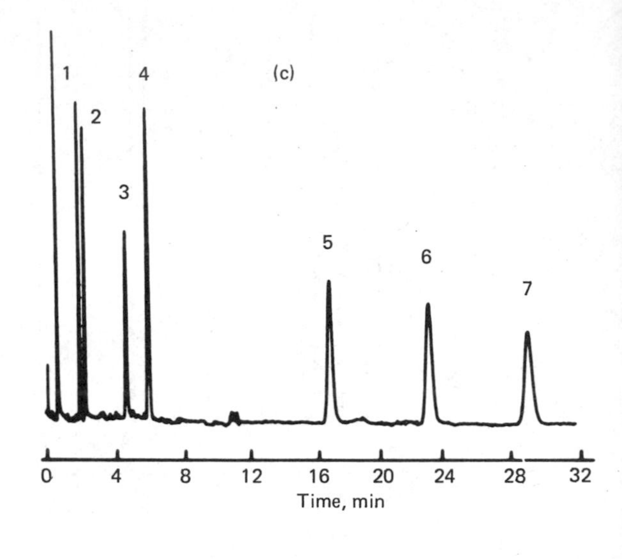

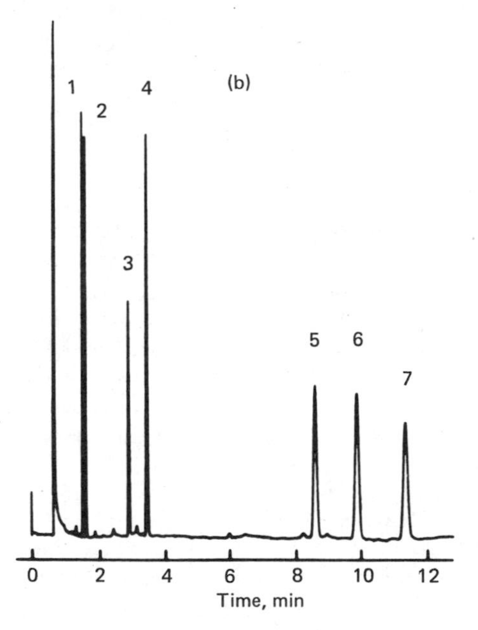

Figure 6. Chromatograms of (1) phenanthrene, (2) anthracene, (3) fluoranthene, (4) pyrene, (5) triphenylene, (6) benz[a]anthracene, and (7) chrysene with columns containing: (a) SE-52; (b) 80% (w/w) SE-52 + 20% BBBT; and (c) BBBT. On request from the authors an error in the original figure (a) has been corrected. (From Ref. 99.)

tion of PAH [70]. This method has, to our knowledge, only been applied to the separation of standard solutions of PAH, but mixtures of BBBT liquid crystal and SE-52 silicone gum coated on glass capillary columns separate triphenylene from chrysene [99], as shown in Fig. 6. This critical separation, which has not been obtained on SE-52, is achieved at the expense of a considerable loss in efficiency.

Grob and Grob [114] stated that "introduction of new phases is justified in only two instances: either the new material, owing to its particular molecular structure, is able to fill a gap between existing phases, or it is able to replace an existing phase because of superior characteristics. . . ." The use of polar phases has—particularly in the field of glass capillary columns—often been restricted to narrow (and low) temperature working ranges. With the synthesis of polyphenyl ether sulfones [115-118] thermally stable polar phases for GC were introduced. Beautiful glass capillary GC/MS data on PAH in coal tar were presented by Borwitzky and Schomburg [119], who were able to coat the polyphenyl ether sulfone Poly S-179 on etched alkali-glass capillaries. The working range of these columns was between 190° and 390°C, which covered the PAH fraction from acenaphthene to coronene/dibenzofluoranthene/dibenzopyrene. It should be noted that this column was able to separate benz[a]anthracene, chrysene, and triphenylene. This separation has been reproduced in our laboratory, as shown in Fig. 7.

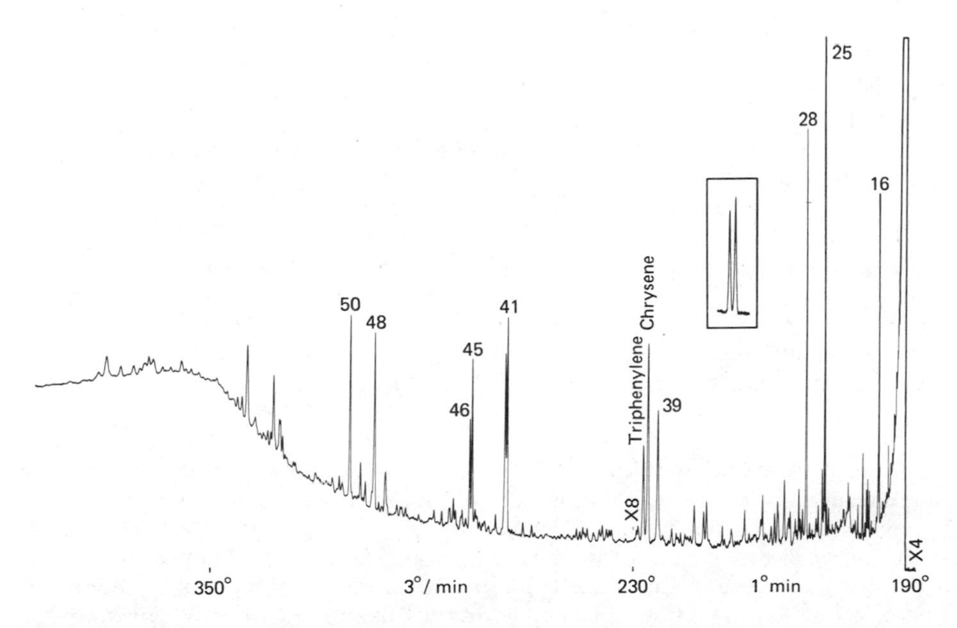

Figure 7. Glass capillary GC separation of PAH from a lake sediment sample. Column 17.9 m × 0.20 mm i.d. was etched dynamically at 450°C with gaseous hydrogen chloride generated by the reaction of concentrated sulfuric acid on sodium chloride. Further pretreatment was as described elsewhere [119]. Dynamically coated with 20% (w/w) Poly S-179 in dichloromethane. Hydrogen carrier 0.7 kg/cm^2, "hot-needle" split injection of 1 μl at 190°C. For peak identification, see Table 2. Insert: Separation of a standard mixture of chrysene and triphenylene under similar conditions.

Table 2. Identification of Compounds in the PAH Fraction[a] of a Coal Tar Sample[b]

Peak No.	Compound	Peak No.	Compound
1	Indene	27	Benzo[def]dibenzothiophene ?
2	Naphthalene	28	Pyrene
3	Benzo[b]thiophene	29	Ethylmethylenephenanthrene ?
4	2-Methylnaphthalene	30	Benzo[a]fluorene
5	1-Methylnaphthalene	31	Benzo[b]fluorene
6	Biphenyl	32	4-Methylpyrene
7	1,3-Dimethylnaphthalene	33	Methylfluoranthene and/or 2-methylpyrene
8	1,4- and/or 2,3-Dimethylnaphthalene	34	1-Methylpyrene
9	Acenaphthylene	35	Benzothionaphthene ?
10	Acenaphthene	36	Benzo[ghi]fluoranthene
11	Dibenzofuran	37	Benzo[c]phenanthrene
12	Fluorene	38	Benzophenanthridine ?
13	2-Methylfluorene	39	Benz[a]anthracene
14	1-Methylfluorene	40	Chrysene and triphenylene
15	Dibenzothiophene	41	Benzo[b]fluoranthene
16	Phenanthrene	42	Benzo[j]fluoranthene
17	Anthracene	43	Benzo[k]fluoranthene
18	3-Methylphenanthrene	44	Benzo[a]fluoranthene ?
19	2-Methylphenanthrene	45	Benzo[e]pyrene
20	2-Methylanthracene	46	Benzo[a]pyrene
21	4,5-Methylenephenanthrene	47	Perylene
22	4- and/or 9-Methylphenanthrene	48	Indeno[1,2,3-cd]pyrene
23	1-Methylphenanthrene	49	Dibenz[a,c/a,h]anthracenes
24	2-Phenylnaphthalene ?	50	Benzo[ghi]perylene
25	Fluoranthene	51	Anthanthrene
26	Benz[e]acenaphthylene ?	52	Coronene, dibenzofluoranthenes, and dibenzopyrenes

[a]Sample cleanup according to Ref. 144.
[b]Gas chromatogram of sample shown in Fig. 15a.

In special applications, the use of selective stationary phases can give valuable information (e.g., the use of OV-61 for an improved separation of the benzofluoranthene isomers [119]). Giger and Shaffner [47] reported resolution of chrysene from triphenylene and of the several benzofluoranthenes on capillary columns coated with polar phases such as polyphenyl ether, Carbowax, or Silar. They did not, however, use such columns routinely "because of their reduced practical temperature range and because of their lower intrinsic separation efficiency (due to coating on a more intensely roughened support surface). . . ." At the moment there are indications that moderately polar phases, such as OV-17 and OV-225, will be commercially available as gum phases. Already introduced, OV-1701 (86% methyl, 7% phenyl, 7% cyanopropyl) has made possible the production of thermostable moderately polar columns. However, some authors [120] have in connection with their work on packed columns stated that the better way of resolving complex PAH mixtures is an improvement in column efficiency rather than searching for new stationary phases.

IV. Evaluation of Separation Efficiency

For many years the theoretical plate number n was considered the most important parameter when evaluating the performance of a GC column; the higher the n, the better the separation efficiency. Within certain limits n will increase with increasing column length. This has been used extensively for complex mixtures of PAH, particularly in earlier separations on glass capillaries where 50-100 m columns were frequently employed. The high-molecular-weight PAH (with boiling points above ~500°C) and the (by then) relatively low maximum temperature limit of the stationary phase in the region of 250-280°C (owing to insufficient surface treatment of the glass) resulted in time-consuming analyses. The enhanced separations in the early part of the chromatogram were thus made at the expense of bad peak shapes (erroneous quantitative data) and lower sensitivity due to long retention times in the upper isothermal region. It should be noted that the number of theoretical plates will depend on the test compound and the temperature at which the test is made. Both the functional group and the partition ratio of the test compound have an effect on these numbers [121]. It has been pointed out by Grob and Grob [122] that the separation number, TZ [123], shows much less dispersion than the theoretical plate number and should therefore be preferred in the evaluation of a capillary column.

V. Choice of Carrier Gas

Hydrogen has been established as the ideal carrier gas in capillary GC [44,84]. The advantages with hydrogen as compared to nitrogen are the following [44]:

Slightly better resolution
Shorter analysis time
Better sensitivity
Decreased discrimination of sensitive sample components
Longer column lifetime
Better separation efficiency
Lower elution temperatures
Less expense

The explosive hazard of hydrogen is the only major drawback. This problem is reduced by installing hydrogen monitors which detect gas leakage in the oven [124]. Helium would also be a better choice than nitrogen [125].

VI. Sampling Techniques

In the following a brief summary of injection techniques will be given. The discussion will be restricted to sampling systems for narrow-bore capillary columns only.

Common for all sample introduction techniques is the necessity to produce an narrow starting band for solutes. The failure to do so will result in broad peaks and consequently a loss in separation efficiency and sensitivity. As pointed out by Grob and Grob [126], there are three governing principles each of which will give band shortening:

1. Stream splitting (split injection)
2. Cold trapping
3. Solvent effect (splitless injection)

Because of the small amount of PAH usually present in environmental samples and the nonquantitative aspect discussed elsewhere, the split injection technique will not be discussed.

A. Cold Trapping

Cold trapping (also termed cryogenic condensation) is based on "freezing" the solutes that are swept from the vaporizing chamber by the carrier gas (or a direct gaseous sample), in the initial part of the column. The column inlet is usually cooled by immersing it into liquid nitrogen or by blowing expanding carbon dioxide onto it. A compound with a boiling point 150°C higher than the column temperature will generally be cold trapped [127]. Solute migration and subsequent separation is achieved by raising the column temperature. Various uses of precolumns may also be regarded as special applications of cold trapping. This is also the case with the solventless falling-needle injector described by van den Berg and Cox [128]. This technique has been used, among other applications, in the analysis of PAH in foodstuffs [129] and automobile exhaust [130]. The loss of some of the more volatile PAH, such as phenanthene during solvent evaporation in the falling-needle technique, is reduced significantly by coating the needle tip with a thin film of SE-54 silicone [130]. Heat desorption has been a means of semiquantitatively transferring organics in air particulates from an exposed filter onto a capillary column. The filter was placed directly in the vaporizing chamber in a glass liner [131]. A similar effect is achieved using an electrically heated injector for rapid transfer from Tenax adsorbent onto the column, as described by Chesler et al. [132].

The use of precolumns is experiencing a renaissance today, the reasons for which have been described by Kaiser [133]. Several authors have taken advantage of the technique for the analysis of PAH, e.g., Novotny and Farlow [134].

B. Splitless Injection

Splitless injection has proven to be the most widely used sampling technique in capillary GC trace analysis (see Table 1). The technique was first described by Grob and Grob [135,136] in 1969. An extended presentation of current understanding and practical technique in 1978 [126] notes that "splitless injection is based on the solvent effect as a mechanism condensing large vapour clouds down to infinitely shortened bands. . . ." To achieve a solvent effect, proper operation parameters must be selected for the amount and type of solvent (polarity and boiling point) related to the column temperature at the time of injection, the rate of injection, and the nature of the sample components. A theoretical basis for the solvent effect has been given by Jennings

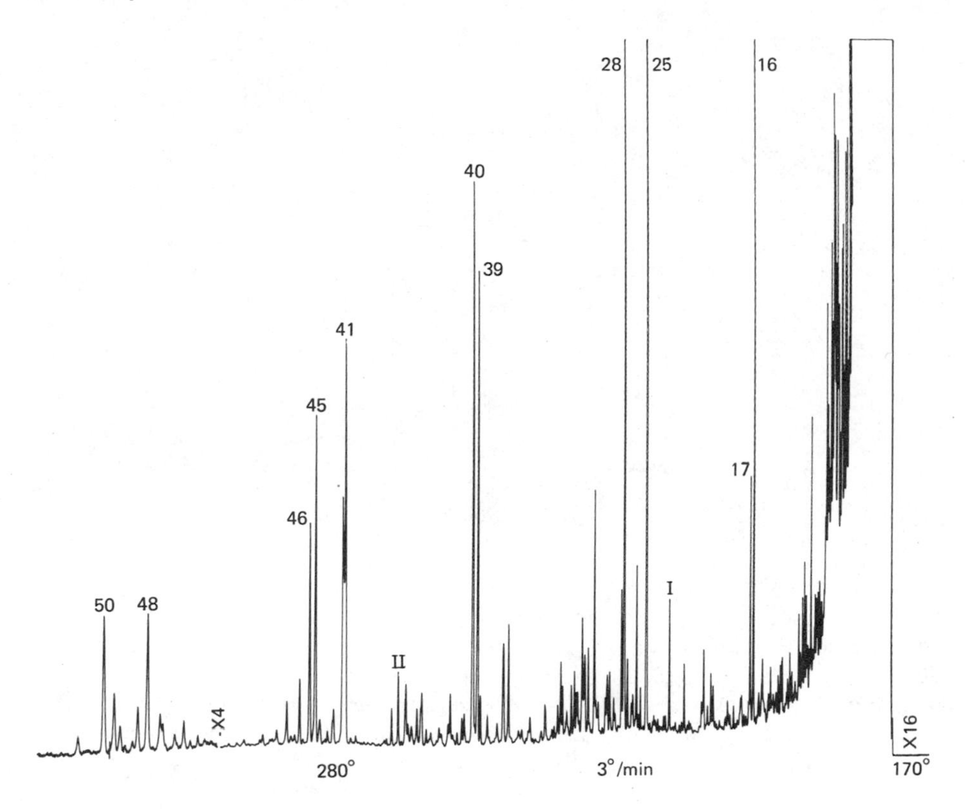

Figure 8. Glass capillary GC separation of PAH from a sample of horse mussel. Column 50 m × 0.34 mm i.d. SE-54. Hydrogen carrier 0.8 kg/cm^2 yielding a deadtime of 2 sec/m for methane at ambient temperature; 1.5 μl injected splitless at 170°C. Temperature program started immediately. Split opened after 30 sec. Solvent: n-dodecane. For peak identification, see Table 2. I and II are internal standards.

et al. [127]. It has been stressed that splitless injection will give a solvent effect even in isothermal analysis at elevated temperatures [137]. This might prove useful as a time-saving factor for the analysis of the higher molecular weight compounds of a PAH fraction. An example of this is shown in Fig. 8. Part of the PAH fraction of contaminated horse mussel (*Modiolus modiolus*) has been separated after splitless injection at 170°C (solvent: n-dodecane). The limitation of splitless injection for PAH analysis is the nonquantitative transfer (and high relative standard deviations) of high boiling compounds onto the column. Quantitations therefore should be restricted to compounds with boiling points below ~525°C.

C. On-Column Injection

The most promising solution to the lack of precision and accuracy of quantitative analysis of high boiling (and thermally unstable) compounds is on-column injection, i.e., direct sampling onto the inlet part of the capillary column.

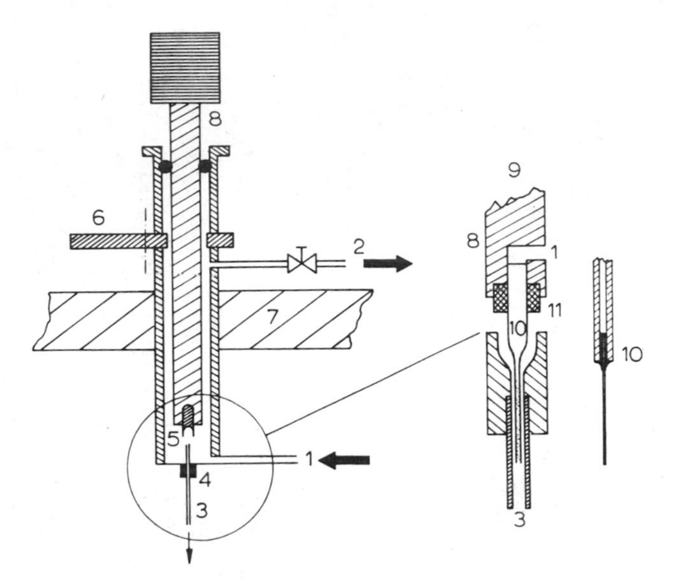

Figure 9. Direct sampling of liquids into capillary columns. Macro- and micro-versions: (1) carrier gas inlet, (2) carrier gas outlet, (3) capillary column, (4) graphite gasket, (5) crucible for sample, (6) sluice valve, (7) insulation of column oven, (8) rod for sample introduction, (9) micro-version of direct sampling, (10) micro-pipette for sampling of nanoliter volumes, (11) silicone rubber gasket. (From Ref. 139.)

Several attempts on on-column injection have been reported with the work of Schomburg and co-workers [138,139] being the first practical application to narrow-bore capillary columns. That work demonstrated a macro- and a micro-version for direct on-column injection of liquid samples, as shown in Fig. 9. Commercially available instruments were introduced following a series of publications by Grob's group on the subject [140-142]. Factors affecting the accuracy and precision of cold on-column (replacing the term on-column) injection in capillary GC were discussed by Grob and Neukom [143]. Cold on-column injection also represents a further application of the solvent effect [126].

The advantages of direct sampling are immediately recognized: no discrimination of compounds upon injection (i.e., the original composition of the sample enters the column). The limiting factor in the analysis of high boiling compounds is no longer the injection but rather the column (i.e., stationary phase and surface treatment). The on-column injector is septum-free; thus, there are no problems with ghost peaks originating from septum material.

The main problem with the on-column injector is connected to routine maintenance. Residue material in the inlet part of the column due to nonvolatile injected material may be removed by rising the first few coils with a proper solvent. This brings the column back to normal. This operation replaces the previous routine operation of cleaning the glass liner of a vaporizing injector.

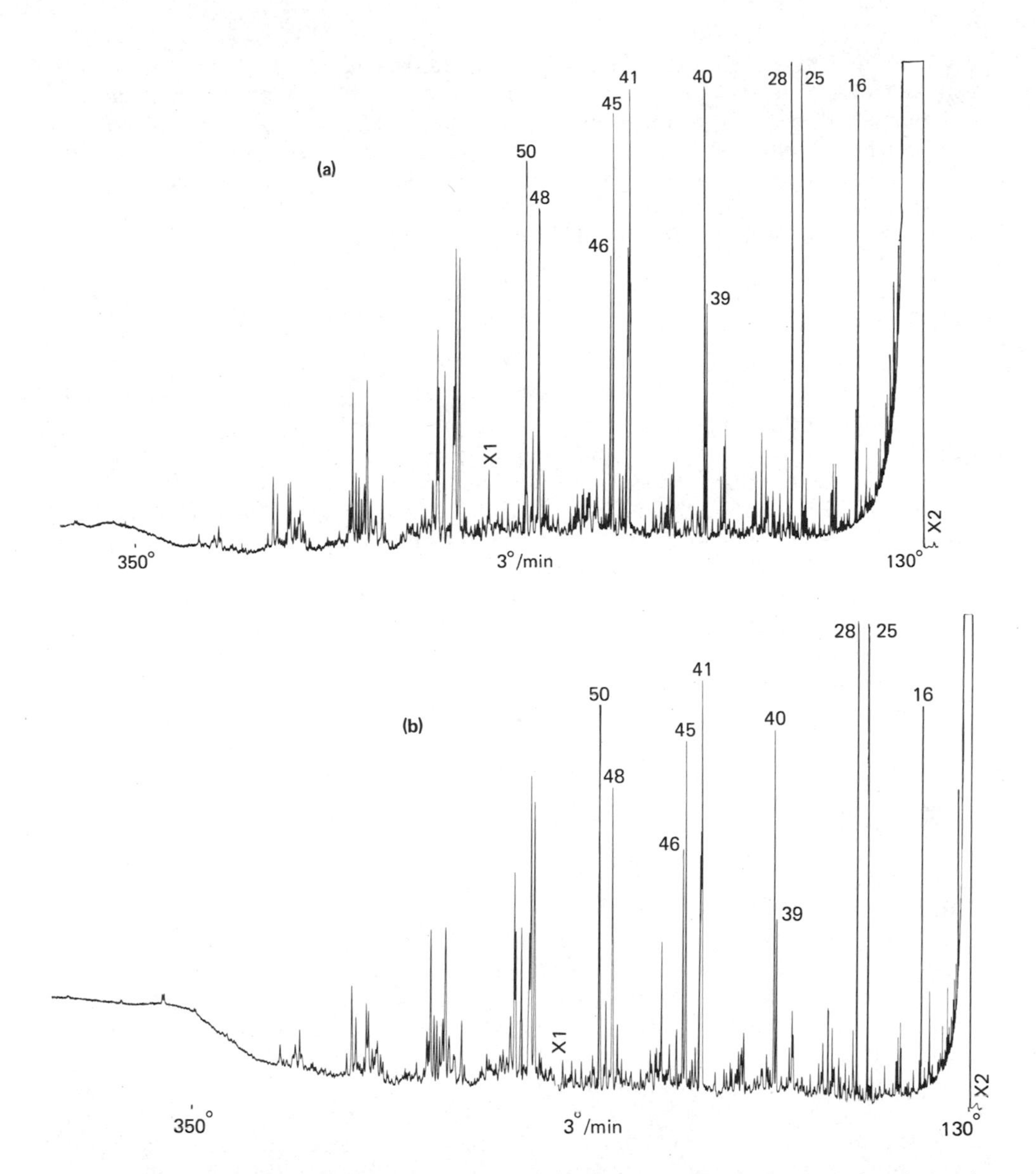

Figure 10. A comparison of splitless injection (a) and cold on-column injection (b) of a sample of PAH from a lake sediment onto the same glass capillary column. Column 30 m × 0.32 mm i.d. persilanized according to Grob (Ammonia test 34%) and coated with 0.05 μm OV-73. Hydrogen carrier: 0.55 kg/cm^2 yielding a deadtime of 2 sec/m for methane at ambient temperature. Detector 375°C. Solvent: cyclohexane. (a) 1.4 μl injected splitless with "hot needle" at ambient temperature. Split opened after 30 sec. (b) 1.6 μl injected on-column at 60°C after 20 sec secondary cooling. For peak identification, see Table 2.

To our knowledge the on-column injector is used in routine analysis of PAH in several laboratories. However, no papers have yet been published that demonstrate the applicability. Figure 10 shows a preliminary comparison between splitless and cold on-column injection. The figures indicate an increase of approximately 70% in the peak height of the last eluting compounds compared to benzo[ghi]perylene (peak 50). There is no doubt that the technique will replace cold trapping and splitless injection for many types of analyses and especially for gas chromatography of PAH.

VII. Detectors

The most frequently used detector for GC analysis of PAH is the flame-ionization detector (FID). Its general response character makes it ideal for several classes of compounds but necessitates an extensive cleanup procedure prior to GC to eliminate possible interfering compounds. An example is shown in Fig. 11. Urban dust from St. Louis, Mo. (sample kindly supplied by Dr. R. Merrill), was analyzed according to the procedure described by Bjørseth [144] (chromatogram A). As can be seen, parts of the chromatogram are obscured by interfering components even after thorough selective partition steps by liquid/liquid extractions. A further purification by chromatography on silica gel [22] resulted in chromatogram B. A remark should be made on the peak eluting between the two benzopyrenes in chromatogram A. This peak is assumably the polycyclic ketone benzo[cd]pyrene-6-one (MW 254), which has been reported present in samples of chip and peat smoke, automobile exhaust, and airborne particulates [145]. The ketone is not quantitatively eluted with cyclohexane from a silica gel column deactivated with 10% of water. The remark is made for one to bear in mind that the presence of the ketone might give rise to erroneous results for benzo[e]pyrene or benzo[a]pyrene due to overlapping depending on the separation efficiency of the column used. This is also the case with the compound eluting between benz[a]anthracene and chrysene/triphenylene in A. The molecular weight is 226, which points to cyclopenteno[cd]pyrene. Chromatography on silica gel partly retains this compound, simplifying quantitation of the aforementioned compounds. We have not been able to confirm this due to the lack of a commercially available standard. (Synthesis of cyclopenteno[cd]pyrene has been described by Jacob and Grimmer [146].)

FID has several advantages, however: the high linearity range makes it ideal for quantitative work based on internal standards; there is usually no need for calibrations, i.e., response factors, in practical routine analysis; and the detector is reliable and easy to maintain. Response factors should be considered for high-boiling compounds, for poorly deactivated columns, or when the age of the column seems to influence the performance.

A few applications with the electron capture detector (ECD) have been reported [10,11]. An early contribution is shown in Fig. 12. Faltusz [29] applied the detector for the analysis of PAH in water. Burchill and coworkers [37] used an ECD in conjunction with OV-1, OV-25, and PMPE columns. They concluded that the method should be used with caution as "the presence of minor components with very high ECD responses eluting at the same time as major PAH may invalidate the method. . . ." The simultaneous detection with FID and ECD has been proposed by some authors. By splitting the effluent

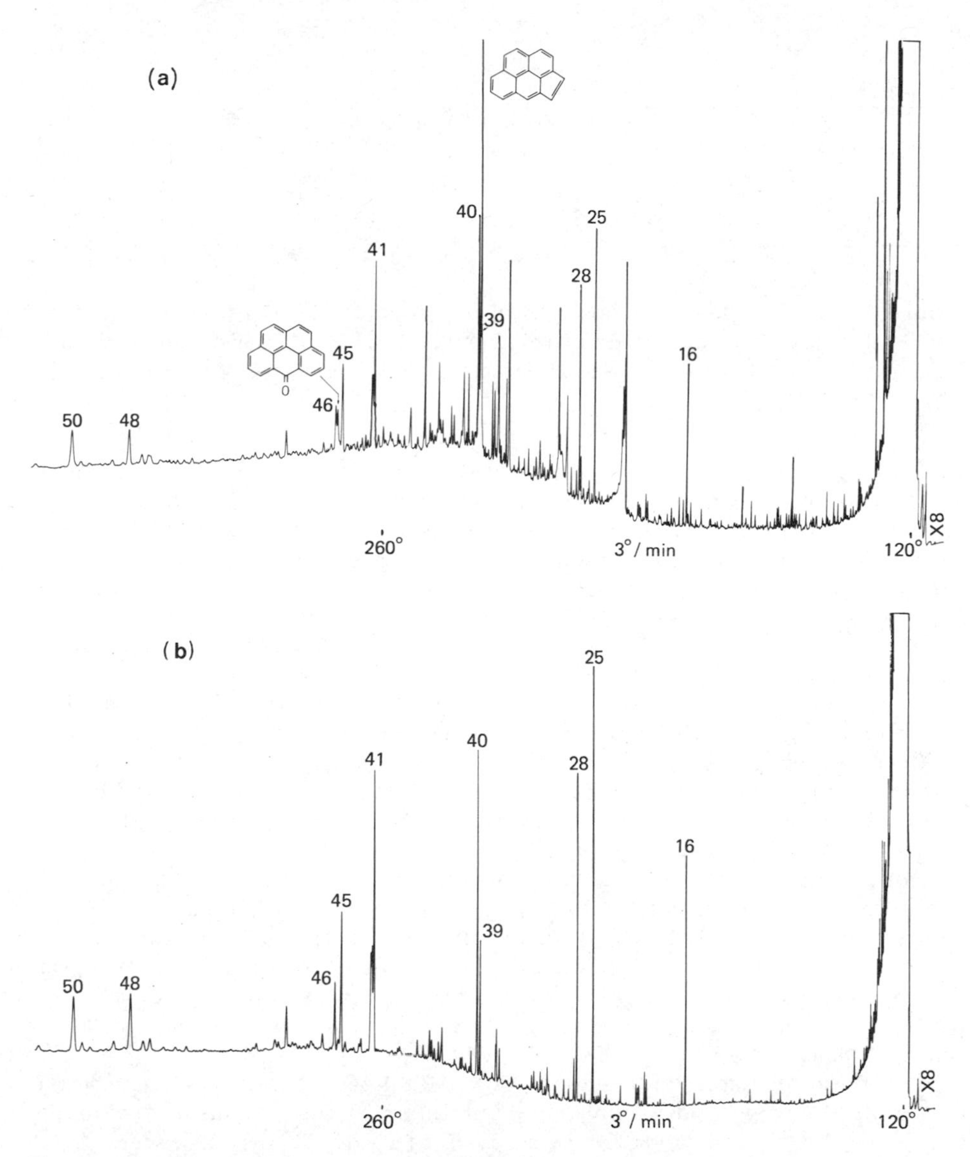

Figure 11. Sample of urban airborne particulates prior to (a) and after (b) cleanup on silica gel. Column 50 m × 0.34 mm i.d. SE-54 glass capillary. Hydrogen carrier: 0.8 kg/cm^2 yielding a deadtime of 2 sec/m for methane at ambient temperature; 0.5 μl injected splitless at ambient temperature. Split opened after 45 sec. Solvent: cyclohexane. For peak identification, see text and Table 2.

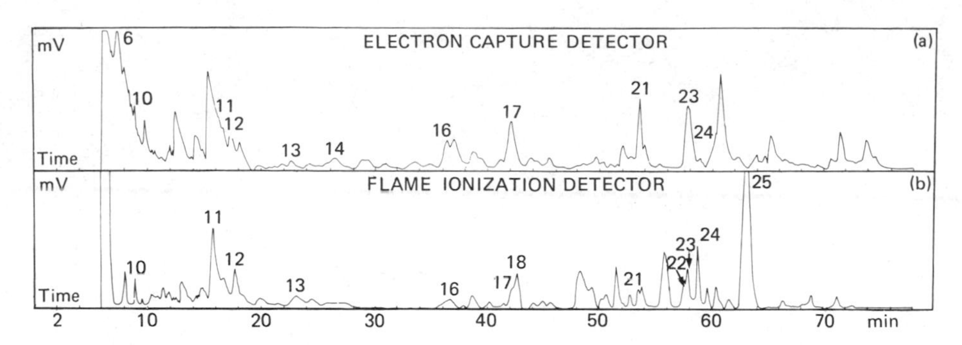

Figure 12. Early glass capillary gas chromatograms of a sample of an atmospheric dust extract, obtained with ECD and FID. Programmed temperature from 150° to 230°C/min on a SE-52 column. (From Ref. 10.)

from the column to two separate detectors a lot of information may be gained on the composition of the sample. Different stream-splitting devices for high-resolution glass capillary columns have been shown [147-150]. Some GC manufacturers now supply detectors where the FID is coupled in series with the ECD. The nondestructive property of the latter is then taken advantage of [151]. The presence of PAH and polychlorinated biphenyls in rain water was recorded simultaneously by Giger and co-workers [152]. Bjørseth and Eklund [153] determined the ECD/FID ratios for 46 polynuclear aromatic hydrocarbons and demonstrated the application to an analysis of urban air particulates; an example of this work is shown in Fig. 13.

Grimsrud et al. [154] studied the alteration of the ECD response to five PAH by oxygen doping of the carrier gas. This would, of course, only be useful for special applications since oxygen would have disastrous effects on most stationary phases.

The disadvantages and problems encountered with the ECD are mainly the necessity to determine response factors for each PAH compound, low linear range (the amount of internal standard must match that of the sample components), and baseline irregularities when the oven temperature is programmed for capillary column work. Flow programming in glass capillary column-electron capture GC has been demonstrated as a substitute to avoid baseline drift during temperature programming [155].

The flame photometric detector (FPD), which was developed for selective detection of sulfur- and phosphorus-containing pesticides and herbicides, serves no direct application in the analysis of PAH. It has, however, and quite accidentally, led to the discovery of impurities in commercially available standards of PAH. Aue and Flinn [156] found sulfur-containing PAH in 40% of the compounds they tested, and the standard with the largest contamination (phenanthrene) contained 0.8% sulfur. Fryčka [157] later confirmed their findings and described some of the contaminants.

Depending on the cleanup procedure, a PAH fraction usually contains sulfur compounds (e.g., benzo[b]thiophene and dibenzothiophene) which may overlap with certain PAH causing erroneous quantitative results when a FID is employed. Ševčik [63] described a problem of separating benzo[b]thiophene

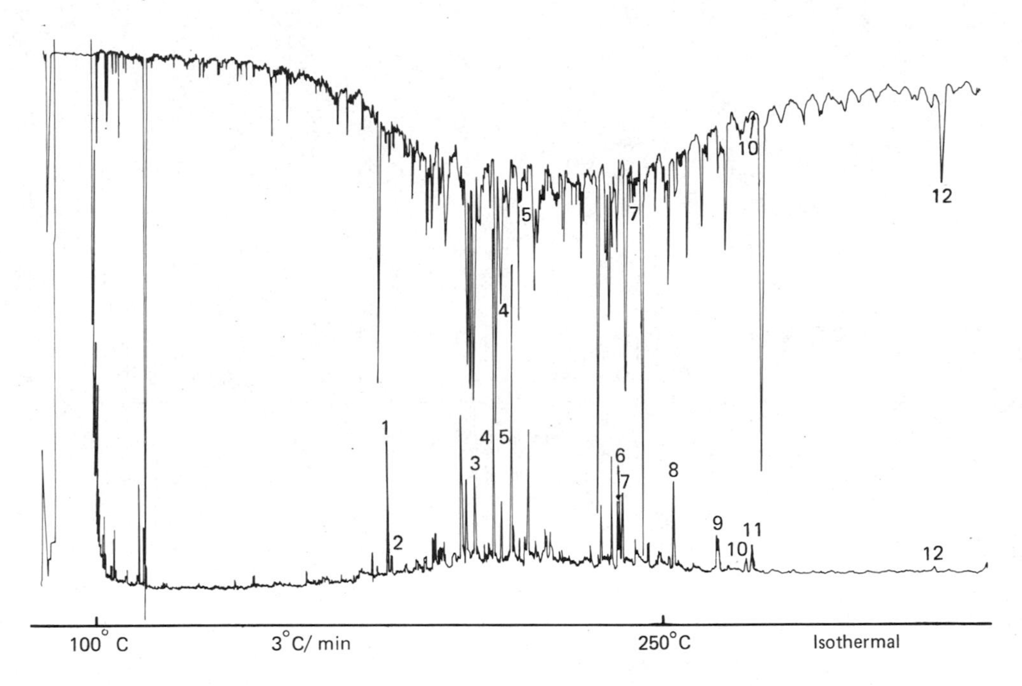

Figure 13. Glass capillary gas chromatogram of PAH in urban air particulates. Upper trace, ECD; lower trace, FID: (1) phenanthrene, (2) anthracene, (3) internal standard = 3,6-dimethylphenanthrene, (4) fluoranthene, (5) pyrene, (6) benz[a]anthracene, (7) chrysene and triphenylene, (8) internal standard = 2,2'-binaphthyl, (9) benzo[b/j/k]fluoranthenes, (10) benzo[e]pyrene, (11) benzo[a]pyrene, (12) indeno[1,2,3-cd]pyrene. (From Ref. 153.)

from naphthalene that would easily have been solved with a FP detector. It should also be noted that several sulfur-containing PAH in coal tar [119] and carbon black [158] have been characterized by glass capillary GC-MS. Lee et al. [159] identified sulfur heterocycles in coal gasification tar by dual detection with FID/FPD. They argued that the FPD is useful for preliminary screening for sulfur compounds in aromatic fractions but pointed out several drawbacks: "The FPD seldom gives linear response, and . . . quenching is a serious problem when the sulfur compounds are present as trace components. Furthermore, . . . it is a poor identification tool, and positive identification must be confirmed by mass spectrometry. . . ."

A nitrogen-selective detector (ND) was used by Lee et al. [51] to enhance "fingerprinting" of the PAH fraction of engine oils. The differences among three samples were shown to be "considerably more dramatic than those observed with [the] flame ionization detector. . . ." A year earlier Hartigan and co-workers [160] had recorded both the FID and ND profiles of nitromethane extracts of urban particulate matter. Schomburg and coworkers [85] split the column effluent between a FID and a ND and recorded both the "general picture" and the nitrogen fraction of coal tar (Fig. 14). The response and multitude of nitrogen-containing compounds will depend on the sample, the cleanup procedure, and the location of the nitrogen in the molecule.

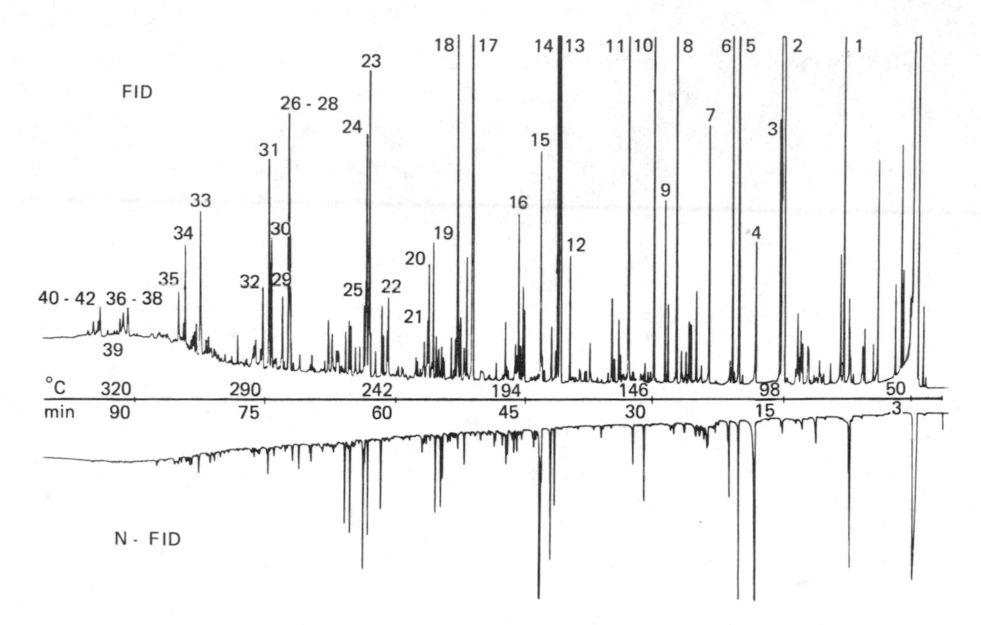

Figure 14. Glass capillary gas chromatogram of coal tar. Simultaneous detection with FID and N-FID. Column 50 m × 0.27 mm i.d. coated with 0.086 μm Dexsil 400. Helium carrier 33 cm/sec at 320°C. (From Ref. 85. Peak numbers refer to identification in this reference.)

Schwende and co-workers [161] tuned an UV gas-phase detector to 200 nm and detected PAH after separation on a glass capillary column. Selective detection of aromatic molecules in a crude coal tar sample was also shown at 250 nm. With proper calibrations and optimization of wavelengths, this form of detection will certainly play an important role in the future analysis of PAH.

The extreme sensitivity and selectivity of fluorescence has also been applied to the detection of PAH in combination with GC [162]. A direct gas-phase isolation and injection system coupled with packed column GC and subsequent detection by electron capture and spectrophotofluorometry was applied to the analysis of PAH in air particulates by Burchfield et al. [66]. A great future has been predicted for the combination of GC and fluorescence detection.

Ettre in his article "Selective Detection in Column Chromatography" [163] has given a brief summary of the characteristics of the most important detectors.

VIII. Identification of PAH

The identification of individual PAH in a gas chromatogram will depend on the sample complexity, sample cleanup, separation, and method/sensitivity of detection. In routine analyses of PAH from well-known sources, the identifications are usually based on recognition of a typical pattern. Identification by means of retention parameters has been discussed by several authors [164,

165]. Most frequently used have been the Kovats retention indices for isothermal runs and modifications thereof for temperature-programmed analyses [166]. The indices have been calculated relative to n-alkane standards. Beernaert [167] calculated indices of 70 PAH relative to alkanes in temperature-programmed GC by improved linear interpolation. The influence of several GC parameters (column length, film thickness, gas flow rate, temperature-programming rate, and the injection system) on the separation and the reproducibility of the retention indices was studied. Aue and Paramasigamani [168] had already studied whether it is mainly a change in the retention of the solutes or mainly a change in the retention of the alkanes that brings about the increasing indices of the solutes on phases of increasing polarities. The authors found that it is the alkane standards whose retention changes drastically with the polarity of the liquid phase; the retention of the solute remains more or less even through the polarity range. Hence, it is essentially alkane retention that determines a liquid phase's standing in the polarity scale. Identification of PAH on this basis should therefore not be regarded as sufficient evidence.

Lee et al. [169] demonstrated the lack of good statistical reliability for retention indices of PAH evaluated with n-alkanes as internal standards. They therefore defined a new retention index based on four PAH compounds as internal standards, namely, naphthalene, phenanthrene, chrysene, and picene. The retention indices of more than 200 PAH were determined using this system. A relationship was found between the connectivity indices and retention indices of PAH by Kaliszan and Lamparczyk [170]. This system has been described in Chap. 13.

Sample cleanup may often be insufficient or difficult. A GC/FID chromatogram will then be difficult to interpret owing to the general response character of the detector. The use of a selective detector such as the ECD, UV, or fluorescence detectors may then prove useful.

The combination of GC/MS is today the most powerful tool in the identification of PAH. This subject is treated in Chap. 7, but some applications are found in Table 1.

The aspect of chemical reactions coupled with gas chromatography has not been comprehensively treated for identification of PAH. Used together with, e.g., retention data it can give additional evidence as to the presence of certain PAH. The following analysis (which is far from optimized) may serve as an example: sulfonation of PAH in concentrated sulfuric acid is a function of the temperature. Sulfonation of pyrene proceeds rapidly at room temperature to produce a mixture of disulfonic acids [171]. The solubility of PAH in cold concentrated sulfuric acid varies, and the colors of the solutions, ranging from yellow to wine-red, indicate possible complex formations [172, 173]. Figure 15 shows the chromatograms of a cyclohexane solution of the PAH fraction of tar prior to and after extraction with concentrated sulfuric acid at room temperature. Equal volumes of tar solution and acid were extracted for 30 sec on a Whirlie mixer and centrifuged. The organic layer was removed, washed twice with water, dried over anhydrous sodium sulfate, and chromatographed on a high-resolution glass capillary column. Several compounds have disappeared from the chromatogram, while others are only partly affected or not affected at all. An optimization of the analytical parameters will undoubtedly add valuable information to the identification of PAH.

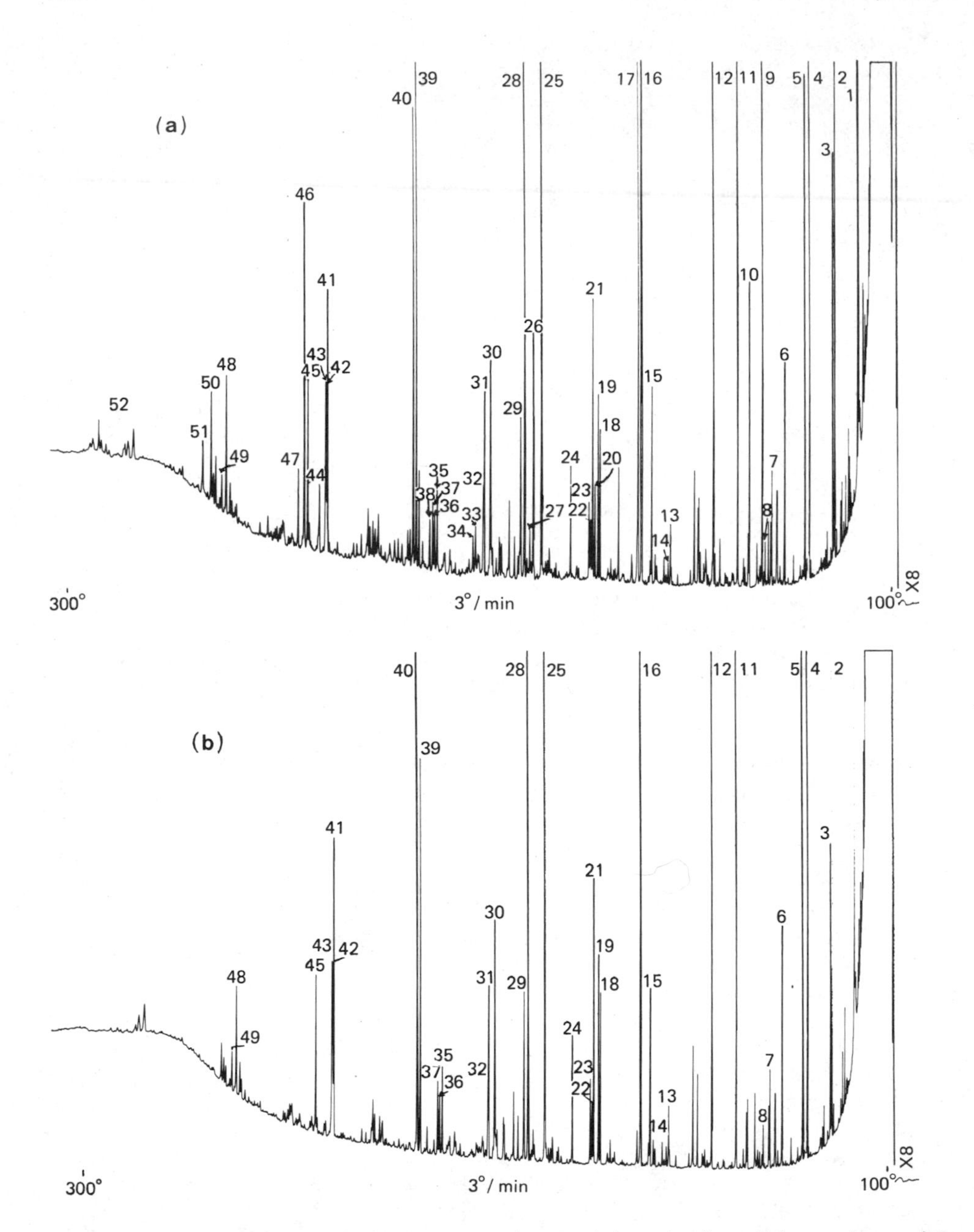

Figure 15. PAH fraction of coal tar prior to (a) and after (b) reaction with concentrated sulfuric acid. Glass capillary column 52 m × 0.31 mm i.d. persilanized according to Grob and coated with 0.1 μm OV-73. Hydrogen carrier 0.83 kg/cm^2 yielding at deadtime of 2 sec/m for methane at ambient temperature; 0.5 μl injected splitless with "hot needle" at ambient temperature. Split opened after 30 sec. For peak identification, see Table 2.

IX. Quantitative GC

A. Aspects to Be Considered for Capillary Columns

Reduced reproducibility has often been a point of criticism for quantitative work on capillary columns. This weakness is mainly related to split injection methods and vaporizing injectors. Discrimination of sample components take place both in the syringe needle upon injection and in the injector, i.e., the composition of the part of the sample which finally reaches the column is different from the original sample. A lot of work has been done to modify vaporizing injectors. Langlais et al. [174] increased the accuracy of split ratio measurements, thus improving the reproducibility of quantitative determinations, by the introduction of a rapid-action gate-valve in the appropriate gas line. A highly linear all-glass splitter based on a large buffer volume was described by Jennings [175]. Schomburg et al. [139] compared different vaporizing tubes for homogenization of sample vapor-carrier gas mixture with split sampling. The standard deviations of repeatability of peak area ratios ranged from 34 to less than 1% with the best results for a glass liner with a long and tight glass-wool plug. Grob and Neukom [176] were not able to reproduce these results in their work on the influence of the syringe needle on the precision and accuracy of vaporizing GC injections. Their conclusion was to pull back the sample into the barrel of the syringe and allow the needle to warm up in the injector before the sample was transferred into it ("hot-needle" injection). Newly developed sampling techniques such as the splitless and on-column injection techniques overcome most of these deficiencies.

Apart from the errors encountered with the syringe needle and injector, the column itself may contribute to quantitative anomalies mainly due to adsorption. Well-shaped peaks and high separation efficiency were often the prevailing criteria for column quality but provide no information about the solutes eluting from the column. Grob et al. [177] have developed a comprehensive, standardized quality test for glass capillary columns which gives information on their quantitative efficiency. The test should be performed on new columns and regularly during prolonged use of the columns.

B. The Internal Standard Method

The importance of using internal standards when analyzing PAH must be stressed. Apart from the mere advantage of a direct measurement of peak area (or height) of solute to peak area of internal standard,* the method eases the sample cleanup; there is no need to know sample volumes or amount injected; and what is lost of the sample component during cleanup is also lost of the internal standard(s). The requirements for a substance which is to be used as an internal standard have been given by Willis [178]:

Must yield a completely resolved peak
Should elute close to the component(s) being measured
Should not be present in the original sample

---

*We here define the internal standard as a compound being added to the sample prior to sampling or prior to cleanup. The internal standard is chemically similar to the compounds to be quantitated.

Must no react chemically with the sample
Should be present in about the same concentration as components measured

As the PAH fraction which is available to GC analysis covers a boiling point range of more than 300°C, there is a need for more than one internal standard (to correct for evaporation losses and peak broadening during a chromatographic run). Grimmer et al. [22] used 3,6-dimethylphenanthrene and benzo[b]chrysene, respectively, as internal standards in their two PAH fractions which were analyzed isothermally by GC. Bjørseth et al. [179,180] routinely use the two compounds 3,6-dimethylphenanthrene and 2,2'-binaphthyl, each of which covers one part of the programmed chromatographic run.

With the exception of the FID, all other detectors will require a response factor evaluation due to the different behaviors of each PAH in the detector. As already mentioned, the concentration of the internal standard becomes particularly critical with detectors showing a short linearity response range. A recovery test should be performed to establish the number of internal standards necessary and to determine which range is best covered by which internal standard in the chromatogram. To detect minor compounds in the chromatogram, it is often necessary to let major peaks go off scale on the recorder. This necessitates the use of an electronic integrator. Of the two possible methods available, measurement of the peak height or integration of the peak area, the latter has been used most frequently. In quantifying minor peaks it should also be kept in mind that adsorption phenomena affect these more than the larger peaks.

GC analysis may sometimes be disturbed by overlapping peaks. Quantitative GC analysis in such cases may be achieved using procedures described by Janik [181].

X. Multiple Column Systems

As stated by Bertsch [182] there are several ways to enhance the information from a GC run. More than one set of retention data will add information, as will splitting the column effluent to several detectors or column-switching techniques. Many approaches for achieving particular separations have been made that take advantage of the different polarities of available stationary phases. Several techniques have been devised ranging from double- or multiple-column chromatography either with capillary or packed precolumns and intermediate trapping followed by a second separation on a glass capillary column [183-185] or with solute switching, which by means of regulated pressure changes the direction of flow between columns [186,187], to multiple-column/multiple-temperature chromatography [188]. The latter method was further developed into the SECAT-technique, i.e., stepless fine tuning of capillary selectivity by serial column temperature optimization [189]. A series of papers by Bertsch [182,190,191], Miller et al. [192], and Anderson et al. [193] give an excellent up-to-date review of these methods which frequently are called two-dimensional GC.

The application of two-dimensional GC to the analysis of PAH are few. Ševčik [63] (see also note by Borwitzky et al. [194]) demonstrated the use of a multidimensional switching system for the separation of benzo[b]thiophene from naphthalene. With an impressive setup Grob [187] enhanced the separa-

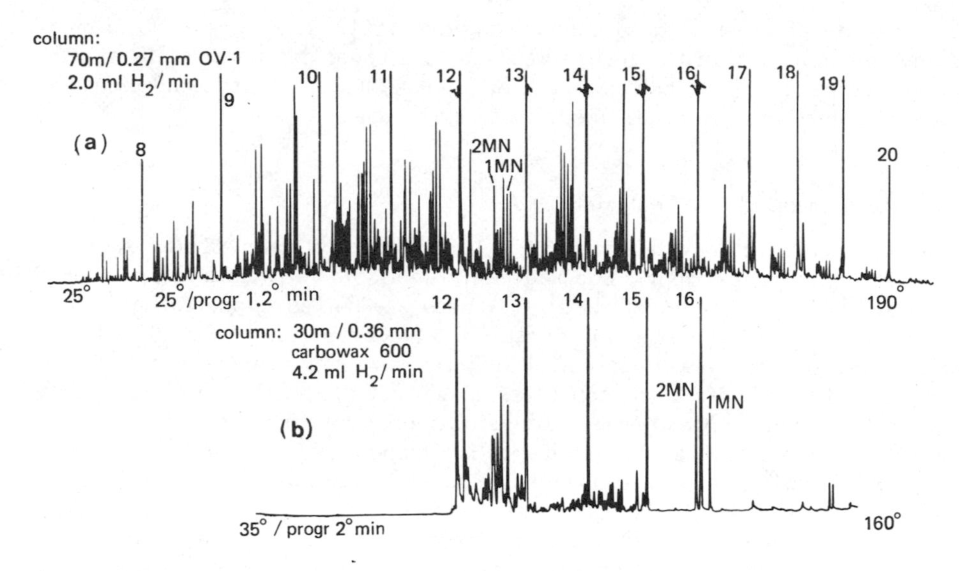

Figure 16. Analysis of diesel oil, partially transferred on and separately eluted from second column. Arrow down means start and arrow up means end of transfer to second column. Key: 2MN = 2-methylnaphthalene; 1MN = 1-methylnaphthalene. (From Ref. 187.)

tion of the two isomers of methylnaphthalene in diesel oil (Fig. 16). The immediate use of two-dimensional GC, apart from its utility in difficult separations, is for preparative GC. There is a lack of available standards of the many isomeric PAH either due to synthesis problems or high cost of production. The suspected health effect of several trace compounds may enforce, e.g., biological tests to establish mutagenic or carcinogenic properties. Glass capillary columns have the power to isolate a majority of these compounds. The drawback of low sample capacity is overcome with preparative GC with glass capillary columns. An elegant, automated system for this purpose has been described by Roeraade and Enzell [195]. After repetitive sampling, sufficient amounts of material can be obtained, say, for NMR spectroscopy. There is most likely a future for this system.

XI. Cleanup Procedures

This chapter has so far described many of the advantages with capillary GC, including resolution and sensitivity. Applications of the method on different complex mixtures have also been demonstrated. However, as pointed out by Jennings [196], "no degree of sophistication in the final analysis can compensate for or correct compositional changes caused by sampling errors. . . ." Sampling, cleanup, preconcentration (by, e.g., evaporation of excess solvent), etc., must therefore be considered carefully. The addition of internal standards (discussed above) will correct for losses during cleanup, and contamination from solvents may be avoided by alternative extraction methods such as

Soxhlet extraction with liquid carbon dioxide [196]. High-pressure liquid chromatography will in the future be used to a great extent as a cleanup procedure prior to GC analysis (see, e.g., Ref. 197). Further applications are well documented in other chapters of this book.

XII. Recommended GC Analysis of PAH

Several considerations will govern the choice of analytical method. Many laboratories are dependent on commercially available columns and are therefore not able to tailormake them for PAH. Others are not equipped to perform high-resolution GC analysis. The latter is, however, fully competent in analyzing PAH provided it knows its limitations. In the following we will also recognize the difference between (1) analysis with the aim of obtaining as much information as possible from one single chromatographic run and (2) routine analysis where a limited number of compounds are measured.

In the following, we have given some recommendations concerning GC analysis of PAH. These recommendations are the results of experience from hundreds of PAH analyses in our laboratory. The fact that our own GC protocol is recommended is, however, not to be interpreted as disapproval of other methods, nor that we have considered all published methods. We simply know by experience that the method works.

Our recommendations follow:

1. Cold on-column injection (alternatively splitless injection, which restricts quantitative work to compounds up to coronene)
    a. WCOT 50 m × 0.25*-0.35 mm ID persilylated or polysiloxane deactivated glass/fused silica capillary column coated with a 0.05-0.15 μm film of OV-73 or immobilized/crosslinked SE-54 (alternatively a commercial high temperature glass/fused silica capillary coated with a thin film of silicone gum, or Poly S-179 prepared according to Ref. 119)
    b. Hydrogen carrier gas
    c. Temperature programming 1-4°C/min, preferably up to 350°C
    d. FID (alternatively MS or flow-splitting to FID/MS)
    e. Electronic area integration (peak broadening correction should be programmable)
    f. A minimum of two internal standards
2. Splitless injection (or split injection according to Ref. 52 in the case of concentrated samples)
    a. WCOT 15-25 m × 0.25-0.35 mm ID glass/fused silica capillary (or 4 m × 0.20-0.35 mm ID for split injection) coated with a silicone gum
    b. Temperature programming 6-10°C/min
    c. FID
    d. Electronic area integration
    e. Two internal standards

Packed columns with their large amount of stationary phase may play a role in preparative GC for the isolation of micrograms of unknown PAH for UV and fluorescence analysis.

---

*The limiting factor for on-column injection is the syringe outer diameter.

References

1. World Health Organization, *International Standards for Drinking Water*, 3rd ed., WHO, Geneva, 1971.
2. C. D. Kalfadelis, E. M. Magee, and T. D. Searl, in Evaluation of pollution control in fossil fuel conversion processes. Exxon Research and Engineering Co., Linden, N.J. EPA contract No. 68-02-0629, Publ. No. 650/2-74-009-1 (1975).
3. E. M. Smith and R. L. Levins, in *Polynuclear Aromatic Hydrocarbons: Chemistry and Biological Effects*, A. Bjørseth and A. J. Dennis (Eds.), Battelle Press, Columbus, Ohio, 1980, pp. 973-982.
4. F. Dupire, *Z. Anal. Chem. 170*:317 (1959).
5. N. Carugno, *Natl. Cancer Inst. Monogr. 9*:171 (1962).
6. E. Sawicki, F. T. Fox, and W. Elbert, *Amer. Ind. Hyg. Assoc. J. 23*:482 (1962).
7. W. Lijinsky, I. Domsky, G. Mason, H. Y. Ramahi, and I. Safari, *Anal. Chem. 35*:952 (1963).
8. J. R. Wilmshurst, *J. Chromatogr. 17*:50 (1965).
9. A. Liberti, G. P. Cartoni, and V. Cantuti, *J. Chromatogr. 15*:141 (1964).
10. V. Cantuti, G. P. Cartoni, A. Liberti, and A. G. Torri, *J. Chromatogr. 17*:60 (1965).
11. N. Carugno and S. Rossi, *J. Gas Chromatogr.* 5:103 (1967).
12. G. P. Cartoni, *Estrat. Cron. Chim. No. 22* (Dec. 1968).
13. R. C. Lao, R. S. Thomas, H. Oja, and L. Dubois, *Anal. Chem. 45*:908 (1973).
14. E. D. John and G. Nickless, *J. Chromatogr. 138*:399 (1977).
15. R. S. Thomas, R. C. Lao, D. T. Wang, D. Robinson, and T. Sakuma, in *Carcinogenesis*, Vol. 3: *Polynuclear Aromatic Hydrocarbons*, P. W. Jones and R. I. Freudenthal (Eds.), Raven Press, New York, 1978, pp. 9-19.
16. W. Cautreels and K. van Cauwenberghe, *Atmos. Environ. 10*:447 (1976).
17. H. Tausch and G. Stehlik, *Chromatographia 10*:350 (1977).
18. A. Bjørseth and G. Lunde, *Amer. Ind. Hyg. Assoc. J. 38*:224 (1977).
19. A. Candeli, G. Morozzi, A. Paolacci, and L. Zoccolillo, *Atmos. Environ. 9*:843 (1975).
20. K. Adachi, *Bull. Environ. Contam. Toxicol. 25*:416 (1980).
21. H. J. Neumann and D. T. Kaschani, *Wasser Luft Betrieb 21*:648 (1977).
22. G. Grimmer and H. Böhnke, *J. Assoc. Offic. Anal. Chem. 58*:725 (1975).
23. G. Grimmer and H. Böhnke, *Cancer Lett. 1*:75 (1975).
24. Y. L. Tan, *J. Chromatogr. 176*:319 (1979).
25. G. Locati, A. Fantuzzi, G. Consonni, I. Li Gotti, and G. Bonomi, *Amer. Ind. Hyg. Assoc. J. 40*:644 (1979).
26. Supelco, Catalog 16, Supelco, Inc., Bellefonte, Pa., p. 22 (1980).
27. B. D. Crittenden and R. Long, *Combust. Flame 20*:359 (1973).
28. F. M. Benoit, G. L. Lebel, and D. T. Williams, *Intern. J. Environ. Anal. Chem. 6*:277 (1979).
29. E. Faltusz, *Z. Anal. Chem. 294*:385 (1979).
30. R. C. Lao, R. S. Thomas, and J. L. Monkman, *J. Chromatogr. 112*:681 (1975).
31. R. F. Severson, M. E. Snook, H. C. Higman, O. T. Chortyk, and F. J. Akin, in *Carcinogenesis*, Vol. 1: *Polynuclear Aromatic Hydrocarbons:*

Chemistry, Metabolism, and Carcinogenesis, R. I. Freudenthal and P. W. Jones (Eds.), Raven Press, New York, 1976, pp. 253-270.
32. M. E. Snook, R. F. Severson, H. C. Higman, R. F. Arrendale, and O. T. Chortyk, *Beitr. Tabakforsch. 8*:250 (1976).
33. G. Grimmer, in *Luftqualitätskriterien für ansgewählte polyzyklische aromatische Kohlenwasserstoffe*, Ber. 1/79, Umwelt Bundes Amt, Schmidt Verlag, Berlin, 1979, pp. 70-76.
34. C. Vidal-Madjar and G. Guiochon, *Nature 215*:1372 (1967).
35. A. Zane, *J. Chromatogr. 17*:130 (1968).
36. J. Fryčka, *Chromatographia 8*:413 (1975).
37. P. Burchill, A. A. Herod, and R. G. James, in *Carcinogenesis*, Vol. 3: *Polynuclear Aromatic Hydrocarbons*, P. W. Jones and R. I. Freudenthal (Eds.), Raven Press, New York, 1978, pp. 35-45.
38. M. J. E. Golay, in *Gas Chromatography 1958*, D. H. Desty (Ed.), Academic Press, New York, 1958, pp. 36-55.
39. D. H. Desty, J. N. Haresnape, and B. H. F. Whyman, *Anal. Chem. 32*:302 (1960).
40. G. Grimmer, H. Böhnke, and A. Glaser, *Erdöl Kohle Erdgas. Petrochem. 30*:411 (1977).
41. T. H. Gouw, I. M. Whittemore, and R. E. Jentoft, *Anal. Chem. 42*:1394 (1970).
42. T. Doran and N. G. Taggart, *J. Chromatogr. Sci. 12*:715 (1974).
43. K. Grob and G. Grob, in *Identification and Analysis of Organic Pollutants in Water*, L. H. Keith (Ed.), Ann Arbor Science Publs., Ann Arbor, Mich., 1976, pp. 75-85.
44. K. Grob and G. Grob: *HRC & CC 3*:109 (1979).
45. G. Schomburg, *HRC & CC 7*:461 (1979).
46. M. Novotny, *Anal. Chem. 50*:16A (1978).
47. W. Giger and C. Schaffner, *Anal. Chem. 50*:243 (1978).
48. P. J. Blokzijl and R. Guicherit, *TNO-Nieuws* p. 653 (Nov. 1972).
49. A. Bjørseth, in *Polynuclear Aromatic Hydrocarbons*, P. W. Jones and P. Leber (Eds.), Ann Arbor Science Publs., Ann Arbor, Mich., 1979, pp. 371-381.
50. M. L. Lee, M. Novotny, and K. D. Bartle, *Anal. Chem. 48*:405 (1976).
51. M. L. Lee, K. D. Bartle, and M. V. Novotny, *Anal. Chem. 47*:540 (1975).
52. B. W. Wright and M. L. Lee, *HRC & CC 3*:352 (1980).
53. E. Sawicki, T. W. Stanley, S. McPherson, and M. Morgan, *Talanta 13*:619 (1966).
54. L. DeMaio and M. Corn, *Anal. Chem. 38*:131 (1966).
55. W. H. Griest, H. Kubota, and M. R. Guerin, *Anal. Lett. 8*(12):949 (1975).
56. D. F. S. Natusch and B. A. Tomkins, *Anal. Chem. 50*:1429 (1978).
57. K. A. Schulte, D. J. Larsen, R. W. Hornung, and J. V. Crable, *J. Amer. Ind. Hyg. Assoc. 36*:131 (1975).
58. W. Cautreels and K. Cauwenberghe, *Water Air Soil Pollut. 6*:103 (1976).
59. B. D. Crittenden and R. Long, in *Carcinogenesis*, Vol. 1: *Polynuclear Aromatic Hydrocarbons: Chemistry, Metabolism, and Carcinogenesis*, R. I. Freudenthal and P. W. Jones (Eds.), Raven Press, New York, 1976, pp. 209-223.

60. P. E. Strup, R. D. Giammar, T. B. Stanford, and P. W. Jones, in *Carcinogenesis*, Vol. 1: *Polynuclear Aromatic Hydrocarbons: Chemistry, Metabolism, and Carcinogenesis*, R. I. Freudenthal and P. W. Jones (Eds.), Raven Press, New York, 1976, pp. 241-251.
61. A. Hase, P. H. Lin, and R. A. Hites, in *Carcinogenesis*, Vol. 1: *Polynuclear Aromatic Hydrocarbons: Chemistry, Metabolism, and Carcinogenesis*, R. I. Freudenthal and P. W. Jones (Eds.), Raven Press, New York, 1976, pp. 435-442.
62. R. C. Lao and R. S. Thomas, in *Polynuclear Aromatic Hydrocarbons*, P. W. Jones and P. Leber (Eds.), Ann Arbor Science Publs., Ann Arbor, Mich., 1979, pp. 429-452.
63. J. Ševčik, *HRC & CC 1*:25 (1980).
64. M. R. Guerin, J. L. Epler, W. H. Griest, B. R. Clark, and T. K. Rao, in *Carcinogenesis*, Vol. 3: *Polynuclear Aromatic Hydrocarbons*, P. W. Jones and R. I. Freudenthal (Eds.), Raven Press, New York, 1978, pp. 21-33.
65. K. H. Palmork, S. Wilhelmsen, and T. Neppelberg, in *Report on the Contribution of PAH to the Marine Environment from Different Industries*. International Council for the Exploration of the Sea, Fisheries Improvement Committee, C.M. (1973), W:33.
66. H. P. Burchfield, E. E. Green, R. J. Wheeler, and S. M. Billedeau, *J. Chromatogr. 99*:697 (1974).
67. E. Sawicki, R. C. Corey, A. E. Dooley, J. B. Gisclard, J. L. Monkman, R. E. Neligan, and L. A. Ripperton, *Health Lab. Sci. 7*:56 (1970).
68. J. L. Glajch, J. A. Lubkowitz, and L. B. Rogers, *J. Chromatogr. 168*:355 (1979).
69. N. A. Goeckner and W. H. Griest, *Sci. Total Environ. 8*:192 (1977).
70. R. J. Laub and W. L. Roberts, in *Polynuclear Aromatic Hydrocarbons: Chemistry and Biological Effects*, A. Bjørseth and A. J. Dennis (Eds.), Battelle Press, Columbus, Ohio, 1980, pp. 25-58.
71. F. Belsito, L. Boniforti, R. Dommarco, and G. Laguzzi, *Quant. Mass Spectrom. Life Sci. Eng. J. 2*:431 (1978).
72. E. Sawicki, *Health Lab. Sci. 11*:218 (1974).
73. E. Sawicki, *Health Lab. Sci. 11*:228 (1974).
74. A. F. Shushunova, P. E. Shkodich, N. G. Shkanakin, and L. Z. H. Lembik, *Gig. Sanit. 8*:61 (1975).
75. B. B. Chakraborty and R. Long, *Environ. Sci. Technol. 1*:828 (1967).
76. G. Chatot, R. Dangy-Caye, and R. Fontanges, *J. Chromatogr. 72*:202 (1972).
77. K. Alben, *Environ. Sci. Technol. 14*:468 (1980).
78. K. Grob, *Chromatographia 10*:625 (1977).
79. Chrompack News No. 24, Chrompack, Middelburg, The Netherlands (1979).
80. G. Becher, A. Bjørseth, and B. Sortland Olufsen, in *Chemical Hazards in the Workplace. Measurement and Control*, G. Choudhary (Ed.), American Chemical Society, Washington, D.C., 1981, pp. 369-381.
81. M. Blumer, *Sci. Amer. 234*:35 (1976).
82. P. Sandra, M. Verstappe, M. Verzele, and J. Verzele, *Intern. Lab.* p. 57 (July/Aug. 1980).

83. K. Grob, Jr., and G. Grob, *Chromatographia 10*:250 (1977).
84. G. Schomburg, R. Dielmann, H. Borwitzky, and H. Husmann, *J. Chromatogr. 167*:337 (1978).
85. G. Schomburg, H. Husmann, and H. Borwitzky, *Chromatographia 12*:651 (1979).
86. K. Grob, G. Grob, and K. Grob, Jr., *HRC & CC 1*:31 (1979).
87. K. Grob, G. Grob, and K. Grob, Jr., *HRC & CC 11*:677 (1979).
88. K. Grob and G. Grob, *HRC & CC 3*:197 (1980).
89. L. Blomberg and T. Wännman, *J. Chromatogr. 148*:379 (1978).
90. L. Blomberg, J. Buijten, J. Gawdzik, and T. Wännman, *Chromatographia 11*:521 (1978).
91. L. Blomberg and T. Wännman, *J. Chromatogr. 168*:81 (1979).
92. L. Blomberg and T. Wännman, *J. Chromatogr. 186*:159 (1979).
93. R. Dandeneau, P. Bente, T. Rooney, and R. Hiskes, *Intern. Lab.* p. 69 (Nov./Dec. 1979).
94. K. Sato, M. Matsui, and N. Ikekawa, *Bunseki Kagaku 17*:639 (1968).
95. H. D. Sauerland and M. Zander, *Erdöl Kohle Erdgas Petrochem. 25*:526 (1972).
96. G. M. Janini, K. Johnston, and W. L. Zielinsky, Jr., Anal. Chem. 47:670 (1975).
97. G. M. Janini, G. M. Muschik, and W. L. Zielinsky, Jr., *Anal. Chem. 48*:809 (1976).
98. G. M. Janini, G. M. Muschik, J. A. Schroer, and W. L. Zielinski, Jr., *Anal. Chem. 48*:1879 (1976).
99. R. J. Laub, W. L. Roberts, and C. A. Smith, *HRC & CC 7*:355 (1980).
100. W. L. Zielinski, Jr., *Analabs* p. 2 (Sept. 1977).
101. W. L. Zielinski, Jr., *Ind. Res./Develop.* p. 178 (Feb. 1980).
102. S. Wasik and S. Chesler, *J. Chromatogr. 122*:451 (1976).
103. *Gas-Chrom Newslett. 18*, 8 (1977).
104. J. W. Strand and A. W. Andren, *Anal. Chem. 50*:1508 (1978).
105. W. L. Zielinsky, Jr., and G. M. Janini, *J. Chromatogr. 186*:237 (1979).
106. Chrompack News No. 25, p. 3, Chrompack, Middelburg, The Netherlands (1979).
107. V. Hložek and H. Gutwillinger, *Chromatographia 13*:234 (1980).
108. Supelco, GC Reporter, Supelco, Inc., Bellefonte, Pa., p. 4 (Apr. 1977).
109. G. M. Janini B. Shaikh, and W. L. Zielinski, Jr., *J. Chromatogr. 132*:136 (1977).
110. A. Radecki, H. Lamparczyk, J. Grzybowski, and J. Halkiewicz, *J. Chromatogr. 150*:527 (1978).
111. R. J. Laub and J. H. Purnell, *J. Chromatogr. 112*:71 (1975).
112. R. J. Laub and J. H. Purnell, *Anal. Chem. 48*:799 (1976).
113. R. J. Laub, J. H. Purnell, and P. S. Williams, *J. Chromatogr. 134*:249 (1977).
114. K. Grob, Jr., and K. Grob, *J. Chromatogr. 140*:257 (1977).
115. R. G. Mathews, R. D. Schwartz, C. D. Pfaffenberger, S.-N. Lin, and E. C. Horning, *J. Chromatogr. 99*:51 (1974).
116. P. van Hout, J. Szafranek, C. D. Pfaffenberger, and E. C. Horning, *J. Chromatogr. 99*:103 (1974).
117. S.-N. Lin, C. D. Pfaffenberger, and E. C. Horning, *J. Chromatogr. 104*:319 (1975).

118. R. D. Schwartz, R. G. Mathews, S. Ramachandran, R. S. Henly, and J. E. Doyle, *J. Chromatogr. 112*:111 (1975).
119. H. Borwitzky and G. Schomburg, *J. Chromatogr. 170*:99 (1979).
120. L. S. Lysyuk and A. N. Korol, *Chromatographia 10*:712 (1977).
121 W. Jennings, *Gas Chromatography with Glass Capillary Columns*, Academic Press, New York, 1978, p. 68.
122. K. Grob and G. Grob, *J. Chromatogr. Sci.* 7:515 (1969).
123. R. Kaiser, *Z. Anal. Chem. 189*:1 (1962).
124. B. Sortland Olufsen, *HRC & CC 10*:578 (1979).
125. M. L. Lee and B. W. Wright, *J. Chromatogr. Sci. 18*:345 (1980).
126. K. Grob and K. Grob, Jr., *HRC & CC 1*:57 (1978).
127. W. G. Jennings, R. R. Freeman, and T. A. Rooney, *HRC & CC* 5:275 (1978).
128. P. M. J. van den Berg and T. P. H. Cox, *Chromatographia* 5:301 (1972).
129. B. Larsson, personal communication (1980).
130. F. S.-C. Lee, T. J. Prater, and F. Ferris, in *Polynuclear Aromatic Hydrocarbons*, P. W. Jones and P. Leber (Eds.), Ann Arbor Science Publs., Ann Arbor, Mich., 1979, pp. 83-110.
131. I. Wauters, P. Sandra, and M. Verzele, *J. Chromatogr. 170*:125 (1979).
132. S. N. Chesler, F. R. Guenther, and R. G. Christensen, *HRC & CC 7*: 351 (1980).
133. R. E. Kaiser, *HRC & CC 2*:95 (1979).
134. M. Novotny and R. Farlow, *J. Chromatogr. 103*:1 (1975).
135. K. Grob and G. Grob, *J. Chromatogr. Sci.* 7:584 (1969).
136. K. Grob and G. Grob, *J. Chromatogr. Sci.* 7:587 (1969).
137. K. Grob and K. Grob, Jr., *J. Chromatogr. 94*:53 (1974).
138. G. Schomburg, H. Husmann, and F. Weeke, *Chromatographia 10*:580 (1977).
139. G. Schomburg, H. Behlau, R. Dielmann, F. Weeke, and H. Husmann, *J. Chromatogr. 142*:87 (1977).
140. K. Grob and K. Grob, Jr., *J. Chromatogr. 151*:311 (1978).
141. K. Grob, *HRC & CC* 5:163 (1978).
142. M. Galli, S. Trestianu, and K. Grob, Jr., *HRC & CC 6*:366 (1979).
143. K. Grob, Jr., and H. P. Neukom, *J. Chromatogr. 189*:109 (1980).
144. A. Bjørseth, *Anal. Chim. Acta 94*:21 (1977).
145. T. Alsberg and U. Stenberg, *Chemosphere* 7:487 (1979).
146. J. Jakob and G. Grimmer, *Zbl. Bakteriol. Hyg., I. Abt. Orig. B165*: 305 (1977).
147. F. Etzweiler and N. Neuner-Jehle, *Chromatographia 6*:503 (1973).
148. K. Bächmann, W. Emig, J. Rudolph, and D. Tsotsos, *Chromatographia 10*:684 (1977).
149. Hewlett-Packard Operating Note No. 5985A, H-P Scientific Instruments Div., Palo Alto, Calif. (1978).
150. E. L. Anderson and W. Bertsch, *HRC & CC 1*:13 (1978).
151. F. Poy, *HRC & CC* 5:243 (1979).
152. W. Giger, M. Reinhard, C. Schaffner, and F. Zürcher, in *Identification and Analysis of Organic Pollutants in Water*, L. H. Keith (Ed.), Ann Arbor Science Publs., Ann Arbor, Mich., 1976, pp. 433-452.
153. A. Bjørseth and G. Eklund, *HRC & CC 1*:22 (1979).

154. E. P. Grimsrud, D. A. Miller, R. G. Stebbins, and S. H. Kim, *J. Chromatogr.* *197*:51 (1980).
155. S. Nygren and P. E. Mattson, *J. Chromatogr.* *123*:101 (1976).
156. W. A. Aue and C. G. Flinn, *J. Chromatogr.* *153*:305 (1978).
157. J. Fryčka, *J. Chromatogr.* *174*:488 (1979).
158. M. L. Lee and R. A. Hites, *Anal. Chem.* *48*:1890 (1976).
159. M. L. Lee, C. Willey, R. N. Castle, and C. M. White, in *Polynuclear Aromatic Hydrocarbons: Chemistry and Biological Effects*, A. Bjørseth and A. J. Dennis (Eds.), Battelle Press, Columbus, Ohio, 1980, pp. 59-73.
160. M. J. Hartigan, J. E. Purcell, M. Novotny, M. L. McConnell, and M. L. Lee, *J. Chromatogr.* *99*:339 (1974).
161. F. J. Schwende, M. Novotny, and J. E. Purcell, *Chromatogr. Newslett.* *8*:1 (1980).
162. H. P. Burchfield, R. J. Wheeler, and J. B. Bernos, *Anal. Chem.* *43*: 1976 (1971).
163. L. S. Ettre, *J. Chromatogr. Sci.* *16*:396 (1978).
164. E. Kováts, *Helv. Chim. Acta* *41*:1915 (1958).
165. G. Schomburg and G. Dielmann, *J. Chromatogr. Sci.* *11*:151 (1973).
166. H. van den Dool and P. D. Kratz, *J. Chromatogr.* *11*:463 (1963).
167. H. Beernaert, *J. Chromatogr.* *173*:109 (1979).
168. W. A. Aue and V. Paramasigamani, *J. Chromatogr.* *166*:253 (1978).
169. M. L. Lee, D. L. Vassilaros, C. M. White, and M. Novotny, *Anal. Chem.* *51*:768 (1979).
170. R. Kaliszan and H. Lamparczyk, *J. Chromatogr. Sci.* *16*:246 (1978).
171. H. Vollmann, H. Becker, M. Corell, and H. Streeck, *Ann. Chem.* *531*:1 (1937).
172. R. S. Mulliken, *J. Am. Chem. Soc.* *74*:811 (1952).
173. H. C. Brown and J. D. Brady, *J. Am. Chem. Soc.* *74*:3570 (1952).
174. R. Langlais, R. Schlenkermann, and M. Weinberg, *Chromatographia* *9*: 601 (1974).
175. W. G. Jennings, *J. Chromatogr. Sci.* *13*:185 (1975).
176. K. Grob, Jr., and H. P. Neukom, *HRC & CC* *1*:15 (1979).
177. K. Grob, Jr., G. Grob, and K. Grob, *J. Chromatogr.* *156*:1 (1978).
178. D. E. Willis, *Chromatographia* *5*:42 (1972).
179. A. Bjørseth, J. Knutzen, and J. Skei, *Sci. Total Environ.* *13*:71 (1979).
180. A. Bjørseth, G. Lunde, and A. Lindskog, *Atmos. Environ.* *13*:45 (1979).
181. A. Janik, *Chromatographia* *8*:563 (1975).
182. W. Bertsch, *HRC & CC* *2*:85 (1978).
183. G. Schomburg, H. Husmann, and F. Weeke, *J. Chromatogr.* *99*:63 (1974).
184. G. Schomburg, H. Husmann, and F. Weeke, *J. Chromatogr.* *112*:205 (1975).
185. W. Blass, K. Riegner, and H. Hulpke, *J. Chromatogr.* *172*:67 (1979).
186. D. R. Deans, *Chromatographia* *1*:18 (1968).
187. K. Grob, *Chromatographia* *8*:423 (1975).
188. R. E. Kaiser and R. I. Rieder, *HRC & CC* *4*:201 (1978).
189. R. E. Kaiser and R. I. Rieder, *HRC & CC* *7*:416 (1979).
190. W. Bertsch, *HRC & CC* *4*:187 (1978).
191. W. Bertsch, *HRC & CC* *6*:289 (1978).
192. R. J. Miller, S. D. Stearns, and R. R. Freeman, *HRC & CC* *2*:55 (1979).

193. E. L. Anderson, M. M. Thomason, H. T. Mayfield, and W. Bertsch, *HRC & CC 6*:335 (1979).
194. H. Borwitzky, G. Schomburg, and H. Husmann, *HRC & CC 4*:206 (1980).
195. J. Roeraade and C. R. Enzell, *HRC & CC 3*:123 (1979).
196. W. G. Jennings, *HRC & CC 5*:221 (1979).
197. F. P. DiSanzo, P. C. Uden, and S. Siggia, *Anal. Chem. 52*:906 (1980).
198. D. T. Kaschani, *Erdöl Kohle Erdgas Petrochem. 32*:572 (1979).
199. Chrompack Special News, Chrompack, Middelburg, The Netherlands (1980).
200. R. L. Hanson, R. E. Royer, R. L. Carpenter, and G. J. Newton, in *Polynuclear Aromatic Hydrocarbons*, P. W. Jones and P. Leber (Eds.), Ann Arbor Science Publs., Ann Arbor, Mich., 1979, pp. 3-19.
201. W. H. Griest, B. A. Tomkins, J. L. Epler, and T. K. Rao, in *Polynuclear Aromatic Hydrocarbons*, P. W. Jones and P. Leber (Eds.), Ann Arbor Science Publs., Ann Arbor, Mich., 1979, pp. 395-409.
202. U. Stenberg, T. Alsberg, L. Blomberg, and T. Wännman, in *Polynuclear Aromatic Hydrocarbons*, P. W. Jones and P. Leber (Eds.), Ann Arbor Science Publs., Ann Arbor, Mich., 1979, pp. 313-326.
203. U. Rapp, U. Schröder, S. Meier, and H. Elmenhorst, *Chromatographia 8*:474 (1975).
204. G. Lunde, *Ambio 5*:207 (1976).
205. G. Lunde and A. Bjørseth, *Nature 268*:518 (1977).
206. A. Bjørseth, O. Bjørseth, and P. E. Fjeldstad, *Scand. J. Work Environ. Health 4*:212 (1978).
207. F. I. Onuska and M. E. Comba, *J. Chromatogr. 126*:133 (1976).
208. G. Grimmer, K.-W. Naujack, and D. Schneider, in *Polynuclear Aromatic Hydrocarbons: Chemistry and Biological Effects*, A. Bjørseth and A. J. Dennis (Eds.), Battelle Press, Columbus, Ohio, 1980, pp. 107-125.
209. G. Grimmer, Chap. 4 in this handbook.
210. W. Giger, M. Reinhard, and C. Schaffner, *Vom Wasser 43*:343 (1974).
211. K. Grob, *Chromatographia 7*:94 (1974).
212. W. Bertsch, E. Anderson, and G. Holzer, *J. Chromatogr. 126*:213 (1976).
213. R. Herrmann, *Catena 5*:165 (1978).
214. J. Eyem, *Chromatographia 8*:456 (1975).
215. R. W. Schimberg, P. Pfäffli, and A. Tossavainen, *Staub-Reinhaltung Luft 7*:273 (1978).
216. Hewlett-Packard, Publ. No. 43-5953-1470 (D), Hewlett-Packard Co., Avondale, Pa. (1980).
217. M. Novotny, M. L. Lee, and K. D. Bartle, *J. Chromatogr. Sci. 12*:606 (1974).
218. M. L. Lee, M. Novotny, and K. D. Bartle, *Anal. Chem. 48*:1566 (1976).
219. F. I. Onuska, A. W. Wolkoff, M. E. Comba, R. H. Larose, M. Novotny, and M. L. Lee, *Anal. Lett. 9*:451 (1976).
220. K. D. Bartle, M. L. Lee, and M. Novotny, *Proc. Anal. Div. Chem. Soc.* p. 304 (Oct. 1976).
221. M. L. Lee, G. P. Prado, J. B. Howard, and R. A. Hites, *Biomed. Mass Spectr. 4*:183 (1977).
222. L. S. Sheldon and R. A. Hites, *Environ. Sci. Technol. 12*:1188 (1978).
223. R. H. Bieri, M. Kent Cueman, C. L. Smith, and C.-W. Su, *Intern. J. Environ. Anal. Chem. 5*:293 (1978).

224. R. E. LaFlamme and R. A. Hites, *Geochim. Cosmochim. Acta 42*:289 (1978).
225. M. L. Lee, D. L. Vassilaros, W. S. Pipkin, and W. L. Sorensen, in *Trace Organic Analysis: A New Frontier in Analytical Chemistry*, National Bureau of Standards Spec. Publ. No. 519, U.S. Government Printing Office, Washington, D.C., 1979, pp. 731-738.
226. C. M. White, A. G. Sharkey, Jr., M. L. Lee, and D. L. Vassilaros, in *Polynuclear Aromatic Hydrocarbons*, P. W. Jones and P. Leber (Eds.), Ann Arbor Science Publs., Ann Arbor, Mich., 1979, pp. 261-275.
227. R. V. Schultz, J. W. Jorgenson, M. P. Maskarinec, M. Novotny, and L. J. Todd, *Fuel 58*:783 (1979).
228. J. M. Meuser, F. R. Moore, P. E. Strup, J. E. Wilkinson, and A. Bjørseth, in *Polynuclear Aromatic Hydrocarbons: Chemistry and Biological Effects*, A. Bjørseth and A. J. Dennis (Eds.), Battelle Press, Columbus, Ohio, 1980, pp. 405-415.
229. S. G. Wakeham, C. Schaffner, and W. Giger, *Geochim. Cosmochim. Acta 44*:403 (1980).
230. S. G. Wakeham, C. Schaffner, and W. Giger, *Geochim. Cosmochim. Acta 44*:415 (1980).
231. B. Olufsen and M. Skogland, *Kjemi 8*:42 (1977) (in Norwegian).
232. E. Winkler, A. Buchele, and O. Müller, *J. Chromatogr. 138*:151 (1977).
233. P. E. Strup, J. E. Wilkinson, and P. W. Jones, in *Carcinogenesis*, Vol. 3: *Polynuclear Aromatic Hydrocarbons*, P. W. Jones and R. I. Freudenthal (Eds.), Raven Press, New York, 1978, pp. 131-138.
234. A. Bjørseth, in *Carcinogenesis*, Vol. 3: *Polynuclear Aromatic Hydrocarbons*, P. W. Jones and R. I. Freudenthal (Eds.), Raven Press, New York, 1978, pp. 75-83.
235. A. Bjørseth, O. Bjørseth, and P. E. Fjeldstad, *Scand. J. Work Environ. Health 4*:224 (1978).
236. A. Bjørseth and G. Eklund, *Anal. Chim. Acta 105*:119 (1979).
237. M. E. Snook, R. F. Severson, H. C. Higman, R. F. Arrendale, and O. T. Chortyk, in *Polynuclear Aromatic Hydrocarbons*, P. W. Jones and P. Leber (Eds.), Ann Arbor Science Publs., Ann Arbor, Mich., 1979, pp. 231-260.
238. B. Olufsen, in *Polynuclear Aromatic Hydrocarbons: Chemistry and Biological Effects*, A. Bjørseth and A. J. Dennis (Eds.), Battelle Press, Columbus, Ohio, 1980, pp. 333-343.
239. J. Knutzen and B. Sortland, *Water Research 16*:421 (1982).
240. L. Sucre, W. Jennings, G. L. Fisher, O. G. Raabe, and J. Olechno, in *Trace Organic Analysis: A New Frontier in Analytical Chemistry*, National Bureau of Standards Spec. Publ. 519. U.S. Government Printing Office, Washington, D.C., 1979, pp. 109-120.
241. Hewlett-Packard, *Peak No. 2* (1980).
242. G. Grimmer, H. Böhnke, and A. Hildebrandt, *Z. Anal. Chem. 279*:139 (1976).

7

Mass Spectrometric Analysis of Polycyclic Aromatic Hydrocarbons

BJÖRN JOSEFSSON / Chalmers University of Technology, University of Göteborg, Göteborg, Sweden

---

I. Introduction

Mass spectrometry (MS), especially in combination with gas chromatography (GC), is the most powerful analytical tool available today in the identification of individual components in complex mixtures. The high resolving capability of GC together with the MS molecular structure information often leads to complete identification of volatile PAH components in complex mixtures. The main drawback of MS is the difficulty in separating positional isomers, a common problem also in PAH analysis.

The analytical procedure for PAH can generally be divided into (1) isolation and separation and (2) identification and quantitation. Different MS techniques may be applied depending on the sample working-up procedure. There will be a compromise between obtainable MS structure information and the sophistication of prior isolation methods. Thus, identification of hundreds of PAH components in trace amounts in a crude extract demands either a time-consuming pretreatment, at the risk of a change or loss of components, or advanced MS methods.

The first part of this chapter will discuss more or less direct MS methods, while the second will deal with the outstanding tool of high-resolution glass capillary column separation combined with MS.

II. Ionization Techniques

Ion production is the central point of MS because it basically determines what kind of information is measured by the spectrometer. Several techniques have been used to attain molecular ionization of PAH compounds, the most common of which is electron impact ionization (EI). Other methods of increasing importance are chemical ionization (CI) and field or field desorption ionization (FI and FD, respectively).

The designs of ion sources vary; however, the basic elements are similar. The need for high ionization efficiency and easy transmission of ions to the analyzer is favored by differential pumping of the ion source and the analyzer. The mean free path of ions in the analyzer is then essentially independent of the ion source pressure, which will result in more stable conditions.

Field desorption ionization can be practiced only with special sampling devices. The technique is not feasible for continuous monitoring of column effluents. However, this is the only ionization principle that can be applied to completely nonvolatile compounds.

A. Electron Impact Ionization

Electron impact ionization is the most common method, used mainly because of stability, ease of operation, and narrow kinetic energy spread in the ions formed. The ionization is performed with an electron gun, which produces electrons with a certain energy that can be varied between 5 and 100 eV. These electrons bombard the sample molecules and may impact sufficient energy to remove an electron from the sample molecule. The result is a molecular ion and in most cases decomposition into smaller fragment ions. The fragmentation depends on the energy of the bombarding electron and will result in a mass spectral pattern typical of the compound studied. An EI source operating at the conventional 70 eV provides sufficient energy to ionize and cause the characteristic fragmentation of most molecules. The fragmentation does not change significantly above 25-30 eV of ionization energy. With a low voltage such as 15 eV the spectrum will be considerably simplified because of decreased fragmentation. Ionization potentials of complex molecules can be measured by studying the decrease in the molecular ion signal as a function of ionization voltage. Gallegos [1] reported the ionization energies for different PAH to be in the range of 6.55-8.63 eV.

The positive ions produced in the source are extracted from the ionization chamber by a positively charged repeller electrode. The ions are there accelerated toward the analyzer with a positive field, whose strength varies, e.g., from 8000 to 800 V for a magnetic sector instrument or from 20 to 10 V for quadrupole instruments.

B. Chemical Ionization

Many compounds do not give a molecular ion upon EI. In those cases CI techniques will provide important information concerning molecular weights. CI will yield less fragmentation; however, the ionization conditions may be varied to a greater extent compared with EI.

CI occurs via ion-molecule reactions between neutral sample molecules and a high pressure (0.5-5.0 torr) reagent gas ion plasma. The plasma is produced by electron bombardment to the reagent gas with ionizing electrons (100-500 eV). Ionization and collisions of the reagent gas molecules produce ionic species such as CH_5^+ and $C_2H_5^+$ in case of methane. These ions will be present in excess and in turn react with sample molecules as either Brönsted acids or hydride abstractors, depending on the sample molecule. For a molecule XH species such as so-called quasi-molecular ions, XH_2^+ or X^+ may be produced. A great number of sample molecules can be ionized in this way. EI techniques, on the other hand, will yield 10 times lower ionized species. As a relative comparison of sensitivities, it can be mentioned that the CI mode yields between one and two orders of magnitude higher sensitivity.

The most important parameter of the CI source is the reagent gas. The type of reagent gas will determine, e.g., the amount of fragmentation. It is also possible to vary the type of ionization utilizing proton transfer, charge exchange, or EI type of ionization. Some reagent gases, e.g., ammonia, nitric oxide, or gas mixtures, will specifically ionize a particular functional group. Deuterium oxide can be used to determine the presence of active hydrogen. Argon-water has been found to give both a quasi-molecular ion and a fragmentation pattern characteristic of the corresponding EI spectra. Generally, the amount of fragmentation is dependent on the proton affinity of the reagent gas: The lower the proton affinity of the reagent gas, the greater the amount of energy which is transferred upon ionization. This greater amount of excess energy results in a greater amount of fragmentation.

Lee and Hites [2] used a reagent gas mixture (10% methane in argon) which yields a mixed charge exchange upon CI of PAH. The methane produces a protonated molecular ion $M^+ + 1$, while argon participates in charge-exchange reactions to give a fragmentation similar to EI spectra, i.e., a large M^+ peak. The relative rates of these two competing reactions determine the intensity ratio of the two ions. This effect may be used to differentiate PAH isomers, e.g., anthracene and phenanthrene, which may not be done with EI [3].

CI/MS systems with methane as the reagent gas have been used to confirm the formation of nitrated PAH upon exposure to nitrogen dioxide and nitric acid [4].

C. Field Ionization

Like CI, FI is a kind of soft ionization method, producing mainly molecular ions with reduced fragmentation. A strong electrical field, typically 10,000 V, is established between a sharp point or edge and a second electrode. In this field the sample molecules in the gas phase are subjected to electron removal by a quantum mechanical tunnel effect. The greatest difference between EI and FI is that in the EI case a large amount of excitation energy is transferred to the molecules whereas only very small amounts of excitation energy are transferred in the FI case. The excess energy is on the same order as for CI; however, the FI produces an odd electron molecule ion for most molecules rather than a quasi-molecular ion. The sensitivity is lower compared with that of EI. The FI technique may be best suited for the analysis of multi-component mixtures of great structural complexity without preseparation. Severin [5] and Scheppele et al. [6] have carried out preliminary investigations on PAH.

D. Field Desorption Ionization

For thermally labile compounds or nonvolatile organic compounds, which cannot be run by direct probe EI/MS, a modified FI can be used. The FD, EI, CI, and FI all require that the sample molecules exist in the vapor state for ionization. In FD, however, the sample is coated directly onto a removable emitter, a tungsten wire with benzonitrile. Then, the applied sample is ionized as in FI, but directly from the solid state. Quantitative determinations of certain PAH constituents have been performed by Barofsky et al. [7] and Pfeifer et al. [8].

E. Atmospheric Pressure Ionization

An atmospheric pressure ionization (API) source is external to the mass analyzer vacuum system. The sample is introduced by either direct injection GC or high-performance liquid chromatography (HPLC), together with a preheated inert gas such as nitrogen or argon at atmosphere pressure. The initial ionization is produced by a β-ray-emitting ^{63}Ni foil or by high-voltage corona discharge. In the weak plasma a series of ion molecule reactions occur, resulting in ions by, e.g., proton transfer or charge transfer. The ionization of trace impurities is very high, giving a high sensitivity [parts per trillion (ppt) level]. The ions then enter through a pinhole aperture into the high-vacuum mass spectrometer region. Sampled ions are separated from neutral molecules in the high-vacuum region by a series of ion lenses. The mass spectrometer can be optimized for either positive or negative ion operation. The principle was first presented by Horning et al. [9] and has been described in detail by Siegel and Fite [10].

F. Combination of Ionization Sources

Modern instruments provide the capability of different ionization modes within one source. Most common is the combined CI/EI source. There are devices which allow rapid change from EI to CI mode during the elution of a GC peak (2-3 sec wide). The system often utilizes the carrier gas as a reagent gas and requires a high capacity pumping system (200 liters/sec in the ion source).

III. Mass Analyzers

Three different types of mass analyzers are common, especially with GC/MS combinations. These are magnetic deflection (single or double focusing), quadrupole, and time-of-flight mass spectrometers.

The most important parameter determined by the analyzer is the mass resolution. The widely used definition is based on the "resolution of 10% valley." The resolution R is defined as $M/\Delta M$, where M is the first peak in a doublet and ΔM is the difference in masses of the two peaks. The resolution of an instrument is then equal to the highest m/z value for which an ion can be separated from the nearest higher ion with the same abundance. Besides, the valley between the ions should be 10% of the height. When the sample has a molecular weight of 200, R should be at least 200.

There are several instrumental factors influencing the resolving power, most of which are built into the instrument and can therefore not be adjusted by the operator. With a single focusing magnetic deflection instrument a resolution of up to 3000 is attainable. Double-focusing mass spectrometers which also accommodate an energy divergence filter have a considerably higher resolution capacity, up to 150,000. High resolution means that unit masses are not separated and determined at the numerical value of the resolving power but peaks are separated only millimass units apart at much lower nominal masses. Accurate mass measurements can be carried out on a low-resolution instrument by the method referred to as peak matching. The instrument is focused on two peaks—a known standard and the unknown peak to be measured at a constant magnetic field—but at different accelerating voltages. The ratio of the accelerating voltages multiplied by the mass of the known standard yields the accurate mass of the unknown. This technique can be computerized and used on rather rapidly eluted GC/MS peaks. In magnetic instruments the resolution is dependent on slit widths, which means that narrower slits give higher resolution at the expense of a corresponding decrease in sensitivity. Also, for all mass spectrometers including quadrupole types, the resolution is decreased as the scan rate is increased.

The quadrupole type of mass spectrometers is very different from magnetic deflection instruments. The ions are separated on the basis of their m/z solely by means of electric fields. There are some advantages with this technique since the path is to a smaller extent dependent on the kinetic energy or angular divergence of the incoming ions resulting in high transmissions. The scanning is performed linearly by a change in voltage, which can be done very rapidly. With the linear scan function, each peak has the same width over the entire mass range. The drawbacks of quadrupoles are the limited resolution (commonly not exceeding 1000 amu, with a resolution of 700-800) and considerable mass discrimination of ions of higher mass. The latter effect may to some extent be reduced by chemical ionization. Quadrupole mass filters are attractive combined with capillary column GC because of the possibility of rapid scanning.

IV. Sample Handling and Introduction Systems

Mass spectrometers can be coupled to different inlet systems, such as batch inlets for gases and liquids and direct inlet probes for thermally unstable samples. The most efficient inlet system is without doubt the gas chromatograph, which will be discussed in Sec. V.

PAH in samples may exist in the liquid or the vapor phase. In addition, they may occur in the solid phase as an integral constituent of particulate matter. The identification and quantitation of PAH constituents are hard to make directly, since the samples may contain a great number of other associated organic materials. Therefore, the isolation is often performed in several steps, using extraction and chromatographic methods. However, a simpler, more direct analytical performance is sometimes preferred. One reason is that labile compounds such as highly substituted PAH and their nitrogen and sulfur analogs are difficult to preserve during separation. Furthermore, large polycyclic aromatic ring systems containing more than seven benzene rings have low vapor pressure for successful GC analysis. There are some methods which to a greater or lesser extent eliminate working up of samples before the MS analysis.

A. Analysis of Complex Mixtures by Probe Distillation

Probe distillation of crude PAH extracts has been reported by Giger and Blumer [11]. The separation is accomplished in high vacuum from a probe glass capillary with a restricted opening (0.5 mm) and 8 mm in length. The probes are filled with a syringe, and the solvent is evaporated before the probes are inserted through the vacuum lock system of the MS. Gradual temperature programming permits a successive distillation of the PAH constituents. The vapor composition in the MS source changes more slowly than in GC, yielding longer integration time and thus higher sensitivity. A low ionization voltage or an effective electron beam energy at 12 eV essentially eliminates fragments of contaminating material. If the repeller voltage is higher than zero, the electron beam voltage should be decreased to keep the effective energy constant. A potential on the repeller may increase the sensitivity and repeatability [12].

The probe distillation method has been especially successful in the analysis of aza-arenes in sediment samples [13,14]. After isolation of the aza-arene fraction, the sample is introduced to the MS, which is run in the repetitive scanning mode. The resulting spectra are summed to give the composite mass spectrum for certain boiling point intervals or the total sample (Fig. 1). The total sample composition is not dependent on a changing vapor composition during distillation. The odd number (odd number of nitrogen) ions are easily identified. Information about higher molecular weight components as well as the relative abundance of alkyl homologs of aza-arenes is thus obtained.

Schuetzle et al. [15] have used a similar technique to determine the PAH constituents in gaseous and particulate air samples. The PAH is collected on a gas adsorber (Chromosorb 102), which in turn is connected to a specially designed inlet system on the MS. The sample is cooled to 0°C before the air is pumped off. The pumping removes water from the polymer, but hydrocarbons above C_4 are not lost. Thereafter, a gradual heating from 0° to 150°C desorbs organics directly into the spectrometer. The magnitude of each ion current peak varies as a function of time or temperature and is called a mass thermogram. The mass thermogram shows similarities to gas chromatograms. With this technique it is possible to separate isomers such as anthracene and phenanthrene.

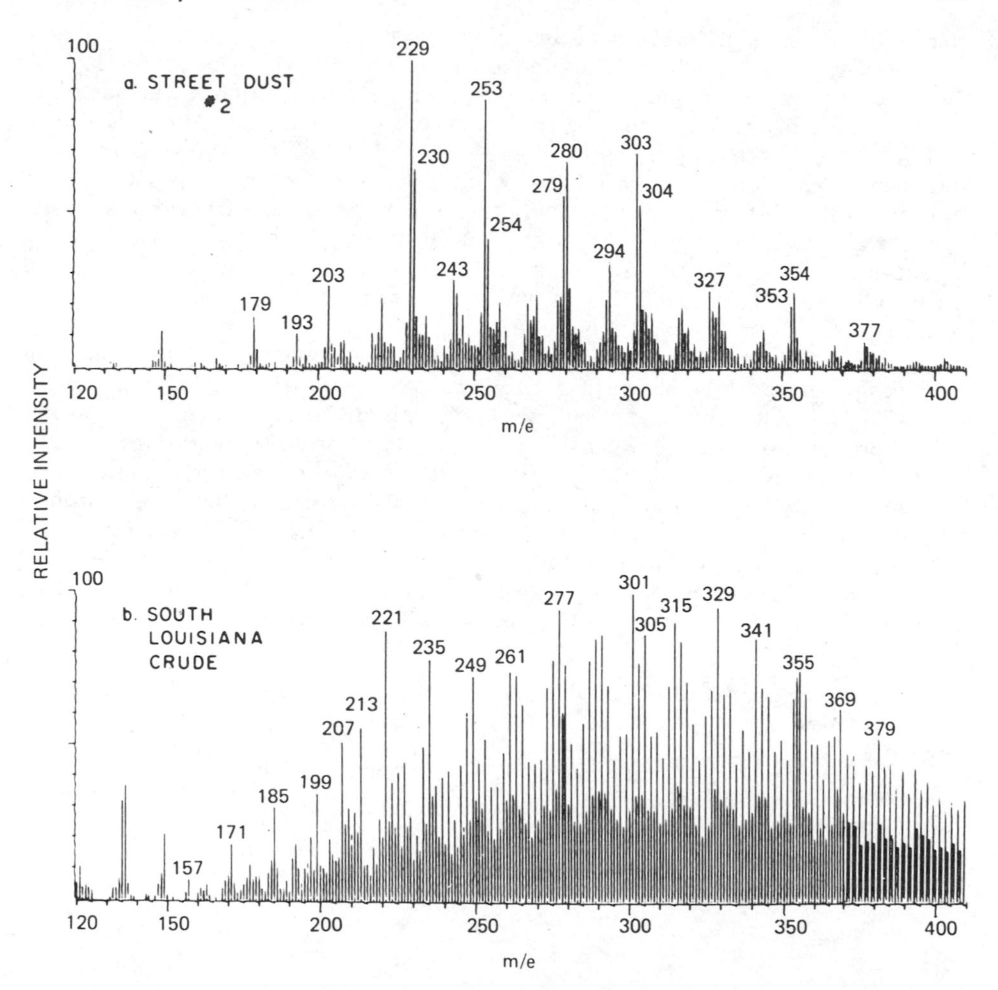

Figure 1. Mass spectra obtained by probe distillation/low voltage MS of the aza-arenes fractions. The respective spectra constitute the composite mass spectrum for total sample as a result of a number of spectra which have been summed. (Reprinted with permission from S. G. Wakeham, Azaarenes in recent lake sediments. *Environ. Sci. Technol. 13*:1122. Copyright 1979, American Chemical Society.)

B. High-Resolution Mass Spectrometry of Complex Mixtures Without Prior Separation

The combination of low EI ionization voltages (7-10 eV) and high-resolution MS (HRMS) can be used directly on crude extracts for specific aromatic group analysis. The sample is introduced by, e.g., a direct probe into the ion source, where the aromatics are ionized at low energies yielding just the molecular ions. Saturated hydrocarbons will not interfere. The elemental composition of the different molecular ions are determined from precise mass measurement (R > 10,000). The idea is to further simplify the evaluation by determining groups of similar compounds from Z values. The Z number is a measure of hydrogen deficiency in hydrocarbons according to the formula C_nH_{2n+z}. This technique has been especially valuable in the petroleum industry. When environmental samples are analyzed, the method gives initial information, which may provide guidance in selecting and optimizing the subsequent analysis [16-19].

Hites and Biemann [20] analyzed sediment extracts, which were introduced directly into an HRMS with R = 20,000. The sample was vaporized at a continually increasing temperature and ionized at 70 eV. The separated ions were registered on photographic plates. The plates were developed and read on a computerized comparator regarding exact masses with their intensities (roughly linear). The precise masses were converted to elemental compositions and corresponding intensities. A semiquantitative indication of relative abundance of PAH was possible by a special arrangement of data. The total carbon number vs. ring and double bonds as well as third-dimension intensity were arranged in tabular form. With this technique, Hites and Biemann were able to find, e.g., that pyrene and fluoranthene (m/z 202, $C_{16}H_{10}$) were the most abundant group of isomers. The method is especially useful for finding the distribution of alkylated species within a PAH mixture.

C. Atmospheric Pressure Chemical Ionization/Mass Spectrometry

Lane et al. [21] have developed an interesting real-time MS technique to determine PAH in ambient air without sample preparations, using a Trace Atmospheric Gas Analyzer (TAGA; see Fig. 2). The technique is based on CI with a reagent gas added to the air sample and ionization by corona dis-

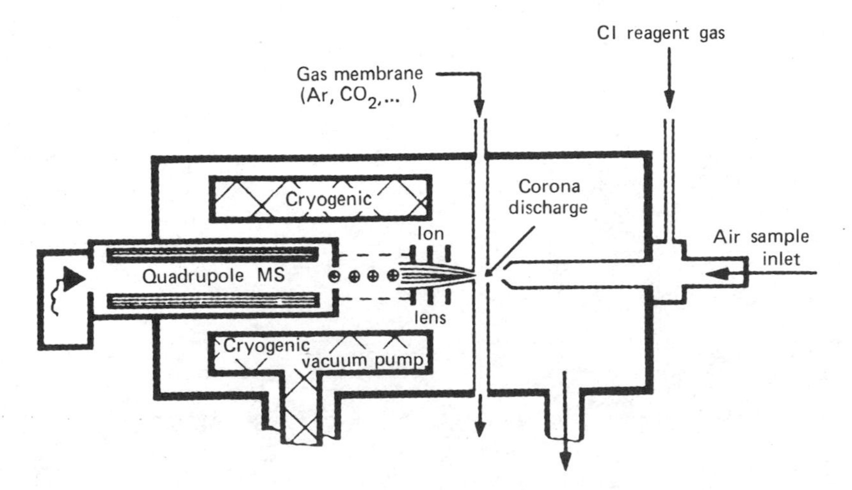

Figure 2. Schematic view of the Trace Atmospheric Gas Analyzer (TAGA) instrumentation. (From Ref. 21.)

charge. High sensitivity is obtained at atmospheric pressure, where a high frequency of ion-molecule collisions forming product ions occurs. The air sample and reagent gas are directed together into the corona discharge region in the ion source in such a way that wall or memory effects are kept low. The ion products formed are extracted through an orifice after having passed a gas membrane. The gas membrane rejects remaining air molecules and particulates, etc., to keep the MS clean and to prevent clogging of the orifice. Finally, the ions are focused by electrostatic lenses into a quadropole spectrometer. The large pressure difference, i.e., 760 torr in the ion source and 10^{-5} torr in the mass analyzer, requires pumps with very high capacity.

The CI may be performed without any reagent gas being added. The water naturally occurring in the air forms hydrated protons, which in turn cause proton transfers to compounds with higher proton affinity than water. The proton affinity increases as the reagent gas benzene (1 ppm) will cause charge transfer reactions, yielding molecular ions of PAH. A problem may be the variation of water content in the air. Different concentrations of water may influence the chemical ionization yield.

Lane et al. [21] showed that the APCI/MS technique is very sensitive, i.e., PAH could be detected in the lower ppt range in air.

V. High-Resolution Gas Chromatography/Mass Spectrometry

The combination of high-resolution glass capillary column GC separation and mass spectrometry (GCGC/MS) is the most powerful analytical instrumentation for obtaining information about constituents in complex mixtures, typically PAH samples. In this chapter only glass capillary column separation will be treated since this techniquc exhibits certain advantages in combination with MS.

Compared with packed columns, the capillary columns, besides higher resolution, yield substantially lower contamination of the ion source. Modern glass capillary columns can be made very stable; thus, the stationary phase bleeding to the MS is low, resulting in low background noise. Properly deactivated columns with low catalytic effects result in a further decrease of the background noise. The loadability, i.e., the injected sample size, may be quite large when splitless or on-column injection techniques are used. Moreover, the sensitivity is much better since the GC^2 peaks are eluted in a more concentrated band. Details of GCGC techniques are discussed in Chapter 6.

A. Interface of GCGC/MS

Efficiency of GCGC/MS is dependent on the ability of the system to maintain the GC separation power during the passage of the sample constituents into the ion source of the MS. The interface between the GC column and the ion source determines the overall performance and constitutes the most critical part. The interface should therefore meet the following requirements:

1. The GC performance should not be altered.
2. The chemical structure of each component should not be affected as it passes the interface.
3. The total GC effluent must enter the ion source in such a way that the MS performance remains optimal.

The GC effluent should be rapidly transmitted and as directly as possible to eliminate extra column band broadening by dead volumes, exposure to active surfaces, cold spots, etc. The inertness of the interface is very important to avoid catalytic and adsorption effects on sample components. The GCGC effluent (range 0.5-5 ml/min) should be evacuated by the pumping system to maintain the necessary vacuum level ($\leq 10^{-4}$ torr) in the ion source. Furthermore, the optimal conditions regarding MS sensitivity and resolution should not be changed during the GC^2 run as a result of altered carrier gas flow characteristics during temperature programming.

Several types of GC/MS separator interfaces are reviewed in textbooks; e.g., see McFadden [22]. As to GCGC, there are a few interfaces of interest, which are not based on separators. In recent years different direct connection modes have gained in popularity, and these have been found to be the most adequate for full sample transfer. Different direct connections need not to be complicated. The availability of large-capacity pumps (200 liters/sec) for the ion source makes it possible to design interfaces ideal for the accommodation of the carrier gas flow typical for GCGC. Three different interfaces of current interest will be discussed: direct connection with restrictor, direct connection without restrictor, and open split connection.

1. Direct Connection with Restrictor

The GC capillary column is connected to the ion source via a restrictor in the transfer line. The restrictor may be a small-bore capillary tubing (typically 30-40 cm × 0.010 cm i.d.) of glass or platinum/iridium materials. The pressure drop from 1 atm to 10^{-4} torr occurs over the restrictor; thus the GC column is working under normal conditions. Cold spot and eddy cavity effects should be avoided. An all-glass system is to be preferred since the glass surfaces can be deactivated in the same way as the GC column. Different practical arrangements have been described, e.g., by Leferink and Leclercq [23] (glass), Grob and Jaeggi [24] and Grob [25] (glass, platinum), and Neuner-Jehle et al. [26] (platinum).

2. Direct Connection without Restrictor

The glass capillary column is inserted into the ion source. The GC column is working under partial vacuum. From a theoretical point of view, the separation efficiency should be better when the column is run under reduced pressure, according to early work by Giddings [27]; Schulze and Kaiser [28] could practically show maintained separation efficiency with a direct connection interface. In a recent work, Cramers et al. [29] concluded that the GC peak will be narrower, and thus higher, which results in a lowering of the detection limits.

In the author's laboratory an all-glass deactivated direct connection arrangement has been used since 1976 [30,31]. Our experience shows that during temperature-programmed runs the focusing in the ion source will change as a result of the changed conductance of the carrier gas flow through the capillary column. The ion source chamber is tight, and the preset optimal conditions for the sensitivity will change slightly over the temperature programmed run. Another drawback is a small distortion of the MS total ion current recording compared with the GCGC/FID chromatogram. However, these

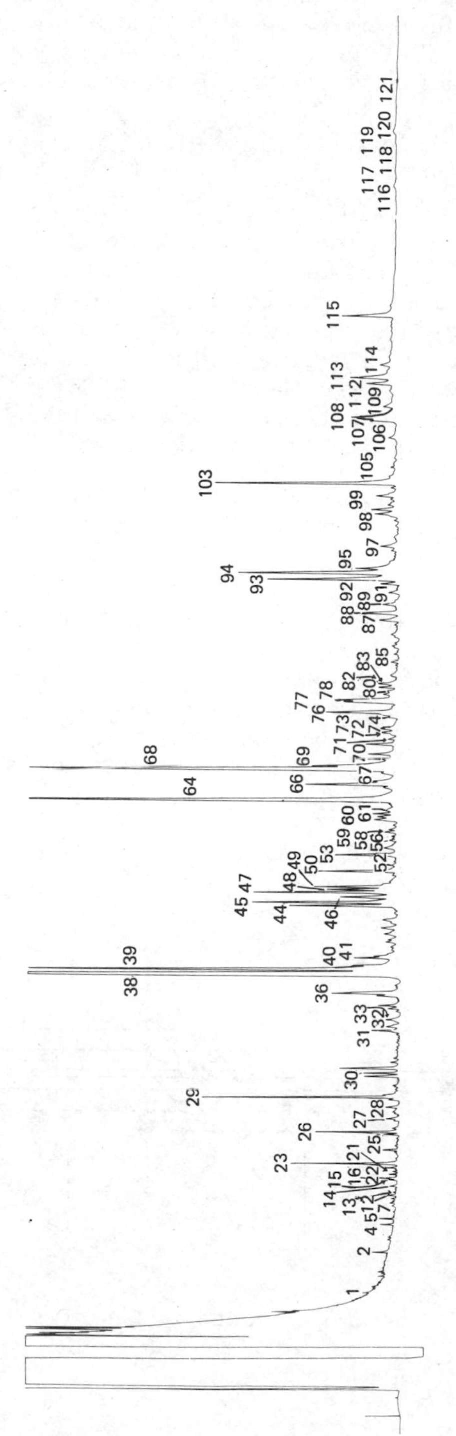

Figure 3. The recording is the total ion current trace obtained from a GCGC/MS run of PAH from the working atmosphere in a coke plant. (From Ref. 31.)

problems are small compared with the attained high overall efficiency of the system. As shown in Fig. 3, more than a hundred PAH constituents could be identified because of the maintained GCGC resolving power. The system has also a very high sensitivity, making it possible to obtain good spectra at the 1 ng level.

3. *Open-Split Connection*

Open-split connection works with a restrictor on the MS side. The GC column effluent reaches a manifold at atmospheric pressure; thus undisturbed chromatographic conditions are maintained. The manifold is flushed with a makeup gas, e.g., helium, and the effluent from the GC column together with a part of helium is sucked at constant flow into the MS. The excess makeup gas is vented from the manifold. The makeup gas may also be used for chemical ionization in such a way that the outlet of the manifold is throttled. The main advantage of an open-split connection is the rapid and easy interchangeability of GC columns. Henneberg et al. [32] have reported an all-glass open-split interface. Schmid et al. [33] presented a system which has high versatility with respect to column changing (Fig. 4).

B. MS as a GC Detector

There are several ways of monitoring the GC effluent in a GC/MS run. The total ion current (TIC) may be registered by an electrode positioned between the ion source and the magnetic sector to collect a fixed portion of the ion beam. Another method of TIC monitoring is based on a separate EI ion source behind the main source, which receives a portion of the GC effluent. The auxiliary ion source operates at 20 eV and thus does not respond to the helium carrier gas. Actually, this detector is to be compared with a flame-ionization detector (FID).

The TIC detectors mentioned cannot be used together with CI because of the large amount of reagent gas ions produced. In this case electronic integra-

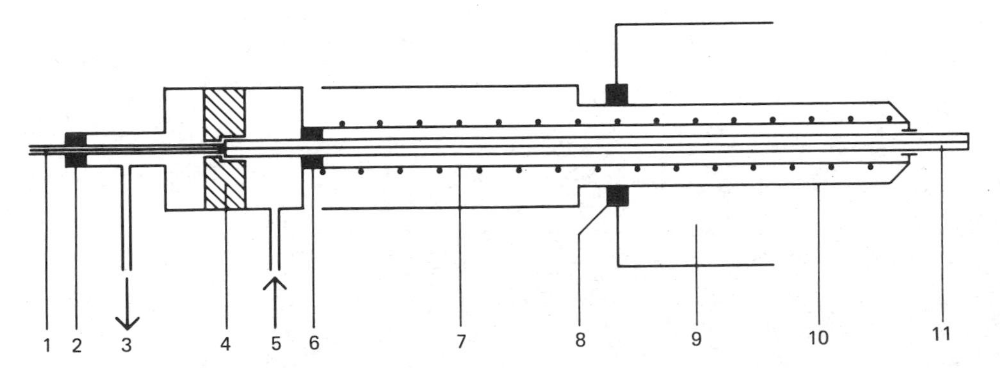

Figure 4. Schematic view of an open split interface: (1) GCGC; (3) makeup gas outlet; (4) device for positioning of the two capillaries; (5) makeup gas inlet; (7) heating device for the interface capillary; (9) high-vacuum region; (11) interface capillary. (From Ref. 33.)

tion is used to integrate the entire mass spectrum signal. Quadrupole instruments register the TIC by shutting off the DC potential on the rods, allowing all the ions to pass through to the detector. The mass scanning is then performed with alternative DC on and off. The mass range is determined by the RF voltage on the rods. By presetting a proper RF voltage, ions above a certain mass will be transmitted; thus, for example, reagent gas ions (methane, m/z 29) are discriminated.

Modern instruments are often provided with combined CI/EI ion sources. Different ionization modes may be applied by rapid changeover in the course of the GC run.

1. *Cyclic Scanning Mode*

The speed of scanning is of great importance in GC effluent analysis. Scan speeds must be fast enough to achieve representative mass spectra when rapid peaks pass from GCGC columns. Normally, a scan speed of 1 sec/mass decade is sufficient with respect to sensitivity and accuracy for low-resolution mass spectra. With modern computer technology the cyclic scan mode has become a feasible arrangement. A repetitive scan computer system collects and stores all information on, e.g., a disk after processing the signals as normalization and background subtraction. The stored information is available for different outputs. The computer can be used to plot a reconstructed gas chromatogram of the ion intensities from each scan vs. the spectrum number (time) yielding a TIC chromatogram. Computer regenerated gas chromatograms for ions with certain masses are called mass chromatograms (see Fig. 5).

2. *Selective Ion Detection*

The MS may be used as a specific detector for the GC if it is tuned to detect ions with specific masses. The selective ion detection (SID) technique is characterized by an extreme sensitivity (10^{-12} g level) combined with high selectivity. The selectivity may be raised by using more than one ion for monitoring at the sacrifice of sensitivity. The sensitivity is dependent on operational variables such as slit width (magnetic sector) or peak width (quadrupole) ionization, and dwell time when counting specific ions. On a magnetic sector instrument the switching between selected ions is best accomplished by a gross change of the acceleration voltage combined with fine tuning of the electromagnetic field. With quadrupoles the tuning is accomplished by varying the electric and the radio frequency field. Because these voltages can be varied rapidly and reproducibly, the quadrupole is ideally suited for SID analyses. In contrast to magnetic instruments, the quadrupole is not limited in mass range.

Quantitative measurements may be properly carried out by using isotopically labeled internal standards, since the MS can distinguish between such compounds. The labeled standards should be as close as possible to the selective mass analyzed in order to avoid the influence of mass discrimination. Nowadays, commercially available ^{13}C-labeled standards are to be preferred to deuterated substances since carbon will not undergo isotopic exchange in the cleanup procedure.

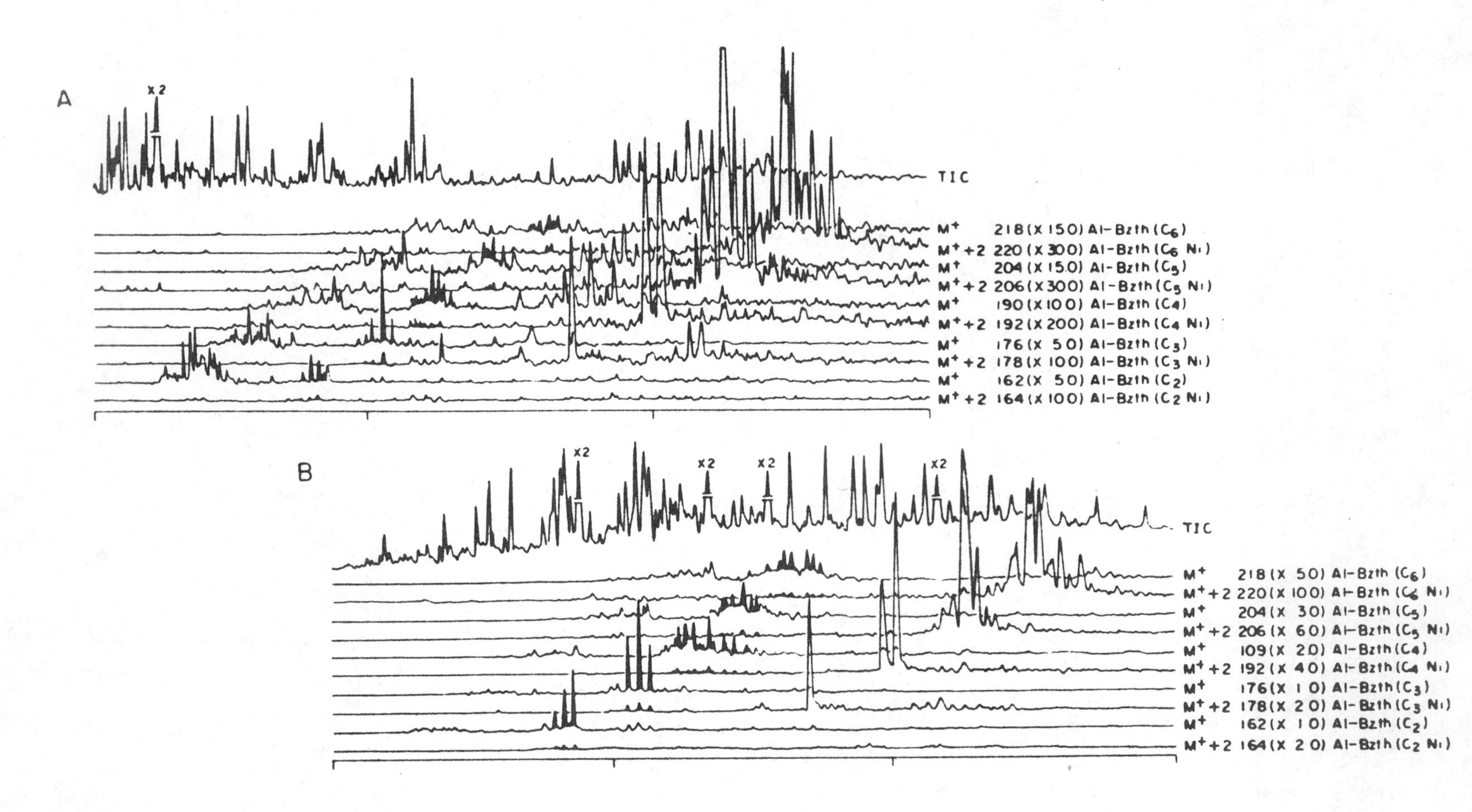

A
x2
TIC
M+ 218(x 150) Al-Bzth (C6)
M+ +2 220(x 300) Al-Bzth (C6 Ni)
M+ 204(x 150) Al-Bzth (C5)
M+ +2 206(x 300) Al-Bzth (C5 Ni)
M+ 190(x 100) Al-Bzth (C4)
M+ +2 192(x 200) Al-Bzth (C4 Ni)
M+ 176(x 50) Al-Bzth (C3)
M+ +2 178(x 100) Al-Bzth (C3 Ni)
M+ 162(x 50) Al-Bzth (C2)
M+ +2 164(x 100) Al-Bzth (C2 Ni)
B
x2
x2
x2
x2
TIC
M+ 218(x 50) Al-Bzth (C6)
M+ +2 220(x 100) Al-Bzth (C6 Ni)
M+ 204(x 30) Al-Bzth (C5)
M+ +2 206(x 60) Al-Bzth (C5 Ni)
M+ 109(x 20) Al-Bzth (C4)
M+ +2 192(x 40) Al-Bzth (C4 Ni)
M+ 176(x 10) Al-Bzth (C3)
M+ +2 178(x 20) Al-Bzth (C3 Ni)
M+ 162(x 10) Al-Bzth (C2)
M+ +2 164(x 20) Al-Bzth (C2 Ni)

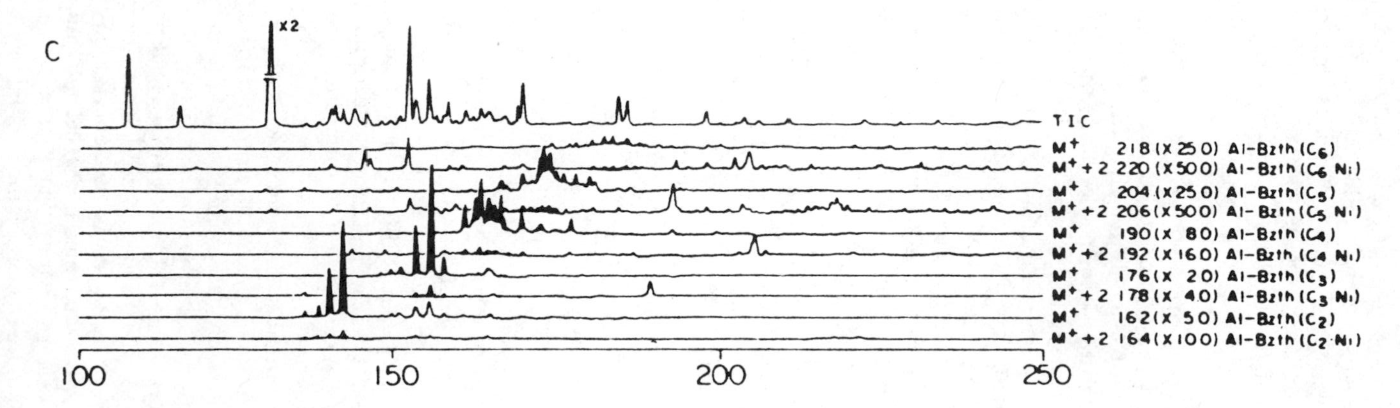

Figure 5. Mass chromatograms of alkyl benzothiophenes in (a) crude oil; (b) soft bodies of clams; and (c) eel flesh. (From Ref. 47, by permission of Preston Publications, Inc.)

C. Elemental Composition Determinations

The most advanced GCGC/MS system incorporates HRMS, which permits the assignment of elemental compositions to each resolved, measured exact mass in each high-resolution mass spectrum. This information will facilitate structure elucidation of unknown compounds. The most important requirement is the maintenance of accurate mass measurements during rapid scan cycle operation with adequate sensitivity. The limiting part in such a system is the magnetic field, which has to be very stable upon rapid changes. Besides, the computer system capacity should be large.

Recently, Meili et al. [34] presented a GCGC/HRMS/computer system with an open-split interface of the type proposed by Henneberg et al. [32]. The scan rate was in the range of 6-8 sec/decade with a precision of ±15 ppm of the exact mass. The sensitivity was in the range of 5-100 ng of the injected amount (urinary organic acids).

Powers et al. [35] used a low-resolution MS for accurate mass determination by improving the reproducibility of the cyclic scan function. They used a special magnetic field sensor to control the field upon rapid scanning. Applied to seven PAH compounds, separated on a 20-m OV-101 capillary column, the parent ion masses were measured with about 10 ppm precision. As reference they used ions from tetraiodoethylene, and the sensitivity was in the nanogram range.

Rapp et al. [36] used a medium resolution MS (R = 3000) interfaced with a glass capillary column via a platinum restrictor. With this system they were able to obtain a precision of 10 ppm for accurate mass measurements in the 3-10 ng range of tobacco smoke constituents. It should be noted they tested the dead volume and the inertness of the GCGC/MS coupling device with PAH standards. Coronene, a compound which is normally difficult to separate with symmetric peaks, showed some tailing on the TIC trace.

D. Interpretation and Identification

Mass spectra obtained from an optimally working GCGC/MS system may be directly identified by comparison with those of a standard MS library [37,38]. High GC resolution of sample constituents as they pass through the ion source vouches for certainty in the identification. However, some PAH are identical with respect to elemental composition. Actually, there are a number of isomers resulting in the same parent peak. Since aromatic compounds exhibit relatively intense parent peaks and little fragmentation, additional information may be necessary. In the GCGC/MS case, the GC retention time of standards may confirm the identity.

Of great help in the identification of unknown compounds are accurate mass measurements to determine elemental compositions of ions. The formula is calculated with algorithms based on the noninteger masses of all the atomic weights of all the elements plus isotopes other than ^{12}C. In the case of PAH the deviations from integer values of a measured mass are used to calculate the number of hydrogen atoms. The deviation from integer masses will be entirely due to the ^{1}H (1.0078246) atoms present, since ^{12}C and all its multiples have integer masses with multiples of 12.0000. Thus, pyrene ($C_{16}H_{10}$) will have the value $16 \times 12 + 10 + 10 \times 0.0078246 = 202.078246$. Medium- and high-resolution work is handled with computer systems. This is especially necessary when heteroelements are involved. The results may be presented as elemental maps for different ions.

There are several common fragmentation pathways for most PAH. Mechanistic studies of the decomposition of PAH are discussed by Budzikiewicz et al. [39]. Unsubstituted, monomethyl or heteroatom PAH show the molecular ion as the most abundant ionic species. In general, the isomers are difficult to distinguish. The major fragmentation of alkylated aromatics, e.g., 2-methylnaphthalene is the β-cleavage with the loss of a hydrogen atom, yielding a resonance stabilized phenyltropylium ion, m/z 141, which undergoes the elimination of acetylene to yield the indene ion m/z 115. Unsubstituted PAH shows hydrogen cleavage of the molecular ion, yielding M^{+}-1, M^{+}-2, and M^{+}-3 fragments. A common fragmentation is the expulsion of C_2H_2 from the parent ion as well as from the hydrogen cleavage products. Sometimes doubly charged ions M^{+}/2z and M^{+}-26/2z are observed. A detailed discussion of PAH fragmentation was given by Safe and Hutzinger [40].

E. Applications

The development of efficient GCGC/MS systems for PAH analyses has to a great extent been governed by glass capillary column specialists. At an early stage they became aware of the fact that GC separation of elemental composition isomers simplifies mass spectrometric identification.

Giger et al. [41] and Wakeham et al. [42,43], of the EAWAG group in Zürich, have carried out comprehensive studies of PAH in the environment, especially in sediments. With Grob-type glass capillary columns, connected via platinum capillary to a quadrupole mass spectrometer, they were able to determine 31 different constituents in sediment samples, using cyclic scanning recording mass chromatography and mass ion detection (MID) monitoring. In studies of recent lake sediments, Wakeham et al. [42,43] were able to distinguish between PAH compounds of anthropogenic origin and compounds derived from biogenic precursors.

The determination of petroleum-derived compounds discharged into the sea is of great concern. Grahl-Nielsen et al. [44] have developed a MID method to determine naphthalene, phenanthrene, dibenzothiophene, and some of their alkylated homologs typical of oil spills. The instrumental setup was similar to that of the EAWAG group. By selecting the parent ion masses (naphthalene m/z 128, etc.) and some typical fragment ions (M-15), these investigator's recorded so-called mass fragmentograms (MID). The precision varied between ±2 and ±7% for different specific compounds using internal standards.

Borwitzky and Schomburg [45] used a polar stationary phase polyphenyl ether sulfone Poly S-179 coated capillary column connected to a quadrupole mass spectrometer by a platinum capillary. The MS background of the stationary phase bleeding was low at 250°C. It should be noticed that polar temperature stable stationary phases are very difficult to coat on glass capillary columns. The system was used for analyzing PAH in coal tar.

Lee et al. [46] have studied PAH in tobacco and marijuana smoke condensates as well as on airborne particulates. The glass capillary column was connected directly to the ion source of a dodecapole MS.

Bjørseth and Eklund [31] analyzed PAH in trace quantities in working atmospheres, with an all-glass interface system without a restrictor and a computerized double-focusing MS. The interface was deactivated with Carbowax 20M to obtain high inertness. Mass spectra were taken in the mass range m/z 50-500 (R 700) at a scan rate of 1 sec/decade with automatic repetitive scanning. A spectrum was stored every 1.5 sec. The TIC trace of PAH from a coke plant sample is shown in Fig. 3. The system was very sensitive, i.e., spectra at the 1 ng level were enough for structure elucidation. Thus PAH constituents could be identified at less than 0.5 ppm in the final extract. By comparing GC retention times with standards combined with mass spectral information, they detected and identified more than a hundred compounds.

Mass chromatography of sulfur-containing oil constituents such as benzothiophenes, dibenzothiophenes, and their alkylated homologs in clams was carried out by Ogata and Miyake [47]. They utilized the natural isotope ratio of ^{32}S and ^{34}S (approximate abundance: $M^+/M^+ + 2$; 95/4) for confirmation (Fig. 5). The electron impact energy was 20 eV, and the repeating scan speed was 3 sec.

VI. Concluding Remarks

The combination of high-resolution GC and computerized HRMS is a very useful means for PAH analysis as far as identifications are concerned. However, it still constitutes a costly and complicated alternative to be used only by specialists. In recent years simpler GCGC/MS instruments with adequate capacity have become commercially available and gained wide acceptance. Well-performed integration of glass capillary GC columns to ordinary MS has proved to be a remarkably useful analytical tool.

In the future, the combination of HPLC and MS may play an important role in PAH analysis. Then it might complement the GCGC/MS. The problems involved with the development of a practical HPLC/MS system are still much more complicated than those of GCGC/MS. Dack et al. [48], however, carried out some preliminary studies on PAH in liquefied coal products using HPLC/MS. The HPLC column was coupled with a commercially available belt transport LC/MS interface to a quadrupole CI/MS. Bjørseth [49] used the same MS instrument setup interfaced with a straight-phase HPLC column to study PAH on particulates from an aluminum plant. The eluent from the column was split, allowing it to be simultaneously monitored by an ultraviolet (UV) detector as well as the MS. Peaks appearing late in the chromatogram arose from the more polar compounds which are often difficult to analyze by GCGC. Among these, different carbazoles, acridines, and oxygen-containing compounds were identified by their mass spectra.

Recently, a multistage mass spectrometer technique has been introduced [50,51]. The technique is not dependent on chromatographic separation; thus both separation and identification of the constituents of a mixture are performed mass spectrometrically. First, the components in a sample are ionized and separated in one mass spectrometer; then a second mass spectrometer coupled to the first produces mass spectra of each component.

Kondrat et al. [52] as well as Lafferty and Bockhoff [51] constructed MS/MS instruments by reversing the geometry of conventional double-focusing mass spectrometers. They put the magnetic analyzer first where, e.g., the quasi-molecular ions formed from CI are separated from a mixture. Then the ions are subjected to fragmentation in a collision cell with an inert gas, say, helium. The collisions excite the ions, causing them to fragment. The kinetic energy of each fragment is measured by the electrostatic analyzer, which produces a mass spectrum pattern similar to conventional EI mass spectra. Instead of the double-focusing mass spectrometer with reversal geometry, Hunt et al. [53] used a triple quadrupole mass spectrometer for mixture analysis, e.g., of PAH.

Much larger and more polar molecules can be analyzed by this MS/MS than by GC/MS because the technique works for compounds with much lower vapor pressures. Furthermore, the method is sensitive and rapid, which increases its potential for the future.

Acknowledgments

The author thanks Drs. Göran Eklund and Claes Roos for valuable discussions.

References

1. E. J. Gallegos, *J. Phys. Chem. 72*:3452 (1968).
2. M. L. Lee and R. A. Hites, *J. Am. Chem. Soc. 99*:2008 (1977).
3. M. L. Lee, D. L. Vassilaros, W. S. Pipkin, and W. L. Sorensen, *Natl. Bur. Standards Spec. Publ. 519*:731 (1979).
4. J. N. Pitts, L. A. Van Cauwenberge, D. Grosjean, J. P. Schmid, D R. Fitz, W. L. Belser, G. B. Knudson, and P. M. Hynds, *Science 202*:515 (1978).
5. D. Severin, *Erdöl Kohle 25*:514 (1972).
6. S. E. Scheppele, P. L. Grizzle, C. J. Greenwood, T. D. Marriott, and N. B. Perreira, *Anal. Chem. 48*:2105 (1976).
7. D. F. Barofsky, E. Barofsky, and R. Held-Aigner, *Adv. Mass Spectrom. 7A*:109 (1978).
8. S. Pfeifer, H. D. Beckey, and H. R. Schulten, *Z. Anal. Chem. 284*:193 (1977).
9. E. C. Horning, M. G. Horning, D. I. Carroll, I. Dzidic, and R. N. Stillwell, *Anal. Chem. 45*:936 (1973).
10. M. W. Siegel and W. L. Fite, *J. Phys. Chem. 80*:2871 (1976).
11. W. Giger and M. Blumer, *Anal. Chem. 46*:1663 (1974).
12. B. H. Johnson and T. Aczel, *Anal. Chem. 39*:682 (1967).
13. M. Blumer and T. Dorsey, *Science 195*:283 (1977).
14. S. G. Wakeham, *Environ. Sci. Technol. 13*:1118 (1979).
15. D. Schuetzle, D. Cronn, A. L. Crittenden, and R. J. Charlson, *Environ. Sci. Technol. 9*:838 (1975).
16. I. P. Fisher and P. Fisher, *Talanta 21*:867 (1974).
17. A. Herlan, *Erdöl Kohle 27*:138 (174).
18. I. P. Fisher and A. Johnson, *Anal. Chem. 47*:59 (1975).
19. J. L. Stauffer, P. L. Levins, and J. E. Oberholzer, in *Carcinogenesis*, Vol. 3: *Polynuclear Aromatic Hydrocarbons*, P. W. Jones and R. I. Freudenthal (Eds.), Raven Press, New York, 1978, p. 89.
20. R. A. Hites and W. G. Biemann, in *Adv. Chem. Ser. 147: Analytical Methods in Oceanography*, T. R. P. Gibbs (Ed.), American Chemical Society, Washington, D.C., 1975.

21. D. A. Lane, T. Sakuma, and E. S. K. Quan in *Polynuclear Aromatic Hydrocarbons: Chemistry and Biological Effects*, A. Bjørseth and A. J. Dennis (Eds.), Battelle Press, Columbus, Ohio, 1980.
22. W. H. McFadden, *Techniques of Combined Gas Chromatography/Mass Spectrometry: Applications in Organic Analysis*, Wiley, New York, 1973.
23. J. G. Leferink and P. A. Leclercq, *J. Chromatogr. 91*:385 (1974).
24. K. Grob and H. Jaeggi, *Anal. Chem. 45*:1788 (1973).
25. K. Grob, *Chromatographia 9*:509 (1976).
26. N. Neuner-Jehle, F. Etzweiler, and G. Zarske, *Chromatographia 6*:211 (1973).
27. J. C. Giddings, *Anal. Chem. 36*:741 (1964).
28. P. Schultze and K. H. Kaiser, *Chromatographia 4*:381 (1971).
29. C. A. Cramers, G. I. Scherpenzeel, and P. A. Leclercq, *J. Chromatogr. 203*:207 (1981).
30. E. Eklund, B. Josefsson, and A. Bjørseth, *J. Chromatogr. 150*:161 (1977).
31. A. Bjørseth and G. Eklund, *Anal. Chim. Acta 105*:119 (1979).
32. D. Henneberg, U. Henrichs, H. Husmann, and G. Schomburg, *J. Chromatogr. 167*:139 (1978).
33. P. P. Schmid, M. D. Müller, and W. Simon, *HRC & CC 2*:225 (1979).
34. J. Meili, F. C. Walls, R. McPherron, and A. L. Burlingame, *J. Chromatogr. Sci. 17*:29 (1979).
35. D. Powers, P. H. D'Arcy, J. C. Bill, and M. J. Wallington, *Proc. 26th Annual Conference on Mass Spectrometry and Allied Topics*, American Society for Mass Spectrometry, St. Louis, 1978, p. 480.
36. U. Rapp, U. Schröder, S. Meier, and H. Elmenhorst, *Chromatographia 8*:474 (1975).
37. E. Stenhagen, S. Abrahamson, and F. W. McLafferty (Eds.), *Registry of Mass Spectral Data*, Wiley (Interscience), New York, 1974.
38. A. Cornu and R. Massot, *Compilation of Mass Spectral Data*, Heyden, London, 1975.
39. H. Budzikiewicz, C. Djerassi, and D. H. Williams, *Interpretation of Mass Spectra of Organic Compounds*, Holden-Day, San Francisco, 1965, Chaps. 9 and 11.
40. S. Safe and O. Hutzinger, *Mass Spectrometry of Pesticides and Pollutants*, CRC Press, Cleveland, 1973, Chap. 6.
41. W. Giger, M. Reinhard, and C. Schaffner, *Vom Wasser 43*:343 (1975).
42. S. G. Wakeham, C. Schaffner, and W. Giger, *Geochim. Cosmochim. Acta 44*:403 (1980).
43. S. G. Wakeham, C. Schaffner, and W. Giger, *Geochim. Cosmochim. Acta 44*:415 (1980).
44. O. Grahl-Nielsen, J. T. Staveland, and S. Wilhelmsen, *J. Fisheries Res. Board Can. 35*:615 (1978).
45. H. Borwitzky and G. Schomburg, *J. Chromatogr. 170*:99 (1979).
46. M. L. Lee, M. Novotny, and K. D. Bartle, *Anal. Chem. 48*:405 (1976).
47. M. Ogata and Y. Miyake, *J. Chromatogr. Sci. 18*:594 (1980).
48. W. A. Dack, W. H. McFadden, and D. L. Bradford, *J. Chromatogr. Sci. 15*:454 (1977).
49. A. Bjørseth, *VDI-Ber. 358*:81 (1980).
50. R. W. Kondrat and R. G. Cooks, *Anal. Chem. 50*:81A (1978).
51. F. W. McLafferty and F. M. Bockhoff, *Anal. Chem. 50*:69 (1978).

52. R. W. Kondrat, G. A. McClusky, and R. G. Cooks, *Anal. Chem. 50*:2017 (1978).
53. D. F. Hunt, J. Shabanowitz, and A. B. Giordani, *Anal. Chem. 52*:386 (1980).

8

Optical Spectrometric Techniques for Determination of Polycyclic Aromatic Hydrocarbons

E. L. WEHRY / Department of Chemistry, University of Tennessee, Knoxville, Tennessee

I. Introduction

This chapter considers the techniques of optical spectrometry and their application to identification and quantitative determination of polycyclic aromatic hydrocarbons (PAH) and their derivatives. Encyclopedic coverage of the voluminous literature in the field has not been attempted; rather, following some general remarks pertaining to the important analytical characteristics of the various methods, a few typical applications are treated in some detail and may, in effect, be regarded as "case studies." In addition, this chapter points to newer techniques and extensions of present methods which have not yet been used extensively in PAH analysis but appear to offer substantial promise for future applications. Much research remains to be performed in the analytical chemistry of PAH by optical spectrometry, and one of the purposes of this chapter is to point out areas where additional research effort appears to be needed.

II. Ultraviolet-Visible Absorption Spectrophotometry

A. Introduction and Quantitative Considerations

Ultraviolet (UV) and visible absorption spectrophotometry is a well-established analytical technique used almost exclusively for quantitative analyses. In this region of the electromagnetic spectrum (200-1000 nm wavelength range), many molecules exhibit transitions involving promotion of valence electrons from normally occupied to normally unoccupied molecular orbitals; the theoretical principles of molecular electronic spectra are discussed in considerable detail elsewhere [1,2]. The "information content" of the electronic absorption spectra of large molecules is relatively small (compared with that of infrared, nuclear magnetic resonance, or mass spectra), and is not easily related a priori to the finer details of molecular structure. Hence, while the UV or visible spectrum of a PAH can occasionally be employed to confirm the identity of the compound (by comparison with a spectrum of an authentic sample of the compound in question obtained under similar conditions), the vast majority of reported applications of UV-visible spectrophotometry to PAH analyses are quantitative rather than qualitative.

The basis of quantitative absorption spectrometry in *any* region of the electromagnetic spectrum is the Beer-Lambert-Bouger law (more commonly known simply as Beer's law):

$$A = \log\left(\frac{P_0}{P}\right) = \varepsilon bc \qquad (1)$$

In Eq. (1), A is the absorbance of the sample, P_0 is the optical power (often misleadingly called "intensity") incident upon the sample, P is the optical power transmitted by the sample, ε is the absorptivity of the absorbing compound, b is the optical path length of the sample, and c is the concentration of the absorbing substance.

While we cannot provide a detailed discussion of quantitative UV-visible absorption spectrophotometry here, several aspects of Beer's law merit specific comment:

1. The absorptivity ε of a compound is a fundamental physical property of that compound. It represents, in essence, the probability that a given compound will absorb light of a given energy (or wavelength). In addition to the wavelength of the incident light, the value of ε for a particular compound depends upon the identity of the solvent (or other matrix within which

the substance is present at the time of measurement), the temperature of the sample, the units in which b and c are expressed, and whether the logarithm in Eq. (1) is taken to base 10 or base e. The accepted practice is to use base-10 logs and to express b and c in centimeters and molarity (M, moles per liter), respectively; thus, ε usually has the units $M^{-1}\ cm^{-1}$. Under these specific circumstances, ε is termed the *molar absorptivity* of the compound in question. It is conventional to tabulate ε at the absorption maxima for a particular compound (ε_{max}).

2. According to Beer's law, ε for a particular compound is independent of concentration. Hence, a plot of A vs. c is linear, having a slope equal to εb. Such a plot is termed a *Beer's law plot*.

3. Beer's law is obeyed rigorously only under a moderately stringent set of measurement conditions, discussed in detail elsewhere [3,4]. Instrumental deviations from Beer's law usually are negligible if reasonably high-quality spectrophotometric equipment is used. It should, however, be noted that ε varies with wavelength. Thus, the effective value of ε for a particular compound at a given nominal wavelength will be slightly different for different spectrophotometers, owing to variations in the wavelength accuracy and monochromaticity of the incident light ("spectral bandpass" [5]) from one instrument to another. Therefore, although tabulated ε_{max} values in common solvents are widely available for PAHs, it is preferable in accurate work to determine ε for the analyte, using the same spectrophotometer to be employed for the actual analyses, by obtaining a Beer's law plot from standard solutions of a pure sample of the analyte.

"Chemical" deviations from Beer's law also can occur, especially for compounds containing acidic or basic functional groups [6].

4. If a sample contains more than one compound which absorbs at a particular wavelength, the absorbance of the sample at that wavelength should be equal to the sum of the absorbances due to each of the individual absorbing substances. Hence, quantitative analyses for n absorbing compounds can in principle be accomplished by measuring the absorbance at n wavelengths and solving n simultaneous Beer's law equations (assuming that ε is known for each compound at each wavelength). Such a procedure assumes that ε for each analyte is unaffected by the presence of the others (i.e., that the compounds in question do not interact appreciably with each other at the concentrations present in the sample). In practice, procedures of this type are easily amenable to implementation by computer but tend to produce substantial errors unless the number of components with overlapping spectra is fairly small (five or fewer).

5. Measurements of very small (<0.10) or very large (>1.5) absorbances usually are subject to large inherent errors and should be avoided in accurate quantitative work [7].

6. Most of the quantities in Eq. (1) appear in the literature under a variety of names, not all of which are particularly descriptive. For example, absorbance (A) appears often, especially in the biological literature, under the misleading term "optical density." Recommended nomenclature and symbolism for terms important in spectrometric analysis are summarized annually in *Analytical Chemistry* [8].

The design and operating principles for UV-visible spectrometric instrumentation, routine analytical procedures, and data-handling techniques are discussed in textbooks on instrumental analysis [3-6,9,10] and in specialized monographs [11].

B. The UV-Visible Absorption Spectra of PAH

Virtually all organic molecules which absorb strongly in the UV and/or visible are unsaturated molecules; the electronic transitions in question involve promotion of an electron from a bonding to an antibonding π orbital ($\pi \rightarrow \pi^*$ transition). Indeed, the main qualitative use of UV-visible spectrophotometric analysis is simply to ascertain whether an unknown molecule appears to possess significant unsaturation [12]. Virtually all PAH and their substituted and heteroatom derivatives exhibit intense absorption in the UV; some large PAH (five or more rings) also absorb strongly in the visible. A typical electronic absorption spectrum of a PAH (pyrene), measured in liquid solution using conventional absorption spectrometric techniques, is shown in Fig. 1 [13].

The following characteristics of the UV absorption spectra of PAH and their derivatives are relevant to analytical applications:

1. Absorption spectra of unsubstituted PAH usually possess considerable vibrational fine structure (e.g., Fig. 1). In favorable cases, this characteristic facilitates UV absorption spectrometric analysis of multicomponent samples (assuming that the different analytes exhibit different λ_{max} values).

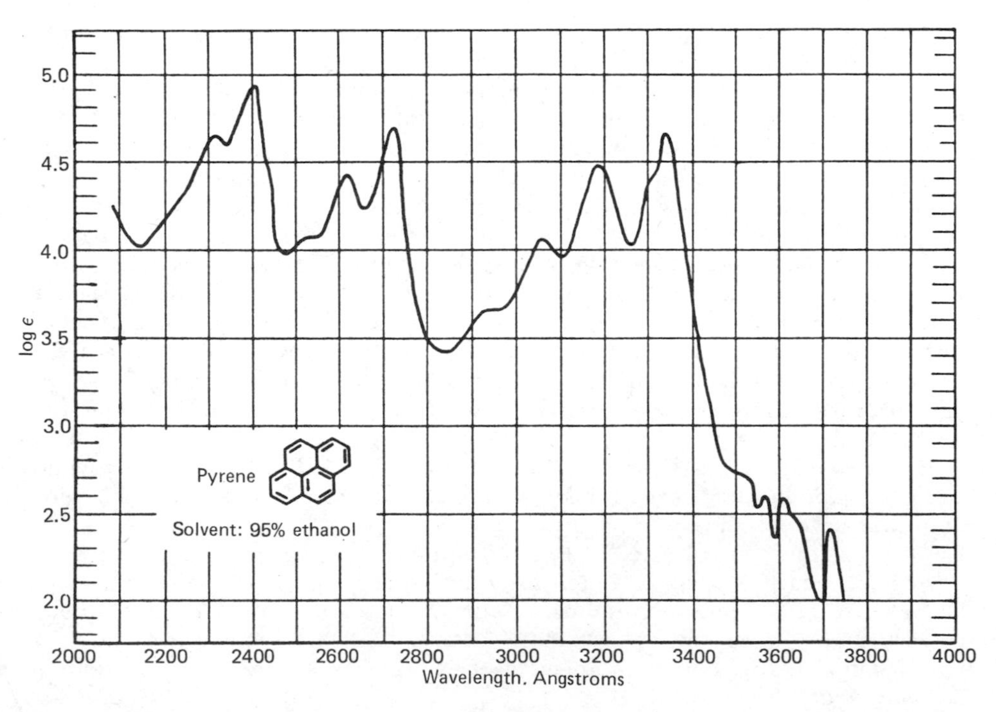

Figure 1. UV absorption spectrum of pyrene in ethanolic solution. (From Ref. 13.)

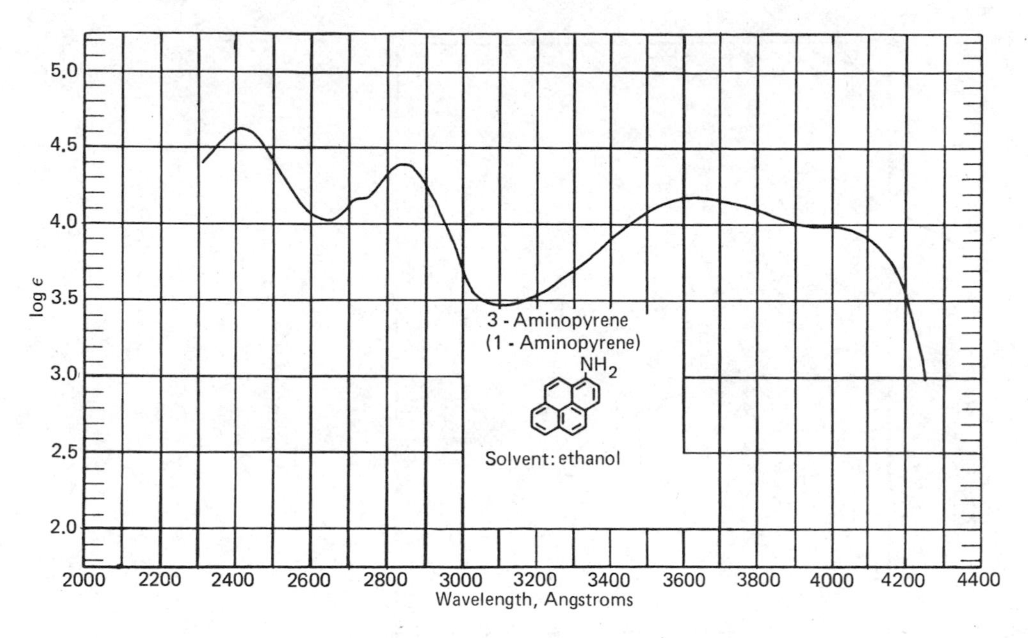

Figure 2. UV absorption spectrum of 3-aminopyrene in ethanolic solution. (From Ref. 13.)

2. The absorption spectra of polar derivatives of PAH usually exhibit much less well-resolved fine structure, especially in polar solvents, than do the parent compounds. Compare, for example, the spectrum of 3-aminopyrene (Fig. 2) [13] to that of pyrene (Fig. 1). Another example is afforded by Fig. 3 [14], which compares the absorption spectra of chrysene [1] and a nitrogen heteroaromatic derivative thereof, naphtho[2,1-f]quinoline [2].

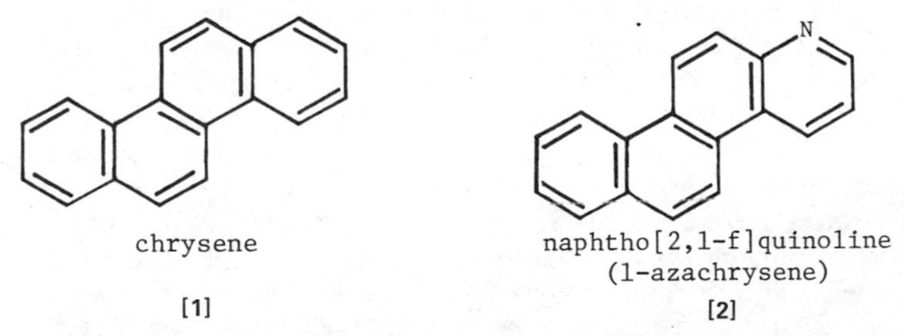

3. Substituent groups and heteroatoms frequently have little effect upon the spectral region in which absorption is observed; as noted in Figs. 1-3, the substituted and/or heteroatom derivatives exhibit virtually the same λ_{max} values as the parent PAH. This is particularly true in nonpolar solvents (e.g., saturated hydrocarbons).

4. Wavelengths of maximum absorption (λ_{max}) are dependent upon the solvent. For unsubstituted PAH, small shifts to lower wavelength (greater energy) of spectral bands are observed as the "polarity" of the solvent is

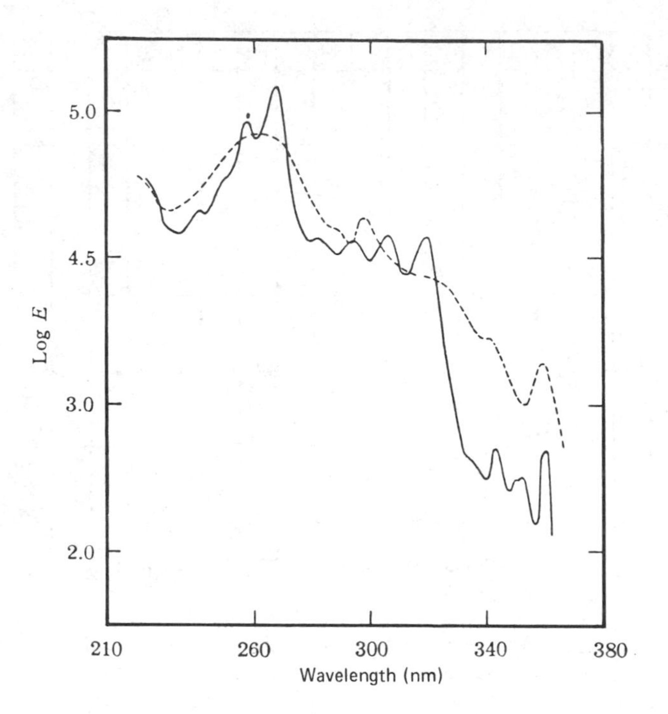

Figure 3. UV absorption spectra of chrysene (—) and naphtho[2,1-f]quinoline (---) in ethanolic solution. (Reprinted with permission from W. S. Johnson, E. Woroch, and F. J. Mathews, Cyclization studies in the benzoquinoline and naphthoquinoline series. *J. Amer. Chem. Soc. 69*:569. Copyright 1947, American Chemical Society.)

decreased; the dielectric constant is a reasonably satisfactory estimate of solvent "polarity" in this regard. Somewhat larger shifts in the same direction usually are observed as a function of solvent polarity for aromatic compounds having polar substituent groups ($-NO_2$, -CN, -OH, etc.). Such solvent shifts are typical of the behavior of molecules exhibiting $\pi \rightarrow \pi^*$ transitions [15]; the data shown for anthracene in Table 1 are representative.

For some heteroaromatics and for carbonyl compounds, solvent effects can be more complex, and occasionally very dramatic. The electronic spectroscopy of many nitrogen, sulfur, and oxygen heterocycles, as well as that of ketones, aldehydes, carboxylic acids, and esters, differs from that of the parent hydrocarbons due to the presence of an unshared electron pair on the heteroatom or carbonyl oxygen. This electron pair occupies a nonbonding (n) orbital, which often is higher in energy than any of the occupied π orbitals. Promotion of one of the nonbonding electrons to an antibonding π molecular orbital, though formally "forbidden" by spectroscopic selection rules, can occur, and this $n \rightarrow \pi^*$ transition often is lower in energy (i.e., occurs at longer wavelength) than any of the $\pi \rightarrow \pi^*$ transitions.

Table 1. Solvent Effects on Position of Long-Wavelength UV Absorption Band of Anthracene

Solvent	Dielectric constant at 25°C	λ_{max} (nm)
N-Methylformamide	182.4	392
Formamide	109.5	387
Acetonitrile	38.0	381
Chlorobenzene	5.7	380
Toluene	2.4	379
Dioxane	2.2	377
n-Hexane	1.9	375

The characteristics of $n \rightarrow \pi^*$ transitions have been reviewed by Kasha [16]. For analytical purposes, two characteristics of $n \rightarrow \pi^*$ transitions are especially relevant. First, they are usually much less intense than $\pi \rightarrow \pi^*$ transitions; $\pi \rightarrow \pi^*$ transitions can exhibit ε_{max} values of 10^5 M^{-1} cm^{-1} or greater (e.g., Fig. 1), whereas ε_{max} values for $n \rightarrow \pi^*$ transitions seldom exceed 500. Often, therefore, an $n \rightarrow \pi^*$ transition will be buried in the $\pi \rightarrow \pi^*$ absorption region of a heterocycle or carbonyl compound. Second, λ_{max} for an $n \rightarrow \pi^*$ transition is very solvent sensitive, with the transition shifting to lower wavelength (higher energy) as the hydrogen-bond-donating power of the solvent increases [17]. In many cases, $n \rightarrow \pi^*$ and $\pi \rightarrow \pi^*$ absorption bands shift in opposite directions as the solvent is changed (for an example of this effect, see Fig. 4 [18]). This phenomenon can sometimes be used as a diagnostic criterion to distinguish between a parent hydrocarbon and a heterocyclic derivative thereof [16].

5. Many polar derivatives of PAH (e.g., phenols, carboxylic acids, nitrogen heterocycles, or primary and secondary amines) undergo acid-base dissociation in polar solvents, and their absorption spectra accordingly are a function of pH. The absorption spectra of an acid and its conjugate base often are dramatically different both in λ_{max} and ε_{max}. Variation of pH therefore can be a useful (and simple) technique for determining acidic or basic compounds in the presence of other absorbing species whose spectra do not change with pH.

6. The larger the number of conjugated aromatic rings in a PAH, the greater the complexity of its absorption spectrum and the longer the wavelength of its lowest-energy absorption band. Figure 5 illustrates this trend for the "linear" PAH through pentacene.

7. Many PAH are intensely fluorescent (see Sec. III). The fluorescence excited by the source in a UV-visible spectrophotometer may be sufficiently intense to be detected. Whenever this happens, the measured absorbance will be too small. This problem can be especially serious for strongly absorbing samples; it is recommended that samples be diluted such that the absorbance at the wavelength of measurement is less than 1.0. If a PAH exhibits

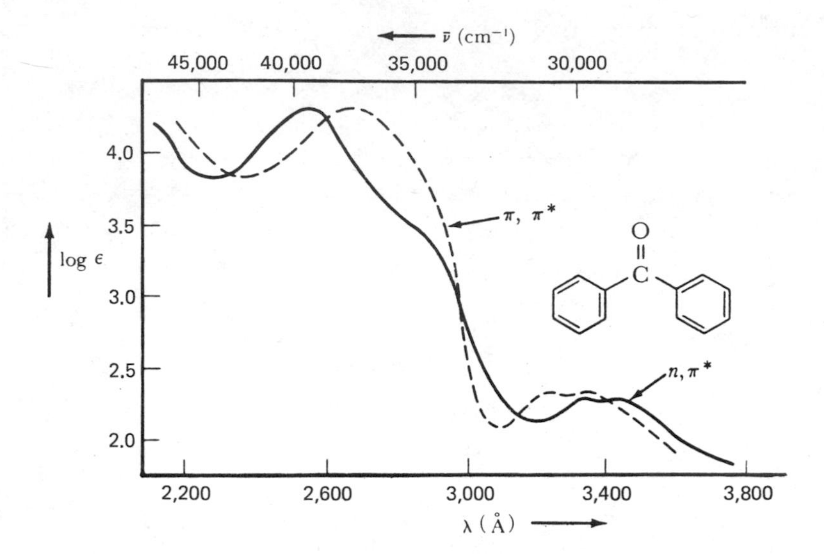

Figure 4. UV absorption spectra of benzophenone in cyclohexane (-) and ethanol (---). Note that the $\pi \rightarrow \pi^*$ absorption band shifts to longer wavelength but the much weaker $n \rightarrow \pi^*$ band shifts in the opposite direction as the solvent is changed from cyclohexane to ethanol. (Reproduced from N. J. Turro, *Molecular Photochemistry*, 1965, with permission of the publisher, Benjamin/Cummings, Inc., Reading, Mass.)

fluorescence which is sufficiently intense to produce appreciable absorbance errors, the spectrum of the compound will be distorted, and the extent to which this occurs will increase with increasing analyte concentration. Invariance of the absorption spectrum as a function of analyte concentration usually is a satisfactory indication that, over the absorbance range studied, fluorescence is a negligible source of error.

C. Sources of Spectral Data

One of the important analytical advantages of UV-visible absorption spectrophotometry is the large number of sources of electronic spectra of PAH. One can therefore often ascertain rather readily whether a given quantitative analysis is feasible by absorption spectrometry, simply by referring to published spectra of the analyte and suspected interfering sample constituents. Below are listed a number of the principal compilations of such data (in decreasing order of general utility for PAH analyses):

1. The compilation of spectra by Friedel and Orchin [13] is extremely convenient to use because full spectra are plotted in easy-to-read format. While not comprehensive, this collection includes spectra of most of the PAH of greatest analytical interest.

2. The treatise on polycyclic hydrocarbons by Clar [19], which will already be familiar to many readers of the present book, presents a large number of UV-visible spectra, including many of compounds for which the

data cannot be readily located elsewhere. The format of printed spectra often is not as useful as that employed by Friedel and Orchin [13].

3. Other collections of plotted spectra include those of the Sadtler Research Labs [20], Plenum Press [21], and Lang [22] compilations. These collections all include a large number of spectra; those of PAH and their derivatives are a small minority in a much larger compilation.

4. The continuing series entitled *Organic Electronic Spectral Data* (OESD), which extracts UV-visible absorption data from the journal literature, is now complete up through the year 1973; the most recent volume appeared in 1979 [23]. Every compound for which UV absorption data have appeared in the literature since 1945 will have an entry somewhere in OESD. Entries in this monumental series are by molecular formula. Only λ_{max} and ε_{max} values are given (in tabular form), but citations to the literature from which the original data were taken are provided. This series is the "court of last resort" for locating UV spectral data for an organic compound. If an entry for the compound in question cannot be found in OESD, either the data in question do not exist in the literature over the span of years covered by OESD or

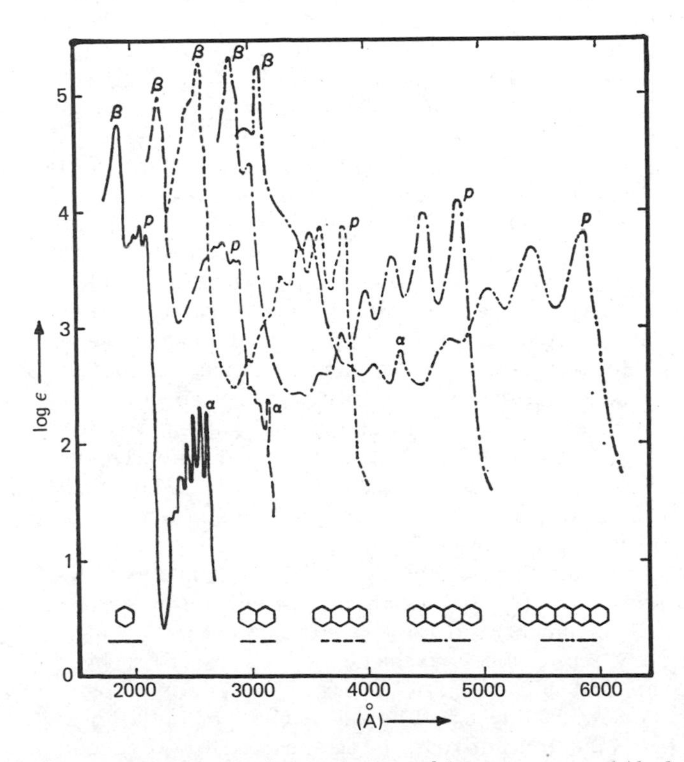

Figure 5. Absorption spectra of benzene, naphthalene, anthracene, tetracene, and pentacene. Note the increasing spectral complexity and shift of the long-wavelength absorption band into the visible region, as a function of increasing number of rings. (Reproduced with permission from E. Clar, *Polycyclic Hydrocarbons*, Vol. 1, 1964. Copyright by Academic Press Inc., New York.)

the person carrying out the search has made a mistake in the molecular formula of the compound!

It need hardly be emphasized that the data included in many of these compilations is of uncertain reliability (owing principally to impurities present in the supposedly authentic samples of the compounds used to obtain the data). As noted previously, reliance upon literature ε values for quantitative analyses should be avoided unless high accuracy is not a requirement of the analytical procedure.

D. Analysis of PAH-Containing Samples by UV Absorption Spectrophotometry: An Example

Analyses of complex PAH-containing samples by UV absorption spectrophotometry proceed by relatively well-defined stepwise patterns, most of the features of which are exemplified in an analytical procedure for alkylphenols in coal liquefaction recycle solvents developed by Schabron et al. [24].

For a complex sample of this type, the actual analytical step (UV absorption measurement) is virtually trivial, both in difficulty and in time required, compared with the sample cleanup steps required to reach the point at which UV analysis may profitably be employed. This comment is especially applicable to UV-visible absorption spectrophotometry, which is not an especially sensitive technique and which (due to the complex and relatively broad spectra obtained) is not readily applicable to the analysis of very complex mixtures of absorbing substances.

In the study under consideration [24], a sample of the solvent in question was diluted with chloroform and chromatographed on a silica gel column to produce three fractions containing (1) PAH, (2) ortho-substituted alkylphenols, and (3) meta- and para-alkylphenols. Fractions 2 and 3 (which were the ones of interest) then were subjected to liquid chromatography on μBondapak NH_2, producing 25 fractions of the o-alkylphenol cut and 30 fractions from the m- and p-alkylphenol cut. The 55 fractions so produced were subjected to a quick UV spectral scan to ascertain which of them appeared to contain aromatic compounds. Those which appeared promising were then subjected to, successively: liquid chromatography on a μBondapak C_{18} column; adsorption chromatography over silica gel; and liquid chromatography successively on μPorasil, μBondapak CN, μBondapak NH_2, and μBondapak C_{18} columns. The intended result of the final step in the chromatographic sequence was production of a solution containing a single alkylphenol in 65% aqueous methanol, suitable for UV determination. Because the phenols are weakly acidic, it was possible to use both λ_{max} values (in both very acidic and very basic media) and the absorbance ratio for a particular compound at a particular wavelength in very acidic and very basic media to confirm the identity of a particular compound and to ensure that the compound whose absorbance was being measured actually was phenolic. The combination of such data with chromatographic retention times is emphasized by Schabron et al. [24] as essential to produce unambiguous identification of particular compounds.

Obviously, a procedure of this type is laborious and time consuming, but the large number of sample cleanup steps is absolutely essential in the UV absorption spectrophotometric analysis of a sample of this complexity. This represents one of the principal disadvantages of UV-visible absorption spectrophotometry in PAH analysis.

E. Summary

Below are summarized some of the more important analytical advantages and disadvantages of UV-visible absorption spectrophotometry in PAH analysis.

1. *Advantages*

a. Readily Available Instrumentation Virtually any laboratory which performs chemical analyses has available a reasonably high-performance UV-visible spectrophotometer. Such instrumentation is readily amenable to operation by technicians, and large numbers of samples can be analyzed per hour once a "standard operating procedure" for a particular analysis has been devised.

b. High Quantitative Accuracy When proper care is exercised (see Sec. II.A), UV-visible molecular absorption spectrometry is capable of substantially greater accuracy and precision than any of the other spectroscopic techniques discussed in this chapter. The reasons for this will become more apparent in later sections; here it should be noted that the instrumentation (particularly sources, detectors, and sample containers) for UV-visible spectroscopy operates in a more nearly "ideal" manner than that for other regions of the spectrum (particularly the infrared).

c. Freedom from Interference by Nonaromatic Compounds The only organic compounds which exhibit appreciable absorption in the accessible UV or visible spectral regions are unsaturated compounds. Hence, aliphatic hydrocarbons and their derivatives, which may be major constituents of fossil fuel samples, do not ordinarily constitute analytical interferences.

2. *Disadvantages*

a. Need for Extensive Sample Cleanup As emphasized in the preceding sections, the complexity and lack of highly resolved fine structure characteristic of the electronic absorption spectra of large molecules limits the selectivity of analytical methods based on UV-visible absorption.* Here it is important to emphasize that losses or contamination problems in the sample cleanup steps may vitiate the high accuracy of UV-visible absorption measurements; i.e., while the analytical measurement itself is inherently very accurate, large errors in an analysis may be introduced in the procedures required to prepare the sample for measurement.

---

*It should be noted in this connection that the utility of absorption spectrometry for mixture analysis often is enhanced by taking derivatives of the spectra. For example, a second-derivative UV absorption technique has been demonstrated applicable to the identification and quantitation of individual PAH in samples of moderate complexity [A. R. Hawthorne and J. H. Thorngate, *Appl. Spectrosc. 33*:301 (1979)]. Commercial instrumentation is available for the performance of derivative UV-visible absorption spectrometry.

b. Lack of "Fingerprinting" Capabilities As emphasized in the preceding discussion, UV-visible spectroscopy normally must be supplemented by more "structure-specific" spectroscopic techniques if the analytical task requires identification of unknown sample constituents.

c. Relatively Low Sensitivity Conventional "absorption" spectroscopy actually involves measurement of the extent to which a sample *transmits* light incident upon it [cf. Eq. (1)]. When the analyte concentration is low, A (absorbance) is small; in that case, P is only slightly smaller than P_0. The measurement problem then becomes one of distinguishing a small difference between two large numbers, which is always a losing proposition. Hence, despite the high effective power of UV-visible sources and the high detectivity of UV-visible detectors (particularly photomultiplier tubes), absorption spectrophotometry is inherently a technique of limited sensitivity; meaningful detection limits are on the order of 10^{-7} M in the most favorable cases. This limitation is not particularly crucial in dealing with samples containing high PAH concentrations (e.g., coal liquids) but is extremely important in any analytical situation requiring determination of PAH at trace levels (many environmental analyses obviously fall in this category). For such analytical problems, recourse to techniques which can detect PAH at lower levels may be essential.

It should be noted that various "absorption" spectroscopic techniques which truly measure the extent to which a sample absorbs light, such as photoacoustic spectroscopy [25] and thermal lensing [26], are under rapid development. However, these techniques have not yet been adapted to the routine analytical characterization of complex organic samples.

3. *Conclusion*

UV-visible absorption spectrophotometry is a mature "workhorse" technique especially suited for quantitative analysis in samples of limited complexity. Its use in PAH analysis is gradually being superseded by luminescence techniques (see Sec. III), which often are orders of magnitude more sensitive. *If* the objective of the analysis is quantitative, *if* one is willing to accommodate to the rather extensive sample cleanup required for complex samples, and *if* very low detection limits are not required, UV-visible absorption spectrophotometry is likely to be superior to any of the other techniques discussed in this chapter.

Finally, it should be noted that developments and new procedures in UV-visible absorption spectrophotometry are reviewed biennially (in even-numbered years) in *Analytical Chemistry* [27].

III. Fluorescence and Phosphorescence Spectrometry

A. Introduction and Quantitative Considerations

Photoluminescence is the term used to denote light emitted by a molecule which had previously been promoted to an electronically excited state by absorption of light. If the electronic transition responsible for the observed emission occurs without a change in the spin quantum number of the emitting molecule, the emission is termed *fluorescence*; if the electronic transition in question occurs with a change in the spin state of the molecule, the luminescence is called *phosphorescence*. The principles of fluorescence and phosphorescence have been discussed thoroughly in monographs [28,29], and we confine our remarks in this section to the most important generalizations.

Fluorescence and phosphorescence obviously can be observed only if the analyte in question absorbs light. Fortunately, as noted in Sec. II, PAH absorb strongly in the UV. Intense absorption is a necessary, but not sufficient, condition for the observation of fluorescence or phosphorescence. Indeed, many strongly absorbing molecules are weakly luminescent, and their emission is of little or no analytical utility. Nonradiative decay processes, and possibly photochemical decomposition, always compete with the luminescence processes, and it is only those molecules for which emission competes effectively with other excited-state decay pathways that analytically useful luminescence is observed [28,29].

Because fluorescence has thus far been used much more widely than phosphorescence in PAH analysis, we will confine most of our attention to fluorescence. The vast majority of aromatic molecules possess ground electronic singlet states (i.e., all electron spins paired), and the intense bands observed in the UV-visible absorption spectra of PAH and their derivatives result from transitions to various excited singlet states (labeled S_1^*, S_2^*, S_3^*, etc. in order of increasing energy) from the ground singlet state (conventionally denoted by the symbol S_0).

A potentially important advantage of fluorescence over UV-visible absorption analysis is the inherent involvement of two wavelengths (excitation and emission) in the fluorescence process. If two compounds absorb at the same wavelength but fluoresce at different wavelengths, then an emission spectrum (signal as a function of emission wavelength at constant excitation wavelength) will differentiate between them. Likewise, two compounds whose fluorescence spectra overlap but which absorb at different wavelengths can be distinguished by the corresponding differences in their fluorescence excitation spectra (signal as a function of excitation wavelength at constant emission wavelength) or, more simply, by setting the excitation wavelength such that the compound of interest is selectively excited and measuring the emission signal at the wavelength of maximum fluorescence.

The probability that a molecule which has been promoted to an excited singlet state will decay back to the ground state by fluorescence is termed the *fluorescence quantum yield* Φ_F. The fluorescence power F produced by a particular molecule is proportional to the rate at which excited states are produced by absorption of light and to the fluorescence quantum yield:

$$F \propto P_{abs}\Phi_F \tag{2}$$

where P_{abs}, the energy per unit time of incident light absorbed by the sample, is given by

$$P_{abs} = P_0 - P \tag{3}$$

where P_0 and P have the same significance as in Beer's law [Eq. (1)]. By Beer's law,

$$P = P_0 10^{-A} = P_0 10^{-\varepsilon bc} \tag{4}$$

Therefore

$$F \propto \Phi_F P_0 (1 - 10^{-\varepsilon bc}) \quad (5)$$

The important fact conveyed by Eq. (5) is that the fluorescence signal is *not* inherently linear in the concentration of the analyte. The problems which can thereby be introduced into quantitative analysis by fluorescence spectrometry are discussed elsewhere [30].

The quantity $1 - 10^{-\varepsilon bc}$ can be expressed by a power series approximation:

$$1 - 10^{-\varepsilon bc} = \varepsilon bc + \frac{(\varepsilon bc)^2}{2!} + \frac{(\varepsilon bc)^3}{3!} + \cdots \quad (6)$$

For small absorbances ($\varepsilon bc < 0.15$ at the exciting wavelength), the higher order terms in Eq. (6) are negligible with respect to εbc. In that case, Eq. (5) simplifies to

$$F = k \Phi_F P_0 \varepsilon bc \quad (7)$$

where k is a proportionality factor the significance of which is considered below. Under certain limiting conditions, therefore, fluorescence signals are linear in concentration. One normally wishes in practical fluorometric analysis to ensure that the sample is dilute enough for the approximation inherent in Eq. (7) to be valid.

By far the most important advantage of fluorescence over absorption methods is that, for molecules which exhibit appreciable fluorescence quantum yields, fluorescence methods are capable of detecting much lower concentrations of analyte species than are absorption techniques. For example, Richardson and Ando [31] have reported limits of detection for PAH on the order of 10^{-11} M (ca. 1 part per trillion) by laser-induced fluorescence; this is an improvement of at least a factor of 10^4 in detection limits over what could be achieved, under the most favorable conditions, by UV absorption spectrophotometry.

This characteristic of fluorometric analysis is sufficiently important that it has become generally known as the "fluorescence advantage" [32]. Its origin is apparent from Eq. (7). Because the signal F is directly proportional to the incident power P_0 (which is of course not true in absorption spectrophotometry), there is a strong incentive in fluorometric analysis to employ very intense sources. The risk in this approach is that the analyte may be destroyed (by photochemical reactions or, less likely, by heating) if one does not proceed with care. It is also obvious that in fluorometry (unlike absorption spectrometry) ideally no optical signal is detected when no analyte is present. Hence, when the analyte concentration is small, the measurement problem in fluorometry (ideally) is to distinguish between a small signal and zero signal rather than to measure a small difference between two large signals (as is the case in absorption spectrometry whenever the absorbance is small). It must, however, be stressed that, in most practical fluorometric analyses, the existence of background fluorescence from contaminants in the solvent [33] or the sample itself limits the ability of fluorescence to detect very small quantities of analytes, especially when very sensitive modern fluorescence instrumentation is used.

It is obvious from Eq. (7) that the analyte must exhibit a reasonably large fluorescence quantum efficiency in order for fluorescence to be a useful analytical method for that analyte. Most PAH are intensely fluorescent; a brief tabulation of Φ_F values for some common PAH is provided in Table 2.

Table 2. Fluorescence Quantum Efficiencies of Common PAH

Compound	Solvent[a]	Φ_F
Naphthalene	Cyclohexane	0.19
Anthracene	Cyclohexane	0.30
Pyrene	Cyclohexane	0.65
Benz[a]anthracene	Cyclohexane	0.19
Chrysene	Cyclohexane	0.12
Perylene	Ethanol	0.87
Fluoranthene	Cyclohexane	0.25

[a]All data obtained in liquid solution at room temperature.
Source: Adapted from J. B. Birks, *Photophysics of Aromatic Molecules*. Copyright 1970. Reprinted by permission of John Wiley & Sons, Ltd., New York.

Perhaps the most serious difficulty in quantitative analysis by fluorometry is the tendency for Φ_F for a given compound to vary as a function of the solvent and (more importantly) in the presence of other solutes. ***Quenching*** is defined as any process wherein the fluorescence quantum efficiency of a particular compound is decreased by interactions with other constituents of a sample. A number of different mechanisms (discussed in detail elsewhere [34,35]) for fluorescence quenching have been elucidated; here we need only note that quenching can proceed by both collisional and "long-range" processes. Molecular oxygen is a notorious fluorescence quencher of many organic molecules, including most PAH, and procedures for quantitative fluorometric analysis of PAH therefore usually include a deoxygenation step.

In some, but not all, cases of quenching, electronic energy transfer occurs, resulting in formation of an electronically excited state of the quenching molecule. This excited molecule may decay by fluorescence. Such a process, wherein fluorescence of a second compound is produced by energy transfer following excitation of the analyte by absorption of light, is termed *sensitization*. Sensitization occasionally is of value in the fluorometric determination of a species which is difficult to excite by direct absorption (either because it is a weak absorber or because it does not absorb at a wavelength convenient for excitation with a particular source), but more often it is a nuisance.

Parker [35] has emphasized that fluorescence quenching can be suppressed simply by diluting the sample until no solute is present at a concentration exceeding 10^{-3} M. Such a procedure is readily applicable if the analyte is a major constituent of the sample but may not be feasible if the analyte is a minor component present in a sample containing potential quenchers at much higher concentrations. The quenching problem constitutes a major headache in applications of fluorescence spectrometry to quantitative analysis in complex samples; it has no analog in absorption spectrophotometry.

Another problem with which one must often contend in fluorometric analysis of multicomponent samples is the *inner-filter effect*, which can take several forms [36]. For example, if a sample contains several different compounds which absorb in the same wavelength region, then the efficiency of excitation of fluorescence from any one compound will be decreased in a manner proportional to the sum of the absorbances of the others. In effect, P_0 [Eq. (7)] is decreased by this form of interference. Alternatively, an interfering sample constituent may absorb in the wavelength region where the analyte fluoresces; again, the result is a decrease in the signal observed for the analyte. These artifacts are distinguishable from quenching because they do not alter the fluorescence quantum yield of the analyte. In principle, inner-filter effects can be reduced to negligible proportions by dilution, but again this approach may fail if the analyte is a minor constituent of a complex sample. Inner-filter effects also can be alleviated by careful consideration of the illumination geometry employed in the fluorescence spectrometer [36].

Finally, we should note that few tabulated values of Φ_F for PAH in analytically realistic solvents are available. The fluorescence quantum yield is a rather difficult quantity to measure accurately, and many of the existing literature values are of uncertain validity [37]. Moreover, even if an accurate literature value for Φ_F can be unearthed, it may not pertain to the analyte in a real sample (due to the ever-present possibility of quenching). For practical purposes, therefore, Φ_F must virtually always be regarded as unknown. Equation (7) contains an additional unknown, k, which includes (among other variables) the fraction of the fluorescence emitted by the analyte which is actually detected. In most samples, the fluorescence is emitted in all directions, and design of an optical system which will collect all (or even a large fraction of) the emitted photons is difficult. Hence, k is usually termed the "instrumental geometry factor," the value of which depends upon the illumination and collection optics used in the fluorescence spectrometer. It is virtually impossible to measure k accurately, so it must also be treated as an unknown (though presumably reproducible) parameter. In effect, Eqs. (5) and (7) reduce, respectively, to

$$F = KP_0(1 - 10^{-\varepsilon bc}) \quad (8)$$

and

$$F = KP_0\varepsilon bc \quad (9)$$

where $K = k\Phi_F$ must be determined empirically.

If one is certain that quenching and inner-filter effects are absent, it is possible to determine K by preparing an analytical calibration curve of F vs. c for solutions of the pure analyte in the particular solvent to be used for the analysis. For most real samples, it is safer to employ some type of empirical standardization technique which does not automatically presume the absence of quenching and inner-filter phenomena; the *method of standard additions* is used most commonly in molecular fluorescence spectrometry. Details of this procedure and the assumptions inherent in its use are discussed elsewhere [38].

It should be evident from the preceding discussion that fluorescence is inherently somewhat less suitable than absorption spectroscopy for quantitative analyses in very complex samples, at least when "conventional" fluorometric measurement techniques are used. A number of new fluorometric procedures, designed to alleviate some or all of the difficulties considered above, are currently being developed (see Sec. III.E).

Instrumentation for conventional molecular fluorescence spectrometry has been reviewed in great detail by Parker [39] and in a more introductory manner by Ellis [40] and Guilbault [41].

B. The Fluorescence Excitation and Emission Spectra of PAH

For most large organic molecules, it is found that excitation into any of the "higher" excited singlet states (S_2^*, S_3^*, etc.) is followed by rapid nonradiative decay to S_1^*. This fact has two consequences. First, fluorescence spectra tend to be simpler than absorption spectra because the former contain only one electronic transition ($S_1^* \rightarrow S_0$) whereas several transitions ($S_0 \rightarrow S_1^*$, $S_0 \rightarrow S_2^*$, etc.) are observed in the absorption spectrum. Thus, the fluorescence spectrum of a PAH has a lower "information content" than the absorption spectrum but also is more suitable for multicomponent sample analysis because of its reduced complexity. Second, the fluorescence quantum efficiencies of most large molecules are independent of the wavelength of excitation. This means that (subject to a very important caveat noted later) the fluorescence excitation spectrum of a PAH should be identical with its UV-visible absorption spectrum in a given solvent. It is then obvious from Eqs. (8) and (9) that the highest sensitivity in fluorometric analysis is achieved when the excitation wavelength is chosen to be that at which the analyte exhibits the highest molar absorptivity, irrespective of which excited singlet state is thereby populated.

While $S_0 \rightarrow S_1^*$ absorption normally involves a transition from the ground vibrational state of S_0 to a vibrationally excited level of S_1^*, fluorescence usually entails a transition from the ground vibrational state of S_1^* to an excited vibrational level of S_0. Consequently, the fluorescence spectrum normally is displaced to longer wavelength (lower energy) relative to the $S_0 \rightarrow S_1^*$ absorption band system, though frequently there is some overlap between the two. Because the vibrational level spacings in S_0 and S_1^* usually are similar, the fluorescence spectrum often is virtually the "mirror image" of the $S_0 \rightarrow S_1^*$ region of the absorption spectrum. These relationships between absorption and fluorescence spectra are illustrated in Fig. 6.

In fluorescence, as for absorption, the spectrum usually is displaced to longer wavelength as the size of the conjugated aromatic ring system is increased. An example of this trend in the fluorescence spectra of "linear" PAH is shown in Fig. 7 [42].

The influence of molecular structure on the fluorescence spectra and quantum yields of organic molecules is treated in detail elsewhere [43-46]. As noted in Sec. III.A, most parent PAH exhibit rather intense fluorescence. While many substituent groups have minor effect on the wavelength or efficiency of PAH fluorescence, some "heavy-atom" (—Br, —I) and other (—NO_2) substituents are notorious for decreasing Φ_F, often (but not invariably) with concomitant enhancement of the phosphorescence efficiency. Heterocycles and carbonyl compounds, which usually have low-lying n-π^* transitions, often fluoresce very weakly in nonpolar solvents but may fluoresce with useful efficiency in polar, hydrogen-bonding media [46].

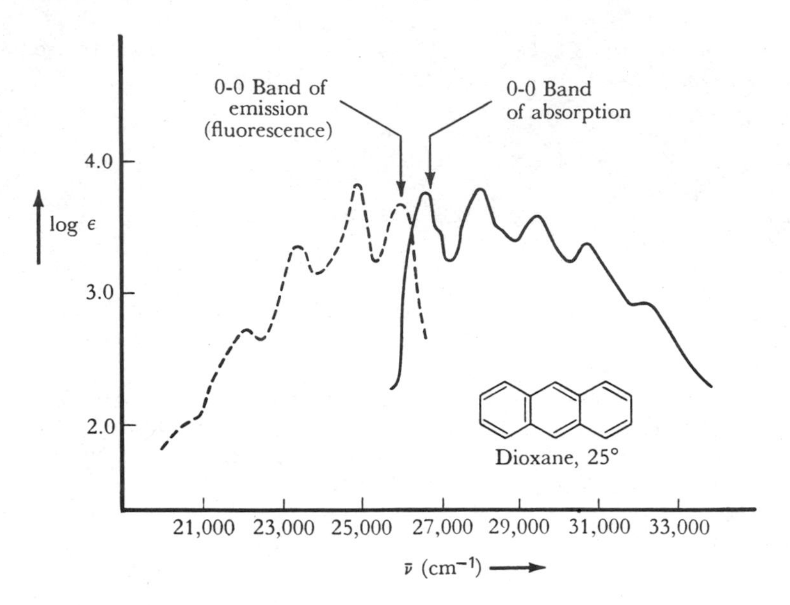

Figure 6. Fluorescence (---) and absorption (—) spectra of anthracene in dioxane solution. Note that the fluorescence appears at lower frequency (longer wavelength) than the absorption and that the fluorescence and absorption spectra are virtually "mirror images" of one another. (Reproduced from N. J. Turro, *Molecular Photochemistry*, 1965, with permission of the publisher, Benjamin/Cummings, Inc., Reading, Mass.)

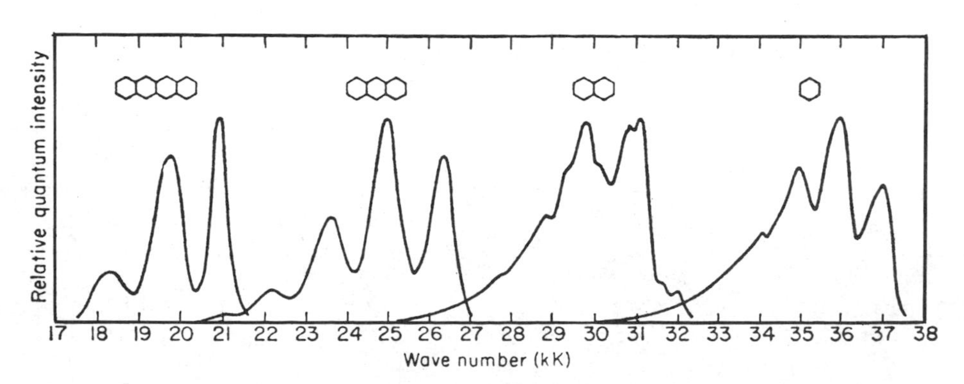

Figure 7. Fluorescence spectra of benzene, naphthalene, anthracene, and tetracene in cyclohexane solution at room temperature. The horizontal scale is frequency in kK (1 kK = 1000 cm^{-1}). (From J. B. Birks, *Photophysics of Aromatic Molecules*. Copyright 1970. Reprinted by permission of John Wiley & Sons, Ltd., New York.)

We emphasize once again that only those compounds which absorb in the accessible UV-visible spectral region (i.e., at wavelengths greater than about 220 nm) can exhibit analytically useful fluorescence. This is a necessary but not sufficient condition, and many aromatic molecules which are strong UV-visible absorbers fluoresce only weakly. Fluorescence is accordingly a less "universally" applicable analytical technique than UV-visible absorption. On the other hand, "spectral interferences" in multicomponent samples may be less severe in fluorescence than in absorption, for three reasons: First, as already noted, one has two wavelengths (excitation and emission) which can be adjusted to achieve selectivity. Second, as just discussed, absorption spectra usually are more complex than fluorescence spectra because more than one electronic transition commonly is observed in absorption, thus increasing the probability of spectral overlaps in the absorption spectra of complex mixtures. Finally, potentially interfering sample constituents may be nonfluorescent (though their presence may nonetheless be manifested in the form of quenching and inner-filter effects).

Both the excitation and emission spectra, as well as the fluorescence quantum yield, of aromatic compounds are influenced by changes in the solvent. Such effects are especially pronounced for PAH derivatives containing polar substituent groups. These phenomena are discussed in detail elsewhere [47-49].

C. Sources of Spectral Data

In contrast to UV-visible absorption spectrophotometry, few published compilations of fluorescence spectra exist. Probably the most useful for PAH analysis is the book by Berlman [50], which provides full plots of excitation and fluorescence spectra for many PAH in liquid solution, as well as fluorescence decay times and quantum efficiencies; unfortunately, the quantum yields in this collection appear to be uniformly too high (by a factor of ca. 1.2 [51]). Sadtler has issued a collection of fluorescence spectra [52] in their conventional format; a number of PAH and PAH derivatives are included in that collection (which can be obtained in computer-searchable files as well as conventional hard copy). The extensive Landolt-Börnstein tabulation of physical properties includes a volume on organic molecular luminescence [53]. This compilation is rather old (1967) and, for the most part, includes only tables of $\tilde{\nu}_{max}$, Φ_F, and decay time values rather than plotted spectra. However, its coverage of the literature up to the time of its publication was quite thorough; therefore, this can serve as a much more comprehensive data base than the two previously cited. Finally, for a number of years Passwater edited a *Guidebook to Fluorescence Literature* [54], consisting of tabulated journal citations pertaining to fluorescence, with a separate index provided of the principal compounds dealt with in the cited papers. To some extent, this series served the same purpose in fluorometry as OESD in absorption spectrometry. However, it appears that the "fluorescence guide" series terminated with Vol. 3 (1974).

Those compilations of fluorescence spectra which do exist must be used with considerable care. UV absorption spectra of the same sample obtained with two or more instruments of comparable spectral resolution will (in the absence of instrumental malfunctions) be virtually identical (in terms of band widths and λ_{max} and ε_{max} values). In contrast, fluorescence spectra obtained for the same sample with two different fluorometers of comparable design, or even with the same instrument at different times, may exhibit drastic differences in λ_{max} and apparent fluorescence intensity. The reasons for this situation were first discussed clearly by Hercules [55] and have subsequently been treated in great detail by Parker and Rees [56,57]. The problem is that, whereas virtually all recording UV-visible absorption spectrophotometers employ a double-beam optical configuration, the majority of commercial fluorescence spectrometers presently are of single-beam design. Hence, no provision is made for correction of spectra for variations in source brightness, detector response, or monochromator throughput as a function of wavelength. A change of one detector for another, even if they have the same nominal response curve, may bring about significant changes in the overall spectral response factor for the instrument. The problem also arises for excitation spectra measured with a single-beam fluorometer. An excitation spectrum will be identical with the absorption spectrum (see Sec. III.A) only if the appropriate corrections are made or if the spectrum is measured with a self-correcting spectrometer.

Fluorescence spectrometers which correct automatically for variations in source brightness and detector responsivity with wavelength have been designed [58-61] and are now commercially available. However, the vast majority of fluorescence and excitation spectral data reported in the literature were obtained with "noncorrecting" instruments. Manual correction of spectra obtained with such an instrument is possible but laborious [57]. Caution must therefore be exercised in the use of literature compilations of uncorrected fluorescence spectra.

D. Analysis of PAH-Containing Samples by Fluorometry: An Example

A useful "case study" of the analysis of a complex PAH-containing sample by fluorometry is offered by a series of very thorough investigations by Katz's group of the PAH content of atmospheric particulate matter [62-64]. The particles were collected on fiberglass filters by conventional high-volume air sampling and were then subjected to a lengthy (8-12 hr) Soxhlet extraction with dichloromethane or benzene. The resulting extract was filtered, the filtrate was evaporated to dryness, and the residue was dissolved in a small volume of toluene.

Because the solution so obtained contained many PAH, it was necessary to subject it to separation prior to fluorometric analysis. Thin-layer chromatography (TLC) was chosen for this purpose. Use of commercial precoated alumina plates and a pentane/diethyl ether (19:1) mobile phase produced a separation of PAH into four distinct zones. Each of these four areas on the alumina TLC plate was individually removed by scribing, and the PAH present in each zone were removed from the scribed region of the plate by elution with dichloromethane. The dichloromethane was evaporated, and the resulting residue was dissolved in a small volume of toluene.

Each of the four individual PAH fractions so obtained was then subjected to a second stage of TLC, using cellulose of different degrees of acetylation as adsorbent. The mobile phase was n-propanol/acetone/water (2:1:1).

This second TLC step separated most of the PAH of interest, particularly if two-dimensional development of the plate was carried out (see Fig. 8 for an example).

Each individual spot was removed from the plate by scribing and was placed in a weighing bottle in contact with diethyl ether (which served to elute PAH from the adsorbent). The diethyl ether was separated from the adsorbent and was evaporated; the residue was dissolved in n-pentane, and the resulting solution was transferred to cells for fluorometric analysis. Prior to measurement of fluorescence, the cells were purged of oxygen by bubbling with dry N_2 so as to minimize quenching of PAH fluorescence by O_2. Fluorescence was then measured with a commercial fluorometer. For most PAH, the fluorescence emission and excitation spectra obtained from the relevant TLC spot were quite similar to those obtained for pure samples of the corresponding compounds. Figure 9 shows example comparison spectra for one PAH (naphtho-[1,2,3,4-def]chrysene).

Recoveries of PAH in the TLC procedure generally were in the 87-95% range (as determined by studies using pure PAH), with a precision of recovery of ±3-5% relative standard deviation. In this, as in many similar procedures, the accuracy and precision of the final analytical result are determined by those of the sample cleanup and separation steps, rather than by that of the actual fluorescence measurement. On the basis of "extraction curves" (quantity of PAH extracted as a function of duration of extraction), it was presumed that the initial Soxhlet extraction of PAH from the original particulate sample was quantitative. While that conclusion probably was valid in this particular case, it has been observed that exhaustive solvent extraction of PAH adsorbed on other solids (e.g., coal fly ash [65]) often exhibits nonquantitative recovery (see Chap. 3).

The essential aspects of the procedure outlined above are as follows: first, separation of the PAH from other sample constituents and the particulate matrix; second, separation of the PAH into fractions (often according to number of rings); third, resolution of each PAH fraction into pure compounds or very simple mixtures; and, finally, fluorometric analysis of each of the separated PAH. Obviously, such procedures tend to be time consuming and not readily amenable to automation. However, in view of the unsuitability of fluorometry for analysis (especially quantitation) of very complex mixtures, there is little alternative to a separation scheme such as that outlined. Much current research in molecular fluorescence spectrometry is devoted to increasing the applicability of the technique to complex mixture analysis. A number of these new procedures are described briefly in the following section.

E. Recent Innovations in Fluorometric Analysis

In conventional fluorescence spectrometry, the source is a xenon or mercury-xenon arc, the sample is a liquid solution, and the detector is a photomultiplier tube. Diffraction gratings are used in the excitation and emission monochromators (unless a simple filter fluorometer is used), and either one or the other (but not both) of the monochromators is scanned to produce either an excitation or an emission spectrum. A number of innovations in fluorescence instrumentation and technique currently are being developed. Although these new fluorometric procedures generally have not yet become available

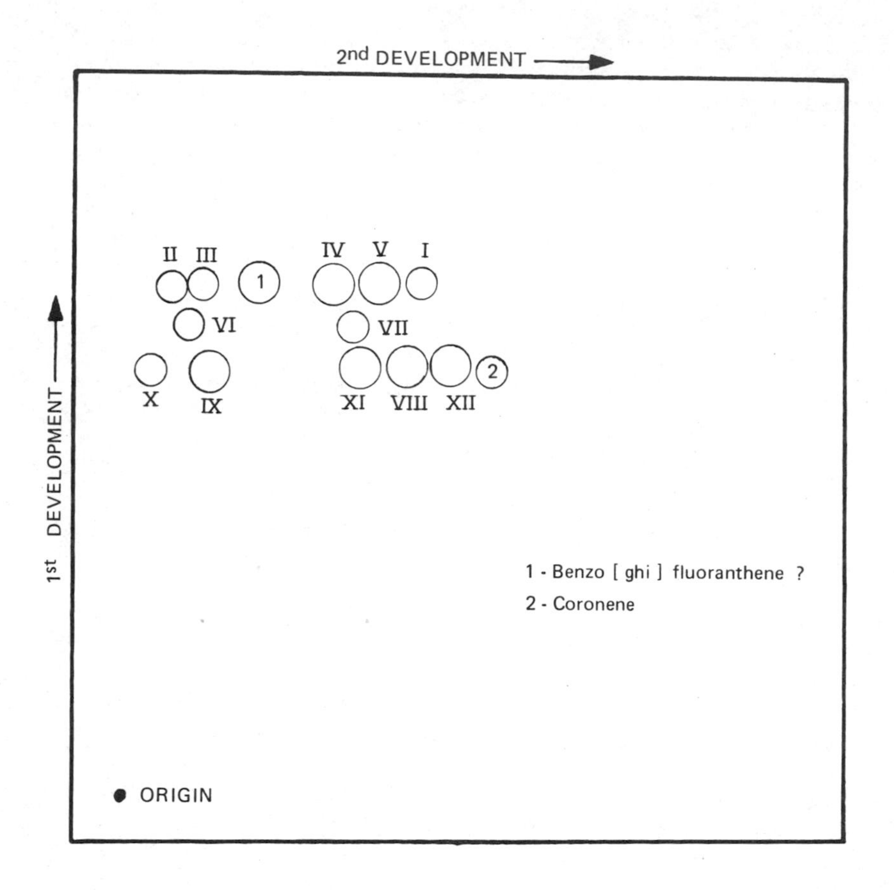

Figure 8. Two-dimensional thin-layer chromatogram (visualized by fluorescence) of five- and six-ring PAH fraction from benzene-soluble fraction of extracted airborne particulate matter. Compounds: I, Benzo[e]pyrene; II, benzo[a]pyrene; III, benzo[b]fluoranthene; IV, benzo[k]fluoranthene; V, pyrene; VI, dibenzo[def,mno]chrysene; VII, benzo[ghi]perylene; VIII, naphtho[1,2,3,4-def]chrysene; IV, benzo[rst] pentaphene; X, dibenzo[b,def]-chrysene; XI, naphtho[2,1,8-gra]naphthacene; XII, dibenzo[def,p]chrysene. Locations of the spots for the individual compounds were established by TLC measurements with authentic samples of the compounds. (Reprinted with permission from R. C. Pierce and M. Katz, Determination of atmospheric isomeric polycyclic arenes by thin-layer chromatography and fluorescence spectrophotometry. *Anal. Chem.* 47:1746. Copyright 1975, American Chemical Society.)

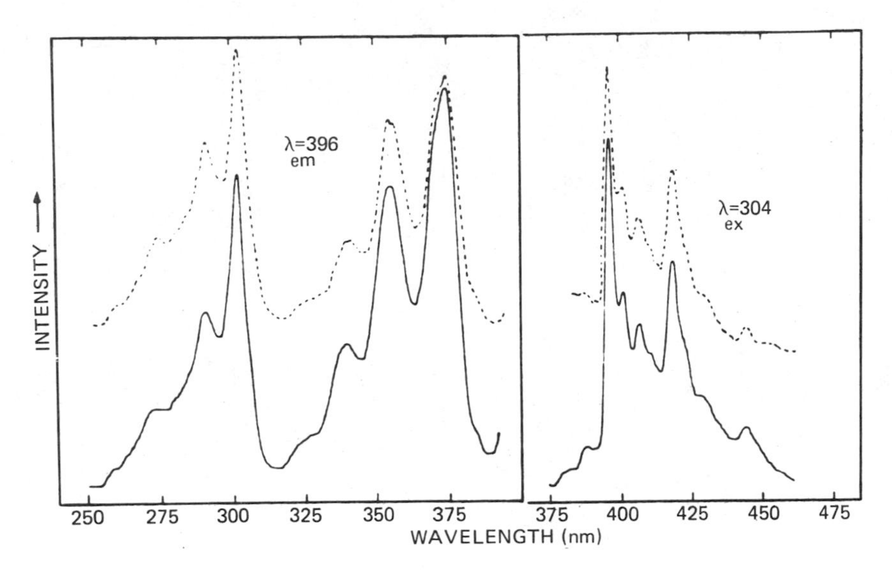

Figure 9. Fluorescence excitation (left) and emission (right) spectra of naphtho[1,2,3,4-def]chrysene. Curves: (—) pure compound; (---) from TLC spot. Solvent: n-pentane. (Reprinted with permission from R. C. Pierce and M. Katz, Determination of atmospheric isomeric polycyclic arenes by thin-layer chromatography and fluorescence spectrophotometry. *Anal. Chem.* *47*:1747. Copyright 1975, American Chemical Society.)

in the form of commercial instrumentation, they promise to greatly enhance the capabilities of fluorometry in PAH analysis. A number of these modifications to conventional fluorescence spectrometry are discussed very briefly below.

1. Laser-Induced Fluorescence

As shown in Eqs. (8) and (9), the signal generated in a fluorescence experiment is directly proportional to the rate of absorption of photons by the sample. Commercial lasers now exist which (in certain wavelength regions) can deliver very much greater photon fluxes per unit wavelength than can conventional arc lamp sources. Moreover, the spatial coherence of laser light makes it possible to excite fluorescence from very tiny samples, or from very small volume elements of larger samples. In favorable cases, these characteristics of lasers can be used to achieve extremely low detection limits for PAHs. For example, Fig. 10 shows an analytical calibration curve for pyrene excited with a nitrogen laser-pumped dye laser [31]; in this case, a detection limit of less than 1 part per trillion of pyrene was achieved. Such high sensitivities are virtually impossible to achieve with arc lamp sources unless very large lamps (3 kW or greater) are used. We emphasize that, in practical analytical situations, achievement of such low limits of detection is possible and meaningful only if scrupulous attention is devoted to purity of reagents and solvents and to avoidance of contamination of the sample in cleanup steps [33]. In

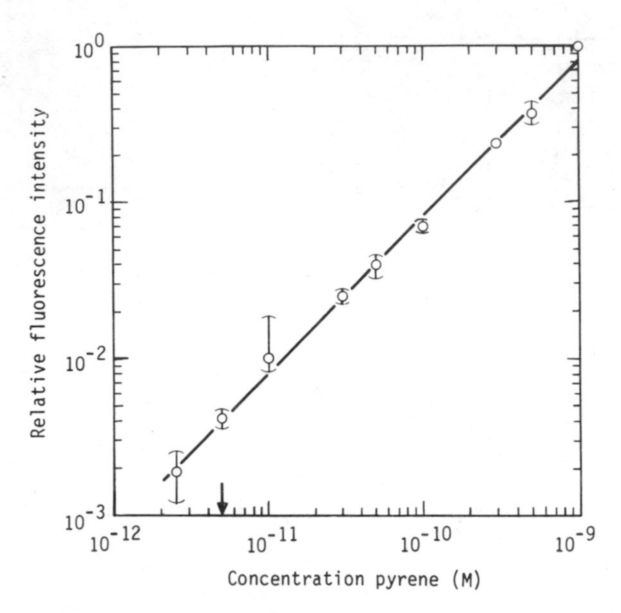

Figure 10. Plot of measured fluorescence signal as a function of concentration for pyrene in aqueous solution. The arrow corresponds to a concentration of 1 part per trillion; the limit of detection was estimated to be 0.5 parts per trillion. The excitation source was a nitrogen-paser-pumped dye laser. (Reprinted with permission from J. H. Richardson and M. E. Ando, Sub-part-per-trillion detection of polycyclic aromatic hydrocarbons by laser-induced molecular fluorescence. *Anal. Chem. 49*: 958. Copyright 1977, American Chemical Society.)

quantitative analysis at very low levels, care must also be exercised to avoid losses by volatilization (particularly of PAH such as naphthalene and anthracene), photodecomposition, and adsorption onto the walls of containers (particularly glass).

Lasers have some shortcomings as sources for fluorometric analysis. Lasers usually are expensive, not amenable to use by unskilled personnel, less than ideally reliable, and relatively inflexible in terms of the number of different types of experiments to which any given laser can be adapted. Continuously tunable output at high power in the UV (where most PAH absorb most strongly) is not yet easily achieved. Continuous tunability over the range of wavelengths which one would like to use in fluorometric analysis of a complex sample is not readily obtained (any given laser dye has a rather limited tuning range). However, the technology of lasers is developing rapidly, and most of these disadvantages (except cost!) should eventually disappear.

It should also be stressed that the high monochromaticity of laser light can be extremely useful for exciting fluorescence from individual PAH in mixtures, as noted in Sec. III.E.4. Indeed, for almost any fluorescence experiment in which highly selective excitation is the goal, laser excitation is virtually indispensable.

Useful overviews of lasers as excitation sources in fluorescence spectrometry have been presented by Richardson [66] and Wright and Wirth [67]; the reader is referred to them for more detailed discussion and references to the original research literature on the subject.

2. Synchronous Fluorescence and Related Procedures

The fact that the phenomenon of photoluminescence involves two distinct spectra (excitation and emission) has long been recognized as an advantage of fluorescence over absorption techniques but is not exploited fully by conventional fluorometric instrumentation. A relatively simple method, termed *synchronous scanning*, was first employed by Lloyd [68] in forensic applications of fluorometry. In the simplest synchronous fluorescence technique, both the excitation and emission monochromators are scanned simultaneously, with a constant wavelength difference $\Delta\lambda$ maintained between the monochromator settings. At any given time, a fluorescence signal is observed in this procedure for a given compound only when both the excitation and emission wavelengths fall within the respective spectra for that compound.

Synchronous fluorescence offers important advantages over conventional fluorometry for both qualitative and quantitative analyses of complex samples. Provided that the proper value of $\Delta\lambda$ is chosen for an analyte, its synchronous spectrum is much simpler than either the conventional excitation or emission spectra of that compound [69]; an example is shown in Fig. 11. Band overlaps obviously are reduced in the spectra of complex mixtures of fluorophores, as shown in Fig. 12. Reports in the literature describe application of synchronous luminescence spectrometry to real samples, including crude oil identification [70,71] and PAH analysis in wastewater from a coal gasifier [72].

Vo-Dinh has reviewed the principles and practice of synchronous luminescence spectrometry [73]; the reader is referred to that review for additional detail.

An extension of the synchronous luminescence concept, wherein a two-dimensional contour map of the information content of the excitation and emission spectra is prepared, has interesting possibilities for fluorometric analysis of complex mixtures. A comparison of a conventional spectral presentation with an "excitation-emission contour plot" is shown in Fig. 13. In the contour map, the excitation wavelength is plotted against the emission wavelength; the contour lines denote regions of equal fluorescence signal. A conventional fluorescence spectrum at a given excitation wavelength is equivalent to a horizontal slice through the contour map at that excitation wavelength. Similarly, a conventional excitation spectrum corresponds to a vertical slice through the contour plot at the emission wavelength in question. The contour map (referred to variously as an "excitation-emission matrix" [74] and a "total luminescence contour map" [75] in the literature) contains all the information present in the excitation and emission spectra of a fluorophore. It can be thought of as containing the information which would be obtained from a large number of synchronous scans at different $\Delta\lambda$ values.

Such contour plots can be much more useful than conventional presentations of emission or excitation spectra for "fingerprinting" individual compounds (as, for example, in oil-spill identification [76]). Computer algorithms for quantitative analysis by this procedure also are under development [77], and the technique promises to have broad applicability to quantitative fluoro-

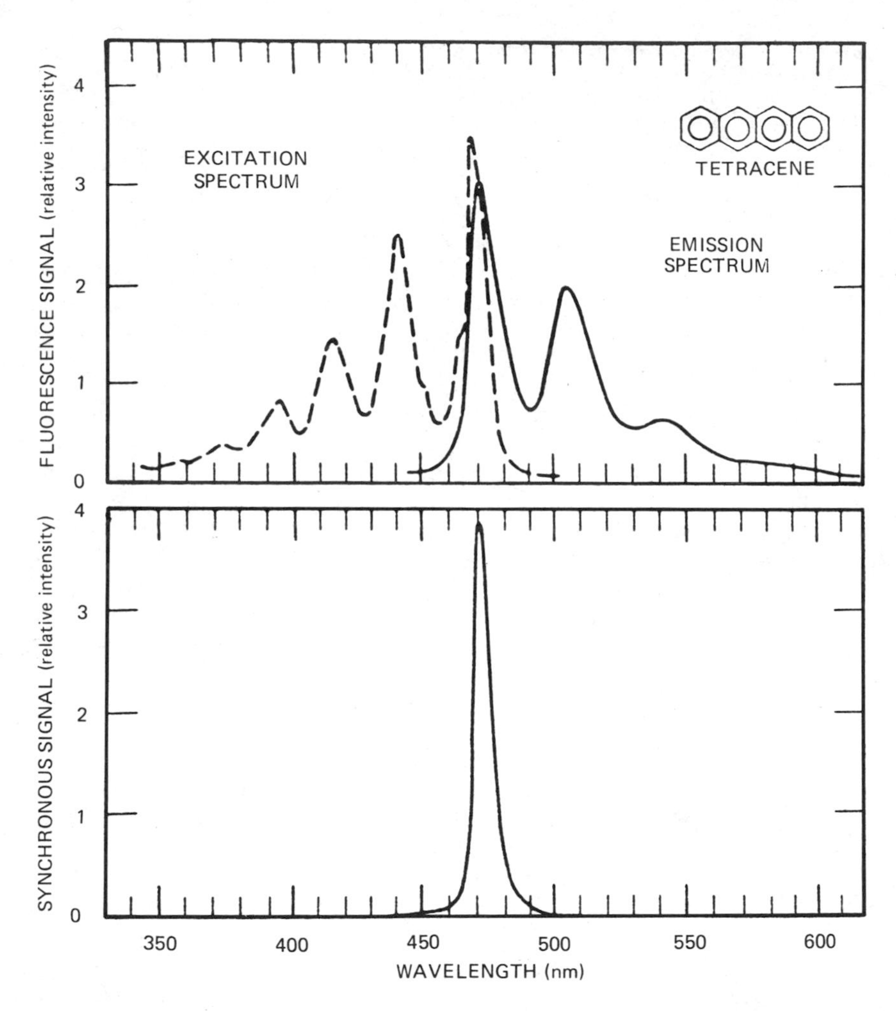

Figure 11. Conventional excitation (top left) and fluorescence (top right) spectra of tetracene in ethanolic solution, compared with the synchronous fluorescence spectrum of the same solution at the "optimum" $\Delta\lambda$ (3 nm). Note the much simpler appearance of the synchronous spectrum. (Reprinted with permission from T. Vo-Dinh, Multicomponent analysis by synchronous luminescence spectrometry. *Anal. Chem. 50*: 397. Copyright 1978, American Chemical Society.)

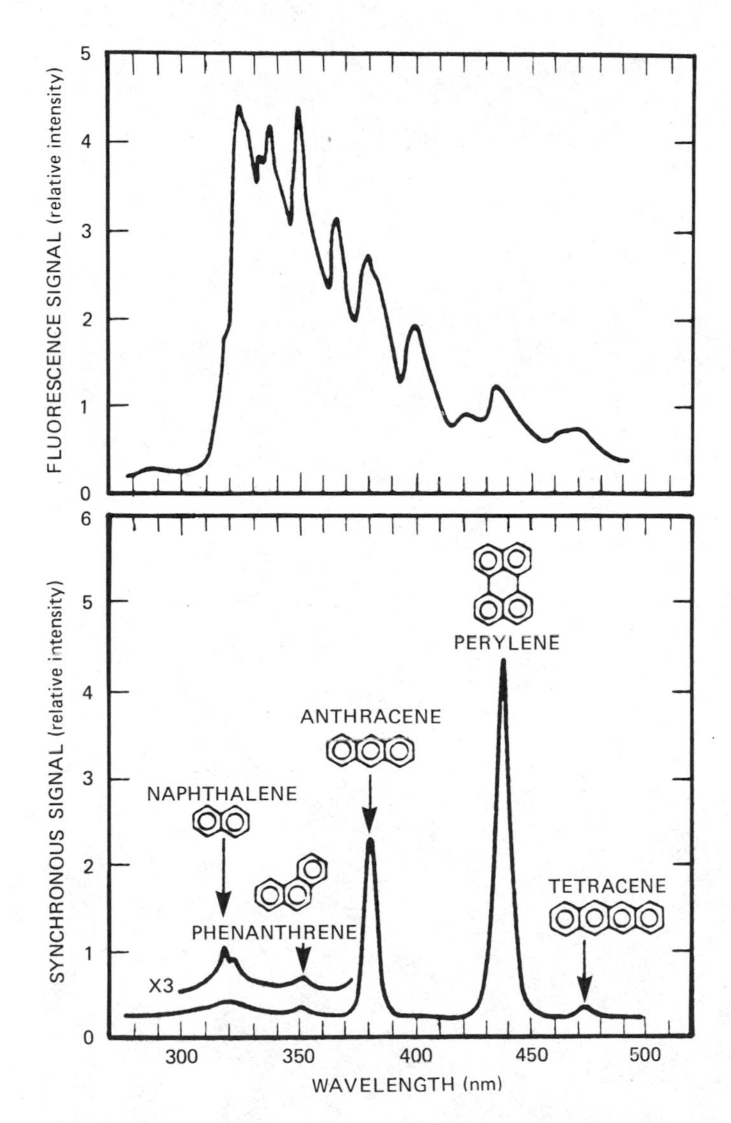

Figure 12. (a) Conventional fluorescence spectrum of a mixture of naphthalene, phenanthrene, anthracene, perylene, and tetracene. (b) Synchronous fluorescence signal for the same mixture ($\Delta\lambda$ = 3 nm). To some extent, the relative magnitude of the weaker features in the synchronous spectrum can be increased by varying $\Delta\lambda$. (Reprinted with permission from T. Vo-Dinh, Multicomponent analysis by synchronous luminescence spectrometry. *Anal. Chem.* *50*:399. Copyright 1978, American Chemical Society.)

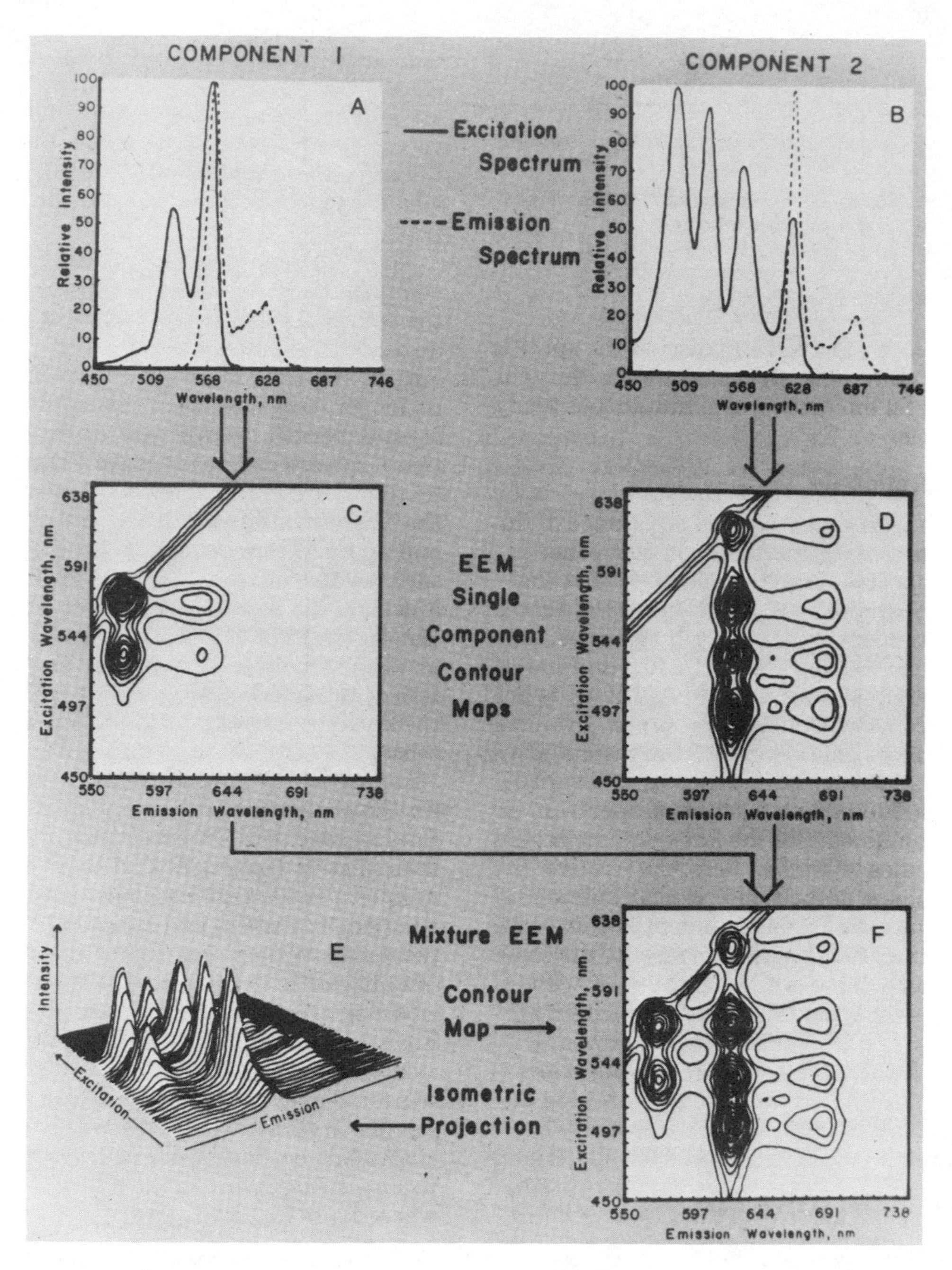

Figure 13. (A,B) Fluorescence excitation and emission spectra of zinc octaethylporphin. (C-F) Excitation-emission contour plots of these spectra. (Reprinted with permission from D. W. Johnson, J. B. Callis, and G. D. Christian, Rapid scanning fluorescence spectroscopy, *Anal. Chem. 49*:747A. Copyright 1977, American Chemical Society.)

metric analysis in mixtures. The procedure recently has been reviewed in depth by Christian and coworkers [78], and the reader is referred to that review for more detailed examination.

3. *Derivative and Wavelength Modulation Techniques*

The problem of band overlap in the fluorescence spectrometry of multicomponent samples can in some cases be alleviated by differentiating the spectra [79,80]. This effect is illustrated in Fig. 14, wherein it is noted that the derivative of a fluorescence spectrum may often be more useful than the spectrum itself for identifying a component which occurs spectrally as a shoulder of a band due to another sample constituent. Because derivatives can easily be taken electronically, this procedure can be implemented for any scanning fluorometer without modifying the optics or scanning hardware.

An extension of this procedure is *wavelength modulation*, wherein, as the excitation or emission wavelength is scanned, it is rapidly and repetitively displaced back and forth over a small interval $\Delta\lambda$. In most cases, the wavelength is modulated with a sinusoidal waveform, so that the operation can be thought of as introducing "AC ripple" to the linear (DC) scan of the monochromator [81-83]. In the simplest case, wherein $\Delta\lambda$ is very small relative to the widths of the spectral bands observed for the sample, wavelength modulation will produce a close approximation to the first derivative of the spectrum in question. The procedure also can be employed to achieve more sophisticated objectives. For example, Fig. 15 compares fluorescence spectra

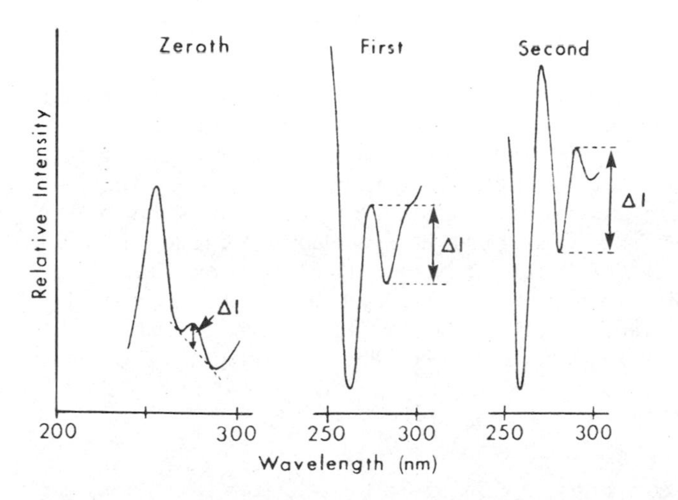

Figure 14. Comparison of fluorescence excitation spectrum of a mixture of 440 ppb pyrene and 360 ppb anthracene (in isopropanol solution) with its first and second derivatives. In the "zeroth derivative," the large band is due to anthracene and the small band to pyrene; note how the relative contribution of pyrene is accentuated by taking derivatives of the spectrum. (Reprinted with permission from G. L. Green and T. C. O'Haver, Derivative luminescence spectrometry. *Anal. Chem.* *46*:2195. Copyright 1974, American Chemical Society.)

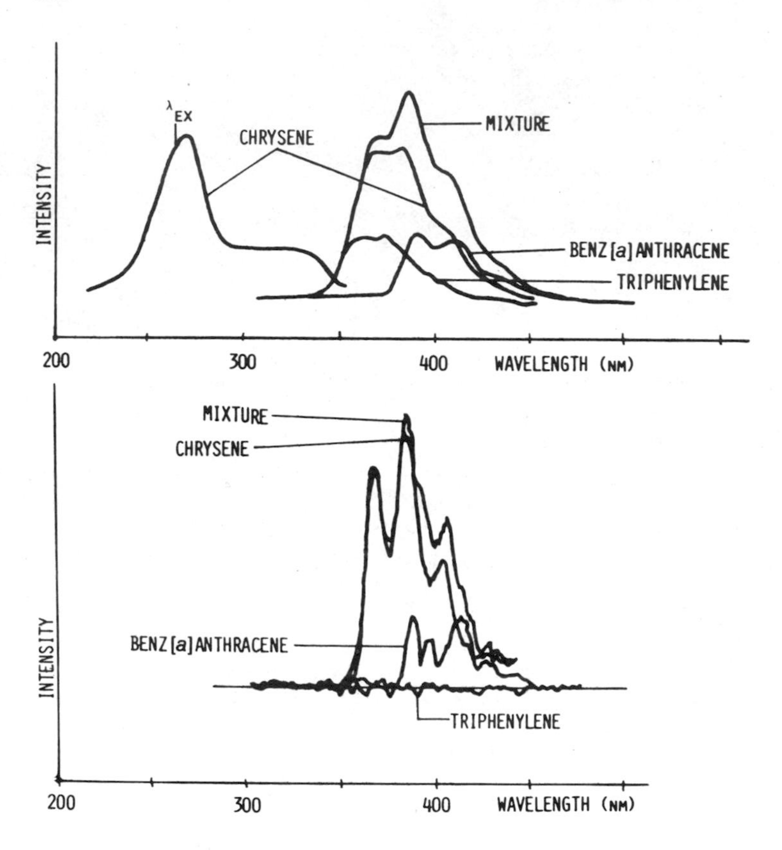

Figure 15. Comparison of selective excitation (top) and "selective modulation" (bottom) of 0.09 ppm chrysene in cyclohexane solution in the presence of equal concentrations of its isomers benz[a]anthracene and triphenylene. Note that, by performing the wavelength modulation experiment in the proper manner, the spectral interference from the other two compounds can be virtually eliminated, whereas this is not possible by selective excitation (due to the degree of overlap in the excitation spectra of the three compounds). (Reprinted with permission from T. C. O'Haver and W. M. Parks, Selective modulation: A new instrumental approach to the fluorometric analysis of mixtures without separation. *Anal. Chem. 46:*1893. Copyright 1974, American Chemical Society.)

for chrysene in a three-component PAH mixture by selective excitation (top) and "selective modulation" [83] (bottom). It is evident that the wavelength modulation technique is much more successful in eliminating the spectral interference than is the attempt at selective excitation. The ability to perform selective modulation requires that the modulation frequency be adjusted to a value which depends upon the positions and widths of the absorption bands of the analyte and the interfering compounds. Thus, spectra of authentic samples of these compounds must already available to effectively implement the procedure. Modification of the scanning hardware of commercial spectrofluorometers to enable performance of wavelength modulation fluorometry is not difficult [83]; the only additional electronic apparatus required beyond that associated with a conventional fluorometer is a lock-in amplifier. The use of this technique for fluorometric analysis of multicomponent samples has been discussed carefully by O'Haver [81-83], to which the reader is referred for additional detail.

4. Low-Temperature Luminescence Spectrometry

Thus far, all fluorometric analyses for PAH considered in this chapter have involved liquid solutions as the sampling medium. It has long been known, however, that important analytical advantages are offered by the use of low-temperature solid matrices in molecular fluorescence spectrometry. The two paramount advantages of low-temperature over liquid-solution samples in fluorometry are the following:

1. Fluorescence excitation and emission spectra in cryogenic solids tend to exhibit much greater resolution of vibrational fine structure than do the corresponding spectra in liquid solution (see Fig. 16 for an example). Thus,

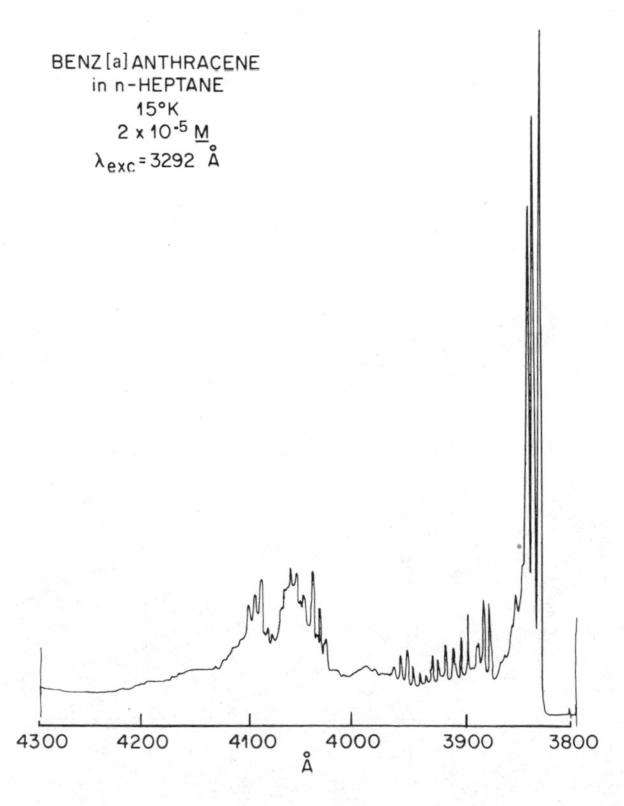

Figure 16. Shpol'skii fluorescence spectrum of benz[a]anthracene in an n-heptane frozen solution at 15 K.

the extent of band overlap in fluorescence spectra of mixtures is reduced by using low-temperature matrices. Moreover, the increased resolution in excitation spectra facilitates the use of selective excitation for fluorometric analysis of mixtures in low-temperature solid samples.

2. The "rigidity" of a cryogenic solid minimizes or completely eliminates fluorescence quenching, provided that aggregation of solute molecules in the low-temperature sample can be avoided. Since quenching ultimately is the factor limiting the accuracy of quantitative fluorometric analysis in complex samples, its elimination greatly enhances the ability of fluorescence to deal with real samples without lengthy preliminary sample cleanup steps.

The subject of low-temperature fluorometric analysis has recently been reviewed in detail by Wehry and Mamantov [84], and only the most salient features of the techniques can be discussed here. Low-temperature fluorometric analysis generally proceeds either by *frozen-solution* or *matrix-isolation* techniques, each of which is considered briefly next.

a. Frozen Solution Fluorometry: The Shpol'skii Effect The easiest way to prepare a low-temperature solid from a liquid solution is simply to freeze the solution, to a temperature of 77 K (liquid nitrogen) or, preferably, lower. Such procedures are routine in phosphorescence analysis (see Sec. III.F) and also have important applications in fluorescence. The most important frozen-solution fluorometric technique is based on a phenomenon commonly termed the *Shpol'skii effect*. It was first observed by Shpol'skii and co-workers that exceptional resolution of fine structure is observed in the low-temperature fluorescence spectra of PAH in certain solvents (usually straight-chain alkanes). This phenomenon has been studied in great detail by spectroscopists, especially in the USSR; these studies have been reviewed in detail elsewhere [85-87].

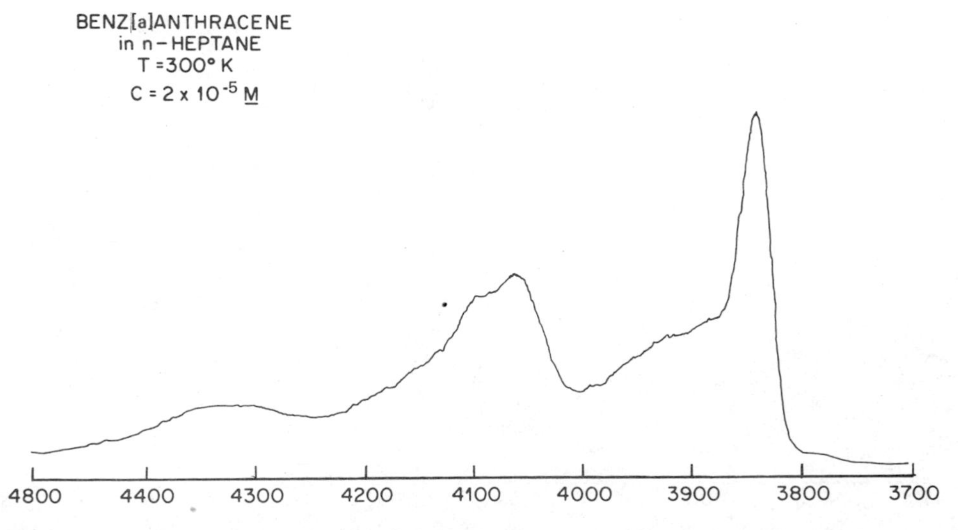

Figure 17. Room-temperature fluorescence spectrum of benz[a]anthracene (concentration = 2×10^{-5} M) in h-heptane.

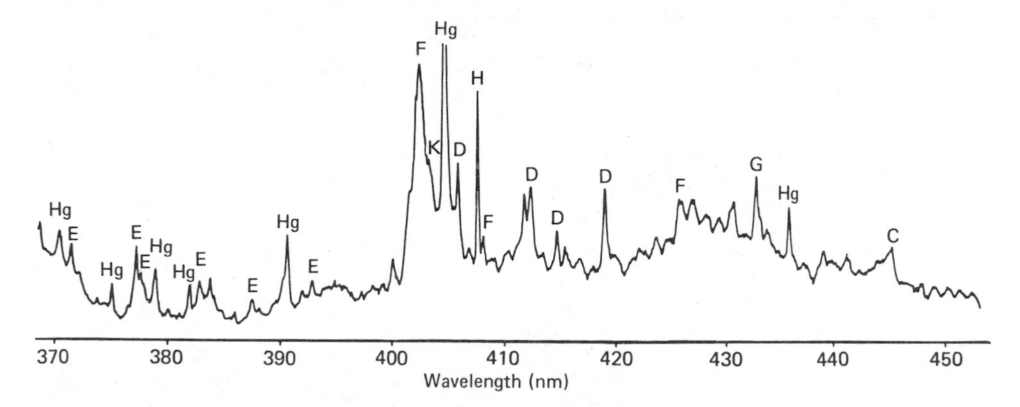

Figure 18. Shpol'skii luminescence spectrum (in n-hexane at 77°K) of a cyclohexane extract from coal-tar pitch. Identified compounds: C, perylene; D, benzo[ghi]perylene; E, pyrene; F, benzo[a]pyrene; G, dibenzo[ai]pyrene. Here, "Hg" denotes stray light (mercury lines from the mercury-xenon lamp used to excite the fluorescence). (From Ref. 88.)

Low-temperature (Shpol'skii) and room-temperature fluorescence spectra of benz[a]anthracene in n-heptane are shown in Figs. 16 and 17, respectively; the increase in spectral resolution produced by freezing the liquid solution is very dramatic. It must be emphasized that such enormous enhancements in resolution occur, for any particular PAH, only in certain very specific solvents. It is believed that the PAH solute occupies substitutional positions in the polycrystalline lattice formed by freezing the liquid solution; the Shpol'skii effect generally is observed only in solvents whose carbon skeleton rather closely matches the longest dimension of the PAH in question [85]. Therefore, different solvents are optimal for PAH of different molecular dimensions, and this fact can cause some difficulty in fluorometric analysis of a complex PAH mixture by Shpol'skii frozen-solution fluorometry. A more detailed discussion of the fundamental origin of the Shpol'skii effect is given elsewhere [84].

The "quasi-linear" spectra obtained in Shpol'skii matrices can be exceedingly useful for fingerprinting individual PAH in complex samples. For example, Fig. 18 shows the Shpol'skii fluorescence spectrum of an extract from coal-tar pitch [88]; five PAH can readily be identified from this fluorescence spectrum obtained at a single excitation wavelength. Examination of Shpol'skii fluorescence spectra of samples of this complexity at a number of different excitation wavelengths usually is necessary to achieve a reasonably "complete" qualitative profile of the PAH present.

It is important to emphasize that use of Shpol'skii low-temperature matrices brings about dramatic sharpening of excitation, as well as fluorescence, spectra. Thus, the possibilities for carrying out selective excitation of fluorescence from individual constituents of complex samples are much greater in Shpol'skii matrices than in liquid solutions. Under practical analytical conditions, full exploitation of the possibilities for selective excitation usually

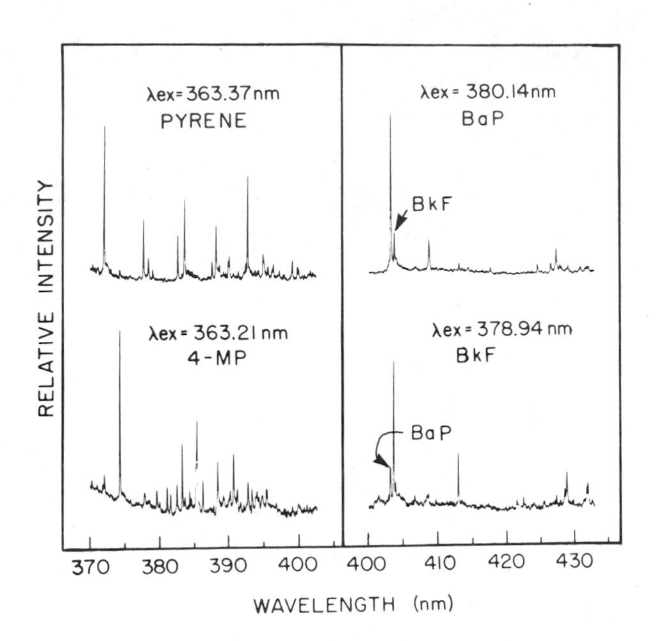

Figure 19. Dye-laser excited fluorescence spectra, at four different excitation wavelengths, of a solvent refined coal liquid sample, in an n-octane matrix at low temperature. The spectra are very similar to "pure compound" spectra of the compounds in question, despite the complexity of the sample. (Reprinted with permission from Y. Yang, A. P. D'Silva, V. A. Fassel, and M. Iles, Direct determination of polynuclear aromatic hydrocarbons in coal liquids and shale oil by laser-excited Shpol'skii spectrometry. *Anal. Chem.* 52:1351. Copyright 1980, American Chemical Society.)

requires that a tunable laser be used as the fluorometric source. For example, Fig. 19 shows Shpol'skii fluorescence spectra, obtained at four different excitation wavelengths (obtained via a dye laser), from a coal-derived material [89]. The spectra obtained under these conditions in the complex sample are very similar in appearance to those observed in the same Shpol'skii matrix for pure samples of the four PAH in question. Though lasers are not without their disadvantages (see Sec. III.E.1), their use in the fluorometric analysis of complex samples in Shpol'skii low-temperature matrices offers intriguing possibilities which have yet to be exploited fully.

Fassel and D'Silva and co-workers have employed x-rays, rather than (conventional) UV or visible photons, to excite luminescence spectra of PAH in Shpol'skii frozen-solution samples [90,91]. It is not feasible to excite luminescence from individual compounds in complex samples by this technique. Instead, the x-ray-excited luminescence spectrum will contain contributions from all, or virtually all, luminescent PAH present in the sample (including compounds for which luminescence may be difficult to excite efficiently with UV or visible photons). Thus, this method is particularly suitable for obtaining qualitative PAH profiles of complex samples in a relatively short time and has been used for this purpose with shale oil, airborne particles, residues from coal liquefaction processes, and other complex PAH-containing samples [91].

The more conventional optically excited Shpol'skii fluorescence method has been used to identify specific PAH in a variety of complex real samples, including coal and coal-tar pitch extracts [88], coal liquefaction products [89], automobile exhaust [92], jet airplane engine emissions [93], and extracts from marine sediments [94].

The use of Shpol'skii fluorometry for quantitative analysis of PAH often has tended to be problematic. As discussed in detail elsewhere [87], the appearance and intensity of a Shpol'skii spectrum may depend upon such factors as the rate at which the solution is frozen, the concentration of the analyte PAH, and the identities and concentrations of other solutes present in the sample. Many PAH are sparingly soluble in organic solvents at room temperature; the solubility decreases further with decreasing temperature. Hence, formation of solute aggregates, or even microcrystallites, may accompany frozen-solution formation. These occurrences usually are accompanied by spectral band broadening and fluorescence quenching, and high quantitative precision and accuracy is particularly difficult to achieve under such conditions [87]. Optimal quantitative results in Shpol'skii fluorometric analyses usually are achieved when the original liquid solution is frozen very rapidly, preferably to a temperature below 50 K, and when the initial liquid solution is very dilute (or else contains analytes which are very soluble in the particular solvent used).

It should be stressed that not all analytical applications of fluorometry in frozen solutions are based on the Shpol'skii effect. An alternative approach, site-selection fluorometry, is discussed in the following section.

b. Matrix-Isolation Fluorometry The technique of matrix isolation represents an entirely different approach to the preparation of samples for low-temperature fluorometric analysis. In matrix isolation, a solid or liquid sample is vaporized (usually under vacuum) and the resulting vapor is mixed thoroughly with a large excess of a diluent gas (termed the "matrix gas") which serves as the "solvent." The gaseous mixture then is deposited as a solid on a surface at very low temperature (usually 20 K or less). Any material which has an appreciable vapor pressure at room temperature can be used as the matrix gas; if the matrix is a liquid at room temperature, it is not required that the sample constituents be soluble in that liquid. Conventional matrices for this form of spectroscopy include the rare gases (especially argon) and nitrogen; organic solvents also can be used as matrix materials.

The technique of matrix isolation, originally developed for the spectroscopic study of transient and reactive entities such as free radicals, has been discussed in detail elsewhere as a sampling procedure for analytical spectroscopy [84,95], and only the principal aspects are considered here. There are several reasons for considering matrix isolation as an alternative to frozen-solution procedures in analytical fluorescence spectrometry. First, as already noted, solubility restrictions no longer are relevant, and the number of possible "solvents" for a particular low-temperature fluorometric analysis therefore is greatly increased. Second, the main objective of matrix isolation, implied by its name, is to ensure that in the final spectroscopic solid sample the solute molecules are thoroughly separated from each other and "see" only matrix molecules as near neighbors. Under these conditions, fluorescence

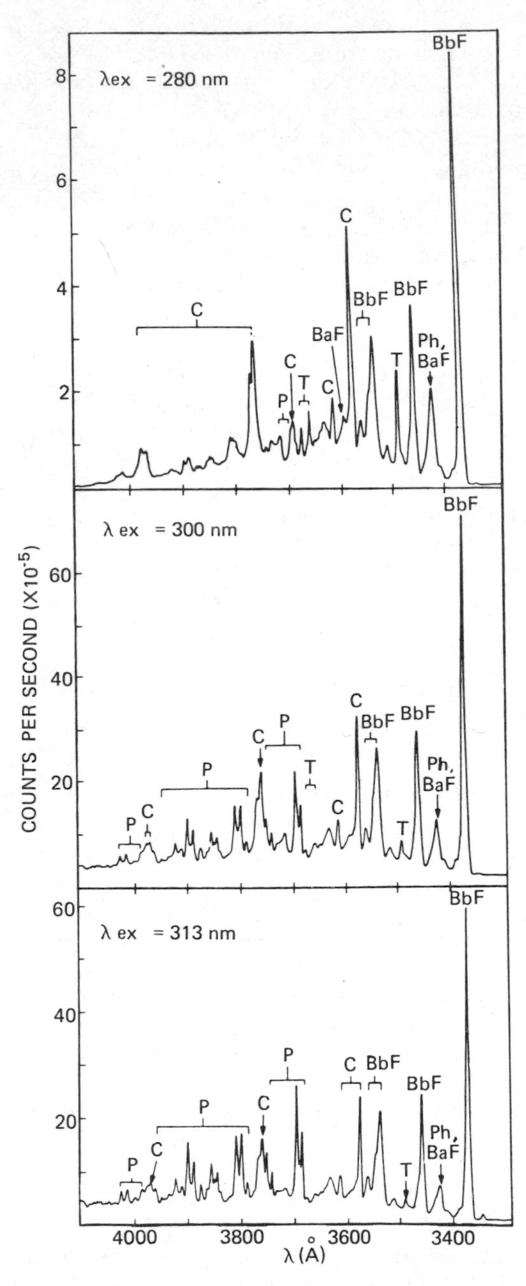

Figure 20. Matrix isolation fluorescence spectra of six-component mixture in a nitrogen matrix at 16 K. Compounds: benzo[b]fluorene (BbF), benzo[a]fluorene (BaF), chrysene (C), pyrene (P), triphenylene (T), and phenanthrene (Ph). (Reprinted with permission from R. C. Stroupe, P. Tokousbalides, R. B. Dickinson, Jr., E. L. Wehry, and G. Mamantov, *Anal. Chem.* *49*:702. Copyright 1977, American Chemical Society.)

quenching can be totally suppressed and quantitative fluorometric analysis can therefore proceed in the absence of quenching or sensitization artifacts. Finally, matrices which are gases at room temperature, such as nitrogen or argon, are more readily purged of luminescent impurities than are liquid solvents; passage through a series of cold traps is very effective in removing fluorescent contaminants from argon or nitrogen. Because background fluorescence from the solvent often limits the quantities of PAH which can be detected fluorometrically, matrix isolation in conventional matrices offers the possibility for enhanced sensitivity. It must be stressed, however, that the highest spectral resolution in matrix-isolation fluorometry often is observed when an organic matrix (such as a Shpol'skii solvent) is used; in that case, the fluorescent impurity problem is no less serious than in conventional fluorometry.

In Fig. 20 are shown matrix-isolation fluorescence spectra of a six-component synthetic mixture of three- and four-ring PAH in a nitrogen matrix, at three different excitation wavelengths [87]. These spectra demonstrate the resolution which can be obtained in conventional matrices, as well as the use of changes in the excitation wavelength to bring out the fluorescence of individual components of mixtures. A matrix-isolation fluorescence spectrum of a real sample (an adsorption chromatography fraction from a coking-plant water sample) in nitrogen is shown in Fig. 21. This sample

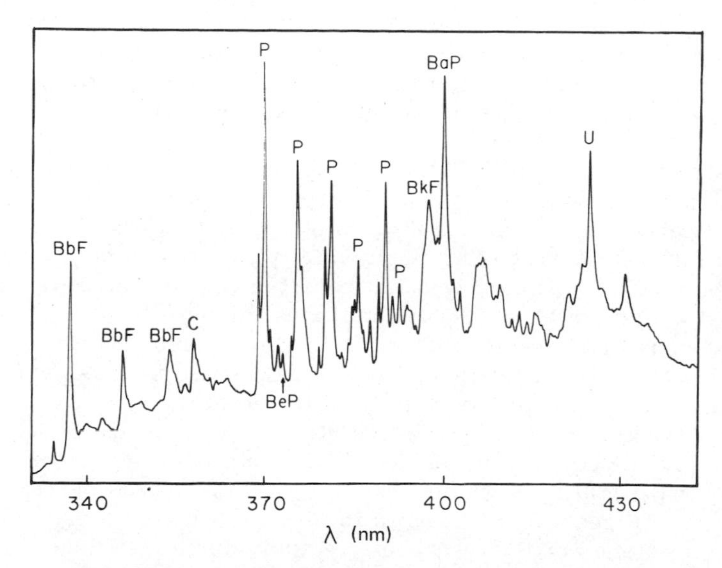

Figure 21. Matrix isolation fluorescence spectrum of a chromatographic fraction from a steel mill coking plant water sample. Matrix: N_2. Excitation wavelength: 313 nm. Excitation source: xenon-mercury lamp. Identified compounds: BbF, benzo[b]fluorene; C, chrysene; BeP, benzo[e]pyrene; P, pyrene; BkF, benzo[k]fluoranthene; BaP, benzo[a]pyrene. U denotes band due to unidentified sample constituent. (Reprinted with permission from E. L. Wehry and G. Mamantov, Matrix isolation spectroscopy. *Anal. Chem.* *51*:649A. Copyright 1979, American Chemical Society.)

contained at least 25 PAH, yet bands assignable to individual PAH are easily identifiable in the spectrum.

The quantitative aspects of matrix-isolation fluorometry have received detailed study [87,95-97]. Because the concentration of analyte in the actual spectroscopic sample is not known (and is in any case of no particular interest), detection limits and linear working ranges in matrix-isolation fluorescence are specified in weight, rather than concentration, units. In favorable cases, subpicogram detection limits have been achieved by matrix isolation, and linear working regions of six decades or more in the amount of an individual PAH have been obtained [96]. Quenching and inner-filter effects on quantitative analyses are eliminated if the experiment is performed properly (i.e., if the mole ratio of matrix gas to sample is sufficiently great that isolation is achieved). For example, Fig. 22 compares analytical calibration curves for four different four-ring PAH in equimolar mixtures. The calibration curves are linear and parallel; moreover, the curve for each individual compound in the mixture is virtually superimposable upon that for the same compound in pure form (e.g., chrysene in Fig. 22).

Comparison of Figs. 16 and 20 shows that the spectral resolution attainable by matrix isolation in conventional matrices is less than that which can be observed in Shpol'skii frozen-solution systems. Two different approaches can be used to circumvent this difficulty. The first is simply to use, as the matrix, the vapor of an organic solvent which is a "Shpol'skii solvent" for the PAH of interest [96,98]. The resulting quasi-linear spectra can be used for analytical purposes in precisely the same ways as described for Shpol'skii frozen-solution spectra. For example, Fig. 23 compares the laser-induced fluorescence spectra of perylene in nitrogen and heptane (vapor-deposited) matrices [96]; the resolution advantages of using a Shpol'skii matrix are obvious.

As in the case of frozen-solution Shpol'skii fluorometry, optimum advantage of the spectral resolution attainable in such matrices can be taken by use of a laser to excite the fluorescence. Figure 24 offers an example in which the fluorescence spectrum of a steel mill coking plant water chromatographic fraction, in an n-heptane matrix using dye-laser excitation, is compared with the spectrum obtained under exactly the same conditions for pure benzo[a]-pyrene [96]. It is obvious that, despite the complexity of the real sample (which contained at least 25 PAH), this technique has succeeded in selectively exciting fluorescence from benzo[a]pyrene without interference from any of the other sample constituents. By tuning the dye laser to the optimum excitation wavelength for a given PAH, similar results were obtained for a number of other PAH in this sample [96]. By way of contrast, a lamp-excited spectrum of the same sample in a nitrogen matrix is shown in Fig. 21, wherein it is obvious that selective excitation of fluorescence from individual constituents of the sample was not achieved.

Of course, the main disadvantage of the Shpol'skii procedure (irrespective of whether frozen solutions or vapor-deposited matrices are employed) is that different solvents are optimal for different PAH. Moreover, compounds which cannot readily be incorporated into substitutional positions within a solid matrix (such as alkylated PAH) may fail to exhibit quasi-linear spectra in any solvent. It would be very nice to be able to achieve resolution comparable to that observed in Shpol'skii matrices for *any* solute in *any* matrix. While

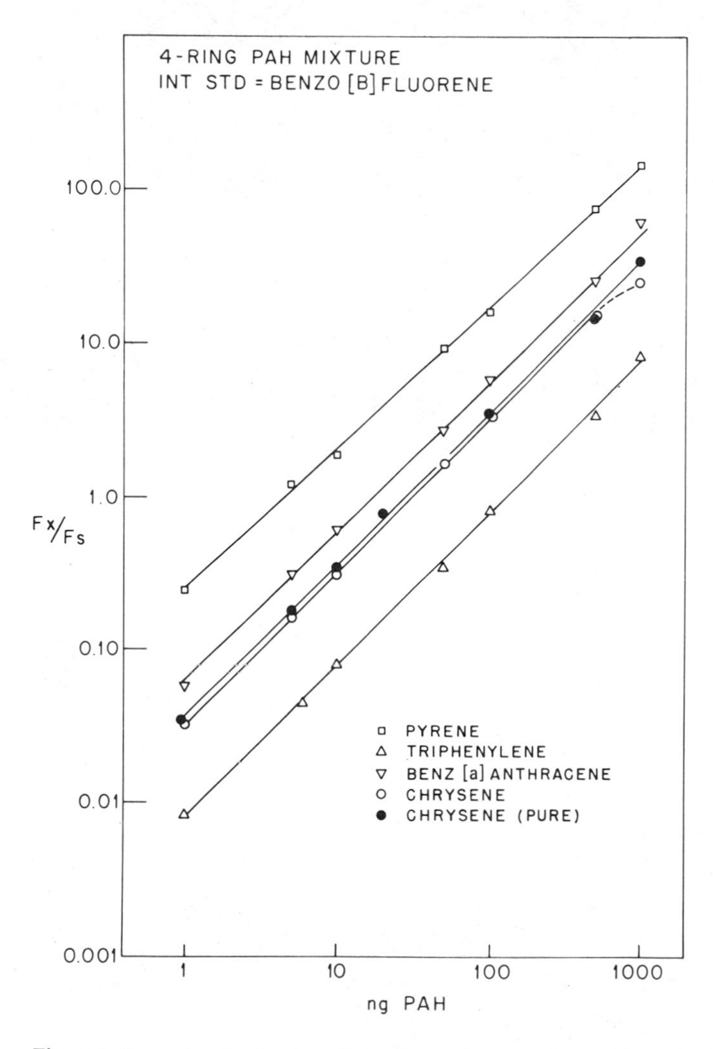

Figure 22. Analytical calibration curves for matrix isolation fluorometric determination of individual PAH in four-component synthetic mixture of four-ring PAH. Each sample contained an equal weight of each of the four components; the fluorescence intensity in each case was ratioed to that of an internal standard, benzo[b]fluorene. Shown for comparison is a working curve, obtained under identical conditions, for pure chrysene; the coincidence of the "pure compound" and "mixture" calibration curves for chrysene over three decades in quantity of PAH demonstrates suppression of quenching and inner-filter artifacts in the mixture by use of the low-temperature matrix. (Reprinted with permission from R. C. Stroupe, P. Tokousbalides, R. B. Dickinson, Jr., E. L. Wehry, and G. Mamantov, Low-temperature fluorescence spectrometric determination of polycyclic aromatic hydrocarbons by matrix isolation. *Anal. Chem.* *49*:703. Copyright 1977, American Chemical Society.)

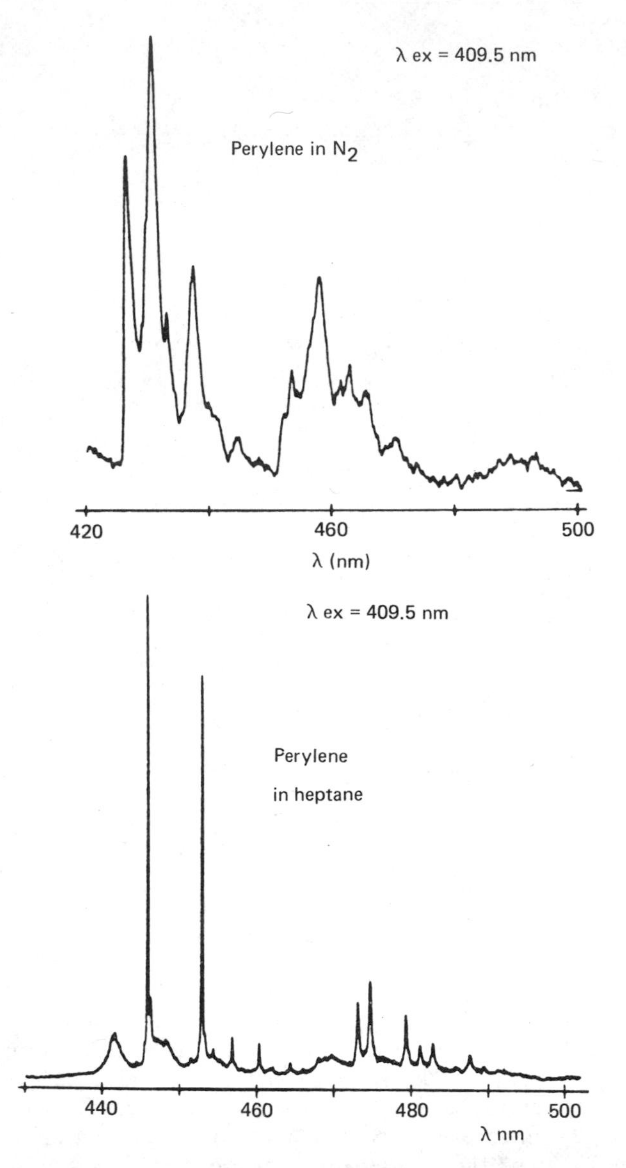

Figure 23. Dye-laser-excited matrix isolation fluorescence spectra of perylene in nitrogen (top) and n-heptane (bottom) matrices at 15 K. (Reprinted with permission from J. R. Maple, E. L. Wehry, and G. Mamantov, Laser-induced fluorescence spectrometry of polycyclic aromatic hydrocarbons isolated in vapor-deposited n-alkane matrices. *Anal. Chem.* *52*: 921. Copyright 1980, American Chemical Society.)

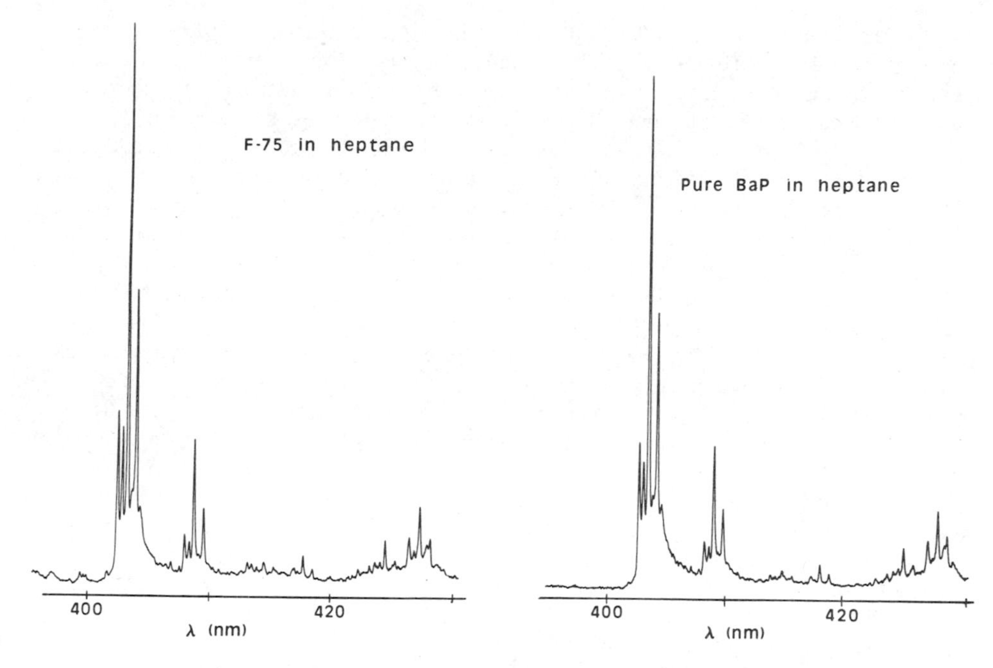

Figure 24. Dye-laser excited fluorescence spectra in n-heptane matrices at 15 K of chromatographic fraction from coking plant water sample (left) and pure benzo[a]pyrene (right). Note the similarity of the spectra. (Reprinted with permission from J. R. Maple, E. L. Wehry, and G. Mamantov, Laser-induced fluorescence spectrometry of polycyclic aromatic hydrocarbons isolated in vapor-deposited n-alkane matrices. *Anal. Chem. 52*:923. Copyright 1980, American Chemical Society.)

this goal as stated is somewhat utopian [99], procedures do exist by which very-high-resolution molecular electronic spectra can be obtained in many different matrices. One of the factors determining the widths of individual features in the fluorescence spectrum of a large organic molecule is the extent to which different molecules of the compound in question experience different microenvironments in the spectroscopic sample [84,99]. In the Shpol'skii effect, the extent of this "heterogeneous broadening" effect is minimized by using, as the matrix, a substance which forms a crystalline lattice into which the solute molecules "fit" in such a manner that they all occupy virtually identical sites. An alternative approach is to tolerate the occupation of a variety of different solute sites within the solid sample but to selectively excite only those molecules (of that particular solute) which occupy virtually identical lattice sites. This procedure, first described by Personov and co-workers [100], has become known variously as site-selection [99-103], fluorescence line narrowing [104], and (probably most accurately) energy-selection [105] spectroscopy. The principle of the method is to use exciting light having a bandwidth much smaller than the width of the inhomogeneously broadened absorption transition in the analyte, so that only solute molecules which presumably occupy virtually identical sites within the solid matrix have the opportunity to fluoresce. To accomplish this process under analytically realistic conditions requires use of a laser as excitation source and may in certain cases require the use of very low sample temperatures (less than 10 K) [106]. Other instrumental requirements of the technique and a more detailed discussion of its underlying principles can be found elsewhere [84,99]. Site selection has been observed both in frozen solutions and vapor-deposited matrices; an example of site selection in frozen solution media is shown in Fig. 25.

Small and co-workers have recently described analytical applications of site-selection fluorometry [104,107]. These authors have used, as solvent, a 1:1 glycerol/water mixture which forms a glass when frozen. Line-narrowed fluorescence spectra have been reported in this glassy solvent at 4.2 K for a variety of PAH (both as pure compounds and in mixtures). Site-selection fluorometric analysis in vapor-deposited matrices also has been reported [108]; for such matrix-isolation experiments, argon or "inert" organic materials such as fluorocarbons appear to be the most suitable matrices. Though a good deal of work remains to be performed before the analytical utility of site-selection fluorometry is fully established, the results reported thus far [104,107,108] are quite promising and it seems highly probable that the technique will emerge as one of very significant utility for identification and quantitation of PAHs in complex samples.

c. Summary We have devoted considerable attention in this chapter to low-temperature fluorometric techniques because they have already been employed widely in PAH analysis and because their use seems destined to increase significantly in the future. As emphasized in the preceding sections, low-temperature fluorescence provides much greater analytical selectivity than can usually be achieved by fluorometry in liquid solutions. Low-temperature fluorescence spectra usually are sufficiently characteristic to serve as fingerprints for individual PAH. Moreover, under proper circumstances, low-temperature matrices eliminate the fluorescence quenching and inner-filter effects which plague room-temperature quantitative fluorometric analyses of complex mixtures.

It must also be stressed that the low-temperature fluorescence methods suffer from several disadvantages. Foremost among these is the expense of the equipment required. To exploit the high spectral resolution achievable by low-temperature fluorometry, one needs a high-resolution emission monochromator. The vast majority of existing commercial spectrofluorometers exhibit inadequate resolution for effective use of low-temperature matrices (though it must be emphasized that some commercial instruments do exist which can provide the desired resolution). For optimum results in low-temperature fluorometry, a tunable laser is required for excitation. A cryostat of some sort also is required. For work at temperatures of 77 K or higher, a simple (and inexpensive) liquid nitrogen immersion Dewar usually is adequate. However, for best results in matrix-isolation or frozen-solution fluorometric analysis, it is generally necessary to employ lower temperatures. Temperatures as low as 15 K can be achieved by use of closed-cycle refrigerators [109] which are available commercially; while these devices are easy to use, they are not inexpensive. More sophisticated experiments, such as some site-selection spectral analyses, may require temperatures as low as

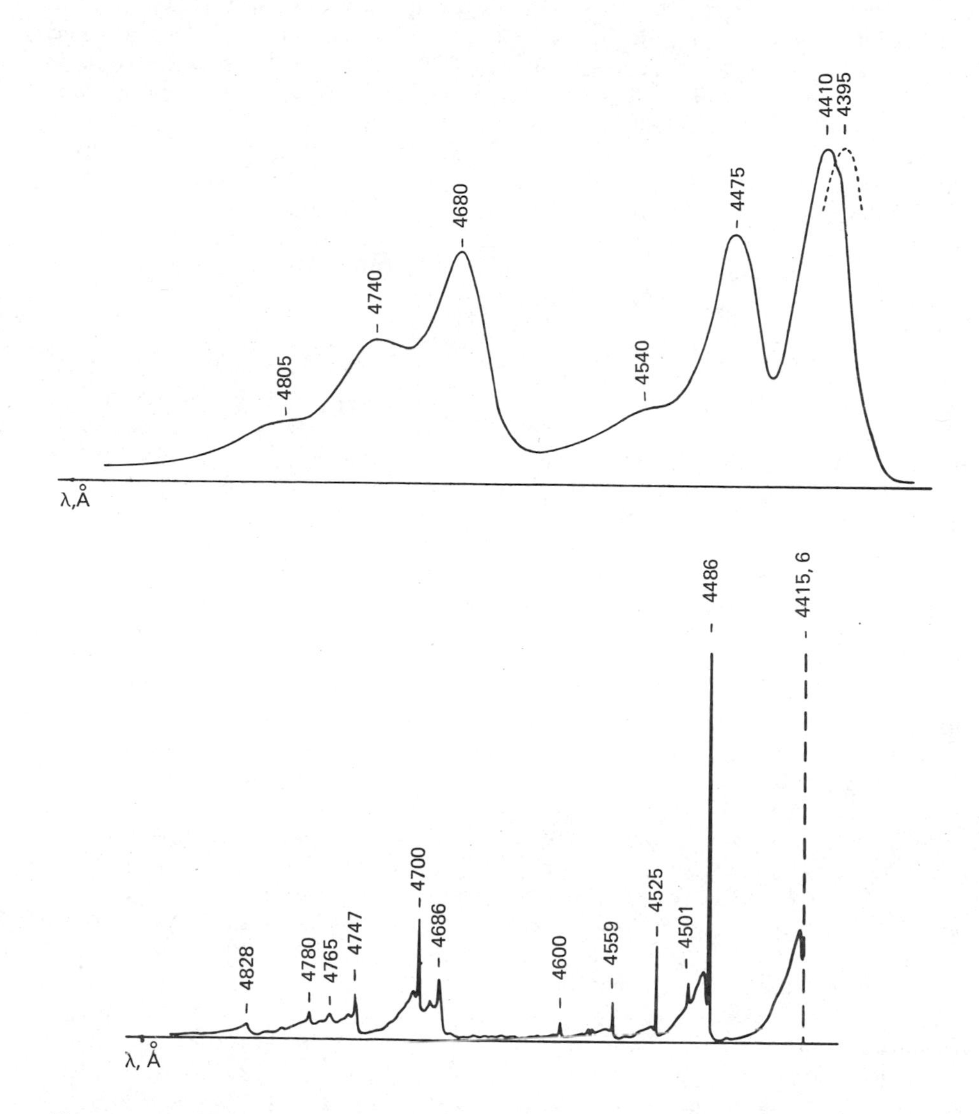

Figure 25. Fluorescence spectra of perylene in ethanolic frozen solutions at 4.2 K: (a) excited by mercury lamp; (b) excited by dye laser (λ_{exc} = 443.0 nm; bandwidth = 0.0125 nm). Note the high spectral resolution obtained with laser site-selective excitation. (From Ref. 100.)

4 K, for which a liquid helium cryostat is required. While these devices are not usually especially expensive, they are somewhat cumbersome to use [110] and the cost of the liquid helium is not inconsequential. The total cost of the apparatus required for detailed analytical studies by low-temperature fluorometry is such that only laboratories which perform large numbers of PAH analyses or which encounter samples of great complexity are likely to make the investment. It must be emphasized that the high-resolution capabilities of low-temperature fluorometry can greatly diminish the amount of sample cleanup required prior to analysis. Therefore, low-temperature fluorometric techniques should be capable of increasing the number of analyses which can be performed per unit time (as compared, for example, with UV-visible absorption or liquid-solution fluorescence). In that event, the initial investment in hardware will pay for itself rather quickly.

5. *Electronic Array Detectors*

In a conventional fluorescence spectrometer, an emission spectrum is scanned by the mechanical action of turning the diffraction grating in the emission monochromator. Within any instant of time, only one "resolution element" (the width of which, in wavelength units, is defined by the resolution characteristics of the monochromator and the width of the extrance and exit slits) is viewed by the detector. Such an instrument commonly is termed a "sequential scanning" spectrometer.

An alternative approach is to use an array of detectors, rather than a single detector, and use each individual detector to view one resolution element of the spectrum. If N detectors are used, N resolution elements of the spectrum can then be viewed at once, rather than in sequence. Such an approach is widely used in atomic emission spectrometry, in the "direct reader" (which uses one phototube per element) and the emission spectrograph (which uses photographic film for detection). Neither of these approaches lends itself to quantitative molecular fluorescence spectrometry. What is ideally needed is an electronic array detector (which may be thought of, rather crudely, as the electronic analog to a photographic plate). This electronic array device should be sensitive, exhibit a wide linear dynamic range, and lend itself readily to quantitative measurements. In fact, electronic array detectors do exist that exhibit at least some of these characteristics, and they are finding more and more use in fluorometric analysis.

Two general classes of electronic array detectors are available. First, various types of television camera pickup tubes, wherein the spectral information is scanned by a rastered electron beam, can serve as spectroscopic detectors. The most common device of this type is the *vidicon tube*, which exists in several variants. The principles and characteristics of vidicon detectors have been reviewed by Talmi [111,112] and Christian et al. [78]. A second class of electronic array detectors comprises a family of solid-state integrated-circuit devices, the most common of which is the self-scanned *photodiode array*. The characteristics of these detectors have been discussed by Horlick and Codding [113], Christian et al. [78], and Talmi and Simpson [114].

In a fluorometer equipped with an array detector, a fluorescence spectrum as dispersed by the emission monochromator is displayed across the face of the detector. No exit slit is used. Scanning the spectrum with such an instrument is an electronic operation (performed by the detector) rather than a mechanical procedure. Such detectors therefore are particularly useful in situations in which rapid acquisition of spectra is desired. For example, in an analysis based on reaction kinetics, in which the rate of change of fluorescence intensity or the nature of changes in a fluorescence spectrum as a function of time provide the data upon which the analysis relies, the use of an array-detector fluorometer can have significant advantages [115]. A second obvious situation requiring fast acquisition of spectra is the measurement of fluorescence of materials eluting from chromatographic columns [116]. Indeed, applications of vidicon tubes as fluorometric detectors in liquid [117, 118] and gas [119] chromatography already have appeared. This particular application of electronic array detectors is certain to increase greatly as such devices become more widely available to analytical chemists.

Array detectors probably will eventually supplant photomultiplier tubes as detectors in fluorometry [78], even when rapid spectral acquisition is not essential. At present, however, these devices are not devoid of shortcomings, the most obvious of which is cost. Moreover, some array detectors are appreciably less sensitive and exhibit a much narrower linear dynamic range than present-day photomultipliers, though the technology of array detectors is still developing. Further, many present-day array detectors have very small active areas; thus, the wavelength range which can be displayed across them without unacceptable loss in resolution may be undesirably small. In some cases, this inherent tradeoff between spectral resolution and range of wavelength coverage can be circumvented by optical modifications to the monochromator [120,121].

6. *Time Resolution*

The excited states responsible for fluorescence have finite lifetimes, usually in the nanosecond time regime. Therefore, in addition to excitation and emission spectra, fluorescence decay times also can be used as parameters for identification of sample constituents and for distinguishing between similar compounds. In time-resolved fluorescence spectrometry, an excitation or emission spectrum is obtained within a narrow "time window" much shorter than the decay time of the excited state in question. The experimental techniques for time-resolved fluorescence spectrometry have been reviewed in detail elsewhere [122-125].

An example application of time resolution in fluorometric PAH analysis is afforded by Fig. 26. As discussed in Sec. III.E.4, low-temperature fluorometric techniques are intended to provide spectral resolution of similar PAH, including isomers. However, that objective is not invariably accomplished. For example, when matrix isolated in nitrogen, the isomeric PAHs benzo[a]-pyrene (BaP) and benzo[k]fluoranthene (BkF) exhibit strong spectral overlap, which becomes particularly obnoxious when BkF is present in substantial excess over BaP (Fig. 26, left). It happens, however, that the fluorescence decay times of BaP and BkF are quite different (78 and 13 nsec, respectively, in nitrogen matrices at 15 K). Therefore, fluorometric determination of BaP in the presence of 100-fold excess quantities of BkF can be accomplished by time resolution, pupulating the excited states of the two compounds by a short pulse of light from a laser and then measuring the fluorescence within a time window of 75 psec in width, delayed 90 nsec with respect to the exciting

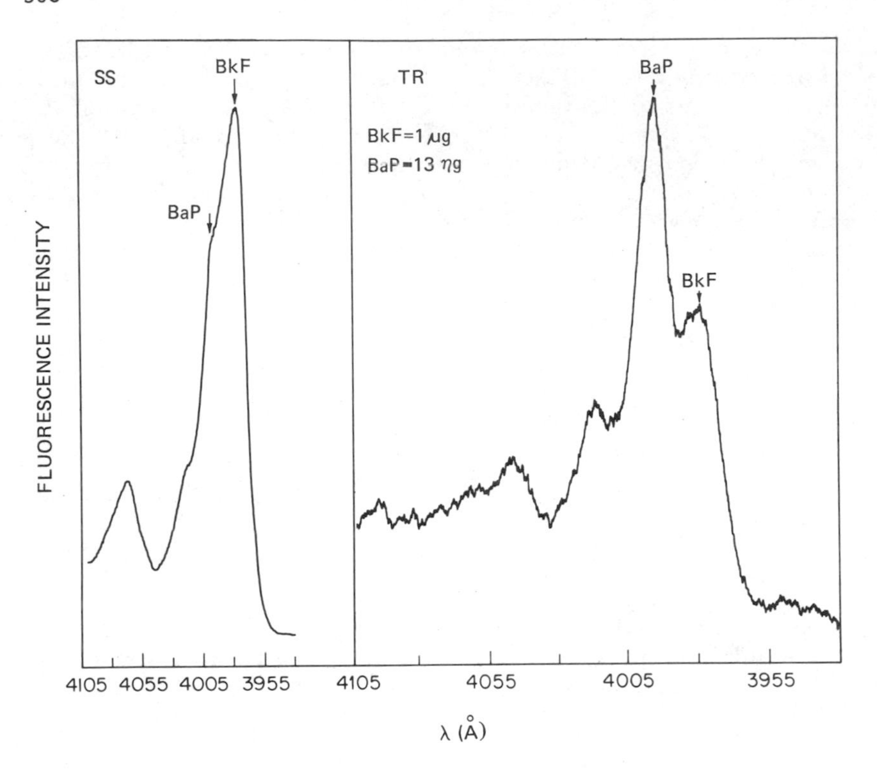

Figure 26. Conventional (left) and time-resolved (right) matrix-isolation fluorescence spectra of a sample containing 13 ng benzo[a]pyrene (BaP) and 1,000 ng benzo[k]fluoranthene(BkF). Because BaP has a much greater fluorescence decay time than BkF, BaP predominates in time-resolved spectra of the mixture at long delays, despite its almost total disappearance in the BkF band under conventional "steady-state" measurement conditions. (Reprinted with permission from R. B. Dickinson, Jr., and E. L. Wehry, Time-resolved matrix-isolation fluorescence spectrometry of mixtures of polycyclic aromatic hydrocarbons. *Anal. Chem. 51*:780. Copyright 1979, American Chemical Society.)

laser pulse [126]. Under those conditions, the fluorescence of BkF decays away virtually completely prior to measurement, and one is left with the signal from the much longer-lived BaP [126] (see Fig. 26, right).

Time resolution has thus far experienced relatively little application in analytical fluorescence spectrometry, for two reasons. First, the apparatus required for time-resolved fluorometry is complex and expensive, though recent innovations in measurement concepts and instrumentation [125,127-130] promise to alleviate that problem. Second, and more fundamentally significant, many compounds having very similar fluorescence and/or excitation spectra also exhibit virtually identical fluorescence decay times. Hence, in such cases temporal resolution may be more difficult to achieve than spectral resolution. However, it must be stressed that time resolution also has another important use in fluorometric analysis: distinguishing the fluorescence of the analyte from background fluorescence (e.g., of sample or solvent contaminants) or background scattering (Rayleigh or Raman) [131,132]. As time-resolution instrumentation becomes more mature and easier to use in practical analytical situations, its use certainly will expand.

7. *Two-Photon-Induced Fluorescence*

With the advent of high-power lasers as spectroscopic sources, it is possible to produce electronically excited states by "simultaneous" absorption of two photons (neither of which by itself is sufficiently energetic to effect excitation). Because the cross-section for two-photon excitation is very small, some of the "sensitivity" advantage normally associated with fluorometric analysis is lost whenever two-photon excitation is employed. However, two other important advantages may be derived by using two-photon excitation. First, the selection rules for one-photon and two-photon absorption processes are different, and transitions forbidden for one-photon absorption may be allowed for two-photon excitation [133]. In certain cases, therefore, excitation of fluorescence by two-photon absorption actually may be more efficient (in terms of the number of excited states produced by a given photon flux from the source) than conventional one-photon processes. Moreover, excitation at wavelengths for which high-power tunable output is not now readily available (e.g., in the deep UV) can be avoided by use of the two-photon approach (which obviously can proceed which photons of lower energy than one-photon absorption).

A second, and more important, advantage of two-photon excitation is that the fluorescence is always observed at higher energy (lower wavelength) than the absorption. This fact can be very important for samples which exhibit a very strong Rayleigh or Raman scatter background (the Stokes lines of the latter always are at longer wavelength than the incident light). In two-photon fluorescence, the fluorescence is well separated from and at shorter wavelength than the exciting light. Also, for samples exhibiting severe inner-filter effects (due to absorption by the solvent or interfering sample constituents), two-photon fluorescence can reduce the extent of nonlinearity in analytical calibration curves (which are usually obtained for strongly absorbing samples by conventional one-photon excitation) [133,134]. Therefore, the technique has interesting potential for use in fluorometric analysis of complex samples (particularly if an absorbing solvent is used—which is unthinkable in conventional fluorometry) and for fluorometric detection in liquid chromatography [135]. Its main shortcoming at present is the lack of high-power lasers in suitable wavelength regions, and its applicability to fluorometric PAH analysis will be dependent upon developments in laser technology.

8. *Summary*

This section has presented a quick overview of a number of rather recent developments in fluorometric analysis. While most of these techniques have been applied to PAH analysis, they are for the most part still in the process of development. The vast majority of fluorometric PAH determinations are now performed using conventional sequential scanning fluorometers in room-temperature liquid solution media. A number of newer techniques are now available which increase the analytical information available from fluorescence measurements and/or suppress some or all of the artifacts and interferences encountered in conventional fluorometric techniques. Much work remains to be done in establishing the utility of many of these more "modern" techniques to practical PAH analyses in real samples, but research activity in this area is now rather widespread and these techniques will become increasingly important in the future.

Developments in fluorometric instrumentation and technique are reviewed biennially (in even-numbered years) in *Analytical Chemistry* [136].

F. Phosphorescence Spectrometry

Thus far, we have virtually ignored phosphorescence in our discussion of PAH by luminescence methods. Until recently, virtually all analytical phosphorescence measurements [137] were performed in low-temperature media (usually at 77 K). The experimental difficulties presumed (not always accurately) to be associated with the low-temperature procedures have not generally been offset in any obvious way in the analysis of PAH by other advantages of phosphorescence over fluorescence. For most PAH, quantum yields for fluorescence are comparable to (and in many cases larger than) those for phosphorescence; of the common PAH, only triphenylene is known to exhibit a much larger quantum efficiency for phosphorescence than for fluorescence. Thus, examples in the literature of PAH analysis by phosphorescence in conventional frozen-solution media are rather uncommon. On the other hand, for many PAH derivatives (especially heterocycles, carbonyl compounds, and PAH having heavy-atom substituents such as bromine or iodine), the phosphorescence quantum efficiency may be much greater than that for fluorescence. This is especially likely to be true if the phosphorescence measurement is performed in a heavy-atom matrix [138]. For such compounds, there is an obvious advantage to phosphorescence over fluorescence for trace analysis.

The practice of analytical phosphorescence was greatly affected by the report of Schulman and Walling [139] in 1972 that certain organic molecules exhibited phosphorescence at room temperature when adsorbed on a variety of solid substrates, including alumina, silica gel, and chromatography paper. It has subsequently been reported by a number of investigators that a wide variety of compounds, including PAH, exhibit this phenomenon under the proper conditions. In a relatively short time, a large literature pertaining to this room-temperature phosphorescence (RTP) technique has appeared, much of which has been reviewed thoroughly by Vo-Dinh and Winefordner [140].

The "external heavy-atom effect" [138] operative in conventional frozen-solution phosphorescence also is observed in the RTP of aromatic solutes on solid surfaces [141-143]. Not only does the use of a heavy-atom-treated solid adsorbent (such as filter paper impregnated with an aqueous NaI solution) tend to increase the intensity of RTP, but the magnitude of the enhancement varies from analyte to analyte and from one heavy-atom matrix to another. Because both the absolute and relative RTP signals for different PAH in a mixture vary from one heavy-atom material to another, observation of the RTP spectrum for a multicomponent PAH sample in the presence of several different heavy atoms provides additional analytical selectivity [143]. This effect is particularly important in view of the fact that RTP spectra of most compounds are rather featureless (see Fig. 27 for an example) and spectral overlaps are common in multicomponent samples.

Figure 28 illustrates how the use of several different excitation wavelengths and several heavy-atom matrices can enable the analysis of a mixture to proceed in spite of the relatively featureless spectra which are obtained. While this may initially appear to require a great deal of work, the solid substrates can be specially prepared in batches ahead of time and the acquisition of an RTP spectrum of any given sample on a particular solid is rapid. In fact, the speed and simplicity of RTP (as compared with other spectroscopic techniques) is one of its principal attractions for routine analytical use. Moreover, it is relatively easy to modify commercial spectrofluorometers [144,145] or thin-layer fluorescent scanners [146] for use in RTP analysis.

Experimental procedures in RTP have been discussed thoroughly by Vo-Dinh and Winefordner [140]. Among recent innovations in RTP procedures are the use of synchronous excitation [147], second-derivative techniques

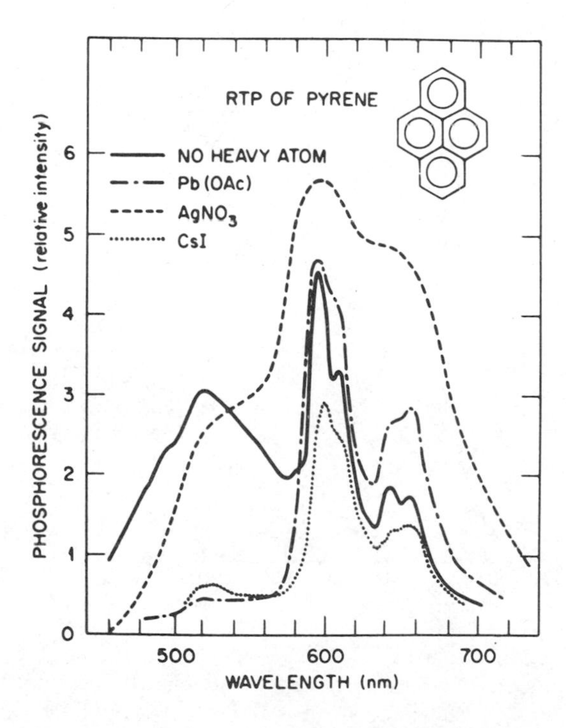

Figure 27. Room-temperature phosphorescence spectra of pyrene on filter paper as a function of the identity of the heavy-atom perturbing species. Note that changes in the heavy-atom species affect both the magnitude of the RTP signal and the nature of the spectrum. (Reprinted with permission from T. Vo-Dinh and J. R. Hooyman, Selective heavy-atom perturbation for analysis of complex mixtures by room-temperature phosphorimetry. *Anal. Chem.* 51:1917. Copyright 1979, American Chemical Society.)

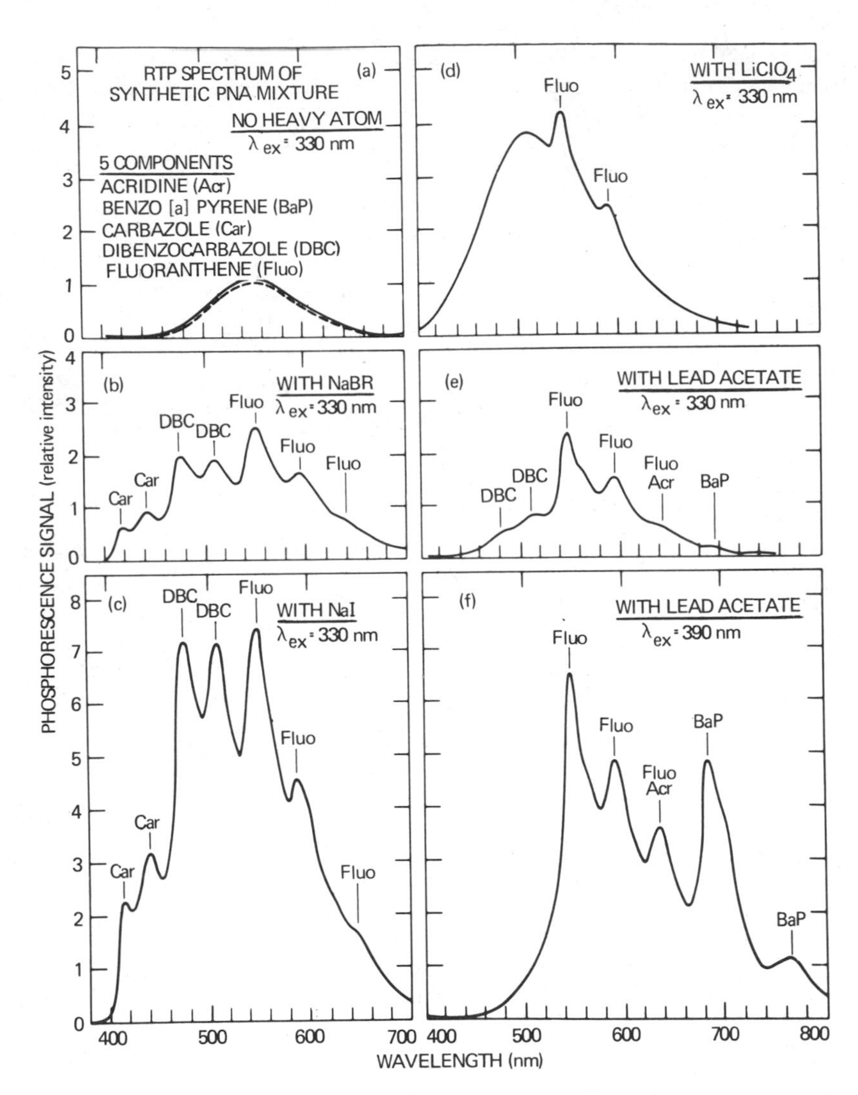

Figure 28. The use of different heavy-atom perturbers and variations in excitation wavelength to achieve selectivity in multicomponent sample analysis by room-temperature phosphorometry. The excitation wavelength and heavy-atom species used in each case is indicated. (Reprinted with permission from T. Vo-Dinh and J. R. Hooyman, Selective heavy-atom perturbation for analysis of complex mixtures by room-temperature phosphorimetry. *Anal. Chem. 51*: 1919. Copyright 1979, American Chemical Society.)

[148], and flow-through adsorbent cells [149]. It is certain that use of RTP for PAH analysis will expand, particularly in analytical situations requiring speed, simplicity of procedure, and low capital cost.

In 1978 it was reported that phosphorescence of organic molecules often can be observed in liquid solution at room temperature in heavy-atom solvents in the presence of micelle-forming surfactants [150]. This phenomenon has more recently been utilized in PAH analysis [151]; detection limits for PAH on the order of 10^{-6} M have been reported. This form of RTP is a liquid-solution technique which can be regarded as directly competitive with conventional solution fluorescence spectrometry. The possibilities for improved selectivity in solution RTP (as compared with solution fluorometry) by selective heavy-atom perturbation and differences in solubility of different PAH in micelles as a function of the identity of the surfactant and counterion are interesting, and further developments in this new RTP technique are anticipated.

One important characteristic of phosphorescence (as compared with fluorescence) is the much greater decay time of triplet states, thus greatly simplifying the instrumentation required to acquire time-resolved spectra [152]. Enhanced selectivity in multicomponent sample analysis by phosphorescence also can be sought by "phase resolution" [153]; in this procedure, an amplitude-modulated excitation source is used and the phase relationships between the time-dependent phosphorescence signal and the modulated incident light depend upon (among other parameters) the phosphorescence decay time of the analyte. Hence, this procedure can be thought of as a special case of time-resolved phosphorometry.

It is apparent that the analytical potential of the various phosphorescence techniques has not yet been thoroughly scrutinized. Interest in phosphorometric analysis of PAH has been rekindled by discovery of the solid-phase and solution RTP phenomena, and increasing analytical use of these techniques appears highly probable.

IV. Vibrational Spectroscopy

A. Introduction

Although the techniques of vibrational spectroscopy have been used widely in organic analysis (principally for qualitative purposes [12]), the literature dealing with application of infrared and Raman spectrometry to characterization of PAH and their derivatives is rather sparse. This situation has resulted in large part from the relatively low sensitivity of infrared and Raman techniques (as compared with UV-visible absorption or, particularly, fluorescence analysis). To quote from the Committee on Biologic Effects of Atmospheric Pollutants of the National Research Council (U.S.) [154]:

> The disadvantages of vibrational spectroscopy are the relatively weak bands, the fact that infrared band strengths are not proportional to concentration, the requirement for a vibrationally transparent medium, and the lack of unique polycyclic spectral features. The disadvantages far outweigh the advantages of these techniques.

This viewpoint, while historically accurate, represents something of an overstatement in view of recent developments in instrumentation and technique in vibrational spectroscopy. In this section, we consider some aspects of the limited literature dealing with applications of infrared and Raman spectrometry to PAH characterization and briefly examine some newer developments which should significantly increase the analytical attractiveness of these techniques.

B. Infrared Absorption Spectrometry

1. *Introductory Remarks*

Infrared (IR) absorption spectrometry is a well-established analytical technique, the fundamental principles and basic techniques of which are treated in detail elsewhere [12,155-158]. Because both UV-visible and IR absorption spectrometry involve the same basic measurement, albeit in different regions of the electromagnetic spectrum, it is useful to begin by comparing the relative advantages and disadvantages of the two techniques.

The most significant advantage of IR over UV-visible absorption spectrometry is the much greater information content of an IR spectrum, as well as the fact that the information is related in a more direct manner to molecular structure than is that of a UV-visible absorption spectrum. IR spectra contain many bands, and (except in special cases, such as intermolecular hydrogen bonding) those bands are quite sharp. Hence, while qualitative analysis is a relatively minor application of UV-visible spectrometry, it is the major analytical use of IR spectrometry.

It must be stressed that the use of IR as a "fingerprinting" technique—with the interpretation of IR spectra in terms of molecular structure—is an empirical procedure [12,157] usually requiring libraries of reference spectra of pure compounds. Although the principal functional groups tend to exhibit characteristic absorption bands within relatively narrow regions of the IR spectrum and can therefore usually be identified by use of functional-group correlation charts, it is rarely possible to assign a "complete" structure to an unknown compound on the basis of its IR spectrum unless a library spectrum of that compound, obtained under similar conditions, is available for comparison [159]. In practice, therefore, IR often is used as one of several tools (the others being mass spectrometry and nuclear magnetic resonance) for "proving" structures of unknown molecules.

For quantitative analysis, the disadvantages of IR (relative to UV-visible) techniques are quite numerous:

1. As noted above, molar absorptivities of IR bands usually are smaller than those of the strongest bands in the UV-visible absorption spectrum of a PAH. This situation exacerbates the sensitivity limitation inherent in any form of absorption spectroscopy (Sec. III.A).

2. Sampling media and sample containers for the IR often are quite "nonideal." Virtually all organic solvents absorb strongly in portions of the IR and effectively obliterate those spectral regions; those solvents which are transparent throughout most of the region (such as CCl_4 and CS_2) tend to be poor solvents for most organic compounds. Most materials sufficiently transparent in the IR to be used as sample containers are susceptible to chemical attack by polar materials, including water. Thus, the path thickness of IR cells for solutions usually cannot be regarded as known or constant and must be redetermined at frequent intervals for quantitative work. Very thin cells (on the order of millimeters) often must be used because of the solvent absorption problem, thus exacerbating the sensitivity problem.

Other IR sampling methods also suffer from significant disadvantages for quantitative applications. KBr disks, by the use of which one can avoid the "solvent problem," are very useful for qualitative analysis but are virtually useless for quantitative measurements. Mulls, such as mineral oil, obliterate portions of the spectrum and also are not well suited to quantitative measurements.

3. Relatively weak sources and insensitive detectors (compared with those used in the UV-visible) result in signal-to-noise performance inferior to that encountered in the UV-visible. These problems are, to a significant extent, overcome by use of Fourier transform techniques in IR spectrometry, as noted below.

4. The use of literature values of molar absorptivities for quantitative analysis by IR absorption is even less advisable than in the case of UV-visible absorption (see Sec. II.A). IR follows the same fundamental quantitative relationship (i.e., Beer's law) as does UV-visible absorption. In this regard, the sharpness characteristic of IR absorption bands, which is exceedingly important for applications of the technique to qualitative analysis, is a serious hindrance for quantitative measurements. The problem is that the instrumental resolution required to avoid distortion of an IR spectrum usually cannot be achieved without unacceptable loss in analytical sensitivity. As a result, IR band intensities are extremely dependent upon the instrument used to measure them. They can be corrected for finite resolution errors if the instrumental resolution (or "instrument line shape function," as it often is termed) is accurately known, but such computations can be rather tedious for routine analytical purposes. Therefore, for most quantitative applications of IR spectrometry, it is absolutely essential that calibration (e.g., acquisition of a Beer's law plot) and measurement of the absorbance of the analyte be performed with the same spectrometer. Such procedures do not necessarily require pure samples of the analyte compounds; methods for quantitative calibration in IR spectrometry which require only several mixtures containing different relative amounts of the compounds in question, rather than pure samples of each, have been devised [160].

A good practical guide to these and other problems in quantitative IR analysis (and techniques to minimize the effects of these difficulties) has been presented by Smith [156]. It must be stressed that new instrumental developments have reduced the severity of many of these problems, as noted in the following subsection.

2. *Applications of Dispersive and Fourier Transform IR Spectrometry to PAH Analysis*

Applications of IR spectrometry to PAH analysis using conventional dispersive (grating or prism) spectrometers have not been numerous; most have involved identification of functional groups or classes of compounds, rather than individual compounds, in complex mixtures. For example, Schweighardt and co-workers have shown that the asphaltene fractions of coal liquids can be qualitatively characterized by IR spectrometry in terms of the presence of hydrogen-bonding functional groups (such as —OH, pyrrolic —NH, and pyridinic —N) [161]. In those studies, confirmatory tests (thin-layer chromatographic separations followed by chemical derivatization) were used to verify the assignments of specific IR bands to functional groups. The ability of IR techniques to distinguish between intra- and intermolecular hydrogen bonding was stressed in this study. Extension of such procedures to quantitative determinations of —OH and pyrrolic —NH functionalities in coal-derived materials have been described [162].

The sensitivity limitations associated with IR spectrometry using dispersive instrumentation have retarded use of the technique in PAH analysis. Considerable interest in IR analysis has, however, been stimulated by the development and commercialization of Fourier transform infrared (FTIR) spectrometers. These instruments operate on an interferometric rather than dispersive basis and are "multiplex" instruments (i.e., all portions of the spectrum are detected simultaneously rather than sequentially, as is the case in a conventional dispersive spectrometer). The principles and practice of FTIR spectrometry have been treated thoroughly in books [163-165] and review articles [166-171], to which the reader is referred for additional detail.

A useful example of the application of FTIR spectrometry to a problem relevant to the subject matter of this book is the use of group frequencies to identify the various functional groups present in chromatographic fractions of solvent-refined coal samples [172]. Functional groups identified in this way included alkyl groups, phenolic —OH, pyrrolic —NH, pyridinic —N, and carbonyl groups (both ketone and ester functionalities). Patterns of substitution of various groups on the aromatic framework of the material were inferred from intensity patterns of the various aromatic C—H out-of-plane bending vibrations. Such samples are difficult to examine by dispersive IR spectrometry because they are very strongly absorbing throughout virtually the entire IR and, when examined as films cast from liquid solution, are optically inhomogeneous.

Because an FTIR spectrometer contains a digital computer as an integral component, the advent of FTIR has stimulated interest in computer-based techniques for obtaining, manipulating, and interpreting IR spectra. Among these procedures are spectral subtraction methods (sometimes termed "spectral stripping"), wherein the spectrum of an interfering sample constituent or of broad background absorption is removed by subtraction from the spectrum of a multicomponent sample [173-175]. Such procedures, if used with care, are helpful both for identification of suspected sample constituents and for quantitative analysis, particularly in cases for which pure samples of the sample components are not available [169,173,174]. Moreover, as emphasized by McDonald [159], IR data originating in a digital instrument and remaining in digital form exhibit superior precision and dynamic range to those converted to analog form and then plotted on chart paper. Thus, both identification of unknown compounds (which is facilitated by very high precision of the wave number or wavelength scale) and quantitative analysis (which requires high precision of the absorbance scale) benefit from use of computerized IR instrumentation. Another important characteristic of an FTIR spectrometer is that the "instrument line shape function" (the equivalent to the "spectral bandpass" of a dispersive spectrometer) is known with high accuracy. Consequently, the computer can be used to correct measured absorbances for "finite bandpass" errors [176,177]; as noted above, these can cause substantial deviations from Beer's law if ignored.

Computer storage of "library" spectra for subsequent comparisons with those of unknowns is an obvious additional use of a computer associated with an IR spectrometer. Moreover, computer techniques for the actual interpretation of IR spectra in terms of molecular structure (such as artificial intelligence and pattern recognition) are currently receiving great attention [178]; as yet, however, these techniques have not displaced people practiced in the science (and art) of IR spectral identification.

Of course, these computer techniques are not inherent in the FTIR method; they can also be implemented with a computerized dispersive IR spectrometer [179], and such instruments are now commercially available (usually at substantially lower cost than FTIR spectrometers).

Because IR spectra are "fingerprints" of individual compounds, there has developed great interest in using IR as a detection method in gas chromatography (GC) [180,181]. Uses of this technique in PAH analysis are beginning to appear in significant numbers. For example, Uden et al. [182] have used a rapid-scanning dispersive IR spectrometer as a detector in the GC separation of basic shale oil fractions; several different nitrogen heterocyclic components of these samples were identified by comparing vapor-phase IR spectra of chromatographically separated oil components with those of authentic samples of the suspected constituents. Erickson and coworkers have reported a GC/FTIR analysis of semivolatile organics (trapped on XAD-2 resin) collected from a coal gasifier [183]. Again, comparison of vapor-phase IR spectra of separated components with those of pure compounds established the identity of many of the sample constituents. Among the 19 components in these samples identified in this manner were benzene, phenol, naphthalene, and a number of alkylbenzenes and alkylnaphthalenes. An example of a vapor-phase IR spectrum (that of 2,4-xylenol) obtained in this study is presented in Fig. 29.

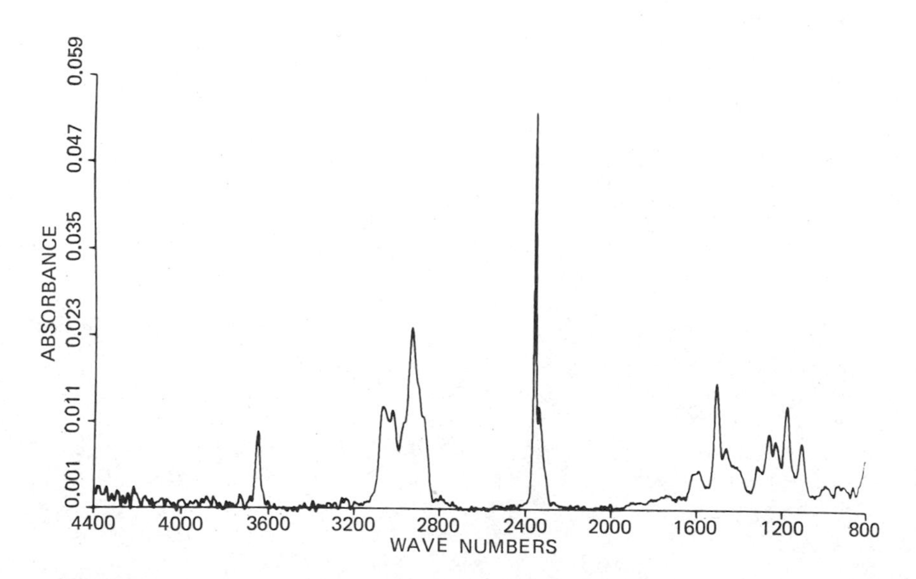

Figure 29. Vapor-phase FTIR spectrum of a component eluted in a GC analysis of semivolatile organic matter sampled from a coal gasifier. On the basis of a comparison with an FTIR spectral library, the compound was identified as 2,4-xylenol. (From Ref. 183.)

Applications of this type should increase rapidly as GC/FTIR interfaces having adequate sensitivities become commercially available. In addition, FTIR is receiving increasing consideration as a detection method in high-performance liquid chromatography [184,185]. It does not appear that use of such systems for difficult PAH analyses has yet been reported, but future developments in LC/FTIR instrumentation could have very substantial implications for PAH analysis.

3. *Matrix Isolation Fourier Transform IR Spectrometry*

One of the principal limitations of IR spectrometry, especially for quantitative applications, is the unsuitability of the various conventional sample-preparation procedures (see Sec. IV.B.1). We have noted previously (Sec. III.E.4) that matrix isolation, wherein a vaporized sample is codeposited with a large excess of a diluent gas on a cold surface, has several advantages over more conventional sample-preparation procedures in molecular fluorescence spectrometry. Many of these same advantages merit consideration for IR spectrometry as well. A particularly noteworthy advantage of matrix isolation (MI) is that, if a matrix such as nitrogen or argon is used, the matrix is transparent throughout the mid-IR (where most organic molecules absorb). The "multiplex advantage" of FTIR spectrometers can be fully exploited for signal-to-noise enhancement only if the "solvent" is nonabsorbing. Another important characteristic of MI as an IR sampling technique is that, in a properly prepared solid matrix, solute-solute interactions should be negligible (as opposed to the situation which commonly pertains in liquid solutions, KBr disks, and mulls). Thus, adherence to Beer's law (Sec. II.A) should be enhanced by use of MI. A third important characteristic of MI is that spectra obtained of solutes in cryogenic solids prepared by deposition from the gas phase are expected to be sharper and less complex than those observed in conventional IR samples [186]. Finally, MI techniques can be adapted readily to preparation of samples which are physically very small (and which therefore contain a relatively small quantity of analyte). Microsampling is currently an area of considerable interest in IR analysis [187-190], and MI is likely to prove useful as an IR microsampling technique.

In our laboratory, we (Mamantov, Wehry, and several colleagues) have described techniques for microsampling in FTIR spectrometry by MI [191] and have described the application of these techniques to the analysis of individual PAH in mixtures [95,97,192-196]. The ability of MI FTIR spectrometry to generate highly resolved spectra suitable for distinguishing between isomeric PAH has been stressed in this work. For example, Fig. 30 compares FTIR spectra of naphthalene in a KBr disk and in a nitrogen matrix at 15 K; the improved resolution characteristic of the low-temperature spectrum is obvious. The indicated region of the spectrum (700-900 cm^{-1}), wherein out-of-plane C—H bending modes predominate, has proven to be most useful for distinguishing between PAH. Other regions of the IR may be more useful for discriminating between substituted PAH or heterocycles; for example, the spectral region between 1100 and 1400 cm^{-1} appears most generally useful for distinguishing between nitrogen heterocycles.

Examples of the ability of MI FTIR spectra to distinguish between isomeric PAH are shown in Fig. 31 and 32. Figure 32 shows that benzo[a]pyrene and benzo[e]pyrene are readily distinguishable; note that bands having half-widths

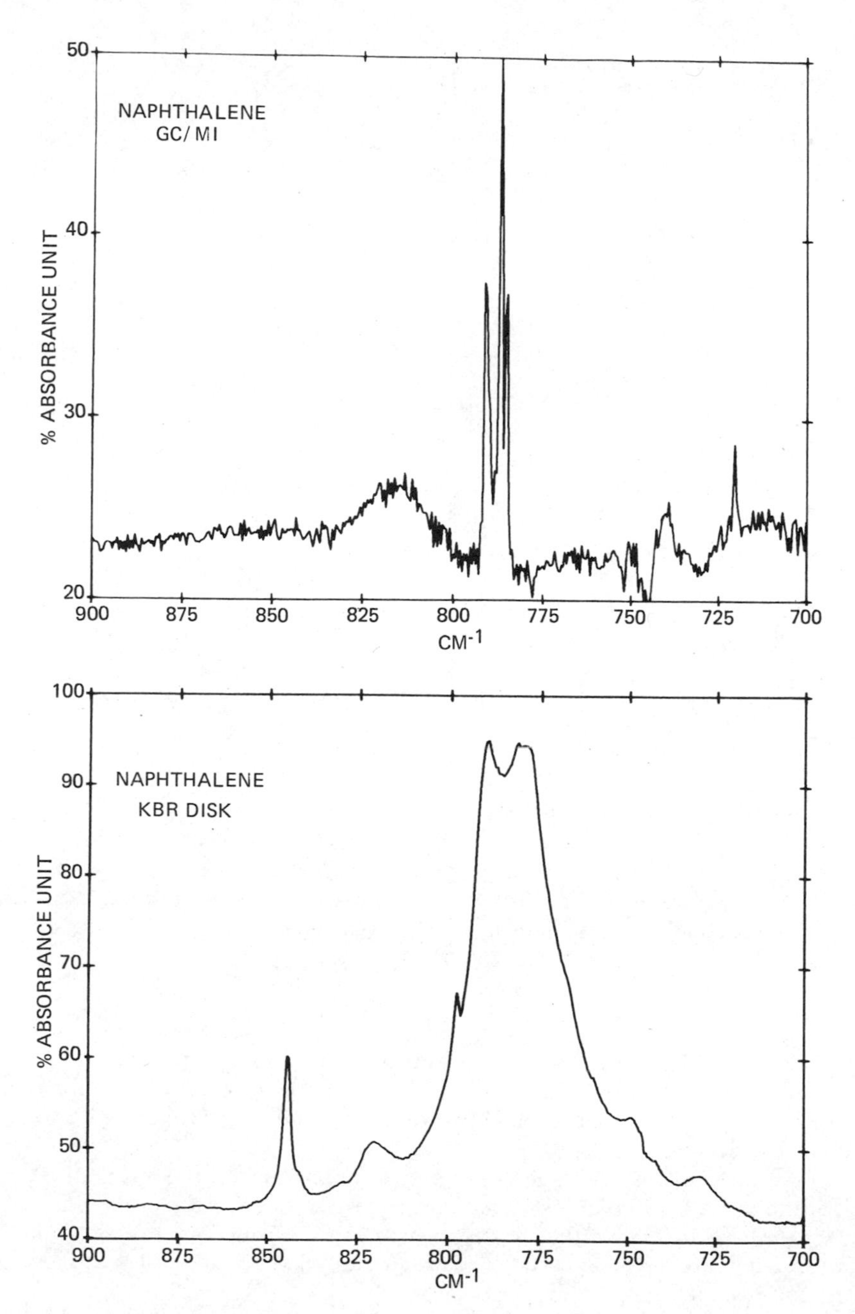

Figure 30. Top: FTIR spectrum of naphthalene matrix isolated in N_2 at 15 K. Bottom: FTIR spectrum of naphthalene in a KBr disk at room temperature.

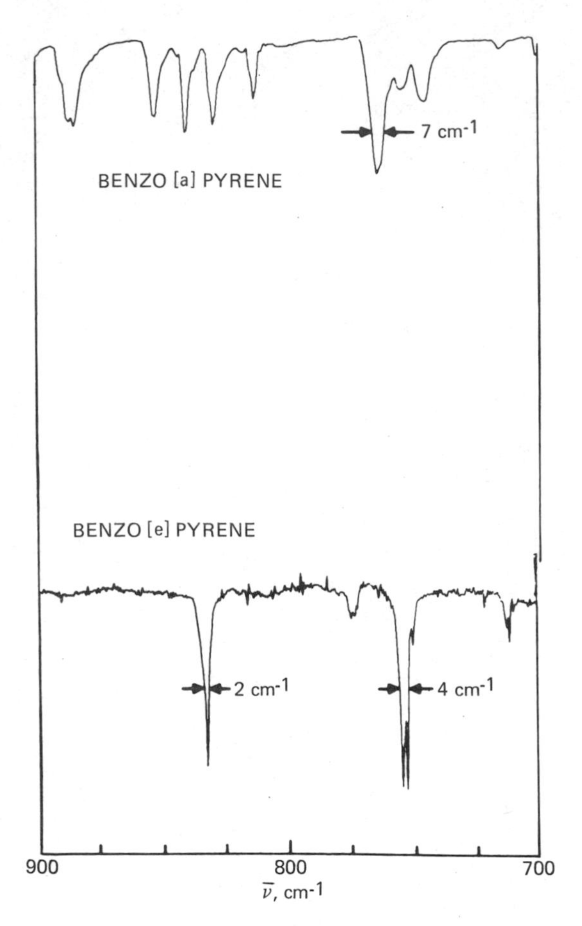

Figure 31. Matrix-isolation FTIR spectrum of benzo[a]pyrene (top) and benzo[e]pyrene (bottom) in nitrogen at 15 K. Bandwidths (full width half maximum) are indicated. (From Ref. 196.)

as small as 2 cm^{-1} can be obtained by MI. Figure 33 compares the MI FTIR spectra of the six isomeric methylchrysenes [97]; again, the results indicate the utility of MI FTIR spectrometry to "fingerprint" specific compounds in mixtures of isomers.

Figure 33 shows the MI FTIR spectrum of a real sample (the coking plant chromatographic fraction whose low-temperature fluorescence spectrum is shown in Fig. 21). The compounds indicated in Fig. 33 were each identified by comparison with "library" spectra of pure PAH obtained under identical MI sampling conditions.

Quantitative analytical techniques in MI FTIR spectrometry also have been described [97,191-193]. For example, Fig. 34 shows a Beer's law plot for the absorbance of naphthalene (matrix isolated in N_2 at 15 K) in the presence of constant amounts of 1- and 2-methylnaphthalene [193]. Despite the presence of the methyl derivatives, the Beer's law plot for the parent compound is linear over two decades in amount of naphthalene. The replicate points in Fig. 34 also provide an indication of the precision of quantitative FTIR analysis by MI; relative standard deviations in the 5-8% range are obtained if an internal standard is used [195]. The characteristics of matrix-isolated samples for quantitation are very favorable by comparison with those of conventional IR sampling methods, and this aspect of MI is likely to receive particular emphasis.

As noted previously, the use of FTIR spectrometry for detection in gas chromatography is attracting considerable attention. Because sample preparation by matrix isolation requires that the sample be vaporized, the MI technique should be amenable to the examination of GC effluents by FTIR. The first experimental system for this purpose was described by Reedy et al. [197]

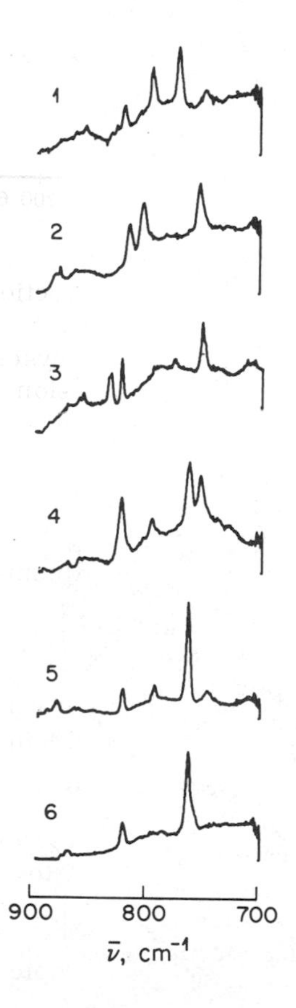

Figure 32. Matrix-isolation FTIR spectra (in N_2) of the six methylchrysenes in the 700-900 cm^{-1} region at 15 K. (Reprinted with permission from P. Tokousbalides, E. R. Hinton, Jr., R. B. Dickinson, Jr., P. V. Billota, E. L. Wehry, and G. Mamantov, Analysis of the isomeric methylchrysenes by matrix isolation fluorescence and Fourier transform infrared spectrometry. *Anal. Chem.* *50*:1191. Copyright 1978, American Chemical Society.)

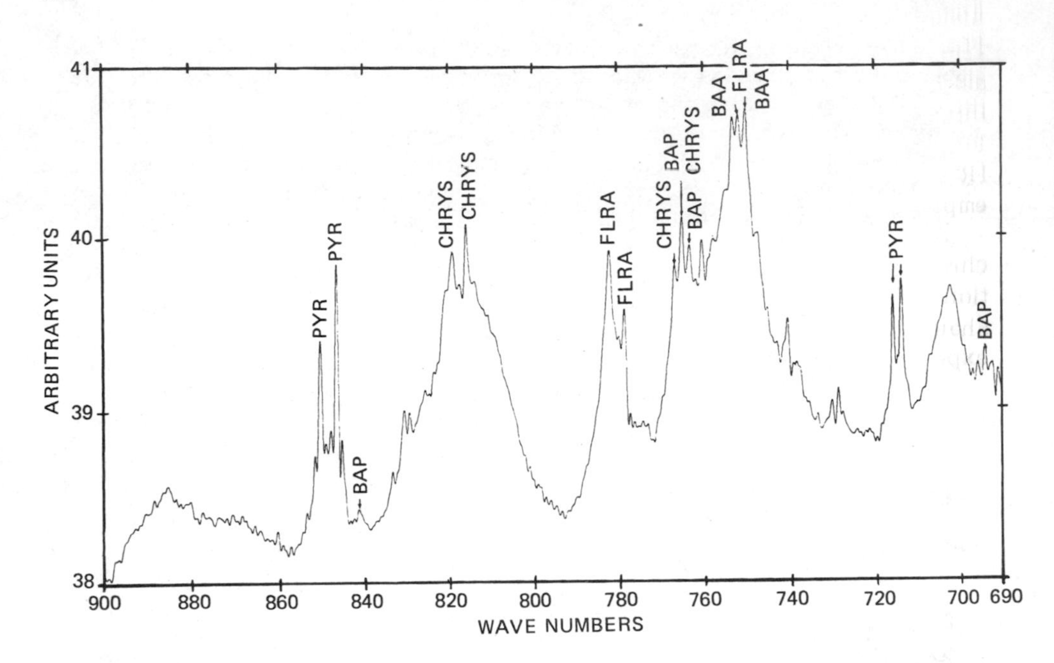

Figure 33. Fourier transform infrared spectrum of a chromatographic fraction from a steel mill coking plant water sample, in a nitrogen matrix at 15 K. Identified compounds: PYR, pyrene; BAP, benzo[a]pyrene; CHRYS, chrysene; FLRA, fluoranthene; BAA, benz[a]anthracene. (Reprinted with permission from E. L. Wehry and G. Mamantov, Matrix isolation spectroscopy. *Anal. Chem.* *51*:654A. Copyright 1979, American Chemical Society.)

and Bourne et al. [198], who demonstrated that FTIR spectra of GC effluents could be achieved under conditions of true matrix isolation, even for compounds (such as phenols) which exhibit strong tendencies to aggregate. Applications of the GC/MI FTIR technique to PAH analysis have recently been examined [199,200]; detailed descriptions of the experimental apparatus and results will be described elsewhere [201]. The use of a high-resolution spectroscopic detector in GC means that the resolution requirements of the separation itself can be relaxed somewhat. For example, if it is possible to differentiate spectroscopically between sets of isomeric PAH that are separable only with much difficulty, then there is no need to separate them chromatographically. The tradeoffs between chromatographic and spectroscopic resolution in analytical systems of this type have yet to be delineated adequately; however, the potential utility of combined separation-spectroscopic methods for PAH analysis appears great.

In addition to matrix isolation, other relatively new sampling techniques for FTIR spectrometry include diffuse reflectance from solid samples dispersed in KCl [202] and photoacoustic FTIR spectrometry of solid samples [203,204]. As yet, neither of these techniques appears to have been utilized in PAH analysis.

4. Sources of Spectral Data

It has been emphasized repeatedly that qualitative analysis by IR spectrometry is dependent upon the availability of libraries of reference spectra of pure compounds. Fortunately, there exist many libraries of IR spectra, though the IR data for many PAH derivatives are unfortunately not included in any existing library. Of the numerous IR data bases available, the enormous Sadtler collection [205], which now contains nearly 60,000 spectra in plotted form (and is available in printed form, microfilm, or microfiche) probably is the most generally useful from the standpoint of PAH analysis. Another large collection of plotted IR spectra is that compiled by the American Petroleum Institute and, subsequently, by the Thermodynamics Research Center of Texas A&M University [206]. This collection, concerned principally with the spectra of petroleum constituents, contains a number of spectra or aromatic hydrocarbons and derivatives thereof. The largest library of IR spectra is that assembled by the American Society for Testing and Materials (ASTM). This library, which is not currently being expanded, contains the major IR absorption bands of some 145,000 substances. The ASTM file is available from Sadtler Research Laboratories (Philadelphia, Pa.) on computer-readable magnetic tape. Unfortunately, this library includes only the most intense IR bands of the compounds included (no relative intensity data are filed), and many of the spectra are rather old.

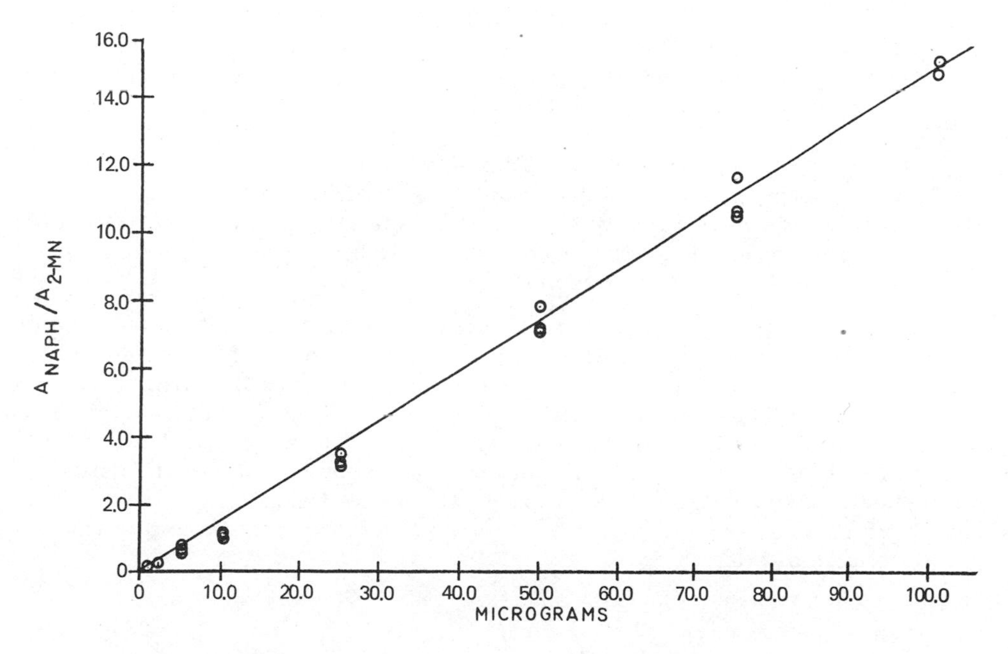

Figure 34. Beer's law plot for absorbance of naphthalene matrix-isolated in N_2 at 15 K in the presence of 25 μg each of 1- and 2-methylnaphthalene. The 2-methylnaphthalene served as an internal standard, and in each case the absorbance of naphthalene was ratioed to that of 2-methylnaphthalene. (From Ref. 193.)

Various other, smaller collections of IR spectra exist; these are listed, and their characteristics compared, by the Coblentz Society (see Craver [207], Fiske et al. [208], and McDonald [159]). Many of the spectra contained in libraries were obtained a number of years ago using instrumentation incapable of the resolution and signal-to-noise performance characteristics of modern spectrometers. Moreover, in many cases the sampling techniques are not described clearly. Hence, duplication of some of the IR spectra found in library files under modern spectrometric and sampling conditions is not possible. The Coblentz Society has issued specifications for "research quality infrared spectra" [209], which unfortunately are not demonstrably satisfied by many of the spectra in existing data bases. All of this does not mean that these files are not useful but rather that they must be used with some care and with a proper recognition of their present limitations. Many investigators (particularly those having computerized IR spectrometers) prefer to assemble their own libraries of pure-compound spectra, using the sampling and spectrometric conditions which pertain to their particular applications.

5. *Summary and Conclusions*

Although IR spectrometry has had little impact on PAH analysis to date, there are reasons to believe that the technique will experience increased application in the future. A number of developments promise to enhance the analytical capabilities of IR spectrometry; these include the following:

a. Continued improvements in sensitive FTIR instruments and microsampling procedures appropriate for use with such instrumentation
b. The development of tunable laser source IR spectrometry [210,211], which will be of use in analytical applications requiring very high spectral resolution
c. Advances in computer techniques for automatic recognition and interpretation of spectra, spectral subtraction, and quantitation
d. Improvements in the quality of files of IR spectra, particularly in computer-compatible form (while this process will certainly not be inexpensive [159], it is destined to occur eventually)

Although there is no possibility that IR will supplant UV-visible absorption and fluorescence techniques in PAH analysis, the developments itemized above (among others) should secure a more important role for IR in PAH characterization than it has had in the past.

It should be noted that developments in IR instrumentation and techniques are reviewed in the biennial fundamental review issue of *Analytical Chemistry* published in even-numbered years [159]. In recent years, these reviews also have provided very useful status reports on IR data bases and automated interpretation of spectra.

C. Raman Spectrometry

The fundamentals and applications of Raman spectrometry have been discussed lucidly in a book by Long [212]. Although Raman spectrometry has certain analytical advantages over IR absorption (in particular, the ability to use sources, detectors, and sampling procedures applicable to the UV-visible region of the spectrum), the technique thus far has received very little consideration for use in PAH analysis. The principal reason for this is that the Raman effect is inherently very weak, and high sensitivity is not an attribute usually associated with conventional Raman techniques.

Considerable enhancement of the strength of a Raman signal usually can be achieved by use of the "resonance Raman effect" (recently reviewed in terms of its analytical potentialities by Morris and Wallan [213]), wherein the incident wavelength is constrained to fall within—or at least close to—that of an allowed electronic transition of the molecule of interest. In favorable cases, detection limits of 10^{-8} M or smaller can be achieved by resonance Raman spectrometry [213,214]. Unfortunately, PAH do not generally represent a "favorable case" for resonance Raman spectrometry. Because of the tendency of most PAH to fluoresce, their resonance Raman signals will tend to be submerged in an intense background fluorescence.

There are several ways to observe Raman spectra (conventional or resonance enhanced) of fluorescent samples. One simple procedure is to add to the sample a large quantity of an efficient fluorescence quencher (such as nitrobenzene dissolved in chloroform), and indeed the resonance Raman spectra of a number of intensely fluorescent PAH have been observed, using an argon-ion laser as source, by precisely this procedure [215]. Other, more instrumentally sophisticated, techniques for fluorescence rejection in Raman spectrometry (including time resolution [216,217], wavelength modulation [218,219], and methods based on differences between the polarization characteristics of fluorescence and Raman scattering [220]) also have been developed, and intensely fluorescent PAH often have been included among the test cases studied to demonstrate the feasibility of these methods. Nevertheless, the "fluorescence problem" remains a formidable obstacle to use of Raman spectrometry for PAH analysis. Even in conventional (i.e., nonresonance) Raman spectrometry, fluorescence from a trace sample constituent (rather than of the analyte itself) may be sufficiently intense to obliterate the Raman signal of the analyte.

It is therefore not likely that Raman spectrometry will soon achieve a position of widespread importance in the determination of PAH. It is likely to encounter use for samples, or particular analytes, for which other optical spectrometric methods are inapplicable (e.g., nonfluorescent species). Raman has important points of superiority over IR spectrometry in being readily able to deal with aqueous samples. In liquid chromatographic detection, Raman has an important potential advantage over fluorescence in that the same incident wavelength can be used for all solutes in a conventional Raman experiment, but obviously this is not the case in a fluorescence (or resonance Raman) measurement. However, the sensitivity of conventional Raman spectrometry may often be inadequate for application to chromatographic detection under realistic conditions.

Finally, it should be noted that a plethora of new techniques related to the Raman effect, including coherent anti-Stokes Raman spectrometry (CARS) [221,222], photoacoustic Raman spectrometry (PARS) [223], and stimulated Raman gain techniques [224-226], are currently in the process of being developed. While none of these new techniques has yet been developed as a practical analytical method, they all appear to have interesting analytical possibilities; these and other "nonlinear" Raman effects may eventually be shown to have useful applications to analyses of PAH.

V. Addendum

Since completion of the original manuscript for this chapter, a number of developments have occurred which deserve mention. This addendum will serve to point out some useful references which have appeared recently.

A. Reviews

Optical spectroscopic techniques for analysis of PAH, with special reference to the aquatic environment, have been reviewed in great detail by Futoma et al. [227]. A book by Lee et al. [228] and review article by Bartle et al. [229] on the analytical chemistry of PAH includes extensive discussion of optical spectrometric techniques.

B. UV-Visible Absorption Spectrometry

The use of photoacoustic spectrometry (PAS) for determination of PAHs has now been reported, and detection limits for determination of PAH in liquid solution by PAS have been compared with those obtained by fluorescence and photoionization measurements [230]. Although there is little obvious incentive for using PAS to determine intensely fluorescent species such as parent PAH, the technique may be extremely useful for determinations of weakly fluorescent derivatives of PAH.

C. Conventional Fluorescence Techniques

Jurgensen et al. have compiled excitation and fluorescence wavelengths and limits of detection for the fluorometric determination of 60 PAH in liquid solution media [231]. A new book entitled *Standards in Fluorescence Spectrometry* contains authoritative discussions of a great many topics of practical importance in analytical fluorescence spectrometry, including criteria for defining sensitivities of fluorometers, stray-light effects, effects of sample temperature and photochemical decomposition in analytical fluorometry, sample illumination options, corrections for inner-filter effects, and corrections for instrumental distortion of excitation and fluorescence spectra [232].

D. Recent Innovations in Fluorometric Analysis

A review of applications of *lasers* in analytical spectroscopy by Omenetto and Winefordner [232a] includes a useful discussion of the use of laser excitation in fluorometric analysis.

Vo-Dinh and co-workers have described the use of *synchronous fluorescence spectrometry* for identification and quantification of PAH in solvent-refined coal [233] and airborne particulate matter [234] samples. Use of synchronous fluorescence following liquid chromatography for identification of PAH in coal-derived liquids has been reported [235]. Lloyd [236] and Latz et al. [237] have exchanged comments regarding possible difficulties in using synchronous luminescence for quantitative analysis of individual compounds in mixtures.

A major development in *low-temperature luminescence spectrometry* is the use of a supersonic jet to produce very highly resolved fluorescence spectra of PAH and other large organic molecules in the gas phase [238]. (Such supersonic expansions also can be used to achieve highly selective laser-induced photoionization of gas-phase molecules [239], and such phenomena will in due course be applied to determination of PAH.) The use of laser excitation to achieve greatly improved analytical selectivity in the matrix-isolation fluorometric determination of PAH in multicomponent samples has been discussed by Maple and Wehry [240,241] and Wehry et al. [242]. The selectivity of matrix-isolation fluorometry of mixtures can be enhanced further by use of "photoselection" techniques, wherein a polarizer is used to transmit selectively the fluorescence of one component in a mixture of PAHs excited with polarized light from a dye laser [243]. Yang and co-workers have demonstrated that the analytical selectivity of Shpol'skii fluorometry in frozen-solution matrices is greatly enhanced by use of dye-laser excitation [244,245]. A detailed quantitative study of frozen-solution Shpol'skii fluorometric analysis indicates that solute aggregation and site distribution irreproducibility can be minimized by using reproducible freezing conditions and ensuring that very dilute solutions are frozen [244]. Small's group has reported applications of "fluorescence line narrowing" spectrometry of PAH to the observation of highly resolved fluorescence spectra of individual PAH in complex mixtures, using aqueous glycerol frozen solutions as the cryogenic matrix [246,247]. Bykovskaya and co-workers have discussed the principles of such experiments and have demonstrated the detection of individual PAH in mixtures without prior separation by this technique [247a]. Maple and Wehry have employed a similar approach for determination of individual polar derivatives of PAH in mixtures by matrix isolation [241]. Lamp-excited Shpol'skii fluorescence spectrometry has been used to identify PAH in petroleum fractions and marine sediments [248]. The use of "excitation-emission matrices" (or "total luminescence spectroscopy") in low-temperature luminescence measurements has been discussed [249,250].

Christian et al. have reviewed the use of *electronic array detectors* in fluorometric analysis [251]. A review of the fluorometric analysis of complex mixtures by Warner and McGown focuses special attention on the use of array detectors and the "excitation-emission matrix" mathematical formalism [252].

The use of *time-resolved fluorometry* to eliminate errors due to quenching, which is usually a very important source of error in the fluorometric analysis of complex samples in liquid solution, has been discussed [253]. Richardson and coworkers have discussed the use of time resolution to enhance selectivity in the detection of PAH by fluorometry following liquid chromatography [254].

Relevant developments in *two-photon-induced fluorescence* include a discussion of "two-photon polarization spectroscopy" and its possible use in mixture analysis [255], and the use of two-photon excited fluorescence in liquid chromatographic detection [256].

E. Phosphorescence

A book by Hurtubise reviews solid-surface luminescence methods, including room-temperature phosphorescence (RTP) of organic compounds adsorbed on filter paper or other solid supports [257]; Parker and co-workers also have reviewed RTP principles and techniques [258,259]. The use of RTP on filter-paper substrates to determine PAH in coal liquids [233,260] and airborne particulate samples [234] has been described by Vo-Dinh and colleagues.

Additional results have been presented by Cline Love's group pertaining to observation of RTP from PAH solubilized in surfactant micelles and the analytical implications of the phenomenon [261,262]. Fluorescence of PAH likewise may be enhanced when they are solubilized in micelles, and this observation may lead to development of new fluorometric methods for PAH [263].

F. Vibrational Spectroscopy

The use of matrix isolation Fourier transform infrared spectrometry in the detection of components eluting from a capillary gas chromatography column has been reported, and use of the technique for identifying and determining PAH in complex samples has been described [264].

Acknowledgments

Research in this laboratory on matrix isolation Fourier transform infrared and fluorometric analysis of PAH has been supported by contracts with the Electric Power Research Institute. Our investigations of laser-induced matrix isolation fluorometric analysis of PAH has been supported by grants from the National Science Foundation.

References

1. H. H. Jaffé and M. Orchin, *Theory and Applications of Ultraviolet Spectroscopy*, Wiley, New York, 1962.
2. J. N. Murrell, *The Theory of the Electronic Spectra of Organic Molecules*, Wiley, New York, 1963.
3. D. G. Peters, J. M. Hayes, and G. M. Hieftje, *Chemical Separations and Measurements*, Saunders, Philadelphia, 1974, pp. 614-617, 641-642.
4. H. A. Strobel, *Chemical Instrumentation*, 2nd ed., Addison-Wesley, Reading, Mass., 1973, pp. 432-436, 444-449.
5. C. K. Mann, T. J. Vickers, and W. M. Gulick, *Instrumental Analysis*, Harper & Row, New York, 1974, pp. 291-293.
6. E. D. Olsen, *Modern Optical Methods of Analysis*, McGraw-Hill, New York, 1975, pp. 69-70.
7. Ref. 5, pp. 435-436.
8. Spectrometry nomenclature. *Anal. Chem. 52*:221 (1980).
9. H. H. Willard, L. L. Merritt, Jr., and J. A. Dean, *Instrumental Methods of Analysis*, 6th ed., Reinhold Van Nostrand, New York, 1981, Chaps. 2 and 3.
10. H. H. Bauer, G. D. Christian, and J. E. O'Reilly, *Instrumental Analysis*, Allyn & Bacon, Boston, 1978, Chap. 7.
11. R. P. Bauman, *Absorption Spectroscopy*, Wiley, New York, 1962.
12. J. B. Lambert, H. F. Shurvell, L. Verbit, R. G. Cooks, and G. H. Stout, *Organic Structural Analysis*, Macmillan, New York, 1976, pp. 317-342.
13. R. A. Friedel and M. Orchin, *Ultraviolet Spectra of Aromatic Compounds*, Wiley, New York, 1951.

14. W. S. Johnson, E. Woroch, and F. J. Mathews, *J. Amer. Chem. Soc. 69*, 566 (1947).
15. E. L. Wehry, in *Practical Fluorescence: Theory, Methods, and Techniques*, G. G. Guilbault (Ed.), Dekker, New York, 1973, pp. 101-102.
16. M. Kasha, in *Light and Life*, W. D. McElroy and B. Glass (Ed.), Johns Hopkins Univ. Press, Baltimore, 1961, p. 31.
17. Ref. 15, pp. 111-112.
18. N. J. Turro, *Molecular Photochemistry*, Benjamin, New York, 1965, p. 45.
19. E. Clar, *Polycyclic Hydrocarbons*, Academic Press, New York, 1964.
20. *Sadtler Standard Spectra*. Sadtler Research Laboratories, Inc., Philadelphia.
21. *Ultraviolet Atlas of Organic Compounds*, Plenum Press, New York, 1968, 5 vols.
22. L. Lang, *Absorption Spectra in the Ultraviolet and Visible Region*, Academic Press, New York, Vols. 1-20.
23. J. P. Phillips, D. Bates, H. Feuer, B. S. Thyagarajan, *Organic Electronic Spectral Data*, Vol. 15: *1973*, Wiley, New York, 1979.
24. J. F. Schabron, R. J. Hurtubise, and H. F. Silver, *Anal. Chem. 51*: 1426 (1979).
25. Y.-H. Pao, *Optoacoustic Spectroscopy and Detection*, Academic Press, New York, 1977.
26. J. M. Harris and N. J. Dovichi, *Anal. Chem. 52*:695A (1980).
27. L. G. Hargis and J. A. Howell, *Anal. Chem. 52*:306R (1980); *54*:171R (1982).
28. D. M. Hercules, *Fluorescence and Phosphorescence Analysis*, Wiley (Interscience), New York, 1966, Chap. 1.
29. C. A. Parker, *Photoluminescence of Solutions*, Elsevier, New York, 1969, Chaps. 1, 2, and 4.
30. R. B. Lam and J. J. Leary, *Appl. Spectrosc. 33*:17 (1979).
31. J. H. Richardson and M. E. Ando, *Anal. Chem. 49*:955 (1977).
32. T. Hirschfeld, *Appl. Spectrosc. 31*:245 (1977).
33. T. G. Matthews and F. E. Lytle, *Anal. Chem. 51*:583 (1979).
34. J. B. Birks, *Photophysics of Aromatic Molecules*, Wiley, New York, 1970, pp. 90-93.
35. Ref. 29, pp. 72-78, 83-85.
36. Ref. 29, pp. 20-21, 220-234.
37. J. N. Demas and G. A. Crosby, *J. Phys. Chem. 75*:991 (1971).
38. T. C. O'Haver, in *Trace Analysis: Spectroscopic Methods for Elements*, J. D. Winefordner (Ed.), Wiley (Interscience), New York, 1976, pp. 38-43.
39. Ref. 29, pp. 128-245.
40. D. W. Ellis, in *Fluorescence and Phosphorescence Analysis*, D. M. Hercules (Ed.), Wiley (Interscience), 1966, pp. 41-79.
41. G. G. Guilbault, *Practical Fluorescence: Theory, Methods, and Techniques*, Dekker, New York, 1973, Chap. 2.
42. Ref. 34, p. 107.
43. Ref. 29, pp. 428-438.
44. E. L. Wehry and L. B. Rogers, in *Fluorescence and Phosphorescence Analysis*, D. M. Hercules (Ed.), Wiley (Interscience), 1966, pp. 81-99.
45. R. S. Becker, *Theory and Interpretation of Fluorescence and Phosphorescence*, Wiley (Interscience), 1969, pp. 116-181.
46. Ref. 15, pp. 87-99.

47. Ref. 15, pp. 99-131.
48. B. L. Van Duuren, *Chem. Rev. 63*:325 (1963).
49. Ref. 29, pp. 373-385.
50. I. B. Berlman, *Handbook of Fluorescence Spectra of Organic Molecules*, Academic Press, New York, 1965.
51. Ref. 34, pp. 103-104.
52. "Standard Fluorescence Spectra," Sadtler Research Laboratories, Inc., Philadelphia.
53. A. Schmillen and R. Legler, *Lumineszenz Organischer Substanzen* (Landolt-Börnstein New Ser., Vol. 3), Springer-Verlag, Berlin, 1967.
54. R. A. Passwater, *Guide to Fluorescence Literature*, Vol. 3, Plenum Press, New York, 1974.
55. D. M. Hercules, *Science 125*:1242 (1957).
56. C. A. Parker and W. T. Rees, *Analyst 85*:587 (1960).
57. Ref. 29, pp. 246-261.
58. J. E. Wampler and R. J. De Sa, *Rev. Sci. Instrum. 25*:623 (1971).
59. I. Landa and J. C. Kremen, *Anal. Chem. 46*:1694 (1974).
60. T. Vo-Dinh and U. P. Wild, *Appl. Opt. 13*:2899 (1974).
61. J. E. Wampler, in *Modern Fluorescence Spectroscopy*, E. L. Wehry (Ed.), Vol. 1, Plenum Press, New York, 1976, pp. 25-32.
62. R. C. Pierce and M. Katz, *Anal. Chem. 47*:1743 (1975).
63. M. Katz, T. Sakuma, and A. Ho, *Environ. Sci. Technol. 12*:909 (1978).
64. M. Katz and C. Chan, *Environ. Sci. Technol. 14*:838 (1980).
65. W. H. Griest, L. B. Yeatts, Jr., and J. E. Caton, *Anal. Chem. 52*:199 (1980).
66. J. H. Richardson, in *Modern Fluorescence Spectroscopy*, E. L. Wehry (Ed.), Vol. 4, Plenum Press, 1981.
67. J. C. Wright and M. J. Wirth, *Anal. Chem. 52*:988A (1980).
68. J. B. F. Lloyd, *Nature 231*:64 (1971).
69. T. Vo-Dinh, *Anal. Chem. 50*:396 (1978).
70. P. John and I. Soutar, *Anal. Chem. 48*:520 (1976).
71. S. G. Wakeham, *Environ. Sci. Technol. 11*, 272 (1977).
72. T. Vo-Dinh, R. B. Gammage, A. R. Hawthorne, and J. H. Thorngate, *Environ. Sci. Technol. 12*:1297 (1978).
73. T. Vo-Dinh, in *Modern Fluorescence Spectroscopy*, E. L. Wehry (Ed.), Vol. 4, Plenum Press, 1981.
74. D. W. Johnson, J. B. Callis, and G. D. Christian, *Anal. Chem. 49*:747A (1977).
75. L. P. Giering and A. W. Hornig, *Amer. Lab. 9*(11):113 (1977).
76. D. Eastwood, in *Modern Fluorescence Spectroscopy*, E. L. Wehry (Ed.), Vol. 4, Plenum Press, New York, 1981.
77. I. M. Warner, G. D. Christian, E. R. Davidson, and J. B. Callis, *Anal. Chem. 49*:564 (1977).
78. G. D. Christian, J. B. Callis, and E. R. Davidson, in *Modern Fluorescence Spectroscopy*, E. L. Wehry (Ed.), Plenum Press, New York, Vol. 4, 1981.
79. T. C. O'Haver, in *Modern Fluorescence Spectroscopy*, E. L. Wehry (Ed.), Vol. 1, Plenum Press, New York, 1976, pp. 66-73.
80. G. L. Green and T. C. O'Haver, *Anal. Chem. 46*:2191 (1974).
81. Ref. 79, pp. 74-80.

82. T. C. O'Haver, *Anal. Chem. 51*:91A (1979).
83. T. C. O'Haver and W. M. Parks, *Anal. Chem. 46*:1886 (1974).
84. E. L. Wehry and G. Mamantov, in *Modern Fluorescence Spectroscopy*, E. L. Wehry (Ed.), Vol. 4, Plenum Press, New York, 1981.
85. E. V. Shpol'skii and T. N. Bolotnikova, *Pure Anal. Chem. 37*:183 (1974).
86. G. F. Kirkbright and C. G. de Lima, *Analyst 99*:338 (1974).
87. R. C. Stroupe, P. Tokousbalides, R. B. Dickinson, Jr., E. L. Wehry, and G. Mamantov, *Anal. Chem. 49*:701 (1977).
88. J. A. G. Drake, D. W. Jones, B. S. Causey, and G. F. Kirkbright, *Fuel 57*:663 (1978).
89. Y. Yang, A. P. D'Silva, V. A. Fassel, and M. Iles, *Anal. Chem. 52*:1350 (1980).
90. C. S. Woo, A. P. D'Silva, V. A. Fassel, and G. J. Oestreich, *Environ. Sci. Technol. 12*:173 (1978).
91. C. S. Woo, A. P. D'Silva, and V. A. Fassel, *Anal. Chem. 52*:159 (1980).
92. A. Colmsjö and U. Stenberg, *Anal. Chem. 51*:145 (1979).
93. L. M. Shabad and G. A. Smirnov, *Atmos. Environ. 6*:153 (1972).
94. J. Joussot-Dubien (Université de Bordeaux, France), personal communication (April 1980).
95. E. L. Wehry and G. Mamantov, *Anal. Chem. 51*:643A (1979).
96. J. R. Maple, E. L. Wehry, and G. Mamantov, *Anal. Chem. 52*:920 (1980).
97. P. Tokousbalides, E. R. Hinton, Jr., R. B. Dickinson, Jr., P. V. Billota, E. L. Wehry, and G. Mamantov, *Anal. Chem. 50*:1189 (1978).
98. P. Tokousbalides, E. L. Wehry, and G. Mamantov, *J. Phys. Chem. 81*:1769 (1977).
99. B. E. Kohler, in *Chemical and Biochemical Applications of Lasers*, C. B. Moore (Ed.), Vol. 4, Academic Press, New York, pp. 31-53.
100. R. I. Personov, E. I. Al'Shits, and L. A. Bykovskaya, *Opt. Commun. 6*:169 (1972).
101. K. Cunningham, J. M. Morris, J. Fünfschilling, and D. F. Williams, *Chem. Phys. Lett. 32*:581 (1975).
102. W. C. McColgin, A. P. Marchetti, and J. H. Eberly, *J. Amer. Chem. Soc. 100*:5622 (1978).
103. G. Flatscher, K. Fritz, and J. Friedrich, *Z. Naturforsch. 31a*:1220 (1976).
104. J. C. Brown, M. C. Edelson, and G. J. Small, *Anal. Chem. 50*:1394 (1978).
105. M. Labhart, G. Miklos, J. Keller and U. P. Wild, *J. Chem. Phys. 72*:1764 (1980).
106. I. I. Abram, R. A. Auerbach, R. R. Birge, B. E. Kohler, and J. M. Stevenson, *J. Chem. Phys. 63*:2473 (1975).
107. J. C. Brown, J. A. Duncanson, Jr., and G. J. Small, *Anal. Chem. 52*:1711 (1980).
108. J. R. Maple and E. L. Wehry, *Anal. Chem. 53*:266 (1981).
109. H. E. Hallam and G. F. Scrimshaw, in *Vibrational Spectroscopy of Trapped Species*, H. E. Hallam (Ed.), Wiley, New York, 1973, pp. 29-31.
110. B. Meyer, *Low Temperature Spectroscopy*, Elsevier, New York, 1971, pp. 222-227.
111. Y. Talmi, *Anal. Chem. 47*:658A, 699A (1975).
112. Y. Talmi, *Amer. Lab. 10*(3):79 (1978).

113. G. Horlick and E. G. Codding, *Contemp. Topics Anal. Clin. Chem. 1*, 195 (1977).
114. Y. Talmi and R. W. Simpson, *Appl. Opt. 19*:1401 (1980).
115. J. D. Ingle, Jr., and M. A. Ryan, in *Modern Fluorescence Spectroscopy*, E. L. Wehry (Ed.), Vol. 3, Plenum Press, New York, 1981.
116. P. Froehlich and E. L. Wehry, in *Modern Fluorescence Spectroscopy*, E. L. Wehry (Ed.), Vol. 3, Plenum Press, New York, 1981.
117. J. R. Jadamec, W. A. Saner, and Y. Talmi, *Anal. Chem. 49*:1316 (1977).
118. J. R. Jadamec, W. A. Saner, and R. W. Sager, *ACS Symposia Ser. 102*: 115 (1979).
119. R. P. Cooney, T. Vo-Dinh, and J. D. Winefordner, *Anal. Chim. Acta 89*: 9 (1977).
120. R. M. Hoffman and H. L. Pardue, *Anal. Chem. 51*:1267 (1979).
121. K. W. Busch, B. Malloy, and Y. Talmi, *Anal. Chem. 51*:670 (1979).
122. D. Holten and M. W. Windsor, *Ann. Rev. Biophys. Bioenerg. 7*:189 (1978).
123. M. G. Badea and L. Brand, *Methods Enzymol. 61*:378 (1979).
124. F. E. Lytle, *Anal. Chem. 46*:545A, 817A (1974).
125. G. M. Hieftje and E. E. Vogelstein, in *Modern Fluorescence Spectroscopy*, E. L. Wehry (Ed.), Vol. 4, Plenum Press, New York, 1981.
126. R. B. Dickinson, Jr., and E. L. Wehry, *Anal. Chem. 51*:778 (1979).
127. G. M. Hieftje, J. M. Ramsey, and G. R. Haugen, *Advan. Chem. Ser. 55*: 85 (1978).
128. G. M. Hieftje, G. R. Haugen, and J. M. Ramsey, *Appl. Phys. Lett. 30*: 463 (1977).
129. C. C. Dorsey, M. J. Pelletier, and J. M. Harris, *Rev. Sci. Instrum. 50*: 333 (1979).
130. J. M. Ramsey, G. M. Hieftje, and G. R. Haugen, *Appl. Opt. 18*:1913 (1979).
131. N. Strojny and J. A. F. de Silva, *Anal. Chem. 52*:1554 (1980).
132. T. Hirschfeld, *J. Histochem. Cytochem. 27*: 96 (1979).
133. M. J. Wirth and F. E. Lytle, *ACS Symposia Ser. 85*:24 (1978).
134. M. J. Wirth and F. E. Lytle, *Anal. Chem. 49*:2054 (1977).
135. M. J. Sepaniak and E. S. Yeung, *Anal. Chem. 49*:1554 (1977).
136. E. L. Wehry, *Anal. Chem. 52*:75R (1980); *54*:131R (1982).
137. J. J. Aaron and J. D. Winefordner, *Talanta 22*:707 (1975).
138. J. J. Aaron, J. J. Mousa, and J. D. Winefordner, *Talanta 20*:279 (1973).
139. E. M. Schulman and C. Walling, *Science 178*:53 (1972).
140. T. Vo-Dinh and J. D. Winefordner, *Appl. Spectrosc. Rev. 13*:261 (1977).
141. T. Vo-Dinh, E. Yen, and J. D. Winefordner, *Anal. Chem. 48*:1186 (1976).
142. I. M. Jakovljevic, *Anal. Chem. 49*:2049 (1977).
143. T. Vo-Dinh and J. R. Hooyman, *Anal. Chem. 51*:1915 (1979).
144. J. N. Miller, D. L. Phillipps, D. T. Burns, and J. W. Bridges, *Anal. Chem. 50*:613 (1978).
145. E. L. Yen-Bower and J. D. Winefordner, *Appl. Spectrosc. 33*:9 (1979).
146. C. D. Ford and R. J. Hurtubise, *Anal. Chem. 51*:659 (1979).
147. T. Vo-Dinh and R. B. Gammage, *Anal. Chem. 50*:2054 (1978).
148. T. Vo-Dinh and R. B. Gammage, *Anal. Chim. Acta 107*:261 (1979).
149. J. B. F. Lloyd, *Analyst 103*:775 (1978).

150. N. J. Turro, K. C. Liu, M. F. Chow, and P. Lee, *Photochem. Photobiol.* *27*:523 (1978).
151. L. J. C. Love, M. Skrilec, and J. G. Habarta, *Anal. Chem.* *52*:754 (1980).
152. J. D. Winefordner, *Acc. Chem. Res.* *2*:361 (1969).
153. J. J. Mousa and J. D. Winefordner, *Anal. Chem.* *46*:1195 (1974).
154. National Research Council (U.S.), *Particulate Polycyclic Organic Matter*, National Academy of Sciences, Washington, D.C., 1972, p. 298.
155. N. B. Colthup, L. H. Daly, and S. E. Wiberley, *Introduction to Infrared and Raman Spectroscopy*, 2nd ed., Academic Press, New York, 1975.
156. A. L. Smith, *Applied Infrared Spectroscopy: Fundamentals, Techniques, and Analytical Problem Solving*, Wiley, New York, 1979.
157. L. J. Bellamy, *The Infra-Red Spectra of Complex Molecules*, 3rd ed., Chapman & Hall, London, 1975.
158. J. E. Stewart, *Infrared Spectroscopy: Experimental Methods and Techniques*, Dekker, New York, 1970.
159. R. S. McDonald, *Anal. Chem.* *52*:361R (1980); *54*:1250 (1982).
160. J. L. Koenig and D. Kormos, *Appl. Spectrosc.* *33*:349 (1979).
161. F. R. Brown, S. Friedman, L. E. Makovsky, and F. K. Schweighardt, *Appl. Spectrosc.* *31*:241 (1977).
162. I. Schwager and T. F. Yen, *Anal. Chem.* *51*:569 (1979).
163. P. R. Griffiths, *Chemical Infrared Fourier Transform Spectroscopy*, Wiley, New York, 1975.
164. J. R. Ferraro and L. J. Basile, *Fourier Transform Infrared Spectroscopy: Applications to Chemical Systems*, Vols. 1 and 2, Academic Press, New York, 1978 and 1979.
165. P. R. Griffiths, *Transform Techniques in Chemistry*, Plenum Press, New York, 1978.
166. P. R. Griffiths, C. T. Foskett, and R. Curbelo, *Appl. Spectrosc. Rev.* *6*:31 (1972).
167. J. B. Bates, *Science* *191*:31 (1976).
168. G. Horlick, *Appl. Spectrosc.* *22*:617 (1968).
169. J. L. Koenig, *Appl. Spectrosc.* *29*:293 (1975).
170. N. Sheppard, in *Molecular Spectroscopy*, A. R. West (Ed.), Heyden, London, 1976, p. 149.
171. D. W. Green and G. T. Reedy, in *Fourier Transform Infrared Spectroscopy: Applications to Chemical Systems*, J. R. Ferraro and L. J. Basile (Ed.), Vol. 1, Academic Press, 1978, p. 1.
172. P. C. Painter and M. M. Coleman, *Fuel* *58*:301 (1979).
173. T. Hirschfeld, *Anal. Chem.* *48*:721 (1976).
174. R. M. Gendreau, P. R. Griffiths, L. E. Ellis, and J. R. Anfinsen, *Anal. Chem.* *48*:1907 (1976).
175. R. P. Goehner, *Anal. Chem.* *50*:1223 (1978).
176. R. J. Anderson and P. R. Griffiths, *Anal. Chem.* *47*:2339 (1975).
177. K. Scanlon, L. Laux, and J. Overend, *Appl. Spectrosc.* *33*:346 (1979).
178. L. A. Grybov and M. E. Elyashberg, *CRC Crit. Rev. Anal. Chem.* *8*:111 (1979).
179. J. S. Mattson, *Anal. Chem.* *49*:470 (1977).
180. P. R. Griffiths, in *Fourier Transform Infrared Spectroscopy: Applications to Chemical Systems*, J. R. Ferraro and L. J. Basile (Eds.), Vol. 1, Academic Press, New York, 1978, p. 143.

181. M. D. Erickson, *Appl. Spectrosc. Rev. 15*:261 (1979).
182. P. C. Uden, A. P. Carpenter, Jr., H. M. Hackett, D. E. Henderson, and S. Siggia, *Anal. Chem. 51*:38 (1979).
183. M. D. Erickson, S. D. Cooper, C. M. Sparacino, and R. A. Zweidinger, *Appl. Spectrosc. 33*:575 (1979).
184. D. W. Vidrine, in *Fourier Transform Infrared Spectroscopy: Applications to Chemical Systems*, J. R. Ferraro and L. J. Basile (Eds.), Vol. 2, Academic Press, New York, 1979, p. 129.
185. D. T. Kuehl and P. R. Griffiths, *Anal. Chem. 52*:1394 (1980).
186. H. E. Hallam, in *Vibrational Spectroscopy of Trapped Species*, H. E. Hallam (Ed.), Wiley, New York, 1973, p. 1.
187. T. Hirschfeld, *Appl. Opt. 17*:1400 (1978).
188. R. Cournoyer, J. C. Shearer, and D. H. Anderson, *Anal. Chem. 49*: 2275 (1977).
189. H. J. Sloane and R. J. Obremski, *Appl. Spectrosc. 31*:506 (1977).
190. T. Hirschfeld, *Appl. Spectrosc. 31*:550 (1977).
191. D. M. Hembree, E. R. Hinton, Jr., R. R. Kemmerer, G. Mamantov, and E. L. Wehry, *Appl. Spectrosc. 33*:477 (1979).
192. G. Mamantov, E. L. Wehry, R. R. Kemmerer, and E. R. Hinton, *Anal. Chem. 49*:86 (1977).
193. E. R. Hinton, G. Mamantov, and E. L. Wehry, *Anal. Lett. 12*:1347 (1979).
194. E. L. Wehry, G. Mamantov, D. M. Hembree, and J. R. Maple, in *Polynuclear Aromatic Hydrocarbons: Chemistry and Biological Effects*, A. Bjørseth and A. J. Dennis (Eds.), Battelle Press, Columbus, Ohio, 1980, p. 1005.
195. E. R. Hinton, Ph.D. dissertation, University of Tennessee, 1979.
196. G. Mamantov, E. L. Wehry, R. R. Kemmerer, R. C. Stroupe, E. R. Hinton, and G. Goldstein, in *Analytical Chemistry of Liquid Fuel Sources: Tar Sands, Oil Shale, Coal, and Petroleum*, P. C. Uden, S. Siggia, and H. B. Jensen (Eds.), Advances in Chemistry Ser. No. 170, American Chemical Society, Washington, D.C., 1978, p. 106.
197. G. T. Reedy, S. Bourne, and P. T. Cunningham, *Anal. Chem. 51*:1535 (1979).
198. S. Bourne, G. T. Reedy, and P. T. Cunningham, *J. Chromatogr. Sci. 17*:460 (1979).
199. E. L. Wehry, G. Mamantov, D. M. Hembree, and J. R. Maple, in *Polynuclear Aromatic Hydrocarbons: Chemistry and Biological Effects*, A. Bjørseth and A. J. Dennis (Eds.), Battelle Press, Columbus, Ohio, 1980, p. 1005.
200. D. M. Hembree, Ph.D. dissertation, University of Tennessee, 1980.
201. D. M. Hembree, A. A. Garrison, R. A. Crocombe, R. A. Yokley, G. Mamantov, and E. L. Wehry, *Anal. Chem. 53*:1783 (1981).
202. M. P. Fuller and P. R. Griffiths, *Anal. Chem. 50*:1906 (1978).
203. D. W. Vidrine, *Appl. Spectrosc. 34*:314 (1980).
204. M. G. Rockley, *Chem. Phys. Lett. 68*:455 (1979).
205. *Standard Infrared Grating Spectra*. Sadtler Research Laboratories, Inc., Philadelphia.
206. *Selected Infrared Spectral Data*. American Petroleum Institute Research Project No. 44, Thermodynamics Research Center, Texas A&M University, College Station, Tex.

207. C. D. Craver, *The Coblentz Society Desk Book of Infrared Spectra*, Coblentz Society, Kirkwood, Mo., 1977.
208. C. L. Fiske, G. W. A. Milne, and S. R. Heller, *J. Chromatogr. Sci. 17*: 441 (1979).
209. The Coblentz Society specifications for evaluation of research quality infrared spectra (Class II). *Anal. Chem. 47*: 945A (1975).
210. R. S. McDowell, *Advan. Infrared Raman Spectrosc. 5*: 3 (1978).
211. D. H. Whiffen, in *Molecular Spectroscopy*, A. R. West (Ed.), Heyden, London, 1977, p. 169.
212. D. A. Long, *Raman Spectroscopy*, McGraw-Hill, New York, 1977.
213. M. D. Morris and D. J. Wallan, *Anal. Chem. 51*: 182A (1979).
214. R. J. Thibeau, L. Van Haverbeke, and C. W. Brown, *Appl. Spectrosc. 32*: 98 (1978).
215. D. L. Gerrard and W. F. Maddams, *Appl. Spectrosc. 30*: 554 (1976).
216. R. P. Van Duyne, D. L. Jeanmaire, and D. F. Shriver, *Anal. Chem. 46*: 213 (1974).
217. J. M. Harris, R. W. Chrisman, F. E. Lytle, and R. S. Tobias, *Anal. Chem. 48*: 1937 (1976).
218. F. L. Galeener, *Chem. Phys. Lett. 48*: 7 (1977).
219. K. H. Levin and C. L. Tang, *Appl. Phys. Lett. 33*: 817 (1978).
220. C. A. Arguello, G. F. Mendes, and R. C. C. Leite, *Appl. Opt. 13*: 1731 (1974).
221. A. B. Harvey, *Anal. Chem. 50*: 905A (1978).
222. L. B. Rogers, J. D. Stuart, L. P. Goss, T. B. Malloy, Jr., and L. A. Carreira, *Anal. Chem. 49*: 959 (1977).
223. J. J. Barrett and M. J. Berry, *Appl. Phys. Lett. 34*: 144 (1979).
224. A. Owyoung, *IEEE J. Quantum Electron. 14*: 192 (1978).
225. B. F. Levine and C. G. Bethe, *Appl. Phys. Lett. 36*: 245 (1980).
226. J. P. Haushalter, G. P. Ritz, D. J. Wallan, K. Dien, and M. D. Morris, *Appl. Spectrosc. 34*: 144 (1980).
227. D. J. Futoma, S. R. Smith, and J. Tanaka, *CRC Crit. Rev. Anal. Chem. 13*: 117 (1982).
228. M. L. Lee, M. V. Novotny, and K. D. Bartle, *Analytical Chemistry of Polycyclic Aromatic Compounds*, Academic Press, New York, 1981.
229. K. D. Bartle, M. L. Lee, and S. A. Wise, *Chem. Soc. Rev. 10*: 113 (1981).
230. E. Voigtman, A. Jurgensen, and J. D. Winefordner, *Anal. Chem. 53*: 1921 (1981).
231. A. Jurgensen, E. L. Inman, Jr., and J. D. Winefordner, *Anal. Chim. Acta 131*: 187 (1981).
232. J. N. Miller, *Standards in Fluorescence Spectrometry*, Chapman & Hall, London, 1981.
232a. N. Omenetto and J. D. Winefordner, *CRC Crit. Rev. Anal. Chem. 13*: 59 (1981).
233. T. Vo-Dinh and P. R. Martinez, *Anal. Chim. Acta 125*: 13 (1981).
234. T. Vo-Dinh, R. B. Gammage, and P. R. Martinez, *Anal. Chem. 53*: 253 (1981).
235. T. Kato, S. Yokoyama, and S. Yozo, *Fuel 59*: 845 (1980).
236. J. B. F. Lloyd, *Anal. Chem. 52*: 189 (1980).
237. H. W. Latz, A. H. Ullman, and J. D. Winefordner, *Anal. Chem. 52*: 191 (1980).

238. J. A. Warren, J. M. Hayes, and G. J. Small, *Anal. Chem. 54*:138 (1982).
239. T. G. Dietz, M. A. Duncan, M. G. Liverman, and R. E. Smalley, *Chem. Phys. Lett. 70*:246 (1980).
240. J. R. Maple and E. L. Wehry, *Anal. Chem. 52*:920 (1980).
241. J. R. Maple and E. L. Wehry, *Anal. Chem. 53*:266 (1981).
242. E. L. Wehry, R. R. Gore, and R. B. Dickinson, Jr., in *Lasers in Chemical Analysis*, G. M. Hieftje, J. C. Travis, and F. E. Lytle (Eds.), Humana Press, New York, 1981, p. 201.
243. J. R. Maple and E. L. Wehry, *Anal. Chem. 53*:1244 (1981).
244. Y. Yang, A. P. D'Silva, and V. A. Fassel, *Anal. Chem. 53*:894 (1981).
245. Y. Yang, A. P. D'Silva, and V. A. Fassel, *Anal. Chem. 53*:2107 (1981).
246. J. C. Brown, J. M. Hayes, J. A. Warren, and G. J. Small, in *Lasers in Chemical Analysis*, G. M. Hieftje, J. C. Travis, and F. E. Lytle (Eds.), Humana Press, New York, 1981, p. 237.
247. I. Chiang, J. M. Hayes, and G. J. Small, *Anal. Chem. 54*:315 (1982).
247a. L. A. Bykovskaya, R. I. Personov, and Y. V. Romanovskii, *Anal. Chim. Acta 125*:1 (1981).
248. M. Ewald, M. Lamotte, F. Redero, M. J. Tissier, and P. Albrecht, *Phys. Chem. Earth 12*:275 (1980).
249. M. M. Corfield, H. L. Hawkins, P. John, and I. Soutar, *Analyst 106*:188 (1981).
250. C.-N. Ho and I. M. Warner, *Trends Anal. Chem. 1*:159 (1982).
251. G. D. Christian, J. B. Callis, and E. R. Davidson, in *Modern Fluorescence Spectroscopy*, E. L. Wehry (Ed.), Vol. 4, Plenum Press, New York, 1981, p. 111.
252. I. M. Warner and L. B. McGown, *CRC Crit. Rev. Anal. Chem. 13*:155 (1982).
253. G. M. Hieftje and G. R. Haugen, *Anal. Chim. Acta 123*:255 (1981).
254. J. H. Richardson, K. M. Larson, G. R. Haugen, D. C. Johnson, and J. E. Clarkson, *Anal. Chim. Acta 116*:407 (1980).
255. M. J. Wirth, A. Koskelo, and M. J. Sanders, *Appl. Spectrosc. 35*:14 (1981).
256. E. S. Yeung and M. J. Sepaniak, *Anal. Chem. 52*:1465A (1980).
257. R. J. Hurtubise, *Solid Surface Luminescence Analysis: Theory, Instrumentation, and Applications*, Dekker, New York, 1981.
258. R. T. Parker, R. S. Freedlander, and R. B. Dunlap, *Anal. Chim. Acta 119*:189 (1980).
259. R. T. Parker, R. S. Freedlander, and R. B. Dunlap, *Anal. Chim. Acta 120*:1 (1980).
260. T. Vo-Dinh, R. B. Gammage, and P. R. Martinez, *Anal. Chim. Acta 118*:313 (1980).
261. M. Skrilec and L. J. Cline Love, *Anal. Chem. 52*:1559 (1980).
262. L. J. Cline Love, J. G. Habarta, and M. Skrilec, *Anal. Chem. 53*:437 (1981).
263. H. Singh and W. L. Hinze, *Anal. Lett. 15*:221 (1982).
264. D. M. Hembree, A. A. Garrison, R. A. Crocombe, R. A. Yokley, E. L. Wehry, and G. Mamantov, *Anal. Chem. 53*:1783 (1981).

9

Analysis of Polycyclic Aromatic Hydrocarbons by Thin-Layer Chromatography

JOAN M. DAISEY / Institute of Environmental Medicine,
New York University Medical Center, New York, New York

I. Introduction

A. Purpose

The purpose of this review is to present an overview of the applications of thin-layer chromatography (TLC) to the analysis of polycyclic aromatic hydrocarbons (PAH) in a variety of media, with particular emphasis on practical aspects and recent advances in this field. Analytical methods for PAH vary considerably depending upon the sample matrix, e.g., food, water, or airborne particulate matter. For the details of sample preparation prior to PAH analysis by TLC, the reader is referred to the original publications cited. There are also many excellent books available as an introduction to TLC in general [1-4] and a number of older reviews specific to PAH compounds [5-10].

B. General Description of the Technique

In TLC, mixtures are separated by the movement of a solvent or solvent mixture, the moving phase, through a thin layer of adsorbent, the stationery phase. Compounds are separated due to their differing affinities for the moving and stationary phases of the TLC system.

The adsorbent layer is generally 250 or 500 μm thick and is coated on a support such as a glass plate, metal foil, or plastic sheet. The commercially prepared plates which are most widely used are 10 × 20 cm or 20 × 20 cm. Silica gel, alumina, and powdered cellulose are the most widely used adsorbents for PAH analyses. A binder is often incorporated into the adsorbent to make it more cohesive and hold it to the plate so that the plate can be easily handled without dislodging the thin layer. Silica gel, in particular, is rarely used without a binder such as gypsum. The chemical composition of the adsorbent and its water content and surface area will all influence the effectiveness of a TLC system for a given separation.

The solvent used as the moving phase or developer in a TLC system may be polar or nonpolar, with one component or a mixture of two or more. It must have characteristics which make it compatible with the adsorbent and and suitable for the desired separation.

Solutions of PAH mixtures which are to be separated are applied to the adsorbent with a micropipette, microsyringe, or capillary tube about 1.5 cm from one end of the TLC plate as shown in Fig. 1. Care is taken not to disturb the surface of the adsorbent and to keep the spot at the origin or site of application 1-2 mm in diameter. When the solvent has evaporated from the spot(s), the end of the plate closest to the origin is placed in a developing solvent in a tank (cf. Fig. 1) or sandwich chamber. If a tank is used, the atmosphere of the tank must be saturated with solvent vapor. The level of the solvent is kept below the origin, as shown in Fig. 1(a). In ascending chromatography, as shown, the solvent moves up the plate through the adsorbent layer by capillary action. Individual PAH compounds are separated due to differences in their affinities for the adsorbent and solvent which results in characteristic rates of travel up the plate. When the height of the solvent front reaches 10 or 15 cm, the plate is removed from the chamber and the developing solvent is allowed to evaporate in a hood. The location of individual PAH is readily apparent if they are colored compounds. Most PAH are detected by their fluorescence under ultraviolet (UV) light.

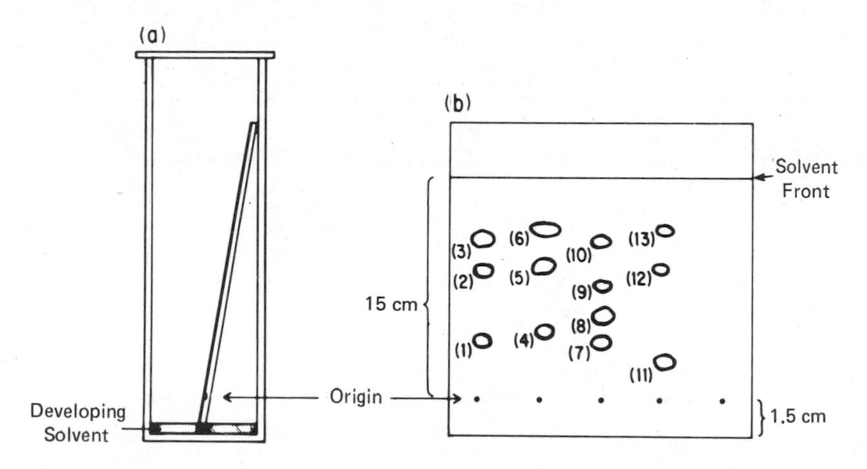

Figure 1. (a) Side view of thin-layer plate in a chromatography tank during development. (b) Front view of a thin-layer chromatographic separation of polycyclic aromatic hydrocarbons on 20% acetylated cellulose with n-propanol-acetone-water (2:1:1): (1) benzo[a]pyrene; (2) benz[a]anthracene; (3) benzo[e]pyrene; (4) anthracene; (5) dibenz[a,h]anthracene; (6) fluoranthene; (7) benzo[b]fluoranthene; (8) chrysene; (9) benzo[k]fluoranthene; (10) benzo[ghi]perylene; (11) triphenylene; (12) perylene; (13) pyrene.

The position of each compound can then be characterized by the R_f or R_B value:

$$R_f = \frac{\text{distance of the compound from the origin}}{\text{distance of the solvent front from the origin}} \tag{1}$$

$$R_B = \frac{\text{distance of the compound from the origin}}{\text{distance of benzo[a]pyrene from the origin}} \tag{2}$$

$$hR_f = 100 \times R_f \tag{3}$$

R_f values are never greater than 1, whereas R_B can be greater than 1. A permanent record of the chromatogram is generally made by sketching the chromatogram in a notebook and recording R_f values. Photocopies or photographs may also be used. The photocopier will record only those PAH which absorb UV radiation at 366 nm.

Individual PAH compounds can be quantitated in situ by manual measurements of the areas of the spots, by transmission or absorption spectrophotometry, or by spectrofluorimetry. Alternatively, the spots can be removed from the plate, eluted from the adsorbent, and then quantitated.

In addition to the simple TLC technique described here, there are a number of special development techniques which can be used to advantage [1-4]. With the two-dimensional development technique, the mixture is applied to the plate a few centimeters from one corner and developed. It is then turned 90° and developed in a second direction, generally with a different solvent, as shown in Fig. 2.

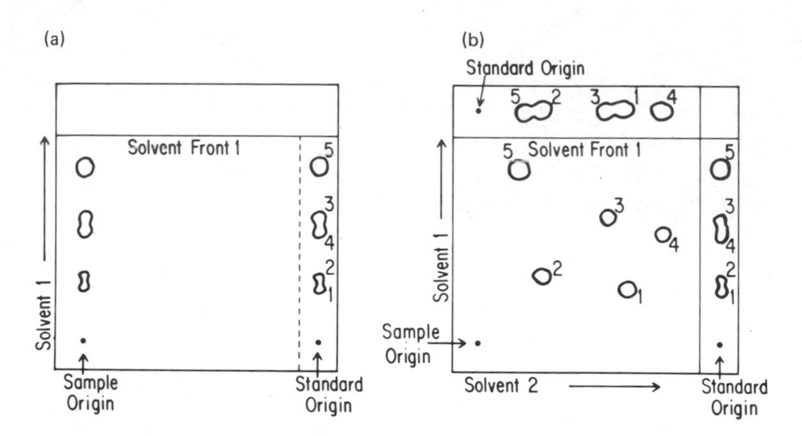

Figure 2. Schematic diagram of a two-dimensional development of a TLC plate to separate five compounds (1-5): (a) following development in one direction; (b) following second development.

There are two other relatively simple techniques which can be used to give increased resolution of individual PAH compounds in mixtures or to reduce background interferences. With the multiple development method, the same solvent is used several times for development. The plate is dried between each development. The R_f value after the nth development, ${}^nR_f$, can be calculated from the relationship:

$$ {}^nR_f = 1 - (1 - R_f)^n \tag{4}$$

Stepwise development employs two or more developing solvents. The plate is developed 5-6 cm with the first solvent, dried, and then developed with the second solvent to 10-12 cm.

C. Advantages and Disadvantages

The principal advantages of TLC analytical methods for PAH compounds are speed, simplicity, low cost, and the small sample size required. Samples can be applied to a plate, chromatographed and quantitated in a matter of hours. The technique, as described above, is simple to use and requires very little investment in equipment. The basic equipment and supplies, such as a developing chamber, commercially prepared plates, and solvents, can be purchased for under $200. Relatively inexpensive equipment is also available for investigators wishing to prepare their own plates. Sample size requirements are minimal. Under optimal conditions, as little as a few nanograms of benzo[a]pyrene (BaP) can be determined using spectrophotofluorometric methods.

In contrast to GC (gas chromatographic), GC/MS (gas chromatographic/mass spectrometric), and HPLC (high-performance liquid chromatographic) methods, samples and standards can be run simultaneously and multiple analyses can be carried out simultaneously. TLC, like HPLC, can be used for all types of compounds (not gaseous), whereas GC and GC/MS are limited to compounds that have appreciable vapor pressures or can be derivatized to more volatile products which do not undergo decomposition at the temperatures used for analysis. TLC can also be used to advantage to quickly and inexpensively screen samples for the presence of PAH compounds or to determine optimal solvent systems for the separation of mixtures by either TLC or HPLC.

There are also some disadvantageous features of TLC analysis. In general, there are fewer theoretical plates for TLC than for many other chromatographic techniques presently available and, therefore, less inherent resolution capability. However, through a careful selection of excitation and emission wavelengths, the specificity of spectrofluorometric analysis can partially compensate for the lower resolution of TLC.

TLC methods have not been as fully automated, with respect to sample application and quantitation, as have GC and HPLC methods. In addition, if individual compounds or fractions must be recovered from the adsorbent for quantitation, the methods are quite tedious and time consuming. R_f values are not generally as reproducible as retention times for GC and HPLC, although the reproducibility can be optimized with appropriate measures.

D. Historical Background

During the 1950s, concern over human exposures to carcinogenic PAH in food, in industrial and ambient atmospheres, in cigarette smoke, and in other media prompted the development of analytical methods for these compounds [11-19]. The most widely used analytical method for PAH compounds in this decade entailed a column chromatographic separation coupled with absorption and/or fluorescence spectrophotometry. This method required relatively large samples and several days for a complete analysis. In addition, problems were sometimes encountered in the resolution of certain isomers.

In parallel with the growing interest in analysis for PAH compounds in the human environment, the technique of TLC was rapidly developing. After 1958, when TLC equipment and materials became commercially available, the theory and applications of TLC expanded enormously. These parallel developments merged in the decade of the 1960s.

In 1962 Wieland and Determann [20] reported a TLC separation of eight PAH standards on acetylated cellulose with methanol-ethanol-water (4:4:1). A similar report from Badger and coworkers [21] followed in 1963, for a slightly different set of PAH standards separated on acetylated cellulose. The use of silica gel and alumina adsorbents for the separation of PAH standards was reported by Kucharczyk and co-workers [22] in 1963. In 1964 the use of adsorbents (silica gel and alumina) impregnated with complexing agents such as caffeine and 2,4,7-trinitrofluorenone was reported for separation of parent PAH compounds [23,24]. Hydro derivatives, synthesized by reduction, were separated from parent PAH compounds by TLC on silica gel impregnated with 2,4,7-trinitrofluorenone or trinitrobenzene by Harvey and Halonen [25]. All of this work was directed toward the separation of PAH compounds in standard mixtures.

One of the first applications of TLC to the analysis of PAH was that of Domsky and co-workers [26], who used TLC (silica gel/benzene-isooctane, 1:19) to separate 7,12-dimethylbenz[a]anthracene (DMBA) from interfering materials in extracts of tissues from mice treated with DMBA. The DMBA was then extracted from the silica gel and quantitated by absorption spectrophotometry.

During the 1960s most of the emphasis was on the development of analytical methods for PAH in airborne pollutants and in food products. In 1964, Sawicki et al. [27] published a key paper in the development of TLC applications to PAH analysis. In it, they described systematic investigations of three different TLC systems for the separation and analysis of PAH in the benzene-soluble organic (BSO) fraction of airborne particulate matter and source emissions. They found that alumina/pentane-ether (95:5) gave the best separation of the BSO extracts but that individual PAH could be better resolved on acetylated cellulose/ethanol-toluene-water (17:4:4) or cellulose/dimethylformamide-water (1:1). Concentrations of BaP could be determined with as little as 0.2 mg of BSO containing 0.05 μg of the BaP, using spectrofluorometric analysis, in only a few hours. The column chromatographic method that was in general use at this time required approximately 20-150 mg of BSO (an urban air sample of about 2000 m^3 or more) and several days to complete an analysis. In numerous publications during the 1960s, Sawicki's group [10,27-47] reported developments in and applications of TLC and ancillary techniques which permitted the resolution, identification, and quantitation of PAH [27-34] aza-arenes [38-40], and other related derivatives [41-47] in source emissions and airborne particulate matter. In 1967, Sawicki and co-workers [34] compared 11 methods for the analysis of BaP in airborne particulate matter, many of which incorporated TLC as part of the analytical scheme. In 1972, the American Public Health Association (see Katz [48]) adopted one of the TLC methods as a tentative method of microanalysis for BaP in airborne particulate matter and source effluents.

For the analysis of PAH in food products, one of the first TLC methods reported was that of Genest and Smith [49] which employed silica gel G/2,2,4-trimethylpentane-benzene (97:3) and multiple development to reduce background interferences in the determination of BaP. Howard and co-workers [50] developed a method for the determination of PAH in foods and modified [51] the method of Genest and Smith for BaP to obtain a lower limit of detection of 0.5 ppb. The general method was later shortened and modified [52] to apply to the determination of PAH in a variety of food products. A collaborative study [53] was conducted of the general method for PAH analysis developed by Howard and co-workers [52]. Average percent recovery at the 10 ppb level for four PAH compounds ranged from 70 to 90%. The standard deviations of the recoveries within and between laboratories did not vary more than 12.7%.

The methods developed by Howard and co-workers were widely adopted. The method for BaP in smoked foods was adopted as an official method by the Association of Official Analytical Chemists (AOAC) in 1972 [54] and accepted by the International Union of Pure and Applied Chemistry (IUPAC) as a recommended method in 1975 [55]. The International Agency for Research on Cancer (IARC) has also included these methods in their publication on selected methods of analysis for environmental carcinogens [56].

The decade of the 1960s was in many respects the "Golden Age" of the development of TLC methods for PAH. During the 1970s, the emphasis shifted to methods which employed GC and HPLC, incorporating automated injectors and integrators as well as interfaces with mass spectrometers.

II. Principles of TLC Separations of PAH

The properties of the adsorbent layer and developing solvent(s) in a TLC system determine the mechanism of separation and, thereby, the effectiveness of that TLC system for a given application. The TLC systems which have been used most frequently in the analysis of PAH are shown in Table 1. R_B values for many PAH compounds for selected TLC systems are shown in Table 2 [25,27,57,58]. These systems may be classified according to mechanisms of separation as adsorption or partition chromatography, although in actuality both mechanisms often contribute to a given separation.

A. Adsorption TLC

In adsorption chromatography, the stationary phase—the adsorbent—is polar in nature and the moving phase is nonpolar. Oxides, hydrated oxides, and salts such as silica gel and alumina are the most frequently used adsorbents. The PAH compounds are repeatedly adsorbed and desorbed from the surface of the adsorbent via surface-induced dipole moments. Differences in the adsorption characteristics of PAH compounds determine differences in rates of migration on the plate and thus effect separations. Hydrogen bonding appears to play a role as well in the case of silica gel since PAH compounds are Lewis bases [59,60]. Both silica gel and alumina tend to separate PAH compounds on the basis of ring size, i.e., those compounds with more condensed rings and more polarizable π electrons interact more strongly with the surface of the absorbent and thus have lower R_f values, as can be seen in Table 2. The planarity of the ring system also affects R_f values.

Adsorption systems are effective in separating PAH from other classes of compounds in a mixture and have often been used as a preseparation or cleanup technique to isolate a PAH fraction or subfractions [27,61-64]. For example, comparing alumina (pentane-ether, 19:1) to two partition chromatographic TLC systems, Sawicki and co-workers [27] noted that the alumina gave the best separation of the organic fraction of airborne particulate matter but the poorest separation of the PAH compounds.

B. Partition TLC

In partition chromatography, the compounds to be separated are partitioned or distributed between a stationary liquid phase which is part of the thin layer and a moving liquid phase. In normal-phase partition TLC, the stationary phase is a polar or hydrophilic liquid, such as water, which is bound as water of hydration to a material such as cellulose. The moving phase is a nonpolar or hydrophobic solvent. In reverse-phase chromatography, the stationary phase is hydrophobic and the solvent is hydrophilic. A paraffin-impregnated thin layer and acetylated cellulose thin layer used with a polar solvent are examples of reverse phase TLC systems. Acetylation of the hydroxyl groups of cellulose increases the hydrophobic or nonpolar nature of cellulose. In many instances (see Table 1), two organic solvents are used in combination with water. The organic solvent of intermediate polarity functions as a homogenizer, making the polar and nonpolar solvents miscible.

In partition chromatography, compounds are separated by being repeatedly partitioned between the stationary and moving phases of the TLC system.

Table 1. Thin-Layer Chromatographic Systems Commonly Used for the Analysis of Polycyclic Aromatic Hydrocarbons[a]

	Adsorbents				
	Silica gel (SG)	Alumina (A)	Cellulose (C)	Acetylated Cellulose (AC)	Silica gel impregnated with a complexing agent (SG/C)
Chemical nature of adsorbent	SiO_2, acidic to neutral pH	Al_2O_3, basic to neutral pH	CH_2OH, O, HO, OH, O, n	CH_2OAc, AcO, OAc, O, n	SiO + complexing agent such as trinitrofluorenone or p-benzoquinone
Typical developing solvents[b]	(1) Benzene (2) Toluene (3) Cyclohexane-benzene 1:1.5 (4) Benzene-heptane 1:4	(1) Pentane-ether 19:1 (2) Hexane-ether (3) Isooctane	(1) Dimethylformamide-water 1:1	(1) Ethanol-toluene-water 17:4:4 (2) Ethanol-dichloromethane-water (20:10:1) (3) n-Propanol-acetone-water (2:1:1)	(1) Benzene-heptane (1:4) (2) Carbon tetrachloride
Predominant mechanism of separation	Adsorption	Adsorption	Partition (normal phase)	Partition (reverse phase)	Complexation
Application	Class separation of PAH	Class separation of PAH, separation of individual PAH	Separation of individual PAH	Separation of individual PAH	Separation of PAH from alkyl and hydro derivatives

[a]Abbreviations for TLC systems are given according to adsorbent (see abbreviations at head of column) and solvent system, designated as 1, 2, 3, etc.

[b]Solvent proportions given by volume.

Table 2. R_B Values of PAH Compounds in Various TLC Systems

Compound	Molecular formula	TLC System[a]						
		SG/4	A/1	C/1	AC/1[b]	AC/2[c]	AC/3[b]	SG/C[d]/1
Phenanthrene	$C_{14}H_{10}$	0.95	1.13	1.99	3.74	—	—	0.73
Anthracene	$C_{14}H_{10}$	0.90	1.14	1.99	3.33	—	—	0.76
Pyrene	$C_{16}H_{10}$	1.21	1.25	1.72	3.16	2.7	3.2	1.13
Fluoranthene	$C_{16}H_{10}$	1.02	1.09	1.89	2.92	2.9	3.2	0.97
Chrysene	$C_{18}H_{12}$	1.00	1.10	1.75	—	1.4	1.6	1.00
Triphenylene	$C_{18}H_{12}$	1.00	1.07	1.49	—	—	0.8	0.96
Benz[a]anthracene	$C_{18}H_{12}$	1.00	1.03	1.47	2.70	2.3	2.5	0.98
7,12-Dimethylbenz[a]anthracene	$C_{20}H_{16}$	0.90	—	—	—	—	2.9	0.89
Benzo[a]pyrene	$C_{20}H_{12}$	1.00	1.00	1.00	1.00	1.00	1.00	1.00
Benzo[e]pyrene	$C_{20}H_{12}$	—	1.04	1.16	2.94	2.8	3.0	—
Perylene	$C_{20}H_{12}$	0.93	0.91	1.14	2.86	—	2.5	0.94
Benzo[k]fluoranthene	$C_{20}H_{12}$	—	0.98	1.03	2.40	2.0	—	—
Benzo[b]fluoranthene	$C_{20}H_{12}$	—	—	—	—	1.7	—	—

(continued)

Table 2 (continued)

Compound	Molecular formula	TLC System[a]						
		SG/4	A/1	C/1	AC/1[b]	AC/2[c]	AC/3[b]	SG/C[d]/1
Anthanthrene	$C_{22}H_{12}$	0.90	0.71	0.70	2.17	3.6	1.4	0.92
Benzo[ghi]perylene	$C_{22}H_{12}$	0.86	0.89	0.69	3.04	2.8	3.0	0.87
Dibenz[ah]anthracene	$C_{22}H_{14}$	0.30	0.74	0.66	2.92	2.3	2.6	0.72
Coronene	$C_{24}H_{12}$	—	0.46	0.37	2.87	3.0	—	—
Dibenzo[a,i]pyrene	$C_{24}H_{12}$	0.32	0.68	0.45	2.41	—	—	0.77
Dibenzo[a,h]pyrene	$C_{24}H_{12}$	0.31	—	—	—	—	—	0.75
From Reference No.:		[25]	[27]	[27]	[27]	[57]	[58]	[25]

[a]Systems designated by adsorbent and solvent following abbreviations in Table 1, e.g., SG/3 = silica gel/solvent system No. 3.

[b]20% acetylated cellulose.

[c]30% acetylated cellulose; R_B estimated from Fig. 4 in Ref. 57.

[d]Trinitrofluorenone used as a complexing agent.

The separation of PAH compounds in these systems is thus determined largely by differences in their solubilities in the two phases that affect the rates of migration and R_f values. Both molecular size and shape affect the solubilities of PAH compounds in various solvents [2,65]. As partition chromatographic systems are quite sensitive to changes in molecular size and shape, they are effective in separating certain PAH isomers which are difficult to resolve such as benzo[a]pyrene (BaP), benzo[e]pyrene (BeP), and perylene [27,58,62].

C. Complexation TLC

A third approach to TLC separation of PAH compounds involves the use of alumina or silica gel impregnated with organic compounds which are electron acceptors, such as 2,4,7-trinotrofluorenone or caffeine, which can form electron donor-acceptor or charge-transfer complexes with PAH compounds [23-25,66]. The R_f values of PAH compounds on such plates depend inversely on structural features which enhance complexation, such as the number of condensed rings, their planarity, and the presence of alkyl substituents [25]. Schenk and co-workers [60] have reported evidence of secondary effects in such TLC systems related to the concentration of the electron acceptor in the thin layer. Increased concentrations of the electron acceptor can reduce the number of available hydroxyl sites on a silica gel layer. When the PAH donor interacts more strongly with hydroxyl groups than with the added electron acceptor, the R_f on the impregnated plate is greater than for the silica gel alone due to the reduced number of hydroxyl sites.

To date, complexation TLC systems have not been used for the analysis of environmental samples, although they are potentially useful for the separations of alkyl and hydro derivatives of PAH [25]. Application of such TLC systems has been limited chiefly by the lack of a commercially produced plate.

III. Applications

TLC has been used in the analysis of environmental samples, both as a preparative technique to isolate PAH from other classes of compounds and as a means of resolving individual PAH compounds. TLC is generally used in combination with other separation techniques such as liquid-liquid extraction (LLE), column chromatography (CC), GC, HPLC, and GC/MS. A wide variety of TLC methods have been developed and applied to the analysis of PAH in environmental samples. The choice of method for a given application will depend upon the sample matrix, sample size, time and cost restraints, and available equipment. Applications of TLC to the analysis of particulate matter from ambient air and source emissions were reviewed in 1972 by Sawicki and Sawicki [10]. More recently Herod and James [79] reviewed all types of analytical methods for PAH, with particular reference to coke oven emissions. Amos [9] has recently reviewed chromotographic methods for analysis of heavy distillates, residues, and crude oils.

A. Preparative TLC in the Analysis of PAH

Table 3 presents a summary of analytical methods for PAH in various media which use TLC as a preparative technique [58,61,67-78]. For most types of samples a preliminary extraction and/or concentration step is required.

Table 3. Applications of Preparative TLC

Media	Analytical scheme	TLC system	Comments	Ref.
Airborne particulate matter	TLC → GC	Silica gel G/cyclohexane-benzene (1:1.5)	Isolation of PAH fraction by TLC	61
	TLC → GC	2:1 Alumina-acetylated cellulose/(1) pentane; (2) ethanol-toluene-water (17:4:4)	Isolation of PAH fractions by two dimensional TLC	67
	TLC → TLC → GC	(1) Silica gel/benzene (2) 2:1 Alumina-acetylated cellulose/(1) pentane-ether (98.5:1.5), (2) ethanol-toluene-water (17:4:4)	Two-step TLC; two-dimensional development for the second system	68-70
	TLC → GC	Acetylated cellulose/1-propanol-acetone-water (2:1:1)	TLC isolation of three PAH fractions	58
Sediment	CC[a] → TLC → GC	Kieselgel G/hexane	TLC isolation of PAH fraction followed by GC analysis	71
	TLC	Silica Gel G/hexane with 1% NH_4OH	Separation of aromatics from aliphatics	64
	CC → TLC → HPLC	Kieselguhr G-20% acetylated cellulose/methanol-ether-water (4:4:1)	TLC isolation of 4 PAH fractions	72
Refinery wastewater	TLC → HPLC	Silica gel/(1) trichlorotrifluoroethane; (2) benzene	Two step TLC development used to separate PAH fraction	73

Tar, pitch, bitumen	TLC → HPLC	Silica Gel G-Bentone 34/toluene	TLC isolation of PAH fraction	74
Heavy oils	TLC	Three-stage adsorbent layer-Florisil G, aluminum oxide G and silica gel G/cyclohexane-benzene-ethyl acetate-acetone (380:1:1)	TLC separation into saturated monoaromatic and polyaromatic fraction	75
	IEC[b] → GPC[b] → TLC	2:1 alumina-acetylated cellulose/ether-toluene-cyclohexane (2:1:1)	Two-dimensional development	76
Shale oils	TLC → GC [c]	Alumina or silica gel/hexane	Multiple developments with increasing concentrations of chloroform or ether for improved separation	77
Mineral oils	TLC → MS	Silica gel/light petroleum	Separation of PAH from other fractions	78

[a] CC = column chromatography.
[b] IEC = ion exchange chromatography; GPC = gel permeation chromatography.
[c] GC^2 = glass capillary gas chromatography.

Soxhlet extraction with an organic solvent, for example, is used to separate organic from inorganic materials in airborne particles and in sediments. Details of such preliminary separations are omitted from Table 3 for simplicity but can be found in the original publications cited therein.

Following preparative TLC, the isolated PAH fraction(s) are generally located on the plate by their fluorescence, extracted from the absorbent, and further separated by a second TLC system, by GC or by HPLC.

TLC has been used as a preparative technique in a number of instances [58,67-72] in order to resolve isomers which are difficult to separate on most GC columns and some HPLC columns, e.g., BaP, BeP, and perylene. Although this type of analysis requires considerable time for the GC or HPLC analyses (generally an hour or more for each fraction) many individual PAH compounds can be resolved. As capillary column GC has become more widely available and HPLC columns have continued to improve, the need for this type of preparative technique has decreased.

Silica gel in combination with a nonpolar solvent is the most commonly used type of TLC system for separating PAH from other classes of compounds (see Table 3). This system has been used for airborne particulate matter, sediment, refinery waste water, tars, and oils. Grant and Meiris [74] combined silica gel with Bentone-34, an organo-clay, to separate a PAH fraction for subsequent HPLC analysis. In contrast to silica gel alone, this system resolved many individual PAH compounds and the R_f values were used as additional evidence of identity for individual PAH.

Peurifoy and co-workers [75] used a gradient-layer TLC plate with three stages or adjacent layers for fractionating heavy oils. With development from the Florisil G band to aluminum oxide to silica gel, monoaromatics moved to the top (silica gel) band while the polyaromatics were found in the aluminum oxide.

Martinu and Janak [80] have investigated the use of Porapak T, a copolymer of ethylvinylbenzene and divinylbenzene, as a TLC adsorbent for the separation of PAH compounds and their partially hydrogenated derivatives. With light petroleum used as a developing solvent, PAH compounds were more strongly adsorbed (R_f = 0.06-0.17) than the hydro derivatives (R_f = 0.13-0.71). Such a system appears to have considerable potential as a preparative technique for the separation of hydro derivatives from parent PAH.

B. TLC Analysis of Benzo[a]pyrene

BaP is generally a significant proportion of the PAH fraction in most media. It is a potent carcinogen in animals and can be separated from other PAH using TLC methods. For these reasons, it has often been used as an indicator or index of PAH levels.

In 1967 Sawicki and co-workers [34] compared a number of TLC methods and other methods for the determination of BaP in airborne particulate matter. The best resolution of BaP from other PAH was obtained using a two-dimensional development on a mixed adsorbent layer consisting of 2:1 alumina/cellulose acetate. However, recoveries were quite variable (40-75%), and the two-dimensional development required about 3 hr. A TLC system consisting of alumina/pentane-diethyl ether (19:1) gave recoveries of 95 ± 5% for BaP and required only an hour for development. In this method the spot containing

BaP was extracted with diethyl ether. The ether was evaporated and the residue dissolved in concentrated H_2SO_4 for spectrofluorometric quantitation. Even though BaP was not separated from BeP, benzo[k]fluoranthene, and perylene, with appropriate selection of excitation and emission wavelengths the spectrofluorometric determination was specific for BaP. The lower limit of detection for this assay was 3 ng BaP. The alumina/pentane-diethyl ether (19:1) TLC system was adopted by the American Public Health Association in 1972 as a method of microanalysis for BaP in particulate matter [48]. Lannoye and Greinke [81] investigated a number of variables affecting the accuracy and precision of this method and suggested a number of modifications. Due to the degradation of BaP which was observed, caused by peroxides and residual oxygen in diethyl ether, it was suggested that other solvents be used instead in the development and extraction steps [81]. Lannoye and Greinke [81] also suggested use of 520 nm as the excitation wavelength for spectrofluorometry to give better sensitivity and reduce interferences.

Keefer [82] suggest the use of magnesium hydroxide in place of alumina as an effective TLC adsorbent for the separation of PAH. The principal advantages of this adsorbent are a greater sample capacity and less sensitivity to variations in humidity and extent of activation. Magnesium hydroxide has not been used to any great extent for PAH separations but may be useful for preparative TLC in environmental analysis.

Most TLC methods for the isolation of BaP employ acetylated cellulose as an adsorbent. As noted by Sawicki and co-workers [27] in 1964, this absorbent with appropriate solvent gives a good separation of BaP from its isomers BeP, perylene, and benzo[k]fluoroanthene.

Schultz and co-workers [83] used 30% acetylated cellulose/methanol-ether-water (4:4:1) to isolate BaP from the organic fraction of airborne particulate matter. The organic fraction was isolated by vacuum sublimation of the filter at 300°C. A scanning spectrofluorometer was used to quantitate BaP on the TLC plate at levels as low as 1 ng with a reproducibility of ±10%.

Howard and co-workers [51] also used acetylated cellulose/ethanol-toluene-water (17:4:4) to isolate BaP in a PAH fraction isolated from smoked food. The PAH fraction was isolated by alcoholic KOH digestion combined with LLE and CC on Florisil. Recovery of BaP was 73-100% at the 2 ppb level for the overall analysis. Although the TLC separation required little time, considerable sample cleanup was required prior to the TLC separation.

Seifert and Steinbach [84] have employed a two-dimensional development on acetylated cellulose/95% ethanol-dichloromethane (8:2), followed by pyridine-methanol-water (3:5:2), to separate BaP from the organic fraction of airborne particulate matter. BaP was quantitated by spectrofluorometric scanning of the plate. The two-dimensional development gave a reliable separation of BaP from benzo[k]fluoranthene and anthanthrene.

C. TLC as a Final Separation Step in the Analysis of PAH

A summary of applications of the TLC as a final separation step in the analysis of PAH is presented in Table 4 according to sample type. The initial separation step in the overall separation scheme has been omitted from the table. In the case of airborne particulate matter or emissions from combustion sources, Soxhlet extraction or sonication with an organic solvent such as cyclohexane

Table 4. Applications of TLC as a Final Separation Step in the Analysis of Polycyclic Aromatic Hydrocarbons in Various Types of Samples

Type of sample	Separation scheme[a]	TLC system(s)	Detection and quantitation	Comments	Ref.
Airborne particulate matter, combustion source emissions	LLE → CC → TLC → CC	Acetylated cellulose/ethanol-toluene-water (17:4:4)	Extraction of PAH, UV spectrophotometry	Identities of PAH confirmed by fluorescence spectra; no estimate of time required for analysis	85
	CC → TLC	Acetylated cellulose/ethanol-dichloromethane-water (20:10:1)	Scanning spectrofluorometry, lower limit of detection was 1.1 ng BaP	Four PAH fractions isolated by Sephadex CC (11 hr); PAH fractions separated by TLC	57
	TLC	Acetylated cellulose/ethanol-dichloromethane-water (20:10:1)	Vacuum sublimation, low temperature (77°K) fluorescence in alkane solvent	Special equipment required; spectra highly specific and well resolved; no estimate of time required for analysis	86
	TLC → TLC	(1) Preparative: Alumina/hexane-ether (19:1) (2) Acetylated cellulose/n-propanol-acetone-water (2:1:1)	Extraction of individual PAH, spectrofluorometry	Four isomeric fractions of PAH separated on alumina; individual PAH resolved on acetylated cellulose	62,87
	TLC → MS	Cellulose/dimethylformamide-water (7:3)	Mass spectrometry; lower limit of detection in picogram range	MS more practical used in combination with GC	88

TLC → TLC (2D)	(1) Preparative TLC on silica gel (2) Mixed layer of 2:1 alumina-acetylated cellulose/two-dimensional development (1) pentane-ether (98.5:1.5); (2) ethanol-toluene-water (75:10:15)	Extraction, UV spectrophotometry	11 PAH separated	70,89
TLC → TLC (2D)	(1) Preparative TLC on silica gel (2) Mixed layer of silica gel-acetylated cellulose/two-dimensional development (1) ethanol-toluene-water (17:4:4); (2) cyclohexane-ethyl acetate (95:5)	Extraction; UV spectrophotometry	15 PAH separated and quantitated	90
TLC (2D)	Mixed layer of acetylated cellulose-silica gel-alumina/two-dimensional development (1) n-heptane-benzene (9:1); methanol-ether-water (4:1:1)	Extraction, spectrofluorometry		91

(continued)

Table 4 (continued)

Type of sample	Separation scheme[a]	TLC system(s)	Detection and quantitation	Comments	Ref.
	TLC (2D)	Mixed layer of 2:1 alumina-acetylated cellulose/two-dimensional development (1) hexane-toluene (9:1); (2) methanol-ether-water (4:4:1)	Extraction, spectrofluorometry or scanning spectrofluorometry	Industrial exhaust fumes analyzed; 66 PAH isolated	92
	TLC (2D)	Dual band plate (1) alumina-cellulose band (4 cm wide)/n-hexane-ether (19:1)-15 cm development (2) Acetylated cellulose band (16 cm wide)/methanol-ether-water (4:4:1)-10 cm development	Extraction, spectrofluorometry	76 PAH compounds separated; 31 identified; ~2 hr for TLC development	93, 94
Food products	LLE → CC → LLE → TLC → TLC	(1) Cellulose-dimethylformamide/isooctane (2) Acetylated cellulose/ethanol-toluene-water (17:4:4)	Extraction, UV spectrophotometry; identity confirmed by spectrofluorometry; lower limit of detection 0.02 ppb for BaP, benzo[k]-fluoranthene and perylene, 2 ppb for other PAH	Alcoholic KOH digestion for initial separation; 9 PAH compounds separated and quantitated; recoveries of 75-100% at 2 ppb level	52, 54, 95

Water and wastewater	LLE → CC[b] → TLC (2D)	2:1 alumina-acetylated cellulose/two dimensional development (1) n-hexane-benzene (9:1); (2) methanol-ether-water (4:4:1)	Scanning spectrofluorometry, lower limit of detection is 0.5-1 ng/liter for 2 1 samples	6 PAH compounds separated and used as indicators of total PAH	96
Sediments	LLE → TLC → TLC	(1) Preparative TLC on silica gel/cyclohexane-benzene (4:1) (2) Alumina-acetylated cellulose; two-dimensional development (1) pentane-ether (19:1); (2) ethanol-dichloromethane-water (20:20:1)	Extraction, spectrofluorometry; detection limit: 2.5 ppb for BaP; 1.5 ppb for perylene	7 PAH compounds separated, 2 quantitated; 72 ± 5% recovery for BaP; 30 g samples; at 500 ppb level, 5-7% reproducibility	97
Cigarette smoke condensate	TLC → TLC	(1) Preparative TLC on alumina/isooctane (2) Acetylated cellulose/ethanol-dichloromethane-water (20:10:1)	Scanning spectrofluorometry	Separation and quantitation of BaP, benz[a]anthracene and chrysene possible in 6-8 hr; 91-95% recovery; reproducibility of ±5%; sample size: 10-20 cigarettes	63

(continued)

Table 4 (continued)

Type of sample	Separation scheme[a]	TLC system(s)	Detection and quantitation	Comments	Ref.
Petroleum products for medicinal and cosmetic uses	CC → CC → TLC	Acetylated cellulose/ ethanol-dichloromethane-water (20:10:1)	Extraction, spectrofluorometry	8 PAH fractions from CC separated by TLC; overall recovery of ~60% for BaP at ppb level for 100-g samples; CC steps require about 100 hr	98

[a]LLE = liquid-liquid extraction; CC = column chromatography; TLC = thin-layer chromatography; TLC (2D) = TLC with two-dimensional development; MS = mass spectrometry.
[b]Only required for highly colored extracts.

or dichloromethane is frequently used to separate organic from inorganic material. Some investigators [83,88] have used vacuum sublimation as a first separation.

Alcoholic KOH digestion is commonly used as a first separation for food products. Both food products and sediments generally require a number of cleanup steps to isolate the PAH fraction, as can be seen in Table 4. In contrast to airborne particulate matter in which PAH are present at the parts per thousand level, PAH compounds are generally at the ppm and ppb levels in sediments and food products and thus must be separated from relatively large proportions of other materials. Potthast and Eigner [99] suggested the use of propylene carbonate as part of the initial cleanup in the analysis of PAH in meat products, since PAH are very soluble in this solvent and a fat-free extract of PAH can be obtained in the initial steps of sample treatment, thus reducing the time required for sample preparation.

Polycyclic aromatic hydrocarbons in water and wastewater are initially extracted and concentrated with a nonpolar solvent such as hexane or by adsorption on XAD resin. In the case of wastewater, an additional cleanup by column chromatography may be required [96] prior to TLC separation.

Petroleum and petroleum products present problems in the initial cleanup steps because the character of the bulk sample is very similar to that of the PAH compounds to be separated. TLC (see Table 3) and CC as well as ion exchange and gel permeation chromatography have been used to separate the PAH fraction from other types of hydrocarbons in these complex mixtures.

Two major types of TLC systems have been used to effect a final separation of PAH fraction(s). A one-step separation on acetylated cellulose with a mixed solvent consisting of ethanol, toluene (or dichloromethane), and water has been used frequently in the analysis of airborne particulate matter, source emissions, and cigarette smoke condensate (see Table 4). This system is quite effective in separating the isomers of BaP, BeP, perylene, and benzo[k]fluoranthene. In some instances, the PAH fraction is separated into subfractions by a preparative TLC step [62,87] or by CC [57,63,85,95,98]. This, of course, increases the resolution capabilities of the system.

The second major type of TLC system that has been used employs a mixed layer of aluminum oxide and acetylated cellulose in combination with two-dimensional development. Although only one sample can be chromatographed on a plate, this eliminates the need to extract and concentrate the sample after the first development step and also reduces sample losses. A nonpolar solvent mixture is used in the first development step to separate any other types of compounds which may still be present in the PAH fraction and to give some separation of the PAH compounds. After this, the plate is dried, turned 90°, and developed in a mixed solvent appropriate for acetylated cellulose in order to resolve individual PAH compounds.

Masushita and Suzuki [93] and Masushita et al. [94] have used a dual band plate rather than a mixed layer in order to gain the advantages of both alumina and acetylated cellulose on a single TLC plate. The sample is developed first on a band of alumina mixed with 5% cellulose (4 cm wide × 20 cm high) with hexane-ether (19:1) to a height of 15 cm. The plate is then turned 90° and developed in a second solvent across a band of acetylated cellulose (16 cm × 20 cm). Using this type of plate, 76 spots were separated of which 31 were identified.

Several other TLC systems for the separation of PAH have been investigated in recent years. Shiraishi and co-workers [100] effected a rapid separation of 28 PAH compounds by reverse-phase TLC on Kieselguhr G impregnated with 10% liquid paraffin with acetonitrile-water (13:7) as the moving phase. Polyamide thin layers have been used by Bories [101] with toluene-methanol (4:1) and by Armstrong and McNeely [102] with an aqueous micellar solution of sodium dodecylsulfate as the mobile phase. There appear to have been no applications of these systems to the analysis of PAH in environmental samples. Thus, it is not clear at present whether these systems would be advantageous in such analyses.

The time required for the separations outlined in Table 4 can vary widely depending upon the type of sample and, therefore, the number of preparative steps required prior to the final TLC step. The time required for LLE will depend upon the number of steps and upon whether emulsions are formed. CC separations generally require some 6-10 hr; however, as many as 10-12 columns can be run simultaneously. For TLC, samples can be applied as individual spots at the origin to about 4-6 plates by hand in about an hour. Development in one dimension generally requires 1-2 hr, but a number of plates can be developed simultaneously. The limit for the number of plates to be developed in a given day is determined largely by the final analysis. If individual bands or spots must be removed from the plate and extracted, then 4 or 5 samples are probably the limit for a single day for one person, as the extractions should be completed on the same day as the development. In situ analysis by a scanning spectrofluorometer requires much less time for analysis.

D. TLC in the Analysis of Related Classes of Compounds

TLC separations have been developed for several classes of compounds which are closely related to PAH, e.g., aza-arenes, polycyclic quinones, phenols, and nitro compounds. Interest in these types of compounds has stemmed largely from the fact that some compounds in these classes are carcinogenic in animals [103] and/or mutagenic in the Ames bioassay [104,105]. There has also been some interest in these classes of compounds as they relate to atmospheric chemistry [106], identification of sources of anthropogenic contamination of the environment [107], and formation of filter artifacts during air sampling [105]. Some of the TLC methods which have been developed for these classes of compounds are presented in Table 5.

The aza-arenes are present in airborne particulate matter [32,106,114], source emissions, and coal tar pitch [32], in cigarette smoke [115], in sediments [113,114,116,117], and in crude and synthetic oils [118,119], generally at levels 10 to 100 times lower than those of PAH compounds. The aza-arenes have been identified in synthetic fuels produced from coal and appear to be responsible for a significant proportion of the mutagenic activity of these fuels in the Ames bioassay [119].

TLC has been used to separate aza-arenes from other classes of compounds [110], to rapidly isolate and quantify selected aza-arenes such as benz[c]-acridine [108] or acridine [109], and to separate and characterize aza-arenes in 100 fractions from a CC separation of the basic organic fraction of airborne particulate matter [30,32,107]. In the latter instance, 200 spots were sepa-

Table 5. TLC Systems for the Analysis of Some Related Classes of Compounds

Compound class	Sample type	Separation scheme[a]	TLC system(s)	Detection and quantitation	Comments	Ref.
Aza-arenes	Airborne particulate matter, coal tar pitch, source emissions	LLE → CC → TLC	Cellulose/dimethylformamide-water (35:65)	Scanning spectrofluorometry; quenchofluorometric techniques used for characterization	100 fractions from CC on alumina separated by TLC; 200 spots separated, 25 aza-arenes characterized, 11 identified unequivocally	30,32, 107
	Airborne particulate matter	TLC	Alumina/pentane-ether (19:1)	Extraction; spectrofluorometry in pentane-trifluoroacetic acid	Benz[e]acridine separated and quantitated	108
		TLC (2D)	2:1 Alumina-cellulose/two-dimensional development (a) pentane-ether (19:1); (b) 35% aqueous dimethylformamide	Extraction; spectrofluorometry in pentane-trifluoroacetic acid	Benzo[h]quinoline separated and quantitated	108
	Airborne particulate matter, coal tar pitch	TLC	Glass fiber filter impregnated with silica gel/toluene-methanol-trifluoroacetic acid (93:5:2)	Scanning spectrofluorometry	Acidine separated from other compounds in basic fraction in 18 min	109

(continued)

(Table 5 continued)

Compound class	Sample type	Separation scheme[a]	TLC system(s)	Detection and quantitation	Comments	Ref.
	Airborne particulate matter	TLC → GC^2	Silanized silica gel/benzene-hexane-pyridine (4:5:1)	Extraction, separation by GC equipped with flame-ionization detector	Aza-arenes (R_f = 0.55) separated from PAH (R_f = 0.76) and paraffins (R_f = 0.9) by TLC; 1-2 g particulate matter required	110
Polycyclic quinones	Airborne particulate matter	TLC	1:1 aluminum oxide silica gel/pentane-ether (19:1)	Extraction, spectrofluorometry in trifluoroacetic acid	Benzo[a]pyrene, benz[c]acridine and 7H-benz[de]-anthracen-7-one in BSO isolated by TLC	111
	Airborne particulate matter	TLC	Glass fiber paper impregnated with silica gel/pentane-ether (19:1)	Trifluoroacetic acid used to intensify fluorescence and locate compounds; extraction followed by spectrofluorometric determination in conc. H_2SO_4	7H-benz[de]-anthracen-7-one (R_f = 0.68) and phenalen-1-one (R_f = 0.27) isolated and quantified; detection limits: 5 and 2 ng, respectively	47,112
	Airborne particulate matter	CC → TLC	Polyamide 11/acetic acid-water (19:1)	Extraction, UV spectrophotometry	Identity confirmed by reduction of quinones to parent PAH and fluorescence spectrophotometry; recoveries not reproducible	106

Polycyclic phenols	Airborne particulate matter, coal tar pitch	LLE → TLC → TLC	Silica gel/triethylamine or silica gel/toluene-ethylformate-formic acid (5:4:1) saturated with 0.3 M sodium acetate and several other systems	Colorimetric reagent sprays, no quantitation	Individual phenols isolated by using different sequences of TLC systems; 8 phenols in coal tar pitch identified, 2-fluorenol identified in sample from NYC	46
Polycyclic nitro compounds	Airborne particulate matter	CC → TLC	Silica gel/cyclohexane-chloroform (1:1) or cellulose/N,N-dimethyl-formamide-water (2:3)	Reduction to amine, fluorescence and fluorescence quenching; detection limit = 1-5 ng depending upon compound	100 mg of dichloromethane-soluble organics fractionated by CC on silica gel; benzene fraction further separated by TLC; 3-nitrofluoranthene and 6-nitrobenzo-[a]pyrene identified	113

[a]LLE = liquid-liquid extraction; CC = column chromatography; TLC = thin-layer chromatography; TLC (2D) = TLC with two-dimensional development; GC^2 = glass capillary gas chromatography.

rated, 25 aza-arenes were isolated, and 11 were identified unequivocally using various combinations of TLC systems and fluorescence and quenchofluorometric techniques. In general, the TLC systems used for the aza-arenes are similar to those used for PAH compounds. Glass fiber paper impregnated with silica gel has also been used to effect rapid (15-18 min) separations of acridine and two quinones commonly found in organic extracts of airborne particulate matter with little or no prior cleanup [47].

Two other TLC adsorbents have been used for the separation of aza-arene standards but not for analysis of "real-world" samples. Janák and Kubecová [120] have investigated separations of PAH and related compounds on Porapak Q with various solvents. With ethanol or propanol as a developing solvent PAH compounds generally exhibited R_f values of 0.1-0.3, whereas aza-arenes had R_f values of 0.3-0.6 and hydroxy derivatives were found near the solvent front. Wagner and Lehmann [121] separated N-heterocyclic compounds on silica gel thin layers impregnated with $CuSO_4$. The compounds were detected through fluorescence and through the fluorescence-thermochromism of heterocycle-CuI complexes formed by spraying the developed plates with KI-ascorbic acid solution followed by saturated Na_2CO_3 solution.

TLC separations of 7H-benz[de]anthracen-7-one and phenalen-1-one, which have been found in airborne particulate matter and source effluents, have been reported [47,111,112]. A rapid separation on glass fiber paper impregnated with silica gel was developed and designated as a tentative method by the Intersociety Committee on Methods of Air Sampling and Analysis (see Katz [112]). Pierce and Katz [106] separated and quantitated quinone derivatives of PAH in airborne particulate matter by column chromatography followed by TLC on polyamide 11/acetic acid-water (19:1) in an investigation of atmospheric oxidation of PAH compounds.

Very little work has been done on TLC separations of polycyclic phenols or nitro compounds. Sawicki and co-workers [46] separated a number of phenols in airborne particulate matter using various combinations of TLC systems, but the method was not further developed. Dietz and co-workers [122] reported methods for separating and identifying individual phenols in water samples by means of six one-dimensional TLC systems and various spray reagents. Jäger [113] identified two nitro derivatives of PAH compounds in airborne particulate matter using cellulose/N,N-dimethylformamide-water (2:3).

There has been even less development of TLC systems for alkyl and hydro derivatives of PAH. Separations on Porapak T [80] and on thin layers impregnated with complexing agents [25] have been reported for standard compounds, but applications of these systems to complex samples have not been developed.

IV. Detection, Characterization, and Quantitation in TLC

A. Detection and Characterization

The intense fluorescence of PAH compounds under ultraviolet (UV) light is almost invariably used to locate individual PAH compounds on TLC plates. Estimates of the limit of fluorescence detectability by the human eye for some PAH compounds on 20% acetylated cellulose (250 μm) are presented in Table 6. The limits depend principally upon the structure of the individual PAH com-

Table 6. Estimated Visual Limits of Detectability of Some PAH Compounds Under Long Wavelength Ultraviolet Light[a]

Compound	Fluorescence detection limit (ng)
Benzo[a]pyrene	1-2
Benzo[ghi]perylene	2-3
Fluoranthene	5-10
Pyrene	10-15
Benzo[e]pyrene	20-30
Chrysene	20-30
Dibenz[a,h]anthracene	20-30
Anthracene	30-40
Benz[a]anthracene	30-40

[a]250-μm 20% acetylated cellulose TLC plates; hand-held UV lamp, Model UVL-21, Ultraviolet Products Inc., San Gabriel, Calif.: 400 μW at 366 nm at 15-cm distance.

pounds but also depend upon the observer, size of the spot, thickness of the layer, and other factors. The fluorescent color of PAH and related compounds as well as their R_f values can be useful in characterizing unknown PAH. Differences in fluorescence under acidic, basic, and neutral conditions can also be useful in characterization of unknown compounds [41].

Low-temperature fluorescence and phosphorescence have been used to detect and characterize PAH [35,123]. Compounds which fluoresced at room temperature were usually fluorescent at liquid nitrogen temperatures [35]. Differences in fluorescent color depended upon temperature and the presence or absence of chloroform, concentrated H_2SO_4, trifluoroacetic acid, and other reagents [35]. More significantly, many polar compounds such as quinones and aromatic nitro compounds, which are not fluorescent at room temperature, were found to be intensely fluorescent at liquid nitrogen temperatures [35]. Thus, low-temperature fluorescence offers considerable potential as a detection and quantitation technique for such compounds.

Fluorescence quenching, a technique in which reagents are used to selectively quench the fluorescence of some compounds, has been used to characterize PAH and related compounds on TLC plates [32,42,113,124,125]. Fluorescence quenching reagents, e.g., picric acid, may be sprayed on the chromatogram after development [38,43] or incorporated into the adsorbent [43] or the solvent mixture [38,124]. Both volatile and nonvolatile quenching reagents have been used [38,43]. Use of a volatile reagent permits further characterization with other reagents. Sawicki and co-workers [42,124,125] were the first to apply this technique to analysis for PAH in airborne particulate matter, particularly for aza-arenes [32]. More recently Jäger [113] used aniline, phenylhydrazine, and carbon disulfide as quenchofluorometric reagents to distinguish nitro-aromatic compounds in airborne particulate matter.

Colorimetric reagents have also been used to detect and characterize PAH and related compounds [35,41,126-128]. Such reagents are particularly useful for compounds which are not fluorescent or only weakly fluorescent. These reagents can be sprayed onto the chromatogram after development or incorporated into the adsorbent. Many of the reagents used, such as 7,7,8,8-tetracyanoquinodimethane [127] and 2,4,7-trinitro-9-fluorenone [128] form colored complexes. Other reagents (e.g., H_2SO_4) form highly colored fluorescent and/or phosphorescent cationic or anionic species [35,41].

B. Quantitation

Semiquantitative determinations of the concentrations of PAH can be obtained by chromatographing different amounts of standard PAH on the same plate as the samples and then visually comparing sample and standards [2]. Although this technique is not often used, it can be valuable in an initial screening of samples for which no composition data are available.

Quantitative methods may be classified into two types. In the first type, individual spots or bands on the TLC plate are removed from the plate and the PAH are extracted or volatilized from the adsorbent for quantitation. The extraction solvent is frequently replaced by a solvent such as pentane or concentrated sulfuric acid, which is more suitable for absorbance or fluorescence spectrophotometric quantitation and characterization, the most common methods of quantitation. Spectrofluorometry is 10-1000 times more sensitive than absorbance spectrophotometry for PAH compounds. Concentrations as low as 0.003 μg/ml for BaP in concentrated sulfuric acid can be determined using spectrofluorometry [48]. Background fluorescence, light scattering, and instrument sensitivity, as well as the inherent fluorescence of a given compound, will all affect the lower limit of detection. Spectrofluorometry can also be highly selective. For example, the emission spectrum of BaP in sulfuric acid has been obtained in the presence of 50 other PAH and aza-arenes [28]. However, fluorescence quenching at high concentrations or by other compounds present in solution can also affect quantitation. For this reason absorbance spectrophotometry is sometimes used for quantitation and the identities of the isolated compounds are confirmed by their fluorescence spectra.

Although absorbance and fluorescence spectrophotometry have been used most frequently to quantitate PAH following extraction from the TLC adsorbent, any suitable quantitation method may be used. For example, investigators have used MS [88] and the Shpol'skii effect [86] to quantitate individual PAH isolated by TLC.

In situ quantitation by reflectance or transmission spectrophotometry constitutes the second class of quantitation methods for TLC. In situ spectrofluorometry began to be developed as a technique for quantitation and characterization of PAH on TLC plates in the mid-1960s [36,37,129]. Both transmission and reflectance measurements can be made on a TLC plate. However, Frei [130] evaluated both modes and concluded that reflectance spectrophotometry was superior. In situ spectrofluorometry has the obvious advantage of timesaving as individual compounds or fractions do not have to be extracted from the TLC adsorbent. In addition there are no extraction losses. Although the cost of a scanning spectrofluorometer is considerable, the savings in cost of labor are sufficient to compensate for the capital investment for a laboratory in which many routine analyses are done. In addition, integrators and data-processing systems can be interfaced with such equipment to further reduce the time required for analysis. Excitation and emission spectra can be readily obtained for identification, and individual PAH can be quantified by measuring the areas of the spots on the TLC plate. The lower limit of detection for scanning spectrofluorometers ranges from 0.1 to 0.001 μg depending upon the individual PAH compound, while the linearity of response is as high as 0.2 μg [129]. The reproducibility of in situ spectrofluorometry has been reported to be 5-10% [83,129,130]. Errors in sample volume spotted onto the TLC plate, in the uniformity of the thin layer, in the reproducibility of the chromatographic process, and in positioning the TLC plate in the light beam all contribute to uncertainties in in situ determinations of PAH [131].

V. Some Practical Aspects of TLC for the Analysis of PAH

A. General Comments

As for all analytical methods, good analytical practices in general and a quality-assurance program are required for the analysis of PAH by TLC. These practices include regular maintenance and calibration of instruments, the use of pure reagents, of standards, blanks, and spiked samples (for recovery determinations) and good record keeping.

Exposure of samples and PAH standards in solution or on TLC plates to light and air should be minimized. Many PAH compounds absorb light between 350 and 450 nm and subsequently undergo degradation in the presence of oxygen. It is common practice to install yellow lights ($\lambda > 450$ nm) in laboratories in which such analyses are done or to work only with light admitted through windows.

In order to reduce background interferences and minimize chances of contamination, sources of contamination such as cigarette smoke, dust, and plastic ware should be eliminated from the laboratory.

During analysis, samples must often be concentrated by evaporation of organic solvents. Losses of PAH can be reduced by working at pressures greater than 12 mmHg and water bath temperatures below 45°C [132]. When samples must be reduced to a few milliliters, distillation in a Kuderna-Danish apparatus, evaporation at low temperatures or freeze-drying is recommended. As there is generally some loss of PAH throughout the separations, extractions, and solvent reductions, internal recovery standards or spiked samples should be used to determine the recovery of PAH.

Reproducibility of Rf values and good resolution are prerequisites for good TLC analysis. Many of the factors which influence the reproducibility and resolution in a TLC system are discussed below.

B. Safety

As many of the PAH compounds are potent carcinogens, safety procedures should be established for their use and laboratory personnel should be informed of the hazards and familiarized with the safety procedures. The pure compounds should be handled in a glove box or glove bag. Use of a static eliminator (generally a polonium source) during operations involving pure compounds is strongly recommended as they tend to be charged with static electricity and difficult to control. Protective clothing, disposable gloves, and disposable lab bench pads should be used when working with the standard solutions or PAH fractions. Operations with these materials are most safely conducted in a hood. Work surfaces and hoods should be regularly monitored for contamination using UV light. Gloves, towels, etc. which become contaminated should be incinerated. Contaminated glassware and equipment should be decontaminated by rinsing with solvent and soaking in chromic acid solution. If the adsorbent is removed from the TLC plate for extraction of PAH, use of a TLC recovery tube is strongly recommended because of the hazard of inhalation of PAH-contaminated adsorbent dusts which can be generated during recovery of materials from the plate. These commercially available devices consist of a glass tube with a fritted glass disk. A vacuum line attached to one end of the tube draws the adsorbent into the tube and onto the fritted disk, where it is retained. The PAH can then be extracted in situ from the adsorbent.

C. Reagents and PAH Standards

Reagents used in the analyses should be of sufficiently high purity to give the lowest possible background and should be routinely monitored to determine blank corrections.

PAH purchased from commercial sources frequently require further purification for use as standards. TLC can be used as a convenient and rapid check on the purity of PAH standards. Acetylated cellulose and cellulose used with an appropriate developing solvent (see Table 1) are particularly useful for such determinations, as they are effective in separating individual PAH--particularly isomers which are difficult to separate by other means. Recrystallization and/or preparative TLC, CC, or HPLC may be used to purify PAH standards, if necessary.

D. Thin-Layer Plates

Thin-layer plates may be made in the laboratory or purchased commercially. Multiple band plates and mixed layer plates must be prepared in the laboratory, as most types are not commercially available. Silica gel, alumina, cellulose, and acetylated cellulose coated plates on glass, aluminum foil, or plastic sheets are commercially available. The uniformity of the adsorbent layer of these commercial plates is advantageous in achieving reproducible chromatograms, particularly for the beginner. Aluminum foil-backed thin layers are convenient for subsequent extractions, as the chromatogram may be cut with scissors and extracted without separating the absorbent from the support. Channeled silica gel plates are now produced which are convenient for separation of multiple samples on a single plate. Recently, commercially produced TLC plates which incorporate a diatomaceous earth preadsorbent at the base of the plate have become available. Relatively large amounts (50-60 μl per application) are applied to the preadsorbent portion of the plate without regard to the size of the spot or band. As the developing solvent passes through the preadsorbent, it extracts and concentrates the applied sample and the sample reaches the actual adsorbent thin layer as a very narrow band. This results in increased resolution, increased sensitivity (due to higher concentrations per square millimeter) and more reproducible R_f values. In addition sample application time can be reduced substantially since 50-60 μl can be applied at one time as compared to the usual 2-3 μl per spot.

Predevelopment of TLC plates is often required to reduce background interferences in the analysis [55,64,71]. After predevelopment, the plates should be stored in a desiccator at an appropriate relative humidity to protect them from adsorbing atmospheric contaminants. Plates should not generally be stored for more than a few weeks after predevelopment [48]. In addition, TLC plates often have a shelf life beyond which the absorbent layer tends to separate from the support during development.

Relative humidity is an important consideration for reproducible separations and recoveries of PAH. This is particularly true for adsorption chromatographic systems, e.g., silica gel and alumina, since separations in such systems depend strongly on the surface activity of the adsorbent [81,133-135]. Relative humidities recommended for TLC separations are presented in Table 7. Plates may be stored in constant humidity chambers after activation and/or predevelopment and also prior to development for reconditioning. Silica gel and alumina plates are generally activated by heating in an oven at 100-120°C for 30-45 min [48,62,70,71].

E. Application of Samples and Standards

For maximum resolution and sensitivity samples and standards must be applied as small spots (a few millimeters) or narrow streaks. As the size of the spot on the thin layer is proportional to the volume applied, the simplest and least expensive technique for sample application is to apply the sample in portions of a few microliters about 1.5 cm from the edge of the plate and evaporate the solvent between applications. Syringes or micropipettes may be used for quantitative sample application. Spotting guides or templates are useful in accurately positioning the spots on the plates, particularly when repeated applications are required. Automatic spotting devices which deliver large volumes (100-200 μl) in a short time period and provide small spots of sample are also available. These devices employ automated syringes which slowly and continuously deliver the sample to the surface of the plate while a warm air stream is used to evaporate the solvent.

Table 7. Recommended Relative Humidities for TLC Separations on Various Adsorbents

Adsorbent	Relative humidity (%)	Ref.
Silica gel	40	133
Alumina	40-50	62,81,135
Acetylated cellulose	50	62

Table 8. Approximate Capacities of Various TLC Absorbents[a]

Adsorbent	Milligrams per spot	Type of sample	Ref.
Silica gel	1.0-5.0[b]	Cyclohexane-soluble fraction of airborne particulate matter	61
Alumina	0.1-0.5	Benzene-soluble fraction of airborne particulate matter	27,48,62
Acetylated cellulose	0.1-0.3	Cyclohexane-soluble fraction of airborne particulate matter	58

[a]250-μm-thick layer.
[b]Estimated from data in Ref. 61.

Samples may also be streaked on the TLC plate by manual means, semi-automatic devices, or fully automated streakers [1-4,136]. The fully automated streakers can produce streaks as narrow as 0.5 mm. This results in greater resolution and improved accuracy in in situ scanning spectrofluorometry or densitometry.

Variations in the amount of sample applied and the concentrations of the components in the sample can alter R_f values and resolution. Applications of too large a sample will result in streaking and loss of resolution upon development. Thus, some initial experiments to determine optimal sample size for a given type of sample and a particular TLC system are strongly recommended. Table 8 presents some data on approximate sample size for several TLC systems which may be useful in planning such experiments.

Finally, standards and samples (~1% in concentration) should always be chromatographed on the same TLC plate and the concentrations and masses of the PAH standards should be similar to those of the PAH in the sample. A volatile solvent which does not degrade the TLC adsorbent should be used for the solutions of samples and standards. The solvent should also be as nonpolar as possible to minimize separations at the origin during sample application [137].

F. Development

Both the solvent mixture and the conditions of development must be considered in the development of a TLC plate. The choice of solvent mixture will depend principally upon the adsorbent layer and the type of separation sought, i.e., separation of PAH from other classes of compounds or separation of individual PAH compounds (see Tables 1, 3, and 4).

It is important to carefully measure the volumes of the individual solvents used in a mixture of solvents as small changes in composition can significantly affect separations in a TLC system. This is particularly important for partition TLC systems. For this reason, solvent mixtures should not be reused after development and should not be stored for long periods, as the proportions of the components will change, particularly for volatile components.

The use of glass-distilled solvents or other highly purified solvents is strongly recommended as impurities in a solvent can affect the resolution and stability of the compounds being separated and also increase the blank. If ether is one of the solvents used in development, freshly opened bottles should be used or the ether should be checked for the presence of peroxides. Peroxides tend to form in anhydrous ether which has been in contact with air and can react with the PAH in samples and standards reducing the recovery of the PAH and precision of the analysis.

If TLC plates are developed in a tank, it is extremely important to equilibrate the solvent and vapor above the solvent in the tank prior to development in order to obtain a more uniform solvent front and sample development and a reproducible separation. This may be done by placing Whatman filter paper or blotter paper against one side of the tank, pouring the solvent mixture over it, and allowing the system to equilibrate for 10-15 min. After this, the TLC plate should be placed in the center of the tank, 3-5 cm from the sides of the tank. The level of the solvent should be high enough to supply sufficient solvent for development but must be below the origin so that the solvent does not dissolve the sample. The tank should be kept tightly closed during development and shielded from light.

Alternatively, a "sandwich" chamber may be used for development. This type of chamber consists of two glass plates with a small recess in the inner surface, and the TLC plate is sandwiched between the two glass plates for development. The "sandwich" chamber has a minimum volume and thus tends to saturate rapidly. A presoaked cellulose plate can be used as a cover plate to maximize solvent saturation. The sandwich chamber often provides a more easily reproducible environment for TLC development and thus more reproducible chromatograms.

A number of other types of development chambers have been developed which can enhance TLC resolution [1-4]. These are available commercially from the companies which produce and sell TLC supplies and equipment.

Multiple development and continuous development [1-4] are techniques which can also be used to increase resolution during TLC. In multiple development TLC, the plate is developed in a given solvent, dried, and then developed in the same direction through one or more additional cycles. In continuous development TLC, solvent is evaporated from the top of the TLC plate during development in order to resolve more slowly moving components in the mixture. Vieth and co-workers [138] have described an inexpensive apparatus for continuous development of TLC plates.

After development to a height of 10-15 cm, the TLC plate is removed from the development chamber and placed in a hood to dry. The plate is then quickly examined under a UV lamp (preferably long-wavelength UV), and bands or spots to be removed are marked. Exposure to the UV light should be minimized as it can cause decomposition of the PAH on the plate.

G. Record Keeping

Records of the exact chromatographic conditions must be kept if the work is to be reproduced. Such records should include information on the type and thickness of the adsorbent layer, the commercial source of the plate or adsorbent, activation conditions, predevelopment, storage, relative humidity, temperature, sample application (equipment, concentrations, amounts), type of chamber, saturation method, and the techniques for visualization. After development, the R_f values and colors of the spots should be carefully recorded. It may be appropriate to include a sketch or photograph of the plate and spots as well. For routine work, a standard form is quite useful for recording all information.

H. Quantitation

PAH separated by TLC can be extracted from the TLC adsorbent and quantitated by various techniques such as spectrophotometry or spectrofluorometry or quantitated in situ by scanning densitometers or spectrofluorometers. Quantitative TLC requires that the methods for sample cleanup, for application and separation of the sample on the TLC plate, and for recovery of the PAH be carefully controlled and reproducible [129-131,133,134,139,140].

If the PAH compounds are to be extracted from the adsorbent for quantitation, a recovery as close to 100% as possible with a very low background or blank is desirable. The TLC plate should be prewashed with eluting solvent prior to activation and sample application to reduce background. The PAH should be removed from the plate and extracted on the same day as the chromatographic separation is carried out, as the compounds can decompose if left on the plate. Thin-layers on aluminum foil sheets can greatly facilitate the recovery of PAH, as the spot can be cut from the sheet for elution with the foil backing left in place.

The extraction solvent used to recover PAH should give the maximum recovery possible, must not dissolve the adsorbent, and must give the lowest possible background for the type of quantitation chosen. High-purity solvents are quite critical for the extraction step as the volumes used must frequently be reduced for the final quantitation; thus, any impurities present are concentrated. Certain solvents may not be used with certain adsorbents. Methanol, for example, will dissolve silica gel and dichloromethane tends to break down acetylated cellulose.

Pierce and Katz [62] used ether to extract PAH from acetylated cellulose for subsequent quantitation by spectrofluorometry. Daisey and Leyko [58] found that this type of extraction gave too high a background if GC separation was followed with quantitation by a flame-ionization detector. Lannoye and Greineke [81] have shown that it may be advisable to avoid the use of ether, as any peroxide impurities can lead to decomposition of benzo[a]pyrene.

It should be noted that recoveries of PAH frequently show some dependence on mass [141], with recoveries becoming poorer and less reproducible as the sample mass becomes smaller.

Extracts from TLC absorbents should be carefully filtered to remove any fine particles of adsorbent which might interfere with quantitation [81]. A 6-10 μm fritted-glass filter [81] will generally retain fine particles. Fluorescence spectrofluorometric methods are particularly sensitive to suspended particles and filtration through a Teflon membrane filter, 0.45 μm pore size, may be required.

Filtering and evaporating and extraction solvent under an inert gas (nitrogen or argon) can reduce decomposition of PAH. A positive-pressure filter apparatus for this purpose has been described by White and Howard [142].

The filtering equipment should be all glass. Rubber stoppers are to be avoided. An automated elution system is also available [143] which can extract six spots simultaneously and requires less than 0.2 ml of solvent for extraction of each spot.

It is often necessary to change solvents for quantitation after extraction. If the sample is to be taken to dryness, this should be done at a low temperature under an argon or nitrogen atmosphere and the sample must be taken *just* to dryness. Alternatively, a Kuderna-Danish apparatus can be used if the solvent to be evaporated is more volatile than the replacement solvent.

If fluorescence measurements are to be made, dissolved oxygen should be removed by flushing the solvent with nitrogen prior to measurement [81]. Some investigations have used solvents with low affinities for oxygen as an alternative. Concentrations of samples and standards must also be in the range in which the response of the spectrofluorometer is linear. If concentrations are too high, quenching will occur.

In situ quantitation of PAH compounds on TLC plates by reflectance spectrophotometry requires carefully controlled and reproducible conditions for good accuracy and precision. Major sources of variability are sample application, uniformity of the thin layer, the reproducibility of chromatographic conditions, errors in positioning the plate with respect to the center of the light beam, and errors in the light measurements [129-131,136]. Errors in measurements of peak heights can also contribute to experimental uncertainties [144]. Increasingly automated equipment for in situ reflectance measurements and data analysis is becoming available and will result in greater accuracy and precision in in situ reflectance analysis for PAH compounds as well as reducing the amount of time required per sample for analysis.

VI. Future Trends

During the 1970s, there was relatively little development of TLC as a method for the analysis of PAH, although there were some refinements in the methods. Rather major advances in PAH analyses were seen in HPLC and capillary column GC methods. A number of recent developments in the field of TLC, however, have the potential for revolutionizing PAH analysis by TLC in the decade of the eighties. These include the development of high-performance TLC (HPTLC), development of more automated equipment for sample application and for in situ reflectance spectrophotometry, and data processing and development of a number of new products of TLC.

HPTLC employs thin layers of small particles (5 μm, for example), of an extremely narrow particle size distribution, with pore diameters of the order of 40-60 Å [136,144-149]. The development of HPTLC permits separation of nanogram and even picogram (with fluorescence detection) quantities, effective separations in 3-7 cm in 1-4 min.

R_f values are reproducible up to ±0.01 units [136] and plate heights of the order of 12 μm [146] have been obtained. Smaller spot diameters and increased reflectance due to smaller particle size for HPTLC result in lower limits of detection. HPTLC can be used to advantage as a pilot technique for HPLC as the solvent system is more easily changed [150,151], or it can be used for PAH analysis. Multiple samples and the standards can be run on the same plate at the same time, e.g., 32 samples on a 10 × 10 cm plate.

A variety of new TLC products have become commercially available in the past few years which can greatly facilitate TLC. These include plates with preadsorbents, including HPTLC plates which incorporate preadsorbents, dual band plates, individually prepackaged plates with templates, and reverse-phase chemically bonded C_{18} TLC plates [152-154].

Information on these products and on many types of ancillary equipment which facilitate and improve TLC separations can be obtained in the catalogs from manufacturers. In addition some companies publish technical bulletins that provide useful information on various aspects of TLC. Several recent reviews on advances in HPTLC are also available [155,156].

Abbreviations

A	alumina
AC	acetylated cellulose
BaP	benzo[a]pyrene
BeP	benzo[e]pyrene
C	cellulose
CC	column chromatography
DMBA	7,12-dimethylbenz[a]anthracene
GC	gas chromatography
GC^2	glass capillary gas chromatography
GPC	gel permeation chromatography
HPLC	high-performance liquid chromatography
HPTLC	high-performance thin-layer chromatography
hR_f	$100 \times R_f$
IEC	ion exchange chromatography
LLE	liquid-liquid extraction
MS	mass spectrometry
PAH	polycyclic aromatic hydrocarbons
ppb	parts per billion
ppm	parts per million
R_B	distance of a compound from the origin divided by distance of benzo[a]pyrene form the origin
R_f	distance of a compound from the origin divided by distance of the solvent front from the origin
${}^nR_f$	R_f value of a compound after n multiple developments
SG	silica gel
SG/C	silica gel impregnated with a complexing agent
TLC	thin-layer chromatography
TLC (2D)	thin-layer chromatography with two-dimensional development

Acknowledgments

The assistance and comments of Dr. Robert Hazen, Hunter College of the City University of New York, and of Mrs. Elizabeth McCarthy, New York University Medical Center, in the preparation of this chapter are gratefully acknowledged.

References

1. J. M. Bobbitt, *Thin-Layer Chromatography*, Van Nostrand Reinhold, New York, 1963.
2. K. Randerath, *Thin-Layer Chromatography*, 2nd rev. ed., D. D. Libman (Transl.), Academic Press, New York, 1966.
3. E. Stahl (Ed.), *Thin-Layer Chromatography, A Laboratory Handbook*, M. R. F. Ashworth (Transl.), Springer-Verlag, New York, 1969.
4. J. G. Kirchner, *Thin-Layer Chromatography: Techniques of Chemistry*, E. S. Perry (Ed.), 2nd ed., Vol. 14, Wiley, New York, 1978.
5. A. Van Langmeersch, *Chim. Anal. 50*:3 (1968).
6. D. Halot, *Talanta 17*:729 (1970).
7. R. E. Schaad, *Chromatogr. Rev. 13*:61 (1970).
8. P. L. Gupta and P. Kumar, *Petrol. Hydrocarbons 6*:140 (1971).
9. R. Amos, *Chromatogr. Sci. 11*:329 (1979).
10. C. R. Sawicki and E. Sawicki, in *Progress in Thin-Layer Chromatography and Related Methods*, A. Niederwieser and G. Polaki (Eds.), Vol. 3, Ann Arbor Science Publs., Ann Arbor, Mich., 1972, p. 230.
11. R. E. Waller, *Brit. J. Cancer 6*:8 (1952).
12. M. B. Shimkin, B. K. Koe, and L. Zechmeister, *Science 113*:650 (1951).
13. P. Wedgwood and R. L. Cooper, *Analyst 78*:170 (1953).
14. R. L. Cooper, *Chem. Ind. (London)* p. 1364 (1953).
15. R. L. Cooper, A. J. Lindsey, and R. E. Waller, *Chem. Ind. (London)* p. 1418 (1954).
16. R. L. Cooper, *Analyst 79*:573 (1954).
17. P. Kotin, H. L. Falk, P. Mader, and M. Thomas, *Ind. Hyg. Occup. Med. 9*:153 (1954).
18. E. J. Bailey and N. Dungal, *Brit. J. Cancer 12*:348 (1958).
19. B. L. VanDuuren, *J. Nat. Cancer Inst. 21*:1 (1958).
20. T. Wieland and H. Determann, *Experientia 18*:430 (1962).
21. G. M. Badger, J. K. Donnelly, and T. M. Spotswood, *J. Chromatogr. 10*:397 (1963).
22. N. Kucharczyk, J. Fohl, and J. Vymetal, *J. Chromatogr. 11*:55 (1963).
23. A. Berg and J. Lam, *J. Chromatogr. 16*:157 (1964).
24. M. Franck-Neuman and P. Jossang, *J. Chromatogr. 14*:280 (1964).
25. R. G. Harvey and M. Halonen, *J. Chromatogr. 25*:294 (1966).
26. I. I. Domsky, W. Lijinsky, K. Spencer, and P. Shubik, *Proc. Soc. Exp. Biol. Med. 113*:110 (1963).
27. E. Sawicki, T. W. Staley, W. C. Elbert, and J. D. Pfaff, *Anal. Chem. 36*:497 (1964).
28. E. Sawicki, T. R. Hauser, and T. W. Stanley, *Int. J. Air Pollut. 2*:253 (1960).
29. E. Sawicki, T. R. Stanley, J. D. Pfaff, and W. C. Elbert, *Chemist-Analyst 53*:6 (1964).
30. E. Sawicki, T. W. Stanley, J. D. Pfaff, and W. C. Elbert, *Anal. Chim. Acta 31*:359 (1964).
31. E. Sawicki, J. E. Meeker, and M. J. Morgan, *Air Water Pollut. Int. J. 9*:291 (1965).
32. E. Sawicki, T. W. Stanley, and W. C. Elbert, *J. Chromatogr. 18*:512 (1965).

33. D. F. Bender and E. Sawicki, *Chemist-Analyst 54*:73 (1965).
34. E. Sawicki, T. W. Stanley, W. C. Elbert, J. Meeker, and S. McPherson, *Atmos. Environ. 1*:131 (1967).
35. E. Sawicki and H. Johnson, *Microchem. J. 8*:85 (1964).
36. E. Sawicki, T. W. Stanley, and H. Johnson, *Microchem. J. 8*:257 (1964).
37. E. Sawicki, T. W. Stanley, and W. C. Elbert, *J. Chromatogr. 20*:348 (1965).
38. E. Sawicki, W. C. Elbert, and T. W. Stanley, *J. Chromatogr. 17*:120 (1965).
39. E. Sawicki, T. W. Stanley, S. McPherson, and M. Morgan, *Talanta 13*: 619 (1966).
40. C. R. Engel and E. Sawicki, *J. Chromatogr. 31*:109 (1967).
41. D. F. Bender, E. Sawicki, and R. M. Wilson, Jr., *Anal. Chem. 36*:1011 (1964).
42. E. Sawicki, H. Johnson, and K. Kosinski, *Microchem. J. 10*:72 (1966).
43. E. Sawicki and H. Johnson, *J. Chromatogr. 23*:142 (1966).
44. E. Sawicki, T. W. Stanley, and W. C. Elbert, *Mikrochim. Acta* p. 1110 (1965).
45. E. Sawicki, T. W. Stanley, W. C. Elbert, and M. Morgan, *Talanta 12*:605 (1965).
46. E. Sawicki, M. Guyer, R. Schumacher, W. C. Elbert, and C. R. Engel, *Mikrochim. Acta* p. 1025 (1968).
47. T. W. Stanley, M. J. Morgan, and J. E. Meeker, *Environ. Sci. Technol. 3*:1198 (1969).
48. M. Katz (Ed.), *Methods of Air Sampling and Analysis*, 2nd ed., American Public Health Association, Washington, D.C., 1977, pp. 216, 220 (1st ed., 1972).
49. G. Genest and D. M. Smith, *J. Assoc. Offic. Agr. Chem. 47*:894 (1964).
50. J. W. Howard, R. T. Teague, Jr., R. H. White, and B. E. Fry, Jr., *J. Assoc. Offic. Anal. Chem. 49*:595 (1966).
51. J. W. Howard, R. H. White, B. E. Frey, Jr., and E. W. Turricchi, *J. Assoc. Offic. Anal. Chem. 49*:611 (1966).
52. J. W. Howard, T. Fazio, R. H. White, and B. A. Klimeck, *J. Assoc. Offic. Anal. Chem. 51*:122 (1968).
53. T. Fazio, R. H. White, and J. W. Howard, *J. Assoc. Offic. Anal. Chem. 56*:68 (1973).
54. (a) *Official Methods of Analysis*, 11th ed., AOAC, Washington, D.C., 1970, Secs. 21.C01-21.C09; (b) *J. Assoc. Offic. Anal. Chemists 57*:209 (1974).
55. J. W. Howard, T. Fazio, and R. White, *IUPAC Inform. Bull., Tech. Rept. No. 4*, International Union of Pure and Applied Chemistry, Oxford, England, February 1972.
56. H. Egan (Ed.-in-chief), *Environmental Carcinogens: Selected Methods of Analysis*, Vol. 3: *Analysis of Polycyclic Aromatic Hydrocarbons in Environmental Samples*, IARC Publ. No. 29, International Agency for Research on Cancer, Lyons, France, 1979.
57. R. Tomingas, G. Voltmer, and R. Bodnarik, *Sci. Total Environ. 7*:261 (1977).
58. J. M. Daisey and M. A. Leyko, *Anal. Chem. 51*:24 (1979).
59. L. R. Snyder, *J. Chromatogr. 25*:247 (1966).

60. G. H. Schenk, G. L. Sullivan, and P. A. Fryer, *J. Chromatogr.* *89*:49 (1974).
61. D. Brocco, V. Cantuti, and G. P. Cartoni, *J. Chromatogr.* *49*:66 (1970).
62. R. C. Pierce and M. Katz, *Anal. Chem.* *47*:1743 (1975).
63. H. J. Klimisch and E. Kirchheim, *Chromatographia* *9*:119 (1976).
64. J. N. Gearing, P. J. Gearing, T. F. Lytle, and J. S. Lytle, *Anal. Chem.* *50*:1833 (1978).
65. C. Horvath and W. Melander, *Amer. Lab.* *17*:(1978).
66. G. D. Short and R. Young, *Analyst (London)* *94*:259 (1969).
67. G. Chatot, W. Jecquier, M. Jay, R. Fontanges, and P. Obaton, *J. Chromatogr.* *45*:415 (1969).
68. G. Chatot, M. Castegnaro, J. L. Roche, and R. Fontanges, *Chromatographia* *3*:507 (1970).
69. G. Chatot, M. Jay, W. Jecquier, and R. Fontanges, *Chim. Anal.* *52*:1264 (1970).
70. G. Chatot, R. Dangy-Caye, and R. Fontanges, *J. Chromatogr.* *72*:202 (1972).
71. E. D. John and G. Nickless, *J. Chromatogr.* *138*:399 (1977).
72. K. Takami, H. Ishitani, Y. Kuge, and S. Asada, *Nippon Kagaku Kaishi* p. 223 (1979).
73. H. G. Preston and A. Macaluso, in *Measurement of Organic Pollutants in Water and Wastewater*, C. E. Van Hall (Ed.), ASTM STP 686, American Society for Testing and Materials, Philadelphia, 1979, p. 152.
74. D. W. Grant and R. B. Meiris, *J. Chromatogr.* *142*:339 (1977).
75. P. V. Peurifoy, M. J. O'Neal, and L. A. Woods, *J. Chromatogr.* *51*:227 (1970).
76. J. F. McKay and D. R. Latham, *Anal. Chem.* *45*:1050 (1973).
77. S. Saluste, I. Klesment, and S. Kivirahk, *Eesti NSV Teaduste Akad. Toimetised, Keem. Seeria* *28*:7 (1979).
78. G. A. Gilchrist, A. Lynes, G. Steel, and B. T. Whitman, *Analyst (London)* *97*:880 (1972).
79. A. A. Herod and R. G. James, *J. Inst. Fuel* *51*:164 (1978).
80. V. Martinu and J. Janak, *J. Chromatogr.* *65*:477 (1972).
81. R. A. Lannoye and R. A. Greinke, *Amer. Ind. Hyg. Assoc.* *35*:755 (1974).
82. L. Keefer, *J. Chromatogr.* *31*:390 (1967).
83. M. J. Schultz, R. M. Orheim, and H. H. Bovee, *Amer. Ind. Hyg. Assoc.* *34*:404 (1973).
84. B. Seifert and I. Steinbach, *Z. Anal. Chem.* *287*:264 (1977).
85. K. S. Rhee and L. J. Bratzler, *J. Food Sci.* *33*:626 (1968).
86. A. Colmsjo and U. Stenberg, *J. Chromatogr.* *169*:205 (1979).
87. M. Katz, in *Environmental Carcinogens: Selected Methods of Analysis*, Vol. 3: *Polycyclic Aromatic Hydrocarbons in Environmental Samples*, H. Egan (Ed.-in-chief), IARC Publ. No. 29, International Agency for Research on Cancer, Lyons, France, 1979, p. 193.
88. J. R. Majer, R. Perry, and M. J. Reade, *J. Chromatogr.* *48*:328 (1970).
89. G. Chatot, R. Dangy-Caye, R. Fontanges, and P. Obaton, *Chromatographia* *5*:460 (1972).
90. M. Kertesz-Saringer and Z. Morlin, *Egeszsegtudomany* *16*:392 (1972).
91. G. Heinrich and H. Gusten, *Z. Anal. Chem.* *278*:257 (1976).
92. M. Agirova, *Khig. Zdraveopaz.* *18*:290 (1975).

93. H. Matsushita and Y. Suzuki, *Bull. Chem. Soc. Japan 42*:460 (1969).
94. H. Matsushita, Y. Esumi, and K. Yamada, *Bunseki Kagaku 19*:951 (1970).
95. J. Howard, in *Environmentql Carcinogens: Selected Methods of Analysis*, Vol. 3: *Polycyclic Aromatic Hydrocarbons in Environmental Samples*, H. Egan (Ed.-in-chief), IARC Publ. No. 29, International Agency for Research on Cancer, Lyons, France, 1979, p. 175.
96. J. Borneff and H. Kunte, in *Environmental Carcinogens, Selected Methods of Analysis*, Vol. 3: *Polycyclic Aromatic Hydrocarbons in Environmental Samples*, H. Egan (Ed.-in-chief), IARC Publ. No. 29, International Agency for Research on Cancer, Lyons, France, 1979, p. 129.
97. W. A. Maher, J. Bagg, and J. D. Smith, *Intern. J. Environ. Anal. Chem. 7*:1 (1979).
98. S. Monarca, *Sci. Total Environ. 14*:233 (1980).
99. K. Potthast and G. Eigner, *J. Chromatogr. 103*:173 (1975).
100. Y. Shiraishi, T. Yamashita, and T. Shirotori, *Eisei Kagaku 23*:310 (1977).
101. G. Bories, *J. Chromatogr. 130*:387 (1977).
102. D. W. Armstrong and M. McNeely, *Anal. Lett. 12*:1285 (1979).
103. National Academy of Sciences (U.S.), *Particulate Polycyclic Organic Matter*, Washington, D.C., 1972.
104. J. McCann, E. Choi, E. Yamasaki, and B. N. Ames, *Proc. Nat. Acad. Sci. (U.S.) 72*:5135 (1975).
105. J. N. Pitts, Jr., K. A. VanCauwenberge, D. Grosjean, J. P. Schmid, D. R. Fitz, W. L. Belser, Jr., G. B. Knudson, and P. M. Hynds, *Science 202*:515 (1978).
106. R. C. Pierce and M. Katz, *Environ. Sci. Technol. 10*:45 (1976).
107. E. Sawicki, S. P. McPherson, T. W. Stanley, J. Meeker, and W. C. Elbert, *Intern. J. Air Water Pollut. 9*:515 (1965).
108. E. Sawicki, T. W. Stanley, and W. C. Elbert, *J. Chromatogr. 26*:72 (1967).
109. E. Sawicki and C. R. Engel, *Mikrochim. Acta* p. 91 (1969).
110. D. Brocco, A. Cimmino, and M. Possanzini, *J. Chromatogr. 84*:371 (1973).
111. T. W. Stanley, M. J. Morgan, and E. M. Grisby, *Environ. Sci. Technol. 2*:699 (1968).
112. M. Katz (Ed.), *Methods of Air Sampling and Analysis*, 2nd ed., American Public Health Association, Washington, D.C., 1977, p. 231.
113. J. Jäger, *J. Chromatogr. 152*:575 (1978).
114. M. W. Dong, D. C. Locke, and D. Hoffman, *Environ. Sci. Technol. 11*:612 (1977).
115. B. L. VanDuuren, J. A. Bilbao, and C. A. Joseph, *J. Nat. Cancer Inst. 25*:53 (1960).
116. M. Blumer, T. Dorsey, and J. Sass, *Science 195*:283 (1977).
117. S. G. Wakeham, *Environ. Sci. Technol. 13*:1118 (1979).
118. W. E. Haines and D. R. Latham, *Anal. Chem. 49*:256R (1977).
119. M. R. Guerin, J. L. Epler, W. H. Griest, B. R. Clark, and T. K. Rao, in *Carcinogenesis*, Vol. 3: *Polynuclear Aromatic Hydrocarbons*, P. W. Jones and R. I. Freudenthal (Eds.), Raven Press, New York, 1978, p. 21.
120. J. Janák and V. Kubecová, *J. Chromatogr. 33*:132 (1968).

121. H. Wagner and H. Lehmann, *Z. Anal. Chem. 291*:366 (1978).
122. F. Dietz, J. Traud, P. Koppe, and C. Ruebelt, *Chromatographia 9*:380 (1976).
123. H. P. Raaen, *J. Chromatogr. 53*:600 (1970).
124. E. Sawicki, T. W. Stanley, and W. C. Elbert, *Talanta 11*:1433 (1964).
125. E. Sawicki, T. W. Stanley, and J. Johnson, *Mikrochim. Acta* p. 178 (1965).
126. E. Sawicki and H. Johnson, *Mikrochim. Acta* p. 435 (1964).
127. E. Sawicki, C. R. Engel, and W. C. Elbert, *Talanta 14*:1169 (1967).
128. R. A. Heacock and O. Hutzinger, *Mikrochim. Acta 2*:101 (1975).
129. L. Toth, *J. Chromatogr. 50*:72 (1970).
130. R. W. Frei, *J. Chromatogr. 64*:285 (1972).
131. S. Ebel and E. Glaser, *J. High Resolution Chromatogr. Chromatogr. Commun. 2*:36 (1979).
132. B. T. Commins, in *Symposium on Analysis of Carcinogenic Air Pollutants*, National Cancer Institute Monograph No. 9, U.S. Dept. of Health and Human Services, Washington, D.C., 1962, p. 225.
133. D. Rogers, *Amer. Lab.* p. 77 (May 1979).
134. D. Jänchen, *J. Chromatogr. 7*:129 (1977).
135. Y. Suzuki and H. Matsushita, *Ind. Health (Japan) 5*:65 (1967).
136. H. J. Issaq and E. W. Barr, *Anal. Chem. 49*:83A (1977).
137. H. R. Felton, *Amer. Lab.* p. 105 (May 1980).
138. R. Vieth, D. Fraser, and G. Jones, *Anal. Chem. 50*:2150 (1978).
139. J. C. Touchstone (Ed.), *Quantitative Thin-Layer Chromatography*, Wiley (Interscience), New York, 1973.
140. J. C. Touchstone, S. S. Levin, and T. Murawec, *Anal. Chem. 43*:858 (1971).
141. F. DeWiest, D. Rondia, and H. Della Fiorentina, *J. Chromatogr. 104*:399 (1975).
142. R. H. White and J. W. Howard, *J. Chromatogr. 29*:108 (1967).
143. R. K. Viteck, C. J. Seul, M. Baier, and E. Lau, *Amer. Lab. 6*:109, 112 (1974).
144. S. Ebel and J. Hocke, *J. High Resolution Chromatogr. Chromatogr. Commun. 1*:156 (1978).
145. H. Halpaap and J. Ripphahn, *Chromatographia 10*:613 (1977).
146. J. Ripphahn and H. Halpaap, *J. Chromatogr. Libr. 9*:189 (1977).
147. D. Jäenchen and H. R. Schmutz, GIT *Fachz. Lab. 21*:657 (1977).
148. D. Jäenchen, *J. Chromatogr. Libr. 9*:129 (1977).
149. R. E. Kaiser, *Rev. Anal. Chem. (Proc. 2nd Euroanal. Conf.)* p. 77 (1979).
150. T. Wawrzynowicz, T. Dzido, and J. Kuczmierczyk, *Chem. Anal. (Warsaw) 24*:311 (1979).
151. H. J. Issaq, B. Shaikh, N. J. Pontzer, and E. W. Barr, *J. Liquid Chromatogr. 1*:133 (1978).
152. J. Sherma and M. Latta, *J. Chromatogr. 154*:73 (1978).
153. L. C. Sander and L. R. Field, *J. Chromatogr. Sci. 18*:133 (1980).
154. T. H. Maugh, *Science 216*:161 (1982).
155. D. C. Fenimore and C. M. Davis, *Anal. Chem. 53*:252A (1981).
156. S. A. Borman, *Anal. Chem. 54*:790A (1982).

10

Determination of Polycyclic Aromatic Hydrocarbons in Sediments and Marine Organisms

BRUCE P. DUNN / Department of Medical Genetics, University of British Columbia, Vancouver, British Columbia, Canada

I. Introduction

In recent years there has been extensive research on the contamination of the marine environment by polycyclic aromatic hydrocarbons (PAH). This research has been stimulated, on the one hand, by the presence of PAH in tanker-carried crude oils and in industrial and urban effluents and, on the other hand, by the toxicity of lower-molecular-weight PAH to marine organisms and the carcinogenic properties of higher molecular weight PAH. Neff has recently and extensively reviewed the information available on PAH in the marine and freshwater aquatic environment [1]. The reader is referred to this work for a full discussion of the sources, fates, and biological effects of marine PAH.

It has become apparent that the contamination of marine sediments by PAH is widespread. With the exception of perylene and some unusual substituted PAH, there is little or no evidence that these compounds are biosynthesized in the marine environment or elsewhere. Analysis of dated sediment cores has revealed that, although small amounts of PAH have always been present in the environment, natural sources of PAH such as forest fires are now outweighed by man-made sources of these compounds.

PAH are readily taken up by marine organisms. In some species such as vertebrate fish, PAH are rapidly metabolized and excreted and steady-state levels of PAH in tissues are low or undetectable. However, some organisms, notably shellfish, lack any appreciable capacity to metabolize PAH and therefore accumulate them in their tissues. Such bioaccumulation may pose a risk to human health if carcinogen-contaminated seafoods are harvested and consumed.

The results of any study of PAH in the marine environment are of necessity highly dependent on the analytical methodology employed. In the simplest case, the use of different techniques or the measurement of different PAH isomers may make it very difficult to compare one study with another. In more extreme cases, improper attention to analytical details or (particularly in older studies) incomplete resolution of related PAH isomers may invalidate quantitative and sometimes qualitative data.

The analysis of PAH in marine samples represents one of the more difficult tasks in the field of PAH determination. Concentrations of individual compounds are frequently small; and in some cases, such as organisms or portions of dated sediment cores, amounts of sample are very limited. Especially in organisms, PAH are associated with a complex mix of natural and biogenic hydrocarbons, necessitating extensive sample purification before separation and quantitation of PAH can be attempted. Extensive sample manipulation during purification in turn leads to problems in ensuring adequate and consistent recovery of compounds.

This chapter will review the major procedures which have been employed in the determination of PAH in marine sediments and organisms and will discuss some of their advantages and limitations. No attempt will be made to make an exhaustive review of all reports of the determination of PAH in marine samples. Rather, we will cite representative papers utilizing various techniques, pointing out those papers that illustrate specific points. Table 1 lists some features of a number of schemes which have been utilized for the determination of PAH in marine organisms and sediments. It is evident that there is little or no consensus on the best way of approaching the analysis of marine PAH. For this reason, complete analytical schemes used by different investigators will not be discussed. Instead, the techniques used will be examined under four general headings:

Table 1. Details of Some Representative Analytical Schemes for Analysis of PAH in Marine Sediments and Organisms[a]

Ref.	Authors	Date	Material	Extraction	Purification	Analysis	Compounds measured
[13]	Giger and Blumer	1974	Sediments	Soxhlet (methanol, benzene)	Sulfur removal, Sephadex LH-20 CC, alumina/silica CC, trinitrofluorenone complexation	Chromatography on long alumina column, UV spectrophotometry	12 compounds
[37]	Neff and Anderson	1975	Oysters Clams	Homogenization, blending with hexane	Adsorption with Florisil	UV spectrophotometry	Naphthalene, alkyl-naphthalene
[3]	Dunn	1976	Mussels Sediments	Alcoholic KOH digestion	Florisil CC, DMSO partitioning	TLC on cellulose acetate, fluorimetry	Benzo[a]pyrene
[42]	Warner	1976	Organisms	Aqueous NaOH digestion, ether extraction	Silica CC	GC on packed SE-30, OV-17, or OV-1	Several di- and tri-aromatics
[41]	Pancirov and Brown	1977	Fish Shellfish	Alcoholic KOH digestion	DMSO/phosphoric acid partitioning, alumina CC	GC on unspecified column, peaks trapped and analyzed by UV spectrophotometry	Pyrene, methyl pyrene, benz[a]anthracene, benzo[a]pyrene
[12]	Bieri et al.	1978	Sediments	Freeze-dry, Soxhlet (toluene/methanol)	Saponification, sulfur removal, silica CC	GC on capillary SE-52	11 peaks, some unresolved compounds
[14]	Giger and Schaffner	1978	Sediments	Air- or freeze-dry, Soxhlet (methylene chloride)	Sulfur removal, Sephadex LH-20 CC, silica CC	GC on capillary SE-52	22 peaks, some unresolved compounds

(continued)

Table 1 (continued)

Ref.	Authors	Date	Material	Extraction	Purification	Analysis	Compounds measured
[36]	Hanus et al.	1979	Oysters	Homogenize with acetonitrile, filter, saponify	Silica CC, HPLC on 2 × μStyragel	HPLC on 3 × μBondapac	13 compounds
[43]	Black et al.	1979	Sediments	Alcoholic KOH digestion	Florisil CC	HPLC on various reversed phases	5 compounds
[16]	Maher et al.	1979	Sediment	Air-dry, Soxhlet (cyclohexane)	DMSO partitioning, silica TLC	Two-dimensional TLC on cellulose acetate/alumina, fluorimetry	Perylene, benzo[a]pyrene
[10]	Prahl and Carpenter	1979	Sediments Plankton	Soxhlet (benzene/methanol)	Sephadex LH-20 CC, alumina/silica CC	HPLC on Zorbax ODS, or GC on capillary SE-54	6 compounds (HPLC)
[44]	Dunn and Armour	1980	Sediments Mussels	Alcoholic KOH digestion	Florisil CC, DMSO partitioning, Sephadex LH-20 CC	HPLC on Perkin Elmer HC-ODS	20 compounds, some unresolved

[a] Abbreviations: GC, gas chromatography; CC, column chromatography; TLC, thin-layer chromatography; HPLC, high-performance liquid chromatography; DMSO, dimethyl sulfoxide.

1. Sampling and sample treatment
2. Extraction of PAH
3. Purification of PAH
4. Analysis of PAH

Particular attention will be paid to the first three of these subjects, as they are encountered when analyzing marine samples. Some mention will also be made of procedures utilized for the analysis of PAH which have been extracted and purified from marine samples. Not as much emphasis, however, will be placed on this subject, since other chapters in this book deal at length with such techniques as gas chromatography (Chap. 6), mass spectrometry (Chap. 7), and high-pressure liquid chromatography (Chap. 5). As the problems facing the worker attempting to analyze PAH in marine organisms are essentially identical to the problems faced when analyzing foods, the reader is also directed to the chapter in this book dealing with the determination of PAH in food (Chap. 11).

With the exception of a selective depletion of high-molecular-weight compounds in organisms relative to sediments, the two types of samples generally contain similar arrays of PAH [2]. However, PAH levels in sediments are generally much higher than levels in organisms in the same area [2]. In addition, these higher levels of PAH are associated with lower levels of extractable interfering organic materials. Thus, sample purification and analysis schemes which are employed for analysis of organisms can generally be used for analysis of PAH from sediments. The reverse, however, is not necessarily true, as schemes designed for sediments may not have sufficient purification capability or sufficient sensitivity for use with extracts of marine organisms.

II. Sampling and Sample Treatment

Sedentary organisms such as shellfish have been sampled by simple harvesting at low tide [3-6], by utilizing divers [7,8], or, in the case of oysters, by utilizing commercial dredging equipment [9]. Plankton have been sampled by netting followed by wet seiving to achieve size separations [10]. Sediments have been sampled by mechanical grabs or dredges [11-17], mechanical coring devices [7,10,14,18-21], hand coring devices utilized by divers [17], or simple scoops employed at low tide [22]. Hase and Hites have described the use of a vacuum device for sucking up large quantities of sediments through glass tubing [23], and Prahl and Carpenter have described the use of special traps for collecting newly deposited sedimentary material [10].

Samples of sediments and organisms have generally been stored by freezing. Details of how soon samples have been frozen and how they have been treated until freezing are not generally available, although Chesler et al. have described freezing organisms in the field [24] and Ehrhardt has described refrigerating organisms during transport to the laboratory [25]. Kawakami and Nishimura have described the acidification of sediment samples with HCl in the field in order to prevent bacterial decomposition [15]. Similarly, $HgCl_2$ has been utilized as a metabolic poison to prevent bacterial degradation in anchored sediment traps [10].

In preparation for extraction procedures, both sediments [7,12,14] and organisms [7,10] have been freeze-dried. In addition, sediments have been air-dried at ambient or elevated temperature [11,15,16,21]. Some workers utilizing sample extraction procedures not requiring dry samples have taken subsamples of organisms [5,8] or of sediments [3,11,46] for determination of moisture content by drying. Sediments have also commonly been characterized by determining their organic content, either by commercial carbon analyzers [10,26] or by determining the weight loss on ignition at high temperature [2,7,15,32,33].

III. Sample Extraction

The earliest procedures reported for the extraction of PAH from marine organisms utilized mechanical homogenization of wet tissues with methanol, followed by filtration [4,6,34]. Similar procedures for organisms utilizing acetonitrile have recently been described [35,36]. Neff and Anderson have described the extraction of small tissue samples by homogenization with hexane [37], and Payer et al. have described the extraction of dried algae by refluxing with three changes of benzene [39]. Dried sediments have been extracted by sonication with methanol/methylene chloride [40], stirring with several changes of benzene [15], and refluxing with acetone [18]. Small samples of wet sediments have also been extracted by adding water and vortexing in a sealed tube with hexane [38].

Soxhlet extraction has been widely used to remove PAH from both wet and dried sediments. Wet samples have been extracted utilizing water-miscible solvents such as methanol/benzene mixes [10,21], methanol followed by benzene [13], methanol followed by methanol/benzene [19,20,25], or acetone [46]. Wet sediments have also been extracted with methylene chloride after mixing with sodium sulfate as a water scavenger [22]. Dried sediments have been extracted utilizing benzene [11], methanol/toluene [12], cyclohexane [7,16, 21] and methylene chloride [14,17,23,26]. Extraction times have ranged from 6 to 72 hr, with one to several changes of solvent.

Digestion of samples in aqueous [8,24,42] or alcoholic [3,5,7,9,41,44,45] KOH or NaOH followed by partitioning of hydrocarbons into a non-water-miscible solvent has been widely employed for the extraction of PAH from marine organisms. Marine sediments may also be extracted by digestion, with particulate material removed by decantation or centrifugation before solvent partitioning [3,43-45]. When comparisons have been made between digestion and Soxhlet extraction procedures, recoveries of hydrocarbons have been approximately the same.

For a thorough discussion on extraction procedures, the reader is referred to Chap. 3.

IV. Purification of PAH

In early investigations of PAH in marine samples, there was no sharp distinction made between sample purification and PAH analysis. In modern work, a distinction can usually be made between sample *purification*, in which interfering materials are removed and PAH isolated as a class, and sample *analysis*, in which individual PAH are separated and measured. A variety of sample purification procedures have been employed in the analysis of marine samples, either singly or in combination. The major procedures utilized are discussed in this section.

A. Low-Pressure Column Chromatography on Inorganic Adsorbents

Adsorbents which have been used include alumina [5,11,22,39,41], silica gel [12,14,17-21,26,35,36], alumina/silica gel combinations [10,13,25], and Florisil [3,9,43-46]. The usual practice is to introduce samples onto the column in an aliphatic solvent such as pentane, hexane, or isooctane. Aliphatic hydrocarbons are eluted with the same solvent, whereas PAH, pigments, and a variety of biogenic hydrocarbons are retained on the adsorbent. PAH are then selectively eluted with eluants consisting of or containing stronger solvents such as benzene, methylene chloride, methanol, or acetonitrile. Depending on elution conditions, the PAH may be eluted in one fraction or may be fractionated by ring size during chromatography. As an alternative to stepwise elution, it is possible to pass the sample in a strong solvent through the column in a rapid, one-step elution [39,44,46]. In this case, PAH are rapidly and simply separated from polar interfering materials which are retained on the adsorbent. Aliphatic hydrocarbons, however, are not removed and must be eliminated by a subsequent purification step.

Column chromatography procedures using inorganic adsorbents are easily performed on large samples, use inexpensive equipment and reagents, are relatively rapid, and provide a substantial purification of the sample. The activity of the adsorbent must, however, be carefully controlled to avoid nonreproducible results. Losses of compound may occur through irreversible adsorption to active sites on the column. This appears particularly to be a problem with Florisil [3,33] but can be avoided by utilizing deactivated adsorbent [44].

Column chromatography on inorganic adsorbents has frequently been used as the sole procedure for removing organic interfering materials from extracts of marine samples [5,12,17,21,35,42,43,45]. Giger and Schaffner, however, have reported that chromatography on silica alone provides insufficient purification for extracts of sediments. These authors tentatively identified the interfering materials as long-chain wax esters and utilized chromatography on Sephadex LH-20 prior to chromatography on silica in order to remove these compounds [14].

B. High-Pressure Column Chromatography on Microparticulate Packings

Because of high cost and irreversible degradation of column performance, microparticulate silica columns have not been used for sample purifications. Normal-phase chromatography on bonded-phase columns based on aminosilane with a silica backbone, however, has successfully been used in sample cleanup [24,45]. Similarly, high-pressure chromatography with μStyragel columns and a toluene eluant has been employed for removal of interfering materials [36].

C. Low-Pressure Column Chromatography on Organic Adsorbents

Heit et al. have reported the isolation of the aromatic fraction of mussel extracts by gel permeation chromatography on BioBeads 5-8X [8]. More extensive use has been made of chromatography on Sephadex LH-20, utilizing solvents such as isopropanol [18], benzene/methanol [10,13,14,26], methylene chloride/methanol [23], and toluene/ethanol [44]. Although Hase and Hites have reported utilizing chromatography on Sephadex LH-20 as the sole sample purification step [23], this method of removing interfering materials is commonly employed in combination with other sample purification procedures such as column chromatography on silica. Prahl and Carpenter have noted that chromatography on Sephadex LH-20 by itself was insufficient to purify sediment extracts for analysis by HPLC, and they utilized a silica gel/alumina column in addition to the Sephadex step [10]. Dunn and Armour have reported that with extracts of marine organisms partially purified by chromatography on Florisil, chromatography on Sephadex LH-20 was not as effective as dimethyl sulfoxide (DMSO)/isooctane solvent partitioning as a second purification step [44].

D. Thin-Layer Chromatography

Thin-layer chromatography (TLC) on silica has been utilized by a number of workers for sample purification [16,22,25,40]. Separations are generally similar to those which can be performed by column chromatography but chromatographic resolution is higher. The added resolution, however, is generally of little value in a typical purification, where only crude class separations are being made. Sample manipulation with TLC is more time consuming than with column chromatography, and sample capacity is less time consuming. Unless great care is taken, losses of PAH may be higher than with column chromatography because of the extensive sample evaporation needed before application of the sample to plates, because of oxidation of sensitive compounds on plates, and because of the difficulties of recovering the adsorbent from the TLC plate and the PAH from the adsorbent.

E. Solvent Partitioning

PAH are selectively extracted from aliphatic solvents by reagents such as DMSO, dimethylformamide, and nitromethane. Extracts of marine samples have been purified by partitioning between DMSO and isooctane [3,9,33,44], DMSO/phosphoric acid and isooctane [11,41], DMSO and cyclohexane [16], nitromethane and cyclohexane [19,20,39], and dimethylformamide/water and cyclohexane [7,18]. The chemical basis and the selectivity of partitioning between DMSO and aliphatic solvents has been described by Natusch and Tomkins [47]. Partitioning procedures are simple and relatively rapid, and do not utilize expensive equipment or reagents. They have generally been used in combination with other purification procedures such as chromatography on inorganic adsorbents. Little information is available on the relative effectiveness of the different partitioning procedures—although it has been reported that for extracts initially purified by Florisil chromatography, partitioning between DMSO and isooctane is a more effective purification procedure than partitioning between dimethylformamide and isooctane [3].

F. Charge Transfer Complex Formation

PAH may be purified by adding trinitrofluorene to the extract, which forms an insoluble charge transfer complex with the aromatic compounds. The precipitate is washed, and then the complex is split by careful chromatography on silica [13]. This procedure is described as being able to remove interfering materials not removed by chromatography on alumina or on Sephadex LH-20 [13]. However, the procedure is time consuming and difficulties may be encountered with the exacting chromatography necessary to remove the trinitrofluorene from the sample.

G. Removal of Sulfur for Sediment Extracts

If elemental sulfur is present in sediment samples, it may be extracted and carried through normal purification procedures. Sulfur has been removed from extracts by using columns of activated copper eluted with methylene chloride [14,26], hexane [12], or benzene/pentane [13]. As a majority of analyses of sediment have been performed skipping this sulfur-removal step, it is probably not needed as a routine procedure.

V. Analysis

A. Specialized Procedure for Naphthalene and Alkylated Naphthalenes

Naphthalene and alkylated naphthalenes are present in high concentration in some crude oils and are responsible for much of the toxicity of spilled petroleum to marine organisms. Because of the importance of these compounds, Neff and Anderson developed a specialized, rapid procedure for their measurement [37]. Extracts of tissues are purified by adsorbing interfering materials onto Florisil, and naphthalene and alkylated naphthalenes are measured by an UV spectrophotometric method.

B. Resolution of PAH by Column Chromatography

The earliest analyses of marine organisms for PAH were carried out with large samples. Extracts were subjected to repeated fractionation by column chromatography, and individual PAH were isolated by recrystallization, with confirmation of identity by UV absorption spectra [4,6,34]. Similar lengthy separations by column chromatography have been utilized for extracts of sediments, with quantitation by UV absorption measurement of selected peaks in the absorption spectra of individual PAH [13]. Such procedures, although workable, have a limited sensitivity and selectivity and are very labor intensive. They have now been completely replaced by modern high-resolution chromatographic methods.

C. Thin-Layer Chromatography

Most early studies on the occurrence of PAH in the environment utilized TLC for the separation of various PAH isomers, followed by fluorescence analysis. Such procedures were extensively developed by workers investigating PAH in air-pollution particulates and, under certain circumstances, can still be useful today in the analysis of PAH in marine samples. Fossato has reported the determination of benzo[a]pyrene (BaP) and perylene in mussels by fluorescence spectrophotometry after separation of PAH on silica TLC plates [5]. Chromatography on silica or alumina alone gives only limited resolution of various PAH—in particular, BaP is not separated from some compounds such as benzo[k]fluoranthene which have a similar fluorescence spectrum. The values for BaP reported by Fossato may thus overestimate the amount of this isomer in samples. TLC on acetylated cellulose is capable of resolving BaP from all other PAH and has therefore been utilized with fluorometric detection as a specialized separation tool for the measurement of this compound in marine samples [3,15,33]. Even better (although more tedious) separations can be achieved by two-dimensional chromatography on TLC plates with a mixed adsorbent of cellulose acetate and alumina [16,39]. Some 5-10 major PAH are cleanly separated and may be measured by fluorometry. TLC procedures have the advantage of utilizing simple and inexpensive chromatography equipment, although a spectrophotofluorometer is needed for the quantitation of isolated PAH. Since the entire sample may be chromatographed at once and fluorometry is a very sensitive detection method, limits of detectability for selected PAH may be extremely good. TLC procedures, however, are usually labor intensive and frequently require a considerable degree of operator skill. Discussion of TLC analysis of PAH is given in Chap. 9.

D. Gas Chromatography

Gas chromatography, utilizing either flame-ionization detectors (FID) or coupled to MS systems, has been more widely used than any other analytical method for the determination of PAH in purified extracts of marine samples. A wide variety of stationary phases have been used, including Dexsil 300 [22,45], Dexsil 400 [46], OV-1 [40,42], OV-101 [18], OV-7 [28], OV-17 [19,20,42], SE-30 [42], SE-52 [12,14,20,26], and SE-54 [7,10,17]. GC procedures have inherent difficulty in separating certain groups of PAH isomers, such as phenanthrene/anthracene, benz[a]anthracene/chrysene/triphenylene, and benzo[a]pyrene/benzo[e]pyrene. When using short packed columns, overlapping of certain key peaks is inevitable, no matter what the stationary phase. These problems have partially been eliminated by the use of long packed columns or capillary columns ranging in length from 10 to 50 m. However, some isomeric PAH such as benzofluoranthenes and dibenzanthracenes remain hard to separate. An alternative approach to high-resolution columns has been to trap selected peaks as they come from the GC column and measure specific compounds by selective UV absorption procedures [11,41]. An example of a glass capillary gas chromatogram of a sediment sample is given in Fig. 1.

Coupled GC/MS has proved highly useful in assigning structures to peaks, and a number of workers who routinely quantitate PAH by GC with flame-ionization detection utilize GC/MS with selected samples in order to determine or confirm peak identity. For further information on the current practice of GC and coupled GC/MS, the reader is referred to Chaps. 6 and 7 in this handbook.

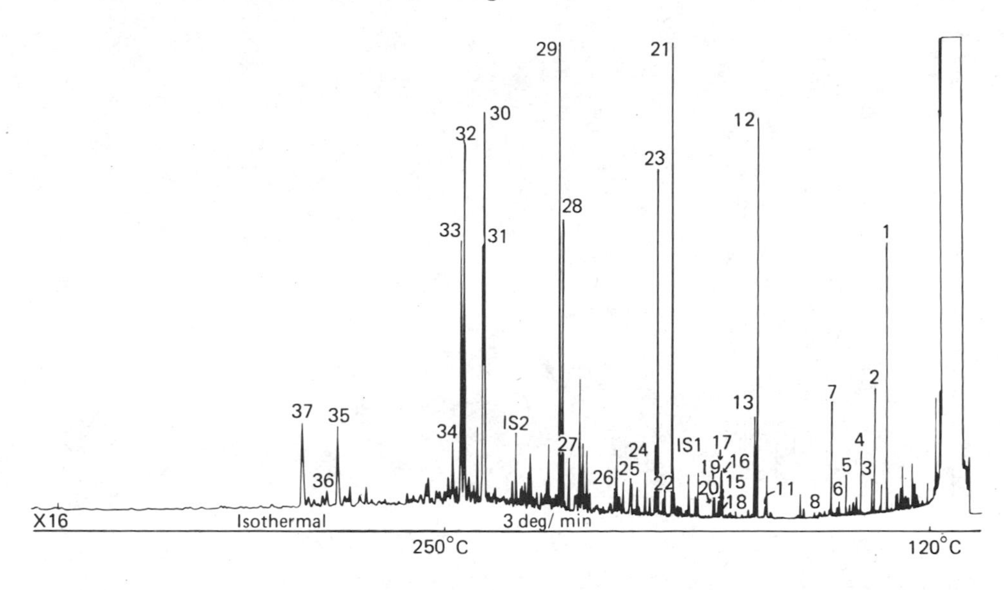

Figure 1. Glass capillary gas chromatogram of PAH in a sediment sample. The peak identities are (1) naphthalene; (2) 2-methylnaphthalene; (3) 1-methylnaphthalene; (4) biphenyl; (5) acenaphthylene; (6) acenaphthene; (7) dibenzofuran; (8) fluorene; (11) dibenzothiophene; (12) phenanthrene; (13) anthracene; (15) methylphenanthrene/methylanthracene; (16) methylphenanthrene/methylanthracene; (17) 2-methylanthracene; (18) 4,5-methylenephenanthrene; (19) methylphenanthrene/methylanthracene; (20) 1-methylphenanthrene; (21) fluoranthene; (22) dihydrobenzo[a/b]fluorenes; (23) pyrene; (24) benzo[a]fluorene; (25) benzo[b]fluorene/4-methylpyrene; (26) 1-methylpyrene; (27) benzo[c]phenanthrene; (28) benz[a]anthracene; (29) chrysene/triphenylene; (30) benzo[b]fluoranthene; (31) benzo[j/k]-fluoranthenes; (32) benzo[e]pyrene; (33) benzo[a]pyrene; (34) perylene; (35) o-phenylenepyrene; (36) dibenz[a,c/a,h]anthracenes; (37) benzo[ghi]-perylene. IS is internal standard. (From Ref. 7. For chromatographic conditions, see that reference.)

E. High-Pressure Liquid Chromatography

The use of high-pressure liquid chromatography on reversed phase microparticulate columns has recently challenged the predominance of GC for the analysis of major parent PAH in marine samples. Compounds are detected by their UV absorption and/or fluorescence. Chromatographic packings which have been utilized include Waters μBondapac [24,35,36,43], Dupont Zorbax [10,43], Altex RP-2 [38], Vydac 201 TP Reversed Phase [44], and Perkin-Elmer HC-ODS [44]. The last two column packings named are more effective than other commercial columns for the separation of a number of important four- and five-ring PAH [44]. Applications of the HC-ODS column are shown in Fig. 2. One company now offers commercially a high-quality,

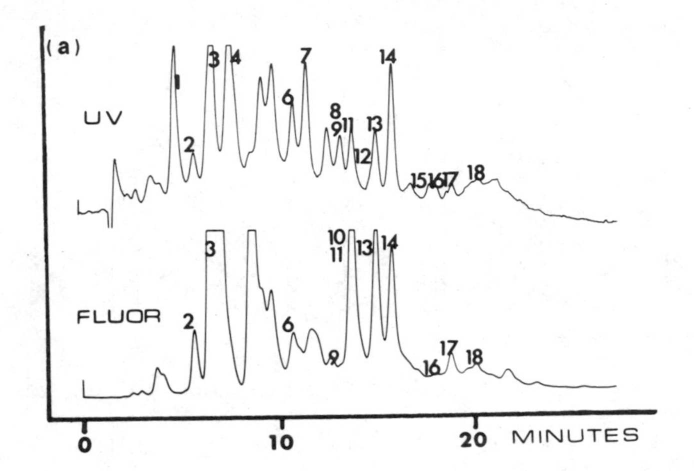
(a)
UV
FLUOR
MINUTES
0
10
20

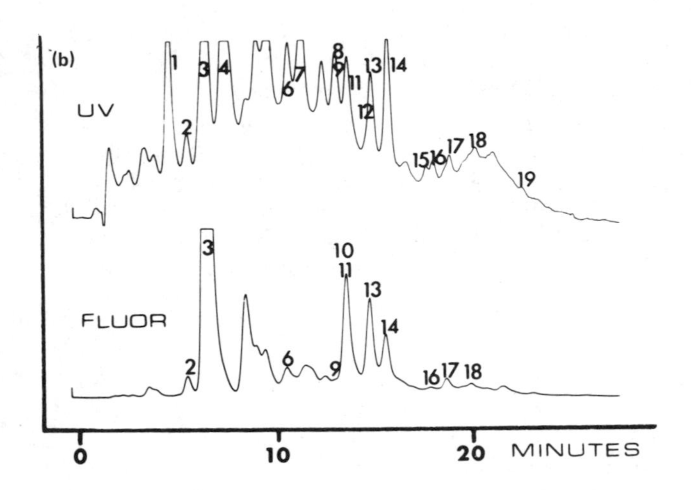
(b)
UV
FLUOR
MINUTES
0
10
20

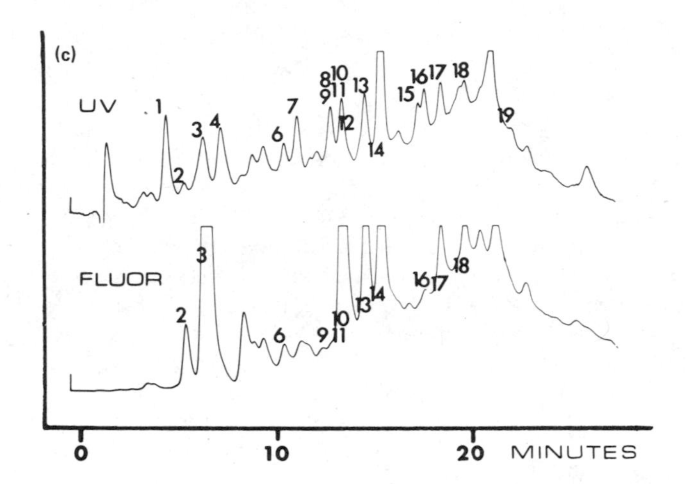

Figure 2. High-pressure liquid chromatogram of purified extract of marine samples. Chromatographic conditions are described in Ref. 44. The samples are (a) creosoted wood from a piling; (b) sediments from the base of the piling; (c) mussels (*Mytilus edulis*) from concrete bridge abutment 3 m from the piling. The peak identities are (1) phenanthrene; (2) anthracene; (3) fluoranthene; (4) pyrene; (6) benz[a]anthracene; (7) chrysene; (8) benzo[e]pyrene; (9) benzo[j]fluoranthene; (10) perylene; (11) benzo[b]-fluoranthene; (12) dibenz[a,c]anthracene (shoulder); (13) benzo[k]fluoranthene; (14) benzo[a]pyrene; (15) dibenz[a,h]anthracene; (16) benzo[ghi]-perylene; (17) indeno[1,2,3-cd]pyrene; (18) benzo[b]chrysene; (19) coronene. (Reprinted with permission from B. P. Dunn and R. J. Armour, Sample extraction and purification for determination of polycyclic aromatic hydrocarbons by reversed phase chromatography. *Anal. Chem.* 52:2027. Copyright 1980, American Chemical Society.)

pretested reversed phase column specifically prepared for the analysis of PAH. The art of high-pressure liquid chromatography of PAH is rapidly advancing at present. The reader is referred to Chap. 5 for the most current information on this topic.

VI. Recommendations

A. Sampling

Samples of sediments or organisms should be taken well away from obvious sources of pollution such as outfalls and wharf areas, unless pollution from these sources is specifically to be studied. Around point sources of pollution, concentration gradients are frequently steep and may be influenced by prevailing currents [27,28]. Care must be taken in these areas to obtain a sufficient number of samples to adequately define the area. Attention should be paid to the possibility in estuarine areas of the layering of pollutant-containing fresh or brackish water over cleaner saline layers; this may result in a vertical gradient of pollution evident in organisms taken from different depths. Several workers have reported seasonal variations in the amounts of PAH in organisms: levels in the winter may be as much as five times summer levels [5,28]. These variations may be related to seasonal influxes of PAH into the water column [10,28], increased photodegradation in summer, or other undefined factors. For comparative purposes, it is important that organism samples be taken at the same time of year, unless only order-of-magnitude comparisons are necessary.

Sampling devices for sediments must not use oil or grease on working parts and must be designed so that they may be easily cleaned. Lowering cables must be oil-free wire or synthetic fiber rope. Particular attention should be paid to the possibility of contaminating samples with surface oil slicks from the sampling boat. Any form of engine which discharges its exhaust underwater (i.e., outboard motors) should be shut off if possible, as should bilge pumps which may discharge oily wastes. Both used crankcase oil and the effluent from outboard motors are rich in PAH [48]. If possible, sampling should be done from the upwind or upcurrent side of an anchored boat so that any petroleum pollution from the sampling vessel is swept away from the area of water through which the sampling device is lowered and recovered. During sampling, the precautions employed should be appropriate for the level of contamination to be expected. Elaborate precautions which may be necessary when one is sampling pristine areas may be entirely a waste of time when sampling in the middle of a busy harbor. Since organisms are generally dissected before analysis and only the internal tissues analyzed, external contamination is not such a problem as with sediments. Obvious exceptions are samples such as seaweeks, where precautions similar to those described for sediments are in order.

Sediment samples and organisms may be frozen in the field by utilizing dry ice. However, it is generally easier to bring samples to the laboratory and process them for storage there. During transport to the laboratory, organisms that have not been frozen should either be refrigerated to prevent bacterial decomposition or should be transported alive. Mussels kept alive for 3 days out of water have shown no loss of BaP [29]. Refrigeration during transport is probably not essential for sediment samples, as the process of bacterial degradation of PAH in sediments is very slow [10,38]. If possible, organisms should be dissected before freezing, as freezing and thawing tend to degrade tissues and make dissection more difficult. If whole organisms must be frozen in the field or in the laboratory, an effort should be made to freeze them as rapidly as possible.

Marine mussels are bioaccumulators of PAH and offer many advantages as monitoring organisms. Recommended procedures for sampling mussels and similar bivalves for hydrocarbon analysis have been described [30]. Because of their relatively small size, mussels are generally analyzed whole, without any attempt to dissect out different internal parts. Byssal threads (the attachment threads secreted by mussels), however, should not be included in the material analyzed. They are largely external to the organism and may be contaminated with exogenous contaminants such as bits of tar or slivers of creosoted wood. Fresh mussels and other bivalves frequently contain seawater inside the closed shell. This should be discarded at the time of shucking. Frozen bivalves and other frozen tissue often exude juices on thawing: if it is clear that the liquid has come entirely from the tissue, it may be included with the sample; otherwise it should be discarded.

For storage by freezing, samples are best kept in cleaned glass containers with a tight lid. Bottles with polyethylene snap caps are convenient. If screw caps are used, particular attention must be paid to the suitability of lid liners. Teflon sheet or cleaned foil may be utilized as an inner liner for screw caps. Whole organisms may be frozen in plastic bags or wrapped in foil that has been solvent washed to remove rolling oil arising from the manufacturing process. Care must be taken that containers and wrappings are airtight in order to prevent inadvertent freeze-drying of the samples during prolonged storage. If there is any doubt about the suitability of a packaging material for contact with stored samples, a sample of the material may be extracted and the extract analyzed by the normal method used for PAH to test for the possibility of interfering materials.

Sediment samples contain varying amounts of water, depending on their physical nature, the type of sampling apparatus, and whether they have had a chance to drain. For ease in making comparisons, it is therefore recommended that analytical results be expressed on the basis of dry sediment weight. If wet sediment samples are extracted, a subsample should be dried for the determination of water content. Sedimentary particles vary considerably in their ability to adsorb PAH [31], and it has been reported that the PAH content of sediments may be positively correlated with their organic content [32,33]. For comparative purposes it is therefore highly recommended that the organic content of the sediments be determined. This may be done by ashing dried subsamples of sediment at 500°C and determining the weight loss. This loss, the "ash-free dry weight," ranges from less than 1% of the dry weight for gravel and coarse sands to greater than 50% for organic muds. The percentage of the dry weight represented by the ash-free dry weight should be reported along with PAH contamination data based on dry weight. Alternatively, contamination data may be reported as nanograms PAH per gram ash-free dry weight. This latter figure gives a measure of the contamination of the organic fraction of the sample, independent of the sample's content of non-PAH-bearing mineral material such as sand and pebbles.

B. Extraction

One of the major problems in the analysis of PAH in marine samples is loss of compounds during complex sample handling procedures. It is strongly recommended that an internal standard of some form be added to every sample before extraction and measured at the end of sample purification procedures. The percentage recovery figures can then be used to correct for losses of PAH and to monitor the performance of the sample purification scheme. It is important to emphasize, however, that internal standards utilized in this way cannot account for inefficient extraction of PAH from the original sample matrix.

Ideally, the standard should match the volatility and solubility properties of those compounds which it is desired to measure. Perdeuterated PAH have been used by Cretney et al. in a procedure using GC/MS to separately measure the standard and nondeuterated PAH present in the sample [45]. Radiolabeled BaP and benz[a]anthracene are both available commercially and have been used as internal standards [3,11,33,44,46]. A typical internal standard might consist of 25,000 dpm [^{3}H]BaP (ca. 0.1 ng at a specific activity of 25 Ci/mmol) or 5000 dpm [^{14}C]BaP (ca. 50 ng at a specific activity of 10 mCi/mmol). In either case, if 10% of the purified sample is taken for the determination of radioactivity by scintillation counting, the level of radioactivity can accurately be measured using a counting time of only a few minutes.

For routine use with samples containing substantial amounts of PAH, the use of ^{14}C-labeled BaP is recommended. In such cases, the amount of PAH contributed by the isotopically labeled standard (typically a minimum of approximately 50-100 ng of PAH) can often be ignored. With smaller or less contaminated samples, a correction will have to be made for the contribution of the standard to the total amount of BaP measured. When analyzing very clean or very small samples, however, the amount of internal standard may be considerably more than the amount of BaP originally present in the sample, and correction may be inaccurate. In this case, [^{3}H]BaP may be used, as it has a much higher specific activity and a typical standard represents only ca. 0.1 ng of PAH. Care must be taken when one is utilizing [^{3}H]PAH as standards, as it has been reported that radiolabel may be lost from the standard by proton exchange catalyzed by active sites on silica or alumina during chromatography or during prolonged refluxing with sediments [46]. Tritiated standards may, however, be used with Florisil chromatography [3,33,44,46], chromatography with Biobeads [46], or presumably chromatography with Sephadex LH-20.

A number of workers have utilized extraction procedures which involve mechanical blending or mixing of sample with an extracting solvent. Such procedures appear useful for multiple small samples where the extraction may be performed in sealed tubes and particulate material removed by centrifugation. Such procedures may also be used with large sediment samples, although the necessity of repeatedly changing the extracting solvent in order to obtain complete extraction may make the technique more labor intensive than Soxhlet extraction. Blending procedures are not recommended for use with large tissue samples, where considerable difficulty may be encountered in efficiently separating the extracting solvent from the sample.

Sediments may be extracted either by Soxhlet procedures or by alkaline digestion. Soxhlet procedures take longer to complete than digestion procedures and may require a considerable inventory of equipment if multiple samples are to be extracted. However, if dry samples and non-water-miscible solvents are used in Soxhlet procedures, fewer man-hours may be needed than with digestion procedures. The latter require labor for the separation of sediment particles from the alkaline digest and for partition of the hydrocarbons into a nonaqueous solvent. This apparent advantage of Soxhlet procedures may be lost if much labor is needed for the drying of samples or if wet samples and water-miscible solvents are utilized (thus necessitating a partitioning step). The Soxhlet procedure of Giger and Schaffner [14] is recommended as being simple and fast and giving good results. Sediments are freeze-dried, or air-dried at 40°C, and then Soxhlet extracted for 6 hr with methylene chloride. This short extraction is stated to be nearly as efficient as longer extractions and provides an extract in an easily evaporated, relatively nontoxic solvent.

Although Soxhlet extraction of freeze-dried tissue is possible, digestion of tissues in alcohol and KOH [3,5,7,9,41,44,45] followed by partitioning into a hydrocarbon solvent is recommended as a better procedure: there is no need to dry the sample; digestion gives complete physical breakdown of the tissue without the need for grinding or homogenization; and saponification of sample lipids is done concurrently with the extraction. Digestions may also be performed in aqueous alkali, followed by extraction with ether [8,42] or pentane [24]. The precautions necessary when working with preservative-free ether [42] and the possibility of poor recoveries because of the low solubility of higher-molecular-weight PAH in the aqueous phase, however, make digestion with alcoholic KOH the first choice. For extracting multiple small samples by digestion procedures, centrifuge tubes with Teflon-lined screw caps may prove highly useful [42].

C. Purification

For both sediments and organisms, it is recommended that the initial purification consist of low-pressure column chromatography. This may consist of chromatography on silica, alumina, Sephadex LH-20, or Florisil. These procedures provide substantial purification, are relatively rapid, and do not require expensive equipment. Unfortunately, there is little or no comparative information available on the efficiency of purification achieved by chromatography using these different adsorbents. Similarly, no comparative studies have been performed on the best solvents and solvent programs to be utilized with each adsorbent. However, as a general rule, procedures should be designed to elute the PAH as a class, without fractionation. Once the bulk of the interfering material has been removed by these procedures, further fractionation may be done if desired. Benzene should not be utilized in any procedure because of its toxicity and carcinogenicity; suitable substitutes may be toluene or methylene chloride.

TLC and high-pressure liquid chromatography on aminosilane columns can also be utilized as initial purification steps. However, the complexities of sample application and recovery for TLC and the high cost of high-pressure liquid chromatography equipment make these procedures less attractive than low-pressure column chromatography. The additional resolution of these chromatographic procedures does not appear to be of much practical advantage in initial purifications, where the aim is only to make crude class separations between PAH and interfering materials.

For extracts of sediments with a substantial degree of PAH contamination, only one purification step may be necessary. However, for organisms or extracts of sediments which contain only small amounts of PAH or which contain high amounts of interfering materials, a second purification step is generally necessary. Partitioning of the sample between an aliphatic solvent and DMSO is recommended as a second purification. The procedure is simple and rapid, provides a substantial purification, requires a minimum of equipment, and gives good recoveries of PAH. Partitioning of the sample between DMSO and an aliphatic solvent is recommended as a final rather than an initial purification step. When used as an initial purification procedure, relatively large volumes must be employed [11,41], and problems may be encountered with the formation of emulsions [44].

D. Analysis

Because of the limited number of compounds which can be quantitated and the labor-intensive nature of the analysis, TLC is not recommended as a first choice in analyzing PAH from marine samples. However, it may still be useful for the analysis of selected compounds in laboratories which do not have access to GC or HPLC equipment. The choice between GC and high-pressure liquid chromatography as a primary analytical tool is more complex and is influenced by a number of factors such as the nature of the sample, the compounds to be quantitated, and the sensitivity required.

In theory, liquid chromatography is highly attractive for the separation of PAH from marine samples for the following reasons:

1. Chromatography does not depend on volatility, so high-molecular-weight compounds can be chromatographed with the same ease as lower-molecular-weight compounds.
2. Injection of samples into the chromatographic system is much easier with HPLC than with GC, and a much larger portion of the sample (10-50%) may be chromatographed at any one time, aiding detection limits.
3. By using appropriate wavelength settings, fluorescence and UV absorption detectors may be made partially or wholly selective for the measurement of specific compounds.
4. Detection is nondestructive, and the effluent may be collected and subjected to further analysis.
5. Chromatography may be stopped for short periods in the middle of a run, and UV absorption or fluorescence spectra may be obtained of peaks trapped within the detector cell.
6. Highly sensitive and selective fluorescence detectors may be used. Usable data may be obtained with peaks containing less than 0.1 ng of many PAH, and nonfluorescent impurities in the sample are ignored by the detector.
7. Chromatographic separations generally take somewhat less time than separations by gas chromatography.

8. Reversed phase columns effectively separate some important carcinogenic and noncarcinogenic PAH isomers which are difficult to separate by gas chromatography.

The apparent advantages of HPLC are counterbalanced by a number of limitations of the technique relative to gas chromatography:

1. Currently available HPLC columns have only a fraction of the resolving power of capillary or long packed GC columns. This does not present a problem in measuring most major PAH present in samples. However, minor components, such as various alkylated derivatives of PAH, do not generally chromatograph as distinct, separate peaks and are lost in the background of unresolved compounds.
2. Response factors in UV and fluorescence detectors vary widely, so it is necessary to have a standard reference material for each compound to be quantitated. This problem does not arise in GC with flame-ionization detection, where within reasonable limits response factors are identical and detector response may be taken to be equivalent to mass.
3. Because of the high inherent selectivity of fluorescence detectors, detection limits for different compounds separated by HPLC may vary widely. Even at compromise wavelength settings, two or more chromatographic runs at different wavelengths may be needed to quantitate all compounds. Alternatively, wavelength settings must be changed in midrun.
4. The powerful technique of on-line MS is not currently available with reversed phase HPLC as it is with GC. This presents severe problems in attempting to identify unknown chromatographic peaks in HPLC.
5. High pressure liquid chromatography apparatus is somewhat more expensive than GC apparatus.

In general, GC would be the analytical technique of choice if minor as well as major PAH must be quantitated (determination of alkylated PAH), if unknown PAH had to be identified and measured (GC/MS), or if a routine method was needed for major, parent PAH and a laboratory possessed GC but not HPLC apparatus. If, however, only the major nonalkylated parent PAH (which predominate in most marine samples [1]) are to be measured, HPLC has two major advantages. By using fluorescence as a detection method, PAH can be quantitated in the presence of other nonfluorescent hydrocarbons. This may permit sample preparation and purification to be less complex than for gas chromatography. Secondly, detection limits for HPLC are better than with GC, aiding the measurement of PAH in less contaminated samples or in very small samples. Wakeham et al. [28] have reported a detection limit for PAH by capillary GC and FID of 1-2 ng/g sediment for samples as large as 50 g. Using HPLC and fluorescence detection, detection limits for important compounds such as BaP, benzofluoranthenes, and perylene are better than 0.01 ng/g sample for similar sized samples, assuming minimum detectability of 0.1 ng PAH, 25% of sample chromatographed at one time, and 80% recovery of PAH from sample.

References

1. J. M. Neff, *Polycyclic Aromatic Hydrocarbons in the Aquatic Environment: Sources, Fates, and Biological Effects*. Applied Science Publ., London, 1979.
2. B. P. Dunn, Polycyclic aromatic hydrocarbons in marine sediments, bivalves, and seaweeds: Analysis by high pressure liquid chromatography, in *Polynuclear Aromatic Hydrocarbons: Chemistry and Biological Effects*, A. Bjørseth and A. J. Dennis (Eds.), Battelle Press, Columbus, Ohio, 1980, pp. 367-377.
3. B. P. Dunn, *Environ. Sci. Technol. 10*:1018 (1976).
4. B. K. Koe and L. Zechmeister, *Arch. Biochem. Biophys. 41*:396 (1952).
5. V. U. Fossato, C. Nasci, and F. Dolci, *Marine Environ. Res. 2*:47 (1979).
6. L. Zechmeister and B. K. Koe, *Arch. Biochem. Biophys. 35*:1 (1952).
7. A. Bjørseth, J. Knutzen, and J. Skei, *Sci. Total Environ. 13*:71 (1979).
8. M. Heit, C. S. Klusek, and K. M. Miller, *Environ. Sci. Technol. 14*:465 (1980).
9. T. Fazio, Analysis of oyster samples for polycyclic hydrocarbons, in *Proceedings of the 7th National Shellfish Sanitation Workshop*, U.S. Dept. of Health and Human Services, Publ. No. 74-2005 (1971).
10. F. G. Prahl and R. Carpenter, *Geochim. Cosmochim. Acta 43*:1959 (1979).
11. R. A. Brown and P. K. Starnes, *Marine Pollut. Bull. 9*:162 (1978).
12. R. H. Bieri, M. K. Cueman, C. L. Smith, and C. W. Su, *Intern. J. Environ. Anal. Chem. 5*:293 (1978).
13. W. Giger and M. Blumer, *Anal. Chem. 46*:1663 (1974).
14. W. Giger and C. Schaffner, *Anal. Chem. 50*:243 (1978).
15. Y. Kawakami and J. Nishimura, *J. Oceanogr. Soc. Japan 32*:175 (1976).
16. W. A. Maher, J. Bagg, and J. D. Smith, *Int. J. Environ. Anal. Chem. 7*:1 (1979).
17. J. L. Lake, C. Norwood, C. Dimock, and R. Bowen, *Geochim. Cosmochim. Acta 43*:1847 (1979).
18. G. Grimmer and H. Böhnke, *Cancer Lett. 1*:75 (1975).
19. R. A. Hites, R. E. LaFlamme, and H. W. Farrington, *Science 198*:829 (1977).
20. R. E. LaFlamme and R. A. Hites, *Geochim. Cosmochim. Acta 42*:289 (1978).
21. M. Heit, *Water Air Soil Pollut. 11*:447 (1979).
22. E. D. John, M. Cooke, and G. Nickless, *Bull. Environ. Contam. Toxicol. 22*:653 (1979).
23. A. Hase and R. A. Hites, *Geochim. Cosmochim. Acta 40*:1141 (1976).
24. S. N. Chesler, B. H. Gump, H. S. Hertz, W. E. May, and S. A. Wise, *Anal. Chem. 50*:805 (1978).
25. M. Ehrhardt, *Environ. Pollut. 3*:257-271 (1972).
26. S. G. Wakeham, C. Schaffner, and W. Giger, *Geochim. Cosmochim. Acta 44*:403 (1980).
27. B. P. Dunn and H. F. Stich, *Proc. Soc. Exp. Biol. Med. 150*:49-51 (1975).
28. B. P. Dunn and H. F. Stich, *J. Fisheries Res. Board Can. 33*:2040 (1976).
29. B. P. Dunn and H. F. Stich, *Bull. Environ. Contam. Toxicol. 15*:398 (1976).

30. National Academy of Sciences (U.S.), *The International Mussel Watch*, Publishing Office, NAS, Washington, D.C., 1980, p. 50.
31. J. C. Means, J. J. Hasset, S. G. Wood, and W. L. Banwart, Sorption properties of energy related pollutants and sediments, in *Polynuclear Aromatic Hydrocarbons*, 2nd ed., P. W. Jones and P. Leber (Eds.), Ann Arbor Science Publs., Ann Arbor, Mich., 1979, pp. 327-340.
32. H. G. Stich and B. P. Dunn, *Arctic 33*:807 (1980).
33. B. P. Dunn, Benzo[a]pyrene in the marine environment: Analytical techniques and results. In *Hydrocarbons and Halogenated Hydrocarbons in the Aquatic Environment*, B. K. Afghan and D. MacKay (Eds.), Plenum Press, New York, 1980, pp. 109-119.
34. H. J. Cahnmann and M. Kuratsune, *Anal. Chem. 29*:1312 (1957).
35. H. Guerrero, E. R. Biehl, and C. T. Kenner, *J. Assoc. Offic. Anal. Chem. 59*:989 (1976).
36. J. P. Hanus, H. Guerrero, E. R. Biehl, and C. T. Kenner, *J. Assoc. Offic. Anal. Chem. 62*:29 (1979).
37. J. M. Neff and J. W. Anderson, *Bull. Environ. Contam. Toxicol. 14*:122 (1975).
38. W. S. Gardner, R. F. Lee, K. R. Tenore, and L. W. Smith, *Water Air Soil Pollut. 11*:339 (1979).
39. H. D. Payer, C. J. Soeder, H. Kunte, P. Karuwanna, R. Nonhof, and W. Graf, *Naturwissenschaften 62*:536 (1975).
40. S. Thompson and G. Eglinton, *Marine Pollut. Bull. 9*:133 (1978).
41. R. J. Pancirov and R. A. Brown, *Environ. Sci. Technol. 11*:989 (1977).
42. J. S. Warner, *Anal. Chem. 48*:578 (1976).
43. J. J. Black, P. P. Dymerski, and W. F. Zapisek, *Bull. Environ. Contam. Toxicol. 22*:278 (1979).
44. B. P. Dunn and R. J. Armour, *Anal. Chem. 52*:2027 (1980).
45. W. J. Cretney, P. A. Christensen, B. W. McIntyre, and B. R. Fowler, Quantification of polycyclic aromatic hydrocarbons in marine environmental samples, in *Hydrocarbons and Halogenated Hydrocarbons in the Aquatic Environment*, B. K. Afghan and D. Mackay (Eds.), Plenum Press, New York, 1980, pp. 315-336.
46. W. H. Griest, Multicomponent polycyclic aromatic hydrocarbon analysis of inland water and sediment, in *Hydrocarbons and Halogenated Hydrocarbons in the Aquatic Environment*, B. K. Afghan and D. Mackay (Eds.), Plenum Press, New York, 1980, pp. 173-183.
47. D. F. S. Natusch and B. A. Tomkins, *Anal. Chem. 50*:1429 (1978).
48. B. Dunn, unpublished results (1978).

11

Polycyclic Aromatic Hydrocarbons in Foods

THOMAS FAZIO AND JOHN W. HOWARD / Food and Drug Administration, Washington, D.C.

I. Introduction

Polycyclic aromatic hydrocarbons (PAH) represent a very important group of chemical carcinogens. Many of these compounds are not carcinogenic; some of them may act as synergists and some as competitive inhibitors. Therefore the broadest possible profile of PAH should be determined in a particular product. The total burden of these compounds has only recently been investigated since the determination of PAH in foods and other complex matrices is very cumbersome and time consuming. For this reason early investigations were limited to the detection and determination of the potent carcinogen benzo[a]pyrene. Emphasis on this compound over the past decades has increased tremendously due to its carcinogenicity, relative ease of analysis, and the belief by other investigators that this compound can serve as an indicator for the presence of other PAH which contaminate the environment. A review article published recently in the *Journal of the Association of Official Analytical Chemists* updates and emphasizes current findings of benzo[a]pyrene in foods and the environment [1].

In this chapter we are attempting to broaden the coverage of the total PAH burden and its presence in our food while being as authoritative as possible. A substantive portion of this chapter has been gathered from the works of the present authors and other researchers in the field. The accuracy and validity of the data presented by some investigators have been questioned due to a lack of information concerning methodology used in their analyses. Although the work and coverage may not be complete, the most frequently observed PAH in foodstuffs have been extensively treated.

Almost a decade ago, Tilgner [2] in his discussion of carcinogens in food stated: "higher [PAH] of which benzo[a]pyrene is but one, have become ubiquitous and may become contaminants under various circumstances." This important class of pollutants has been shown to occur in water, air, food, and soil from such diverse sources as tobacco smoke, automobile and engine exhausts, high-boiling petroleum distillates, carbon black, coal tar, pitch, and rubber tires.

The vast majority of studies conducted in the environment have concentrated on the determination of benzo[a]pyrene, one of the most potent PAH carcinogens. While these investigations demonstrated that this compound is ubiquitous in the environment, Andelman and Suess noted that it constitutes only between 1 and 20% of the total carcinogenic PAH present [3]. In his assessment of the situation, Suess commented that only a few investigators have sought a variety of carcinogenic PAH in the same environmental sample [4]. Furthermore, the sampling and isolation techniques and the analytical methodology of various researchers are not always mentioned. Thus, these critical factors must be considered in assessing carcinogenic PAH in the environment.

Haenni [5] and Tilgner and Daun [6] suggest that only a small fraction of the potential compounds in this class have been recognized, identified, and toxicologically evaluated. In their study of PAH content of sediments taken from the coastal waters at Buzzards Bay, Mass., Giger and Blumer concluded that the complexity of the isolated aromatic fraction suggested the presence of tens of thousands of aromatic compounds, mostly of unknown structure and biological activity [7].

With the development of reliable analytical procedures in the 1960s, a large volume of data on the presence of PAH in foodstuffs has been accumulated on a worldwide basis. However, as discussed above, much of the work has been concerned only with the determination of benzo[a]pyrene. PAH have been reported in smoked fish and meats, grilled and roasted foods, root and leaf vegetables, vegetable oils, grains, plants, fruits, seafoods, whiskies, etc. The sources of such contamination include curing smokes, contaminated soils, polluted air and water, modes of cooking or preparation of foods, food additives, food processing, and endogenic or biosynthesis by plants and microorganisms. Aside from smoked-cured foods, charcoal-broiled meats, and possibly environmental pollution, Haenni indicated that the most common sources of PAH as potential food contaminants were food additives of petroleum origin [4]. However, in a subsequent review several years later, Haenni and Fischbach [8] offered the opinion that more attention should be directed toward contamination of food crops from soils, air, ground waters, etc. Suess, in his assessment of the environmental load and cycle of PAH, concluded that there is prevailing evidence for the natural occurrence of these compounds, which he attributes to endogenic synthesis (by microorganisms, phytoplankton, algae, and highly developed plants), volcanic activity, open burning of forests, and prairies (not ignited by man), and natural seepage of petroleum [4]. He points out, however, that such contributions of PAH are small compared to man-induced and/or man-controlled combustion processes which include the burning of coal, production of coke in the iron and steel industry, catalytic cracking of petroleum, heating and power generation, emissions from transportation vehicles, asphalt paving, and coal tar pitch. In addition, contamination of waterways by oil spills, marine transportation, and industrial wastes is another factor, although its contribution is considered insignificant compared to land-based high-temperature processes. The author states, however, that the impact of such contamination on the aquatic environment should not be underestimated since the chemical half-life of PAH may be much longer in water than in the atmosphere. Tilgner commented that evidence is mounting that environmental contamination of agricultural raw materials is being caused by contaminated air, water, and soil [9]. He concluded that air pollution seems to be the most potent source of food contamination.

As indicated above, the occurrence of PAH in food should be viewed as a part of the much larger problem of the presence of these compounds in the environment. It is also apparent that the findings of PAH in air, water, and soil are being given much more emphasis with respect to their contamination of the food supply. The purpose of this report is to update and to consolidate the findings on PAH in foods and other products of food-additive significance which have been reviewed previously [2-6,9-15].

II. Formation, General Properties, and Origin

PAH represent a very important group of chemical compounds, some of which are potent carcinogens. PAH, also known as polynuclear aromatic hydrocarbons, can be defined as organic compounds containing two or more fused benzene rings which may or may not have substituent groups attached to one or more rings. Some 100 PAH have been identified in foods and in the environment (see Andelman and Suess [3], Tilgner and Daun [6], and Gunther and Buzzetti [10]). The structure, relative activity, and general occurrence of some of these compounds are shown in Table 1.

The PAH found in foods, apart from minute amounts of geochemical or biosynthetic origin, are formed during the pyrolysis of organic matter. Organic substances containing carbon and hydrogen yield PAH during incomplete combustion or pyrolysis or during the formation of petroleum and coal. The more completely an organic material is burned to carbon dioxide and water, the fewer PAH are formed. Since material compositions and conditions of combustion vary, so do the PAH profiles.

The possible sources of PAH contamination of foodstuffs in the modern human environment are numerous, varied and extremely widespread. A review of the literature reveals that curing smokes, contaminated soils, polluted air and water, modes of cooking, food additives, food processing, and endogenous sources have been considered. However, many of the investigations were conducted at random, and the origin of the contaminants and extent of contamination have not been fully established or evaluated. Aside from smoke-cured foods, charcoal-broiled meats, and environmental pollution,

Table 1. Occurrence and Carcinogenicity of Some PAH

Compound	Molecular formula[a]	Carcino-genicity	Occurrence[b]
Anthracene	$C_{14}H_{10}$	-	E,F,S
Benz[a]anthracene (1,2-benzanthracene)	$C_{18}H_{12}$	+	E,F,S
7,12-Dimethylbenz[a]anthracene (9,10-dimethyl-1,2-benzanthracene)	$C_{20}H_{16}$	++++	E,S
Dibenz[a,h]anthracene (1,2-5,6-dibenzanthracene)	$C_{22}H_{14}$	+++	E,F
Phenanthrene	$C_{14}H_{10}$	-	E,F,S
Benzo[c]phenanthrene (3,4-benzphenanthrene)	$C_{18}H_{12}$	+++	E
Fluorene	$C_{13}H_{10}$	-	E,F,S
Dibenzo[a,h]fluorene (1,2-6,7-dibenzfluorene)	$C_{21}H_{14}$	+ -	E
Fluoranthene	$C_{16}H_{10}$	-	E,F,S
Benzo[b]fluoranthene (2,3-benzofluoranthene)	$C_{20}H_{12}$	++	E,F
Benzo[k]fluoranthene (8,9-benzfluoranthene)	$C_{20}H_{12}$	-	E,F
Aceanthrylene	$C_{16}H_{12}$	-	E
Benz[j]aceanthrylene (cholanthrene)	$C_{20}H_{14}$	++	E,S
3-Methylcholanthrene	$C_{21}H_{16}$	++++	E

(continued)

Compound	Molecular formula[a]	Carcino-genicity	Occurrence[b]
Naphthacene (benz[b]anthracene)	$C_{18}H_{12}$	-	E
Pyrene	$C_{16}H_{10}$	-	E,F,S
Benzo[a]pyrene (1,2-benzpyrene) (3,4-benzpyrene)	$C_{20}H_{12}$	+++	E,F,S
Benzo[e]pyrene (4,5-benzopyrene) (1,2-benzpyrene)	$C_{20}H_{12}$	-	E,F,S
Dibenzo[a,h]pyrene (1,2-6,7-dibenzpyrene) (3,4-8,9-dibenzpyrene)	$C_{24}H_{14}$	+++	E,F,S
Indeno[1,2,3-cd]pyrene (o-phenylenepyrene)	$C_{22}H_{12}$	+	E,F
Chrysene (1,2-benzophenanthrene)	$C_{18}H_{12}$	±	E,F,S
Perylene	$C_{20}H_{12}$	-	E,F,S
Benzo[ghi]perylene	$C_{22}H_{12}$	-	E,F,S
Coronene	$C_{24}H_{12}$	-	E,F,S

[a]Molecular structures are shown in the Appendix to this Handbook.
[b]Key: E = environment (water, air, tobacco smoke, gasoline, and diesel exhaust); F = foods; S = curing smoke; +++ or ++ = strongly carcinogenic; + = carcinogenic; - = not carcinogenic.

Table 2. PAH Appearing in the Environment

Compound	Source
1. Benzo[a]pyrene[a]	Smoked foods Oysters Cigarettes Oils (vegetable, mineral, and cooking) Barbecued beef and pork Wax Charcoal-broiled steaks Roasted coffee Polluted H_2O Fruits and vegetables
2. Benzo[e]pyrene[a]	Oysters Smoked foods Vegetable oils Roasted coffee Wax Charcoal-broiled steaks Fruits and vegetables
3. Fluoranthene[a]	Oysters Smoked foods Liquid smoke Vegetable oils Wax Charcoal-broiled steaks Cigarette smoke Roasted coffee
4. Benzo[b]fluoranthene	Oysters
5. Benzo[k]fluoranthene	Oysters Roasted coffee
6. Anthracene	Smoked foods Liquid smoke Charcoal broiled steaks
7. Pyrene[a]	Oysters Wax Smoked foods Vegetable oils Charcoal-broiled steaks Cigarette smoke Roasted coffee
8. Anthanthrene	Charcoal broiled steaks

(continued)

Compound	Source
9. Chrysene[a]	Oysters Wax Smoked foods Roasted coffee
10. Perylene	Oysters Mineral oil Smoked foods Roasted coffee Charcoal-broiled steaks
11. Phenanthrene	Oysters Liquid smoke Smoked foods Charcoal-broiled steaks
12. 4-Methylpyrene	Smoked food Cigarette smoke Wax
13. 4-Methylbenzo[a]pyrene	Liquid smoke
14. 2-Methylpyrene 1-Methylpyrene	Wax Wax
15. Benz[a]anthracene[a]	Roasted coffee Smoked food Charcoal-broiled steaks
16. Acenaphthylene	Smoked foods
17. Triphenylene	Liquid smoke
18. 5-Methylchrysene	Mineral oil
19. 6-Methylchrysene	Smoked foods Wax Mineral oil
20. 1-Methylchrysene	Wax
21. Fluorene	Smoked foods
22. 4-Methylbenzo[a]pyrene	Liquid smoke
23. Benzo[ghi]perylene[a] (1:12 Benzoperylene)	Vegetable oils Charcoal-broiled steaks Roasted coffee Smoked foods
24. Benzo[g,h]perylene	Smoked foods

(continued)

(Table 2 continued)

Compound	Source
25. Benzo[g]chrysene (5:6 Benzchrysene)	Mineral oil
26. Benzo[k]phenanthrene (9:10 Benzophenanthrene)	Mineral oil
27. Benzo[k]phenanthrene:	Mineral oil
4-methyl	Mineral oil
5,8-dimethyl	Mineral oil
28. Chrysene: 5,6-dimethyl	Mineral oil
29. Benzo[b]chrysene	Charcoal-broiled steaks
30. Coronene	Charcoal-broiled steaks Fruits and vegetables
31. Alkyl benzanthracene (probably ethyl or methyl 1:2 benzanthracene)	Charcoal-broiled steaks
32. Benz[a]anthracene:	
7-methyl	Mineral oil
12-methyl	Mineral oil
7,11-dimethyl	Mineral oil
7,12-dimethyl	Mineral oil
33. Anthracene: 9,10-dimethyl	Mineral oil
34. Benzo[a]pyrene: 6-methyl	Mineral oil
35. Cholanthrene	Mineral oil
Cholanthrene: 3-methyl	Mineral oil
36. Dibenzo[a,b]anthracene	Mineral oil
37. Dibenzo[g,h]anthracene	Charcoal-broiled steaks

[a]Most frequently found.

the most common sources of PAH as potential food contaminants are food additives and food packaging of petroleum origin. Some of these sources of PAH appearing in the environment are summarized in Table 2.

III. Methods of Analysis

Haenni in 1968 comprehensively discussed the development of analytical aids for the control of potential PAH contaminants in food additives and in foods by the use of ultraviolet (UV) specification within specific wavelength ranges [5]. Such specifications are currently employed to control the content of these contaminants in products of petroleum origin such as waxes, mineral oils, petrolatums, and other products of food additive significance. The basic principles of analytical control as developed for U.S. regulatory requirements have been adopted in ensuing years by a number of other nations [8].

With respect to the determination of individual PAH in environmental samples, there have been a number of publications of methods and findings of these compounds in air, water, soil, etc. While this information may be relevant to the contamination of food, it should be recognized that the substrates involved are very different from the complex compositions of foodstuffs. Thus, the use of these methods for trace analysis of foods is in question until their application is established. In 1970, Schaad reviewed various chromatographic separation procedures including column, paper, thin-layer, and gas chromatography [16].

As Haenni [5] noted, PAH most often occur in environmental samples as a minute fraction of a complex mixture of hydrocarbons, and it is the exception rather than the rule to find only one of these compounds or only two occurring together. It is therefore of major importance that the methodology employed in such analyses include separation techniques for the determination and identification of individual PAH. This is particularly true in regulatory work where the analyst must be able to separate and unequivocally identify the carcinogenic from the noncarcinogenic types. For example, one of the most significant difficulties is the resolution of the so-called benzpyrene fraction consisting of the carcinogen benzo[a]pyrene, its isomer benzo[e]pyrene, benzo[k]fluoranthene, and perylene. (Benzo[b]fluoranthene, which also has been found in food, should be included with the aforementioned compounds.) At a joint meeting in 1968 of the International Union Against Cancer (UICC) and International Agency for Research on Cancer (IARC) on environmental carcinogens, the Joint Working Group specified that an acceptable method should be capable of separating at least benzo[a]anthracene, benzo[a]-pyrene, benzo[e]pyrene, benzo[ghi]perylene, pyrene, benzo[k]fluoranthene, and coronene [17,18]. Based on more recent findings of PAH in environmental samples, this listing would probably be extended appreciably if such a working group were convened today.

Most of the studies conducted on environmental samples during the 1960s utilized UV and fluorescence techniques to estimate the PAH content. In an early review, Gunther and Buzzetti [10] discussed the analytical problems associated with the isolation and characterization of these compounds. In subsequent work, further refinement of these isolative procedures and the aforementioned determinative techniques resulted in the development of analytical procedures which were capable of reliably measuring a wide variety of

PAH in various products including foods as discussed below. With respect to fluorescence, Shabad reported that, in the USSR, Dikun employs a luminescence technique which is carried out at low temperatures and gives the fine structure of the PAH under analysis [11]. Dikun has collaborated with the present authors in several studies on the analysis of PAH in foods with excellent agreement. However, exact details of the isolation and the technique were not made available.

An assessment of the literature reveals that only a few methods for determining PAH in foods have been subjected to collaborative study and accepted on a national and/or international basis. Collaborative studies of a method specific for benzo[a]pyrene and a general procedure for PAH have been conducted under the auspices of the Association of Official Analytical Chemists (AOAC) and the International Union of Pure and Applied Chemistry (IUPAC) [19,20]. Very briefly, these procedures involve an initial saponification of the product in ethanolic potassium hydroxide, followed by a partition step between dimethyl sulfoxide and an aliphatic solvent and column chromatography on pretreated Florisil. Thin-layer chromatography (TLC) on cellulose (immobile phase, 20% dimethylformamide in ethyl ether; mobile phase, isooctane) and cellulose acetate (21% acetylated; ethanol-toluene-water (17:4:4, v/v/v)) is then employed as the separative technique with UV and fluorescence spectrophotometric procedures being used for determination of the hydrocarbons. In the study of the benzo[a]pyrene method (with cellulose acetate TLC used alone), smoked ham and fish samples were fortified at 4 and 10 μg/kg. Coefficients of variation obtained with the two determinative techniques ranged from 3.15 to 13.6%. The method was adopted as an AOAC official first action method in 1968 and was accepted as a recommended method by IUPAC in 1972 [19,21,22]. The study of the general method for PAH was conducted on ham samples fortified with benzo[a]pyrene, benzo[e]pyrene, benz[a]anthracene, and benzo[ghi]perylene at a level of 10 μg/kg. Statistical evaluation of the data obtained from collaborators in Canada, the United Kingdom, the Federal Republic of Germany, and the United States showed coefficients of variation between laboratories ranging from 7.4 to 12.7%. The procedure was adopted as an official AOAC method in 1972 and has been accepted by the IUPAC Commission at its Madrid 1975 meeting as a recommended method ([23]; see also Refs. 20 and 30).

Grimmer and Böhnke in 1975 described a method for determining PAH in high-protein foods, oils, and fats [24]. The protein-rich foods (meat, poultry, fish, yeast) are initially saponified in methanolic potassium hydroxide, whereas oil and fatty products soluble in methanol or cyclohexane (without a residue) are processed directly. The PAH are concentrated by liquid-liquid extraction (methanol-water-cyclohexane, N,N-dimethylformamide-water-cyclohexane) and by column chromatography on Sephadex LH-20. The compounds are separated and determined by gas-liquid chromatography (GLC) (5% OV-101 on Gas-Chrom Q) using a flame-ionization detector. This latter method has been subjected to collaborative study, and a report has been published [23]. Sunflower oil and meat were fortified at levels of approximately 10 ug/kg with each of the following compounds: chrysene, benzo[b]-fluoranthene, benzo[a]pyrene, benzo[e]pyrene (not used in the sunflower oil study), perylene, dibenz[a,j]anthracene, indeno[1,2,3-cd]pyrene, benzo[ghi]perylene. The coefficients of variation obtained on statistical evaluation of the data ranged from 9.4 to 24.5% for the sunflower oil and from 7.1 to 27% for the meat product. The Commission on Food Additives, IUPAC, considered it a useful screening procedure, and it was so accepted as a recommended method at the Paris meeting in 1976 [23]. As noted by the Commission, the procedure depends solely on relative retention times for identification of the contaminants and thus findings must be regarded as tentative until confirmation by independent adequate means such as mass, UV, and/or fluorescence spectroscopy. Grimmer and Böhnke have pointed out that the OV-101 column employed is not effective in the separation of some important PAH, such as benz[a]anthracene from chrysene and triphenylene, and the benzofluoranthenes [24]. The authors stated that the use of OV-17 packing and capillary columns will improve the resolution of some of these compounds. An additional item of import in the analyses of PAH in food mentioned in the above report is that "to isolate these compounds quantitatively from insoluble samples (meat, fish, etc.), hydrolysis is an absolute necessity." Studies by the aforementioned authors indicated that only about 30% benzo[a]pyrene and other PAH were extracted from fish with methanol, whereas an alkaline hydrolysis of the fish protein yielded an additional 60% of the PAH compounds. It was concluded that the compounds were linked adsorptively to high-molecular-weight structures. The linkages are not destroyed by methanol alone.

Janini et al. in 1975 reviewed the applications of capillary GLC to the analysis of PAH and related compounds in cigarette smoke and in air and automobile exhausts [25-29]. Various other workers have investigated and/or utilized GLC for analysis of environmental samples; however, as stated by Janini, the technique has been used with varying degrees of success in which columns for specific narrow ranges of PAH have been developed [30-35]. Janini also noted that no liquid phase has been reported which wholly meets the recommendations of the UICC/IARC Joint Working Group. With the exception of the aforementioned Grimmer and Böhnke study [24], none of these procedures have been applied to food.

Gouw et al. [36] studied the separation of 44 PAH using short (10-m) glass capillary columns coated with SE-52. Problems were encountered in the resolution of the "benzpyrene fraction," the benzofluoranthenes, and the chrysene-triphenylene-benz[a]anthracene types. Janini and his associates described the results of their GLC studies of PAH with nematic liquid crystals using a flame-ionization detector (FID) [25,37,38].

In a recent publication, Janini et al. report synthesizing N,N'-bis(p-phenylbenzylidene)-α,α'-bi-p-toluidine (BPhBT) and N,N'-bis(p-hexyloxybenzylidene)-α,α'-bitoluidine (BHXBT) [38]. Both of these liquid crystals as well as a 1:1 mixture of the two products were found to yield optimum separations of the aforementioned PAH compounds. The authors state that the low bleed levels and high efficiency characteristics observed for BPhBT and 1:1 BPhBT-BHXBT packed columns have warranted their application as liquid phases in gas chromatographic/mass spectrometric (GC/MS) systems. Burchill et al. [39] also reported the results of their work with these liquid crystals. They state that limited column life was experienced during their work on PAH in coal gases.

Winkler et al. [40] described a method for the determination of PAH in maize. After initial Soxhlet extraction and separations on silica gel and Sephadex LH-20 column, the PAH are determined by capillary gas chromatography using SE-54 and glass columns 50 m long. Further discussion of GC analysis of PAH is given in Chap. 6.

The technique of high-pressure liquid chromatography (HPLC) offers promise as an effective tool for separation and analysis of PAH. As Krstulovic et al. [41] pointed out, nonvolatile, thermally labile compounds can readily be analyzed by this technique without derivatization. Furthermore, samples are not destroyed in the analysis, and it is possible to collect fractions for subsequent analyses by other techniques. In addition, the very sensitive fluorescence and UV characteristics of these compounds can be utilized with HPLC.

A review of the literature reveals that the majority of studies conducted thus far with HPLC have been concerned with the separation of standard compounds, with applications to the analyses of air particulates on various columnar materials. These analyses include investigation of HPLC using reverse phase [42-44], adsorption [45], and complexation [46-49] packing materials with UV or fluorescence detectors. During the past several years, workers in this area have tended to use chemically bonded phase or reverse phases such as octadecylsilane (ODS) [50,51].

Other workers have reported on the separation of PAH using Durapak OPN and cellulose acetate [52-54]. For further discussion of HPLC analysis, see Chap. 5.

At the present time there are only a few published articles on the application of HPLC to PAH analysis of foods. In 1976 Guerrero et al. [55] employed HPLC (μBondapak C_{18} column; acetonitrile-water, 75:25) for the determination of benzo[a]pyrene and benzo[ghi]perylene in clams taken from waters where oil spills had occurred. No data were given in the report as to separation of other PAH from the compounds determined with the μBondapak column. In 1977 Hunt et al. [51] compared two siloxane bonded phases, 2-phthalimidopropyltrichlorosilane (PPS) and ODS-Partisil 5, in their study of the PAH content of mussels. The authors indicate that some useful separations (16 compounds were studied) were obtained with the PPS column, which was found superior to the Partisil material. Based on the relative retention times and chromatograms, it would appear that the benzo[a]pyrene fraction (benzo[a]pyrene, benzo[e]pyrene, benzo[k]fluoranthene and perylene) was not completely resolved. Panalaks [56] has published his findings of PAH in smoked and charcoal-broiled food using HPLC as the determinative method. A Vydac ODS packing was employed with a mobile phase of 87% (v/v) methanol-water. Review of the chromatograms included indicated that difficulties were encountered in the resolution of various PAH including benz[a]anthracene and members of the benzo[a]pyrene fraction.

The majority of the investigations conducted thus far with HPLC suffer from the fact that the authors did not use sufficient compounds or were not selective enough in choosing the more important compounds for their work. Thus it is difficult in some cases to assess the capabilities of the column materials employed. Novotny and colleagues, in their assessment of various chemically bonded phases [5], point out "that many PAH isomers are still unresolved" and calls for further approaches that should be designed toward the development of detailed analytical knowledge. It is obvious, then, that a more concerted effort needs to be made in this area, which offers some specific advantages with respect to PAH analysis.

During the last several years, MS and the combination of GC/MS have been applied extensively in detailed and in some cases quantitative analyses of PAH in air particulates, petroleum products, marijuana and tobacco smoke condensates, marine sediments, etc. [7,57-62]. The use of high-resolution (capillary) columns in some of these studies along with the development of effective isolative techniques has had a significant impact on the separation and identification of hitherto unresolved toxicologically important isomers. For example, Lee et al. [61] reported the isolation of over 150 PAH compounds from marijuana and tobacco smoke condensates using combined capillary GC/MS. Giger and Blumer [7] developed isolative procedures applicable to the analyses of PAH in marine sediments. A combination of UV, visible, and MS techniques have demonstrated the complexity of the PAH fraction in the marine environment.

Up to the present time MS techniques (primarily probe) have been used on an occasional basis in the analysis of PAH in foods, primarily to aid in the identification of a specific carcinogen isolated from a product. Pancirov and Brown [63] modified the procedure of Howard et al. [19] and applied it to the analysis of various marine tissues. GC, UV spectrophotometry, and MS were employed in this study; however, information on the column packings used is not included.

Grimmer and Böhnke used GC/MS in their profile analyses of the PAH content of various foodstuffs [24]. The authors indicate that all of the samples contained more than 100 PAH; however, detailed information on the confirmation of these compounds by MS is not described. As a result of research and demonstrated applications, it can be expected that capillary GC/MS will find much greater use in the analyses of PAH in foods in the near future. These specialized techniques are needed to confirm the identities of reported carcinogens in the environment and are just as applicable to the definitive characterization of PAH contaminants in the food supply.

IV. Occurrence and Levels Found in Foods and the Environment

A. Smoked Foods

Numerous studies have been conducted on the PAH content of curing smokes, and the reader is referred to reviews by Tilgner [64], Draudt [65], Sikorski and Tilgner [66], Tilgner and Daun [6], and Lo and Sandi [15]. Much of the research has been concerned with attempts to control or eliminate the PAH content. Tilgner and Daun [6] stated that over 25 PAH have been detected in curing smokes and approximately 40 others which have not been characterized nor identified.

Investigations have been conducted on a worldwide basis with respect to determination of PAH in smoked foodstuffs. As shown in Table 3, various levels of benzo[a]pyrene have been reported in a wide variety of smoked products [67-91]. The differences in hydrocarbon content may be ascribed to the many variables involved in the smoking process, including the type of generator, combustion temperature, and degree of smoking [65]. With respect to surface deposition of the smoke constituents, Tilgner [64] and Gorelova et al. [92] stated that the PAH compounds will migrate into the

Table 3. Levels of Benzo[a]pyrene Found in Smoked Foods

Food product	Levels (μg/kg)	Reference	Country
Sausage and fish	1.7-10.5	Dobes et al. [67]	Czechoslovakia
Fish and mutton	0.3-2.1	Bailey and Dungal [68]	Iceland
Fish	1.7-53	Voitelovich et al. [69]	USSR
Fish	7	Gorelova [70]	USSR
Sausage and fish	0.1-1.4	Gorelova and Dikun [71]	USSR
Fish	4.2-60	Petrun and Rubenchik [72]	USSR
Fish (salmon)	2.6-3.0	Manelli [73]	Italy
Wurstel	0.4-1.0	Manelli [73]	Italy
Bacon	1.6-4.0	Manelli [73]	Italy
Salami	2.0-2.8	Manelli [73]	Italy
Sardines	1.8	Manelli [73]	Italy
Provola (cheese)	4.1-6.2	Manelli [73]	Italy
Haddock, salmon	0.3,1.0	Lijinsky and Shubik [74]	U.S.
Herring, sturgeon	1.0,0.8	Howard et al. [75,76]	U.S.
Ham	3.2	Howard et al. [75,76]	U.S.
Ham, belly fat	1.0-14	Toth [77]	Germany
Fish	0.05-5.7	Dikun et al. [78]	USSR
Katsuobushi (bonito), Sabushi (mackerel), Urumegushi (sardines)	2-37	Masuda and Kuratsune [79]	Japan
Katsuobushi (bonito)	8.7-27.2	Shiratori [80]	Japan
Nori (seaweed food)	7.4-31.3	Shiratori [80]	Japan
Meats	9-55	Toth [77] Toth and Blaas [81]	Yugoslavia
Fish	11.5	Wierzchowski and Gajelwska [82]	Poland
Fish, mutton (commercially smoked)	1.0	Thorsteinsson [83]	Iceland

(continued)

Food product	Levels (μg/kg)	Reference	Country
Mutton (home smoked)	23.0	Thorsteinsson [83]	Iceland
Hot sausage	0.8	Malanoski et al. [84]	U.S.
Ham	1.0	Malanoski et al. [84]	U.S.
Turkey fat	2.1	Malanoski et al. [84]	U.S.
Whitefish	4.3	Malanoski et al. [84]	U.S.
Whiting	6.9	Malanoski et al. [84]	U.S.
Chubs	1.3	Malanoski et al. [84]	U.S.
Cod	4.5	Malanoski et al. [84]	U.S.
Meats (bologna, frankfurters, salami, pepperoni, sausages, hams, bacon, beef, pork)	0.2-2.0	Panalaks [84]	Canada
Fish (herring, canned, oysters)	0.5-15	Panalaks [85]	Canada
Cheese, Gouda	0.5	Panalaks [85]	Canada
Bacon	1.2-3.6	Rhee and Bratzler [86]	U.S.
Cured meats, sausage, fish, cheese	<5	Lucisano et al. [87]	Italy
Sausages (mutton), bologna	0-0.15	Fretheim [88]	Norway
Fish, oysters	0-9	Swallow [89]	New Zealand
Meat	<0.5	Swallow [89]	New Zealand
Beef	18.8-24.1	Doremire et al. [90]	U.S.
Pork	25.8-31.6	Doremire et al. [90]	U.S.
Lamb	8.8-12.3	Doremire et al. [90]	U.S.
Turkey	ND	Doremire et al. [90]	U.S.
Frankfurters	ND	Joe et al. [91]	U.S.
Salmon steak	ND	Joe et al. [91]	U.S.

interior of the food, the extent of migration being dependent upon the character of the product and its storage time.

The majority of the earlier findings were reported by European and Russian investigators. These initial studies were no doubt prompted by the hypothesis that a correlation might exist between a high incidence of stomach cancer in some population groups (Icelandic and Baltic fishermen) and the presence of PAH compounds in smoked foods. Both Dungal [93] and Voitelovich et al. [69] claimed that such a correlation did exist; however, some investigators felt that the data were not sufficient to warrant such a conclusion [94].

Some of the highest levels of benzo[a]pyrene have been found in smoked fish as reported by Russian and Japanese investigators. For example, Voitelovich et al. [69] and Petrun and Rubenchik [72] reported findings ranging from 1.7 to 60 μg/kg. Shabad [11] in his review of studies in the USSR did not comment on these data; however, he stated that the technology of smoking had been altered to use special smoke liquids which were free of benzo[a]pyrene and did not induce tumors when tested on animals. Masuda and Kuratsune in 1971 reported finding levels of benzo[a]pyrene of up to 37 μg/kg [79] in their analyses of Japanese smoked dried fish products. The fish products included Katsuobushi, Sababushi, and Urumebushi—made, respectively, from bonito, mackerel, and sardines. Katsuobushi is prepared from broiled boned bonito flesh by smoking through intermittent exposure to wood smoke for some 1 or 2 weeks and drying in the sun. The product is stored in wooden boxes for about a week to permit mold growth and the drying and molding processes are then repeated a few times. The Sababushi and Uremebushi are prepared from flesh of mackerel and sardines by a similar process but with less smoking and no storing for mold production. These authors found 16 PAH including 12 with four or five condensed rings and a number of carcinogenic types including benzo[a]pyrene as mentioned above. They believe that these products probably are the most heavily PAH contaminated foods in the world. The relationship between the high PAH content of these foods (consumed frequently but not in large quantities) and the high incidence of gastric cancer in Japan was considered by the authors, who judged that it is inconclusive in the absence of epidemiological studies. Shirotori [80] confirmed Masuda and Kuratsune's findings in Katsuobushi and also reported levels of benzo[a]pyrene of 7.4-31.3 μg/kg in nori, a seaweed food.

Toth [77] and Toth and Blaas [81] reported levels of 9-55 μg/kg of benzo[a]pyrene and 13 other PAH in six Yugoslav smoked meats. However, the authors state that from their experimental sausage-making tests the benzo[a]pyrene content could be reduced to not over 1 μg/kg by reduced time or different generator conditions. As would be expected, products smoked for long periods to a black color contained higher levels of the PAH.

Thorsteinsson [83] compared the PAH content of foods after traditional home smoking and commercial smoking in Iceland. The home-smoked meat (mutton) contained as much as 23 μg benzo[a]pyrene/kg and correspondingly high amounts of other PAH, including benz[a]anthracene. Samples of meat hung just above the stove showed as high as 107 μg benzo[a]pyrene/kg and 115 μg benz[a]anthracene/kg. In contrast, levels of the latter compound did not exceed 1 μg/kg in meat and fish smoked commercially. This author also concluded that 60-75% of the benzo[a]pyrene occurred in the superficial layers of the meat products and that protective coverings, including loose cotton fabric and cellophane, reduced the PAH content of the smoked meats significantly. The latter covering provided almost full protection against benzo[a]pyrene penetration but did not eliminate the smoked flavor and preservative constituents.

These studies are in agreement with the findings of Rhee and Bratzler [86], who studied the formation and distribution of benzo[a]pyrene in smoked bologna and bacon. They also found that cellulose casing significantly reduced the benzo[a]pyrene content in the bologna and that with or without casing the penetration of the compound did not exceed 1.4-1.6 mm from the surface. With respect to the cooking of the smoked bacon, at least half of the PAH was found in the fat drippings.

In comparison with the numerous studies conducted in Europe, data on smoked foods in North America are meager. Genest and Smith [95] in Canada analyzed various smoked fish, frankfurters, and cheeses but benzo[a]pyrene was not detected; however, the detection limits of the method employed ranged only from 10 to 50 μg/kg. Lijinsky and Shubik [74] reported the presence of benzo[a]pyrene at levels of 0.3 and 1.0 μg/kg and other PAH in two samples of smoked fish [72].

In 1966 the U.S. Food and Drug Administration (FDA) developed methods (detection limits of 2 μg/kg and below) for determination of PAH in smoked foods and applied them to various smoked and unsmoked foodstuffs as shown in Table 4 [75,76]. Levels of benzo[a]pyrene of up to 3.2 μg/kg (smoked ham) were reported. Pyrene and fluoranthene were found in all of the products examined including the unsmoked samples. In follow-up investigations the FDA and U.S. Department of Agriculture developed a cooperative

Table 4. PAH Found in Smoked Food Products (μg/kg)[a]

Product	BaA	BaP	BeP	BghiP	FL	P	4 MP
Beef, chipped	0.4	—	—	—	0.6	0.5	—
Cheese, Gouda	—	—	—	—	2.8	2.6	—
Fish:							
Herring	—	—	—	—	—	—	—
Herring, dried	1.7	1.0	1.2	1.0	1.8	1.8	—
Salmon	0.5	—	0.4	—	3.2	2.0	—
Sturgeon	—	0.8	—	—	2.4	4.4	—
White	—	—	—	—	4.6	4.0	—
Frankfurters	—	—	—	—	6.4	3.8	—
Ham	2.8	3.2	1.2	1.4	14.0	11.2	2.0
Pork roll	—	—	—	—	3.1	2.5	—

[a]Key: BaA = benz[a]anthracene; BaP = benzo[a]pyrene; BeP = benzo[e]pyrene; BghiP = benzo[ghi]perylene; FL = fluoranthene; P = pyrene; 4 MP = 4-methylpyrene.

Source: Refs. 75 and 76.

Table 5. PAH in Smoked Foods (μg/kg)

Compound	Ham	Kippered cod	Smoked whiting	Barbecued beef	Hot sausage
Benz[a]anthracene	1.3,9.6	—	—	13.2	0.5
Benzo[a]pyrene	0.7,0.7	4.0	6.6	3.3	0.4
Benzo[e]pyrene	—	—	—	1.7	—
Benzo[ghi]perylene	—	2.2	2.4	4.3	—
Fluoranthene	0.6,2.9	—	—	2.0	—
Pyrene	0.2,0.9	0.6	<0.5	3.2	1.5
4-Methylpyrene	—	—	—	—	1.9
Chrysene	0.5,2.6	1.4	—	9.6	1.0
Perylene	—	0.4	0.7	—	—
6-Methylchrysene	—	—	—	—	0.5

Source: Ref. 84.

program [84]. Sixty assorted foods and related samples were analyzed for benzo[a]pyrene. Thirty-two of these products contained the hydrocarbon. Levels reported for 21 samples were below 1 μg/kg; greater amounts were isolated from the remaining 11 samples. In the smoked products, levels of benzo[a]pyrene did not exceed 7.0 ppb (smoked whiting), which was well below some of the values reported in the literature by Russian and European workers. These data were not unexpected at the time since Draudt had stated that lightly smoked products are common in the United States in contrast to some countries in which foods are heavily smoked for preservative purposes [65]. Table 5 shows other PAH levels observed, ranging from 0.2 to 9.6 μg/kg.

In a recent study, Panalaks [85], using HPLC, completed the analyses of 70 smoked food products commercially available in Canada. PAH were detected in 70% of the samples. Levels of benzo[a]pyrene were from 0.2 to 15 μg/kg. It is also noteworthy that the reported benz[a]anthracene content ranged from 0.2 to 30 μg/kg. The latter compound was found in both smoked and unsmoked oysters. With respect to the Panalaks study, it appears that some follow-up studies should be conducted, particularly with respect to the reports that 7,12-dimethylbenz[a]anthracene is present in smoked and unsmoked oysters (25-38 μg/kg) and that several other PAH, not previously found in foods, are present in the various products analyzed. Further confirmation of identity would be desirable. A summary of Panalaks's findings (levels of PAH ranging from 0.2 to 50 μg/kg found in smoked and charcoal broiled foods) is presented in Table 6.

Doremire et al. [90] examined charcoal-grilled meats for benzo[a]pyrene fluorometrically and found levels ranging from not detectable (ND) to 133 μg/kg. Concentrations of benzo[a]pyrene appear to be proportional to the

Table 6. PAH Found in Smoked and Charcoal-Broiled Food Products (μg/kg)

Foods	BbFL	BaA	BeP	BaP	DBahP	IP	PR	CR
Smoked:								
Bologna	—	—	5.0	2.0	0.5	—	—	—
Frankfurters	—	1.5	—	2.0	—	—	—	2.0
Salami	—	1.5	—	2.0	—	—	—	2.0
Pepperoni	—	0.2	2.0	—	—	1.0	—	—
Various sausages	—	0.5	2.5	1.0	8.0	1.0	—	—
Ham	—	0.2	0.2	2.0	1.0	0.5	—	—
Wesphalian ham	—	—	5.0	2.0	—	—	—	—
Bacon	—	0.5	—	0.5	1.0	—	—	4.0
Smoked beef	—	8.0	—	—	—	0.5	—	—
Smoked pork	—	—	—	0.2-0.3	—	—	—	—
Smoked herrings	—	20.0	2.0	15.0	—	0.2-9.0	—	8.0
Various smoked fish	—	1.0	1.5	0.5	5.0	—	—	10.0
Canned smoked fish	—	—	1.0	—	—	—	1.0	—
Canned (smoked) oysters	25.0	30.0	16.0	2.0	—	5.0	20.0	—
Canned (nonsmoked) oysters	—	15.0-30.0	1.0	—	—	2.0	—	—
Charcoal-broiled:								
Porterhouse steak	—	1.0	4.0	3.0	—	—	2.0	—
Barbecued chicken	—	—	2.0	—	—	10.0	2.0	—
Hamburger	30.0	50.0	20.0	20.0	—	—	—	—
Frankfurter	2.0	2.0	5.0	5.0	—	—	—	—

[a]Key: BbFL = benzo[b]fluoranthene; BaA = benz[a]anthracene; BeP = benzo[e]pyrene; BaP = benzo[a]pyrene; DBahP = dibenzo[a,h]pyrene; IP = indeno[1,2,3-cd]pyrene; PR = perylene; CR = coronene.

Source: Ref. 85.

fat content of the meat product. Lijinsky and Ross [96] explained this phenomenon by the theory that the rendered fat falls on the hot coals and is pyrolyzed, forming benzo[a]pyrene, which is then volatilized and deposited on the meat surface.

B. Fats and Oils

D'Arrigo [14] did a tabulated review on the presence of PAH in vegetable oils and fats, animal oils and fats, and by-products of oils such as margarines and soaps. The author surveyed the variations in PAH content and attributed the differences in findings to technological treatments of oils and fats, heating, usage of solvents, refining, and oxidation.

The detection of PAH, including pyrene, benzo[e]pyrene, and benzo[a]-pyrene, in edible vegetable oils was first reported by Jung and Morand in the early 1960s [97-99]. Subsequent studies by Ciusa et al. indicated the presence of phenanthrene, pyrene, fluoranthene, benz[a]anthracene, chrysene, and perylene in pressed and refined olive oils [100]. Higher levels were detected in the rectified products; however, no benzo[a]pyrene was isolated. In 1966, Borneff and Fabian published a method for isolation of PAH in fats and oils [101]. Estimated levels of up to 1 μg benzo[a]pyrene/kg were reported for vegetable oils [up to 10 μg total PAH (carcinogenic types) per kilogram]. The authors indicated that heating destroyed about 70% of the PAH originally present in the oils. No PAH were found in pork fat. Howard et al. [102] described a quantitative method and its application to the analyses of soybean, cottonseed, corn, olive, safflower, and peanut oils processed in the United States. With the exception of safflower oil (only three samples) trace quantities of PAH, including benzo[a]pyrene, were found in at least one of each type of oil analyzed. Levels of benzo[a]pyrene ranged from 0.4 to 1.5 μg/kg. The concentrations of the PAH found in the various oils analyzed are shown in Table 7. The highest total PAH content was found in peanut oil, soybean oil, and corn oil in that descending order, but the highest concentration of benzo[a]pyrene was observed in soybean oil (1.5 μg/kg). In subsequent investigations, Howard et al. [103] attempted to establish the source of the contamination. Analyses of 15 commercial hexanes used in the solvent extraction of edible oils were conducted on the thesis that they were contributing the residues. Although nine of the samples contained polycyclics, no carcinogens were detected. However, analysis of various crude oil samples not subjected to solvent extraction did reveal the presence of benzo[a]pyrene. The authors concluded that the results of the studies indicated that the contamination occurred in the initial processing operation or was present in the original starting material, as suggested by reports in the literature.

Subsequent reports by Grimmer and Hildebrandt [104] and Biernoth and Rost [105] in Germany confirmed the presence of PAH compounds in both refined and crude oils. In examination of crude rapeseed, sunflower, palm kernel, palm, peanut, cottonseed, soybean, linseed, and coconut oils for polycyclics, the former authors found the highest concentrations (43.7 μg/kg) of benzo[a]pyrene in oil from smoke-dried coconut, whereas sunflower seed and palm kernel oils contained 10.6 and 4.1 μg/kg, respectively [104]. It is noteworthy that other carcinogenic types, i.e., benz[a]anthracene and

Table 7. PAH Found in Vegetable Oils (μg/kg)

Compound	Soybean	Cottonseed	Corn	Olive	Peanut
Benzo[a]pyrene	1.4[a] (1.2-1.5)[b]	0.4 (0.4-0.4)	0.7 (0.4-1.0)	0.5 (0.4-0.5)	0.6
Benzo[ghi]perylene	1.0 (0.5-1.5)	0.5 (0.4-0.6)	0.6 (0.5-0.7)	0.9 (0.5-1.5)	0.9
Benz[a]anthracene	0.9 (0.4-1.3)	—	0.8 (0.4-1.2)	1.0 (0.6-1.4)	1.1
Benzo[e]pyrene	1.6 (0.6-2.6)	0.5	0.7 (0.8-1.2)	0.4	—
Pyrene	1.6 (0.5-3.4)	—	3.1 (0.4-5.8)	2.6 (2.1-3.3)	2.9
Fluoranthene	1.3 (0.6-2.6)	—	—	3.2 (2.2-4.4)	3.3

[a]The single number is the average.
[b]The numbers in parentheses show the range.
Source: Ref. 102.

dibenz[a,h]anthracene, were also isolated from the aforementioned products. Grimmer and Hildebrandt also concluded that hexane solvents were not the source of the contamination. Benzo[a]pyrene (0.2-29 μg/kg) and other PAH including benz[a]anthracene and dibenz[a,h]anthracene were found by Biernoth and Rost in their study of refined coconut oils [105]. These authors also indicated that although the PAH content of the oils was reduced to only a small extent by bleaching earths, treatment with activated charcoal and subsequent deodorization was an effective means of removing the PAH from the oils.

Fabian analyzed fats and oils of vegetable and animal origin, with findings as high as 20-100 μg of carcinogenic types/kg (3-18 μg/kg benzo[a]pyrene) in various margarines and coconut oil, but these compounds were not found in butter or lard [106-108]. The author stated that levels could be reduced to 1-4 μg/kg by treatment with steam and activated carbon. In a study of margarine and mayonnaise, Fritz [109] and Franzke and Fritz [110] found 0.2-0.6 μg of benzo[a]pyrene and six other PAH per kilogram; they also reported 11 PAH including benzo[a]pyrene, 1.9 and 0.5 μg/kg, in crude and refined safflower oils, respectively. Similar findings were published by Grigorenko et al. [111], who found 1-5 μg/kg in sunflower oils. In subsequent work, levels of 0.5-2 μg of benzo[a]pyrene/kg of soybean, cottonseed, and sunflower oils were reported by the latter authors [112].

In a more recent study [113] in New Zealand, Swallow analyzed various vegetable oils and animal fats. Variable concentrations of PAH were found in all of the seven vegetable oils examined: the highest levels of benzo[a]-pyrene (9 μg/kg) were present in a peanut oil product. Concentrations up to 15 μg of the same hydrocarbon/kg were reported in used beef drippings,

Table 8. PAH in Fats and Oils

Compound	Carcinogen[a]
1. Benz[a]anthracene	+
2. Benzo[a]pyrene	+
3. Benzo[e]pyrene	CC
4. Benzo[b]fluoranthene	++
5. Benzo[k]fluoranthene	-
6. Benzo[ghi]perylene	CC
7. Chrysene	+
8. Pyrene	CC
9. Fluoranthene	CC
10. Coronene	-
11. Anthanthrene	-
12. Indeno[1,2,3-cd]pyrene	+
13. Perylene	-
14. Dibenz[a,h]anthracene	++

[a]CC = cocarcinogen.
Source: Refs. 91,100,102-106.

which is the main fat used for commercial frying in New Zealand. The author stated that the amounts of PAH in butter and margarine were minimal.

In his assessment, Tilgner [9] stated that it is most probable "that the contamination of vegetable fats occurs either through reabsorption by plants grown in contaminated soils or by surface contamination from polluted air—both being contributory factors in oilseeds. The contaminants may find their way from the contaminated vegetable raw material into crude and refined oils. The contamination does not originate during the solvent extraction and processing operations." As discussed in Sec. IV.C, below, there is considerable disagreement on the source(s) of such contamination.

A review of the literature indicates the presence of 14 PAH with three to seven rings in fats and oils (in addition to anthracene and phenanthrene), as shown in Table 8.

C. Plants

The presence of PAH has been demonstrated in a wide variety of plants from diverse sources. In 1969 Guddal first reported the isolation of anthracene, pyrene, and fluoranthene from chrysanthemum roots grown in contaminated soil near a gasworks [114]. The author concluded that the PAH were absorbed by the plant since follow-up investigations of roots grown in uncontaminated soils and not exposed to smoke from the factory were not found to be contaminated with they hydrocarbons. These findings stimulated interest in the PAH content of soils and plants as well as their possible endogenous formation in plants and microorganisms. Andelman and Suess have noted that the question of endogenous formation arose because of the hydrocarbon's apparent ubiquity in the environment, particularly in a wide variety of materials which were not likely to have been associated with pyrolytic processes [3]. Blumer, in 1961, reported levels of benzo[a]pyrene of from 40 to 1300 μg/kg and other PAH in soil taken from Connecticut and Massachusetts forests located away from cities and industrial complexes [115]. The author postulated that the contaminants were derived from the pyrolysis of wood or alternatively from organisms which contribute their organic matter to soils. Borneff and Fischer [116] described findings of 100 μg of fluorescent PAH/kg of lake phytoplankton; 13 compounds including benzo[a]pyrene were identified. The concept of bacterial synthesis of PAH and subsequent translocation to plants was seconded by Mallet [117] in France, who found varying levels of benzo[a]pyrene in tree leaves and in decaying matter under trees.

In a series of articles during the 1960s, various German workers presented evidence that PAH are absorbed and synthesized by plants. Doerr, in 1965, reported that barley roots in soil or water cultures absorbed benzo[a]pyrene with subsequent transferral to the shoots [118]. Graf and Diehl [119] isolated eight PAH from various plant leaves (including salad greens, cauliflower, potatoes, carrots, apples, apricots, edible mushrooms, and wheat and rye grains) at levels of 8 to 40 μg/kg. In the same paper, the authors also described the results of their studies with wheat and rye grown hydroponically in carcinogen-free nutrient solutions in the presence and absence of light. The seedlings contained 10-20 μg of benzo[a]pyrene/kg of dried material, in contrast to the seeds in which only traces of the carcinogen were found. The investigators concluded that the hydrocarbon was synthesized by the plants both in the presence and absence of light. In another series of experiments the authors fertilized five different plant varieties with 10 μg of benzo[a]pyrene/liter and observed a distinct increase in vegetation. Borneff et al. [120,121] also concluded that the endogenous formation of carcinogens in plants does occur. This conclusion was based on the results observed in laboratory culture studies of freshwater algae in which the extracted algae contained 10-15 μg carcinogenic PAH/kg. Knoor and Schenk, in 1968, discussed their laboratory studies of various bacteria, indicating that benzo[a]-pyrene had been accumulated through synthesis in amounts of 2-6 μg/kg of dried material [122].

Based on their own investigations and an analysis of the results of others, Schmidt and Fritz [123] concluded that benzo[a]pyrene and other PAH occur in edible plants to a considerable extent. These authors cited Graf and Diehl's aforementioned experiments indicating that PAH are synthesized by plants as essential materials in their metabolic processes; they suggested that these findings indicate that foodstuffs of plant origin which are consumed in great quantities may in fact be the primary source of ingested PAH, rather than smoked, roasted, or heated products, which had been the most suspect sources.

Hancock et al. [124] conducted studies on leaves of plants grown near a railroad station and another site close to an airport. A comparison of pyrene to benzo[a]pyrene ratios suggested that most of the PAH found on leaves were products of plant biochemical synthesis. In their review of the "Environmental Load and Cycle of PAH," Andelman and Suess [3] concluded that

> the existence of a natural background concentration of PAH has now been well established. It consists of PAH biosynthesized on a worldwide scale by plants and microorganisms on land and in the water and formed during open burning of forests and prairies not ignited by man. Volcanic activity is an additional source.

Tilgner [9] noted that a strong relationship appears to exist between air pollution and the occurrence of benzo[a]pyrene in grain and vegetables. For example, grain samples from the heavily industrialized Ruhr district in Germany were found by Grimmer and Hildebrandt [125] to contain approximately 10 times more PAH than samples taken from Lower Saxony and the Holstein district remote from industry.

Grimmer and Hildebrandt [126] also conducted studies on four different types of vegetables grown simultaneously in the same field. The benzo[a]-pyrene content varied considerably, e.g., tomatoes, 0.22 μg/kg; leeks, 6.6 μg/kg; spinach, 7.4 μg/kg; and kale, 20 μg/kg. Salad greens grown close to Hamburg contained levels of PAH five to six times greater than those grown in a suburban area. In subsequent work, Grimmer [127] was of the opinion that "in the appraisal of the amount of carcinogenic hydrocarbons ingested by man, quite new criteria are apparent. Neither smoked foods nor grilled meat but vegetables and salads contain the largest amounts of PAH."

Gunther et al. [128] reported the presence of about 25 mg/kg of anthracene and six unidentified PAH in the rinds of oranges grown in atmosphere-polluted areas in the United States but not in fruit harvested in uncontaminated locations. Bolling [129] studied the effects of location and of drying with combustion gases on the benzo[a]pyrene content of cereals. Wheat, corn, oats, and barley grown in industrial areas showed a four- to tenfold higher contamination than crops from more remote areas. Drying with combustion gases increased the contamination of the grain three- to tenfold; coke as fuel caused much less contamination than oil.

Hutt et al. [130] reported that PAH are deposited on grains dried by direct heating with fuel oil. The use of fuel oils produced levels between 140 and 250% of the initial benzo[a]pyrene content. Levels of benzo[a]pyrene found ranged from 3 to 18 ppb (and sometimes two- to threefold higher) in dried grains as a result of the fuel oil used in the drying process and the completeness of the combustion flame.

Various Russian investigators reported the contamination of plants by PAH. Shabad et al. [131] stated that, on the basis of available data, there are at least three routes of PAH passage into plants: air deposition, adsorption from soil, and synthesis. The results of their studies have led them to believe that air deposition is the principal route of contamination. According to the authors the PAH "penetrates into the soil mainly from air spread across the layers into the water, and passes into plants, fodder, and finally into human food." With respect to the minimal "background" quantities of PAH, Shabad et al. believe that natural synthesis cannot be ruled out; however, the problem requires further study under well-controlled and fully airtight conditions. Shcherbak [132,133] also stated that pollution of plants may occur from sedimentation of atmospheric dust and soot or by migration of the carcinogens into plants from polluted soils. Shabad and Cohan [134] concluded that the main source of contamination of soil is from air particulates. They indicated that migration or resorption of PAH into plants was dependent on the PAH level in the soil and the type of plant. Less benzo[a]pyrene is found in grain plants (e.g., wheat).

Several other investigators have refuted the biosynthesis of PAH by plants. By careful exclusion of contaminated air from developing plants (lettuce, soybean, rye, and tobacco), Grimmer and Duevel [135] demonstrated that they were free of the PAH found in plants from the same lots of seeds exposed to the atmosphere in fields or greenhouses. Schamp and Van Wassenhove [136] stressed the necessity for extreme precautions in isolation of benzo[a]-pyrene from extracts of plants and bitumens, for otherwise false positives can occur. In their analyses, Wagner and Siddiqi [137] found that an increase in benzo[k]fluoranthene in summer wheat and rye paralleled an increase in its concentration in the soil. These workers also reported 4.8-8.6 μg of benzo[a]pyrene/kg and 24.6-76.8 μg of benzo[k]fluoranthene (dry basis)/kg in young wheat plants. Hertel et al. [138] has reported 0.12-0.54 mg benzo[a]pyrene/kg in commercial wheat.

A summary of PAH found in fruits and plant products was compiled by the IARC in 1973 and the degrees of contamination of these commodities are shown in Table 9 [139]. In these investigations levels of PAH such as benz[a]-anthracene, benzo[a]pyrene, benzo[e]pyrene, benzo[b]fluoranthene, chrysene, and dibenz[a,h]anthracene were also observed for food products such as dried yeast (1.8-203 ppb); peanuts (0.01-0.9 ppb); roasted coffee (1.2-43 ppb); tea (3.9-21.3 ppb); tea extracts (1.9-22 ppb); and whiskey (0.03-0.08 ppb).

D. Seafoods ("Marine Life")

Kraybill [140] stated that PAH are ubiquitous in the aquatic environment and may present the greatest carcinogenic insult. Andelman and Suess [3] surveyed the literature on the incidence and significance of PAH in the water

Table 9. PAH in Fruit and Plant Products (μg/kg)[a]

Food	BaA	BaP	BeP	CH	DBahA
Spinach	16	7.4	6.9	28	0.3
Salad	4.6-15.4	2.8-5.3	3.7-14.7	5.7-26.5	0.6-1.0
Kale	43-230	12.6-48.1	1.1-76.2	58-395	0.1-2.6
Apples	—	0.1-0.5	—	—	—
Tomatoes					
Prunes (dried)	—	0.2-1.5	—	—	—
Other fruits	—	2-8	—	—	—
Soybean	—	3.1	4.3	—	—
Cereals	0.4-6.8	0.2-4.1	0.3-4.9	0.8-14	0.1-0.6

[a]Key: BaA = benz[a]anthracene; BaP = benzo[a]pyrene; BeP = benzo[e]pyrene; CH = chrysene; DBahA = dibenz[a,h]anthracene.

environment. The above reviewers [3,4] also discussed sources of contamination of the waterways including industrial and domestic effluents, atmospheric particulates, oil spills, marine transportation, and biosynthesis by plants and microorganisms.

Most of the early studies on PAH in the aquatic environment were reviewed by two Russian workers, Ilnitsky and Varshavskaya [141]. They covered some of the literature up to 1952 with respect to water contamination with benzo[a]pyrene. These authors concluded that the benzo[a]pyrene was increasingly polluting natural waters. Borneff's group in Germany have investigated the presence of PAH in fresh water as well as the effectiveness of various treatments for removal of the contaminants [142-147]. These reports give the total PAH contents found at the time (1960s) in samples taken from various rivers and lakes in Germany. PAH found in surface waters were 0.065-3 μg/liter; in plants and sediments, 700 μg/kg for phytoplankton and up to 55 mg/kg for suspended solids; in groundwaters, 0.045-0.140 μg/liter; and in pure waters, up to 0.025 μg/liter. According to these investigators, it is through the discharge of urban and domestic sewage, the release of industrial wastes, rainwater, and leaching of the pollutants of industrial origin from vegetation and soils that the compounds enter natural waters and thereby contaminate public water supplies. In experimental studies Saccini-Cicatelli [148,149] has observed that *Tubifex* worms placed in benzo[a]pyrene-contaminated water took up as much as 88.2 mg of the hydrocarbon per kilogram and retained up to 350 μg/kg when placed in pure water. As discussed in Sec. IV.C, Borneff et al. [120,121] and Knorr and Schenk [122] reported the synthesis of benzo[a]pyrene by bacteria and algae, respectively [120-122]. Andelman and Suess [3] reported finding PAH in phytoplankton, in river sediments and suspended solids, and in worms, and pointed out that these aqueous biota serve as food for edible fish. They also indicated that benzo[a]-pyrene has been detected in fish taken from European rivers in which these biota were reported to be contaminated.

The majority of the early studies of PAH in the marine environment were performed by Mallet's group in France [150-157]. The results of these investigations and other early studies conducted off the coasts of France, Italy, Greenland, and the United States indicated the presence of benzo[a]pyrene at varying levels in flora and fish as summarized in Table 10 [3,156,158-161]. According to Mallet and Sardou [152], plankton may be able to fix PAH from exogenous sources and marine fauna may be contaminated regardless of whether in polluted or unpolluted locations. With respect to sources of PAH in marine sediments, Andelman and Suess [3] included surface effluents, ships, volcanic debris, and activity of organisms (including bacteria). The same authors also noted that the benzo[a]pyrene contamination of marine life off the unpopulated coast of Greenland is of the same order as that found off French coasts, thus indicating the ubiquity of benzo[a]pyrene in the oceans. They suggested that endogenous synthesis in flora may be the source of contamination in remote areas.

In 1976, Suess concluded that the ability of marine organisms to concentrate PAH had been demonstrated; however, there is still some controversy with respect to their capacity to degrade the hydrocarbons [4]. The author has stressed that "while some aquatic microorganisms will degrade PAH, their actual existence and bioactivity depend on their surrounding environmental

Table 10. Levels of Benzo[a]pyrene Found in Marine Life at Various Geographic Locations

Source	Marine life	Level of benzo[a]pyrene, μg/kg dry wt	Reference
Greenland	Plankton	5.5	Mallet et al. [151]
	Algae	60	Mallet et al. [151]
	Codfish	15	Mallet et al. [151]
	Mollusk	60	Mallet et al. [151]
	Mussel	18 (shell) 55 (body)	Mallet et al. [151]
French coasts	Plankton	5-400	Mallet and Sardou [152] Mallet and Lami [153]
	Shrimp, oyster, mussel mollusk, crab	1.5-90	Mallet [154]
	Oyster, lower shell	70	
	upper shell	112	Mallet and Schneider [155]
	Mussel	16-22	Greffard and Meury [158]
Italian coasts	Plankton	6.1-21.2	Boucart and Mallet [156]
	Algae	2.2	Boucart and Mallet [156]
	Mussel	11 (shell) 130,540 (body)	Boucart and Mallet [156]
	Mollusk	2.4	Boucart and Mallet [156]
	Sardine	65	Boucart and Mallet [156]
United States:			
California	Thatched barnacles	Present	Shimkin et al. [159] Koe and Zechmeister [160]
	Goose barnacle	Present	Koe and Zechmeister [160]
Virginia	Oyster	2-6	Cahnmann and Kuratsune [161]
Alabama	Oyster (shell)	24	Mallet and Schneider [155]
Maine	Clam	3	Joe et al. [91]

conditions. Changes in salinity, temperature, sunlight and wave action directly affect growth rate and metabolism." Suess also noted that the more highly developed aquatic fauna may contribute to such degradation in that some of the mammalian metabolic pathways were found to involve oxidases of other enzymes needed for degrading of the PAH [4]. However, there is insufficient information as to how widespread these systems are. Thus, while a number of investigations have shown a significant degradation of PAH in some marine fish and invertebrates, other researchers were unable to demonstrate their oxidation in some benthic marine invertebrates, phytoplankton, and some zooplankton within a period of one month. Zobell [162] summarized information on sources and biodegradation related to marine pollution.

The PAH contamination of marine life in U.S. coastal waters (Table 10) was first investigated by Shimkin [159] and Koe and Zechmeister [160], who isolated benzo[a]pyrene from thatched barnacles off the coast of California. Cahnmann and Kuratsune [161] reported estimated levels of 2-6 μg of benzo[a]-pyrene and other PAH per kilogram (estimated total level about 1200 μg/kg) in oysters collected from a marine area off Norfolk, Virginia, that was moderately contaminated with petroleum oils. These workers assumed that the quantities of hydrocarbons in oysters vary with their habitat, based on findings of other investigators who showed variations in the levels of the compounds and in composition of the hydrocarbon mixture, depending upon the environment [159,160,163].

During the 1970s, considerable attention was focused on the occurrence of oil spills and the effects on the marine environment. For detailed information the reader is referred to the report of the workshop on the "Input, Fate and Effects of Petroleum in the Marine Environment" sponsored by the National Academy of Sciences (NAS) in 1973 in Airlie, Virginia [164]. This workshop estimated that approximately 6 million tons of petroleum hydrocarbons enter the oceans annually, the major contributors being marine transportation and urban runoff, some 2.1 and 1.9 million tons/annum, respectively. Coastal refineries, industrial and domestic together, were estimated to contribute 0.8 million tons/annum; natural seepage and atmospheric fallout each add another 0.6 million tons/annum. On the basis of the reported benzo[a]pyrene concentration in crude oil being about 1 mg/kg, NAS (1975) concluded that about 6 tons of petroleum-linked benzo[a]pyrene enters the oceans each year. In assessing this contamination, Suess [4] noted that the contribution to the total environmental load is insignificant in comparison with that from land-based high-temperature processes and constitutes only about 0.1%. Ketchum [165] stated that although oil spills are spectacular events and attract the most public attention, they constitute only about 10% of the total amount of oil entering the marine environment. According to this author the other 90% originates from the normal operation of oil-carrying tankers, other ships, offshore production, refinery operations, and the disposal of oil-waste materials.

Blumer et al. [166] and Ehrhardt [167] suggested that marine organisms nonselectively accumulate petroleum hydrocarbons in amounts present in the water and that, once accumulated, the hydrocarbons (in particular the aromatic hydrocarbons) are retained in the tissues for long periods of time, being depurated only very slowly. Anderson [168], on the other hand, in controlled studies of oysters and clams placed in water-fuel oil mixtures followed by depuration in clean waters, reported that both aromatic and saturated hydrocarbons are released from tissues relatively rapidly. For example, maintenance of the shellfish in clean water for periods from 24 to 52 days was sufficient to cleanse the tissues of detectable levels of hydrocarbons. Lee et al. [169] have exposed mussels to mineral oil and various tagged hydrocarbons such as [^{14}C]heptadecane, [^{14}C]naphthalene, and [^{3}H]benzo[a]pyrene. It was found that the mussels rapidly accumulated the hydrocarbons but when they were transferred to uncontaminated water 80-90% of the hydrocarbons were depurated from the mussels in a 2-week period.

Blumer et al. [166] discussed the extent of the pollution problem associated with an oil spill in Buzzards Bay, Massachusetts. Contamination of edible shellfish with oil resulted and persisted for a number of months after the incident, as did the pollutant in the marine sediments. The GC technique employed by the authors did not permit separation and identification of specific PAH; however, the chromatograms showed clear evidence of higher PAH in the oysters and scallops. As was mentioned earlier, Giger and Blumer [7] analyzed sediments from Buzzards Bay and found 12 PAH; however, MS analyses of the remaining complex mixture indicated that many more condensed ring compounds of unknown structure and biological activity were present.

Fazio [170] obtained data on PAH levels in oysters in Aransas Bay and Galveston Bay, Texas. The Galveston Bay area was suspected of being contaminated with petroleum hydrocarbon residues. The results are summarized in Table 11. Samples were taken as follows: 5 samples of suspected contamination from closed areas in Galveston Bay, 13 samples from approved areas in Galveston Bay, and two samples designated as controls from Aransas Bay approximately 150 miles outside of the suspected contamination area. No benzo[a]pyrene was found in any of the samples; however, slightly higher levels of PAH were found in the oysters taken from the closed area. With few exceptions the hydrocarbon types isolated from the three harvested areas were essentially the same.

Table 11. Levels of PAH (μg/kg)[a] Found in Oysters Taken from U.S. Waters

Compounds	Aransas Bay	Galveston Bay	
		Approved	Closed
Benzo[b]fluoranthene	0.3	1.2	2.2
Benzo[k]fluoranthene	–[b]	0.1	0.4
Benzo[e]pyrene	0.2	1.2	2.1
Chrysene	0.5	0.2	0.6
Fluoranthene	1.7	3.0	7.8
Pyrene	0.9	1.8	6.5
Perylene	–[b]	0.7	1.0
Phenanthrene	–[b]	–[b]	2.2

[a]Average values.
[b]Not found.

Fazio and Howard in 1975 also conducted, under contract, a survey of retail market and growing area oysters for the presence of PAH. The sampling consisted of 25 eastern oysters (*Crassostrea virginica*) from the Chesapeake Bay area, 25 from the Baltimore retail market, 25 from the Galveston Bay area, and 25 from the Galveston retail market. The results of the analyses showed no detectable levels of benzo[a]pyrene in the Chesapeake Bay area or retail market samples (Baltimore and Galveston). The oil-polluted areas of Galveston Bay showed levels of benzo[a]pyrene as high as 9.4 μg/kg, whereas oysters taken from approved (noncontaminated) areas did not contain any detectable levels. Pyrene and fluoranthene were detected in all of the samples from all sources at levels up to 157 μg/kg.

In Canada, Dunn and Stick [171] described their findings of elevated levels of benzo[a]pyrene (up to 21.5 μg/kg) in mussels growing near and on creosoted timbers in Vancouver (British Columbia) coastal waters. The same authors [172] also studied the release of the above hydrocarbon from environmentally contaminated mussels transferred from a polluted area into clean circulating water. The amount of benzo[a]pyrene, initially 45 μg/kg wet weight, declined approximately exponentially over the 6 weeks of the experiment, with a reported overall half-life of 16 days. The investigators concluded that short depuration periods of 1-3 days commonly used to eliminate bacterial contamination from edible shellfish before marketing would have little effect on the tissue content of benzo[a]pyrene in mussels.

In a study of 19 mainland and 6 island stations situated throughout Southern California Bight, Dunn and Young [173] found that even in a heavily populated coastal area, baseline levels of benzo[a]pyrene in mussels are at or near the limits of detectability (0.1 μg/kg) of the analytical method employed. These researchers have pointed out that their data are in contrast with the results reported by European workers who found substantial contamination of marine organisms (biosynthesis) taken from remote and presumably unpolluted areas. The former authors concluded that the geographic distribution and low baseline levels of benzo[a]pyrene in mussels support a human rather than a biogenic origin of benzo[a]pyrene contamination of marine organisms.

Pancirov and Brown [63] investigated the extent of PAH contamination in various samples of shellfish and finfish along the U.S. eastern seaboard and various other locations in the United States and Canada. Ten PAH including benzo[a]pyrene, benz[a]anthracene, and chrysene were isolated from the tissues of edible marine animals. Only the oyster and crab samples taken from Long Island Sound and Raritan Bay waters which were exposed to municipal and industrial wastes were found to contain levels of the PAH above 2 μg/kg. The authors noted that in comparison with other foodstuffs neither shellfish nor finfish show unusually high amounts of PAH. Based on their findings that pyrene had much higher relative concentrations than its methyl isomers, the authors offered the opinion that the hydrocarbon contamination is not of petroleum origin but rather is due to combustion sources. This opinion is based on published data on the PAH content of petroleum which show that the methyl derivatives of pyrene significantly outnumber the parent compound. However, it is pointed out that this line of reasoning does not take into account the possibility of preferential metabolism of the hydrocarbons with side chains or the difference in the rate and extent of accumulation of these compounds in marine tissues.

Brown and Pancirov [174] reported baseline levels of PAH for five species of fish (flounder, scup, black sea bass, butterfish, red hake) and one species of shellfish (sea scallops) obtained from the Baltimore Canyon area. With the start of exploration and possible production of oil and/or gas off the east coast of the United States there was concern that environmental damage might result. Levels of benz[a]anthracene and benzo[a]pyrene ranged from 0.3 to 20 μg/kg and <1 to 11 μg/kg, respectively.

E. Yeasts

Much of the analytical work conducted on yeasts can be ascribed to the interest in production of single cell proteins on petroleum substrates. Although such products are produced in Europe and other countries, production in the United States has been slow to develop. McGinnis and Norris [175] have attributed this situation to stringent regulations which require the food industry to demonstrate the absence of traces of carcinogenic PAH in the food product derived from the petroleum substrate. The point out that Takata [176] demonstrated the absence of carcinogenic properties of petroleum-grown yeast in his animal-feeding studies. Scrimshaw [177] has also dismissed the view that such products constitute a potential hazard.

Shabad [11] reported the finding of 10-20 μg benzo[a]pyrene/kg in hydrolyzed yeasts (grown in a special cellulose medium) employed as a cattle feed in Russia. Further investigations revealed that technical ammonium sulfate salts produced by the treatment of coal gas with sulfuric acid had been added to the nutrient medium for cultivation of the hydrolyzed yeast. Analyses of these salts indicated the presence of benzo[a]pyrene at 1000-1400 μg/kg.

McGinnis and Norris [175] discussed their analytical studies of yeast grown on n-hydrocarbons and dextrose. The dextrose-grown yeast was not found to contain any highly condensed ring aromatics apart from pyrene and fluoranthene. However, benzo[a]pyrene (range of 1.1-5.9 μg/kg) and other PAH such as benz[a]anthracene were found in three of the four n-hydrocarbon-grown yeasts analyzed. (The exception was an n-hydrocarbon sample pretreated with silica gel.) In a subsequent study McGinnis [178] analyzed the n-hydrocarbon substrates used for fermentation; however, pyrene, fluoranthene, and some substituted phenanthrenes and pyrenes were the only compounds isolated. The authors concluded that the contamination of the yeast with the aforementioned carcinogenic PAH could not be traced to the n-hydrocarbon source.

In their report of the development of a GC method for PAH in foods, Grimmer and Böhnke [24] discussed the application of their procedure to yeast, meat, smoked fish, and unrefined sunflower oil. The authors stated that all of the samples were found to contain more than 100 PAH (characterized by MS) of which only the main components were determined: phenanthrene, anthracene, fluorene, fluoranthene, pyrene, benz[a]anthracene, chrysene, benzofluoranthenes, benzo[a]pyrene, benzo[e]pyrene, perylene, dibenzanthracenes, indeno[1,2,3-cd]pyrene, benzo[ghi]perylene, anthranthrene, and coronene.

Truhaut and Ferrando [179] have compared the levels of three PAH found in various European commercial yeasts and two samples grown experimentally on petroleum substrates. The content of the commercial products ranged from 0 to 13.2 μg benzo[a]pyrene/kg and from 0 to 9.7 μg benzo[ghi]perylene/kg. No dibenz[a,h]anthracene was detected. With respect to the two alkane yeasts, the results obtained were 0.6-2.5 μg benzo[a]pyrene/kg, 0-1.3 benzo[ghi]perylene/kg and 0.1-0.5 dibenz[a,h]anthracene/kg. The authors indicate, however, that these compounds were not detected in subsequent analysis of yeast samples (petroleum substrate) taken from actual commercial production.

Santoro et al. analyzed yeasts grown on n-paraffins and molasses by GLC with a column containing liquid crystals as stationary phase [180]. Yeasts grown on n-paraffins showed traces of benz[a]anthracene and chrysene (1-10 μg/kg), and yeasts grown on molasses showed levels of 15.2 ppb benz[a]-anthracene, 24.8 ppb of benzo[a]pyrene, and 41.0 ppb of chrysene. The authors claim that the presence of PAH in molasses is probably due to the manner in which this product is processed.

F. Liquid Smoke Flavors

Smoke flavors are used in this country and others to impart a smoked flavor to foods. The advantages of the use of these products over conventional or direct smoking methods have been discussed by various workers [6,65,181]. Tilgner and Daun [6] noted the reduction in processing costs; however, they also remarked that there are over 40 patents in various countries that are mostly based upon wood distillation procedures yielding so-called liquid smoke with typical and unacceptable aroma and flavor. The authors recommended more fundamental studies to clarify the complex interaction phenomena occurring in the food after addition of the concentrate. Hollenbeck [182,183] described his manufacturing process in detail and stated that the major benefit to be derived is flavor reproducibility.

Various investigators in the United States and abroad have analyzed smoke flavors for PAH. Lijinsky and Shubik [74], in the analyses of two aqueous flavors, isolated various hydrocarbons including pyrene, fluoranthene, benzo[ghi]perylene, chrysene, benz[a]anthracene, carbazole, and an unidentified compound with UV and fluorescence spectra almost identical to those of benzo[a]pyrene. White et al. [184] determined PAH in liquid smoke flavors and the resinous condensates that settle out of the aqueous products on standing. Of seven aqueous flavors analyzed, four were free of PAH; the other three contained from 2 μg of pyrene/kg to 35 μg of phenanthrene/kg, with intermediate levels of anthracene, fluoranthene, and triphenylene. One of these flavors contained the apparent benzo[a]pyrene (2 μg/kg) reported by Lijinsky and Shubik [74]; it was shown to be 4-methylbenzo[a]pyrene, the only carcinogen found. The four resinous condensates contained relatively high levels of benzo[a]pyrene (25-3800 μg/kg) plus a number of uncharacterized fluorescent compounds. White and associates concluded that there was a need to assure the efficient removal of the resinous condensates from the aqueous flavors before they were used in foods.

In his 1967 review, Shabad discussed the work of his Russian colleagues Gorelova et al. [185] and Prokofieva [186] in this area [11]. It is stated that the technology of food smoking was altered and that special smoking liquids were in use which had been found to be free of benzo[a]pyrene and did not induce tumors when fed to animals. In a later publication, Gorbatov et al. [187], also of the USSR, compared liquid smokes for use in cured meats with respect to chemical composition and quality. These workers mentioned the desirability of removing the carcinogenic hydrocarbons and residual tars from the products before use. These hydrocarbons and tars may be removed by distillation or possibly by filtration through cellulose pulp as described by Hollenbeck. With reference to the latter process, Gorbatov et al. noted that Hollenbeck [182,183] admitted that the pulp is not effective for the removal of all carcinogens.

Gorbatov et al. [187] also indicated that "the proper procedure for the production of liquid smokes from condensates, including purification steps, depends on the use to which the liquid smoke is to be put. Thus, liquid smokes intended for surface treatment only, do not require the removal of certain ingredients which must be eliminated if the liquid smoke is to be incorporated internally." As an example, the authors state that liquid smoke for the surface treatment of certain sausages must contain those substances that provide the desired product color (tars and certain carbohydrate compounds). The authors concluded that "this new technology . . . will require additional research and development work in order to perfect it for commercial use." This research would include improvement of separation and purification techniques for the flavoring and components from wood pyrolyzates.

Toth and Blaas [81] in Germany reported the examination of 15 commercial liquid smoke preparations for benzo[a]pyrene. The results of this study in indicate that 12 of the products contained <1 μg/kg; however, the 3 remaining preparations contained 8-15 μg/kg. The authors concluded that the majority of the flavors would impart less benzo[a]pyrene to meat than the conventional smoking process and thus would not present a health risk. No information is presented on the resinous or the oil-soluble content of these products.

G. Beverages

Masuda et al. [188] discussed the contamination of whiskies with PAH. They point out that Scotch whiskey is made from malted barley prepared by exposure of sprouted barley to smoke generated by burning peat, in contrast with American bourbon whiskey aged in white oak barrels charred internally. Five brands of bourbon, eight brands of Scotch, and two brands of Japanese whiskey were analyzed. The presence of phenanthrene, pyrene, or fluoranthene was detected in all of the products. Benz[a]anthracene and chrysene were detected in two kinds of whiskey, a Japanese and a Scotch, while benzo[a]pyrene, benzo[b]fluoranthene, and benzo[e]pyrene were found in one Scotch. Concentrations of the hydrocarbons were extremely low, ranging from 0.03 to 0.08 μg/liter.

Malanoski et al. [84] also analyzed samples of bourbon and Scotch; however, benzo[a]pyrene was not detected.

V. Effects of Cooking and Heating of Foods

The formation of PAH in food as the result of the preparation procedures employed has also received attention. Kuratsune [189] found benzo[a]pyrene in the charred material scraped from biscuits heated over a gas burner, whereas the compound was not detected in the char taken from broiled sardines heated in the same manner. PAH have also been identified in coffee soots, a by-product produced in the commercial roasting of coffee beans, and in roasted coffee as shown in Table 12 [190,191].

According to Fritz [192] the greater part of benzo[a]pyrene together with other PAH such as anthranthrene, coronene, indenopyrene, and various benzofluoranthenes is found in the coffee membrane and tar. In normal roasted beans, the levels of benzo[a]pyrene ranged from 0.3 to 0.5 μg/kg. This compound was also found at levels of 0.4-0.7 μg/kg in malt and barley, the raw materials for malt-coffee and coffee substitutes. Roasting of malt in a coal-fired furnace resulted in its formation at approximately 15.8 μg/kg (mean value), whereas the roasting of malt-coffee and a coffee substitute in a gas-fired roaster produced 0.9 and 1.0 μg/kg, respectively. Masuda et al. [193] reported the presence of 18 PAH in the smoke and scorches from fish broiled in either a gas or electric broiler; the gas-broiled fish contained more PAH than those cooked in the electric broiler, as shown in Table 13. In the same report these authors also indicated that roasted barley used in the preparation of Mugicha (a Japanese drink prepared by infusing roasted unhulled barley in hot water) contained several PAH but not carcinogenic types.

Various investigators have studied the pyrolytic behavior of carbohydrates, fats, and proteins and the resultant formation of PAH [194-196]. Davies and Wilmshurst heated starch in the absence of air and at atmospheric pressure with the following results: at 370-390°C benzo[a]pyrene was found at a level of 0.7 μg/kg in the distillation residue; at 650°C, 17 μg/kg was formed [194]. Similar studies were conducted by Masuda et al. [195] on various carbohydrates, amino acids, and fatty acids. No PAH were found at 300°C, but 19 hydrocarbons including benzo[a]pyrene were detected in the above

Table 12. PAH in Roast Coffee and By-products (μg/kg)[a]

Compound	Dark	Very dark	Soots
Phenanthrene	—	—	130-300
Pyrene	2-8	ND-17	260-720
Fluoranthene	1-7	ND-15	340-1000
Chrysene	—	ND-0.5	530-670
Benz[a]anthracene	—	ND-2	16-150
Perylene	—	ND-0.8	280-660
Benzo[e]pyrene	—	ND-3	190-370
Benzo[a]pyrene	—	ND-4	200-400
Benzo[k]fluoranthene	—	ND-0.8	70-140
Benzo[ghi]perylene	—	ND-4	100-140

[a]ND - none detected.
Source: Refs. 190 and 191.

Table 13. PAH in Japanese Horse Mackerel (μg/kg)[a]

Compound	Electric broiler		Gas broiler	
	Smoke	Scorch	Smoke	Scorch
Naphthalene	–	–	167	–
Acenaphthylene	–	–	22	–
Fluorene	2.6	–	8.2	3.2
Anthracene	1.9	0.2	2.3	2.0
Phenanthrene	9.0	1.0	11.0	8.0
Pyrene	3.6	0.2	4.0	7.0
Fluoranthene	5.2	0.2	3.6	7.0
Benzo[a]fluorene	1.0	–	0.6	1.7
Benz[a]anthracene	1.2	–	0.6	2.9
Chrysene	4.3	–	0.4	2.1
Perylene	tr	–	tr	0.2
Benzo[a]pyrene	0.2	–	0.3	0.9
Benzo[e]pyrene	0.5	–	0.2	1.2
Benzo[b]fluoranthene	0.2	–	0.1	1.2
Benzo[j]fluoranthene	0.5	–	tr	0.5
Benzo[k]fluoranthene	0.2	–	–	0.2
Benzo[ghi]perylene	0.2	–	0.3	2.2
Coronene	–	–	tr	tr

[a]tr = trace.
Source: Ref. 193.

products at 500 and 700°C. Halaby and Fagerson [196] pyrolyzed a number of lipids and minor constituents such as carotene and cholesterol in a tube furnace at 400-700°C under nitrogen. At 700°C, pyrolysis of lipids produced approximately 100 μg/kg concentrations of PAH. Benzo[a]pyrene was produced in all samples heated at the above temperature. At 400°C, the relatively low levels made identification difficult, but in general those compounds found at 700°C were present in sufficient quantities for identification. The authors stated that about 10 times as much PAH is produced by pyrolysis of cholesterol as from other lipid materials, but lower levels are found in β-carotene after pyrolysis.

The production of PAH in the grilling of meat has also received attention. Seppilli and Scasellati-Sforzolini [197] analyzed grilled beef and noted the presence of various PAH. Lijinsky and Shubik [198] reported levels of 5-8 μg benzo[a]pyrene/kg in charcoal-broiled steaks and 10.5 μg/kg in barbecued

Table 14. Levels of PAH (μg/kg) Found in Steaks and Ribs[a]

Compounds	Charcoal-broiled steaks	Barbecued ribs
Anthanthrene	0.6 (2)	1.1
Anthracene	– (4.5)	7.1
Benz[a]anthracene	1.4 (4.5)	3.6
Benzo[b]chrysene	– (0.5)	–
Benzo[ghi]perylene	6.7 (4.5)	4.7
Benzo[a]pyrene	5.8 (8)	10.5
Benzo[e]pyrene	5.5 (6)	7.5
Chrysene	0.6 (1.4)	2.2
Coronene	3.2 (2.3)	4.2
Dibenz[a,h]anthracene	– (0.2)	–
Fluoranthene	43 (20)	49
Phenanthrene	21 (11)	58
Pyrene	0.9 (18)	1.4
Perylene	3.5 (2)	42

[a]Key: - = none detected. Numbers in parentheses indicate PAH levels in laboratory-prepared steaks.
Source: Ref. 198.

ribs. Other carcinogenic compounds of interest found in this work included benz[a]anthracene and dibenz[a,h]anthracene, and their estimated levels are shown in Table 14. The authors concluded that the fat or other carbon-hydrogen-oxygen-containing compounds in the meat were the probable pyrolysis sources of the hydrocarbons. Malanoski et al. [84] have assayed barbecued pork and beef and reused cooking oil with findings of 1.4-4.5 μg benzo[a]pyrene/kg.

The effect of variations in methods of cooking on the content of benzo[a]-pyrene and other PAH in meat was discussed by Lijinsky and Ross [199]. This investigation confirmed that the production of PAH in charcoal broiling was dependent on the fat content and the proximity of the food to the heat source. For example, levels of benzo[a]pyrene as high as 50 μg/kg were found in thick T-bone steaks close to the coals for long periods, whereas concentrations were considerably reduced in samples prepared (to the same end point) at a greater distance from the heat source. It was the conclusions of the authors that if the production of carcinogens is to be minimized the method of cooking should avoid contact of the food with the flames, the food should be cooked for longer periods at lower temperatures, and the meat used should have a minimum of fat.

In his study of cooked foods, Fritz [200] found that roasting, frying, or deep frying resulted in negligible quantities of endogenous carcinogens, whereas exogenous treatment with flue gas increased the hydrocarbons, especially benzo[a]pyrene. The findings support those of Halaby and Fagerson [196] on the pyrolysis of fats at 400°C mentioned earlier. Fritz, in other studies of heated foodstuffs, reported the following levels of benzo[a]pyrene: (1) <0.5 μg/kg in baked bread and biscuits and burnt crust; (2) <1 μg/kg in malt coffee and barley coffee substitutes roasted in a gas-fired roaster, but >15.8 μg/kg when the products were roasted in a coal-fired roaster; (3) 0.3-0.5 μg/kg in normal roasted coffee; and (4) none in oils and lard when cooked at normal temperatures [201-203].

Ballschmieter [204] examined roasted peanuts and concluded that there was no significant problem since levels were <1 μg of the eight PAH determined per kilogram. In a comparison of drying of wheat and rye in indirect vs. direct driers (in the latter the combustion gases are in contact with the grain), Fornal et al. [205] reported levels of PAH to be severalfold higher in the direct dried cereals as compared to those treated with the indirect process. Rohrlich and Suckow [206] found that the drying of wheat over a light fuel oil flame increased the benzo[a]pyrene deposition from approximately 6- to 130-fold depending on the degree of exposure.

In addition to studies of smoked foods in Iceland cited previously, Thorsteinsson and Thordarson [207] investigated singed foods, such as sheepsheads and seabirds, which are Icelandic dietary items. According to the authors the fuel formerly used was peat, dry sheep manure, scrap wood, or coal, but in recent years diesel oil, propane, and acetylene gas have come into use at least for commercial singeing. With propane or acetylene oxygen fuel, the singed products were essentially free of PAH. When coal or diesel oil was used as fuel, up to 28 μg of benzo[a]pyrene/kg was found in the singed sheepheads. Even higher levels, 99 μg/kg, occurred in the seabirds singed over coal. It should be noted that concentrations of benz[a]anthracene and other PAH were also found in the latter samples. Thorsteinsson and Thordarson pointed out that the district in northern Iceland where consumption of singed birds is particularly high is also among the highest in that country with respect to the incidence of gastric cancer.

VI. Conclusions

The results of the studies cited in this review emphasize the extent of the occurrence of trace quantities of PAH in our environment. These contaminants, some of which possess carcinogenic activity, have been shown to occur in water, air, food, and soil, as well as such diverse sources as tobacco smoke, automobile and engine exhausts, high-boiling petroleum distillates, carbon black, coal tar, pitch, and rubber tires. Assessment of the potential health hazard has been of great concern in recent years but has been hampered by the slow development of rapid and specific multicomponent procedures for determining PAH in complex food and environmental matrices.

During the last decade significant progress has been made in the development of multiresidue procedures for the analysis of benzo[a]pyrene and other PAH in food at the low microgram per kilogram (parts per billion) levels. Some of these methods have been studied collaboratively and with various modifications have been shown to be applicable to the analysis of a wide variety of products. However, the multiresidue methods are lengthy, which tends to preclude their use as effective monitoring tools. Obviously, further efforts should be directed toward the simplification of these procedures. The basic extraction techniques have been developed and proven for a wide variety of foods; however, the intermediate cleanup and final separative procedures should be further assessed in light of recent advances. Many of the reports reviewed in this chapter do not even contain data which would permit an assessment of the reliability (recovery, reproducibility, and specificity) of the reported data. Although the values discussed are those reported, they must be judged in the light of their consistency with the remainder of the literature.

As evidenced in this review, much of the effort has been expended on the accumulation of data on the benzo[a]pyrene content of foods, which constitutes only between 1 and 20% of the total carcinogenic PAH in the environment [4]. Grimmer and Böhnke commented that most investigators have relied on methods that measure only benzo[a]pyrene content in food, although other PAH are also present in higher concentrations [24]. There is an obvious need for further research to develop a more complete picture of the PAH contamination in foods. In conducting such studies it is essential that we include those carcinogenic PAH which have been recently isolated and characterized in the environment. For example, cyclopenta[cd]pyrene has been recently isolated from some furnace carbon blacks, soot, engine exhausts, and coal tar pitch [208,209]. In addition, in the analysis of environmental samples, including foods, more emphasis needs to be placed on the use of confirmatory techniques. This applies not only to compounds tentatively identified as being present, based on retention times, but also to the characterization of other moieties (potential PAH) which are carried through the cleanup procedures and cast off or ignored as impurities or background interferences. As stated by Schechter, "Data reported without application of suitable confirmatory techniques may not only be worthless, but what is worse, incorrect information may be seriously misleading and may be unrectifiable." This statement, made in 1968 about pesticide residues [210], applies even more appropriately to carcinogenic residues, which frequently are determined at levels of a few parts per billion or less.

Although the large body of information on PAH in smoked and grilled foods reported in prior reviews continues to be expanded, there appears to be an increasing interest in environmental and indigenous factors as sources of contamination. Various investigators have reported the presence of benzo[a]-pyrene and other PAH in foodstuffs of plant origin. Since these foods are consumed in great quantity, these workers have suggested that such plants may in fact be the primary source of ingested PAH rather than smoked or heated products. At this time, three routes of PAH contamination have been considered: air deposition, absorption from soil and water, and biochemical synthesis. There appears to be at least general agreement that available data support the first two routes; however, there is considerable controversy concerning the synthesis route. Questions of false positives have also arisen with respect to these plant analyses. Obviously, a concerted effort is needed to provide reliable and conclusive data so that a proper assessment can be made with respect to the PAH contamination of food crops.

New potential sources of PAH contamination have been recognized as marine oil spills and discharges from refineries into estuarine waters. With the increase in the transport of foreign crude oils and off-shore drillings to meet consumption demands, a large portion of our ecosystem has been contaminated by petroleum products from oil spills, refineries, seepages, and other sources. The potential health hazard associated with the consumption of commercial species of shellfish (oysters, mussels, and clams) contaminated by petroleum products must be assessed for the presence of PAH compounds which are carcinogenic. At the present time there is not sufficient information to come to any definitive conclusions as to whether or not a health hazard exists. Studies as to the fate of petroleum in the marine environment and the uptake and depuration of marine organisms are at present limited and controversial. Fish and lobsters have been shown to metabolize most petroleum hydrocarbons within 2 weeks, but metabolism in lower organisms is slower and the pathways are poorly understood. Some organisms such as mussels and oysters have been shown to eliminate most absorbed petroleum hydrocarbons when placed in clean water. The length of time required to depurate fully has been variable. NAS in 1975 reported that, although information was limited, the effect of soil contamination on human health appeared not to be cause for alarm [164]. The NAS Ocean Affairs Board does not recommend eating contaminated seafood, but in most cases, because of the taste factor, not many will be tempted to do so. The present authors do not agree that an organoleptic yardstick should be used as the criterion for estimating low levels of potential carcinogenic PAH. More research is needed on all aspects of the marine pollution problem before we can make reasonably accurate predictions, estimations, and recommendations concerning the dangers of oil pollution to commercially important shellfish industries and concerning the public health hazards arising from the consumption of petroleum-contaminated shellfish.

Even though great strides have been made in identifying potential health hazards, much still needs to be learned about the occurrence or accumulation of PAH in our total environment. Therefore, more research is needed to develop accurate data with respect to the many facets of our total environment, so that the overall public health hazard may be effectively evaluated and controlled.

References

1. J. W. Howard and T. Fazio, *J. Assoc. Offic. Anal. Chem. 63*(5):1077 (1980).
2. D. J. Tilgner, *Food Manuf. 43*(6):37 (1968).
3. J. B. Andelman and M. J. Suess, *Bull. WHO 43*:479 (1970).
4. M. J. Suess, *Sci. Total Environ. 6*:239 (1976).
5. E. O. Haenni, *Residue Rev. 24*:42 (1968).
6. D. J. Tilgner and H. Daun, *Residue Rev. 27*:19 (1969).
7. G. W. Giger and M. Blumer, *Anal. Chem. 46*:1663 (1970).
8. E. O. Haenni and H. Fischbach, *Trace PAH Analysis: The Contribution of Chemistry to Food Supplies*, IUPAC, Butterworth, London, 1974, pp. 209-225.
9. D. J. Tilgner, *Food Manuf. 87*:47 (1970).
10. F. A. Gunther and F. Buzzetti, *Residue Rev. 9*:90 (1965).
11. L. M. Shabad, *Cancer Res. 27*:1132 (1967).
12. J. W. Howard and T. Fazio, *J. Agr. Food Chem. 17*:527 (1969).

13. T. Saito, *Kagaku To Seibutsu 8*(3):178 (1970).
14. V. D'Arrigo, *Quad. Merceol. 10*:151 (1971).
15. M. T. Lo and E. Sandi, *Residue Rev. 69*:35 (1978).
16. R. Schaad, *Chromatogr. Rev. 13*:61 (1970).
17. UICC (International Union Against Cancer) Tech. Rept. Ser., Vol. 4, Lyons, France (1970).
18. IARC Internal Tech. Rept. No. 71/002, Lyons, France (1971).
19. J. W. Howard, T. Fazio, and R. H. White, *J. Assoc. Offic. Anal. Chem. 51*:544 (1968).
20. T. Fazio, R. H. White, and J. W. Howard, *J. Assoc. Offic. Anal. Chem. 56*:68 (1973).
21. *Official Methods of Analysis*, 12th ed., Association of Official Analytical Chemists, 1975.
22. IUPAC Inform. Bull. Tech. Rept. No. 4, Geneva, Switzerland (1972).
23. Recommended Methods for PAH in Food, *Pure Appl. Chem. 50*(11/12) (1978).
24. G. Grimmer and H. Böhnke, *J. Assoc. Offic. Anal. Chem. 58*:725 (1976).
25. G. Janini, K. Johnston, and W. Zielinski, *Anal. Chem. 47*:670 (1975).
26. N. Carugno and S. Rossi, *J. Gas Chromatogr.* 5:103 (1967).
27. K. Grob, *Chem. Ind. (London)* p. 248 (1973).
28. G. Grimmer, A. Hildebrandt, and H. Böhnke, *Erdoel Kohle 25*:443, 531 (1972).
29. G. Grimmer and H. Böhnke, *Z. Anal. Chem. 261*:310 (1972).
30. V. Cantuti, G. Cartoni, A. Liberti, and A. G. Torri, *J. Chromatogr. 17*:60 (1965).
31. L. DeMaio and M. Corn, *Anal. Chem. 38*:131 (1966).
32. K. Bhatia, *Anal. Chem. 43*:609 (1971).
33. J. Frycha, *J. Chromatogr. 65*:341,432 (1972).
34. D. Lane, H. Moe, and M. Katz, *Anal. Chem.* 45:1776 (1973).
35. A. Zane, *J. Chromatogr. 38*:130 (1968).
36. T. Gouw, I. Whittemore, and R. Jentoft, *Anal. Chem. 42*:1394 (1970).
37. G. Janini, G. Muschik, and W. Zielinski, *Anal. Chem. 48*:809 (1976).
38. G. Janini, G. Muschik, J. Schroer, and W. Zielinski, *Anal. Chem. 48*:1879 (1976).
39. P. Burchill, A. Herod, and R. James, *Carcinogenesis 3*:35 (1978).
40. E. Winkler, A. Buchele, and O. Muller, *J. Chromatogr. 138*:151 (1977).
41. A. Krstulovic, D. Rosie, and P. Brown, *Anal. Chem. 48*:1383 (1976).
42. J. Schmit, R. Henry, R. C. Williams, and J. F. Dieckmann, *J. Chromatogr. Sci. 9*:645 (1971).
43. C. Vaughan, B. Wheals, and M. Whitehouse, *J. Chromatogr. 78*:203 (1973).
44. B. Wheals, C. Vaughan, and M. Whitehouse, *J. Chromatogr. 106*:109 (1975).
45. W. Strubert, *Chromatographia 6*:205 (1973).
46. R. Jentoft and T. Gouw, *Anal. Chem. 480*:1787 (1968).
47. B. Karger, M. Martin, J. Loheac, and G. Guiochon, *Anal. Chem. 45*:496 (1973).
48. R. Vivilecchia, M. Thiebaud, and R. Frei, *J. Chromatogr. Sci. 10*:411 (1972).
49. C. Lochmuller and C. W. Amoss, *J. Chromatogr. 108*:85 (1972).

50. M. Dong, D. Locke, and E. Ferrand, *Anal. Chem. 48*:368 (1976).
51. D. C. Hunt, P. Wild, and N. T. Crosby, *J. Chromatogr. 130*:320 (1977).
52. N. Ives and L. Giuffrida, *J. Assoc. Offic. Anal. Chem. 55*:757 (1972).
53. H. J. Klimisch, *Anal. Chem. 45*:1900 (1973).
54. J. J. Kirkland, *J. Chromatogr. Sci. 10*:593 (1972).
55. H. Guerrero, E. Biehl, and C. Kenner, *J. Assoc. Offic. Anal. Chem. 59*: 989 (1976).
56. T. Panalaks, *J. Environ. Sci. Health Bull. 4*:229 (1976).
57. M. Novotny, M. Lee, and K. Bartle, *J. Chromatogr. Sci. 12*:606 (1974).
58. M. Lee, M. Novotny, and K. Bartle, *Anal. Chem. 48*:1566 (1976).
59. K. Bartle, M. Lee, and M. Novotny, *Environ. Anal. Chem. 3*:349 (1974).
60. M. Lee, K. Bartle, and M. Novotny, *Anal. Chem. 47*:540 (1974).
61. M. Lee, M. Novotny, and K. Bartle, *Anal. Chem. 48*:405 (1976).
62. R. Lao, R. Thomas, H. Cja, and L. Dubois, *Anal. Chem. 46*:908 (1973).
63. R. Pancirov and R. Brown, *Environ. Sci. Technol. 11*:989 (1977).
64. D. J. Tilgner, *Fleischwirtschaft 10*:649 (1958).
65. H. N. Draudt, *Food Technol. 17*:85 (1963).
66. Z. E. Sikorski and D. J. Tilgner, *Z. Lebensm. Untersuch. Forsch. 124*: 274 (1964).
67. M. K. Dobes, J. Hopp, and J. Sula, *Cesk. Onkol. 1*:254 (1954); see *Chem. Abstr. 49*:4199 (1955).
68. E. Bailey and N. Dungal, *Brit. J. Cancer 12*:348 (1958).
69. F. Voitelovich, P. P. Dikun, and L. Shabad, *Vopr. Onkol. 3*:351 (1957).
70. N. D. Gorelova, *Vopr. Onkol. 9*(8):77 (1963); see *Chem. Abstr. 59*:15862 (1963).
71. N. D. Gorelova and P. P. Dikun, *Gig. Sanit. 30*(7):120 (1965); see *Chem. Abstr. 63*:10574e (1965).
72. A. S. Petrun and B. L. Rubenchik, *Vrach. Delo. 2*:93 (1966); see *Chem. Abstr. 64*:20527h (1966).
73. G. Mannelli, *Ann. Fac. Econ. Commerc. 4*(2):467 (1966).
74. W. Lijinsky and P. Shubik, *Toxicol. Appl. Pharmacol. 7*:337 (1065).
75. J. W. Howard, R. Teague, R. White, and B. E. Fry, *J. Assoc. Offic. Anal. Chem. 49*:595 (1966).
76. J. W. Howard, R. H. White, B. E. Fry, and E. Turicchi, *J. Assoc. Offic. Anal. Chem. 49*:611 (1966).
77. L. Toth, *Fleischwirtschaft 51*:1069 (1971).
78. P. P. Dikun, N. D. Drasnitskaya, I. A. Shendrikova, O. P. Gretskaya, A. V. Emshanova, and I. I. Lapshin, *Vopr. Onkol. 15*(3):79 (1969).
79. Y. Masuda and M. Kuratsune, *Gann 62*:27 (1971).
80. T. Shirotori, *Tokyo Kasei Daigaku Kenkyu Kiyo 12*:47 (1972).
81. L. Toth and W. Blaas, *Fleischwirtschaft 52*:1122 (1972).
82. J. Wierzchowski and R. Gajelwska, *Bromatol. Chem. Toksykol. 5*:481 (1972).
83. T. Thorsteinsson, *Cancer 23*:455 (1969).
84. A. J. Malanoski, E. L. Greenfield, C. J. Barnes, J. M. Worthington, and F. L. Joe, Jr., *J. Assoc. Offic. Anal. Chem. 51*:114 (1968).
85. T. Panalaks, *J. Environ. Sci. Health Bull. 4*:299 (1976).
86. K. Rhee and L. Bratzler, *J. Food Sci. 35*:146 (1970).
87. A. Lucisano, P. DeBattistis, and F. Marzadori, *Vet. Ital. 24*:232 (1973).
88. K. Fretheim, *J. Agr. Food Chem. 24*:976 (1976).

89. W. Swallow, *New Zealand J. Sci. 19*:407 (1976).
90. M. E. Doremire, C. E. Harmon, and D. E. Pratt, *J. Food Sci. 62*:203 (1979).
91. F. L. Joe, Jr., E. L. Roseboro, and T. Fazio, *J. Assoc. Offic. Anal. Chem. 62*:615 (1979).
92. N. K. Gorelova, P. P. Dikun, V. A. Solinke, and A. V. Emshanova, *Vopr. Onkol. 6*(1):33 (1960); see *Chem. Abstr. 55*:4814h (1961).
93. N. Dungal, *J. Am. Med. Assoc. 178*:789 (1961).
94. H. F. Kraybill, A synopsis of information and data on PAH in the environment, including carcinogenicity assessment. Unpublished report to the National Cancer Institute, 1973.
95. C. Genest and D. Smith, *J. Assoc. Offic. Agr. Chem. 57*:894 (1964).
96. W. Lijinsky and A. E. Ross, *Food Cosmet. Toxicol. 5*:343 (1967).
97. L. Jung and P. Morand, *C. R. Acad. Sci., Paris 254*:1489 (1962).
98. L. Jung and P. Morand, *C. R. Acad. Sci., Paris 257*:1638 (1963).
99. L. Jung and P. Morand, *Ann. Fals. Expert Chim. 57*:17 (1964).
100. W. Ciusa, G. Nebbia, A. Bruccelli, and E. Volpones, *Riv. Ital. Sostanze Grasse 42*:175 (1965); see *Chem. Abstr. 63*:10184b (1965).
101. J. Borneff and B. Fabian, *Arch. Hyg. Bakteriol. 150*:485 (1966).
102. J. W. Howard, E. W. Turicchi, R. H. White, and T. Fazio, *J. Assoc. Offic. Anal. Chem. 49*:1236 (1966).
103. J. W. Howard, T. Fazio, and R. H. White, *J. Agr. Food Chem. 16*:72 (1968).
104. G. Grimmer and A. Hildebrandt, *Chem. Ind. (London)* p. 2000 (1967).
105. G. Biernoth and H. E. Rost, *Chem. Ind. (London)* p. 2002 (1967).
106. B. Fabian, *Arch. Hyg. (Berlin) 152*:231 (1968).
107. B. Fabian, *Arch. Hyg. (Berlin) 152*:251 (1968).
108. B. Fabian, *Arch. Hyg. (Berlin) 153*:21 (1969).
109. W. Fritz, *Nahrung 12*:495 (1968).
110. C. Franzke and W. Fritz, *Fette Seifen Anstrichmittel 71*:23 (1969).
111. L. T. Grigorenko, P. P. Dikun, I. A. Kalinina, A. N. Mironova, and V. P. Rzhehkin, *Prikl. Biokhim. Mikrobiol. 6*(2):142 (1970).
112. L. T. Grigorenko, P. P. Dikun, I. A. Kalinina, and A. N. Mironovia, *Tr. Vses. Nauchn.-Issled. Inst. Zhirov 28*:243 (1971).
113. W. Swallow, *New Zealand J. Sci. 19*:407 (1976).
114. E. Guddal, *Acta Chem. Scand. 13*:834 (1969).
115. M. Blumer, *Science 134*:474 (1961).
116. J. Borneff and R. Fischer, *Arch. Hyg. Bakteriol. 146*:334 (1962).
117. L. Mallet, *Proc. 4th Congr. Expert. Chim. Vol. Spec. Conf. Commun., Athens*, p. 301 (1964); see *Chem. Abstr. 66*:84240b (1964).
118. R. Doerr, *Naturwissenschaften 52*(7):166 (1965); see *Chem. Abstr. 62*:16635a (1967).
119. W. Graf and H. Diehl, *Arch. Hyg. Bakteriol. 150*:249 (1966).
120. J. Borneff, F. Selenka, H. Kunte, and A. Maximos, *Environ. Res. 2*:22 (1968).
121. J. Borneff, F. Selenka, H. Kunte and A. Maximos, *Arch. Hyg. (Berlin) 152*:279 (1968).
122. M. Knorr and D. Schenk, *Arch. Hyg. (Berlin) 152*:282 (1968).
123. F. Schmidt and W. Fritz, *GBK-Mettiel. 5*:15 (1968).
124. J. L. Hancock, H. G. Applegate, and J. D. Dodd, *Atmos. Environ. 4*:363 (1970).

125. G. Grimmer and A. Hildebrandt, *Deut. Lebensm. Rundschau.* *61*:237 (1965).
126. G. Grimmer and A. Hildebrandt, *Z. Krebsforsch.* *67*:272 (1965).
127. G. Grimmer, *Deut. Apotheker Ztg.* *108*:529 (1968).
128. F. A. Gunther, F. Buzzetti, and W. Westlake, *Residue Rev.* *17*:81 (1967).
129. H. Bolling, *Tec. Molitoria 15*(24):137 (1964); see *Chem. Abstr.* *62*: 16878h (1965).
130. W. Hutt, A. Meiering, W. Delschlagerr, and E. Winkler, *Can. Agr. Eng.* *20*(2):103 (1978).
131. L. M. Shabad, Y. Cohan, A. Ilnitsky, A. Klresina, N. Shcherbak, and G. Smirnov, *J. Nat. Cancer Inst.* *47*:1179 (1971).
132. N. P. Shcherbak, *Gig. Sanit.* *7*:93 (1968).
133. N. P. Shcherbak, *Vopr. Onkol.* *15*(4):75 (1969).
134. L. M. Shabad and Y. L. Cohan, *Arch. Geschwulstforsch.* *40*(3):237 (1972).
135. G. Grimmer and D. Duevel, *Z. Naturforsch.* *25*:1171 (1970).
136. N. Schamp and F. Van Wassenhove, *J. Chromatogr.* *69*:421 (1972).
137. K. H. Wagner and I. Siddiqi, *Z. Pflanzenernaehr. Dueng. Bodenk.* *130*(3):242 (1971).
138. W. Hertel, P. Suchow, and M. Rohrlich, *Getreide Mehl 20*(9):65 (1970).
139. International Agency for Research on Cancer: IARC Monographs on the Evaluation of Carcinogenic Risk of the Chemical to Man, Vol. 3, Lyons, France, 1973.
140. H. F. Kraybill, *Progr. Exp. Tumor Res.* *20*:3 (1976).
141. A. P. Ilnitsky and S. N. Varshavskaya, *Gig. Sanit.* *29*:78 (1964).
142. J. Borneff and F. Fischer, *Arch. Hyg. (Berlin) 146*:1 (1962).
143. J. Borneff, *Muench. Med. Wochenschr.* *105*:1237 (1963).
144. J. Borneff and H. Kunte, *Arch. Hyg. (Berlin) 147*:401 (1963).
145. J. Borneff, *Arch. Hyg. (Berlin) 148*:1 (1964).
146. J. Borneff, *Landarzt 40*:109 (1964).
147. J. Borneff and H. Kunte, *Arch. Hyg. (Berlin) 149*:226 (1965).
148. M. Saccini-Cicatelli, *Boll. Pesca. Pisci. Idrobiol.* *20*:245 (1965); see *Chem. Abstr.* *66*:1730n (1967).
149. M. Saccini-Cicatelli, *Boll. Soc. Ital. Biol. Sper.* *42*:957 (1967); see *Chem. Abstr.* *66*:22680z (1967).
150. L. Mallet, A. Perdriau, and J. Perdriau, *Ann. Med. Legale Criminol. Police Sci. Toxicol.* *40*:168 (1960).
151. L. Mallet, A. Perdriau, and J. Perdriau, *C. R. Acad. Sci., Paris 256*: 3487 (1963); see *Chem. Abstr.* *59*:1404h (1963).
152. L. Mallet and J. Sardou, *C. R. Acad. Sci., Paris 258*:5264 (1964); see *Chem. Abstr.* *61*:8666a (1964).
153. L. Mallet and R. Lami, *C. R. Soc. Biol., Paris 158*:2261 (1964); *Chem. Abstr.* *63*:2744a (1965).
154. L. Mallet, *C. R. Acad. Sci., Paris 253*:168 (1961); *Chem. Abstr.* *56*: 4539F (1962).
155. L. Mallet and C. Schneider, *C. R. Acad. Sci., Paris 259*:675 (1964); see *Chem. Abstr.* *61*:14404f (1964).
156. J. Boucart and L. Mallet, *C. R. Acad. Sci., Paris 260*:3729 (1965); see *Chem. Abstr.* *62*:15901b (1965).
157. L. Mallet and M. Priou, *C. R. Acad. Sci., Paris 264*:969 (1967); *Chem. Abstr.* *66*:108064a (1967).

158. J. Greffard and J. Meury, *Cah. Oceanogr. 19*:457 (1967); see *Chem. Abstr. 68*:2369 (1968).
159. M. Shimkin, B. K. Koe, and L. Zechmeister, *Science 113*:650 (1951).
160. B. Koe and L. Zechmeister, *Arch. Biochem. 41*:396 (1952).
161. H. Cahnmann and M. Kuratsune, *Anal. Chem. 29*:1312 (1957).
162. C. E. Zobell, Sources and biodegradation of carcinogenic hydrocarbons, in *Proceedings of Convention on Prevention and Control of Oil Spills, U.S. Coast Guard, American Petroleum Institute, and the Environmental Protection Agency*, American Petroleum Institute, Washington, D.C., 1971.
163. L. Zechmeister and B. Koe, *Arch. Biochem. 35*:1 (1952).
164. National Academy of Sciences (NAS), Inputs, fate and effects of petroleum in the marine environment. A report of the Ocean Affairs Board, NAS, Washington, D.C., 1975.
165. B. H. Ketchum, Oil in the marine environment. Background papers for a Workshop on Inputs, Fates, and Effects on Petroleum in the Marine Environment, Vol. 2, National Academy of Sciences, Washington, D.C., 1973, pp. 709-738.
166. M. Blumer, G. Souza, and J. Sass, *Marine Biol. 5*:195 (1970).
167. M. Ehrhardt, *Environ. Pollut. 3*:257 (1972).
168. J. W. Anderson, Uptake and depuration of specific hydrocarbons from oil by the bivalves. Background papers for a Workshop on Inputs, Fates, and Effects of Petroleum in the Marine Environment, Vol. II, National Academy of Sciences, Washington, D.C., 1973, pp. 609-708.
169. R. F. Lee, R. Sauerkeber, and A. Benson, *Science 177*:344 (1972).
170. T. Fazio, Unpublished data, Food and Drug Administration, Washington, D.C., 1971.
171. B. Dunn and H. Stick, *J. Fisheries Res. Board Can. 33*:2040 (1976).
172. B. Dunn, and H. Stick, *Bull. Environ. Contam. Toxicol. 15*:398 (1976).
173. B. Dunn and D. Young, *Marine Pollut. Bull. 7*:231 (1976).
174. R. A. Brown and R. J. Pancirov, *Environ. Sci. Technol. 13*:878 (1979).
175. E. L. McGinnis and M. S. Norris, *J. Agr. Food Chem. 23*:221 (1975).
176. T. Takata, *Hydrocarbon Process. Petrol. Refiner 48*:99 (1969).
177. N. S. Scrimshaw, Symposium on single-cell proteins for animal feeding, Brussels, Belgium. Protein-Calorie Advisory Group (PAG) of the United Nations System. Summary: PAG ad hoc Working Group, Single Cell Proteins, March 31, 1976.
178. E. L. McGinnis, *J. Agr. Food Chem. 23*:226 (1975).
179. R. Truhaut and R. Ferrando, The toxicological aspects of single cell proteins used in animal feeding, in *Proc. Protein-Calorie Advisory Group (PAG) Symposium, Brussels, Belgium, 1976*, pp. 38-48.
180. A. Santoro, R. Modica, S. Paglialunga, and I. Bartosek, *Toxicol. Lett. 3*:85 (1979).
181. D. L. Tilgner, *Fleischwirtschaft 46*:501 (1966).
182. C. M. Hollenbeck, Aqueous smoke solution for use in foodstuffs and method for producing same. U.S. Patent 3,106,473 (Red Arrow Products Corp.), 1963.
183. C. M. Hollenbeck, *Food Process. 25*:136 (1964).
184. R. H. White, J. W. Howard, and C. J. Barnes, *J. Agr. Food Chem. 19*:143 (1971).

185. N. K. Gorelova, P. P. Dikun, and I. I. Lapshin, *Vopr. Onkol.* *5*:341 (1969).
186. O. G. Prokofieva, *Vopr. Onkol.* *8*:95 (1962).
187. V. Gorbatov, *Food Technol.* *25*:71 (1971).
188. Y. Masuda, K. Mori, T. Hirohata, and M. Kuratsune, *Gann* *57*:549 (1966).
189. M. Kuratsune, *J. Nat. Cancer Inst.* *16*:1485 (1956).
190. M. Kuratsune and W. C. Hueper, *J. Nat. Cancer Inst.* *20*:37 (1958).
191. M. Kuratsune and W. C. Hueper, *J. Nat. Cancer Inst.* *24*:463 (1960).
192. W. Fritz, *Nahrung* *12*:799 (1968).
193. Y. Masuda, K. Mori, and M. Kuratsune, *Gann* *57*:133 (1966).
194. W. Davies and J. R. Wilmshurst, *Brit. J. Cancer* *14*:295 (1960).
195. Y. Masuda, K. Mori, and M. Kuratsune, *Gann* *58*:69 (1967).
196. G. A. Halaby and I. S. Fagerson, PAH in heat-treated foods: Pyrolysis of some lipids, beta-carotene, and cholesterol, in *Proc. 3rd Intern. Symposium on Food Sci. Technol.*, *1970* pp. 820-829.
197. A. Seppilli and G. Scasellati-Sforzolini, *Boll. Soc. Ital. Biol. Sper.* *39*:110 (1963).
198. W. Lijinsky and P. Shubik, *Ind. Med. Surg.* *34*:152 (1965).
199. W. Lijinsky and A. E. Ross, *Food Cosmet. Toxicol.* 5:343 (1967).
200. W. Fritz, *Arch. Geschwulstforsch.* *40*:81 (1972).
201. W. Fritz, *Nahrung* *12*:799 (1968).
202. W. Fritz, *Nahrung* *12*:805 (1968).
203. W. Fritz, *Nahrung* *12*:809 (1968).
204. H. M. B. Ballschmieter, *Fette Seifen Anstrichmittel* *71*:521 (1969).
205. J. Fornal, L. Fornal, and L. Babuchowski, *Przegl. Zbozowo-Mlyn.* *14*:445 (1970).
206. M. Rohrlich and P. Suckow, *Getreide Mehl.* *20*:90 (1970).
207. T. Thorsteinsson and G. Thordarson, *Cancer* *21*:390 (1968).
208. L. Wallcave, *Environ. Sci. Technol.* 3:948 (1969).
209. A. Gold, *Anal. Chem.* *47*:1469 (1975).
210. M. S. Schechter, *Pestic. Monitoring J.* *2*:1 (1968).

12

Long-Range Transport of Polycyclic Aromatic Hydrocarbons

ALF BJØRSETH AND BJØRN SORTLAND OLUFSEN / Central Institute for Industrial Research, Oslo, Norway

I. Introduction

In recent years the problems of long-range transport of air pollutants have been of increasing concern [1-3]. This concern is mainly focused on the potential hazard to man and environment caused by exposure to a wide variety of pollutants. Studies have been performed in Europe, both for transfrontier transport of pollutants and in connection with long-range transport of pollutants from industrialized Central Europe to Scandinavia. Studies have also been performed in North America when pollutants are transported across the Great Lakes basin from the United States to Canada. In particular the long-range transport of SO_2 and the transformation of SO_2 to SO_4^{2-} have been studied extensively [4]. Furthermore a study in Canada has shown that SO_2 in a plume can be detected 400 km from the source [5]. It has also been established that particles are transported over long distances and that these particles may result in black and white episodes [6]. Dark particles causing black episodes are found to be slightly acidic, contain ammonium sulfate and nitrate, and are sometimes accompanied by increased concentrations of sulfur and nitrogen dioxides. These particles are of anthropogenic nature. White episodes are caused by colorless particles, consisting mainly of acidic ammonium sulfate, and are probably originated over the ocean [6].

On the basis of knowledge of long-range transport of inorganic pollutants it is reasonable to expect that organic pollutants emitted simultaneously from the same source also would be subject to long-range transport. Some recent studies support this hypothesis. This chapter reviews most of the relevant literature in the area and discusses the long-range transport in terms of atmospheric reactivity of PAH. Finally, a simple model for calculating emissions from immission data is presented.

II. Determination of Long-Range Transport of PAH

A. Analysis of Soils and Sediments

The first evidence as to the importance of long-range transport of PAH was revealed in connection with analysis of soil and sediments. Blumer, while studying the distribution of PAH in soil, was puzzled by the relative uniformity of PAH and in particular the alkyl distribution pattern in samples from different locations [7]. He pointed out that large quantities of pyrolytic products are formed in forests and prairie fires and are widely dispersed by prevailing winds. The haze of the North Atlantic is attributed in part to such fires. The particles contain elemental carbon and in addition aromatic hydrocarbons in great abundance. As the cooling mass of gas ascends from those fires, the freshly formed active carbon particles can readily pick up the aromatic hydrocarbons and protect them from light-induced oxidation during their movement through the atmosphere. Mixing during transport would explain the uniform composition over a wide area and explain why the organisms and the chemical conditions at the site of deposition are not reflected in the chemical makeup of the hydrocarbon mixture. Finally, the wide range of combustion temperatures in natural fires would explain the observed range of alkyl derivatives. It is reasonable to assume that Blumer's argument is relevant to other uncontrolled burning sources such as open refuse burning and agricultural burning as well. These sources for PAH are of significant importance [8]. Blumer's argument should therefore not be limited to forest and prairie fires.

A very thorough study on the global distribution of PAH in recent sediments was published by LaFlamme and Hites [9]. They reported that PAH and their alkyl homologs are distributed in sediments throughout the world. The qualitative PAH pattern is remarkably constant for most of the locations studied. Furthermore, the quantities of PAH increase with approximity to urban areas. Their conclusions are in concert with Blumer, suggesting that (anthropogenic) combustion is the major source of PAH. It was also pointed out that degradation of vegetation may result in specific PAH compounds and that these processes may influence the level of PAH in sediment samples. Progress in this area is, however, hampered by the fact that it is difficult to find an area without influence from anthropogenic sources. A historical record of the PAH concentrations in recent sediments illustrates this [10].

Evidence for long-range transport of PAH has also been given by Windsor and Hites [11]. Based on analysis of PAH in sediment from the Gulf of Maine and Nova Scotia it was suggested that airborne transport of PAH can be a long-range phenomenon which would account for the rather uniform, low PAH levels found in soils and sediments that are far removed from PAH sources. To support their hypothesis Windsor and Hites made some simple Stoke's law calculations. When particles are produced by different anthropogenic sources, the particles will be transported by the prevailing winds for distances which are a strong function of the particle diameter. Approximate calculations of these transport distances may be obtained from Stoke's law, which gives the settling velocity of a particle as a function of particle diameter, particle density, and kinematic viscosity. The particle size distribution of urban air particulates seems to be bimodal, with one mode occurring at less than 1 μm and the other at sizes greater than 10 μm [12]. The PAH loadings also seem to follow this bimodal distribution [13]. For small particles it is calculated that if this particle is released at the height of 20 m above sea level into a 4-5 m/sec wind it could take the particle from 1000 to 1500 km (or even more) to settle to the surface. If these particles settle on land they are incorporated into the soil, probably in association with humic substances [13]. If these particles settle on the surface of the ocean they can become incorporated into larger fecal pellets, and this results in relatively rapid settling. Based on these calculations the authors suggested that long-range transport of small particles should account for the PAH levels in Nova Scotia soils and in the deep ocean sediments.

A similar calculation indicates that the larger airborne particles (10 μm) will settle in about 15 km, which is about the diameter of the source.

B. Analysis of Rainwater and Snow

Evidence for long-range transport of PAH has also been obtained from analysis of rainwater and snow in Norway. Ten rain and snow samples collected from air masses defined by calculated trajectories and 14 snow and ice samples representing a cross section of the precipitation were analyzed by gas chromatography/mass spectrometry (GC/MS) [14]. Using GC/MS, a wide range of organic micropollutants was identified including alkanes, PAH, phthalic acid esters, fatty acid esters and other chemicals of industrial origin. About 25 PAH compounds were detected in the samples including phenanthrene, anthracene, fluoranthene, pyrene, benz[a]anthracene, chrysene, benzo[e]pyrene, and benzo[a]pyrene (BaP). PAH were detected both in the aqueous phase and adsorbed to particles. The concentration of benzo[a+e]pyrene in precipitation was found as high as 300 ng/liter in some samples.

PAH have also been determined [15] in snow across two mountain profiles in northern Bavaria, Federal Republic of Germany (Steigerwald and Fichtelgebirge). Although the concentration of pollutants was influenced by nearby urban sources, the results also indicate a contribution of long-range transported air pollutants.

C. Analysis of PAH in Air

It is generally accepted that the main route for transport of PAH is through the atmosphere [16]. Hence to study the worldwide distribution of PAH it is of particular interest to analyze the air masses directly. Recently, a study was undertaken to develop practical techniques for measurement of hydrocarbons and oxygenated volatiles in rural areas [17]. Sampling locations were selected in a large forest at a distance of about 55 km from a heavily populated area. However, it turned out that this distance was not sufficient to escape the urban influence and compounds typical of urban activities were found at dilute concentrations, indicating long-range transport of organic pollutants.

Despite the many sources of PAH in the New York City area, recent studies give evidence of long-range transport of PAH to this area [18,19]. Daily variations in the ratios of BaP and benzo[ghi]perylene in New York City during August 1976, suggest that long-range transport may be a significant factor in determining aerosol burdens of PAH under certain meteorological conditions [18]. More specifically, examinations of the daily variations in ambient concentrations of PAH measured during transport episodes and identified in the New York Summer Aerosol Study of 1976 present evidence that PAH compounds were transported into the city during these periods. The origins of the PAH were oil- and coal-burning areas in the Midwest region of the United States, including the Ohio Valley [19].

Evidence of long-range transport of PAH [20] has also been reported from Belgium, where BaP and associated polycyclic residues due to incomplete combustion have been found, especially in the finest suspended airborne particles (<2 μm). The atmospheric pollution is most often observed in the presence of cold anticyclonic continental air masses from the northeast. The polluting air masses have low wind speeds and a very pronounced directional stability—atmospheric circumstances which favor the accumulation and the transport of the airborne pollutants on a synoptic scale [20].

Circumstantial evidence for long-range transport of PAH is found by analysis of samples from remote rural areas [21]. Such measurements performed in the United States (in Grand Canyon National Park, Arizona; Acadia National Park, on the coast of Maine; and Shenandoah National Park, Virginia) demonstrate the presence of BaP. The results also reveal a decreasing trend of the BaP concentration paralleling the trend of urban samples, thus indicating the influence of these sources.

Analyses of air samples from Barrow, Alaska, which is a remote site in the Arctic, have also shown the presence of PAH. Ambient concentrations of extractable particulate organic matter, of PAH, and of ^{210}Pb were determined for the periods March and August 1979. The presence of flyash in the samples collected during the March sampling period, as well as seasonal differences in the concentrations of the organic species and ^{210}Pb and in meteorology indicate that the principal source of PAH was fossil fuel combustion in the midlatitudes [22]. All these studies are strong indications of considerable aerial transport of PAH. Conclusive evidences for long-range PAH transport are given by Lunde and Bjørseth [23] and Bjørseth et al. [24]. Air samples were collected by high-volume samplers on glass fiber filters at Birkenes in southern Norway and Rørvik in western Sweden. About 2500 m^3 air was drawn through the filters per day, and the filters were changed every 24 hr. The samples were analyzed by glass capillary gas chromatography, and the results were related to trajectories calculated from meteorological data. A typical gas chromatogram of PAH in the air masses is shown in Fig. 1. About 20 PAH compounds were identified in the long-range transported aerosols.

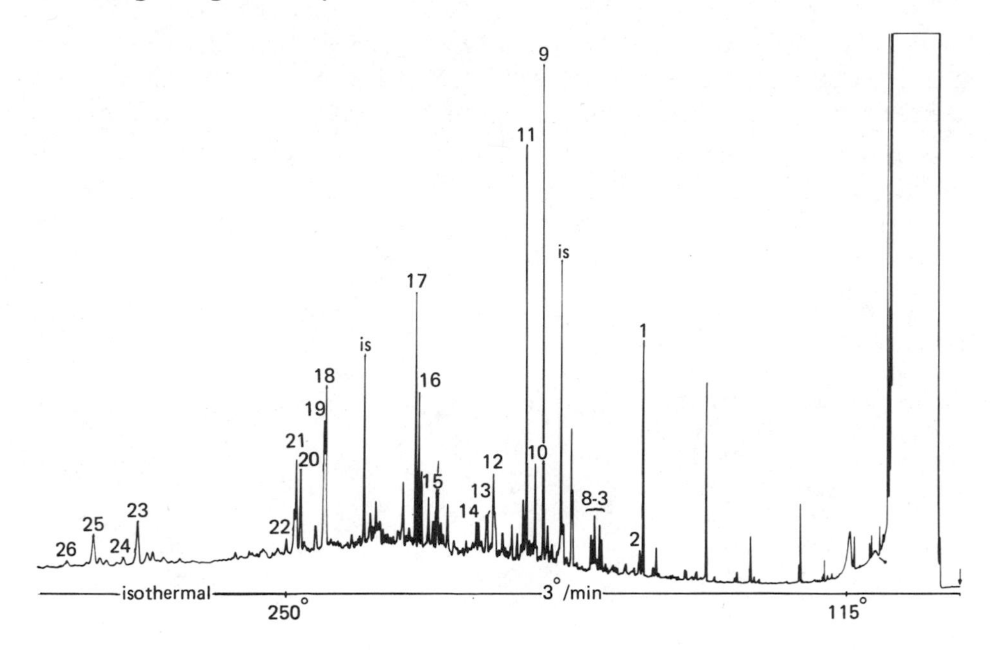

Figure 1. Typical gas chromatogram of an aerosol sample at Birkenes, Norway. The peak identities are (1) phenanthrene; (2) anthracene; (3-8) methylphenanthrene/methylanthracene; (9) fluoranthene; (10)dihydrobenzo-[a/b]fluorene; (11) pyrene; (12) benzo[a]fluorene; (13) benzo[b]fluorene; (14) 1-methylpyrene; (15) benzo[c]phenanthrene; (16) benz[a]anthracene; (17) chrysene/triphenylene; (18) benzo[b]fluoranthene; (19) benzo[j/k]-fluoranthene; (20) benzo[e]pyrene; (21) benzo[a]pyrene; (22) perylene; (23) o-phenylenepyrene; (24) dibenz[a,h]anthracene; (25) benzo[ghi]perylene; and (26) anthanthrene. (From Ref. 24.)

At least four of these compounds (benzo[c]phenanthrene, benz[a]anthracene, BaP, and indeno[1,2,3-cd]pyrene) are suspected carcinogens. The PAH concentrations were up to 20 times higher in air transported from the European continent and Great Britain when compared to air from northern Norway. Under certain meteorological conditions, episodes occur where the concentrations of PAH at remote rural areas reach the same order of magnitude as that for urban atmosphere. Table 1 shows some typical data at the two stations when the wind is from the southwest and the north. These results clearly demonstrate that industrial or urban plumes may be transported over long distances with essentially very low dilution. It is plausible that the values given in Table 1 are somewhat low, since no attempt was made to collect gaseous compounds [25] or to prevent blowoff from already collected particles (see Chap. 4).

Table 1. Concentrations of PAH in Aerosols at the Sampling Stations at Birkenes, Norway, and Rorvik, Sweden (ng/m^3)

Location	Birkenes, Norway					
Sample number	1	2	3	4	5	6
Sampling period	1-3.1.1977	4-9.1.1977	14-17.1.1977	22-24.1.1977	28-30.1.1977	2-4.2.1977
Origin of air	SW	NW-W	S-SE	SW	N-NW	S-SW
Phenanthrene	2.25	–	0.32	0.99	0.51	5.09
Anthracene	0.14	–	0.003	0.04	0.04	0.23
Methylphenanthrene/ methylanthracene	0.31	–	0.05	0.09	0.06	1.13
Methylphenanthrene/ methylanthracene	0.43	–	0.09	0.12	0.09	1.08
2-Methylanthracene	0.05	–	0.02	–	–	0.04
Methylphenanthrene/ methylanthracene	0.42	–	0.08	0.18	0.17	0.66
Methylphenanthrene/ methylanthracene	0.22	–	0.04	0.07	–	0.22
1-Methylphenanthrene	0.30	–	0.05	0.10	0.09	0.49
Fluoranthene	4.10	0.17	0.76	1.62	0.61	6.28
Dihydrobenzo[a/b]-fluorenes	0.74	–	0.13	0.26	0.22	0.99
Pyrene	3.42	0.12	0.68	1.35	0.65	4.89
Benzo[a]fluorene	1.45	–	0.29	0.43	–	0.77
Benzo[b]fluorene	0.10	–	0.02	0.28	–	0.34
1-Methylpyrene	0.35	–	–	0.10	–	0.28
Benzo[c]-phenanthrene	0.20	–	0.04	0.07	0.04	0.92
Benz[a]anthracene	1.39	0.10	0.31	0.49	0.27	1.86
Chrysene/ triphenylene	2.35	0.16	0.51	0.80	0.36	6.18
Benzo[b]fluoranthene	0.83	0.08	0.18	0.25	0.08	5.85
Benzo[j/k]fluoranthenes	0.75	0.07	0.17	0.24	0.07	3.99
Benzo[e]pyrene	1.06	0.11	0.24	0.32	0.13	2.81
Benzo[a]pyrene	0.78	0.06	0.46	0.21	0.18	2.35
Perylene	0.22	0.02	0.05	0.05	0.03	0.49
o-Phenylenepyrene	1.16	0.09	0.26	0.37	0.10	2.67
Dibenz[a,h]-anthracene	0.19	–	0.03	0.04	–	0.52
Benzo[ghi]perylene	0.81	0.07	0.18	0.25	0.07	2.35
Anthanthrene	0.22	–	–	–	–	0.25
Total	24.24	1.05	4.99	8.72	3.77	52.73

Source: Ref. 24.

Rørvik, Sweden						
8 14-16.2.1977	9 19-21.2.1977	10 24-26.2.1977	11 4-6.3.1977	12 13.3.1977	13 19.3.1977	14 21.3.1977
0.52	1.49	0.87	0.27	0.45	0.22	–
0.01	0.18	0.05	0.18	0.12	–	–
0.04	0.52	0.08	–	–	–	–
0.02	0.37	0.09	–	–	–	–
–	–	0.09	–	–	–	–
0.07	0.29	0.09	–	–	–	–
0.03	0.16	0.05	–	–	–	–
0.05	0.29	0.13	–	–	–	–
2.18	3.21	1.03	0.53	1.26	0.42	0.79
0.26	0.50	0.07	–	–	–	0.09
1.41	2.15	0.58	0.21	0.42	0.22	0.60
–	0.95	–	–	–	–	–
–	0.66	–	–	–	–	–
–	–	–	–	–	–	–
0.14	1.13	0.10	0.10	0.24	0.16	–
0.47	1.36	0.23	0.42	0.20	0.12	0.21
1.04	2.79	0.82	2.47	0.90	0.39	0.42
0.27	2.57	2.57	0.37	1.50	0.42	0.35
0.18	0.74	–	–	–	0.16	0.45
0.33	0.99	0.48	0.17	0.31	0.09	0.26
0.39	1.16	0.24	0.13	0.29	0.11	0.24
–	0.09	–	–	–	–	–
0.08	0.82	0.29	–	0.28	–	0.24
–	0.05	0.08	–	–	–	–
0.10	0.84	0.26	–	0.04	–	0.24
–	0.07	–	–	–	–	–

III. Fate of PAH in the Atmosphere

The degree to which airborne PAH will be subject to long-range transport will depend not only on the residence time of the particles in the atmosphere but also on the sensitivity of the adsorbed PAH to various chemical and photo-oxidative reactions. Furthermore, the atmospheric stability of different PAH compounds depends on factors such as molecular structure, intensity, and frequency of incident light and the presence of oxidizing pollutants [8]. In particular, the atmospheric breakdown of selected PAH by chemical reactions with gaseous components can be extremely fast in regions of intense photochemical activity. PAH formed by fossil-fuel combustion will appear in the emitted gases mixed with O_X, SO_X, and NO_X, and reactions between PAH and these species may take place. Such reactions may also take place in ambient air. It has recently been shown [26,27] that urban air particulates have significant mutagenicity in the Ames test and that the total sample mutagenicity was mainly due to acidic and neutral compounds. More than 50% of this mutagenicity required no metabolic activation. The presence of these direct mutagens in ambient particulates may be in part due to the transformation products of PAH [28,29]. Furthermore, because PAH in the atmosphere usually are adsorbed to particulate matter, particle size, porosity of the particle, and adsorption factors are likely to be important in stabilizing these compounds.

In subsequent chapters, some studies of different types of degradation reactions are summarized.

A. Photooxidation

In most studies of PAH photooxidation, the investigators have used light sources that simulate sunlight (>300 nm). However, PAH being photolyzed have appeared in a variety of states, such as gaseous, solutions, adsorbed to different particles, as well as pure solids.

Solutions of PAH in nonpolar solvents are convenient for photolysis studies, but the relevance of solution photochemistry to photolysis of PAH adsorbed by particulate matter is probably remote. Kuratsune and Hirokata [30] photolyzed 13 PAH in cyclohexane and dichloromethane with sunlight and with fluorescent lamps filtered through glass to provide wavelengths longer than 300 nm. They found that naphthacene was the most unstable of the PAH tested, and that BaP and alkyl derivatives of benz[a]anthracene were unstable in light. Further studies by Masuda and Kuratsune [31] showed that 1,6-, 3,6-, and 6,12-benzo[a]pyrene diones are formed by photolysis of BaP with light of wavelengths larger than 280 nm and with the presence of oxygen.

Since PAH in the environment are usually found adsorbed on particles, the study of photolysis of PAH in this state seems more pertinent. Only a few reports on this subject have appeared in the literature, but they indicate that the reactivity of PAH adsorbed to certain particles may be considerably greater than that of hydrocarbons in solution. PAH adsorbed on silica or aluminum oxide may be oxidized very readily [32,33] when exposed to ultraviolet (UV) light. Exposure to ordinary room light or adsorbtion of PAH by cellulose powder or acetylated cellulose gave slower but similar results. Of the 15 PAH listed, no changes were noted for phenanthrene, chrysene, triphenylene, and picene, whereas pronounced changes were detected for anthracene, naphthacene, benz[a]anthracene, dibenz[a,c]anthracene, dibenz[a,h]-anthracene, pyrene, BaP, benzo[e]pyrene, perylene, benzo[ghi]perylene, and coronene [30].

PAH seem to be more stable when adsorbed by natural particulate matter. In perhaps the earliest study of PAH photochemistry, Falk et al. [34] noted that the amount of BaP in ambient air, when compared with other PAH in ambient air, was larger than one would expect on the basis of amounts emitted from known sources. This was attributed to differences in atmospheric stability. They also studied the stability of pure PAH (adsorbed on filter) and adsorbed PAH (combustion particles collected on filters and split into control and exposed samples). The results showed that adsorption of BaP on particles provides protection from photooxidative processes and that adsorption on particulate matter seems to have a stabilizing effect when compared to coating on a filter [34,35]. Furthermore, studies by Lane and Katz [36] indicate that there may be a protective effect of a surface layer of previously oxidized organics. These effects will contribute to relatively high stability of PAH in the form in which they are naturally present in the atmosphere.

Similar conclusions have recently been reached by studies of PAH on blank filters or filters preloaded with rural or highway air particles [37]. It appears that extensive dark oxidation of PAH occurs on most filter types tested and that the filter medium itself assumes a leading role in this degradation [37]. Reactions on filter materials, however, appear to depend on the filter loading [38].

The decrease in concentration of some PAH adsorbed onto fly ashes from an electrostatic precipitator of a coal-fired power plant during their exposure to UV light has been studied [39]. Independent of the PAH identity, intensity and spectral composition of the light used, the decrease of PAH on fly ash without combustible matter during continuous exposure to UV light was rapid for the first hours of irradiation. After a time a point was reached where the decomposition stopped or continued very slowly. The results also showed that the decomposition rate decreases with increasing number of benzene rings in the molecule and also depends on their annihilation. There is practically no decrease of PAH on fly ash with high carbon content after 32 hr of irradiation by light of wavelength 36.5 nm [39].

Other studies indicate that particulate PAH exhibit higher stability than originally estimated in laboratory experiments [40,41] and that PAH may be relatively chemically inert and are removed from air only by rain or slow sedimentation of the particulates [42].

In a recent study of chemical transformations of particulate PAH [43] it was found that pyrene, phenanthrene, fluoranthene, anthracene and BaP adsorbed onto coal fly ash, activated carbon, or graphite particles results in effective stabilization of these compounds against photochemical decomposition. On the other hand, adsorption of fluorene, benzo[a]fluorene, benzo[b]-fluorene, 9,10-dimethylanthracene, and 4-azafluorene onto these same types of particles results in spontaneous dark oxidation to the corresponding quinone or ketone [43].

Further studies of photochemistry of PAH add to the confusion regarding their degradation [8,36,44-47]. It appears that considerable more studies, in particular with respect to PAH on naturally occurring particles, are needed before this subject is clarified.

B. Reactions with Ozone

As indicated in a study in the previous section [8,36], light is not required for atmospheric oxidation of PAH. Ozone reacted readily with PAH [48,49] and several modes of reactions have been identified [8]. For many PAH (excluding BaP) the loss of an individual compound coated on a filter and exposed to air was the same in the dark as in light [34]. Furthermore, BaP reacted readily with low ozone concentrations without irradiation [36], whereas benzo[b+k]fluoranthenes were not nearly as susceptible. It is likely, therefore, that reactivity varies for different structures.

The reactions of nine PAH with ozone in the dark and in simulated sunlight [46] showed that BaP undergoes rapid reactions (Table 2). Pyrene, and especially benzo[b]fluoranthene and benzo[k]fluoranthene, show comparative resistance to dark oxidation.

C. Reactions with SO_X

PAH react readily with SO_2, SO_3, or H_2SO_4, particularly in aerosols or when adsorbed by particles [8,50]. When pyrene and BaP were adsorbed by fly ash and alumina from an acetone solution and irradiated in a quartz tube filled with 10% SO_2, they reacted to yield sulfur-containing compounds including 1-pyrenesulfonic acid and benzo[a]pyrene-4-sulfonic acid [51]. The concentration of SO_2, however, was significantly higher than is normally present in ambient air, and it is unknown if these reactions really occur.

Other studies also prove the facile sulfonation of aromatic hydrocarbons. Gaseous benzene reacts readily with gaseous SO_3 [52]. Pyrene reacts at

Table 2. Half-Lives of PAH under Simulated Atmospheric Conditions (Expressed in Hours)

	Simulated sunlight	Simulated sunlight + ozone (0.2 ppm)	Dark reaction ozone (0.2 ppm)
Anthracene	0.20	0.15	1.23
Benz[a]anthracene	4.20	1.35	2.88
Dibenz[a,h]anthracene	9.60	4.80	2.71
Dibenz[a,c]anthracene	9.20	4.60	3.82
Pyrene	4.20	2.75	15.72
Benzo[a]pyrene	5.30	0.58	0.62
Benzo[e]pyrene	21.10	5.38	7.60
Benzo[b]fluoranthene	8.70	4.20	52.70
Benzo[k]fluoranthene	14.10	3.90	34.90

Source: Ref. 46.

room temperature with concentrated H_2SO_4 to produce a mixture of disulfonic acids [53]. The concentration of H_2SO_4 in aerosols droplets seems to be high enough [54,55] to sulfonate reactive PAH. Sulfur dioxide also reacts with aromatic compounds in smoke, with or without light [44].

The reaction products of PAH with SO_2, SO_3, or H_2SO_4 are sulfinic and sulfonic acids. Since the compounds are water soluble, they will probably not appear in the extracts of particulate matter when common solvents such as cyclohexane and dichloromethane are used. It is likely, therefore, that these compounds are overlooked in most studies of aromatic compounds in atmospheric pollutants.

D. Reactions with NO_x

Nitrogen oxides or dilute HNO_3 have been shown to add to, substitute in, or oxidize PAH [8]. NO_2 adds to and substitute in anthracene [56], and BaP is nitrated in minutes at room temperature with nitric acid diluted by acetic acid and benzene to give a mononitro compound [53]. It has also been shown that NO_2 will react with different PAH (anthracene, BaP, perylene) in laboratory simulations of polluted air to give several products that are directly mutagens in the Ames test [57,58]. In contrast, anthracene and perylene do not show any mutagenic effect, and BaP must be activated by a liver microsomale extract to give a mutagenic effect in the Ames test. The active compounds have been identified as 1-nitro-, 3-nitro-, and 6-nitro-BaP and 3-nitroperylene. These are the compounds that are theoretically most plausible according to molecular orbital calculations and, moreover, those that are formed by classical nitration of PAH in solution.

The occurrence of nitro-PAH in particulate air pollutants still remains to be documented. Even if they are identified in the particulates, their formation may be due to the sampling conditions. J. N. Pitts, Jr. (see Ref. 59) treated a glass fiber filter with BaP and placed it in series with an untreated glass filter. The untreated filter removed particulates from the air, but the treated filter still contained substances that, when extracted, were directly mutagenic in the Ames test. This suggests that some of the mutagens found in air samples collected using glass fiber filters may be caused by reactions on the filter surface.

Jäger and Hanus [60] recently studied the interaction of gaseous NO_2 with solid carrier-adsorbed PAH under laboratory conditions. Carriers utilized included fly ash, silica gel, alumina, and carbon. Special attention was given to factors that might possibly influence the reaction, such as type of carrier, irradiating conditions, NO_2 concentrations, exposure time, and temperature. At a NO_2 concentration of 1.33 ppm, nitroanthracene, nitropyrene, nitrochrysene, two nitro-BaPs, and one dinitro-Bap were detected. The experimental data suggest that the formation of nitro-PAH in the atmospheric environment may occur provided PAH, nitrogen oxides, and a suitable sorbent coexist. Recently, Jäger reported on the detection and characterization of nitroderivatives of PAH in urban air [61]. Airborne particulate matter was collected on glass fiber filters and analyzed by thin-layer chromatography (TLC) and fluorescence quenching. The spots were found on the TLC plate. Two of these were tentatively identified as 3-nitrofluoranthene and 6-nitro-BaP based on R_F values. With the aforementioned reservations, this analysis indicates that nitro-PAH occur in urban atmosphere. These studies are of significance since relatively high levels of PAH and nitrogen oxides coexist in automobile exhaust, plumes from coal-fired power plants, or emissions from other kinds of combustion of organic materials. The high mutagenic activity assigned to some nitro-derivatives of PAH [62,63] stresses the importance of further study of this class of compounds.

IV. Relation between PAH and Other Pollutants

Since PAH are emitted simultaneously with other pollutants from a specific source, there should exist an interrelation between these groups of pollutants. However, relatively few studies which have been published so far demonstrate such interrelations.

Recently, the regional patterns of PAH, pesticide, and trace metal contamination in snow of northeastern Bavaria and their relationship to human influence and orographic effects were investigated [64]. The results showed a regional pattern of PAH that was very similar to that of the trace metals (Zn, Cu, Cd) and the specific conductance measurements. Furthermore, using principal component analysis, the dependencies of the contaminants were shown. The four PAH analyzed, specific conductance, and the trace metals Zn, Cd, and Cu all were highly correlated with each other and in general showed similar regional patterns. The PAH showed no correlation with Pb or with chlorinated pesticides, indicating different sources for these pollutants.

Measurements of PAH and SO_4^{2-} and of PAH and soot have been reported from Birkenes, Norway, and Rørvik, Sweden, respectively. The results are shown in Table 3. At the Birkenes station, high and low values coincide for PAH and SO_4^{2-}, respectively, with a correlation coefficient of 0.76. At the Rørvik station, maximum and minimum values for soot and PAH coincide, and the correlation coefficient between soot and PAH is 0.97.

V. Calculation of PAH Emissions from Long-Range Transport Data

A striking feature of the long-range transport of PAH is the mixing of PAH from different sources, giving a uniform PAH profile. This has been reported in several studies of PAH in sediments [7,9]. The uniform distribution indicates that PAH transported long range have a potential for determining the integrated emissions over a large area [65]. So far, very limited data are available to support this assumption, but a few calculations based on existing data may be performed. The calculations are related to the transport of PAH from the United Kingdom to Norway but will presumably permit us to draw more general conclusions. Until further analytical data are available, the calculations will be limited to a very simple model based on the following assumptions:

a. All substances are stable during the transport.
b. The emitted PAH are all transported with the air.
c. The maximum values measured in Norway and Sweden are representative for the emissions, i.e., lower PAH concentrations are due to fallout, degradation, and vertical mixing in air masses.

Table 3. Concentrations of SO_4^{2-}, Soot and PAH at Birkenes, Norway, and Rørvik, Sweden, January and February 1977

	Birkenes, Norway			Rørvik, Sweden		
Month	Date	SO_4^{2-} ($\mu g\ m^{-3}$)	PAH ($ng\ m^{-3}$)	Date	Soot ($\mu g\ m^{-3}$)	PAH ($ng\ m^{-3}$)
January	1-2	6.3	24.2			
	4-9	0.8	1.1	14-16	14.4	24.2
	22-24	6.4	8.7			
	28-30	1.0	3.8			
Monthly average		3.08	–		14.0	–
February 1977	2-4	7.9	52.7	3-4	41.3	92.4
	8-10	1.9	2.3	5-7	14.5	14.3
	12-14	9.8	23.3	9-10	6.3	6.7
	25-27	0.9	0.8	14-16	12.7	7.6
				19-21	11.2	23.4
				24-26	8.8	8.2
Monthly average		4.86	–		13.6	–

Source: Ref. 24.

The model is based on the assumption that during the sampling period all emissions in the United Kingdom are transported through a plane from Birkenes to Rørvik, as shown in Fig. 2. For simplicity, further assumptions are given in Table 4. Based on these premises it is calculated that the emission in the United Kingdom of BaP and total PAH are 1.2 and 18 tons/day, respectively. This corresponds to an annual emission of 430 and 6480 tons, respectively, provided that the same daily emission level is maintained all year. These numbers must be regarded as rough estimates. However, it is interesting to note that the annual discharge of BaP in the United States was calculated to be 1300 tons/year in 1965 [66]. The annual emissions per capita of BaP according to these calculations are quite comparable: 8.5 g for the United Kingdom and 6.5 g for the United States.

There are several factors that probably will underestimate the emissions:

1. *The assumptions that all emissions are transported with the air masses*—According to the calculations reported by LaFlamme and Hites [9], there will be a continuous segmentation and fallout of particles according to particle size.
2. *Losses of PAH during sampling*—The sampling device used in the studies [23,24] was a simple glass fiber filter with no backup for volatile compounds. Furthermore, rather high sampling velocities were applied. It has been stated that under these circumstances losses of PAH may occur [25]. This may particularly affect the more volatile PAH—and consequently the total PAH value.
3. *The assumption that the substances are not degraded*—As has already been discussed, it is not known precisely to what extent PAH are degraded

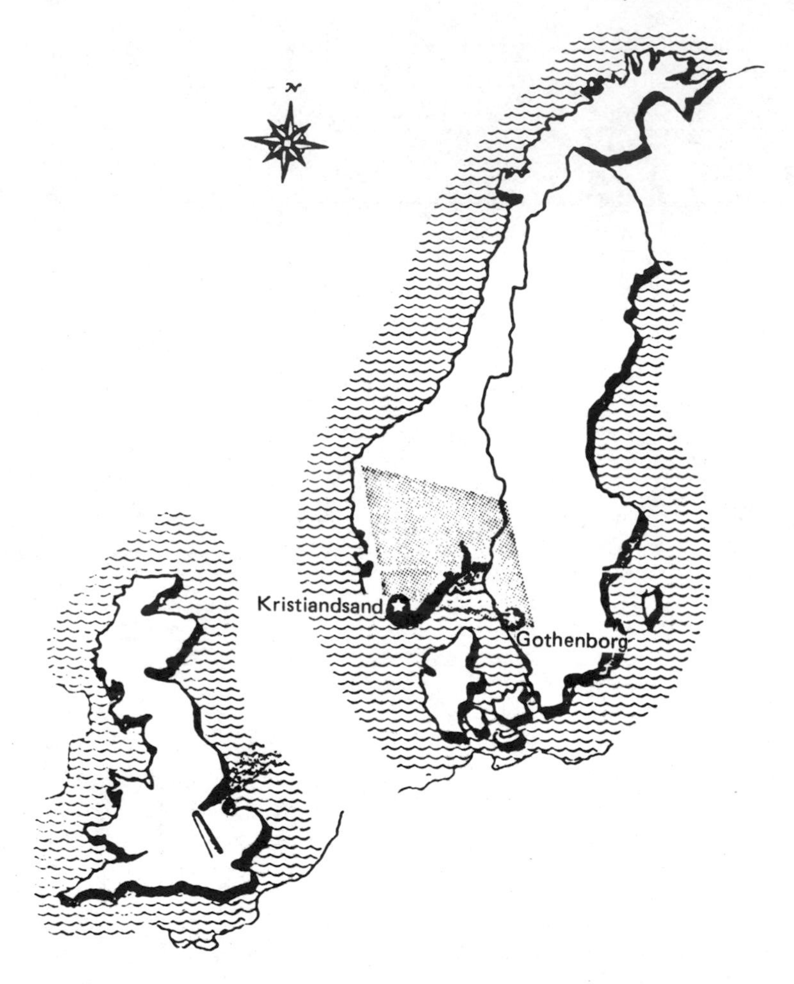

Figure 2. Schematic illustration of long-range transport of PAH from the United Kingdom to Scandinavia.

Table 4. Assumptions for Calculating Long-Range Transport of PAH

Parameter	Value
Average air mass velocity	25 km/hr
Average height of polluted air	1000 m
Width of plane	400 km
Average BaP concentration[a]	5 ng/m^3
Average PAH concentration	75 ng/m^3

[a]Average values for February 3, 1977 [24].

during the air transport. Some results indicate that under real conditions PAH are more stable than results from laboratory studies indicate (see Sec. III). This may be a minor point, since most PAH measurements reported so far were performed during the winter season.

There may also be factors that overestimate the emissions. The most prominent is probably that PAH concentration used is from the winter season. Considerable seasonal variation of PAH concentrations have been reported [21].

The nature of the sources for PAH emissions cannot be described accurately at the present moment. However, some indications may be obtained. The only source that is fairly accurate characterized is gasoline and diesel combustion engines. Emissions are generally in the area 0.10-0.12 g BaP/ton gasoline and 0.08-0.10 g BaP/ton diesel. Consumption in the United Kingdom were 18.3×10^6 tons gasoline and 5.8×10^6 tons diesel (in 1978), corresponding to less than 0.1 g BaP per person per year. Even if it is assumed that all BaP from diesel and gasoline vehicles in the United Kingdom enters the air masses that are transported long range, this explains only a fraction of the total BaP found in the samples. This assumption is, however, not reasonable, since the exhaust from combustion engines is emitted so close to the ground that only to a small extent will it be subject to long-range transport.

Further support for exclusion of combustion engines as sources is evidence that the coronene/BaP ratios in the samples are much lower than usually found in gasoline and diesel exhaust (see Chap. 5). These results therefore indicate that the combustion engine is not a major source of PAH in air masses transported long range.

Blumer [7] has suggested that the relative distribution of individual PAH may be used to characterize the source. The PAH chromatogram presented in Fig. 1 is dominated by the parent PAH. In a few cases methyl substituents are present, but no higher alkyl homologs were detected. This indicates that the source is a medium-to-high temperature process involving carbonaceous material. The relative occurrence of linear vs. angular ring arrangements (i.e., anthracene vs. phenanthrene) indicates that the equilibrium time is long and the reaction temperature relatively high.

VI. Conclusions

Analysis of PAH in soil and sediments, in snow and rainwater, and in air demonstrates the potential of long-range transport of these compounds.

Under certain meteorological conditions, the PAH may be transported over long distances, apparently with no or very low degradation or dilution.

Analyses of well-defined long-range transported air masses with respect to PAH may be used to estimate the integrated emissions over large areas.

References

1. B. Bolin and C. Persson, *Tellus 27*:281 (1975).
2. *The OECD Program on Long-Range Transport of Air Pollutants: Measurements and Findings*, Organization for Economic Cooperation and Development, Paris, 1977.

3. F. H. Braekke (Ed.), Impact of acid precipitation on forest and freshwater ecosystems in Norway: Summary report on the research results from phase I (1972-1975) of the SNSF-project. Agricultural Research Council of Norway and Norwegian Council for Scientific and Industrial Research, Oslo, Norway, 1976.
4. Y. S. Chung, *Atmos. Environ. 12*:1471 (1978).
5. M. M. Millan and Y. S. Chung, *Atmos. Environ. 11*:939 (1977).
6. C. Brosset, *Ambio 5*:157 (1976).
7. M. Blumer, *Sci. Amer. 234*:35 (1976).
8. Committee on Biological Effects, *Particulate Polycyclic Organic Matter*, National Academy of Sciences, Washington, D.C., 1972.
9. R. E. LaFlamme and R. A. Hites, *Geochim. Cosmochim. Acta 42*:289 (1978).
10. R. A. Hites, R. E. LaFlamme, and J. W. Farrington, *Science 198*:829 (1977).
11. J. G. Windsor, Jr., and R. A. Hites, *Geochim. Cosmochim. Acta 43*:27 (1979).
12. K. T. Whitley, R. B. Husar, and B. Y. Liu, *J. Colloid Interface Sci. 39*:177 (1972).
13. M. Katz and R. C. Pierce, in *Carcinogenesis*, R. Freudenthal and P. W. Jones (Eds.), Vol. 1, Raven Press, New York, 1976.
14. G. Lunde, J. Gether, N. Gjøs, and M. B. Støbet Lande, *Atmos. Environ. 11*:1007 (1977).
15. E. Schrimpff, in *International Conference on Ecological Impact of Acid Precipitation*, Sandefjord, Norway, March 11-14, 1980.
16. M. J. Suess, *Sci. Total Environ. 6*:239 (1976).
17. G. Holzer, H. Shanfield, A. Zlatkis, W. Bertsch, P. Juarez, H. Mayfield, and H. M. Liebich, *J. Chromatogr. 142*:755 (1977).
18. J. M. Daisey, M. A. Leyko, and T. J. Kneip, in *Polynuclear Aromatic Hydrocarbons*, P. W. Jones and P. Leber (Eds.), Ann Arbor Science Publs., Ann Arbor, Mich., 1979.
19. J. M. Daisey and P. J. Lioy, *Nature* (in press).
20. F. De Wiest, *Atmos. Environ. 12*:1705 (1978).
21. R. B. Faoro, *J. Air Pollut. Control Assoc. 25*:638 (1975).
22. J. M. Daisey, R. J. McCoffrey, and R. A. Gallagher, *Atmos. Environ.* (in press).
23. G. Lunde and A. Bjørseth, *Nature 268*:518 (1977).
24. A. Bjørseth, G. Lunde, and A. Lindskog, *Atmos. Environ. 13*:45 (1977).
25. C. Pupp, R. C. Lao, J. J. Murray, and R. F. Poffie, *Atmos. Environ. 8*:915 (1974).
26. J. N. Pitts, Jr., K. A. Van Cauwenberghe, D. Grosjean, J. P. Schmid, D. R. Fritz, W. L. Belser, G. B. Knudson, and P. M. Hynds, *Science 202*:515 (1978).
27. M. Møller and I. Alfheim, *Atmos. Environ. 14*:83 (1980).
28. J. N. Pitts, Jr., in *Biological Effects of Environmental Pollutants*, S. D. Lee (Ed.), Ann Arbor Science Publs., Ann Arbor, Mich. 1979.
29. J. N. Pitts, Jr., D. Grosjean, K. A. Van Cauwenberghe, and W. L. Belser, presented at the 175th National Meeting of the Amer. Chem. Soc., Anaheim, Calif., March 12-17, 1978.
30. M. Kuratsune and T. Hirokata, *Nat. Cancer Inst. Monogr. 9*:117 (1962).

31. Y. Masuda and M. Kuratsune, *Air Water Pollut. Inst. J. 10*:805 (1966).
32. M. N. Inscoe, *Anal. Chem. 36*:2505 (1964).
33. G. Kortum and W. Braun, *Ann. Chem. 632*:104 (1960).
34. H. L. Falk, I. Markal, and P. Kotin, *A.M.A. Arch. Ind. Health 13*:113 (1956).
35. H. L. Falk, P. Kotin, and A. Miller, *Intern. J. Air Pollut. 2*:201 (1960).
36. D. A. Lane and M. Katz, in *Fate of Pollutants in the Air and Water Environments*, I. A. Suffet (Ed.), Pt. 2, Wiley (Interscience), New York, 1977.
37. F. S.-C. Lee, W. R. Pierson and J. Ezike, in *Polynuclear Aromatic Hydrocarbons: Chemistry and Biological Effects*, A. Bjørseth and A. J. Dennis (Eds.), Battelle Press, Columbus, Ohio, 1980.
38. R. C. Lao and R. S. Thomas, in *Polynuclear Aromatic Hydrocarbons: Chemistry and Biological Effects*, A. Bjørseth and A. J. Dennis (Eds.), Battelle Press, Columbus, Ohio, 1980.
39. J. Jäger, *Cs. Hyg. 16*:14 (1971).
40. G. Kortum and W. Braun, *Ann. Chem. 632*:104 (1960).
41. R. S. Berry and P. A. Lehman, *Ann. Rev. Phys. Chem. 22*:47 (1971).
42. L. Fishbein in *Chemical Mutagens*, A. Hollaender (Ed.), Vol. 4, Plenum Press, New York, 1976, p. 232.
43. M. M. Hughes, D. F. S. Natusch, D. R. Taylor, and M. V. Zeller, in *Polynuclear Aromatic Hydrocarbons: Chemistry and Biological Effects*, A. Bjørseth and A. J. Dennis (Eds.), Battelle Press, Columbus, Ohio, 1980.
44. B. D. Tebbens, J. F. Thomas, and M. Mukai, *Amer. Ind. Hyg. Assoc. J. 27*:415 (1966).
45. J. F. Thomas, M. Mukai, and B. D. Tebbens, *Environ. Sci. Technol. 2*:33 (1968).
46. M. Katz, C. Chan, H. Tosine, and T. Sakuma, in *Polynuclear Aromatic Hydrocarbons*, P. W. Jones and P. Leber (Eds.), Ann Arbor Science Publs., Ann Arbor, Mich., 1979.
47. W. A. Korfmacher, D. F. S. Natusch, D. R. Taylor, E. L. Wehry, and G. Mamatov, in *Polynuclear Aromatic Hydrocarbons*, P. W. Jones and P. Leber (Eds.), Ann Arbor Science Publs., Ann Arbor, Mich., 1979.
48. P. S. Bailey, *Chem. Rev. 58*:925 (1958).
49. E. J. Moriconi and L. Salce, *Advan. Chem. Ser. 77*:65 (1968).
50. E. E. Gilbert, *Sulfonation and Related Reactions*, Wiley (Interscience), New York, 1965.
51. J. Jäger and M. Rakovic, *J. Hyg. Epidemiol. Microbiol. Immunol. 18*:137 (1974).
52. P. M. Heertjes, H. C. A. van Beck, and G. I. Grimmon, *Rec. Trav. Chim. 80*:82 (1961).
53. E. de B. Barnett, J. W. Cook, and H. H. Graimger, *J. Chem. Soc. 121*:2059 (1922).
54. L. F. Fieser and E. B. Hershberg, *J. Amer. Chem. Soc. 61*:1565 (1939).
55. N. A. Esmen and R. B. Fergus, *Sci. Total Environ. 6*:223 (1976).
56. P. S. Varma and J. L. Das Cupta, *Quart. J. Indian Chem. Soc. 4*:297 (1927).
57. L. Gundel, S.-G. Chang, and T. Novakov, Heterogeneous reactions of polynuclear aromatic hydrocarbons and soot extracts with NO_2. Atmos-

pheric Aerosol Research, Annual Report 1976-77, Energy & Environmental Division, Lawrence Berkeley Laboratory, No. LBL 6819, University of California, Berkeley, 1977, pp. 83-90.

58. J. N. Pitts, Jr., D. Grosjean, K. VanCauwenberghe, and W. L. Belser, Chemical and biological aspects of organic particulates in real and simulated atmospheres, presented at Conference on Carbonaceous Particles in the Atmosphere, Lawrence Berkeley Laboratory, University of California, Berkeley, 1978.
59. Anonymous, *Chem. Eng. News 56(16)*:22 (1976).
60. J. Jäger and V. Hanus, in preparation.
61. J. Jäger, *J. Chromatogr. 152*:575 (1978).
62. H. S. Rosenkranz, E. C. McCoy, D. R. Sanders, M. Butler, D. K. Kiriazides, and R. Mermelstein, *Science 209*:1039 (1980).
63. G. Løfroth, E. Hefner, I. Alfheim, and M. Møller, *Science 209*:1037 (1980).
64. E. Schrimpff, W. Thomas, and R. Herrmann, *Water Air Soil Pollut. 11*: 481 (1979).
65. G. Løfroth, private communication, 1977.
66. R. P. Hangebrauck, D. J. von Lehmden, and J. E. Meeker, *Sources of Polynuclear Hydrocarbons in the Atmosphere*. Public Health Service Publ. 999-AP-33. U.S. Department of Health and Human Services, Cincinnati, Ohio, 1967.

13

Determination of Polycyclic Aromatic Hydrocarbons in Coal-Derived Materials

CURT M. WHITE / Pittsburgh Energy Technology Center,
Bruceton, Pennsylvania

I. Introduction

Coal is the world's major carbonaceous fuel resource. Active investigation of the polycyclic aromatic hydrocarbons (PAH) present in coal and coal-derived materials has been conducted for approximately 70 years [1]. During this period, dozens of analytical techniques have been used to determine PAH in these materials. The purpose of this chapter is to briefly describe those analytical methods that provide the most useful, reliable, and detailed analytical information concerning the occurrence and distribution of PAH in coal and coal-derived materials.

There are many reasons for studying the PAH present in coal. A better understanding of the chemical constitution of coal is important for its increased utilization. Because of the complexity of coal and our inability to nondestructively volatize or solubilize all of it, direct investigations of the chemical structure of coal often cannot be conducted. Therefore, coal extracts have been used to investigate its chemical constitution. The aromatic portion of such extracts is usually the largest fraction, ranging approximately from 40 to 60 weight percent of the extract. Detailed characterization of this fraction is therefore extremely important. Moreover, identification of the PAH in coal can help elucidate the geochemical origins of coal and improve our understanding of the chemistry associated with the coalification process that converted plant debris to coal. An important reason for investigating the occurrence and distribution of PAH in coal is to better understand their possible health effects. Coal workers' pneumoconiosis [2,3], "black lung," can develop after accumulation of respirable coal dust in the lungs. Although there is considerable diversity of opinion concerning the exact cause of this disease, several investigators [2,3] have postulated that the chemical nature of the organic constituents present in coal may conceivably affect the occurrence of the disease. This is particularly true when the adverse health effects of PAH are considered. PAH produce some of the most severe lung irritants [4] when present in combination with inorganic material, which is also a major constituent of coal dust.

Industrial as well as environmental concerns are the main reasons for investigating the PAH in coal-derived materials. The advent of energy shortages has focused attention on increased use of products from the liquefaction and gasification of coal. Since the aromatic portion of coal liquefaction products usually comprises from 20 to 60 weight percent, detailed examination of this fraction has received considerable attention. The composition of the aromatic fraction of coal-derived fuels must be known in detail to determine the reaction mechanisms underlying the conversion process. Then, process variables can be optimized to obtain the desired distribution of products. This is particularly true for processes attempting to make a product suitable as a chemical feedstock [5,6]. Knowledge of PAH in coal-derived liquids is useful for studies of liquefaction kinetics, investigations of reactor design, and determination of the effect of changes in reaction conditions on product distribution.

Another major impetus for detailed characterization of PAH in coal liquefaction, gasification, and coking products has been due to environmental and human health concerns. Many PAH present in coal-derived materials are known or suspected carcinogens. In order to protect the health and safety of workers in these industries, it has been necessary to develop analytical methods for the quantitative collection and analysis of PAH in workplace atmospheres of these plants and in the products themselves. In 1960, researchers at the Union Carbide coal hydrogenation plant at Institute, West Virginia, described some of the industrial hygiene problems associated with coal liquefaction processes [7-10]. The incidence of skin lesions among coal hydrogenation workers was significantly higher than among control populations sampled in the United States. In addition, it is thought that the higher-than-normal incidence of lung cancer in a sister industry, coal coking operations, is due to exposure to PAH in coal tar volatiles [11]. An excellent review, by Herod and James, of the methods for estimating PAH from coke oven emissions has recently appeared [12], and White et al. [13] have described additional analytical aspects of the determination of PAH in workplace atmospheres around coal liquefaction plants.

The remainder of this chapter will be divided into the following sections: (1) a short review of some classical investigations of PAH from coal and coal products; (2) preparative liquid chromatographic (LC) methods applied to coal-derived materials resulting in aromatic fractions; (3) gas chromatographic (GC) and combined gas chromatographic/mass spectrometric (GC/MS) methods for determining PAH; (4) mass spectrometry as a tool for industrial PAH determinations; (5) combined methods for determining sulfur-containing PAH; and (6) new techniques on the horizon for PAH determinations in coal and coal products.

With the exception of a few investigations, discussed in the section on new techniques on the horizon, the most successful studies of PAH in coal and its products have used MS, GC, or combined GC/MS. It has been our experience that LC techniques at present lack sufficient efficiency to adequately separate and identify the wide variety of constituents in the extremely complex mixtures of PAH present in coal or its products. However, LC can be used for the determination of some select PAH when all experimental conditions have been optimized and selective ultraviolet (UV) and fluorescence detectors are used. The recent advent of packed microcapillary LC may increase the usefulness of this technique [14,15]. At present, LC is best used as a means of preparatively separating samples into classes of compounds, resulting in PAH and other fractions.

II. Pioneering Investigations of PAH in Coal and Coal Products

The investigations described in this section are generally considered to be classical, and—although some of them may seem crude by today's standards—they are the cornerstones of our knowledge about PAH in coal and its products, representing enormous effort on the part of the investigators. Each classical investigation reviewed here was (1) a major breakthrough in the development of analytical methods for analysis of PAH and/or (2) a vast improvement in our knowledge of the occurrence and distribution of PAH in coal and coal products. The five investigations reviewed in this section are representative of the excellent work conducted by many scientists in various countries. Some very worthy investigations are not reviewed here due to space limitations.

A. Early Gas Chromatographic Methods

As far as can be determined, the first use of GC for the analysis of PAH from coal products was reported by Dupire and Botquin [16]. Until the results of this classical investigation were published, it was almost impossible to routinely monitor the quality and content of coal tar distillates, other than to measure boiling point distribution, density, and viscosity. The GC instrument used by Dupire and Botquin was completely constructed by these investigators [16]. The chromatographic column (glass, 2.5 m × 6-8 mm) employed He as the carrier gas and was packed with crushed and sieved (60-80 mesh) refractory brick coated with 30% Dow Corning high vacuum silicone grease. The separations were performed isothermally at temperatures between 185° and 295°C, depending on the boiling fraction of the coal tar being analyzed. The chromatographic profiles of the coal tar distillates (200°-230°C, 225°-300°C, and 300°-360°C) appear in Figs. 1, 2, and 3, respectively. Identifications were made by comparing the retention characteristics of the chromatographic peaks in the unknown with those of standards. Dupire and Botquin [16] also noticed that there was a linear relationship between the log of the

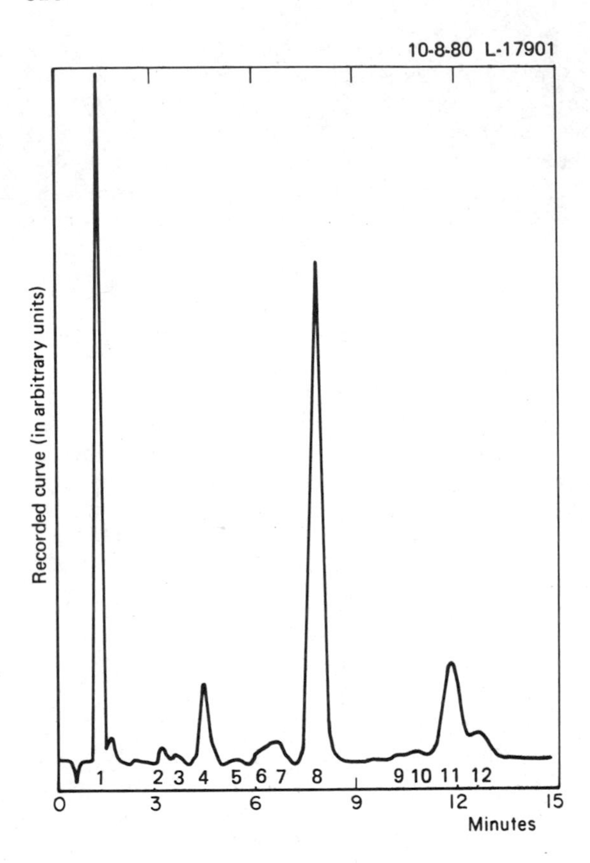

Figure 1. Packed column gas chromatogram of coal tar distillate boiling between 200° and 230°C. Numbered peaks are (1) solvent; (2) pseudocumene; (3) hemimellitene?; (4) indene; (5) durene; (6) o- and m-cresols?; (7) methylindene?; (8) naphthalene; (9) pseudocumenol? (10) oxycoumarone? (11) 2-methylnaphthalene; (12) 1-methylnaphthalene. (From Ref. 16.)

retention time of the PAH and its boiling point under the chromatographic conditions used (see Fig. 4). The results of this investigation were so promising that, in the years that followed, hundreds of investigators have used GC for PAH analysis of samples derived from coal and other products and a new frontier in the analytical chemistry of PAH was opened.

B. Initial Application of High-Resolution Gas Chromatography

In 1968, Helmut Pichler and his colleagues published a paper which represents a milestone in analysis of coal carbonization products [17]. Although Liberti et al. [18] were the first to use capillary GC to separate PAH, Pichler et al. [17] were, to my knowledge, the first to apply high-resolution capillary GC to the analysis of PAH from a coal product. Pichler vacuum-distilled carbonization tars from bituminous coal into several fractions and distillates were

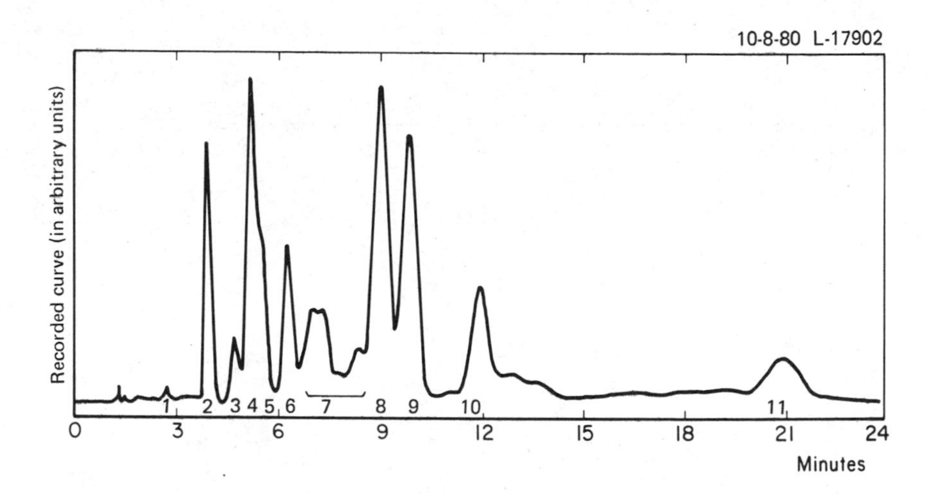

Figure 2. Packed column gas chromatogram of a coal tar distillate boiling between 225° and 300°C. Numbered chromatographic peaks are (1) indene; (2) naphthalene; (3) not identified; (4) 2-methylnaphthalene; (5) 1-methylnaphthalene; (6) biphenyl; (7) dimenthylnaphthalenes; (8) acenaphthene; (9) dibenzofuran; (10) fluorene; (11) phananthrene and anthracene. (From Ref. 16.)

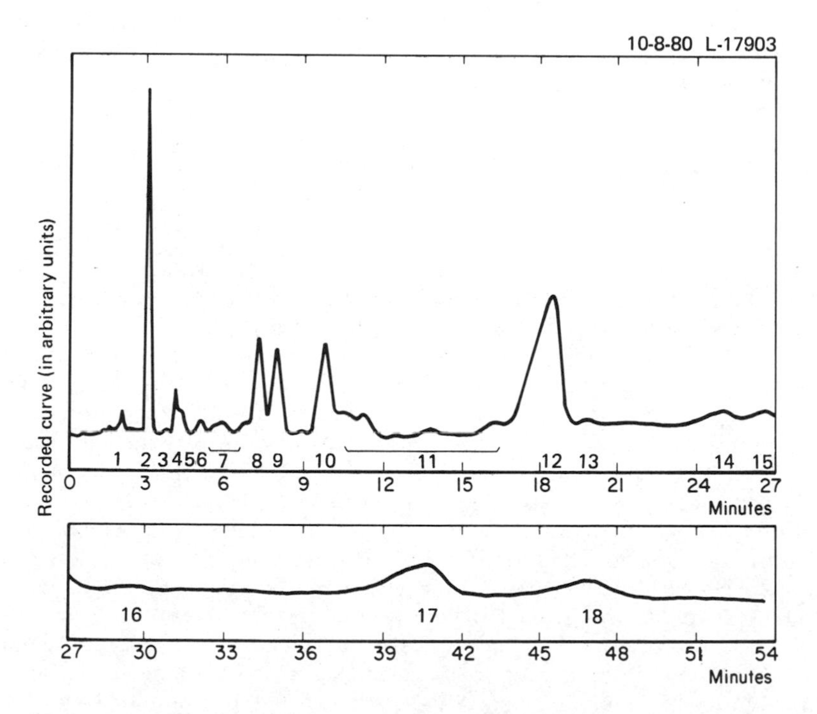

Figure 3. Packed column gas chromatogram of a coal tar distillate boiling between 300° and 360°C. Numbered chromatographic peaks are (1) indene; (2) naphthalene; (3) not identified; (4) 2-methylnaphthalene; (5) 1-methylnaphthalene; (6) biphenyl; (7) dimethylnaphthalene; (8) acenaphthene; (9) dibenzofuran; (10) fluorene; (11) not identified; (12) phenanthrene and anthracene; (13) carbazole?; (14) not identified; (15) methylanthracene?; (16) methylcarbazole? (17) fluoranthene; (18) pyrene. (From Ref. 16.)

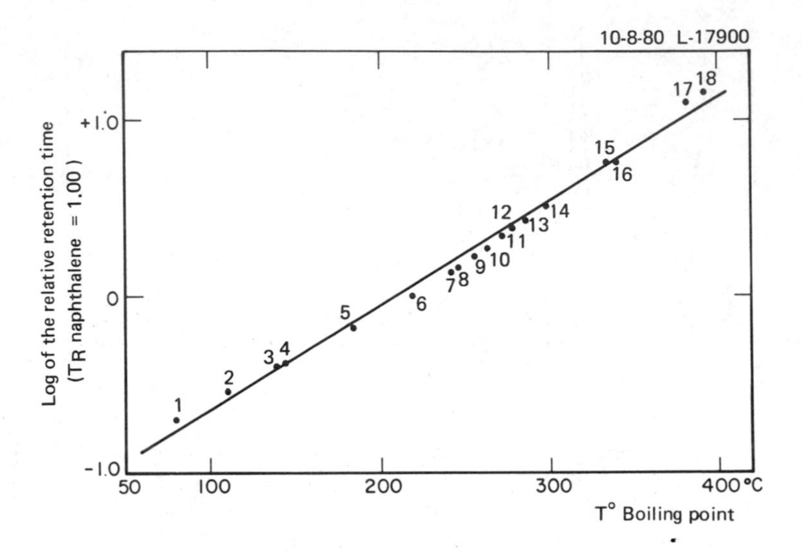

Figure 4. Relationship between boiling point and the log of the relative retention times of some aromatic hydrocarbons, as shown below. (From Ref. 16.)

(1) benzene	(7) 2-methylnaphthalene	(13) dibenzofuran
(2) toluene	(8) 1-methylnaphthalene	(14) fluorene
(3) m- and p-xylene	(9) biphenyl	(15) phenanthrene
(4) o-xylene	(10) 2,6-dimethylnaphthalene	(16) anthracene
(5) indene	(11) 1,2-dimethylnaphthalene	(17) fluoranthene
(6) naphthalene	(12) acenaphthene	(18) pyrene

then separated into classes of compounds using various extraction and preparative LC methods. Each fraction was subsequently analyzed by GC employing wall-coated open-tubular columns (WCOT) and packed columns. Individual constituents in the mixtures were identified using cochromatography with authentic reference compounds. In cases where reference compounds were not commercially available, they were synthesized. The Kovats retention indices were calculated for all of the identified compounds [19]. In addition, verification of these identifications was obtained by preparatively collecting the compounds and comparing their UV spectra with those of standards. The GC profile of the aromatic fraction from a fluidized-bed degasification process, boiling between 80°C and 270°C, is illustrated in Fig. 5. The numbered chromatographic peaks, along with their Kovats retention indices, are in Table 1. The separation was achieved using a 100 m × 0.25 mm i.d. column coated with polypropylene glycol, a N_2 carrier gas, and temperature programming. Although high-molecular-weight PAH were not present in this distillate aromatic fraction, volatile PAH, including naphthalene, methylnaphthalenes, dimethylnaphthalenes, and acenapthylene, were present.

An aromatic fraction from the same degasification process, boiling between 270°C and 410°C, was also analyzed. The chromatographic profile of this

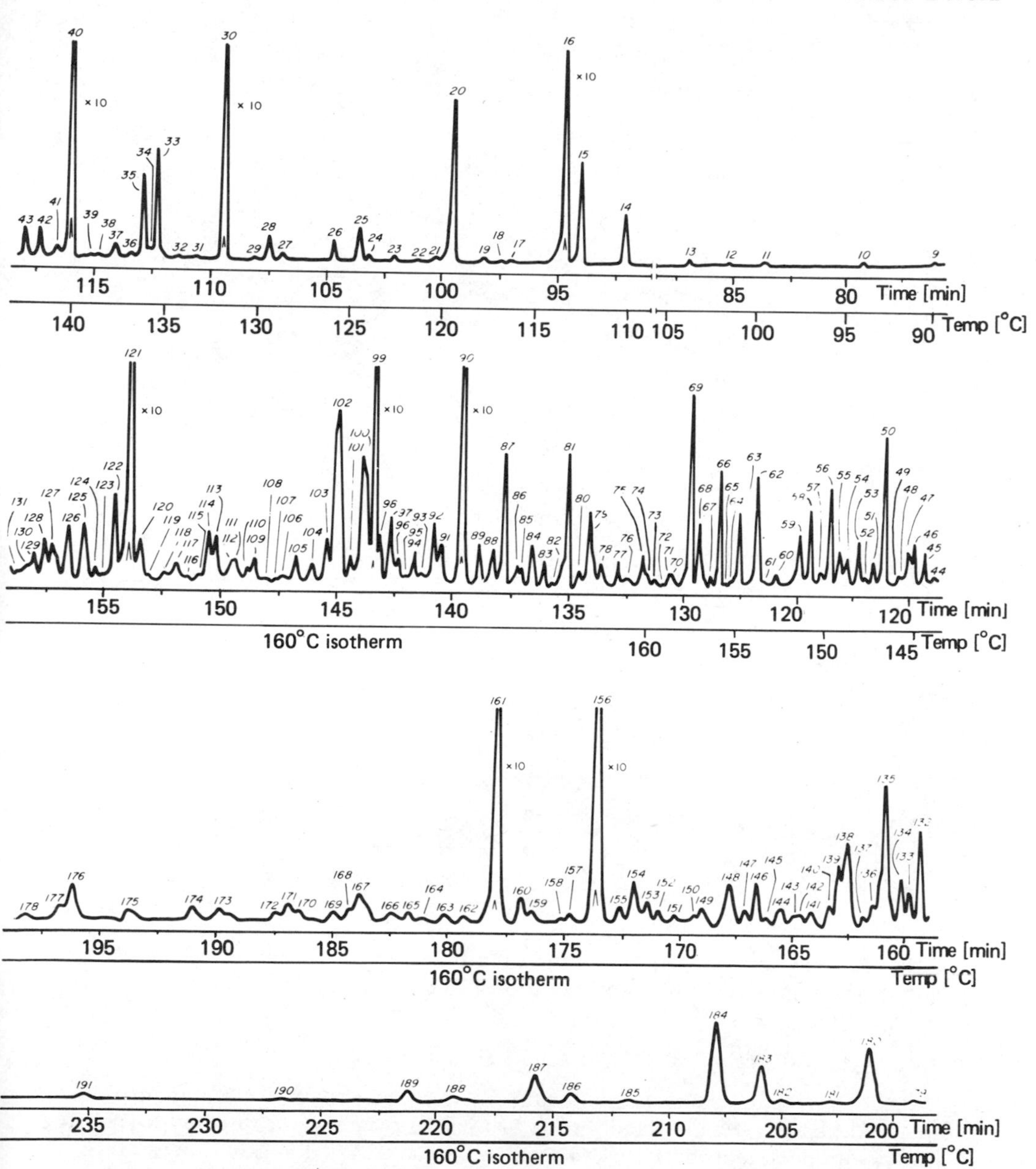

Figure 5. Capillary column gas chromatogram of the aromatic fraction from a distillate boiling between 80° and 270°C of a fluidized-bed degasification tar. Chromatographic conditions: 100 m × 0.25 mm i.d. column coated with polypropylene glycol. Temperature programmed from 90° to 160°C at 1.25°C/min. Numbered chromatographic peaks are identified in Table 1. (From Ref. 17.)

Table 1. Identification of Chromatographic Peaks Due to PAH in Fig. 5

Peak number	Compound	Kovats retention index
69	Indene	1189
102	Methylindene	1305
121	Naphthalene	1373
128	Benzothiophene	1393
129	1-Ethylindene	1396
156	2-Methylnaphthalene	1480
161	1-Methylnaphthalene	1499
175	2-Methylbiphenyl	1556
176	Biphenyl- and 2-ethylnaphthalene	1566
177	1-Ethylnaphthalene	1567
180	2,6-Dimethylnaphthalene and 2,7-dimethylnaphthalene	1583
183	1,7-Dimethylnaphthalene	1598
184	1,6-Dimethylnaphthalene and 1,3-dimethylnaphthalene	1604
186	1,4-Dimethylnaphthalene	1623
187	2,3-Dimethylnaphthalene and 1,5-dimethylnaphthalene	1627
189	1,2-Dimethylnaphthalene	1642
191	Acenaphthylene	1674

Source: Ref. 17.

fraction is shown in Fig. 6, while the identification of the numbered chromatographic peaks and their Kovats retention indices appear in Table 2.

C. Early Investigation of PAH in Coal Extracts

In 1963, Ōuchi and Imuta of the Resources Research Institute in Japan published two papers [20,21] which were to become cornerstones of our knowledge of the occurrence and distribution of PAH in coal. Until these papers were published, very little was known with certainty about the PAH content of coal. The first describes the benzene extraction of Yūbari coal and the preparative LC fractionation of the extract. The fractions were subjected to successive recrystallizations, and two crystalline substances were isolated. The UV and IR (infrared) spectra of these substances showed that one of them was 4,11-dimethylpicene, which had been previously isolated from Yūbari

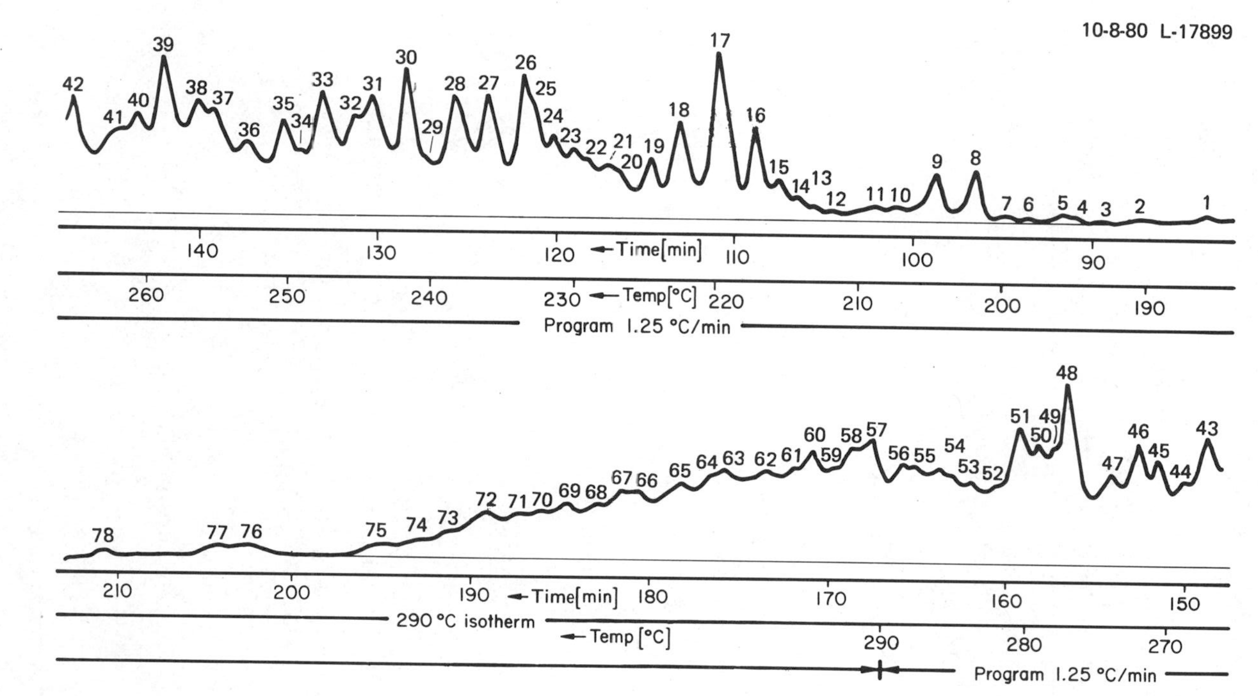

Figure 6. Packed column gas chromatogram of the aromatic fraction from a distillate boiling between 270° and 410°C of a fluidized-bed degasification tar. Chromatographic conditions: 10 m × 4 mm i.d. column packed with Kieselguhr coated with 10% SE-52. Temperature programmed from 190° to 290°C at 1.25°C/min. Numbered chromatographic peaks are identified in Table 2. (From Ref. 17.)

Table 2. Identification of Chromatographic Peaks in Fig. 6

Peak number	Compound	Kovats retention index
1	Naphthalene	1224
8	2-Methylnaphthalene	1342
9	1-Methylnaphthalene	1358
13	Biphenyl	1426
15	1- and 2-Ethylnaphthalene	1446
16	2,6-Dimethylnaphthalene and 2,7-dimethylnaphthalene	1472
17	1,6-Dimethylnaphthalene, 1,7-dimethylnaphthalene, 1,4-dimethylnaphthalene, and 1,3-dimethylnaphthalene	
18	2,3-Dimethylnaphthalene and 1,5-dimethylnaphthalene	1492
19	Acenaphthylene and 2-methylbiphenyl	1518 1512
20	1,2-Dimethylnaphthalene	1522
21	4-Methylbiphenyl	1542
22	Acenaphthene	1549
23	1,2,7-Trimethylnaphthalene	1559
24	1,3,6-Trimethylnaphthalene	1572
25	1,3,7-Trimethylnaphthalene	1589
26	Dibenzofuran	1592
27	2,3,6-Trimethylnaphthalene	1614
28	2,3,5-Trimethylnaphthalene	1627
29	Trimethylnaphthalene	1654
30	Fluorene	1663
31	4-Methyldibenzofuran and 9-methylfluorene	1689
32	Methyldibenzofuran	1695
33	Methyldibenzofuran	1714
34	Xanthene	1729
35	Methyldibenzofuran	1739
37	2-Methylfluorene	1783
38	1-Methylfluorene	1796
39	Dimethyldibenzofuran	1819
40	Dimethyldibenzofuran	1839
41	Dimethyldibenzofuran	1857

Peak number	Compound	Kovats retention index
42	Phenanthrene	1890
43	Anthracene	1905
45	Trimethyldibenzofuran	1943
46	Trimethyldibenzofuran	1957
48	3-Methylphenanthrene	2006
49	2-Methylphenanthrene	2016
50	Methylanthracene	2029
51	Methylanthracene and 1-methylphenanthrene and 4,5-methylenephenanthrene	2042
53	2-Phenylnaphthalene	2079
54	Dimethylphenanthrene and 9-methylanthracene	2092
55	3,6-Dimethylphenanthrene	2113
56	Dimethylphenanthrene	2130
57	Dimethylanthracene and dimethylphenanthrene	2146
58	Dimethylanthracene and dimethylphenanthrene	2158
59	Dimethylanthracene and dimethylphenanthrene	2170
60	Fluoranthene	2185
62	Dimethylanthracene	2236
63	Pyrene	2260
65	p-Terphenyl	2299
66	peri-1,9-Benzoxanthene	2333
68	1,2-Benzofluorene	2366
69	3,4-Benzofluorene and 2,3-Benzofluorene	2386
70	4-Methylpyrene	2399
72	1-Methylpyrene	2432
73	Methylpyrene	2458
76	3,4-Benzophenanthrene	2542
77	2,13-Benzofluoranthene	2559
78	Tetraphenyl	2628

Source: Ref. 17.

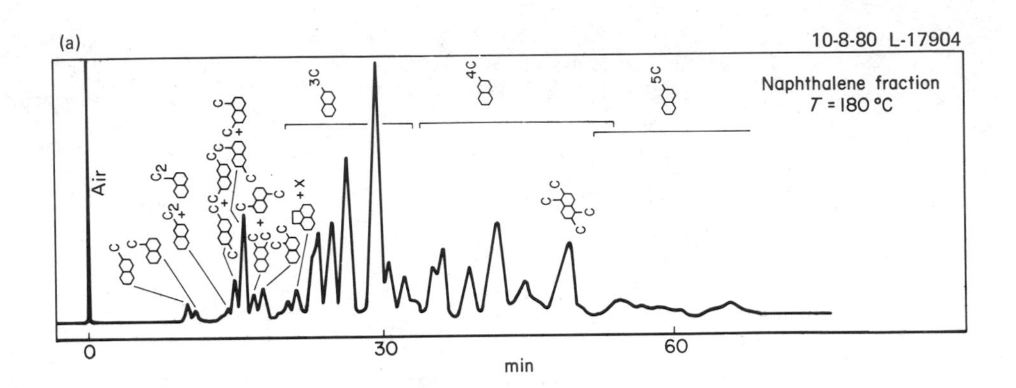
(a)
10-8-80 L-17904
Naphthalene fraction
T = 180 °C
Air
3C
4C
5C
0
30
60
min

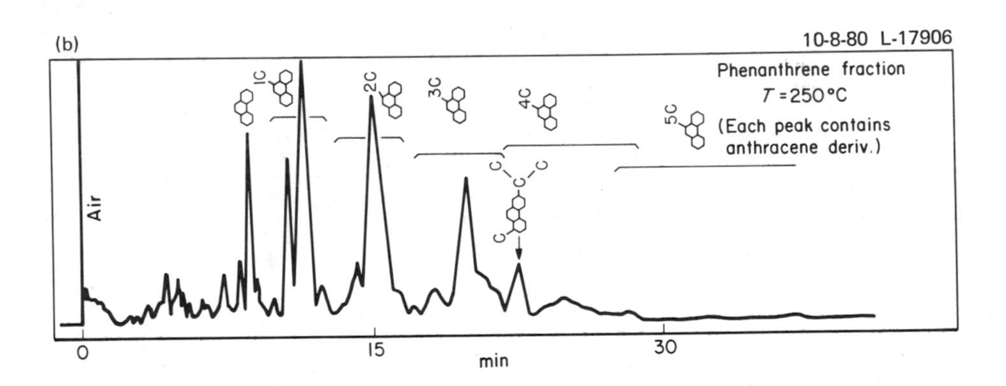
(b)
10-8-80 L-17906
Phenanthrene fraction
T = 250 °C
(Each peak contains
anthracene deriv.)
Air
1C
2C
3C
4C
5C
0
15
30
min

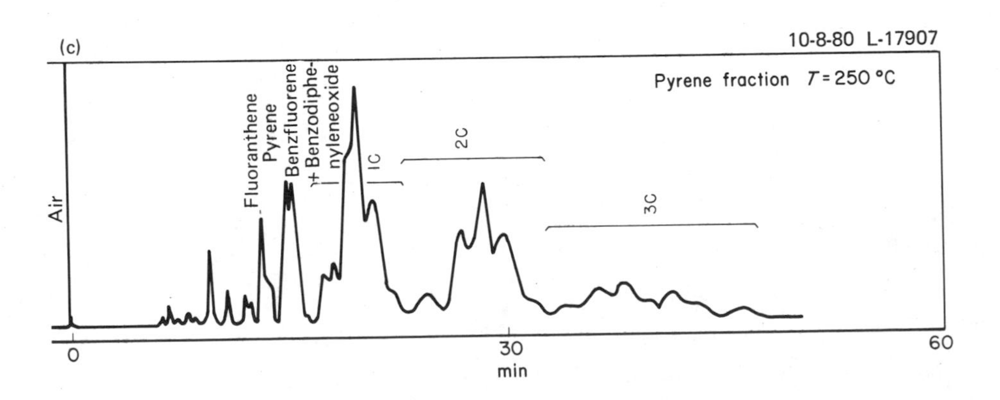
(c)
10-8-80 L-17907
Pyrene fraction T = 250 °C
Air
Fluoranthene
Pyrene
Benzfluorene
+ Benzodiphe-
nyleneoxide
1C
2C
3C
0
30
60
min

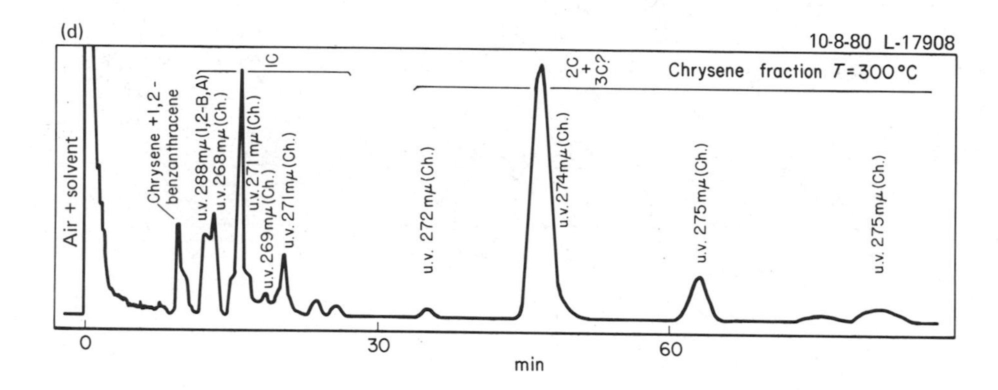

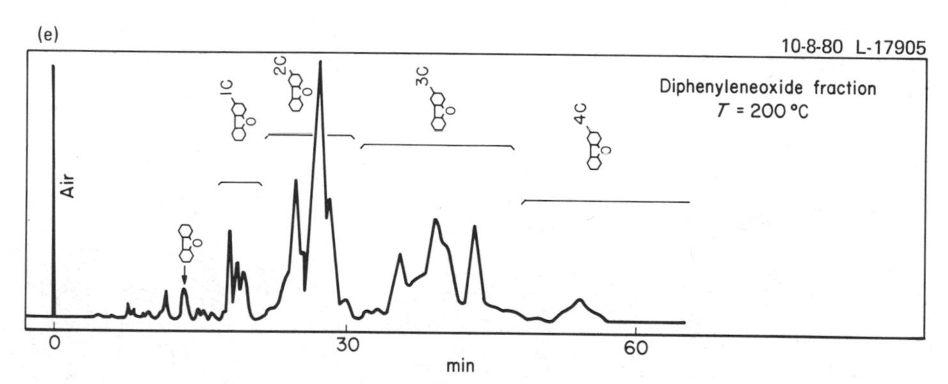

Figure 7. Packed column gas chromatograms of various aromatic fractions from a benzene extract of a Yūbari coal obtained using a 13.5 m × 4 mm i.d. glass column containing 30% high-vacuum silicone grease on 30-60 mesh firebrick. (From Ref. 21.)

coal by Sakabe and Sassa [22], and the second compound was a dimethylchrysene. The second paper describes the detailed analysis of the aromatic and other fractions from the coal extract. The analysis of the aromatic fraction was performed using preparative packed-column GC, followed by UV and IR spectroscopic identification of the isolated components. The GC profiles of the aromatic fractions are given in Fig. 7a-e. This investigation used state-of-the-art GC techniques for that time, and where specific isomeric assignments of PAH are made they are correct. Some of the compounds identified by Ōuchi and Imuta had been previously identified in coals [23-26].

Table 3. Some Aromatic Compounds Isolated from Bohemian Brown Coal

Compound	Formula	Structure
(1) Octahydro-2,2,4a,9-tetramethylpicene	$C_{26}H_{30}$	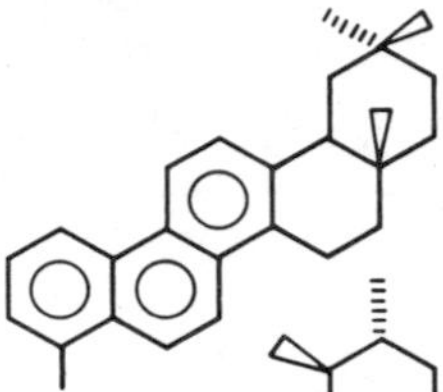
(2) Tetrahydro-1,2,9-trimethylpicene	$C_{25}H_{24}$	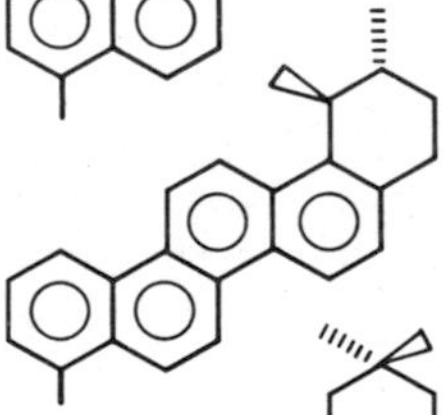
(3) Tetrahydro-2,2,9-trimethylpicene	$C_{25}H_{24}$	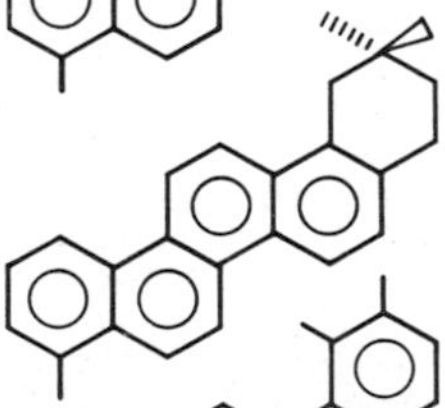
(4) 1,2,9-Trimethylpicene	$C_{25}H_{20}$	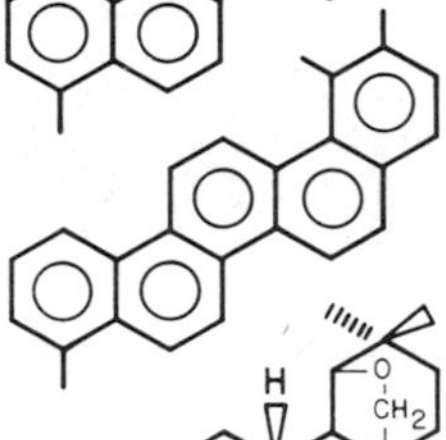
(5) 23,25-Bisnormethyl-2-desoxyallobetul-1,3,5-triene[a]	$C_{28}H_{40}O$	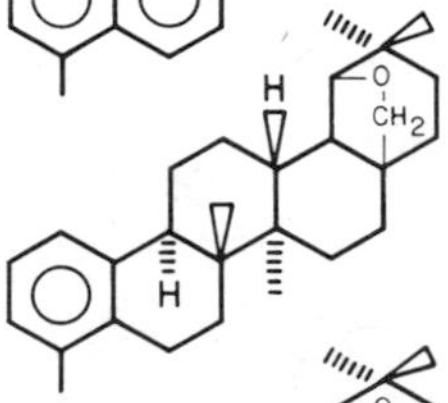
(6) 23,24,25,26,27-Pentanormethyl-2-desoxyallobetul-1,3,5,7,9,11,13-heptaene[a]	$C_{26}H_{28}O$	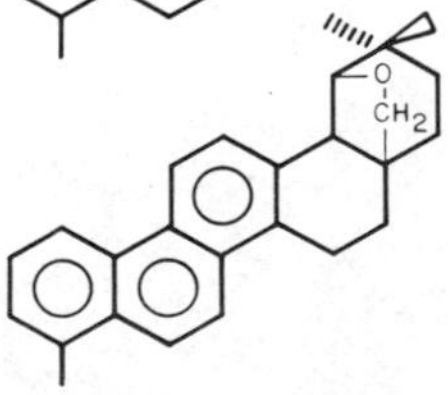

[a]Although the naming system used for these compounds is obscure, these are the names used by Jarolim and co-workers.

Source: Adapted from Refs. 27-32.

However, in this investigation many PAH were positively identified for the first time, and the following major conclusions were drawn: (1) A considerable portion of the organically bound oxygen in coal is present as furanic species. (2) Phenanthrene and alkylphenanthrenes were present in coal in 5 to 10 times greater concentrations than those of anthracene and alkylanthracenes. (3) Kata-condensed PAH were common in coal, whereas peri-condensed PAH were rare.

Jarolim and co-workers of the Institute of Organic Chemistry and Biochemistry in Prague, Czechoslovakia, using various chromatographic and crystallization methods, isolated more than 30 compounds from the benzene extract of North Bohemian brown coal, some of which were PAH [27-32]. The structures of the isolated compounds were assigned based on the results of IR, UV, nuclear magnetic resonance (NMR), and MS analyses of the pure compounds isolated. The PAH identified by these methods are listed in Table 3.

In some cases, isolation of the pure compounds and analysis by a variety of spectral techniques did not provide sufficient information to positively identify the compounds. The investigation conducted by Jarolim and coworkers is truly classical and is part of the foundation of our understanding of PAH from coal. Furthermore, their investigation has helped to elucidate the chemical reactions that occur during the coalification process that converted plant debris into coal. Based on the results of this investigation, the accompanying reaction sequences (1 and 2) have been postulated to occur during coalification. All of the reactants, intermediates, and products depicted in reaction sequences 1 and 2 were isolated from the same coal, thus providing strong evidence that the reactions occurred.

L-17893

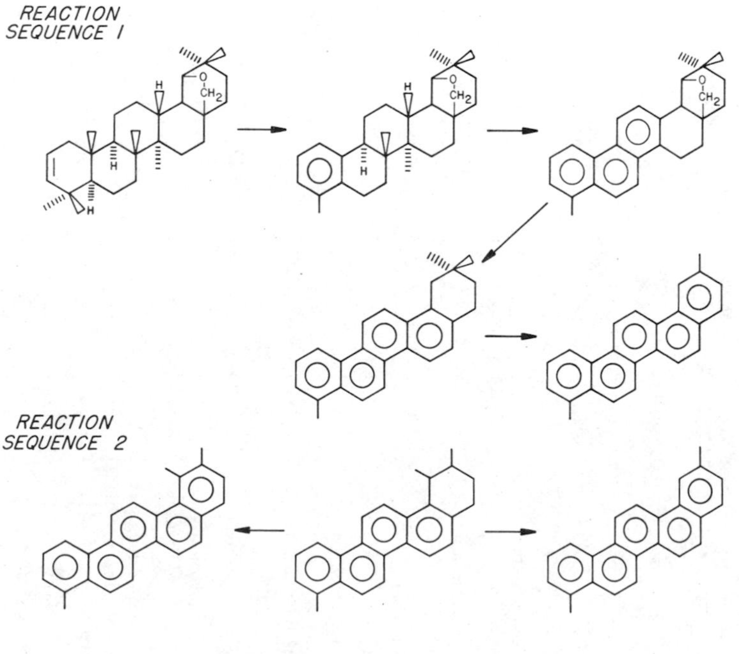

REACTION SEQUENCE 1

H O CH_2 H

REACTION SEQUENCE 2

Table 4. Carbon Number Distribution of Aromatic Hydrocarbons in Coal-Tar Pitch (Softening Point 80°-85°C) from High-Temperature Carbonization (weight percent)

Structural types including alkyl derivatives	C_nH_{2n-x} x	Compound No.[a]	Mol. wt. of 1st member	Carbon number																		
				12	13	14	15	16	17	18	19	20	21	22	23	24	25	26	27	28	29	Total
Indenes	10	188	116			0.1	0.1	0.1														0.3
Naphthalenes	12	214	128		0.1																	0.1
Acenaphthenes; biphenyls	14	251,179	154	0.3	0.3	0.4	0.3	0.2	0.2	0.1												1.8
Acenaphthylenes; fluorenes	16	250,225	152		0.4	0.4	0.4	0.1														1.3
Anthracenes; phenanthrenes	18	262,278	178			4.5	1.4	0.8	0.5	0.2	0.1											7.5
Methylenephenanthrenes; phenylnaphthalenes	20	292	190				0.5	1.2	1.7	1.1	0.7	0.4	0.2	0.2	0.1							6.1
4-Ring aromatics, peri-condensed	22	293,299	202					5.9	2.2	1.2	0.7	0.2										10.2
4-Ring aromatics, cata-condensed	24	309,298	228							3.9	1.5	0.8	0.4	0.1								6.7
Methylenechrysenes; phenylanthracenes; benzo[ghi]-fluoranthenes	26	314	226							0.3	0.6	0.8	1.1	0.7	0.3	0.2	0.1					4.1
5-Ring aromatics, peri-condensed	28	313,320	252									5.2	1.5	0.8	0.4	0.2						8.1

5-Ring aromatics, cata-condensed; methylene-perylenes	30	324	264	0.2	1.6	0.8	0.5	0.2	0.2				3.5
6-Ring aromatics (benzo-[ghi]perylenes)	32	334	276		1.9	0.7	0.6	0.4	0.2	0.1			3.9
6-Ring aromatics, peri-condensed	34	333	302				1.7	0.5	0.2	0.1	0.1		2.6
6-Ring aromatics, cata-condensed	36	335	328						0.2	0.2	0.1		0.5
7-Ring aromatics (coronenes)	36	339	300				0.2	0.2	0.1				0.5
7-Ring aromatics (dibenzo[ghi]perylenes)	38		326						0.6	0.1	0.1		0.8
7-Ring aromatics (dibenzoperylenes)	40		352								0.1	0.1	0.2
Oxygen compounds													2.3
Sulfur compounds													2.7
Nitrogen compounds													7.7
Residue (not volatile at 300°C)													29.1

[a]Compound number is assigned to the structure in H. C. Anderson and W. R. K. Wu, Properties of compounds in coal-carbonization products. *U.S. Bur. Mines Bull. No. 606* (1963).
Source: J. L. Shultz, R. A. Friedel, and A. G. Sharkey, Jr., *Fuel 44*:55 (1965).

Table 5. PAH Present in a Coal Tar Pitch as Determined by HRMS

Previously identified[a]		High-molecular-weight components detected this investigation					
			Precise mass (amu)			Possible origin of component	
Compound	m/e (amu)	Formula	Observed	Theoretical	Δ	Radical from original structure	Addition required[b]
Benzo[kl]xanthene	218.0732	$C_{20}H_{12}O$	268.0894	268.0888	.0006	$C_{16}H_8O$	C_4H_4
		$C_{24}H_{14}O$	318.1052	318.1045	.0007	$C_{16}H_6O$	$2(C_4H_4)$
Naphtho[2,1,8,7-klmn]xanthene	242.0732	$C_{22}H_{12}O$	292.0880	292.0888	.0008	$C_{18}H_8O$	C_4H_4
Benzo[b]naphtho-[2,1-d]thiophene	234.0503	$C_{20}H_{12}S$	284.0657	284.0660	.0003	$C_{16}H_8S$	C_4H_4
Dibenzo[b,def]-chrysene	302.1095	$C_{28}H_{16}$	352.1161	352.1252	.0091	$C_{24}H_{12}$	C_4H_4
5H-Benzo[b]-carbazole	217.0891	$C_{20}H_{13}N$	267.1079	267.1048	.0031	$C_{16}H_9N$	C_4H_4
		$C_{24}H_{15}N$	317.1250	317.1204	.0046	$C_{16}H_7N$	$2(C_4H_4)$
4H-Benzo[def]-carbazole	191.0735	$C_{22}H_{13}N$	291.1077	291.1048	.0029	$C_{14}H_5N$	$2(C_4H_4)$
Acridine	179.0735	$C_{21}H_{13}N$	279.1086	279.1048	.0038	$C_{13}H_5N$	$2(C_6H_5)$
Phenanthrene	178.0783	$C_{20}H_{14}$	254.1056	254.1095	.0039	$C_{14}H_9$	C_6H_5
Chrysene	228.0939	$C_{24}H_{16}$	304.1205	304.1252	.0047	$C_{18}H_{11}$	C_6H_5
Benzo[ghi]perylene	276.0939	$C_{26}H_{14}$	326.1116	326.1095	.0021	$C_{22}H_{10}$	C_4H_4
Coronene	300.0939	$C_{28}H_{14}$	350.1122	350.1095	.0027	$C_{24}H_{10}$	C_4H_4

[a]Listed in Properties of compounds in coal-carbonization products, H. C. Anderson and W. R. K. Wu, *Bur. Mines Bull. No. 606* (1963).

[b]m/e of added radical: C_4H_4 - 52; C_6H_5 - 77.

Source: Ref. 39. Reprinted with permission.

D. Early Mass Spectrometric Methods

In the period between the late 1950s to the mid-1960s, the mass spectrometry research group headed by A. G. Sharkey, Jr., at what was then the U.S. Bureau of Mines Pittsburgh Coal Research Center made many significant contributions to the development and application of MS techniques for the analysis of PAH in coal and coal products. This group pioneered the development of low-voltage mass spectrometric (LVMS) techniques employing low-resolution spectrometers for the analysis of PAH in coal and coal products. The LVMS technique was initially developed by Field and Hastings [33], as well as Lumpkin [34], for analysis of aromatics and other unsaturates in petroleum. Sharkey's group immediately applied and improved the technique for investigations of PAH and other compounds in coal products [35,36]. Previous to the development of the LVMS technique, little information concerning the types of compounds present in coal and its products could be obtained from mass spectra obtained using 70-eV ionizing energies. Analysis of samples by MS using 70-eV ionization resulted in spectra where adjacent mass peaks had approximately the same intensity, not varying even by a factor of 2. In addition, the tungsten electron emitters used in the commercially available spectrometers suffered from a phenomenon known as gas sensitivity. During repeated analyses of samples containing large amounts of aromatics, the nature of the spectra were dependent upon the kinds of samples analyzed previously. Sharkey introduced the use of rhenium filaments, which overcame these difficulties [36]. The use of low ionizing voltages, usually between 9-13 eV, results in selective ionization of aromatic and unsaturated species, producing spectra which consist of one major peak, the molecular ion, for each compound ionized. The Bureau of Mines research group applied the LVMS technique to a wide variety of coal-derived materials, resulting in what was considered at the time to be very detailed analysis of the PAH in coal-derived materials. Some representative results of an investigation of PAH in coal tar pitch are found in Table 4. Previous to the development of the LVMS and GC methods of Dupire and Botquin [16], analysis of coal-derived materials for PAH depended on laborious, time-consuming isolation techniques.

Sharkey's group also pioneered the use of high-resolution mass spectrometry (HRMS) employing 70-eV ionizing potentials as an analytical tool for the analysis of complex mixtures obtained from coal. Previously, Beynon [37] and Biemann [38] had applied HRMS to the solution of structure elucidation problems associated with pure compounds and natural products. Sharkey's group was the first to use HRMS to study PAH in coal and coal-derived materials. In one of their initial investigations of PAH in coal products by HRMS, Sharkey and co-workers studied a Koppers softening point pitch [39]. The sample was introduced to the spectrometer's ion source using a variable-temperature, direct-insertion probe. High-resolution mass spectra were acquired using an ionization energy of 70 eV, at probe temperatures of 100°, 175°, 250°, 290°, and 325°C. Some of the results obtained which include PAH and polycyclics containing N, S, and O appear in Table 5.

III. Preparative LC Separation of Coal Extracts and Coal-Derived Materials Resulting in Aromatic Fractions

As stated previously, to date, the main use of LC in analysis of PAH in coal and coal-derived materials is to provide clean aromatic fractions, free of interferences from other classes of compounds. A large number of methods exist for separation of coal-derived materials into aromatic fractions. Many stationary phases, including silica gel, alumina, silicic acid, Sephadex LH-20, and Florisil, have been used in open-column LC fractionations. Alumina and silica gel are the most commonly used phases, and the methods differ primarily in the eluting solvents and the number of stages. Mobile phases have included a very wide variety of solvents. It is beyond the scope of this chapter to review all the methods, stationary phases, and solvent sequences used; however, those most commonly used will be discussed.

Jewell et al. [40,41] and Ruberto et al. [42] developed a LC separation scheme for the fractionation of petroleum products that has also found widespread use for the separation of coal extracts and products (the SARA separation, depicted in Fig. 8). The method results in the isolation of a PAH fraction as well as other fractions, including saturates, bases, acids, and neutral nitrogen-containing species. A coal-derived heavy oil (pentane solubles) is separated by first removing the acidic components with Amberlyst-A-29 anion-exchange resins, followed by removal of the basic components

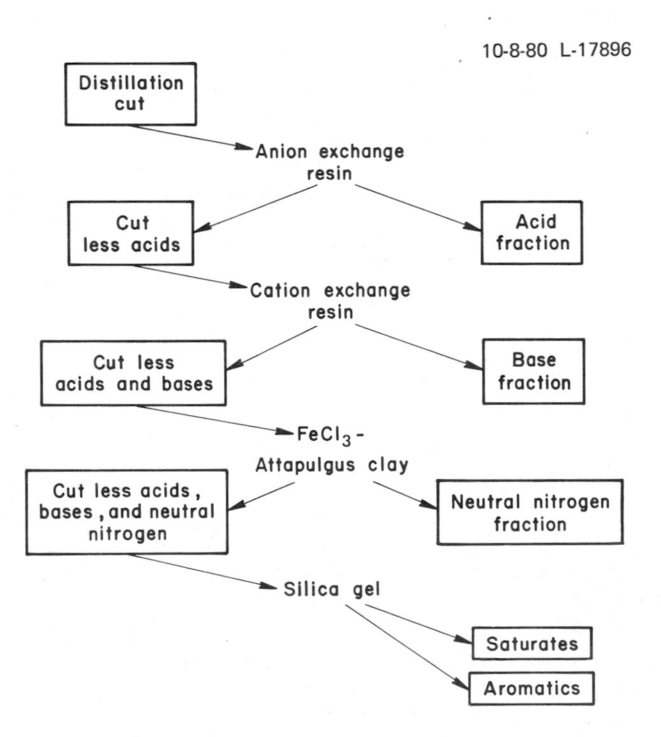

Figure 8. SARA separation scheme for coal-derived materials and petroleum distillates. (From Ref. 41; reprinted with permission.)

using Amberlyst 15 cation-exchange resins. Neutral nitrogen-containing compounds are removed by complexation with $FeCl_3$ deposited on Attapulgus clay. After removal of the acidic, basic, and neutral nitrogen-containing components of the heavy oil, the remaining hydrocarbons are separated on silica gel into saturates and aromatics. The eluting solvent in each chromatographic step is n-pentane. The aromatic fraction can then be further separated on alumina into three additional fractions: one containing aromatic hydrocarbons with one aromatic ring, a second fraction containing aromatic hydrocarbons with two and three aromatic rings, and a third fraction containing hydrocarbons with four or more aromatic rings. Similar separation schemes that were originally developed for chemical class separation of petroleum [43] have been used by Dooley's group (see Holmes et al. [44] and Sturm et al. [45]) to separate coal liquefaction products into corresponding aromatic fractions.

Another separation scheme used for the isolation of an aromatic fraction from coal liquids is that developed by Schiller and Mathiason [46]. This method has proved to be one of the easiest to perform, consisting of pre-

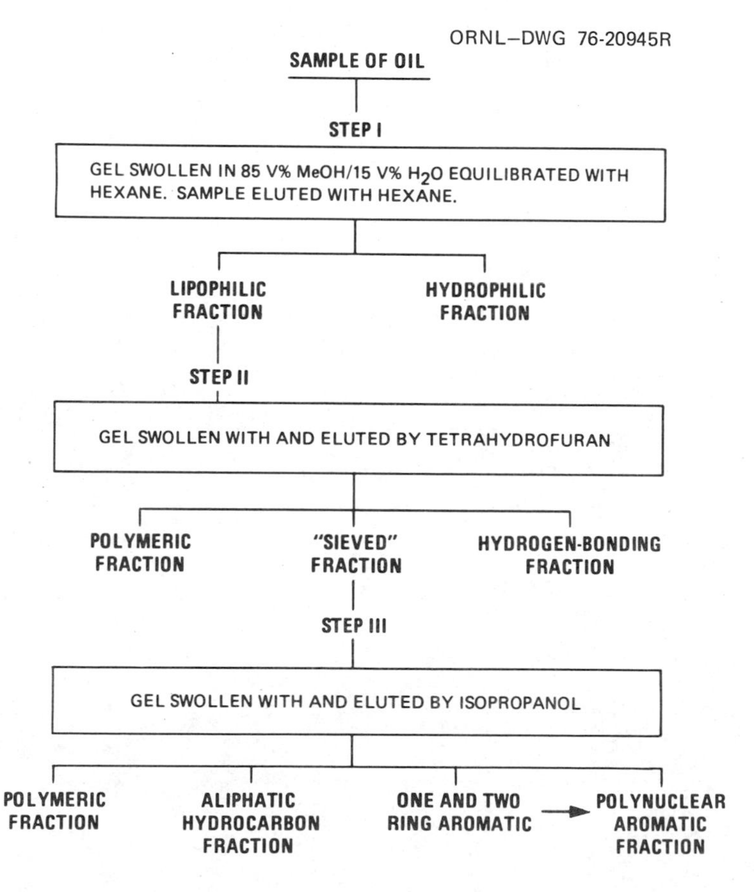

Figure 9. Sephadex LH-20 separation scheme for synthetic crude oils. (From Ref. 47.)

adsorption of the sample on neutral alumina. The alumina with sample is added to the top of a chromatography column (500 mm × 11 mm) packed with alumina, and the fractions are eluted with hexane, toluene, chloroform (two fractions), and 9:1 tetrahydrofuran-ethanol. The PAH are eluted in the toluene fraction.

Jones et al. [47] have developed a separation scheme for chemical class fractionation of syncrudes using Sephadex LH-20. The technique is used primarily to generate fractions for biological testing and is followed by chemical analysis of the biologically active fractions. The separation scheme is outlined in Fig. 9.

Novotny's group (see Schultz et al. [48]) at Indiana University has developed a solvent partitioning scheme that is followed by LC on silicic acid to obtain a PAH fraction from coal products. Solvent partitioning is carried out as described in Fig. 10. PAH fractions are further purified on silicic acid powder by elution with n-hexane. The fraction eluting between the elution volume of naphthalene and coronene (as measured by previous chromatography of the standards) is collected. This fraction is defined as the PAH fraction.

Griest et al. [49] have recently applied a separation scheme, originally developed by Snook et al. [50] and Severson et al. [51] for the separation of PAH from cigarette smoke condensate, to fractionate PAH from a coal liquefaction product into two fractions. The separation of parent and simple alkylated PAH from multialkylated PAH is achieved using gel filtration on

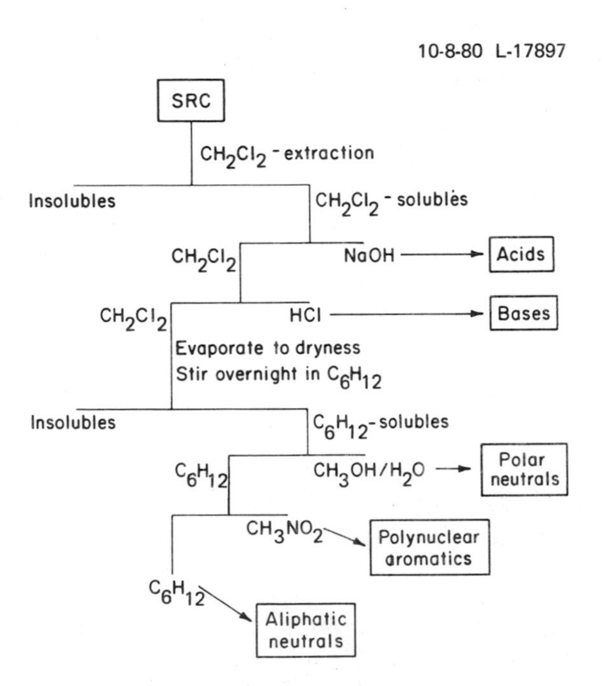

Figure 10. Solvent partition scheme used to obtain a crude PAH fraction from SRC. (From Ref. 48.)

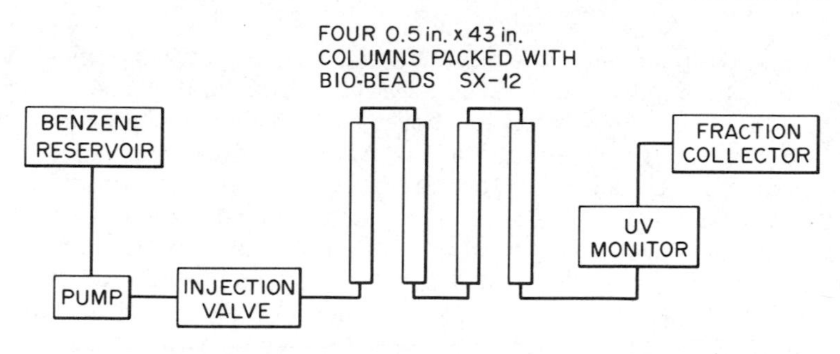

Figure 11. Bio-Beads SX-12 gel filtration apparatus used to separate parent and simple alkylated PAH from multialkylated PAH found in coal liquefaction products. (From Ref. 49; after Severson et al. [51].)

Bio-Beads SX-12, by elution with benzene. A schematic representation of the equipment used appears in Fig. 11, while a representative chromatogram appears in Fig. 12. The Bio-Beads columns are standardized using 2,3,5-trimethylnaphthalene and fluoranthene. The 2,3,5-trimethylnaphthalene/fluoranthene cut point was employed to separate the two fractions.

Other widely used separation schemes that result in clean aromatic fractions from coal and coal products include those described by Farcasiu [52] and Galya [53].

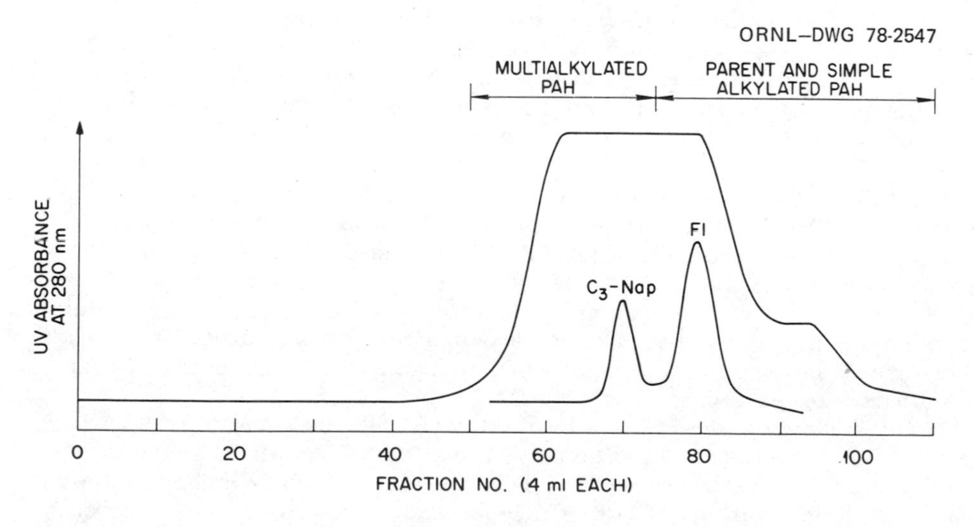

Figure 12. UV absorbance (280 nm) of SX-12 gel filtration eluate. (From Ref. 49.)

IV. GC and GC/MS Analyses of PAH from Coal and Coal Products

Two papers have recently been published concerning the use of high-resolution glass capillary GC for the separation of complex mixtures of PAH. The paper by Lee and Wright [54] is a review article, whereas Borwitzky and Schomburg [55] have described the qualitative results of an investigation of PAH from coal tar. This section will discuss high-resolution GC as well as combined high-resolution GC/MS for the separation and identification of PAH in complex mixtures from coal and coal products. Techniques for the identification of chromatographic peaks are presented, and a reliable method for quantitation of individual isomeric PAH is described. In some places, the present author's opinions are presented.

Franck [56] has estimated the number of compounds present in bituminous coal tars to be 10,000. Mixtures of similar complexity are found in coal extracts and in liquefaction and gasification products and by-products. It is therefore imperative that prior to detailed qualitative analysis by GC or GC/MS the sample be fractionated to provide a clean aromatic fraction. When necessary, the aromatic fractions should be further separated on alumina into three additional fractions—(1) mono-, (2) di- and tri-, and (3) polyaromatic ring systems. Because the aromatic fraction of coal extracts and other coal products consist of hundreds, possibly thousands, of compounds having a wide range of volatility, it is necessary that the GC columns used be of the highest possible efficiency, inertness, selectivity, and thermal stability. These criteria are best met by wall-coated open-tubular (WCOT) glass capillary column GC. When general characterization of a wide variety of PAH is desired, the columns used should have approximately 3000 plates per meter and be capable of near baseline separation of phenanthrene/anthracene, benz[a]anthracene/chrysene, and benz[e]pyrene/benz[a]pyrene. In addition, it should be possible to elute coronene from the column before or shortly (several minutes) after the column reaches its upper temperature limit during a temperature-programmed experiment. A comprehensive discussion of the effect of column dimensions and stationary phase film thickness on efficiency, capacity, and elution temperatures of PAH appears in Lee's review article [54]. Within the last several years, it has been possible to prepare and use glass capillary columns up to 350°C [57], and some even to 390°C [55], for analysis of PAH from coal and coal products. Although the columns may be thermally stable at these temperatures, some PAH may not be. Coal and coal products, particularly those made by hydrogenation of coal, contain partially hydrogenated aromatics, such as dihydrophenanthrene and dihydropyrene. The thermal instability of these compounds is well documented; when heated, they undergo loss of hydrogen and aromatize [58]. This should be taken into account when analyzing PAH from coal and coal products at temperatures above 300°C.

Many stationary phases have been employed for the analysis of PAH from coal and coal products. Traditionally, the silicone gum phases have been used most often because of their high temperature stability and the relative ease with which columns of high efficiency and inertness can be prepared. Lysyuk and Korol [59] have shown that there is little difference in the selectivity of SE-30, SE-52, OV-17, Dexsil 300, and several other phases for PAH. Nematic liquid crystal phases, which Lysyuk and Korol did not investigate,

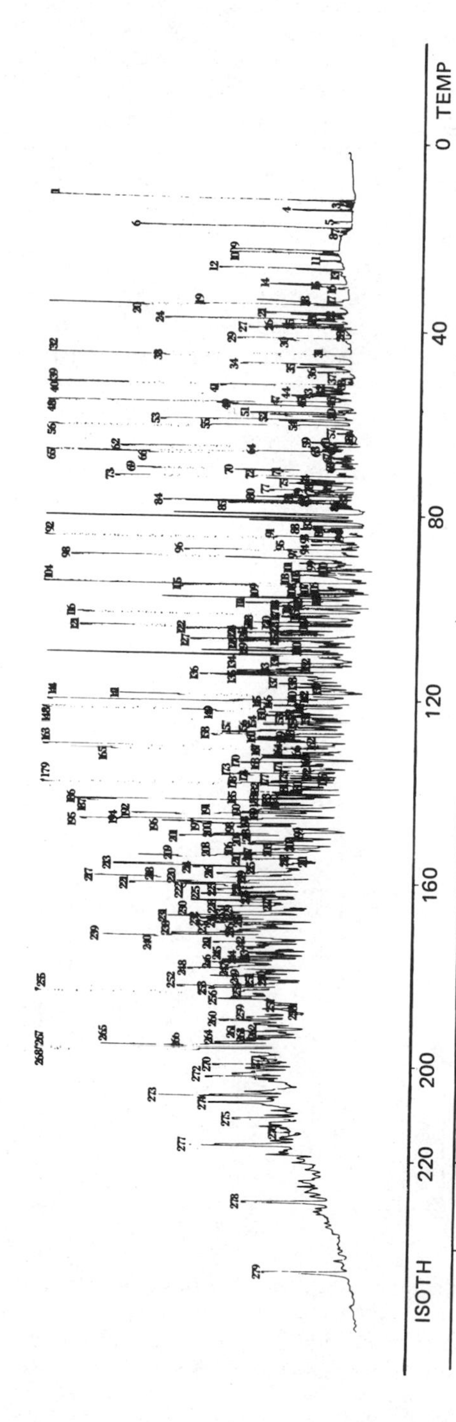

Figure 13. Total ion current monitor gas chromatogram of a coal fluid derived from western Kentucky coal. Chromatographic conditions: glass capillary column, 45 m × 0.35 mm i.d. coated with OV-101, employing a He flow rate of 3.0 ml/min and temperature programming from 0° to 220°C at 2°C/min. Chromatographic peaks are identified in Table 6. (From Ref. 60.)

show a pronounced selectivity for PAH; however the efficiencies and thermal stabilities of columns prepared with the liquid crystal phases have been poor. Recently, glass WCOT columns have been prepared with a liquid crystal phase, but there have not as yet been any published reports of their use for the separation of PAH derived from coal. Our experience is that the silicone gum phases have about the same selectivities for PAH. When columns coated with silicone gum phases are properly deactivated by silanization, they are stable to 350°C for short times, and these phases produce columns of roughly equivalent efficiencies. Therefore, selection of a stationary phase should be based on the amount of chromatographic retention information available about the PAH of interest on the phases (see Chap. 6). Chromatographic profiles of PAH from coal and coal products, obtained by several chromatographers [48, 55,60-64] using different stationary phases, appear in Figs. 13 to 20. The numbered chromatographic peaks in these figures are identified in Tables 6 to 12.

The gas chromatographic retention characteristics of many PAH on different stationary phases are available. Borwitzky and Schomburg [55] have shown that the order of elution of PAH is independent of the polarity of any single phase among the phases they studied. Columns similar to those used in the aforementioned investigations of PAH from coal and coal products are commer-

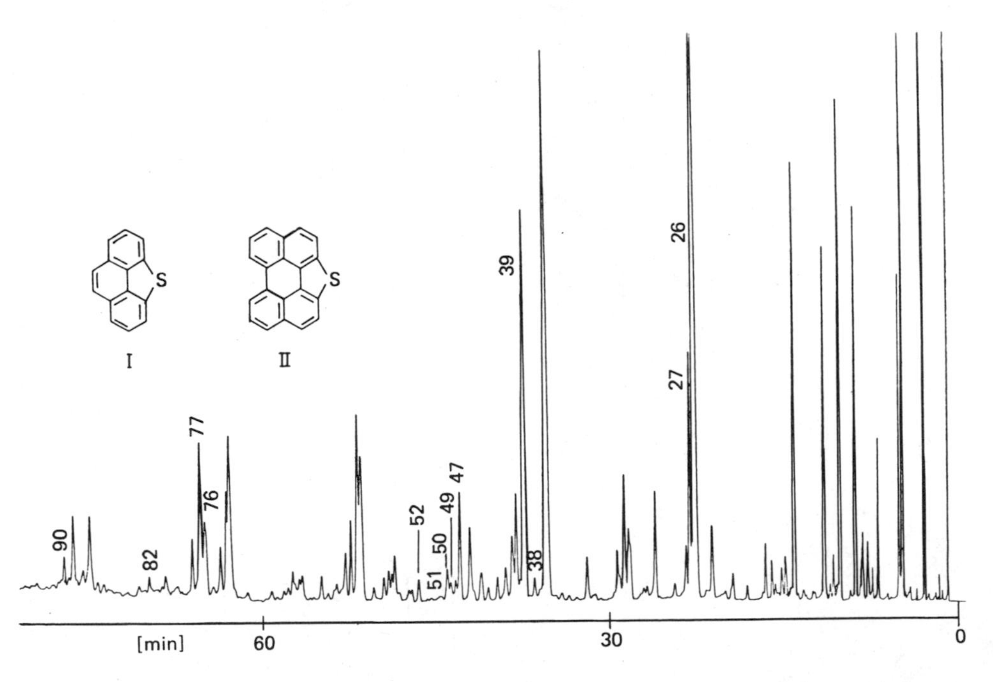

Figure 14. Capillary column gas chromatogram of a distillate from a high-temperature coal tar boiling from 200-600°C/760 atm; 18-m OV 101 capillary, 0.27 mm i.d. Temperature program: 90-300°C (2°C/min). Carrier gas: nitrogen 0.44 bars. Numbered chromatographic peaks are identified in Table 7. (From Ref. 61.)

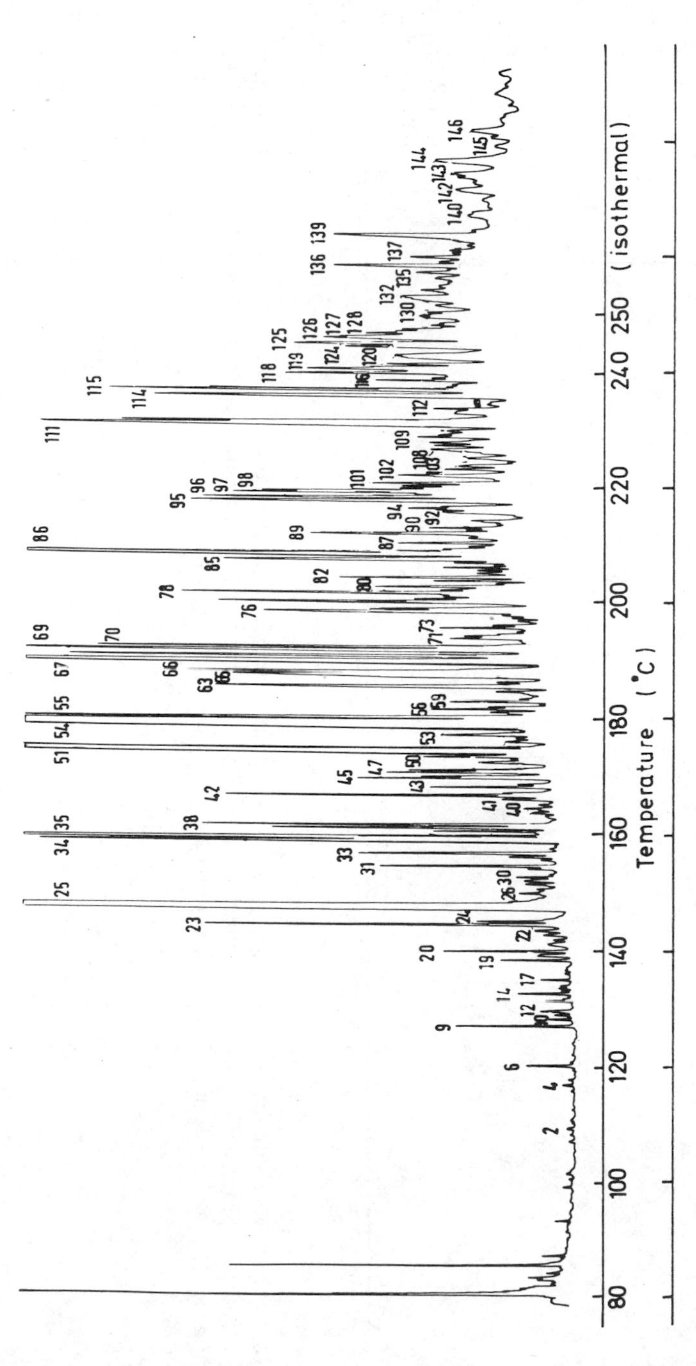

Figure 15. Capillary column gas chromatogram of the PAH fraction of SRC. Chromatographic conditions: 20 m × 0.25 mm i.d. glass capillary column coated with SE-52, employing temperature programming from 80° to 250°C at 2°C/min. The numbered chromatographic peaks are identified in Table 8. (From Ref. 48.)

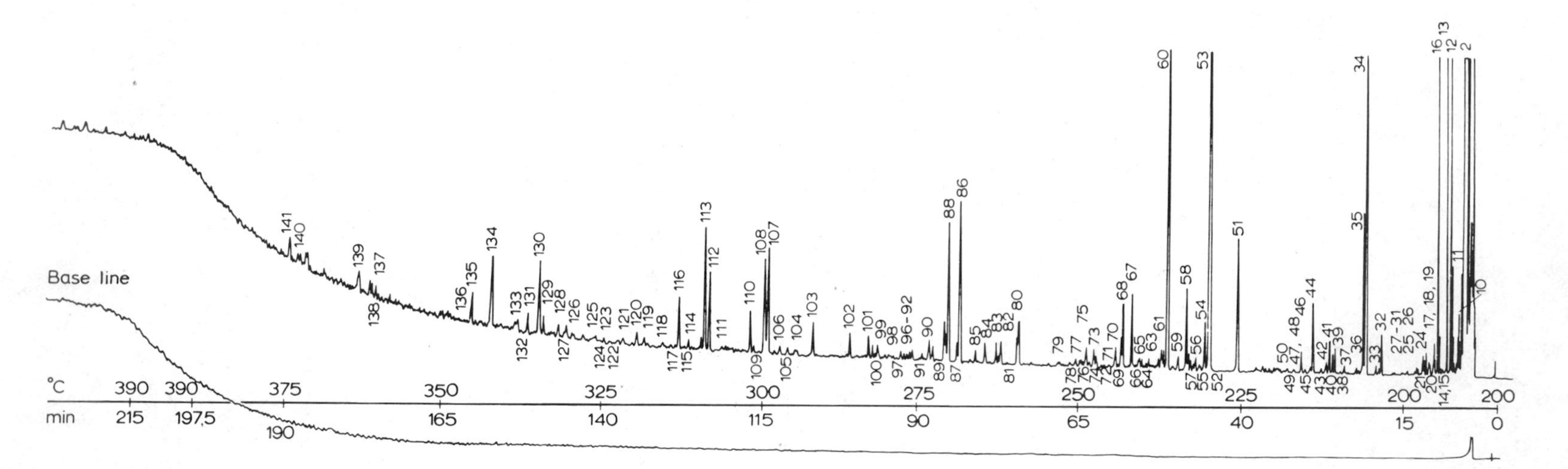

Figure 16. Capillary column gas chromatogram of coal tar. Chromatographic conditions: 92 m × 0.27 mm i.d. glass capillary column coated with Poly S 179, using H_2 carrier at 0.35 m/sec and temperature programmed from 200° to 390°C at 1°C/min. The numbered chromatographic peaks are identified in Table 9. (From Ref. 55.)

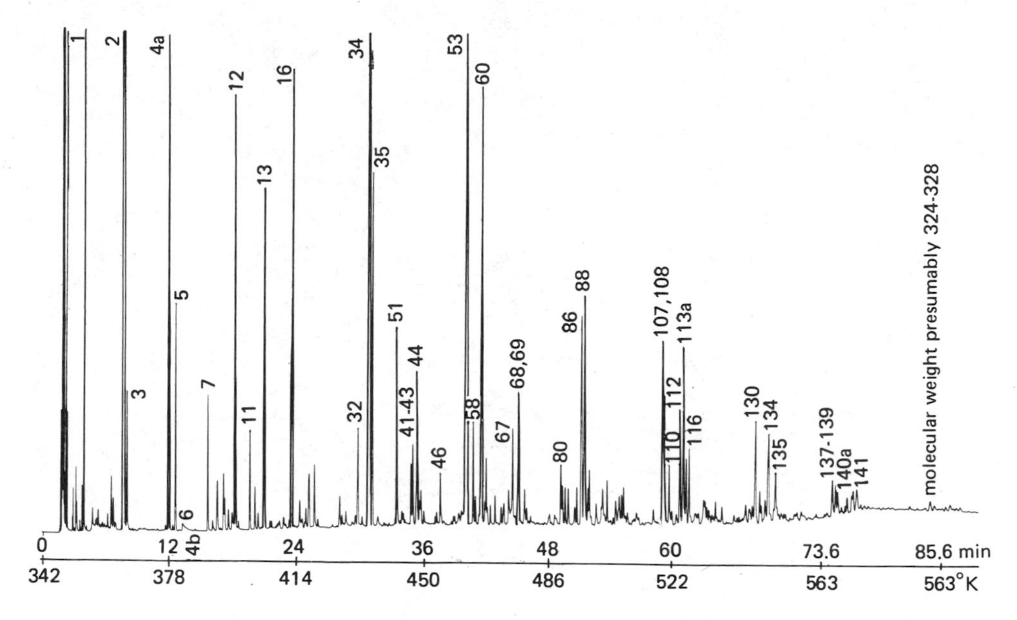

Figure 17. Capillary column gas chromatogram of coal tar. Chromatographic conditions: 48 m × 0.27 mm i.d. glass capillary column coated with SE-54, using H_2 carrier at 0.35 m/sec and temperature programmed from 69° to 290°C at 3°C/min. Numbered chromatographic peaks are identified in Table 9. (From Ref. 55.)

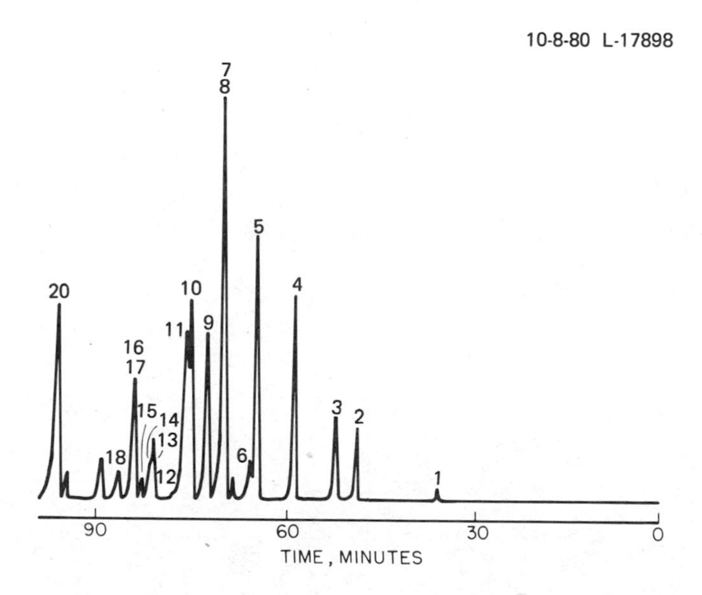

Figure 18. Capillary column gas chromatogram of a distillate from black coal tar. Chromatographic conditions: 100 m × 0.25 mm i.d. stainless steel capillary coated with Apiezon L, operated isothermally at 180°C using N_2 carrier at 0.32 ml/min. Numbered chromatographic peaks are identified in Table 10.

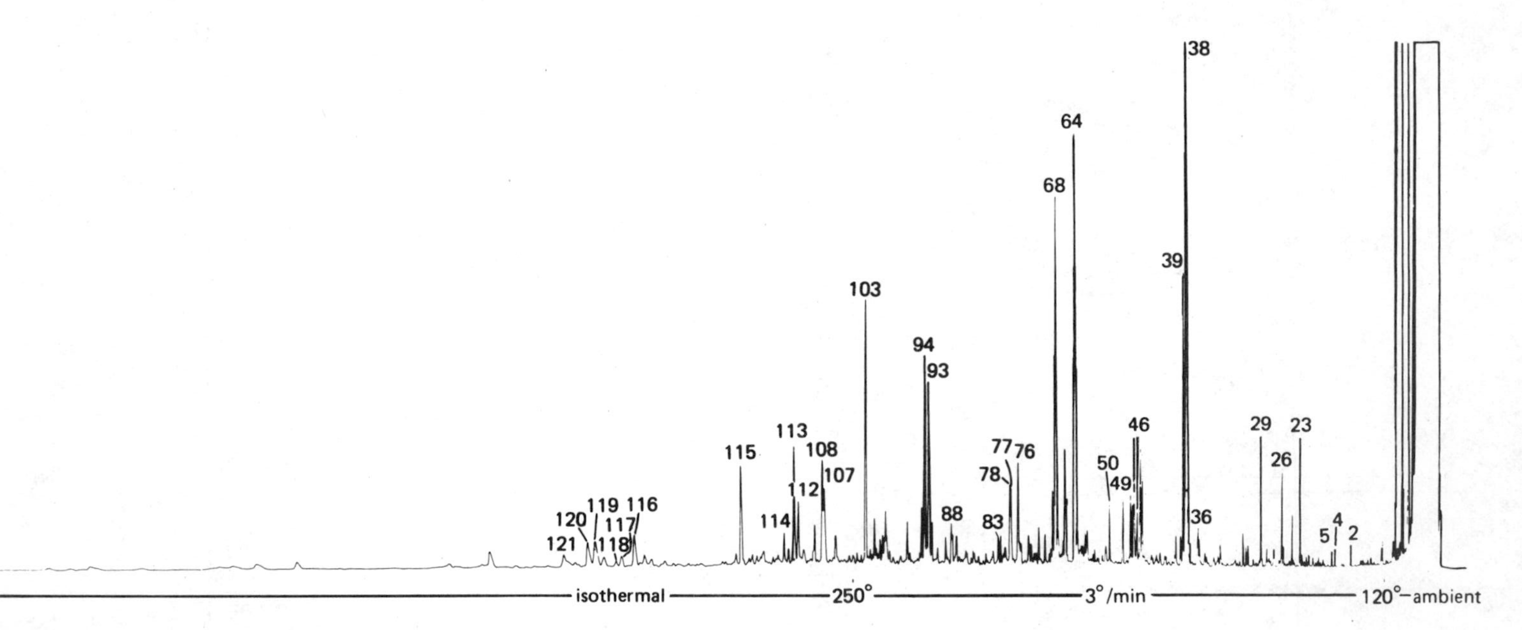

Figure 19. Capillary column gas chromatogram of the PAH fraction of particulate matter collected from the working atmosphere of a coal coking plant. Chromatographic conditions: 50 m × 0.34 mm i.d. glass capillary coated with SE-54, using H_2 carrier at 3 ml/min and temperature programmed from 115° to 260°C at 3°C/min. Numbered chromatographic peaks are identified in Table 11. (From Ref. 63.)

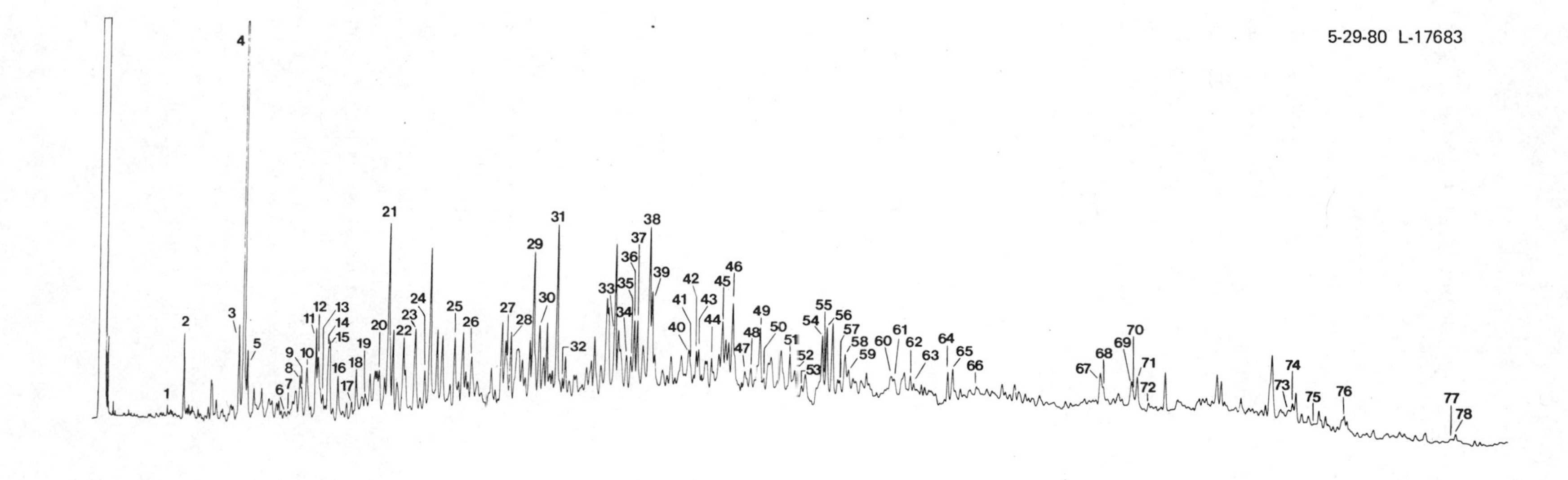

Figure 20. Capillary column gas chromatogram of the aromatic fraction of a Homestead, Kentucky, coal extract. Chromatographic conditions: 19 m × 0.29 mm i.d. glass capillary coated with SE-52, using a He flow rate of 1.9 ml/min and temperature programmed between 70° and 300°C at 2°C/min. Numbered chromatographic peaks are identified in Table 12. See Fig. 29 for the dual FID/FPD chromatograms of the same coal extract. (From Ref. 64.)

Table 6. Identification of Some Chromatographic Peaks in Fig. 13

Peak number	Compound
154	Naphthalene
187	2-Methylnaphthalene
190	1-Methylnaphthalene
213	C_2-Alkylnaphthalene
215	C_2-Alkylnaphthalene
216	C_2-Alkylnaphthalene
219	C_2-Alkylnaphthalene
220	C_2-Alkylnaphthalene
221	C_2-Alkylnaphthalene
225	C_3-Alkylnaphthalene
237	C_3-Alkylnaphthalene
243	C_3-Alkylnaphthalene
247	C_3-Alkylnaphthalene
249	Fluorene
263	Methylfluorene
264	Methylfluorene
269	Methylfluorene
274	Dimethylfluorene

Source: Ref. 60.

cially available or can be prepared in-house. When purchasing columns, caution is advised. Some vendors deactivate the glass with Carbowax, by the Aue [65] or similar methods. Although this deactivation technique produces highly inert neutral columns, it can cause several problems. Carbowax deactivated columns cannot generally be used above 280°C because Carbowax is thermally unstable above that temperature. Moreover, when the desired stationary phase is coated on the glass after Carbowax deactivation, the end result is a mixed stationary phase of unknown polarity, which depends on the amount of Carbowax remaining after deactivation. Exactly what effect this causes on the retention characteristics, order of elution, and/or PAH retention indices is unknown. However, it is known that the retention characteristics of compounds can be different when mixed phases are used [66]. Until the exact effect the presence of deactivating Carbowax has on the retention characteristics of PAH is established, it is the present author's opinion that such columns should not be used for PAH analysis. This is particularly true when the analyst will be depending on published retention characteristics of PAH measured on columns that were not deactivated with Carbowax.

Recently, several supposed disadvantages and limitations of high-resolution GC have been mentioned in the literature [67-69]. It has been stated that chrysene and triphenylene are not adequately resolved by GC. However, several years ago, Blomberg's group developed a chemically bonded, nonextractable silicone gum and nonextractable Carbowax 20M stationary phases that resolve chrysene and triphenylene very nicely [70,71]; the columns are thermally stable to 300°C with very low stationary phase bleed (see also Chap. 6). It has also been stated that gas chromatography is incapable of separating the isomeric methylchrysenes. All six methylchrysenes are separable by high-resolution glass capillary GC, although, using short SE-52 columns, two isomers show some overlap. Where the real difficulty arises is not with the separation of the methylchrysenes from one another but with the possible interferences caused by methylbenz[a]anthracenes and methyltriphenylenes. Even this problem is partially overcome by the use of derivatization techniques [72]. The interfering methylbenz[a]anthracenes are derivatized using a Diels-Alder reaction. Goeckner and Griest successfully used this technique to identify and quantitate the methylchrysenes in a coal liquefaction product [72]. The packed column chromatogram they obtained is shown in Fig. 21.

Identification of chromatographic peaks obtained from samples of coal or coal products is a complicated and controversial issue. It is necessary to differentiate between tentative and positive identification. GC peaks can be tentatively identified by using one of the following identification parameters,

Table 7. Identification of Numbered Chromatographic Peaks in Fig. 14

Peak number	Compound
26	Phenanthrene
27	Anthracene
38	Phenanthro[4,5-bcd]thiophene
39	Pyrene
47	MW = 230
49	MW = 232
50	MW = 230
51	MW = 230
52	MW = 232
76	Benzo[e]pyrene
77	Benzo[a]pyrene
82	MW = 264
90	Anthanthrene (MW = 276)

Table 8. Components Identified from Fig. 15 in the PAH Fraction of Solvent-Refined Coal (SRC)

Peak number	Molecular weight	Formula	Concentration (parts per thousand)	Possible compound types	Peak number	Molecular weight	Formula	Concentration (parts per thousand)	Possible compound types
1	128	$C_{10}H_8$		Naphthalene	28	198	$C_{13}H_{10}S$	0.27	Methyldibenzothiophene
2	142	$C_{11}H_{10}$		2-Methylnaphthalene	29	198	$C_{13}H_{10}S$	0.32	Methyldibenzothiophene
3	142	$C_{11}H_{10}$		1-Methylnaphthalene	30	194	$C_{15}H_{14}$	0.25	Ethylfluorene[b]
4	154	$C_{12}H_{10}$	0.10	Biphenyl	31	198	$C_{13}H_{14}$	0.88	Methyldibenzothiophene
5	156	$C_{12}H_{12}$		2-Ethylnaphthalene	32	204	$C_{16}H_{12}$	0.30	1-Phenylnaphthalene
6	156	$C_{12}H_{12}$		2,6-Dimethylnaphthalene	33	198	$C_{13}H_{10}S$	1.14	Methyldibenzothiophene
	170	$C_{13}H_{14}$		and propylnaphthalene	34	194	$C_{15}H_{12}$	4.30	3-Methylphenanthrene
7	154	$C_{12}H_{10}$		Acenaphthene	35	192	$C_{15}H_{12}$	6.40	2-Methylphenanthrene
8	168	$C_{12}H_{10}O$	0.28	Dibenzofuran	36	190	$C_{15}H_{12}$	0.69	4H-Cyclopenta[def]-
9	168	$C_{13}H_{10}$	0.56	Fluorene					phenanthrene
10	168	$C_{13}H_{10}$	0.16	Allylnaphthalene[c]	37	192	$C_{15}H_{12}$	1.55	9-Methylphenanthrene
11	168	$C_{13}H_{12}$	0.10	Allylnaphthalene[c]	38	192	$C_{15}H_{12}$	1.79	1-Methylphenanthrene
12	180	$C_{14}H_{12}$	0.27	9-Methylfluorene	39	212	$C_{14}H_{12}S$	0.17	Ethyldibenzothiophene[b]
13	182	$C_{13}H_{10}O$	0.21	Methyldibenzofuran					or ethylnaphthothiophene[b]
14	182	$C_{13}H_{10}O$	0.28	Methyldibenzofuran	40	212	$C_{14}H_{12}S$	0.24	Ethyldibenzothiophene[b]
15	182	$C_{13}H_{10}$	0.08	Methyldibenzofuran					or ethylnaphthothiophene[b]
16	182	$C_{14}H_{14}$	0.09	1,1-Diphenylethane	41	212	$C_{14}H_{12}S$	0.28	Ethyldibenzothiophene[b]

17	182		0.15	2,2'-Dimethylbiphenyl,					or ethylnaphthothiophene[b]
				2,3'-dimethylbiphenyl,	42	212	$C_{14}H_{12}S$	2.11	Ethyldibenzothiophene[b]
				2,4'-dimethylbiphenyl,					or ethylnaphthothiophene[b]
				2-methyldiphenylmethane or	43	206	$C_{16}H_{14}$	0.58	Ethylanthracene[b]
				4-methyldiphenylmethane					or ethylphenanthrene[b]
18	180	$C_{14}H_{12}$	0.34	9,10-Dihydroanthracene	44	212	$C_{14}H_{12}S$	0.22	Ethyldibenzothiophene[b]
19	180	$C_{14}H_{12}$	0.30	2-Methylfluorene					or ethylnaphthothiophene[b]
20	180	$C_{14}H_{12}$	0.63	1-Methylfluorene	45	206	$C_{16}H_{14}$	0.99	9-Ethylphenanthrene
21	182	$C_{14}H_{14}$	0.31	3,4'-Dimethylbiphenyl	46	206	$C_{16}H_{14}$	0.64	Ethylphenanthrene
22	196	$C_{14}H_{12}$	0.29	Methylxanthene	47	206	$C_{16}H_{14}$	0.89	3,6-Dimethylphenanthrene
23	184	$C_{12}H_8S$	2.12	Dibenzothiophene	48	206	$C_{16}H_{14}$	0.80	2,7-Dimethylphenanthrene
24	182	$C_{14}H_{14}$	0.73	Ethylbiphenyl[b]	49	206	$C_{16}H_{14}$	0.65	Ethylanthracene[b]
25	178	$C_{14}H_{10}$	27.00	Phenanthrene					or ethylphenanthrene[b]
26	194	$C_{15}H_{14}$	0.47	Ethylfluorene[b]	50	206	$C_{16}H_{14}$	0.96	Ethylanthracene[b]
27	194	$C_{15}H_{14}$	0.18	Ethylfluorene[b]					or ethylphenanthrene[b]
51	202	$C_{16}H_{10}$	17.20	Fluoranthene	99	242	$C_{19}H_{14}$	0.67	6-Methylchrysene
52	205	$C_{15}H_{11}N$	0.32						or 4-methylchrysene
53	208	$C_{16}H_{16}$	0.99	2-Phenyl-1,2,3,4-	100	254	$C_{20}H_{14}$	0.53	Phenylphenanthrene
				tetrahydronaphthalene	101	256	$C_{20}H_{16}$	0.81	Ethyl-$C_{18}H_{12}$[b,e]
54	202	$C_{16}H_{16}$	32.00	Pyrene	102	254	$C_{20}H_{14}$		2,2'-Binaphthyl
55	204	$C_{16}H_{12}$	3.60	Dibenzoheptafulvene	103	254	$C_{20}H_{14}$	1.06	Phenylphenanthrene

(continued)

Table 8 (continued)

Peak number	Molecular weight	Formula	Concentration (parts per thousand)	Possible compound types	Peak number	Molecular weight	Formula	Concentration (parts per thousand)	Possible compound types
56	220	$C_{17}H_{16}$	0.60	Propylphenanthrene[a] or	104	256	$C_{20}H_{16}$	0.46	Ethyl-$C_{18}H_{12}$[b,e]
				propylanthracene[a]	105	256	$C_{20}H_{16}$	0.46	Ethyl-$C_{18}H_{12}$[b,e]
57	218	$C_{17}H_{14}$	0.26	Ethyl-$C_{16}H_{12}$[b,d]	106	256	$C_{20}H_{16}$	0.52	Ethyl-$C_{18}H_{12}$[b,e]
58	218	$C_{17}H_{14}$	0.02	Ethyl-$C_{16}H_{12}$[b,d]	107	256	$C_{20}H_{16}$	1.30	Ethyl-$C_{18}H_{12}$[b,e]
59	218	$C_{17}H_{14}$	0.51	Ethyl-$C_{16}H_{12}$[b,d]	108	256	$C_{20}H_{16}$	0.63	Ethyl-$C_{18}H_{12}$[b,e]
60	220	$C_{17}H_{16}$	0.23	9-Methyl-10-	109	256	$C_{20}H_{16}$	0.91	Ethyl-$C_{18}H_{12}$[b,e]
				ethylphenanthrene	110	256	$C_{20}H_{16}$	0.64	Ethyl-$C_{18}H_{12}$[b,e]
61	218	$C_{17}H_{14}$		Ethyl-$C_{16}H_{12}$[b,d]	111	252	$C_{20}H_{12}$	5.70	Benzo[j]fluoranthene
62	218	$C_{17}H_{14}$	0.52	Ethyl-$C_{16}H_{12}$[b,d]	112	252	$C_{20}H_{12}$	0.75	Benzofluoranthene
63	216	$C_{17}H_{12}$	2.88	Benzo[a]fluorene	113	268	$C_{21}H_{16}$	0.40	Methylbinaphthylene
64	220	$C_{17}H_{16}$	0.32	Propylphenanthrene[a] or	114	252	$C_{20}H_{12}$	3.60	Benzo[e]pyrene
				propylanthracene[a]	115	252	$C_{20}H_{12}$	3.40	Benzo[a]pyrene
65	216	$C_{17}H_{12}$	4.20	Benzo[b]fluorene	116	252	$C_{20}H_{12}$	0.91	Perylene
66	216	$C_{17}H_{12}$	2.98	Benzofluorene	117	266	$C_{21}H_{14}$	0.48	Methyl-$C_{20}H_{12}$[f]
67	216	$C_{17}H_{12}$	8.10	4-Methylpyrene	118	266+	$C_{21}H_{14}$	1.82	Methyl-$C_{20}H_{12}$[f]
68	218	$C_{17}H_{14}$	3.40	Propylphenanthrene[a] or		280	$C_{22}H_{16}$		Ethyl- $C_{20}H_{12}$[b,f]
				propylanthracene[a]	119	266	$C_{21}H_{14}$	2.31	Methyl-$C_{20}H_{12}$[f]

69	216	$C_{17}H_{12}$	4.30	2-Methylpyrene	120	266	$C_{21}H_{14}$	1.96	Methyl-$C_{20}H_{12}$[f]
70	216	$C_{17}H_{12}$	2.72	1-Methylpyrene	121	266	$C_{21}H_{14}$	1.11	Methyl-$C_{20}H_{12}$[f]
71	232	$C_{18}H_{16}$	0.86	Propyl-$C_{16}H_{12}$[a,d]	122	266	$C_{21}H_{14}$	1.77	Methyl-$C_{20}H_{12}$[f]
72	232	$C_{18}H_{16}$	0.67	Propyl-$C_{16}H_{12}$[a,d]	123	266	$C_{21}H_{14}$		Methyl-$C_{20}H_{12}$[f]
73	232	$C_{18}H_{16}$	0.61	Propyl-$C_{16}H_{12}$[a,d]	124	266	$C_{21}H_{14}$	2.47	Methyl-$C_{20}H_{12}$[f]
74	230	$C_{18}H_{14}$	0.17	5,12-Dihydronaphthacene	125	268	$C_{21}H_{16}$	0.62	Methylcholanthrene
75	230	$C_{18}H_{14}$	0.24	Ethylpyrene[b] or ethyl-	126	266	$C_{21}H_{14}$	0.60	Methyl-$C_{20}H_{12}$[f]
				fluoranthene[b]	127	266	$C_{21}H_{14}$	0.54	Methyl-$C_{20}H_{12}$[f]
76	230	$C_{18}H_{14}$	2.61	Ethylpyrene[b] or ethyl-	128	282	$C_{22}H_{18}$	0.44	
				fluoranthene[b]	129	280	$C_{22}H_{16}$		Ethyl-$C_{20}H_{12}$[b,f]
77	230	$C_{18}H_{14}$	1.09	1-Ethylpyrene	130	280	$C_{22}H_{16}$		Ethyl-$C_{20}H_{12}$[b,f]
78	230	$C_{18}H_{14}$	2.62	Ethylpyrene[b] or ethyl-	131	280	$C_{22}H_{16}$	1.49	Ethyl-$C_{20}H_{12}$[b,f]
				fluoranthene[b]	132	280	$C_{22}H_{16}$	0.67	Ethyl-$C_{20}H_{12}$[b,f]
79	230	$C_{18}H_{14}$	0.66	Ethylpyrene[b] or ethyl-	133	280	$C_{22}H_{16}$	0.39	Ethyl-$C_{20}H_{12}$[b,f]
				fluoranthene[b]	134	280	$C_{22}H_{16}$	0.46	Ethyl-$C_{20}H_{12}$[b,f]
80	234	$C_{16}H_{10}S$	0.91	Benzo[b]naphtho[2,1-d]	135	280	$C_{22}H_{16}$	0.52	Ethyl-$C_{20}H_{12}$[b,f]
				thiophene	136	280	$C_{22}H_{14}$	1.12	Dibenzanthracene
81	234	$C_{16}H_{10}S$	0.50	Benzonaphthothiophene	137	268	$C_{22}H_{12}$	0.40	—[g]
82	232	$C_{18}H_{16}$	1.10	Propyl-$C_{16}H_{12}$[a,d]	138	278	$C_{22}H_{14}$		Benzo[b]chrysene
83	230	$C_{16}H_{10}S$	0.32	Benzonaphthothiophene	139	278	$C_{22}H_{14}$	2.23	Picene
84	232	$C_{18}H_{16}$	0.31	Propyl-$C_{16}H_{12}$[a,d]	140	276	$C_{22}H_{12}$	0.67	—[g]
85	228	$C_{18}H_{12}$	2.43	Benz[a]anthracene	141	276	$C_{22}H_{12}$	0.279	Methyl-$C_{22}H_{14}$[h]

(continued)

Table 8 (continued)

Peak number	Molecular weight	Formula	Concentration (parts per thousand)	Possible compound types	Peak number	Molecular weight	Formula	Concentration (parts per thousand)	Possible compound types
86	228	$C_{18}H_{12}$	9.60	Triphenylene and chrysene	142	292	$C_{23}H_{16}$	0.89	Methyl-$C_{22}H_{14}$[h]
87	248	$C_{17}H_{12}S$	0.61	Methylnaphthobenzothiophene	143	292	$C_{23}H_{16}$	0.84	Methyl-$C_{22}H_{14}$[h]
88	242	$C_{19}H_{14}$	0.24	Propylpyrene[a] or	144	292	$C_{23}H_{16}$	1.81	Methyl-$C_{22}H_{14}$[h]
				propylfluoranthene[a]	145	292	$C_{23}H_{16}$		Methyl-$C_{22}H_{14}$[h]
89	242	$C_{19}H_{14}$	1.87	2-Methyltriphenylene	146	290	$C_{23}H_{14}$		Methyl-$C_{22}H_{12}$[h]
90	258	$C_{20}H_{18}$	0.56						
91	248	$C_{17}H_{12}S$	0.43	Methylnaphthobenzothiophene					
92	242	$C_{19}H_{14}$	0.78	11-Methylbenz[a]anthracene					
93	242	$C_{19}H_{14}$	0.43	2-Methylbenz[a]anthracene					
94	242	$C_{19}H_{14}$	0.63	1-Methylbenz[a]anthracene					

95	242	$C_{19}H_{14}$	3.10	1-Methyltriphenylene
96	242	$C_{19}H_{14}$	2.21	3-Methylchrysene
97	242	$C_{19}H_{14}$	1.91	12-Methylbenz[a]anthracene
98	242	$C_{19}H_{14}$	1.49	4-Methylbenz[a]anthracene or 5-methylchrysene

[a]May also be trimethyl or ethylmethyl.
[b]May also be dimethyl.
[c]Good visual fit was shown to standard spectrum of allylnaphthalene. However, methylacenaphthene cannot be ruled out in the absence of standard spectrum.
[d]Possible isomers of $C_{16}H_{12}$ include the following: 4H-cyclopenta[def]phenanthrene, benzacenaphthene, or phenylnaphthalene.
[e]Possible isomers of $C_{18}H_{12}$ include: chrysene, benz[a]anthracene, and triphenylene.
[f]Possible isomers of $C_{20}H_{12}$ include: benzofluoranthene, benzo[e]pyrene, benzo[a]pyrene, and perylene.
[g]Compounds with MW 276 could be isomers of the following: indenopyrene, indenofluoranthene, aceperylene, phenanthrofluorene, acenaphthacenaphthylene, dibenzofluoranthene, benzoaceperylene, or acefluoranthylene.
[h]Possible isomers of $C_{22}H_{14}$ include: dibenzanthracene, benzo[b]chrysene, and picene.
Source: Ref. 48.

Table 9. Assignment of Peaks from Figs. 16 and 17 Based on GC/MS and Comparison of Retention Times on Poly S 179 and OV-7

Peak number	Molecular weight	Assignment	Remarks
1	116	Indene	
2	128	Naphthalene	
3	134	Benzo[b]thiophene	
4a	142	2-Methylnaphthalene	
4b	128	Azulene	
5	142	1-Methylnaphthalene	
6	129	Quinoline	
7	154	Biphenyl	
8	129	Isoquinoline	
9	156	Dimethylnaphthalene	
10	168	Methylbiphenyl	
11	154	Acenaphthene	
12	152	Acenaphthylene	
13	168	Dibenzofuran	
14	131	Methylindole	Or methylindolizine
15a	168	Methylbiphenyl	
15b	180	?	No diphenylethene
16	166	Fluorene	
17	153	Azaacenaphthylene	Or naphthonitrile
18	182	Methyldibenzofuran	
19	182	Methyldibenzofuran	
20a	166	Methylacenaphthylene	Or naphthocyclopentadiene
20b	183	?	
21	166	Methylacenaphthylene	Or naphthocyclopentadiene
22	153	Azaacenaphthylene	Or naphthonitrile
23	180	1,1'-Diphenylethene?	
24a	180	Methylfluorene	
24b	196	?	
25a	180	Methylfluorene	

Peak number	Molecular weight	Assignment	Remarks
25b	196	?	
26	180	9(?)-Methylfluorene	
27	196	?	
28	167	Azafluorene	Or methylcyanonaphthalene
29	180	Methylfluorene	
30	167	Azafluorene	Or methylcyanonaphthalene
31	144	1(?)-Naphthol	
32	184	Dibenzothiophene	
33	184	Naphthothiophene	
34	178	Phenanthrene	
35	178	Anthracene	
36	198	Methyldibenzothiophene	
37	179	Benzo[h]quinoline	
38	184	Naphthothiophene	
39	192	Methylphenanthrene, -anthracene	
40	179	Acridine	
41	192	Methylphenanthrene, -anthracene	
42	192	Methylphenanthrene, anthracene	
43	192	Methylphenanthrenc, -anthracene	
44	190	4H-Cyclopenta[def]-phenanthrene	
45	179	Benzo[f]quinoline	
46	204	2-Phenylnaphthalene	
47	179	Azaphenanthrene, anthracene	
48	206	Dimethylphenanthrene, -anthracene	

(continued)

Table 9 (continued)

Peak number	Molec-ular weight	Assignment	Remarks
49	193	Phenylindole	Or methylbenzoquinoline
50	206	Dimethylphenanthrene, -anthracene	
51	167	Carbazole	
52	181	Methylcarbazole	
53	202	Fluoranthene	
54	218	Benzonaphthofuran	
55	206	9,10-Dimethylanthracene	
56	204	Dihydropyrene?	No 1-phenylnaphthalene
57	208	Thiopheno[def]phenanthrene	Lee and Hites [37]
58	202	Acephenanthrylene-, anthrylene	
59a	218	Benzonaphthofuran	
59b	167	Naphthopyrrole	
60	202	Pyrene	
61	216	Methylfluoranthene, -pyrene	
62	216	Methylfluoranthene, -pyrene	
63	218	Benzonaphthofuran	
64	204	4H-Cyclopenta[def]phen-anthren-4-one	According to Gold
65	216	Methylfluoranthene, -pyrene	
66	216	Methylfluoranthene, -pyrene	
67	216	Benzo[a]fluorene	
68	216	Benzo[b]fluorene	
69	216	Benzo[c]fluorene	
70	216	Methylfluoranthene,	
71	232	Tetrahydo derivative of benzo[c]phenanthrene, benzo[a]anthracene, chrysene, triphenylene	

Peak number	Molecular weight	Assignment	Remarks
72	216	Methylfluoranthene, -pyrene	
73	203	Azafluoranthene, -pyrene	Or phenanthrene or anthracene nitrile
74	230	Dimethylfluoranthene, -pyrene	Dihydro derivatives of benzo[c]phenanthrene, benzo[a]anthracene, chrysene, triphenylene possible
75	203	Azafluoranthene, -pyrene	
76	230	As peak 74	
77	217	Azabenzofluorene	No benzocarbazole
78	230	As peak 74	
79	203	Azafluoranthene, -pyrene	Or phenanthrene or anthracene nitrile
80	228	Benzo[c]phenanthrene	
81	254	1,2'-Binaphthyl	Or phenylphenanthrene, -anthracene
82	191	Pyrrolo[def]phenanthrene	
83a	226	Benzo[mno]fluoranthene	Or cyclopenta[cd]pyrene
83b	234	Benzonaphthothiophene	
84	229	Aza derivative of benzo[c]phenanthrene (presumably)	
85	234	Benzonaphthothiophene	
86	228	Benzo[a]anthracene	
87	228	?	
88	228	Chrysene	
89	228	Triphenylene	
90	229	Aza derivative of benzo[c]phenanthrene, benzo[a]anthracene, chrysene, triphenylene	
91	242	Methyl derivative of benzo[c]anthracene or isomers	
92	242	Methyl derivative of benzo[c]anthracene or isomers	

(continued)

Table 9 (continued)

Peak number	Molecular weight	Assignment	Remarks
93	242	Methyl derivative of benzo[c]anthracene or isomers	
94	242	Methyl derivative of benzo[c]anthracene or isomers	
95	234	Phenylphenanthrene, -anthracene, no binaphthyl	
96	242	Methyl derivative of benzo[a]anthracene or isomers	
97	242	Methyl derivative of benzo[a]anthracene or isomers	
98a	242	Methyl derivative of benzo[a]anthracene or isomers	
98b	254	2,2'-Binaphthyl	
99	240	4H-Cyclopenta[def]chrysene or 4H-cyclopenta[def]triphenylene	
100	240	4H-Benzo[fg]pyrene or dibenzo[def,i]fluorene	
101	240	As peak 100	
102	230	Benzanthrone	
103	217	Benzocarbazole	
104	217	Benzocarbazole	
105a	217	Benzocarbazole	
105b	268	Dinaphthofuran	
106	217	Naphthoindole	
107	252	Benzo[b]fluoranthene	
108a	252	Benzo[k]fluoranthene	
108b	252	Benzo[j]fluoranthene	
109	253	Azabenzofluoranthene or benzo[c]phenanthrene, benzo[a]anthracene, chrysene, triphenylene nitrile	
110	252	Isomer of benzofluoranthene, -pyrene	
111	266	Methylbenzofluoranthene	
112	252	Benzo[e]pyrene	

Table 9 (continued)

Peak number	Molecular weight	Assignment	Remarks
113a	252	Benzo[a]pyrene	
113b	252	?	
114	266	Methylbenzofluoranthene, -pyrene	
115	266	Methylbenzofluoranthene, -pyrene	
116	252	Perylene	
117a	266	Methylbenzofluoranthene, -pyrene	
117b	278	Dibenzophenanthrene, -anthracene	
118	266	Methylbenzofluoranthene, -pyrene, methylperylene	
119a	264	Methylenebenzofluoranthene, -pyrene, methyleneperylene	
119b	253	Aza derivative of $C_{20}H_{12}$ PNA or nitrile of $C_{18}H_{12}$ PNA	
120	254	6H-Benzo[cd]pyrene-6-one or isomer	Gold
121a	284	Dinaphthothiophene	
121b	264	Methylene derivative of benzofluoranthenes, -benzpyrenes, perylene	
122	279	Dibenzoacridine or isomer	
123	278	Dibenzophenanthrene, -anthracene	
124	284	Dinaphthothiophene	
125a	278	Dibenzophenanthrene, -anthracene	
125b	284	Dinaphthothiophene	
126	279	Dibenzoacridine or isomer	
127	278	Dibenzophenanthrene, -anthracene	
128	276	Indenofluoranthene	
129	278	Dibenzo[a,h]- and/or dibenzo[a,c]anthracene	

(continued)

Table 9 (continued)

Peak number	Molecular weight	Assignment	Remarks
130	276	Indeno[a,2,3-cd]pyrene	
131	278	Benzo[b]chrysene	
132	267	Dibenzocarbazole	
133	278	Picene	
134	276	Benzo[ghi]perylene	
135	276	Anthanthrene	
136	290	Methyl derivative of indenofluoranthene, -pyrene, benzo[ghi]-perylene, anthanthrene	Methylene derivative of dibenzophenanthrenes or -anthracenes possible
137	302	Dibenzofluoranthene, -pyrene	
138	302	Dibenzofluoranthene, -pyrene	
139	302	Dibenzofluoranthene, -pyrene	
140a	300	Coronene	
140b	300	?	
141	302	Dibenzofluoranthene, -pyrene	

Source: Ref. 55.

Table 10. Relative Retention Times of Peaks in Fig. 18

Peak No.	Compound	Boiling point	Apiezon L at 180°C
1	Naphthalene	217.96	1.00
2	2-Methylnaphthalene	241.14	1.60
3	1-Methylnaphthalene	244.18	1.77
4	Biphenyl	255.0	2.10
5	2-Ethylnaphthalene	257.9	2.38
6	1-Ethylnaphthalene	258.67	2.43
7	2,6-Dimethylnaphthalene	262.0	2.63
8	2,7-Dimethylnaphthalene	262.0	2.63
9	1,7-Dimethylnaphthalene	262.9	2.67
10	1,3-Dimethylnaphthalene	265.0	2.87
11	1,6-Dimethylnaphthalene	265.5	2.92
12	2-Methylbiphenyl	260.0	2.96
13	2,3-Dimethylnaphthalene	268.0	3.16
14	1,4-Dimethylnaphthalene	268.5	3.20
15	1,5-Dimethylnaphthalene	270.1	3.26
16	1,2-Dimethylnaphthalene	271.1	3.32
17	4-Methylbiphenyl	270.0	3.32
18	3-Methylbiphenyl	272.7	3.42
19	1,8-Dimethylnaphthalene	276.5[a]	3.89
20	Acenaphthene	277.2	3.95

[a]Estimated on the basis of relative retention.
Source: Ref. 62.

Table 11. Identification of Numbered Chromatographic Peaks in Fig. 19
PAH Identified in Particulate Matter from Working Atomsphere of a Coke Plant

Peak number	Compound	Identification: coke plant GC/MS	GC
1	Propylbenzene	+	
2	Naphthalene	+	+
3	Quinoline		
4	2-Methylnaphthalene	+	+
5	1-Methylnaphthalene	+	+
6	MW = 142		
7	Methylquinoline	+	
8	Methylquinoline		
9	Methylquinoline		
10	Methylquinoline		
11	Methylquinoline		
12	Biphenyl	+	+
13	Methylquinoline	+	
14	Dimethylnaphthalene	+	
15	Dimethylnaphthalene	+	
16	Dimethylquinoline	+	
17	Dimethylquinoline		
18	Dimethylquinoline		
19	Dimethylquinoline		
20	Dimethylquinoline		
21	Acenaphthene	+	+
22	Dimethylquinoline	+	
23	Acenaphthylene	+	+
24	Dimethylquinoline		
25	Naphthylnitrile	+	
26	Dibenzofuran	+	+
27	Naphthylnitrile	+	+
28	Methylacenaphthylene	+	
29	Fluorene	+	+
30	Methylacenaphthylene	+	

Peak number	Compound	Identification: coke plant	
		GC/MS	GC
31	9-Methylfluorene	+	+
32	2-Methylfluorene	+	+
33	1-Methylfluorene	+	+
34	Fluorenone		+
35	Methylbenzothiophene		
36	Dibenzothiophene	+	+
37	Methyldibenzofuran		
38	Phenanthrene	+	+
39	Anthracene	+	+
40	Benzoquinoline	+	
41	Acridine	+	+
42	Cyclopenta[def]phenanthrene		
43	Methyldibenzothiophene		
44	Methylanthracene/methylphenanthrene	+	
45	Methylanthracene/methylphenanthrene	+	
46	2-Methylanthracene	+	+
47	4,5-Methylenephenanthrene	+	
48	Methylanthracene/methylphenanthrene	+	
49	1-Methylphenanthrene	+	+
50	Carbazole	+	+
51	Methylbenzoquinoline?		
52	Methylbenzoquinoline?	+	
53	MW = 204	+	
54	MW = 206		
55	Dimethylnaphthothiophene		
56	Methylcarbazole	+	
57	3,6-Dimethylphenanthrene (IS)[a]		
58	Dimethylphenanthrene	+	
59	MW = 209	+	

(continued)

Table 11 (continued)

Peak number	Compound	Identification: coke plant	
		GC/MS	GC
60	Dimethylphenanthrene/methylcarbazole	+	
61	Dimethylphenanthrene/methylcarbazole	+	
62	Methylcarbazole		
63	Methylcarbazole		
64	Fluoranthene	+	+
65	Benzylnaphthalene		
66	Benz[e]acenaphthylene	+	
67	Benzo[def]dibenzothiophene	+	
68	Pyrene	+	+
69	Ethylmethylenephenanthrene	+	
70	Ethylmethylenephenanthrene	+	
71	MW = 203	+	
72	Dihydrobenzofluorene	+	
73	Methylpyrene	+	+
74	Methylpyrene	+	
75	Methylpyrene		
76	Benzo[a]fluorene	+	+
77	Benzo[b]fluorene	+	+
78	4-Methylpyrene	+	+
79	MW = 232		
80	Benzocarbazole	+	
81	Methylpyrene		
82	Methylpyrene	+	
83	1-Methylpyrene	+	
84	Fluorenenitrile		
85	MW = 244	+	
86	Styrylnaphthalene		
87	Benzothionaphthene	+	
88	Benzo[c]phenanthrene	+	+
89	Benzophenanthridine	+	
90	Benzodibenzothiophene		
91	Benzo[a]dibenzothiophene	+	

Peak number	Compound	Identification: coke plant	
		GC/MS	GC
92	Benzo[ghi]fluoranthene	+	
93	Benz[a]anthracene	+	+
94	Chrysene, triphenylene	+	+
95	MW = 227	+	
96	Methylbenz[a]anthracene/methylchrysene		
97	Benzanthrone	+	
98	Methylbenz[a]anthracene/methylchrysene	+	
99	Benzocarbazole	+	
100	MW = 240		
101	Methylchrysene		
102	Benzocarbazole		
103	β,β-Binaphthyl (IS)	+	+
104	Binaphthyl		
105	Benzocarbazole	+	
106	Ethylchrysene/dimethylchrysene	+	
107	Benzo[b]fluoranthene	+	+
108	Benzo[j+k]fluoranthene	+	+
109	Benzofluoranthene	+	
110	MW = 258		
111	MW = 258		
112	Benzo[e]pyrene	+	+
113	Benzo[a]pyrene	+	+
114	Perylene	+	+
115	m,m'-Tetraphenylene (IS)	+	+
116	o-Phenylenepyrene	+	+
117	Dibenz[a,h]anthracene/dibenz[a,c]anthracene	+	+
118	Dibenzanthracene	+	
119	Benzo[b]chrysene	+	
120	Benzo[ghi]perylene	+	+
121	Anthanthrene	+	+

[a]IS = internal standard.
Source: Ref. 63.

Table 12. Identification of Chromatographic Peaks in Fig. 20

Peak number[a]	Compound	Method of identification[b]
1	Tetralin	RI, MS
2	Naphthalene	RI, MS
3	2-Methylnaphthalene	RI, MS
4	Cyclohexylcyclohexane[c]	MS
5	1-Methylnaphthalene	RI, MS
6	1,2,2a,3,4,5-Hexahydroacenaphthylene	RI, MS
7	Biphenyl	RI, MS
8	2-Ethylnaphthalene	RI, MS
9	1-Ethylnaphthalene	RI, MS
10	2,6- and/or 2,7-Dimethylnaphthalene	RI, MS
11	1,3- and/or 1,7-Dimethylnaphthalene	RI, MS
12	1,6-Dimethylnaphthalene	RI, MS
13	C_2-Benzo[b]thiophene	MS, FPD
14	2,3- and/or 1,4-Dimethylnaphthalene	RI, MS
15	1,5-Dimethylnaphthalene	RI, MS
16	1,2-Dimethylnaphthalene	RI, MS
17	1,8-Dimethylnaphthalene	RI, MS
18	Acenaphthene	RI, MS
19	3- and/or 4-Methylbiphenyl	RI, MS
20	Dibenzofuran	RI, MS
21	3,3,5,7-Tetramethyl-1-indanone	MS
22	2,3,6-Trimethylnaphthalene	RI, MS
23	2,3,5-Trimethylnaphthalene	RI, MS
24	Fluorene	RI, MS
25	4,4'-Dimethylbiphenyl	RI, MS
26	Xanthene	RI, MS
27	2-Methylfluorene	RI, MS
28	1-Methylfluorene	RI, MS
29	1-Methyl-7-isopropylnaphthalene (eudalene)	MS
30	Dibenzothiophene	RI, MS, FPD
31	Phenanthrene	RI, MS

Peak number[a]	Compound	Method of identification[b]
32	Anthracene	RI, MS
33	Methyldibenzothiophene	MS, FPD
34	Methyldibenzothiophene	MS, FPD
35	Methyldibenzothiophene	MS, FPD
36	3-Methylphenanthrene	RI, MS
37	2-Methylphenanthrene	RI, MS
38	9- and/or 4-Methylphenanthrene	RI, MS
39	1-Methylphenanthrene	RI, MS
40	2-Phenylnaphthalene	RI, MS
41	S_8	CC, MS, FPD
42	9-Ethylphenanthrene	RI, MS
43	2-Ethylphenanthrene and/or 3,6-dimethylphenanthrene	RI, MS
44	2,7-Dimethylphenanthrene	RI, MS
45	1,7-Dimethylphenanthrene (pimanthrene)	MS
46	Fluoranthene	RI, MS
47	1,8-Dimethylphenanthrene	RI, MS
48	Phenanthro[4,5-bcd]thiophene (benzo[def]dibenzothiophene)	MS, FPD
49	Pyrene	RI, MS
50	1,2,3,4-Tetrahydro-1-methyl-7-isopropyl-phenanthrene (1,2,3,4-tetrahydroretene)	MS
51	Simonellite	MS
52	9-Methyl-10-ethylphenanthrene	RI, MS
53	Benzo[k,l]xanthene	RI, MS
54	Benzo[a]fluorene	RI, MS
55	11-Methylbenzo[a]fluorene	RI, MS
56	1-Methyl-7-isopropylphenanthrene (retene)	RI, MS
57	4,5,6-Trihydrobenz[de]anthracene	RI, MS
58	2-Methylpyrene	RI, MS
59	1-Methylpyrene	RI, MS

(continued)

Table 12 (continued)

Peak number[a]	Compound	Method of identification[b]
60	1-Ethylpyrene	RI, MS
61	2,7-Dimethylpyrene	RI, MS
62	Benzo[ghi]fluoranthene	RI, MS
63	Benzo[c]phenanthrene	RI, MS
64	Benz[a]anthracene	RI, MS
65	Chrysene and/or triphenylene	RI, MS
66	9-Phenylphenanthrene	RI, MS
67	Benzo[b]fluoranthene and/or benzo[j]-fluoranthene	RI, MS
68	Benzo[k]fluoranthene	RI, MS
69	Benzo[e]pyrene	RI, MS
70	Dibenzo[c,kl]xanthene	RI, MS
71	Benzo[a]pyrene	RI, MS
72	Perylene	RI, MS
73	Dibenz[a,c]anthracene	RI, MS
74	Dibenz[a,h]anthracene	RI, MS
75	Benzo[ghi]perylene	RI, MS
76	7-Methyl-3'-ethyl-1,2-cyclopentenochrysene	MS
77	Coronene	MS
78	Dibenzopyrene	MS

[a]Peak numbers refer to labeled chromatographic peaks in Fig. 18.
[b]RI = retention index; MS = mass spectrometry; FPD = flame photometric detection; CC = cochromatography.
[c]The presence of cyclohexylcyclohexane is not clearly understood. It represents the only nonaromatic hydrocarbon found in this fraction. It could be an impurity or artifact from the cyclohexane used in the solvent separation of the coal extract or the cyclohexane used in the clay-gel percolation compound class separation.
Source: Ref. 64.

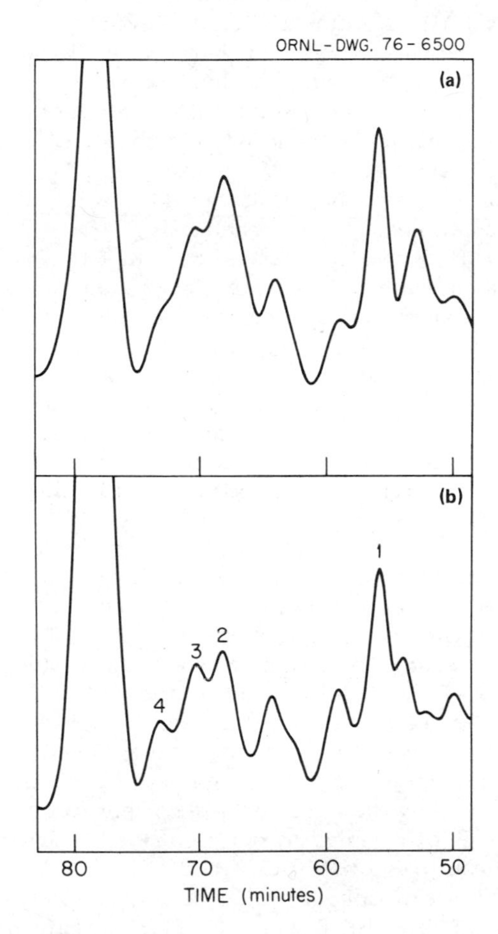

Figure 21. Packed column gas chromatograms of methylchrysene subfractions, unpurified (a) and purified (b), isolated from a coal liquefaction product. Chromatographic conditions: 6.1 m × 3.2 mm o.d. glass column packed with 3.2% OV-22 coated on 80/100 mesh Chromosorb G, using He carrier at 5 ml/min, isothermally at 280°C. Numbered chromatographic peaks are (1) chrysene; (2) 3-methylchrysene; (3) 2-methylchrysene; (4) 6-methylchrysene. (From Ref. 72.)

i.e., matching retention times during cochromatography of the unknown with authentic reference standards of PAH, matching retention indices, or mass spectral data obtained during combined GC/MS experiments. There are varying degrees of reliability of identifications. Clearly, tentative identifications based on cochromatography with authentic standards are more reliable than identifications based solely on matching PAH retention indices. Further, it can be convincingly argued that, since the mass spectra of isomeric PAH are similar to the point of being indistinguishable from one another, tentative identifications based solely on cochromatography or matching PAH retention indices are more reliable than tentative identifications based solely on MS. Qualitative analysis should be performed on samples that represent clean aromatic fractions, free of other classes of compounds that can interfere. If this is not done, the results may be meaningless. Even when using the highest-efficiency WCOT columns available for analysis of crude unfractionated coal products, very few, if any, chromatographic peaks are clearly separated from all other compounds. Invariably, there are shoulders on both sides of each chromatographic peak. Combined GC/MS investigation of the gas chromatographic peaks from samples that have not undergone previous fractionation shows that in almost every case there is evidence for at least two components in each chromatographic peak. Thus, the interferences present in unfractionated samples prevent reliable tentative identification by cochromatography or matching PAH retention indices. Tentative identifications based solely on these chromatographic methods are reliable only on samples that represent clean aromatic fractions.

Identification of chromatographic peaks is best accomplished using cochromatography of clean aromatic fractions with authentic PAH standards, followed by mass spectral information on the same chromatographic peaks to confirm the assignments. Unfortunately, many laboratories lack the necessary reference standards of PAH for analysis of samples by cochromatography. This problem has been partially solved by the development of the PAH retention index system (PAHRIS) [13,64,73]. The PAHRIS is based on the use of four PAH—naphthalene, phenanthrene, chrysene, and picene—as bracketing standards (Table 13), rather than normal paraffins. The development of the PAHRIS and its application to the analysis of complex mixtures of PAH isolated from a coal liquefaction product have been previously described [13]. An example of the use of the PAHRIS for identification of unknown chromatographic peaks appears in Fig. 22 and Table 14. The PAHRIS was developed for linear-

Table 13. Bracketing Standards Used in the PAH Retention Index System (PAHRIS)

Compound	Defined index
Naphthalene	200
Phenanthrene	300
Chrysene	400
Picene	500

Source: Ref. 13.

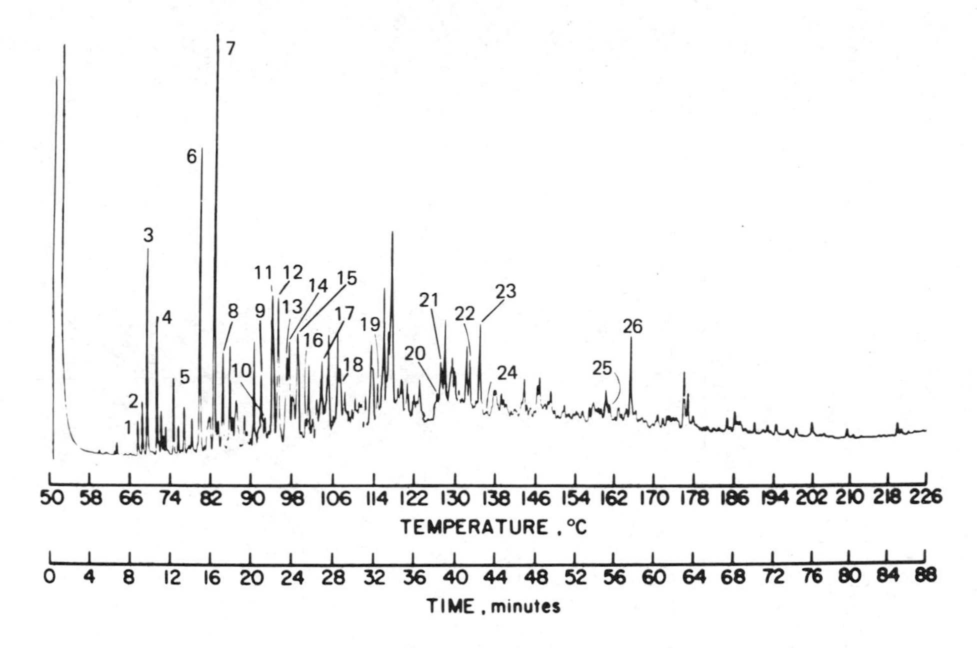

Figure 22. Capillary column gas chromatogram of the aromatic fraction of Synthoil. Chromatographic conditions: 12 m × 0.30 mm glass capillary coated with SE-52, using a He carrier at 2.0 ml/min and temperature programmed from 50° to 250°C at 2°C/min. Numbered chromatographic peaks and their PAH retention indices are identified in Table 14. (From Ref. 13.)

programmed-temperature, glass capillary analysis of PAH using SE-52 as the stationary phase and employing the equation initially suggested by Van Den Dool and Kratz [74].

$$I = 100 \left(n + \frac{X - M_n}{M_{n+1} - M_n} \right) \tag{1}$$

Retention indices were calculated for over 200 PAH standard compounds based on their comparison with these standards using the above equation. In Eq. (1), I is the retention index and n refers to the number of rings in the standard compound eluting just prior to the compound of interest; X, M_n, and M_{n+1} are the measured retention times of the compound of interest and the standard aromatic that elutes just prior to and after the compound of interest, respectively. The average 95% confidence limits for three to eight measurements for each of 12 PAH ranging from naphthalene to perylene was ±0.25 index units [73].

The Kovats retention index system failed to be statistically reliable for PAH analysis. The Kovats retention indices for the same 12 PAH varied widely with the temperature program rate, film thickness, and condition of the column (average 95% confidence limits for six to ten measurements for each PAH was ±4.2 index units). The reasons for the failure of the Kovats retention index system to provide statistically acceptable data for PAH under

Table 14. GC/MS Results from the Analysis of the Aromatic Fraction of SYNTHOIL (See Fig. 22)

GC peak number	Compound	Measured retention index of chromatographic peak from sample (average of three determinations $\bar{x} \pm S$)	Measured retention index using standards ($\bar{x} \pm S$)
1	Methylindane	194.04 ± 0.09	—[a]
2	Methylindane	195.48 ± 0.09	—[a]
3	Tetralin	197.17 ± 0.20	197.04 ± 0.05
4	Naphthalene	200.00	200.00
5	Methyltetralin	205.07 ± 0.33	—[a]
6	Methyltetralin or C_2-indane	213.61 ± 0.10	—[a]
7	2-Methylnaphthalene	218.31 ± 0.24	218.14 ± 0.28
8	1-Methylnaphthalene	220.90 ± 0.04	221.04 ± 0.25
9	1,2,2a,3,4,5-Hexahydroacenaphthylene	232.63 ± 0.06	232.70[b]
10	Biphenyl	233.90 ± 0.04	233.96 ± 0.24
11	2-Ethylnaphthalene	236.02 ± 0.16	236.08 ± 0.16
12	2,6- and 2,7-Dimethylnaphthalenes	237.75 ± 0.10	237.58 ± 0.17, 237.71 ± 0.07
13	1,3- and 1,7-Dimethylnaphthalenes	240.52 ± 0.12	240.25 ± 0.16, 240.66 ± 0.25
14	1,6-Dimethylnaphthalene	240.87 ± 0.19	240.72 ± 0.09
15	1,5-Dimethylnaphthalene	243.83 ± 0.12	243.98 ± 0.30

16	1,2-Dimethylnaphthalene	246.29 ± 0.11	246.49 ± 0.30
17	Acenaphthene	251.17 ± 0.08	251.29 ± 0.14
18	Dibenzofuran	256.94 ± 0.20	257.17 ± 0.05
19	Fluorene	268.30 ± 0.10	268.17 ± 0.15
20	9,10-Dihydrophenanthrene	286.86 ± 0.29	287.09 ± 0.16
21	1,2,3,4,5,6,7,8-Octahydroanthracene	287.76 ± 0.12	287.69 ± 0.20
22	1,2,3,4-Tetrahydrophenanthrene	296.93 ± 0.21	297.21[b]
23	Phenanthrene	300.00	300.00
24	Anthracene	301.84 ± 0.19	301.69 ± 0.08
25	Fluoranthene	344.12 ± 0.25	344.01 ± 0.16
26	Pyrene	351.38 ± 0.29	351.22 ± 0.08

[a]The retention index of these compounds were not determined because standards were not available.
[b]Only two determinations were made; therefore no standard deviation is available.
Source: Ref. 13.

temperature programming conditions have been previously documented [73]. The statistical unreliability of the Kovats retention index system for PAH analysis under temperature programming conditions have been independently confirmed by Beernaert [75]. The PAHRIS has been in use in our laboratories for several years and has been working well. It should be mentioned that when using the PAHRIS, the temperature program *must* be initiated at a temperature low enough to condense naphthalene and other PAH at the front of the column. This usually is 30° to 40°C less than the elution temperature of naphthalene. If temperature programming is started at a higher initial temperature, the retention indices obtained will not coincide with those published [13,73]. The PAH retention indices consist of five numbers. For example, the retention index of fluorene is 268.17. The first number, 2, is descriptive only and indicates that fluorene elutes between the first two bracketing standards, naphthalene and phenanthrene. The remaining numbers, 68.17, are significant. In order to correctly calculate the retention index of fluorene to four significant figures, we must know the retention times of naphthalene, fluorene, and phenanthrene to four significant figures. Since most chromatographic data systems commonly in use report retention times to 0.01 of a minute, the retention time of naphthalene must be 10.00 min or greater in order to provide four significant figures. Although it is apparent from the foregoing discussion, it is worth mentioning that the retention times of the GC peaks must be measured by a chromatographic data system that reports retention times to at least 0.01 of a minute. Under no circumstances should the retention indices be calculated using distances measured from injection point to elution point of peaks on a chromatogram. A small measurement error can cause the retention indices calculated for chromatographic peaks to fail to coincide with those published.

One final comment should be made on the identification of individual PAH—or compounds of any class—from coal or coal-derived products. It is regrettable that compounds are occasionally misidentified in the literature. This is particularly true for identification of constituents not previously reported in the samples being analyzed. Misidentification of PAH not previously reported as present in a sample can lead to wrong conclusions and mislead fuel chemists and chromatographers for decades before the mistakes are corrected. Compounds previously unreported in a sample should not be identified in a publication unless *very* strong evidence can be offered which supports their identification. In most cases, mass spectral evidence alone, obtained from combined GC/MS experiments, is not sufficient to identify a previously unreported compound in publication.

Sample introduction techniques can be very important in analyzing PAH from coal or coal-derived products when using GC. In general, splitless injection is preferred to splitting, since many samples contain a very wide boiling range of components. Grob [76] has described the distillation effects encountered when syringes are used with vaporizing injectors and has developed the on-column capillary injector [77], which eliminates sample discrimination due to differences in volatility, concentration, and polarity. This injector is currently available from only a few manufacturers, and its use for analysis of PAH from coal or coal-derived products has not been reported. Our experience has been that traces of PAH of lesser volatility than chrysene may go undetected unless splitless injection is used because a substantial portion remains in the syringe. Satisfactory precision of peak areas can be obtained using splitless injection by hand. For example, a standard solution of 14 PAH, ranging in concentration from 23 to 49 mg/liter, was hand-injected and analyzed four times, while the peak areas were measured using a chromatographic data system. Using the airplug injection technique [76], 2.0 μl of the standard solution of 14 PAH, ranging in volatility from naphthalene to p-quaterphenyl, was splitlessly injected by hand. The standard deviations of the peak areas for each peak were determined, and from these values the average relative standard deviation of all the peaks was calculated to be 3.39%.

Not only is GC useful as a means of separating complex mixtures of PAH and identifying individual isomeric components, but it can also be used to obtain quantitative information on these compounds. Quantitative analysis is usually performed after the sample has undergone detailed qualitative analysis using the previously described techniques. Obtaining reliable quantitative measurements of good precision and accuracy on a wide variety of PAH of differing volatilities during a single analysis is not an easy task. The relative and absolute peak areas of individual compounds can change as a function of injection technique, and when split injection is used traces of high boiling PAH can fail to be detected because they do not distill from the syringe. It is critical that correct splitless injection techniques [78,79] be used when one is attempting to quantitate a large number of PAH of different volatilities in the same sample at trace amounts. The analyst faces additional problems when attempting to obtain quantitative information. Preseparation of the sample is needed to give a clean aromatic fraction, free of interferences; however, every procedure that the sample is subjected to causes losses of PAH. These losses can be due to evaporation of volatile components, incomplete recovery of individual PAH from LC stationary phases used during sample cleanup, or incomplete extraction. Such losses increase the error in the final quantitative measurements. Furthermore, when using a nonspecific detector, such as a flame-ionization detector (FID), the components of interest must elute from the chromatographic system as pure components free of interferences, which rarely occurs when the sample has not been preseparated to give a clean aromatic fraction. Unfortunately, even when the samples have undergone extensive preseparation, the complexity of the aromatic fractions from coal and coal products can be so great that all PAH are not completely separated from one another, making it difficult to quantitate individual components using an FID.

Many of these problems are overcome through the use of the combined GC/MS technique known as mass fragmentography to quantify a large number of individual PAH from a single sample that has *not* undergone preseparation. Quantification of individual PAH should not be attempted until qualitative analysis has determined their presence in the sample. Mass fragmentography uses a mass spectrometer as an extremely sensitive and specific detector. The sensitivity of common low resolution mass spectrometers is at the low picogram level for individual PAH and is sufficiently specific to provide clean, interference-free chromatographic peaks for most PAH when used with high-efficiency glass capillary columns. The specificity of the mass spectrometer is even greater when precise masses can be monitored using a high-resolution mass spectrometer. The linear dynamic range of the entire instrumental system, including the chromatograph, column, interface, and spectrometer, should be determined for all the compounds of interest before analysis is begun. Because the PAH to be quantitated are in the presence of thousands of other compounds, matrix effects can be large. Therefore, relative response factors as well as internal and external standards should be avoided wherever possible when quantitating individual PAH. Individual PAH can be reliably quantitated by combining mass fragmentography with the method of standard addition [80-83]. Although the method of standard addition does not eliminate matrix effects, it reliably compensates for them while avoiding the use of relative response factors.

A coal-derived material can be quantitatively analyzed for PAH by first dissolving the material in an appropriate solvent that provides a good solvent effect upon splitless injection. An accurately weighed amount, about 200 mg, of the coal-derived material is added to a 25.00-ml volumetric flask and diluted to the mark with solvent. A 2.0-5.0-μl aliquot of this sample is splitlessly injected into the chromatographic system of the GC/MS. The mass spectrometer is set to monitor the molecular ions and/or large fragment ions of the PAH of interest for several minutes before and after their known retention times. Either electron-impact or chemical-ionization techniques can be used. Other compounds may coelute with the PAH; however, they do not interfere with the PAH determinations as long as they do not have fragments or molecular ions of coincident nominal mass with those PAH ions being monitored. The areas of the mass fragmentographic peaks are determined, usually by electronic integration using a computer, and checked to ensure that they are well within the linear dynamic range of the instrumentation. A standard addition consisting of a known amount of the PAH to be quantitated is then added to the original 25.00-ml portion of the sample, and the mass fragmentographic experiment is repeated under identical conditions. The areas of the mass fragmentographic peaks due to the PAH of interest are again determined and checked to ensure that they remain in the linear dynamic range. A second standard addition is made and the mass fragmentographic experiment repeated again; peaks corresponding to the PAH of interest are again checked to ensure they remain in the linear dynamic range. After each standard addition, the mass fragmentographic peaks due to the PAH of interest should enlarge to an amount directly proportional to the amount of the compound added. The concentration of individual PAH can be calculated using the following equation:

$$\frac{x}{x + a} = \frac{A_x}{A_1}$$

where x is the unknown concentration of the PAH, a is the amount of the PAH added by standard addition, A_x is the peak area of the PAH before standard addition, and A is the peak of the PAH after standard addition. Appropriate corrections should be made for background signal. Furthermore, the result should be checked by making a second standard addition. The combined methods of mass fragmentography and standard additions are very powerful and lead to precise and accurate measurements. When a large number of PAH are being measured in a group of samples, every fourth or fifth sample should be analyzed repetitively to determine the precision of the measurements. Before the first addition is made, the unknown should be injected at least three successive times and analyzed using fragmentography. The peak areas for individual PAH from the three experiments should be very

Table 15. Concentration of PAH in SRC-II

Compound	Concentration (mg/g) PETC[a] $\bar{x} \pm s$	NBS $\bar{x} \pm s$
Dibenzothiophene	1.18 ± 0.07	1.02 ± 0.07
Pyrene	6.02 ± 0.31	6.00 ± 0.2

[a]Average of five determinations.

similar. Our experience has been that a relative standard deviation of peak areas of 5-6% is common when more than about 150 pg of PAH have been injected, although this seems to be somewhat dependent upon the volatility of the PAH. When less than this amount of PAH is introduced to the system, the relative standard deviation may still be 5-6% but has been as high as 18%. Three repetitive analyses should also be performed on the sample after the first and second standard additions. It should be mentioned that the amount of PAH added during the first and second standard additions is unimportant as long as the following requirements are met: (1) the total amount of any PAH remains within the linear dynamic range of the instrumentation for that compound; (2) the volume of the total sample does not change more than a few percent when the standard additions are made; and (3) the amount of any single PAH added is enough to cause a real difference in the peak size. For example, if the unknown gives a naphthalene peak area of 90,000 ± 5,000 electronic counts and an amount of naphthalene is added that causes the peak to grow by only 2000 counts, too little naphthalene has been added. It is recommended that enough standard be added to make the peak grow by about 40-80% for each standard addition.

Using these methods, two PAH in a SRC-II coal liquefaction product were determined in our laboratories. This same sample was analyzed for these PAH by the National Bureau of Standards (NBS), as well as by other laboratories. The results obtained by the Pittsburgh Energy Technology Center (PETC) and NBS appear in Table 15.

V. MS Analysis of PAH in Coal and Coal-Derived Materials

MS analysis of PAH in coal and coal-derived materials has traditionally been an extremely important technique. From the late 1950s to the present, MS has been used to obtain information concerning the nature and distribution of PAH in coal and its products. For a brief description of the development of MS as a tool for analysis of complex mixtures of PAH, see Sec. II.D and Chap. 7. Since MS by itself is not capable of differentiating between isomers, it is used to obtain more general information, namely, to determine the total amounts of structural isomers such as phenanthrene and anthracene present in coal extracts or coal products. Today, this general information may not be sufficiently specific to satisfy environmentalists or coal scientists interested in the structure and origins of coal; however, it is ideal for engineers or chemists wishing to rapidly compare the nature and distribution of PAH in a large number of coals or coal-derived samples. In such cases, it may not be important to know exactly which structural isomers are changing in concentration, but only that the concentrations of phenanthrene and anthracene are decreasing relative to the concentrations of dihydrophenanthrene and dihydroanthracene. This kind of information is often required during investigations of the effect of coal liquefaction plant operating conditions (temperature, pressure, flow rate, catalyst) on PAH distribution in the product.

High- and low-resolution magnetic sector mass spectrometers have been used for such analyses. A key feature of these spectrometers is their capability of producing spectra both at low voltages, approximately 12 eV, and at high voltages, 70 eV. The development of the low voltage ionizing technique to selectively ionize aromatics in the presence of other species and to produce spectra that consist mainly of molecular ions has been an important feature of MS analysis of PAH in coal and coal products. However, analytical information of equal importance can be obtained using 70-eV ionizing potentials with high-resolution spectrometers. The choice of ionizing technique and spectrometer resolution should be based on the kind and quality of information needed, as well as the nature of the samples. Low-resolution spectrometers produce nominal mass spectra, which are generally incapable of mass separating ions that have the same nominal masses but different precise masses. Thus both naphthalene and nonane, when present in the same mixture and analyzed at 70 eV with a low-resolution spectrometer, have ions that contribute to the total peak height at mass 128. Alternatively, high-resolution spectrometers are capable of mass separating these ions, thus eliminating the interference at the naphthalene peak at 128.0625 from the nonane peak at 128.1560. Of course, the nonane interference can be eliminated by carefully preseparating the sample to produce a clean aromatic fraction, free of other classes of compounds. Another method of removing the nonane interference at the naphthalene peak at 128 is to use low ionizing voltages. Since naphthalene has a lower ionization potential than nonane, if the applied effective ionization voltage is intermediate between these ionization potentials the interference from nonane is removed because nonane is not ionized and thus not detected, while naphthalene is. In general, the ionization potentials of aromatics and other unsaturated compounds are lower than those of paraffins. There are trade-offs in the use of these techniques. The ionization efficiency is greatly reduced when using low voltages; thus, the sensitivity is greatly reduced and *all* interferences may not be removed completely. When analyzing coal or coal products for PAH by MS alone, best results are obtained using high resolution combined with low-voltage ionization on samples that have been preseparated to yield clean PAH fractions. The confidence that can be placed in the results decreases if the samples have not been preseparated, if 70 eV is used to ionize the sample, or if a low-resolution spectrometer is used.

MS has many uses in the analysis of PAH in coal and coal-derived products. Until the mid-1960s, when combined GC/MS became more common, MS was the major tool for fundamental characterization of coal products. To this day, MS is used as the prime analytical tool in several respected laboratories [84, 85]. More commonly, it is used for PAH analyses needed in engineering, kinetic studies, and general characterization, or for comparison of the kind and distribution of PAH from large numbers of samples and for screening of environmental samples for specific PAH. The remainder of this section will concern itself with some examples of the use of MS for the analysis of complex mixtures of PAH from coal and coal products.

Up to this point, low-voltage, low-resolution and high-voltage (70 eV), high-resolution MS have been discussed. The combination of the two techniques into low-voltage-ionization and high-resolution mass spectrometry (LVHRMS)—as well as its application to analysis of PAH from coal and coal products—was pioneered by Lumpkin [86], Johnson and Aczel [87], and Aczel et al. [88]. The advantages of low-voltage, low-resolution MS spectrometry and high-resolution, high-voltage MS for analyses of PAH in coal and coal products were demonstrated in the late 1950s by Sharkey and co-workers [35,36]. However, it was not until the period between 1967 and 1970 that the two techniques (high-resolution and low-voltage MS) were successfully combined. One of the main road blocks to combining both techniques was that the traditional internal mass standards, such as perfluorokerosene, necessary for high-resolution mass analysis, do not ionize at effective ionizing voltages of 12 eV. (The effective voltage is the sum of the electron beam voltage plus one-third of the repeller voltage. This approximate voltage is attained by charging xylene to the spectrometer and turning down the electron beam voltage until the peak at mass 91 is 1% of the parent peak at mass 106.) This problem was surmounted by Aczel and co-workers [89] by developing a low-voltage mass standard consisting of a variety of halogenated aromatics that ionize at 12 eV and whose negative mass defects and characteristic isotopic peaks are such that they do not interfere with hydrocarbon peaks when operating at resolving powers of 1/10,000. The compounds in the low voltage mass standard and their precise masses are listed in Table 16. At a resolution of 1/10,000, most mass multiplets found in aromatic fractions isolated from coal and coal products are adequately resolved up to about mass 400. Notable exceptions to this are the need for much higher resolutions to separate hydrocarbons from compounds containing sulfur and nitrogen above mass 300.

It is desirable that information concerning the nature and distribution of PAH be quantitative or at least semiquantitative. Since most high-resolution mass spectrometers are qualitative instruments, careful standardization of the instrumental operating conditions is necessary to obtain reliable information concerning the distribution of PAH in samples. Each sample must be analyzed under identical conditions of magnet position, effective voltage, ion beam focusing, slit width, source, and inlet-volume temperatures, etc. The area of a MS peak is directly proportional to the concentration of the component(s) giving rise to it. Quantitative or semiquantitative information is obtained by determining the peak areas using computer techniques described in the literature [89], and then either assuming that all PAH give the same response per unit concentration or applying predetermined sensitivity factors for various components. Either technique can be used when samples from the same source are being analyzed for comparative purposes; however, it is thought that the use of sensitivities results in better quantitative accuracy. There have been several investigations of the parent mass sensitivities of unalkylated, alkylated, and partially hydrogenated PAH. Sensitivity refers to the parent mass peak area or height per unit of compound charged to the spectrometer. Lumpkin [34] and Lumpkin and Aczel [90] showed that at low ionizing voltage the parent mass sensitivity of aromatic hydrocarbons increases rapidly with degree of aromatic ring condensation; upon alkylation, the sensitivity in-

Table 16. Reference Blend for Precise Mass Measurement at Low Voltages

Components	Peaks used			
	1	2	3	4
Pyrrole	67.042197			
Fluorobenzene	96.037525			
Chlorobenzene	112.007976	114.005026		
Chlorofluorobenzene	129.998554	131.995604		
Dichlorobenzene	145.969005	147.966055	149.963105	
Bromobenzene	155.957513	157.955543		
Chloronaphthalene	162.023625	164.020675		
Trichlorobenzene	179.930033	181.927083	183.924133	
Chlorobromobenzene	189.918543		193.912643	
Bromonaphthalene	205.973162	207.971192		
Tetrachlorothiophene	219.847485	221.844396	223.841586	225.838636
Iodochlorobenzene	237.904812	239.901862		
Iodonaphthalene	253.959432			
Perfluoronaphthalene	271.987218			
Perfluoroxylene	285.984022			
Dibromotetrafluorobenzene	305.830389	307.828419	309.826449	
Perfluorodiphenyl	333.984022			
Perfluoroacetophenone	361.978930			
Diiodotetrafluorobenzene	401.802929			
Tetrabromomonofluorophenol	425.672731	427.670761	429.668791	
Octafluorodibromodiphenyl	453.823998	455.822028	457.820058	
Hexabromobenzene	549.506400	551.504430	553.502460	

Source: Ref. 89.

creases initially up to a maximum at C_4 for substituted naphthalenes, and then upon further alkylation the sensitivity decreases regularly. Lumpkin's findings are depicted in Fig. 23. Schiller [91] has found that the sensitivity of mono- and dimethyl homologs of PAH either increases slightly or remains constant relative to the unalkylated PAH. Shultz et al. [92] investigated the parent mass sensitivity of hydroaromatic compounds and developed an equation that can be used to predict the sensitivity of hydroaromatic compounds.

LVHRMS becomes a more useful tool for analysis of PAH from large numbers of samples when the mass spectrometric data acquisition and reduction are performed by a computer. Hundreds of MS peaks need to have their precise masses calculated, a molecular formula assigned, areas measured, and concentration calculated, after appropriate sensitivity factors and other considerations have been applied. Obviously, this cannot be done quickly without the

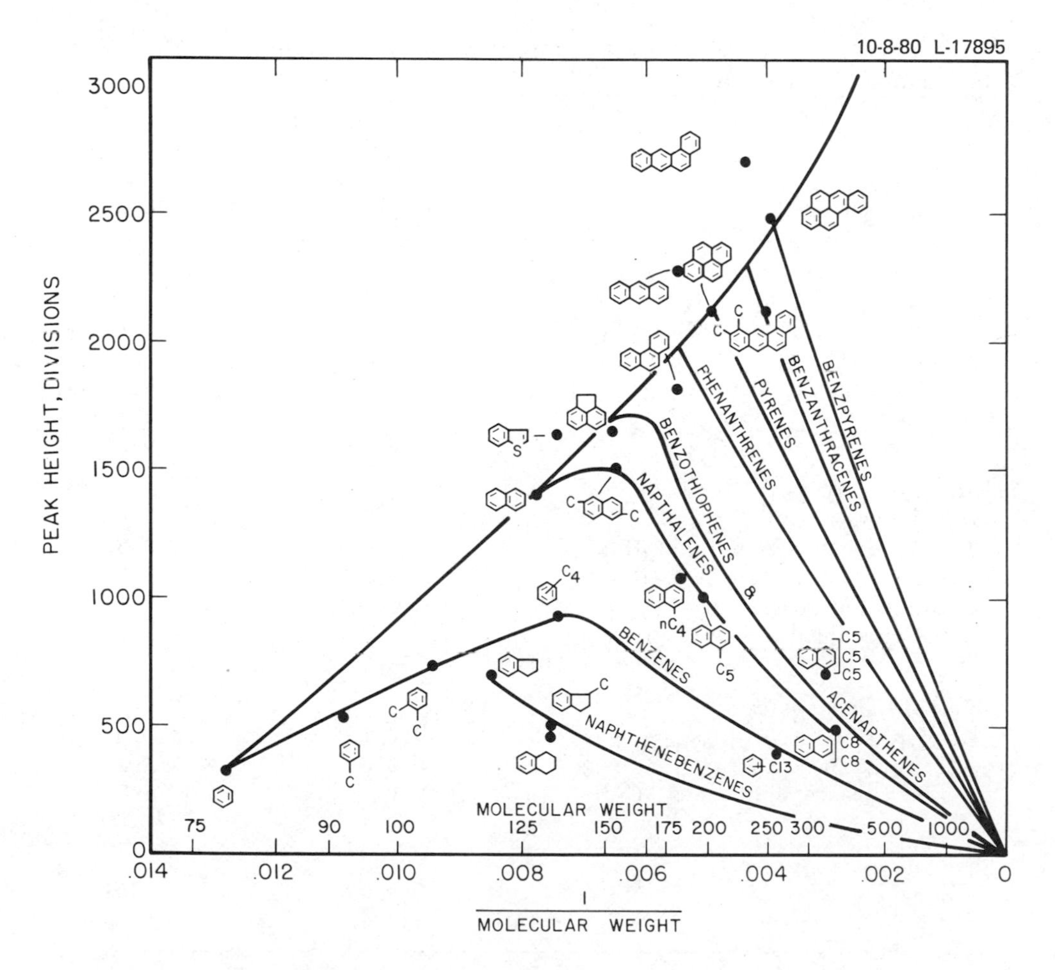

Figure 23. Calibration data for low-voltage aromatic compounds (benzene, 325 divisions). (From Ref. 34.)

aid of a computer, particularly for large numbers of samples. The necessary computer techniques have been developed and described, unfortunately in general terms only, in several of the papers by Aczel and co-workers [89, 93]. Using these computer techniques, it is possible to routinely provide quantitative data for up to 58 compound types and up to 2900 components in each sample in less than 3 hr. This permits a large number of samples to be analyzed for various PAH and is ideal for investigations of the effect of processing conditions on product distributions from bench scale or pilot plant operations.

Aczel and co-workers have applied these methods, primarily developed for the analysis of petroleum products, to the analysis of aromatic components from coal liquefaction products [84]. The analytical method allows determination of PAH as well as of heteroaromatic components ranging in concentration from several percent to parts per million and with molecular weights from 78

Table 17. Concentration of PAH in Coal Liquefaction Products as Determined by LVHRMS

Ring systems	Weight percent: Mildly hydrogenated material	Weight percent: Severely hydrogenated material
Naphthalene system:		
Naphthalenes, C_nH_{2n-12}	14.2	7.9
Dihydronaphthalenes, C_nH_{2n-10}	0.5	0.7
Tetralins, C_nH_{2n-8}	11.5	18.4
Octahydronaphthalenes, C_nH_{2n-4}	0.3	0.4
Decalins, C_nH_{2n-2}	2.0	2.6
Total:	28.5	30.0
Anthracene system:		
Anthracenes, C_nH_{2n-18}	7.2	5.4
Tetrahydroanthracenes, C_nH_{2n-14}	3.0	3.8
Hexahydroanthracenes, C_nH_{2n-12}	1.3	1.5
Octahydroanthracenes, C_nH_{2n-10}	0.7	1.5
Decahydroanthracenes, C_nH_{2n-8}	0.6	1.2
Total:	12.8	13.4
Acenaphthene system:		
Acenaphthylenes, C_nH_{2n-16}	0.1	0.0
Acenaphthenes, C_nH_{2n-14}	6.6	5.3
Tetrahydroacenaphthenes, C_nH_{2n-10}	0.7	1.7
Total:	7.4	7.0
Fluorene system:		
Fluorenes	5.0	5.0
Total:	5.0	5.0

Source: Ref. 88.

to 500. The results obtained are best illustrated by example. Table 17 compares the concentrations of several kinds of PAH, including alkylated homologs and their various hydrogenated forms, in two coal liquefaction products—one mildly hydrogenated, the other severely hydrogenated.

An excellent paper illustrating the use of MS for the analysis of PAH, partially hydrogenated PAH, and other compounds in the pyridine extracts of chemically reduced and untreated coals was published by Kessler et al. [94]. The PAH in the extracts were identified using HRMS at 70 eV and semiquantitated using LVMS. Figure 24 depicts the concentrations of PAH in extracts from both samples. In addition to studying the PAH in both samples, the sulfur-containing PAH were also investigated using the same methods. It should be noted that this is one of the first times, to my knowledge, that the sulfur-containing PAH were actively investigated in coal. The results of the analysis of sulfur-containing PAH are illustrated in Fig. 25.

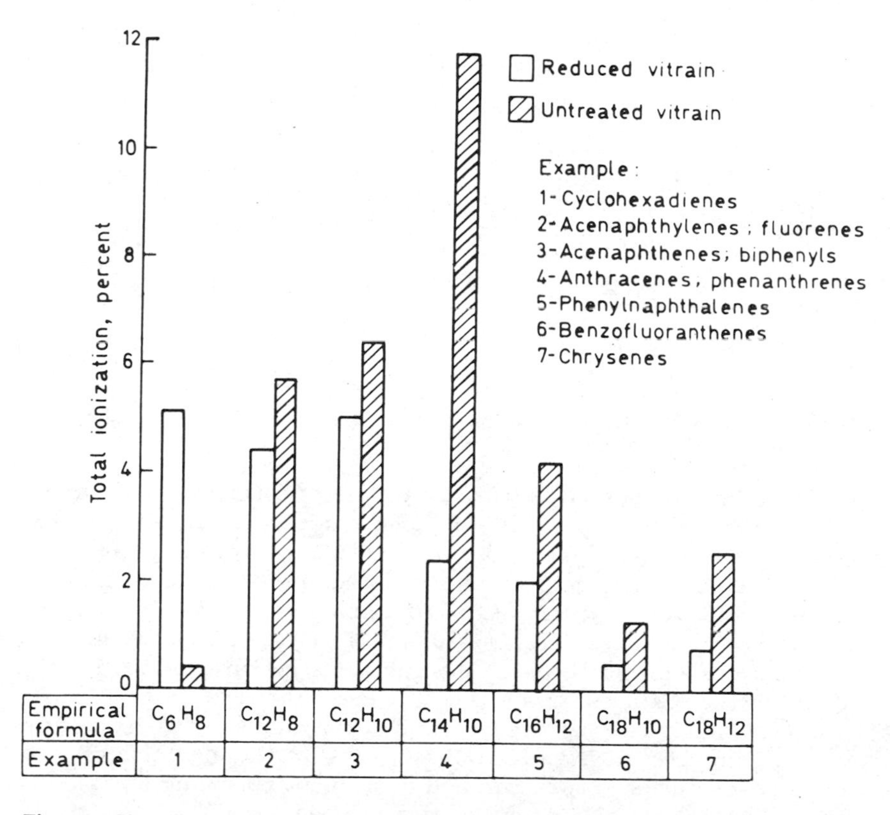

Figure 24. Some aromatic hydrocarbons in the pyridine extracts of reduced and untreated Pittsburgh seam vitrain. (From Ref. 94.)

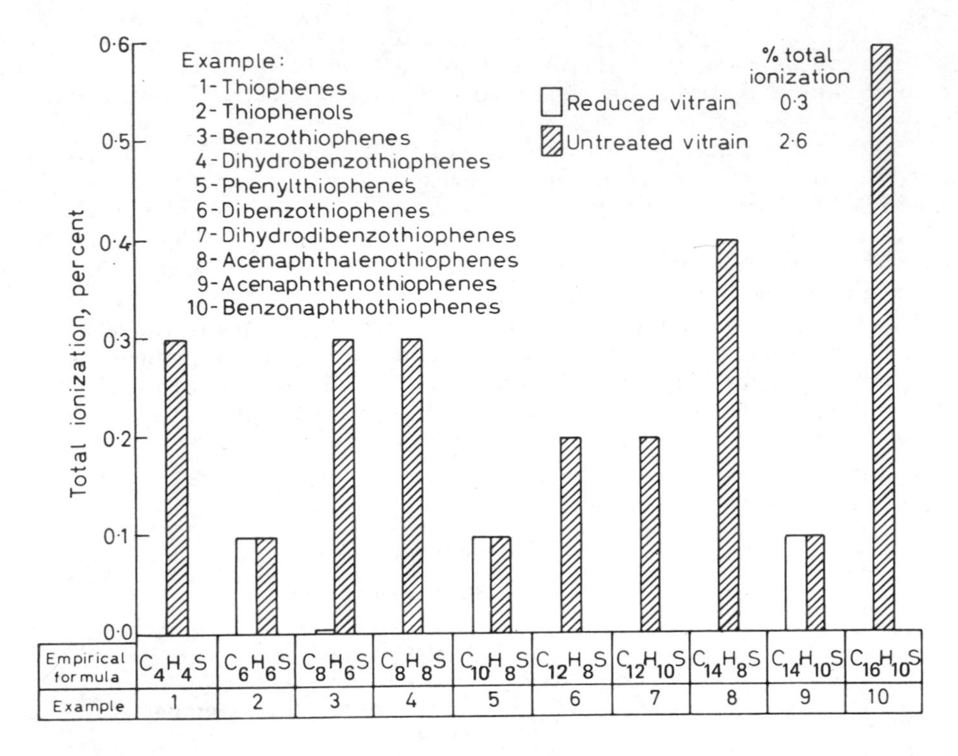

Figure 25. Sulfur compounds in pyridine extracts of reduced and untreated Pittsburgh seam vitrain. (From Ref. 94.)

VI. Combined Methods for Determination of Sulfur-Containing PAH

Sulfur-containing PAH represent an important class of compounds in coal and its products. Active investigation of these compounds in coal-derived materials have been conducted since the early 1920s with varying degrees of success. Kruber [95-98] identified a number of sulfur-containing PAH in coal tars. The detailed characterization of the sulfur-containing compounds in coal-derived products is important for two reasons. First, the human health effects of these products are a direct result of the specific structures of the individual components in the mixtures, as has been found within the class of PAH. Different isomers exhibit varying degrees of carcinogenic activity. Second, the removal of sulfur from fuels is desirable to prevent the formation of noxious sulfur gases during combustion and to prevent poisoning of coal-liquid-upgrading catalysts. The identification of specific sulfur-containing species in coal and coal-derived fuels can provide insights into better methods of sulfur removal. As stated previously, one of the earliest investigations of sulfur-containing PAH in coal was conducted by Kessler et al. [94], using HRMS to identify thiophenic PAH in pyridine extracts of coal. Investigators from this research group also studied the sulfur-containing PAH in coal carbonization tars [99] and in coal liquefaction products [100], again using HRMS. Although HRMS can provide insight into the nature of the sulfur-containing PAH by providing their molecular formulas, it does not by itself identify specific compounds. The investigator must guess at logical structures of sulfur-containing PAH based on the molecular formulas and his knowledge of the sample. Some other investigations of coal products where sulfur-containing PAH have been identified include that of Borwitzky et al. [61], Schultz et al. [48], and Aczel et al. [84].

Investigation of the sulfur-containing PAH in coal and its products is best accomplished through the use of a variety of chromatographic and MS techniques. These techniques include column chromatography, high-resolution glass capillary GC employing sulfur-selective detectors such as the flame-photometric detector (FPD), and combined GC/MS. Helmut Pichler and colleagues [17] were among the first to use GC combined with a sulfur-selective detector to analyze a coal product for sulfur-containing components. White

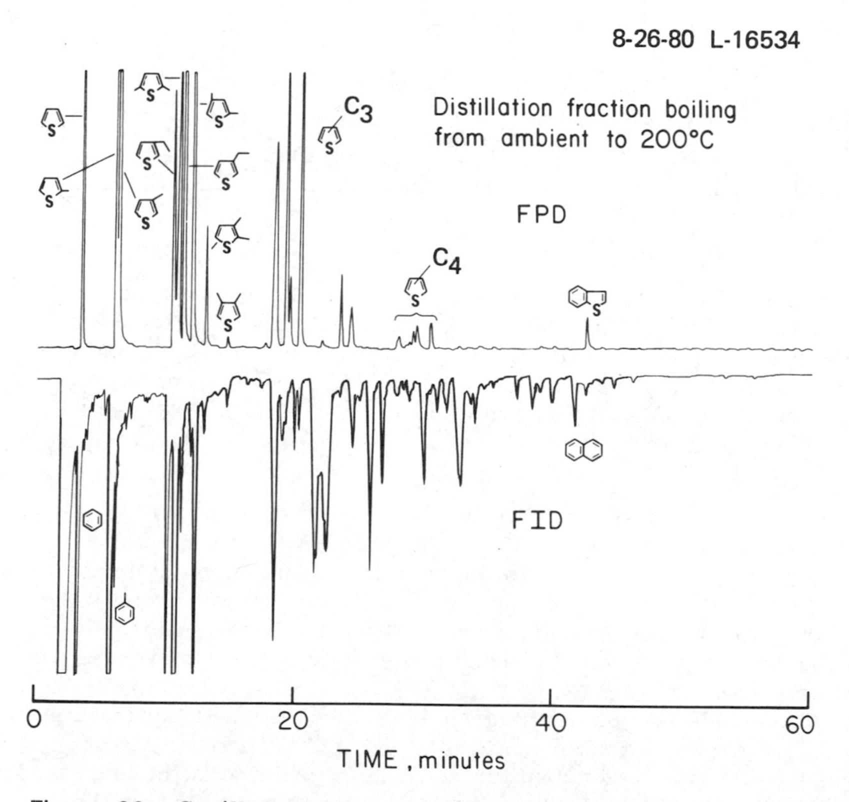

Figure 26. Capillary column gas chromatograms of a coal gasification tar distillate boiling from ambient temperature to 200°C, obtained using a dual sulfur-selective FID/FPD. Chromatographic conditions: 12 m × 0.28 mm i.d. glass capillary column coated with SE-52, using He carrier at 1.5 ml/min and temperature programmed from 50° to 230°C at 2°C/min. (From Ref. 102.)

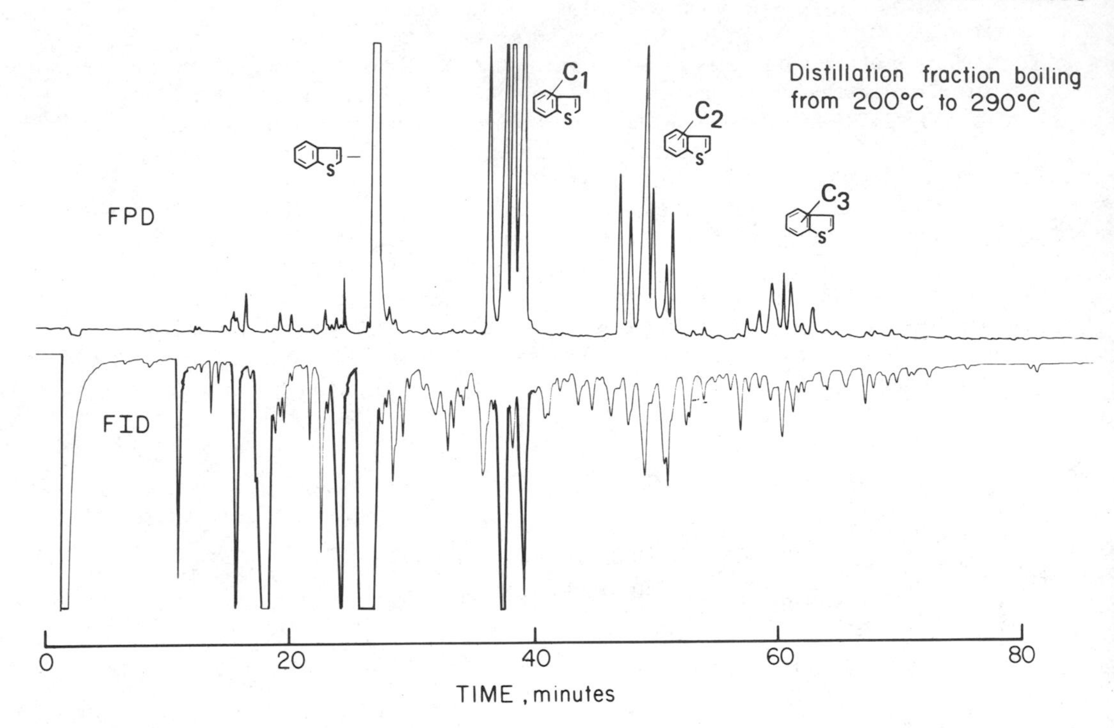

Figure 27. Capillary column gas chromatograms of a coal gasification tar distillate boiling from 200° to 290°C, obtained using a dual sulfur-selective FID/FPD. Chromatographic conditions: 12 m × 0.28 mm i.d. glass capillary column coated with SE-52, using He carrier at 1.5 ml/min and temperature programmed from 90° to 230°C at 2°C/min. (From Ref. 102.)

and Schmidt [101] also used sulfur-selective detection to analyze sulfur-containing components, including a sulfur-containing PAH, in wastewaters from a coal gasification process. GC combined with sulfur-selective detection or, better yet, dual FID/FPD can be used to identify sulfur-containing PAH in coal and its products. Unfortunately, when GC methods are used by themselves, the analyst must rely on cochromatography to identify the individual isomeric sulfur-containing PAH. Since there is a lack of pure sulfur-containing PAH available for cochromatographic studies, when GC is used alone, many sulfur-containing components must go unidentified. For this reason, samples being analyzed for sulfur-containing PAH should be analyzed by combined GC/MS after being analyzed by high-resolution glass capillary GC combined with sulfur-selective detection. The same chromatographic methods used in the GC experiments employing dual detection should be used in the GC/MS experiments. Those components that could not be identified by cochromatography and sulfur-selective detection due to a lack of standards can be tentatively identified by GC/MS. Furthermore, those identifications made by cochromatography can be confirmed by GC/MS. The sulfur-selective detection experiments allow the analyst performing the GC/MS analysis to zero in on those components containing sulfur. Moreover, interpretation of the mass spectral data is easier because the mass spectroscopist knows the compound contains sulfur. The combined techniques of high-resolution GC and sulfur-selective detection have been used to obtain detailed structural information on the sulfur-containing PAH in coal gasification tars [102] and coal liquefaction products. The chromatographic profiles, obtained using a dual FID/FPD, of the sulfur-containing PAH in a gasification tar that had been distilled into three boiling fractions appear in Figs. 26, 27, and 28. White and Lee [64] have also applied these methods to analyze the sulfur-containing PAH in the aromatic fraction of a coal extract (Fig. 29). The results of the investigation of sulfur-containing PAH in a coal extract are detailed in a recent publication [64] and have far-reaching implications concerning the origin of sulfur-containing PAH in coal. It is hypothesized that sulfur-containing PAH are formed by the reaction of PAH with elemental sulfur and/or pyrite. The previously unpublished results of an investigation of the sulfur-containing PAH in a coal liquefaction product appears in Fig. 30.

The analytical methods just described for analysis of sulfur-containing PAH in coal and coal products are very powerful. However, in cases where

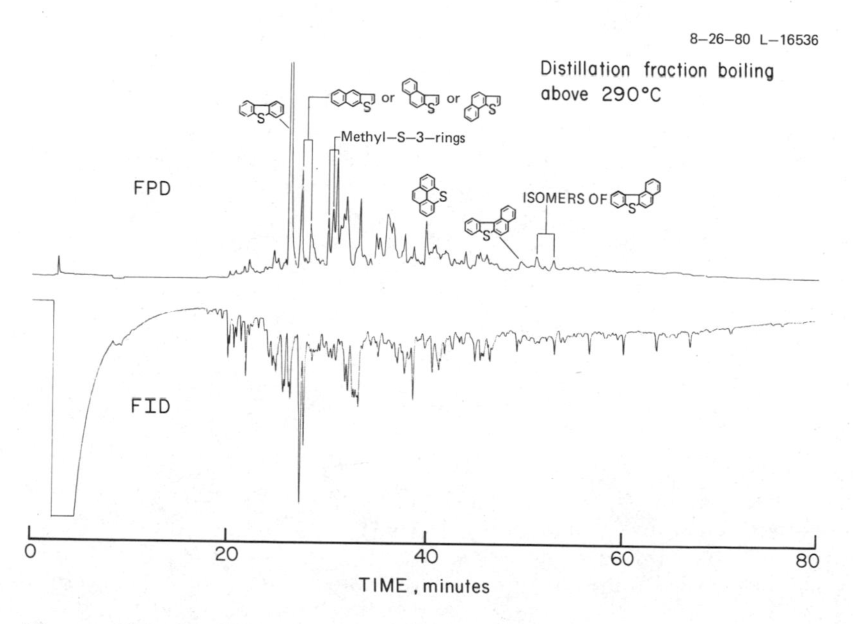

Figure 28. Capillary column gas chromatograms of a coal gasification tar distillate boiling above 290°C, obtained using a dual sulfur-selective FID/FPD. Chromatographic conditions: 12 m × 0.28 mm i.d. glass capillary column coated with SE-52, using He carrier at 1.5 ml/min and temperature programmed from 120° to 230°C at 4°C/min. (From Ref. 102.)

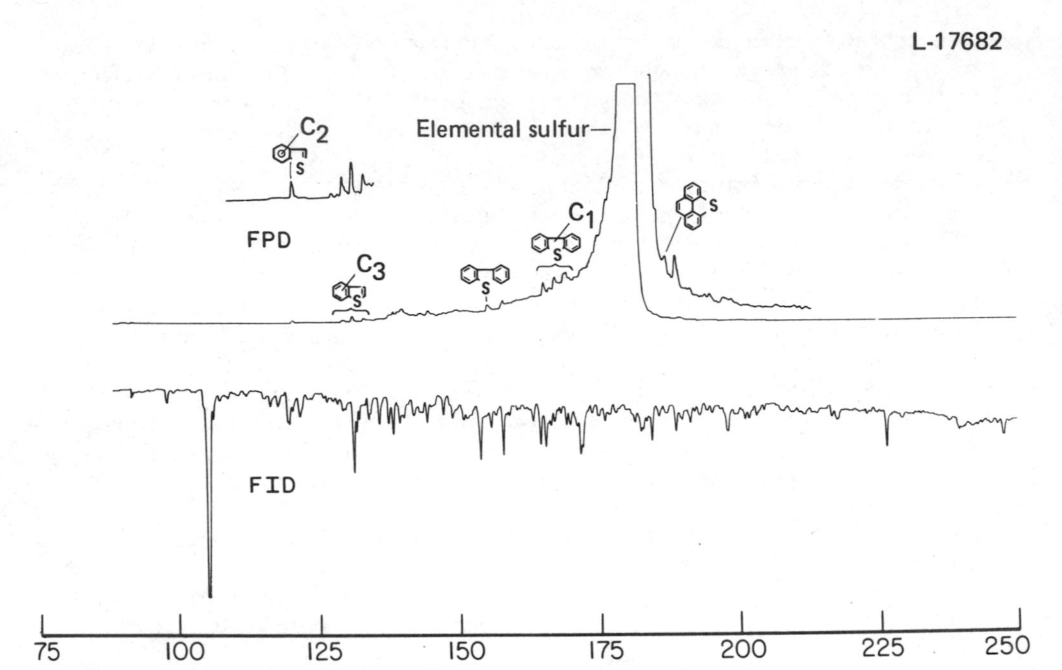

Figure 29. Capillary column gas chromatograms of the aromatic fraction of a Homestead, Kentucky, coal extract obtained using a dual sulfur-selective FID/FPD. Chromatographic conditions: 12 m × 0.28 mm i.d. glass capillary column coated with SE-52, using He carrier at a flow rate of 2.0 ml/min and temperature programmed from 50° to 250°C at a rate of 2°C/min. See Fig. 20 for the FID chromatogram of the same coal extract. (From Ref. 64.)

the sulfur-containing PAH are present in low concentration, obtaining reliable GC/MS data on them can be difficult or impossible. In these cases, it is necessary to isolate a sulfur-containing PAH fraction. This can be achieved by using the separation scheme depicted in Fig. 31, developed by Lee et al. [102]. This separation technique provides a relatively clean sulfur-containing PAH fraction, free of most interferences, that is ideal for combined GC/MS analysis. Chromatographic profiles obtained using an FID on the sulfur-containing PAH fraction isolated from the gasification tar distillates described previously appear in Figs. 32 and 33 (see also Table 18). It should be noted that the reduction step using $LiAlH_4$ can lead to chemical reduction of the thiophenic ring in those sulfur-containing PAH where one side of the thiophenic ring is exposed, not having a fused benzene ring.

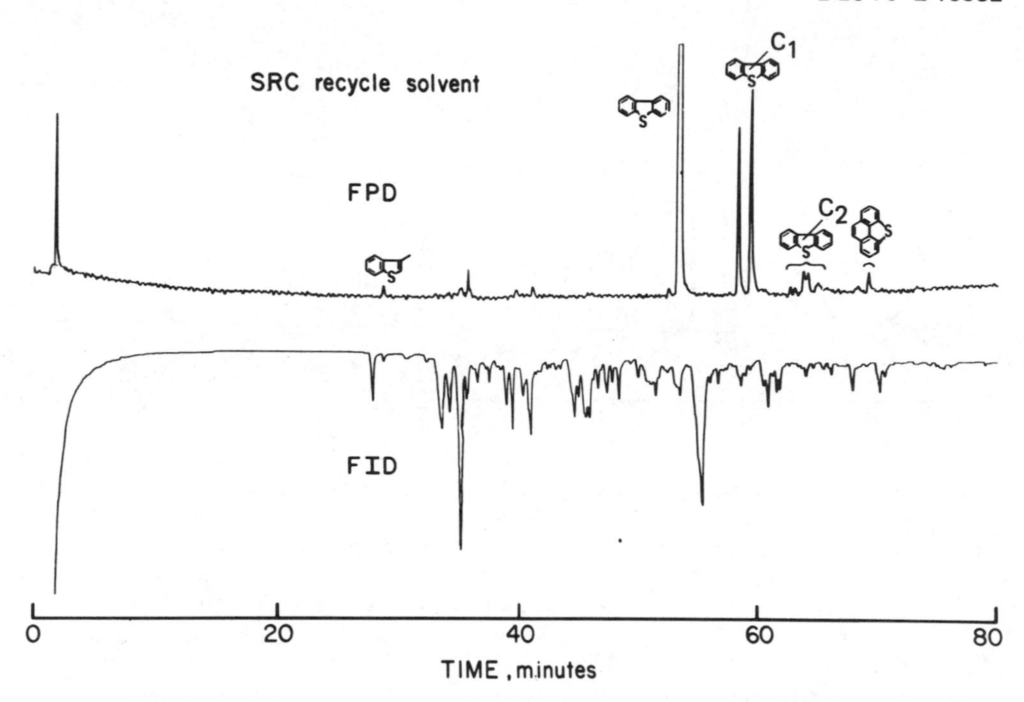

Figure 30. Capillary column gas chromatograms of a coal liquefaction product from the SRC-II process recycle solvent obtained using a dual sulfur-selective FID/FPD. Chromatographic conditions: 18 m × 0.28 mm i.d. glass capillary column coated with SE-52 using He carrier at 1.9 ml/min and temperature programmed from 50° to 210°C at 2°C/min.

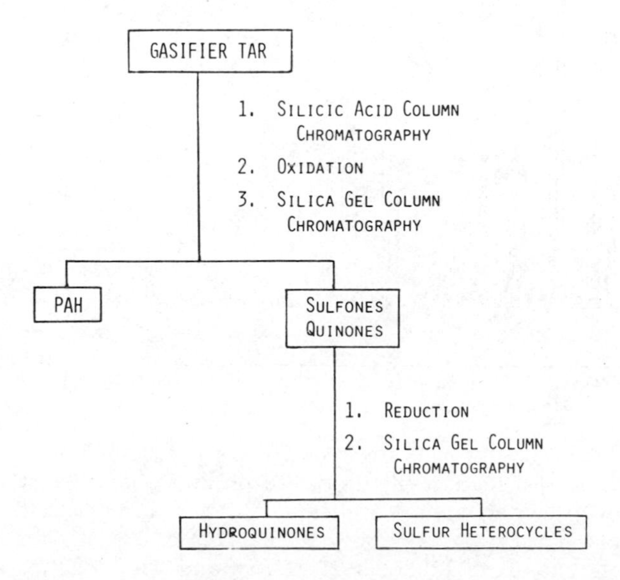

Figure 31. Sulfur heterocycle separation scheme. (From Ref. 102.)

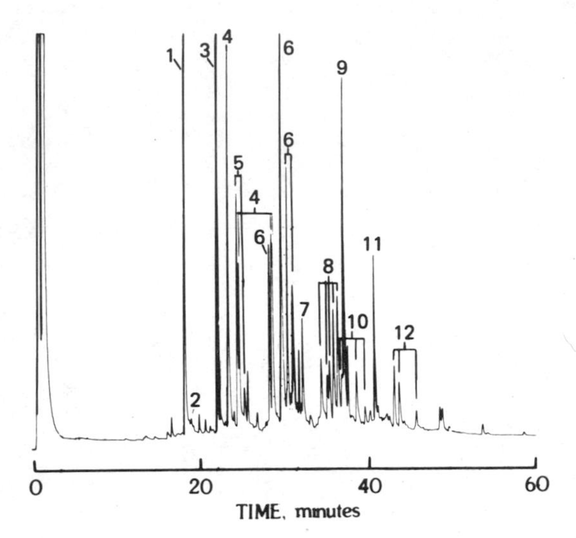

Figure 32. Capillary column gas chromatogram of the sulfur heterocycle fraction isolated from a coal gasification tar distillate boiling between 200° and 290°C, obtained using an FID. Chromatographic conditions: 20 m × 0.28 mm i.d. glass capillary column coated with SE-52, using He carrier at a flow rate of 1.8 ml/min and temperature programmed from 40° to 250°C at 2°C/min. Numbered chromatographic peaks are (1) naphthalene; (2) benzo[b]thiophene; (3) 2,3-dihydrobenzo[b]thiophene; (4) C_1-dihydrobenzothiophene; (5) C_1-benzothiophene; (6) C_2-dihydrobenzothiophene; (7) C_2-benzothiophene; (8) C_3-dihydrobenzothiophene; (9) dibenzofuran; (10) C_3-benzothiophene; (11) fluorene; (12) C_1-dibenzofuran. (From Ref. 102.)

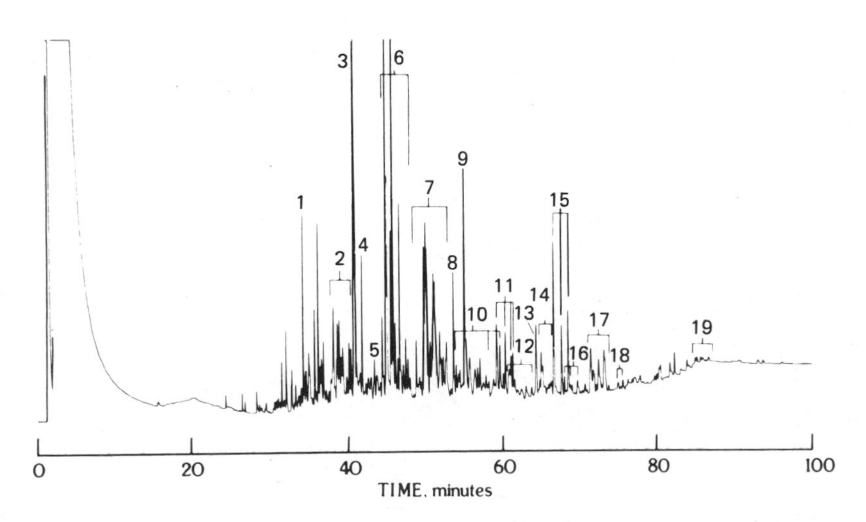

Figure 33. Capillary column gas chromatogram of the sulfur heterocycle fraction isolated from a coal gasification tar distillate boiling above 290°C, obtained using an FID. Chromatographic conditions: 30 m × 0.28 mm i.d. glass capillary column coated with SE-52, using He carrier at 1.8 ml/min and temperature programmed from 40° to 125°C at 10°C/min and then 125° to 250°C at 2°C/min. Numbered chromatographic peaks are identified in Table 18. (From Ref. 102.)

Table 18. Identification of Numbered Chromatographic Peaks from Fig. 33 in the Sulfur Heterocycle Fraction Isolated from the Coal Gasification Tar Distillate Boiling Above 290°C

Peak	Molecular weight	Compound
1	166	Fluorene
2	190 and 204	C_4- and C_5-benzothiophene
3	184	Dibenzothiophene
4	178 and 184	Phenanthrene and naphtho[2,1-b]thiophene
5	184	Naphthothiophene
6	198	C_1-dibenzothiophene
7	212	C_2-dibenzothiophene
8	202	Fluoranthene
9	208	Phenanthro[4,5-bcd]thiophene
10	226	C_3-dibenzothiophene
11	222	C_1-phenanthro[4,5-bcd]thiophene
12	240	C_4-dibenzothiophene
13	254	C_5-dibenzothiophene
14	236	C_2-phenanthro[4,5-bcd]thiophene
15	234	Isomers of benzo[b]naphtho[2,1-d]thiophene
16	250	C_3-phenanthro[4,5-bcd]thiophene
17	248	Isomers of C_1-benzo[b]naphtho[2,1-d]thiophene
18	262	Isomers of C_2-benzo[b]naphtho[2,1-d]thiophene
19	258	Isomers of chryseno[4,5-bcd]thiophene

[a]Numbers refer to peaks in Fig. 33.
Source: Ref. 102.

VII. Analytical Techniques on the Horizon for Determination of PAH in Coal and Coal Products

Over the decades, many analytical methods have been developed for the determination of PAH in coal and coal-derived materials. Since the early 1960s, there has been an amazing improvement in existing analytical techniques and, recently, the development of several revolutionary analytical methods and techniques for qualitative and quantitative analysis of PAH. This section will be concerned with briefly describing some analytical methods currently being developed that have shown great promise in differentiating and thus determining specific isomeric PAH in coal and its products. Among the techniques to be discussed are low-temperature luminescence, room-temperature phosphorimetry, and mixed charge exchange-chemical ionization MS. The following descriptions are not meant to be exhaustive theoretical descriptions of the techniques but rather an introduction to their application for the analysis of PAH in coal and related products.

Electronic spectra of PAH in fluid media are often broad. Thus, spectral overlap of individual PAH is a severe problem in mixture analysis. Fluid media are conducive to energy transfer and intermolecular quenching. These problems are partially overcome at low temperature when some PAH are frozen in crystalline Shpol'skii host solvents (see Chap. 8 and Refs. 103-107).

The technique is not without problems or limitations when used for analysis of complex mixtures of PAH found in coal extracts or products. One of the most difficult problems is microcrystallization or aggregation of PAH from the n-alkanes during freezing. Formation of solids during the freezing process can lead to intermolecular quenching and energy transfer. This problem can be partially solved by using matrix isolation methods [106].

Several variations of the use of the Shpol'skii effect have been developed to characterize PAH in coal extracts, coal maceral extracts, and coal-derived materials. The variations primarily exist in the excitation source used, in the host solvent, and in whether or not the sample was separated into an aromatic fraction prior to analysis.

Drake et al. [107] have applied Shpol'skii luminescence spectroscopy to investigate the PAH content of extracts from the macerals exinite, vitrinite, and inertinite from three British coals. The macerals were extracted with carbon disulfide. The carbon disulfide was removed, and the resulting residue was dissolved in 9:1 hexane/cyclohexane. Shpol'skii luminescence spectra were recorded in the 370-480 nm region at 77°K using either a xenon-arc continuum or mercury-vapor discharge excitation source. The Shpol'skii luminescence spectrum of a vitrinite extract appears in Fig. 34, while the results obtained on the three macerals from three coals and on a pitch extract appear in Table 19. No preseparation of the extract to obtain a PAH fraction was performed.

Woo et al. have developed and applied x-ray-excited optical luminescence (XEOL; i.e., Shpol'skii spectrometry using x-ray excitation) to the analysis of PAH in coal extracts and coal-derived products [67,68]. Seven PAH were tentatively identified in PAH fractions of Iowa and Illinois coals [67]. The coals were extracted, and the PAH fraction was isolated by a 30-hour separation scheme. The resulting solvent-free PAH fractions were dissolved in n-heptane (a Shpol'skii host solvent), solidified at 90K, and irradiated with x-rays. The XEOL spectra of the PAH fractions from the coal extracts appear in Fig. 35, along with the identification of seven PAH. Woo et al. do not specifically state what criteria were used to identify specific PAH; however, it is presumed that the wavelengths of the spectral lines from the PAH of the coal extract were compared with those of pure PAH. The XEOL technique has also been used to tentatively identify a number of PAH in coal liquefaction and gasification products [68]. It should be noted that it is necessary to isolate the PAH fraction of coal products prior to XEOL analysis because of the broad spectral background encountered if this is not done. Figures 36 and 37 illustrate the XEOL spectra of the PAH fractions from solvent refined

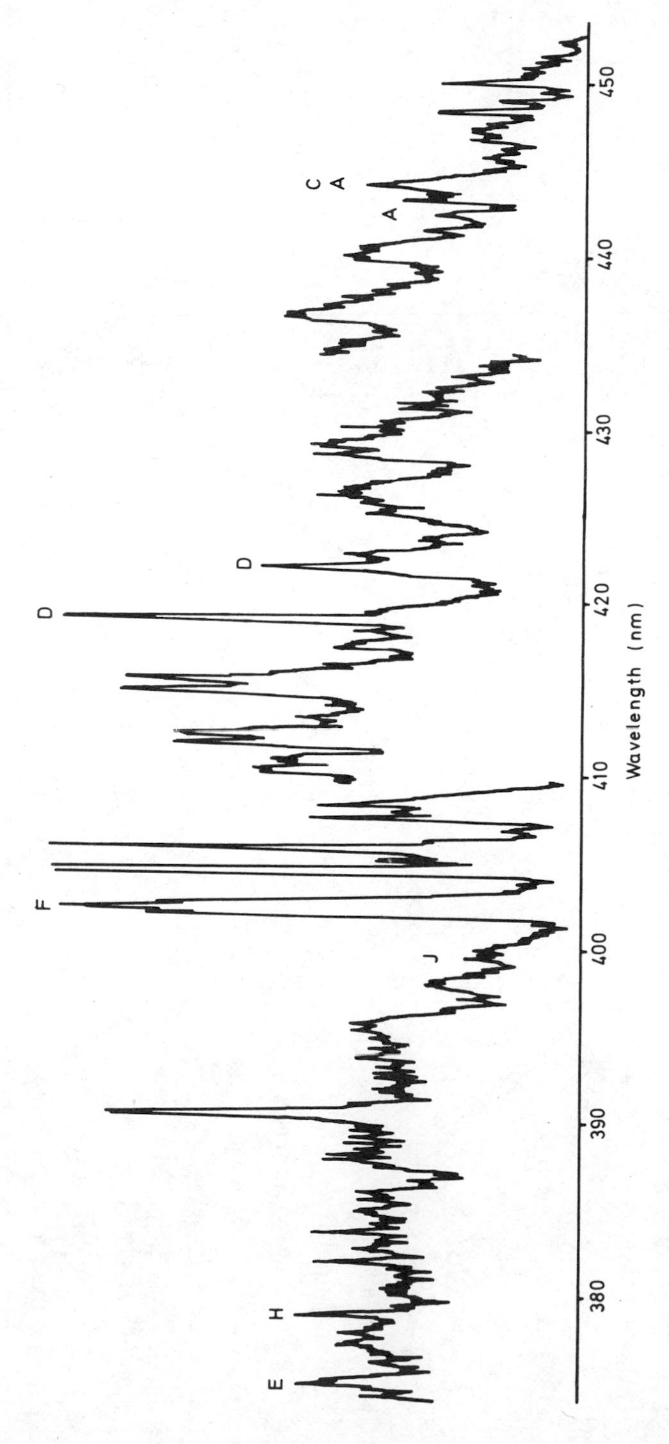

Figure 34. Shpol'skii luminescence spectrum of a CS_2 extract of the vitrinite maceral (Markham Main colliery) in 1:9 cyclohexane/hexane: obtained at 77K with a mercury-vapor lamp. Results on the maceral, macerals from other coals, and a coal tar pitch are in Table 19. (From Ref. 107.)

Table 19. PAH Detected by Shpol'skii Spectroscopy at 77K in Carbon Disulfide Coal and Pitch Extracts (See Fig. 34)[a]

PAH	Chislet Kent) No. 5 Seam			Markham Main (Yorkshire) Barnsley Seam		Cannock Wood (Staffordshire) Shallow Seam		Pitch extract
	85% E	96% V	90% I	89% E	98% V	90% E	98% V	
(A) Coronene	/[b]	/	/	/	/	/	/	–
(B) Ovalene	/	/	/	–	–	–	–	–
(C) Perylene	/	/	/	/	/	/	–	/
(D) Benz[ghi]perylene	/	–	/	/	/	/	–	/
(E) Pyrene	–[c]	–	–	/	/	/	–	/
(F) Benz[a]pyrene	/	–	/	/	/	/	/	/
(G) Dibenz[a,i]pyrene	–	–	–	/	–	/	–	/
(H) Chrysene	–	–	–	–	/	–	–	–
(I) Benz[a]anthracene	–	–	–	/	–	–	–	–
(J) 9,10-Dimethylanthracene	–	–	–	–	/	/	–	–
(K) Benz[b]fluoranthene	–	–	–	–	–	–	–	/
(L) Dibenz[a,h]pyrene	–	–	–	–	–	/	–	/
Ratio of cyclohexane in solvent matrix	1:9	1:9	1:9	1:9	1:9	1:100	1:9	1:50
Amount extracted (wt%)	6.3	0.5	0.3	1.4	0.3	1.1	0.4	–
NCB Coal Rank Code No.	501	601	702		702		902	

[a]Key: E, exinite; V, vitrinite; I, inertinite. Insufficient inertinite was available in the Markham Main and Cannock
[b]/ means detected.
[c]– means not detected.
Source: Ref. 107.

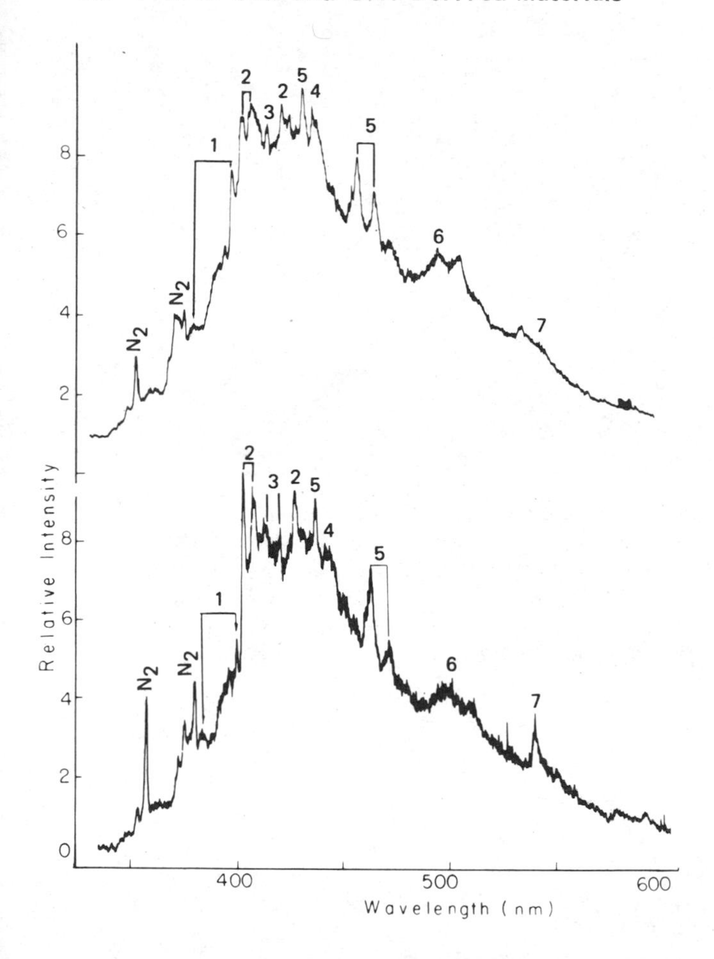

Figure 35. X-ray excited optical luminescence spectra of PAH isolated from coal samples obtained from Iowa (upper) and Illinois (lower). Numbered peaks are (1) 1,2-benzanthracene; (2) 3,4-benzopyrene; (3) benzo[ghi]-perylene; (4) 3,4,8,9-dibenzopyrene; (5) perylene; (6) phenanthrene; (7) 1,2-benzopyrene. (From Ref. 67.)

coal (SRC) process solvent and of a tarry residue from a gasification process. These spectra were obtained as previously described for the coal extracts. Woo and co-workers [68], as well as Drake et al. [107], have claimed that PAH determined by Shpol'skii spectrometry are positively identified. By definition, PAH or any other compounds cannot be positively identified by any single measurement. A minimum of two mutually corroborative identification parameters are needed to positively identify any compound in a sample.

The most impressive example to date of the potential analytical power of Shpol'skii spectrometry to analyze PAH in coal-derived materials has been published by Yang and co-workers [69] at the Ames Laboratory of the U.S.

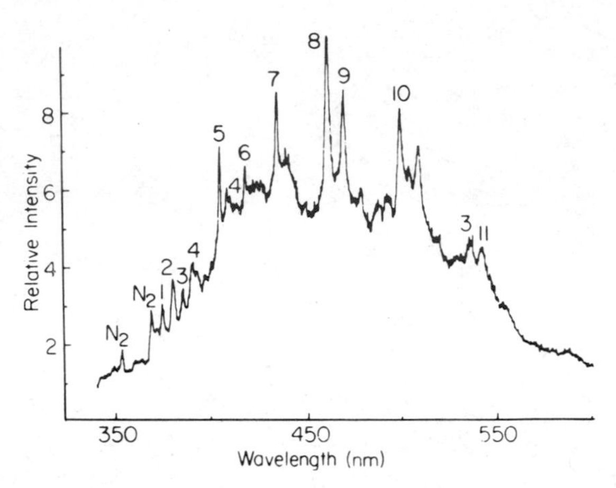

Figure 36. X-ray-excited optical luminescence of PAH extracted from 250 mg of solvent refined coal-process solvent. The spectrum represents PAH present in only 3 mg of the sample: (1) chrysene; (2) benz[a]anthracene; (3) benzo[e]pyrene; (4) 3-methylcholanthrene; (5) benzo[a]pyrene; (6) benzo[ghi]perylene; (7) dibenzo[a,i]pyrene; (8) triphenylene; (9) perylene; (10) phenanthrene; (11) fluoranthene. (From Ref. 68.)

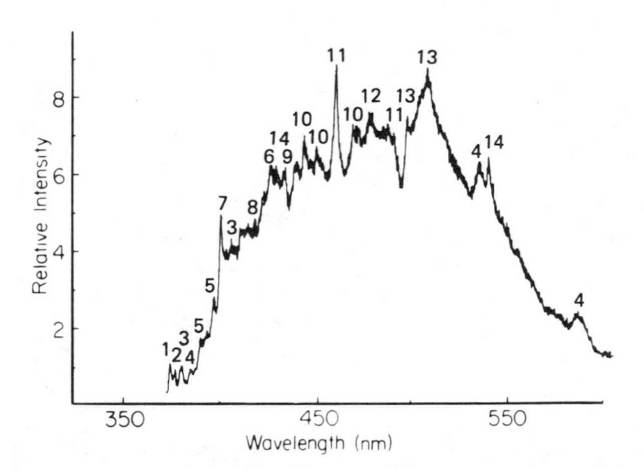

Figure 37. X-ray-excited optical luminescence of PAH extracted from 1.7 mg of a tarry residue obtained during an experiment conducted at the Morgantown Energy Research Center, Morgantown, W.Va. The spectrum represents PAH present in only 1/200 of the processed sample: (1) chrysene; (2) pyrene; (3) benz[a]anthracene; (4) benzo[e]pyrene; (5) dibenz[a,h]anthracene; (6) anthracene; (7) benzo[a]pyrene; (8) benzo[ghi]perylene; (9) dibenzo[a,i]-pyrene; (10) perylene; (11) triphenylene; (12) dibenzo[a,h]pyrene; (13) phenanthrene; (14) fluoranthene. (From Ref. 68.)

Department of Energy. This group has used laser-excited Shpol'skii spectrometry (LESS) employing tunable dye lasers to selectively excite the sample. No preseparation of the sample was performed. A 0.1-g sample of an SRC-II coal liquefaction product was appropriately diluted with purified n-octane. The luminescence spectra were recorded at 15K, while the sample was selectively excited using a tunable dye laser. Some of the spectra obtained and the identifications made from them appear in Fig. 38. The technique can also be used to obtain quantitative data on individual PAH present in coal-derived materials. It is possible that the use of LESS for analysis of PAH from coal and other products will increase dramatically, rivaling and possibly supplanting the use of GC/MS for PAH analysis.

Room-temperature phosphorimetry (RTP) is another relatively new analytical technique that is showing promise of being able to analyze complex mixtures

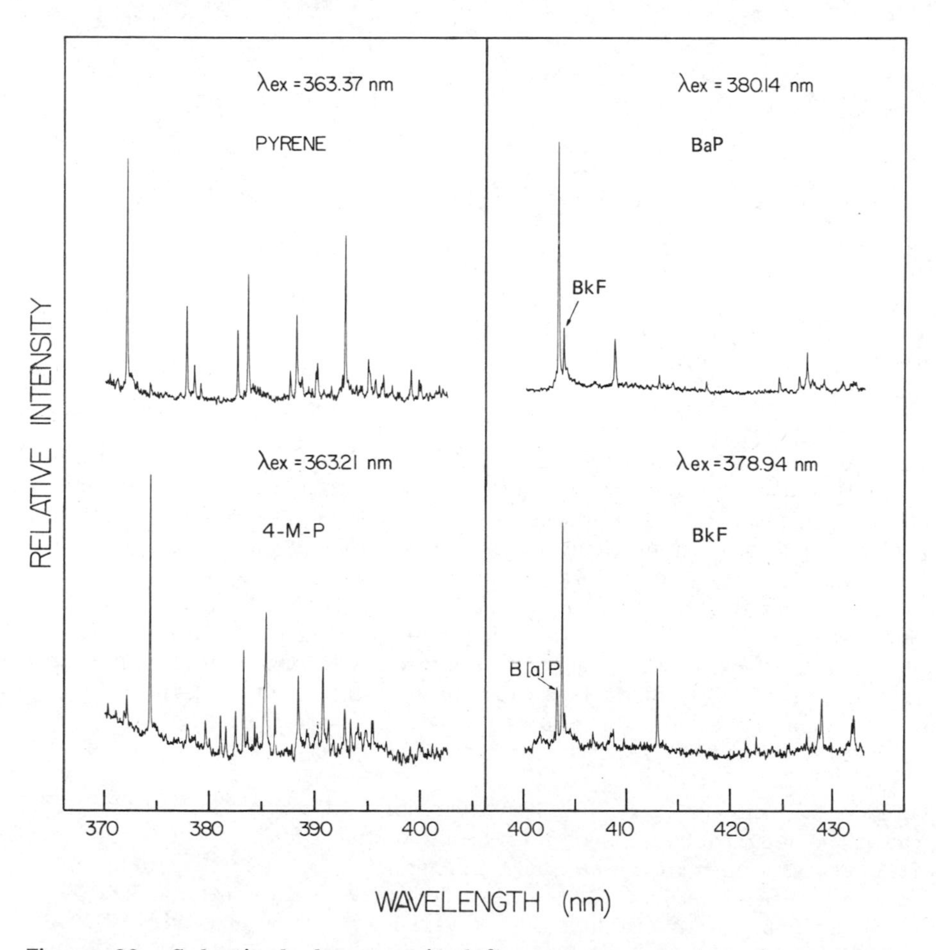

Figure 38. Selectively laser-excited fluorescence spectra of pyrene, 4-methylpyrene (4-M-P), B[a]P, and B[k]F in a solvent-refined coal liquid (SRC-II) sample. (From Ref. 68.)

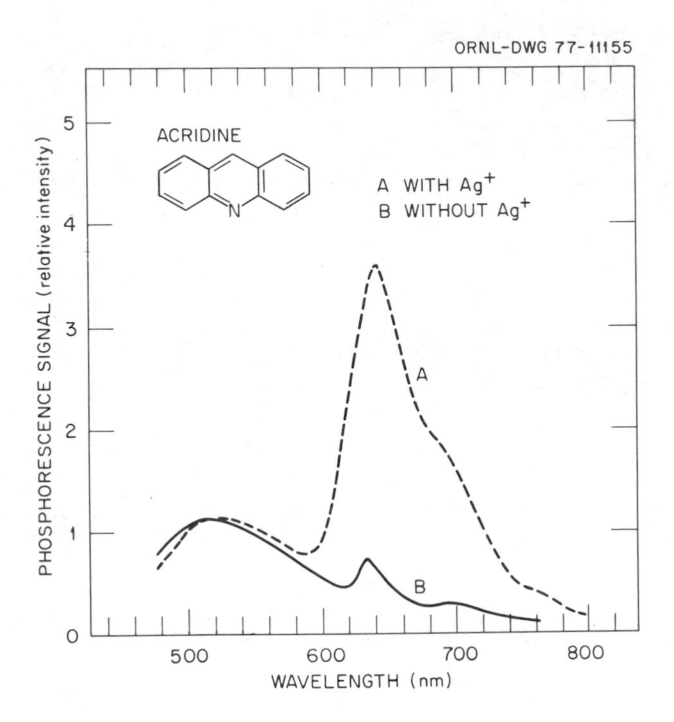

Figure 39. Heavy-atom affect on room temperature phosphorescence of acridine: (a) with silver nitrate (0.25 M), 52-ng sample; (b) without silver nitrate, 210-ng sample. (From Ref. 108.)

of coal-derived materials for specific PAH. Phosphorimetry has conventionally been conducted at cryogenic temperatures so that collisional triplet quenching can be minimized while the sample is in a rigid matrix. This same effect can be achieved at room temperature by adsorption of the sample onto a support such as alumina, silica gel, or filter paper. When this is done, the sample is immobilized by adsorbent-adsorbate forces. Further, the limit of optical detection of some PAH by RTP can be improved through the addition of heavy metal ions to the sample [108,109]. Heavy metal ions encourage population of the triplet state by intersystem crossing and spin orbit coupling, thus enhancing phosphorescence emission by PAH. The effect of the addition of heavy metals on the RTP spectrum of acridine is dramatically depicted in Fig. 39.

A major limitation of RTP for the analyses of multicomponent PAH mixtures has been the occurrence of major interferences due to overlap of other phosphorescent compounds in the sample. RTP spectra of PAH are usually of a broad structureless nature. Thus, the selectivity of conventional RTP is poor with respect to any single PAH. Nevertheless, the selectivity of RTP for individual PAH can be improved by using synchronous excitation, which employs the specificity of energy differences, Δ_{ST}, between the phosphorescence emission band ($T_1 \rightarrow S_0$) and absorption bands ($S_1 \leftarrow S_0$ or $S_N \leftarrow S_0$)

[110]. This technique exploits the singlet-triplet energy difference as a selectivity factor in phosphorimetry. The procedure has the ability to "sort out" selectively an emission signal of one compound present in a mixture. An example of the use of the Δ_{ST} selection approach for a single PAH present in a simple three-component mixture is depicted in Fig. 40. Using a fixed excitation wavelength at 300 nm, the RTP spectrum of the three-component mixture is broad and diffuse with considerable overlap of the spectral bands of the three components. The use of the singlet-triplet energy difference dramatically improves the selectivity. Scanning the phosphorescence spectrum synchronously using λ = 125 nm gives mainly the emission spectrum of fluorene, while the spectra of the other components have been nulled out. The synchronous RTP technique has also been used to analyze a coal liquefaction product for a PAH. It must be remembered that coal products are considerably more complex than a simple three-component mixture. Analysis of a coal liquefaction product for a specific PAH is the acid test. The synchronous

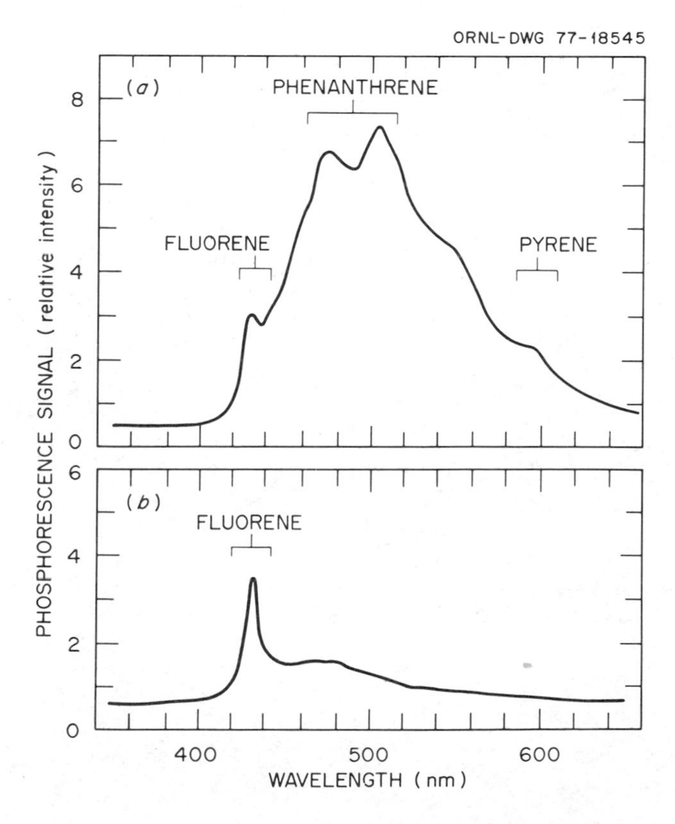

Figure 40. RTP spectra of a mixture of fluorene (20 ng), phenanthrene (50 ng), and pyrene (80 ng): (a) conventional fixed excitation spectrum (λ_{ex} = 300 nm); (b) synchronous spectrum of the same sample ($\Delta\lambda$ = 125 nm). (From Ref. 110.)

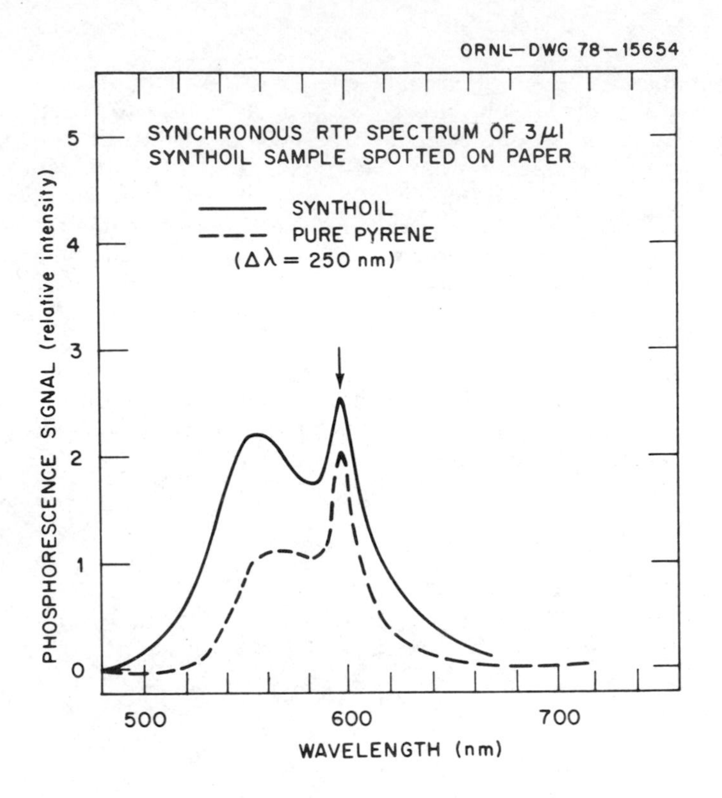

Figure 41. Identification of pyrene in Synthoil by synchronous RTP analysis ($\Delta\lambda$ = 250 nm): spectrum of Synthoil (solid curve); spectrum of pure pyrene (dashed curve). (From Ref. 110.)

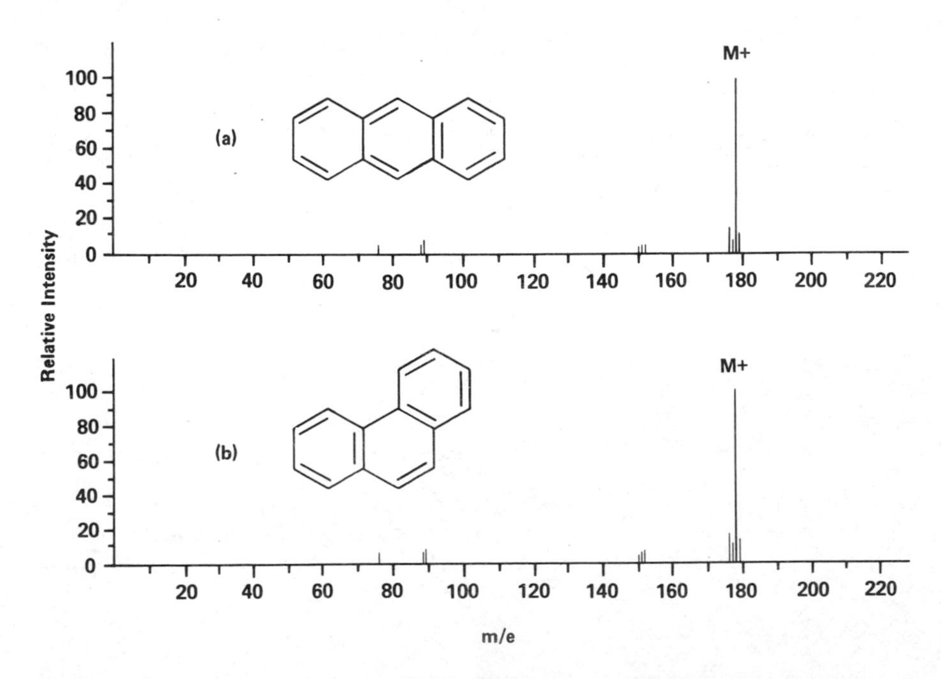

Figure 42. EI mass spectra of (a) anthracene and (b) phenanthrene.

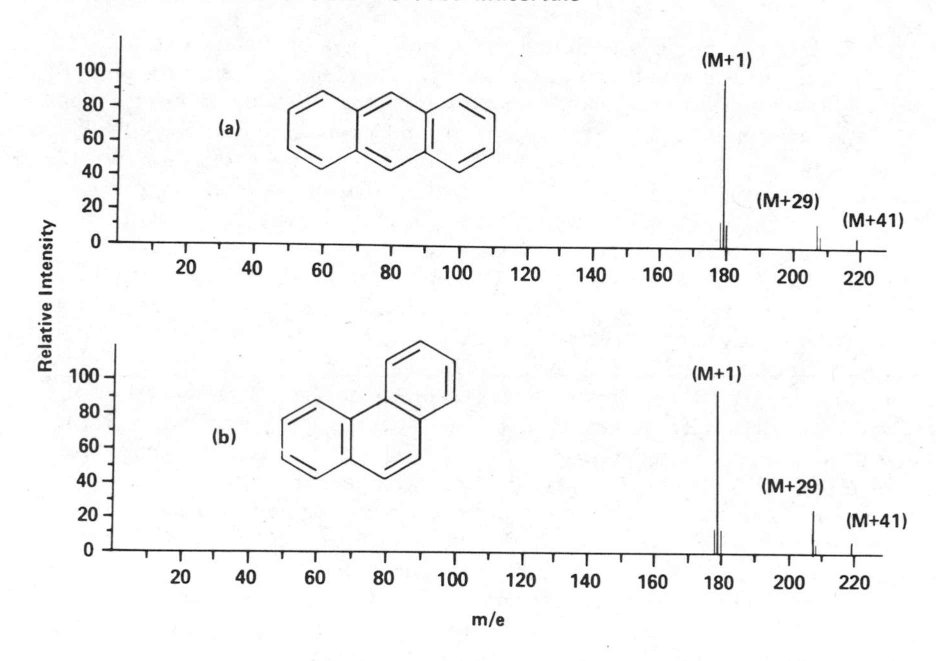

Figure 43. Methane chemical ionization mass spectra of (a) anthracene and (b) phenanthrene.

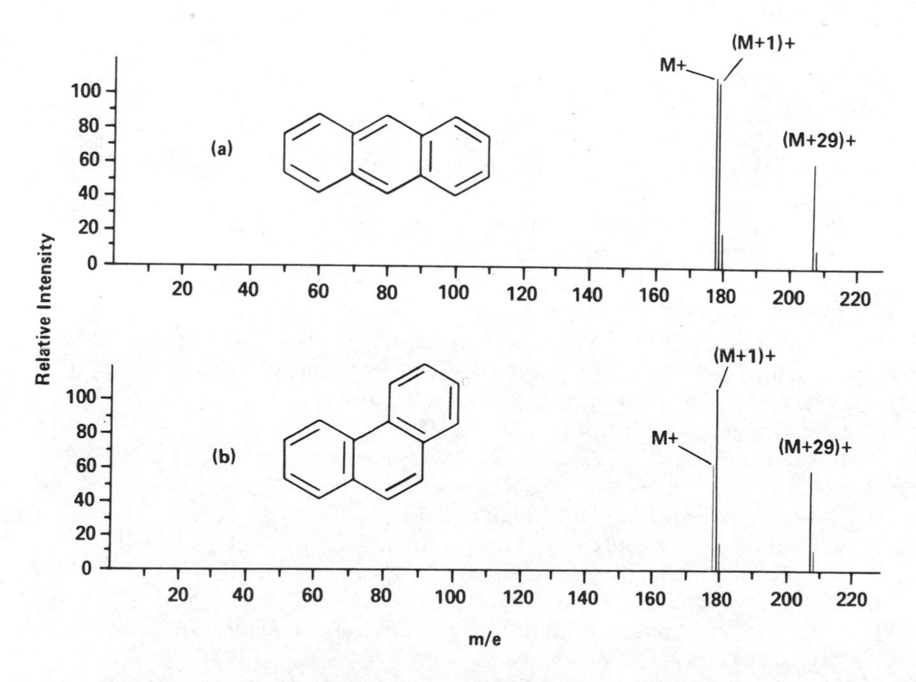

Figure 44. Argon-methane mixed charge exchange-chemical ionization mass spectra of (a) anthracene and (b) phenanthrene.

RTP spectrum of pyrene was obtained on an unseparated crude extract of a coal liquefaction product and is shown in Fig. 41. This is certainly an impressive achievement and indicates that synchronous RTP could become a valuable analytical technique for analysis of coal products for select PAH. It should be noted that synchronous RTP has recently been used to tentatively identify and quantitate pyrene and other PAH in coal liquefaction products.

Another analytical technique that shows promise for differentiating between isomeric PAH in complex mixtures of coal products is the mixed charge exchange/chemical ionization (CE/CI) GC/MS technique being developed by Lee and Hites ([111]; see also Lee et al. [112] and Hites and Dubay [113]). Isomeric PAH usually have electron impact (EI) mass spectra that are indistinguishable from one another. Although in many cases they can be separated from one another by GC, sometimes the gas chromatographic separation of isomeric PAH is difficult. Mixed CE/CI mass spectrometry employing 5-10% methane in argon as the reagent gas has been shown to produce mass spectra of isomeric PAH that are sufficiently different to permit the analyst to recognize a specific isomer based on mass spectral evidence alone. Reasons for the mass spectral differences are discussed elsewhere [111-113]. Figures 42, 43, and 44 depict the mass spectra of anthracene and phenanthrene under EI, methane chemical ionization, and mixed CE/CI. Clearly, the spectra obtained for isomeric anthracene and phenanthrene using mixed CE/CI are different. Thus, in cases where isomeric PAH are not separable from one another by GC, it is sometimes possible to differentiate between such isomers based on differences in their mixed CE/CI mass spectra. To the present author's knowledge, this technique has not been applied to analysis of PAH from coal or its products.

References

1. E. Donath, *Chem.-Ztg. 32*:1271 (1908).
2. R. W. Freedman and A. G. Sharkey, Jr., *Ann. N.Y. Acad. Sci. 200*: 7-16 (1972).
3. R. W. Freedman, in *Analytical Methods for Coal and Coal Products*, C. Karr, Jr. (Ed.), Vol. 2, Academic Press, 1978.
4. V. Saffiotti, F. Cefis, L. H. Kolb, and P. Shubik, *J. Air Pollut. Control Assoc. 15*:23-25 (1965).
5. C. M. White and J. O. H. Newman, Chromatographic and spectrometric investigation of a light oil produced by the synthoil process. PERC/RI 76/3, available from the National Technical Information Center, U.S. Dept. of Commerce, Springfield, Va. (1976).
6. M. J. Finn, G. Fynes, W. R. Ladner, and J. O. H. Newman, *Fuel 59*: 397 (1980).
7. R. D. Sexton, *Arch. Environ. Health 1*:181 (1960).
8. C. S. Weil and N. I. Condra, *Arch. Environ. Health 1*:187 (1960).
9. N. H. Ketcham, and R. W. Norton, *Arch. Environ. Health 1*:194 (1960).
10. R. D. Sexton, *Arch. Environ. Health 1*:208 (1960).
11. N. Fannick, L. T. Gonshor, and J. Shockley, Jr., *Amer. Ind. Hyg. Assoc. J. 33*:461 (1972).
12. A. A. Herod and R. G. James, *Fuel Inst. 51*:164 (1978).

13. C. M. White, A. G. Sharkey, Jr., M. L. Lee, and D. L. Vassilaros, in *Polynuclear Aromatic Hydrocarbons*, 2nd ed., P. W. Jones and P. Leber (Eds.), Ann Arbor Science Publs., Ann Arbor, Mich., 1979.
14. T. Tsuda and M. Novotny, *Anal. Chem. 50*:271 (1978).
15. Y. Hirata, M. Novotny, T. Tsuda, and D. Ishii, *Anal. Chem. 51*:1807 (1979).
16. F. Dupire and G. Botquin, *Anal. Chim. Acta 18*:282 (1958).
17. H. Pichler, P. Hennenberger, and G. Schwarz, *Brennstoff-Chem. 49*: 175 (1968).
18. A. Liberti, G. P. Cartoni, and V. J. Cantuti, *J. Chromatogr. 15*:141 (1964).
19. E. Kovats, *Helv. Chim. Acta 41*:1915 (1958).
20. K. Ōuchi and K. Imuta, *Fuel 42*:25 (1963).
21. K. Ōuchi and K. Imuta, *Fuel 42*:445 (1963).
22. T. Sakabe and R. Sassa, *Bull. Chem. Soc. Japan 25*:353 (1952).
23. I. Gavat and I. Irimescu, *Chem. Ber. 74*:1812 (1941).
24. F. Hofmann and P. Damm, *Brennstoff-Chem. 3*:73 (1922).
25. F. Hofmann and P. Damm, *Brennstoff-Chem. 4*:65 (1923).
26. E. C. Sterling and M. T. Bogert, *J. Org. Chem. 4*:20 (1938).
27. V. Jarolim, M. Streibl, M. Horak, and F. Sorm, *Chem. Ind. (London)* p. 1142 (1958).
28. V. Jarolim, M. Streibl, K. Hejno, and F. Sorm, *Collect. Czech. Chem. Commun. 26*:451 (1961).
29. V. Jarolim, K. Hejno, M. Streibl, M. Horak, and F. Sorm, *Collect. Czech. Chem. Commun. 26*:459 (1961).
30. V. Jarolim, K. Hejno, and F. Sorm, *Collect. Czech. Chem. Commun. 28*: 2318 (1963).
31. V. Jarolim, K. Hejno, and F. Sorm, *Collect. Czech. Chem. Commun. 28*: 2443 (1963).
32. V. Jarolim, K. Hejno, F. Hemmert, and F. Sorm, *Collect. Czech. Chem. Commun. 30*:873 (1965).
33. F. H. Field and S. H. Hastings, *Anal. Chem. 28*:(1956).
34. H. E. Lumpkin, *Anal. Chem. 30*:321 (1958).
35. A. G. Sharkey, Jr., G. Wood, and R. A. Friedel, *Chem. Ind. (London)* p. 833 (1958).
36. A. G. Sharkey, Jr., G. Wood, J. L. Shultz, I. Wender, and R. A. Friedel, *Fuel 38*:315 (1959).
37. J. H. Beynon, *Mass Spectrometry and Its Application to Organic Chemistry*, Elsevier, New York, 1960.
38. K. Biemann, *Ann. Rev. Biochem. 32*:755 (1963).
39. A. G. Sharkey, Jr., J. L. Shultz, T. Kessler, and R. A. Friedel, *Amer. Chem. Soc. Div. Fuel Chem., Preprints 11*:232 (1967).
40. D. M. Jewell, R. G. Ruberto, and B. E. Davis, *Amer. Chem. Soc. Div. Petrol. Chem., Preprints 17*:F81 (1972).
41. D. M. Jewell, J. H. Weber, J. W. Bunger, H. Plancher, and D. R. Latham, *Anal. Chem. 44*:1391 (1972).
42. R. G. Ruberto, D. M. Jewell, R. K. Jensen, and D. C. Cronauer, *Advan. Chem. Ser. 151*:28-47, 1974.
43. W. E. Haines and C. J. Thompson, Separating and characterizing high boiling petroleum distillates: The USBM-API procedure. ERDA LERC/RI-

75/5-BERC/RI-75/2, available from the National Technical Information Service, U.S. Dept. of Commerce, Springfield, Va. (July 1975).
44. S. A. Holmes, P. W. Woodward, G. P. Sturm, Jr., J. W. Vogh, and J. E. Dooley, Characterization of coal liquids derived from the H-coal process. BERC/RI-76/10, available from the National Technical Information Service, U.S. Dept. of Commerce, Springfield, Va. (1976).
45. G. P. Sturm, Jr., P. W. Woodward, J. W. Vogh, S. A. Holmes, and J. E. Dooley, Characterization of a Colstrip coal liquid derived from the zinc chloride hydrocracking process. BERC/RI-78/4, available from the National Technical Information Service, U.S. Dept. of Commerce, Springfield, Va. (May 1978).
46. J. E. Schiller and D. R. Mathiason, *Anal. Chem. 49*:1225 (1977).
47. A. R. Jones, M. R. Guerin, and B. R. Clark, *Anal. Chem. 49*:1776 (1977).
48. R. V. Schultz, J. W. Jorgenson, M. P. Maskarinec, M. Novotny, and L. J. Todd, *Fuel 58*:783 (1979).
49. W. H. Griest, B. A. Tomkins, J. L. Epler, and T. K. Rao, in *Polynuclear Aromatic Hydrocarbons*, 2nd ed., P. W. Jones and P. Leber (Eds.), Ann Arbor Science Publs., Ann Arbor, Mich., 1979.
50. M. E. Snook, R. F. Severson, H. C. Higman, and O. T. Chortyk, in *Polynuclear Aromatic Hydrocarbons*, 2nd ed., P. W. Jones and P. Leber (Eds.), Ann Arbor Science Publs., Ann Arbor, Mich., 1979.
51. R. F. Severson, M. E. Snook, R. F. Arrendale, and O. T. Chortyk, *Anal. Chem. 48*:1866 (1976).
52. M. Farcasiu, *Fuel 56*:9 (1977).
53. L. G. Galya and J. C. Sautoni, *J. Liquid Chromatogr. 3*:229 (1980).
54. M. L. Lee and B. W. Wright, *J. Chromatogr. Sci. 18*:345 (1980).
55. H. Borwitzky and G. Schomburg, *J. Chromatogr. 170*:99 (1979).
56. H. G. Franck, *Angew. Chem. 63*:260 (1951).
57. K. Grob, G. Grob, and K. Grob, Jr., *HRC&CC 1*:31 (1979).
58. J. L. Shultz, *Spectrosc. Lett. 1*:345 (1968).
59. L. S. Lysyuk and A. N. Korol, *Chromatographia 10*:712 (1977).
60. W. Bertsch, E. Anderson, and G. Holzer, *J. Chromatogr. 126*:213 (1976).
61. H. Borwitzky, D. Henneberg, G. Schomburg, H. Sauerland, and M. Zander, *Erdöl Kohle Erdgas Petrochem. 30*:370 (1977).
62. J. Mostecky, M. Popl, and J. Kriz, *Anal. Chem. 42*:1132 (1970).
63. A. Bjorseth and G. Eklund, *Anal. Chim. Acta 105*:119 (1979).
64. C. M. White and M. L. Lee, *Geochim. Cosmochim. Acta 44*:1825 (1980).
65. W. A. Aue and D. R. Younker, *J. Chromatogr. 88*:7 (1974).
66. R. J. Laub and W. L. Roberts, in *Polynuclear Aromatic Hydrocarbons: Chemistry and Biological Effects*, A. Bjørseth and A. J. Dennis (Eds.), Battelle Press, Columbus, Ohio, 1980.
67. C. S. Woo, A. P. D'Silva, V. A. Fassel, and G. J. Oestreich, *Environ. Sci. Technol. 12*:173 (1978).
68. C. S. Woo, A. P. D'Silva, and V. A. Fassel, *Anal. Chem. 52*:159 (1980).
69. Y. Yang, A. P. D'Silva, V. A. Fassel, and M. Iles, *Anal. Chem. 52*:1351 (1980).
70. L. Blomberg and T. Wannman, *J. Chromatogr. 148*:379 (1978).
71. L. Blomberg, J. Buijten, J. Gawdzik, and T. Wannman, *Chromatographia 11*:521 (1978).

72. N. A. Goeckner and W. H. Griest, *Sci. Total Environ. 8*:187 (1977).
73. M. L. Lee, D. L. Vassilaros, C. M. White, and M. Novotny, *Anal. Chem. 51*:768 (1979).
74. H. Van Den Dool and P. D. Kratz, *J. Chromatogr. 11*:463 (1963).
75. H. Beernaert, *J. Chromatogr. 173*:109 (1979).
76. K. Grob, Jr., and H. P. Neukom, *HRC&CC 2*:15 (1979).
77. M. Galli, S. Trestianu, and K. Grob, Jr., *HRC&CC 2*:366 (1979).
78. K. Grob and G. Grob, *J. Chromatogr. Sci. 7*:584 (1969).
79. K. Grob and G. Grob, *J. Chromatogr. Sci. 7*:587 (1969).
80. C. K. Mann, T. J. Vickers, and W. M. Gulick, *Instrumental Analysis*, Harper & Row, New York, 1974.
81. D. A. Skoog and D. M. West, *Principles of Instrumental Analysis*, Holt, Rinehart, and Winston, New York, 1971.
82. I. M. Kolthoff, E. B. Sandell, E. J. Meehan, and S. Bruckenstein, *Quantitative Analytical Chemistry*, Macmillan, New York, 1971.
83. H. H. Willard, L. L. Merritt, and J. A. Dean, *Instrumental Methods of Analysis*, Van Nostrand Reinhold, New York, 1974.
84. T. Aczel, R. B. Williams, R. J. Pancirov, and J. H. Karchmer, Chemical properties of synthoil products and feeds. MERC/RI-8007-1, available from the National Technical Information Service, U.S. Dept. of Commerce, Springfield, Va. (1976).
85. J. T. Swansiger, F. E. Dickson, and H. T. Best, *Anal. Chem. 46*:730 (1974).
86. H. E. Lumpkin, *Anal. Chem. 36*:2399 (1964).
87. B. H. Johnson and T. Aczel, *Anal. Chem. 39*:682 (1967).
88. T. Aczel, J. Q. Foster, and J. H. Karchmer, *Amer. Chem. Soc. Div. Fuel Chem., Preprints 13*:8 (1969).
89. T. Aczel, D. E. Allan, J. H. Harding, and E. A. Knipp, *Anal. Chem. 42*:341 (1970).
90. H. E. Lumpkin and T. Aczel, *Anal. Chem. 36*:181 (1964).
91. J. E. Schiller, *Anal. Chem. 49*:1260 (1977).
92. J. L. Shultz, A. G. Sharkey, Jr., and R. A. Brown, *Anal. Chem. 44*:1486 (1972).
93. T. Aczel, D. E. Allan, J. H. Harding, and E. A. Knipp, presented at 16th Annual Conf. on Mass Spectrometry and Allied Topics, Pittsburgh, Pa., May 12-17, 1968.
94. T. Kessler, R. Raymond, and A. G. Sharkey, Jr., *Fuel 48*:179 (1969).
95. O. Kruber, *Chem. Ber. 53*:1566 (1920).
96. O. Kruber, *Chem. Ber. 86*:366 (1953).
97. O. Kruber, *FIAT Rev. German Sci. 36*:291 (1939-1946).
98. O. Kruber, *Chem. Ber. 73*:1184 (1940).
99. J. L. Shultz, T. Kessler, R. A. Friedel, and A. G. Sharkey, Jr., *Fuel 51*:242 (1972).
100. S. Akhtar, A. G. Sharkey, Jr., J. L. Shultz, and P. M. Yavorsky, *Amer. Chem. Soc. Div. Fuel Chem., Preprints 19*:207 (1974).
101. C. M. White and C. E. Schmidt, *Amer. Chem. Soc., Div. Fuel Chem., Preprints 23*:134 (1978).
102. M. L. Lee, C. Wiley, R. N. Castle, and C. M. White, in *Polynuclear Aromatic Hydrocarbons: Chemistry and Biological Effects*, A. Bjørseth and A. J. Dennis (Eds.), Battelle Press, Columbus, Ohio, 1980.

103. E. V. Shpol'skii, A. A. Il'ina, and L. A. Klimova, *Dokl. Akad. Nauk SSSR 87*:935 (1952).
104. E. V. Shpol'skii, *Soviet Phys. Usp. (English Transl.) 6*:411 (1963).
105. E. V. Shpol'skii, *Pure Appl. Chem. 37*:183 (1974).
106. J. R. Maple, E. L. Wehry, and E. Mamantov, *Anal. Chem. 52*:920 (1980).
107. J. A. Drake, D. W. Jones, B. S. Causey, and G. F. Kirkbright, *Fuel 57*:663 (1978).
108. R. B. Gammage, T. Vo-Dinh, A. R. Hawthorne, J. H. Thorngate, and W. W. Parkinson, in *Polynuclear Aromatic Hydrocarbons*, P. W. Jones and R. I. Freudenthal (Eds.), Raven Press, New York, 1978.
109. T. Vo-Dinh and J. R. Hooyman, *Anal. Chem. 51*:1915 (1979).
110. T. Vo-Dinh and R. B. Gammage, *Anal. Chem. 50*:2054 (1978).
111. M. L. Lee and R. A. Hites, *J. Amer. Chem. Soc. 99*:2008 (1977).
112. M. L. Lee, D. L. Vassilaros, W. S. Pipkin, and W. L. Sorensen, in *Trace Organic Analysis: A New Frontier in Analytical Chemistry*, H. S. Hertz and S. N. Chesler (Eds.), Spec. Publ. 519, National Bureau of Standards (U.S.), Washington, D.C., 1979.
113. R. A. Hites and G. R. Dubay, in *Polynuclear Aromatic Hydrocarbons*, P. W. Jones and R. I. Freudenthal (Eds.), Raven Press, New York, 1978.

14

Analysis of Metabolites of Polycyclic Aromatic Hydrocarbons by GC and GC/MS

JÜRGEN JACOB / Biochemisches Institut für Umweltcarcinogene, Ahrensburg, Federal Republic of Germany

I. Introduction

Since it became evident that polycyclic aromatic hydrocarbons (PAH) are potent and environmentally most relevant carcinogens, tremendous work has been performed in elucidating the metabolic pathways in which these compounds play their carcinogenic role. It soon became obvious that in a primary step PAH are converted into epoxides by monooxygenases of the cytochrome P450 type with reduced nicotinamide adenine dinucleotide phosphate (NADPH) being the cosubstrate:

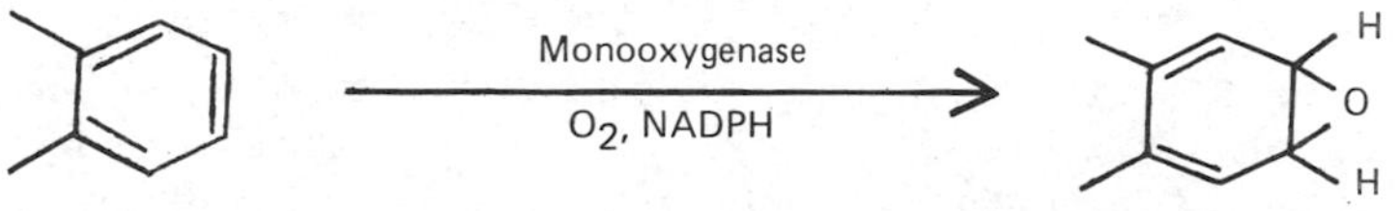

Monooxygenase

O_2, NADPH

H
O
H

Several monooxygenase isoenzymes, which are commonly divided into two large groups, cytochrome P450 and P448, according to their CO-binding spectra, participate in the conversion of PAH into epoxides. Subsequently the epoxides are converted by hydrolases into trans-dihydrodiols, which can be attacked once again by monooxygenases yielding trans-dihydrodiol epoxides, some of which have been shown to operate as ultimate carcinogens (e.g., trans-7,8-dihydroxy-9,10-epoxy-7,8,9,10-tetrahydrobenzo[a]pyrene). Since the balance of these enzymes decides the critical intracellular event in PAH carcinogenesis, it is of great interest to control these initial metabolic steps. This can be achieved by analyzing the profile of PAH metabolites in the cell. In most cases a large number of isomeric phenols, dihydrodiols, and higher oxidized PAH can be expected, so that a very efficient separation technique is necessary to distinguish between the various metabolites.

To overcome this problem in the past, preferentially high-performance liquid chromatography (HPLC) has been used which separates metabolites according to their polarities (for a review, see Ref. 1). Using reverse-phase technique with a methanol/water gradient, dihydrodiols are eluted first, followed by quinones, phenols, and the unconverted PAH. This has been demonstrated for, e.g., benzo[a]pyrene [2,3]. Higher oxidized metabolites can be separated, and even the generally unstable epoxides have been detected. Although HPLC is a rapid technique which does not require derivatization of the metabolites and which does not destroy them during the detection, it exhibits two main disadvantages if compared with the glass capillary gas chromatography (GCGC) technique: (a) the quantitative evaluation requires response factors which may vary from one type of metabolite to the other, and (b) the separation efficiency is poor. Great efforts have been made to overcome the latter by elaborating a high-resolution HPLC using the recycling technique [1]. This, however, is time consuming, and HPLC thus looses one of its greatest initial advantages.

II. GC and GCGC as Detection Methods for PAH Metabolites

The separation and detection of PAH metabolites has repeatedly been achieved by GC. Using 1.8-m packed columns (GasChrom Q coated with 3% OV 17), Stoming and Bresnick [4] were able to separate TMS ethers of trans-11,12-dihydroxy-3-methyl-11,12-dihydrocholanthrene and 11(or 12)-hydroxy-3-methylcholanthrene. They also showed that cis- and trans-dihydrodiols are readily separated under the GC conditions used. Packed columns were also used to detect chloromethyldimethylated trans-dihydrodiols of various PAH by means of an electron-capture GC detector [5]. Recently GCGC has been applied to metabolic studies of PAH and the inducibility of monooxygenases [6-8].

The GC separation of metabolites of 3-methylcholanthrene and benzo[a]-pyrene and their identification by mass spectrometry (MS) has been reported [9,10].

The main advantages of GC and GCGC are as follows:

(a) *The high-resolution efficiency* allows separation of isomeric phenols and dihydrodiols as well as higher oxidized PAH. In case of benz[a]anthracene the trimethylsilyl (TMS) ethers of nine different phenols and five dihydrodiols could be separated on a 50-m glass capillary column coated with Cp(tm)sil 5 (silicone) (Fig. 1) [11]. With highly efficient packed columns of 10 m in length some 20,000-25,000 theoretical plates can be achieved, whereas 70,000 can be reached with 25-m glass capillary columns. Even better separations can be obtained with longer columns.

(b) *The quantification of results*. When one is using a flame-ionization detector (FID), the signal is linear dependent on the carbon content of the compound. Thus no response factors must be determined. Correction factors

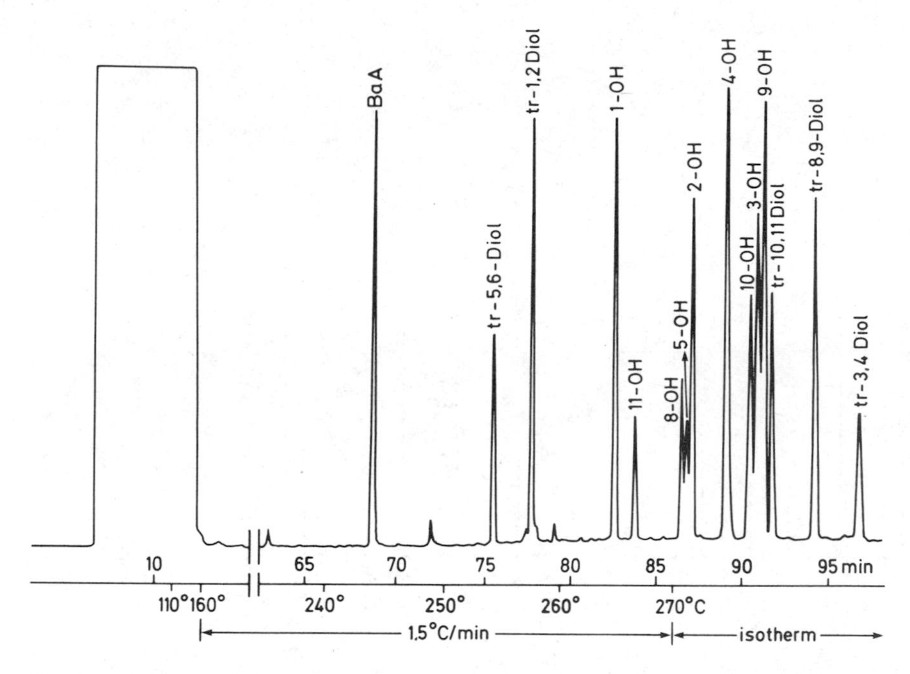

Figure 1. Separation of TMS ethers of nine different benz[a]anthracene phenols and five different dihydrodiols using a 50-m glass capillary column coated with Cp(tm)sil 5 (silicone).

due to the presence of hetero atoms in the molecule can be calculated mathematically.

(c) *Combination with MS*. Coupling GC with MS allows a rapid characterization and in many cases the final identification of metabolites. In the case of benz[a]anthracene metabolites phenols, dihydrodiols, dihydrotriols, and tetrahydrotetrols were identified by GCGC/MS [7,11-14]. Examples are shown in Figs. 2 to 5. Recently, similar results were obtained for pyrene [15]. In all cases more or less intense molecular peaks can be recognized which give some hint as to the degree of oxidation. Moreover, the fragment m/e 147 indicates vicinal trimethylsiloxy (OTMS) groups. From the ratio (m/e 147)/(m/e 191) the presence of a K-region dihydrodiol configuration can be recognized [11].

Disadvantages come from the thermal instability and the insufficient volatility of PAH metabolites which makes derivatization necessary. Silylation which is quantitative within 15-30 min is the recommended method. By this method even tetrols can be converted into volatile derivatives which can be chromatographed without decomposition. Extremely polar compounds, however, may offer problems to GC. Another disadvantage of the GC method must be seen in the fact that the substance is lost to the investigator once it is burned in the FID. On the other hand, GC is a very sensitive technique which enables us to detect only a few nanograms and thus generally requires only a small portion of the original sample.

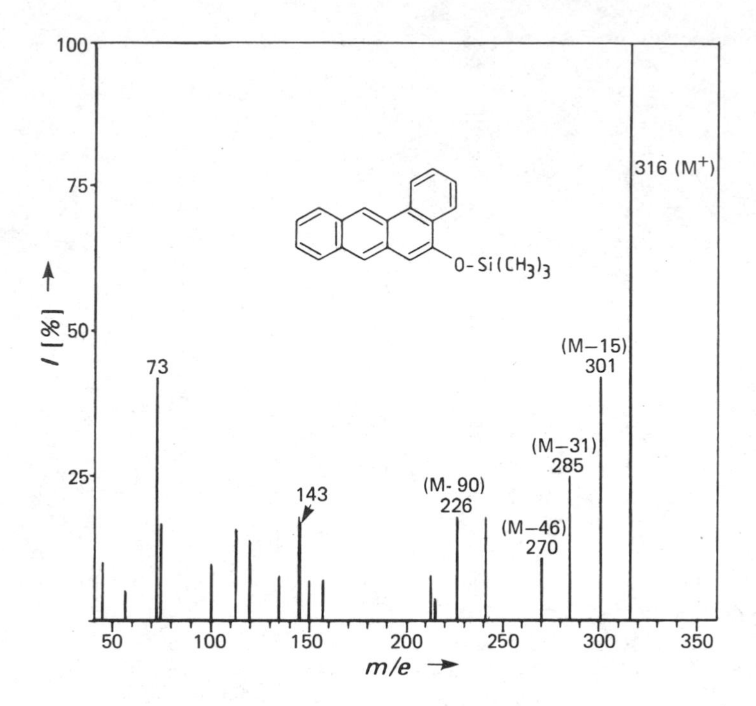

Figure 2. Mass spectrum of the TMS ether of synthetic 5-hydroxybenz[a]-anthracene. (80 eV). (From Ref. 7.)

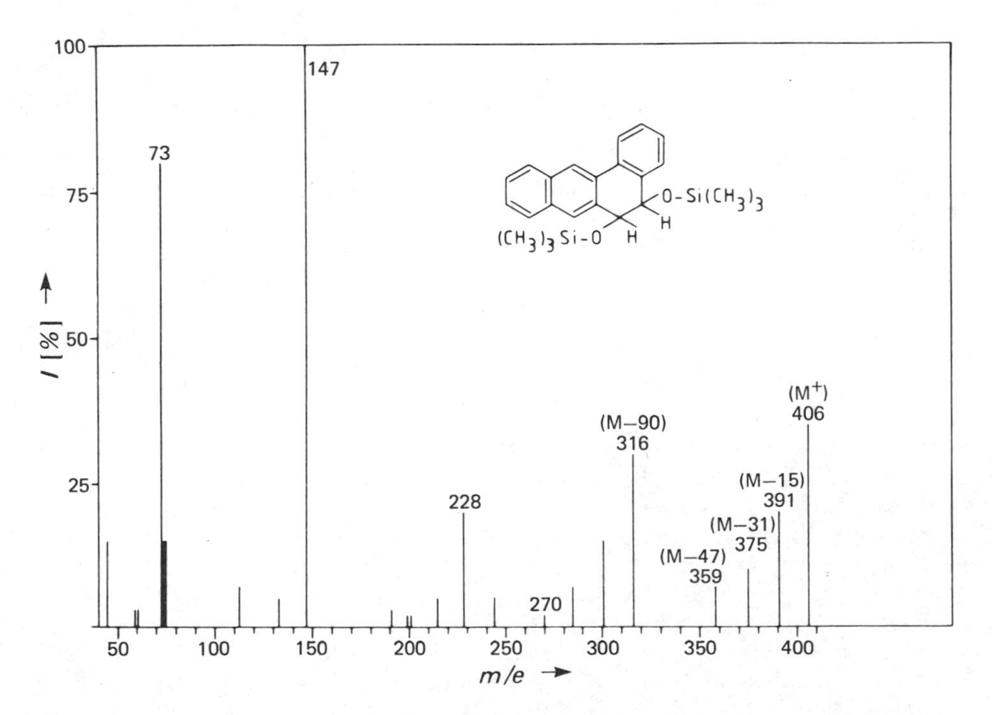

Figure 3. Mass spectrum of the TMS ether of 5,6-dihydroxy-5,6-dihydrobenz[a]anthracene from rat liver microsomes incubation (80 eV). (From Ref. 7.)

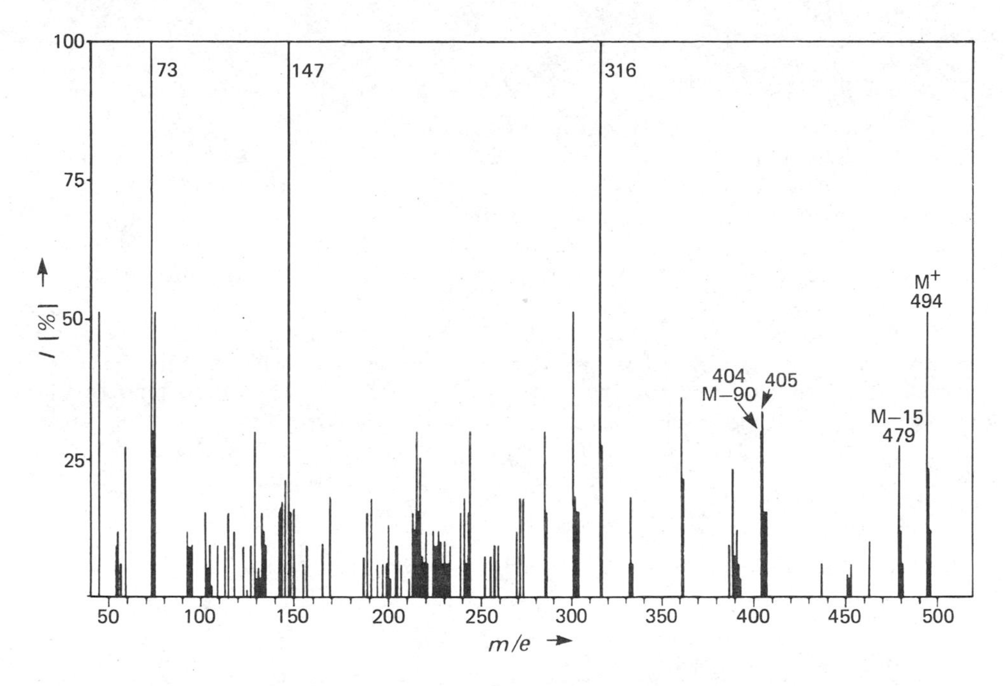

Figure 4. Mass spectrum of the TMS ether of a trihydroxydihydrobenz[a]-anthracene from rat liver microsomes incubation (80 eV). (From Ref. 7.)

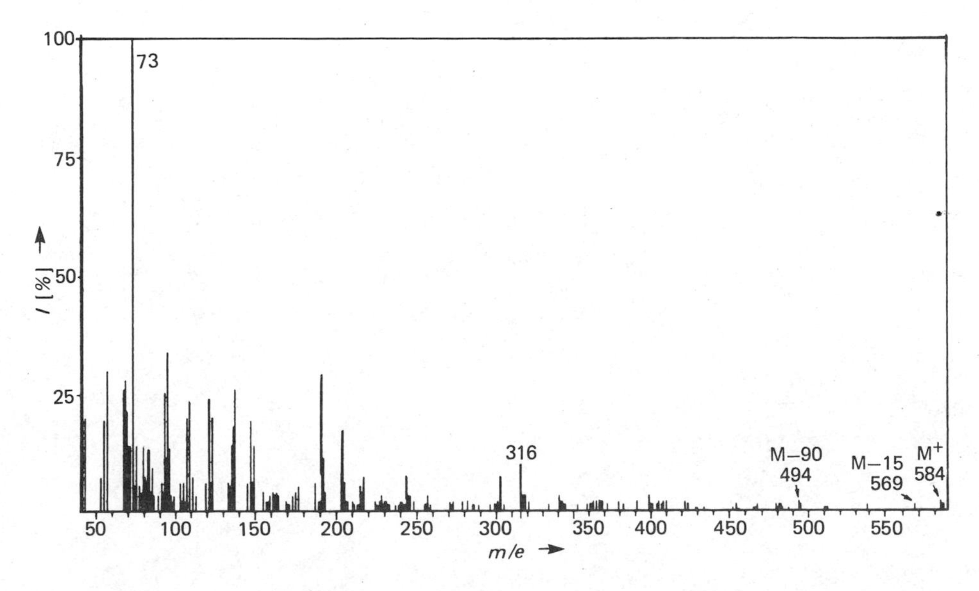

Figure 5. Mass spectrum of the TMS ether of a tetrahydroxytetrahydrobenz-[a]anthracene from rat liver microsomes incubation (80 eV). (From Ref. 7.)

III. Description of the Method

A. Cleanup and Preparation of the Sample for GC

Metabolites and unconverted PAH are extracted with ethyl acetate at pH 3 from rat liver microsomes incubations, cells in culture, or other biological materials, after the enzyme activity is stopped by addition of acetone and after the latter is evaporated under reduced pressure.

After evaporation of the ethyl acetate, the residue is chromatographed on a 10-g Sephadex LH-20 column with isopropanol. The elution volumes of some PAH are given in Table 1.

In the case of benz[a]anthracene, dihydrodiols require 10% larger elution volume and phenols about 30% larger elution volumes. PAH and metabolites are collected in one fraction, and after evaporation of the solvent the residue is silylated. For this purpose the sample is dissolved in 10 μl toluene, and 8 μl of a mixture of Trisil (Pierce) and N,O-bis-(trimethylsilyl)trifluoroacetamide (BSTFA Pierce) (1:1; v/v) is added. Silylation is complete after 30 min at

Table 1. Elution Volumes of Some PAH on a 10-g Sephadex LH-20 Column

PAH	Elution volume (ml)
Aliphatic hydrocarbons	20-35
Phenanthrene	38-50
Anthracene	38-50
Fluoranthene	48-63
Pyrene	48-63
Benzofluorenes (a+b+c)	51-63
Benzo[ghi]fluoranthene	56-71
Cyclopenta[cd]pyrene	60-78
Benz[a]anthracene	60-78
Chrysene	60-78
Benzofluoranthenes (b+j+k)	72-90
Benzo[a]pyrene	73-93
Benzo[e]pyrene	78-99
Perylene	83-106
Indeno[1,2,3-cd]pyrene	84-107
Benzo[b]chrysene	84-118
Anthanthrene	88-113
Benzo[ghi]perylene	90-118
Coronene	104-140

45-50°C. No measurable decomposition of the TMS ethers occurs within 5 days under exclusion of moisture at -10°C. Aliquots of the sample containing 10-30 ng metabolites are injected into the gas chromatograph.

B. GC Conditions

Although 10-m (or 20-m) glass columns packed with GasChrom Q or Supelcoport and coated with 1-5% SE 30, OV 17, OV 101, or comparable impregnations exhibit about 2000 theoretical plates per meter, glass capillary columns are recommended for the separation of TMS ethers of PAH phenols or dihydrodiols. In the case of 10-m packed glass columns (2 mm i.d.), flow rates of 20 ml N_2(He) per minute, injection and column temperatures of 250°C were used. The temperature of the detection bloc should be 10-20°C higher. To avoid decomposition of the sample, contact with hot metal walls must be excluded. Thus all-glass flow ways are to be used, especially the injection, and the connection between column and the FID should be equipped in glass. In the case of 25-m

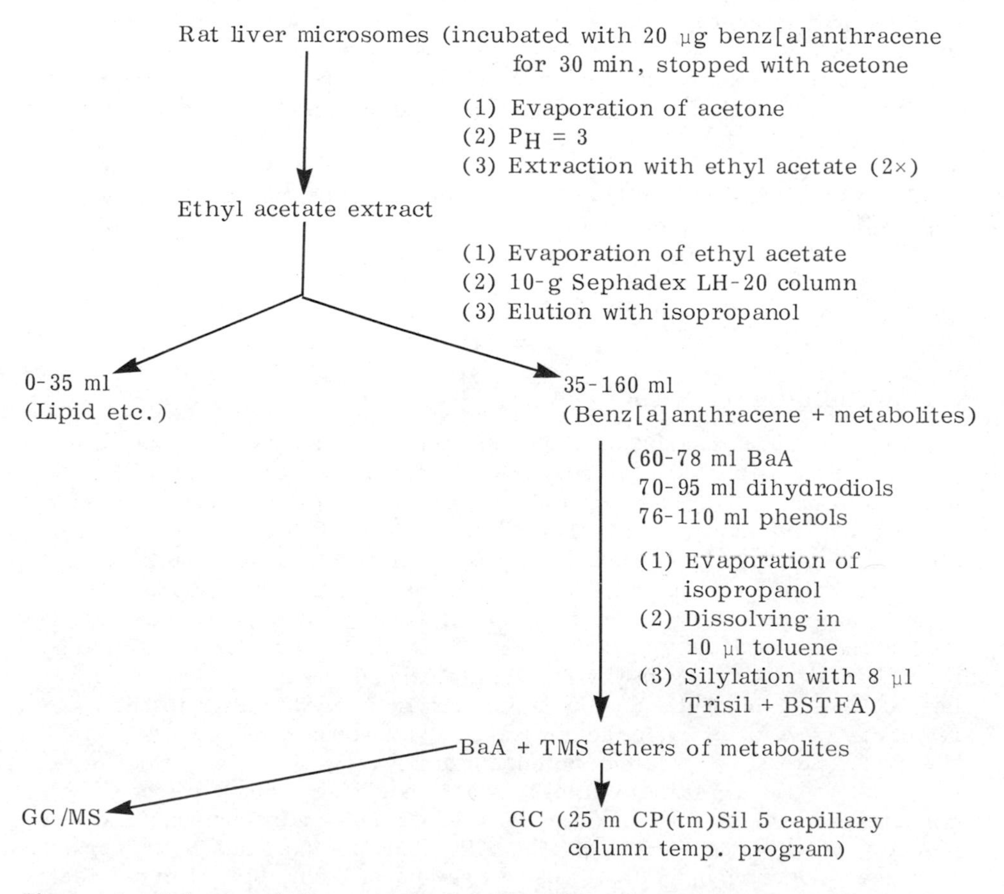

Figure 6. Scheme of the procedure of determination of PAH and metabolites from rat liver microsomal incubations with benz[a]anthracene as substrate.

glass capillary columns (0.27-0.5 mm i.d.), SE 30, OV 101, and CP(tm)sil5 impregnations have been successfully used. Migration rates were about 25 cm/sec. To avoid contamination during GC a poorly bleeding septum and, if available, a septum swinger are recommended. Since splitless injection increases the sensitivity and temperature programming leads to better separations, the following conditions can be recommended: Column temperature during injection 100°C; after 11 min a program from 100° to 260°C with 3°C/min is used. Injection and detection temperatures are 260° and 270°C, respectively. To avoid tailing postcolumn acceleration with argon (25 ml/min) between column outlet and detector is recommended. Since the detector electrode may be coated by the excess of silylation reagents, it is removed for the first 10 min—during which no metabolite peaks occur.

C. MS Conditions

Optimum results have been obtained with double focusing mass spectrometers with open-coupling interface and splitless injection using a septum swinger. All gas ways should be in glass (e.g., from the column outlet into the ion source). Starting temperature of the column is 110°C. The split is closed during the first 10 min to let the solvent pass. One minute after opening the split a temperature program from 110-160°C with 30°C/min followed by a program 160°-270°C with 1.5°C/min is used (injection and interface temperature, 280°C; ion source temperature, 240°C). To avoid burdening the ion source with an excess of solvents and silylation reagents the interface is connected 7 min after injection (blending out). Thus even a large sample dosage does not become critical.

The whole procedure is presented schematically in Fig. 6 for the determination of unconverted PAH and metabolites from a rat liver microsomes incubation with benz[a]anthracene as substrate.

IV. Applications and Results

The aforementioned method has been applied mainly to three problem areas: (1) recording metabolite profiles of various PAH after incubation with rat liver microsomes [7,11-15]; (2) BaA metabolism in fetal hamster lung cells in culture [6,8]; and (3) induction of microsomal rat liver monooxygenases after pretreatment with various inducers (PAH and non-PAH) [6,7,11-13,15].

Table 2 shows that amounts and number of the GC-detectable metabolites vary within a wide range in microsomal rat liver incubation after 30 min.

The induction of monooxygenases has been investigated in detail using the aforementioned method with benz[a]anthracene (BaA) as substrate. It has been found that intraperitoneal application of various PAH to the rat results in very different induction rates of the monooxygenase system in the liver. Some PAH are very potent inducers (e.g., the benzofluoranthenes, chrysene, benzo[a]pyrene); others do not show a significant effect (e.g., benzo[e]pyrene, pyrene). Moreover, there is no relation between inducer potency and the carcinogenicity of a PAH. For example, benzo[k]fluoranthene is either a noncarcinogen or at most a weak carcinogen, but a very potent inducer. In Table 3 the induction rates (i.e., ratio of benz[a]anthracene turnover of induced to that of noninduced animals) calculated from the BaA

Table 2. Turnover and Number of Metabolites of 12 Different PAH after 30-min Incubation with Rat Liver Microsomes[a]

PAH	(nmol/30 min)	Number of GC-detectable metabolites
Phenanthrene	111	2
Fluoranthene	31	6
Pyrene	32	6
Benz[a]anthracene	12	6
Chrysene	0.7	2
Cyclopenta[cd]pyrene	39	6
Benzo[a]pyrene	1.6	2
Benzo[e]pyrene	0.8	2
Benzo[b]fluoranthene	5.2	3
Benzo[k]fluoranthene	—	—
Indeno[1,2,3-cd]pyrene	—	—
Benzo[ghi]perylene	1.6	2

[a]Conditions: 20 μg PAH; 1 mg microsomal protein.
Source: Ref. 7.

Table 3. Ratio of BaA Turnover in Rat Liver Microsomes of Induced to That of Untreated Rats[a]

Inducer	Ratio: $\frac{\text{induced}}{\text{noninduced}}$
None	1.0
5,6-Benzoflavone	4.4
3,3',4,4'-Tetrachlorobiphenyl	5.8
PCB	6.2
Phenobarbital	7.2
Benz[a]anthracene	6.5
Pyrene	0.4
Chrysene	9.1
Benzo[a]pyrene	5.0
Benzo[e]pyrene	1.6
Benzo[b]fluoranthene	4.1
Benzo[j]fluoranthene	6.5
Benzo[k]fluoranthene	3.4

[a]Conditions: 20 μg PAH; 30 min incubation, WISTAR rats; induction with 40 mg PAH/kg rat.
Source: Refs. 7 and 8.

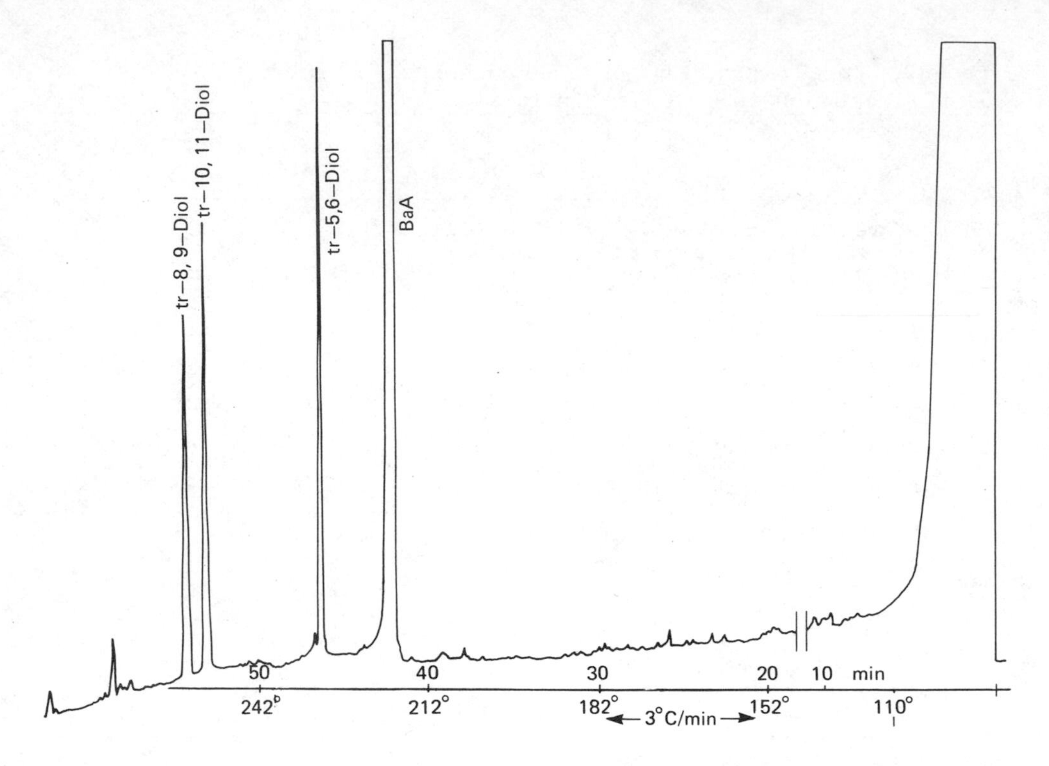

Figure 7. Capillary gas chromatogram of a BaA incubation with liver microsomes of untreated rats.

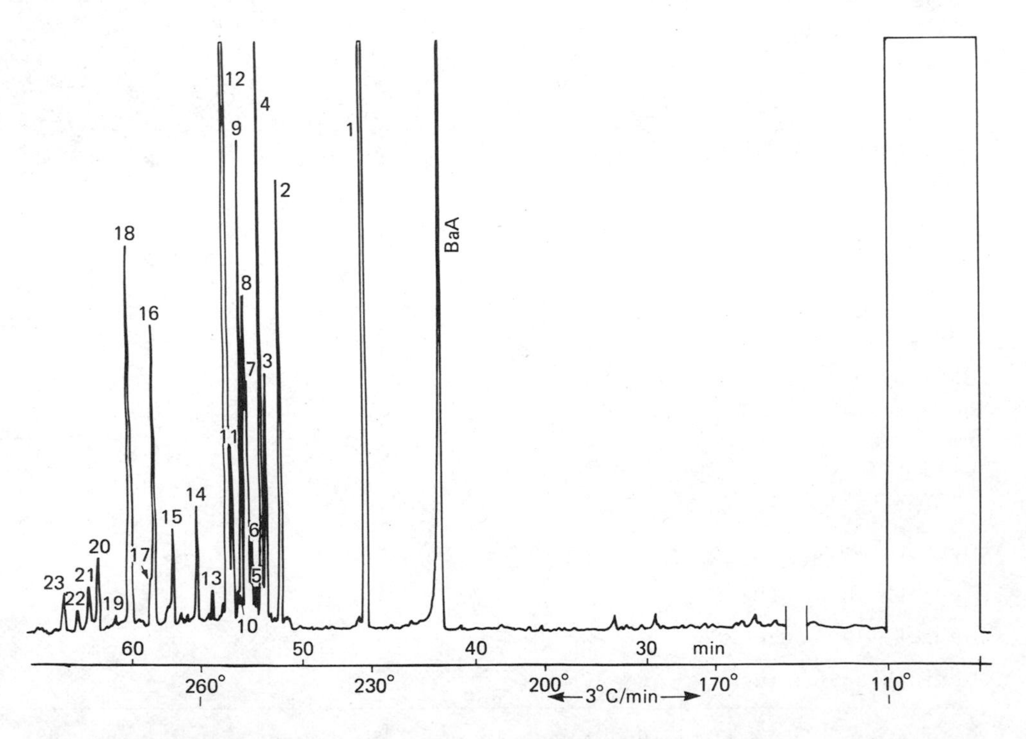

Figure 8. Capillary gas chromatogram of a BaA incubation with liver microsomes of rats pretreated with benzo[b]fluoranthene. (Peak 1 = 5,6-dihydrodiol. Peak 12 = 8,9-dihydrodiol.)

Table 4. Turnover (in Percent of the Incubated PAH) of Various PAH with Rat Liver Microsomes Without and After Pretreatment of the Rat with Benzo[k]fluoranthene as Inducer[a]

Substrate	Turnover without induction (%)	After induction with benzo[k]-fluoranthene (%)	Ratio: Turnover induction / Turnover noninduction
Fluoranthene	21.0	45.0	2.1
Chrysene	0.8	25.0	31.1
Benz[a]anthracene	7.0	23.8	3.4
Pyrene	22.5	47.7	2.1
Benzo[b]fluoranthene	6.6	62.7	9.5
Benzo[j]fluoranthene	2.8	21.0	7.5
Benzo[a]pyrene	2.0	46.2	23.1
Benzo[e]pyrene	1.0	57.3	57.3
Indeno[1,2,3-cd]pyrene	1.0	55.0	55.0
Benzo[ghi]perylene	0	0	–

[a]Conditions as in Table 3; 20 μg substrate.
Source: Ref. 8.

turnover in rat liver microsomes after pretreatment with various inducers are presented.

Noninduced rats exhibit a simple metabolic pattern after rat liver microsomal incubation with BaA in which only the 5,6-, 8,9-, and 10,11-dihydrodiol were detected (Fig. 7). Pretreatment with benzo[b]fluoranthene, however, results in a dramatic change (Fig. 8). Not only the turnover but also the complexity of the pattern is increased; various triols and tetrols and smaller amounts of phenols occur.

It could be demonstrated that a certain inducer such as benzo[k]fluoranthene stimulates the metabolism of various PAH to different extents. For instance, 7.5 times more benzo[j]fluoranthene is metabolized after pretreatment of the rats with benzo[k]fluoranthene under standard conditions in the rat liver microsome model, whereas the induction rate is 57 in case of the substrate benzo[e]pyrene with the same inducer (Table 4).

Acknowledgments

Parts of the author's investigations were supported by the Umweltbundesamt, to which he is greatly indebted. He thanks G. Dettbarn, K. W. Naujack, and G. Raab for their technical assistance.

References

1. J. K. Selkirk, *Advan. Chromatogr. 16*:Chap. 1 (1978).
2. J. K. Selkirk, R. G. Croy, and H. V. Gelboin, *Science 184*:169 (1974).
3. P. Leber, G. Kerchner, and R. I. Freudenthal, A comparison of benzo[a]-pyrene metabolism by primates, rats, and miniature swine, in *Polynuclear Aromatic Hydrocarbons: Chemistry, Metabolism, and Carcinogenesis*, R. L. Freudenthal and P. W. Jones (Eds.), Vol. 1: , Raven Press, New York, 1976, p. 35-43.
4. T. A. Stoming and E. Bresnick, *Science 181*:951 (1973).
5. A. Bettencourt, G. Lhoest, M. Roberfroid, and M. Mercier: *J. Chromatogr. 134*:323 (1977).
6. J. Jacob, G. Grimmer, and A. Schmoldt, Gaschromatographic profile-analysis of PAH-metabolites from rat liver microsome preparations and cells in culture, in *Polynuclear Aromatic Hydrocarbons: Chemistry and Biological Effects*, A. Bjørseth and A. J. Dennis (Eds.), Battelle Press, Columbus, Ohio, 1980, pp. 807-817.
7. J. Jacob, G. Grimmer, and A. Schmoldt, *Z. Physiol. Chem. 360*:1525 (1979).
8. J. Jacob, G. Grimmer, H.-B. Richter-Reichhelm, and E. Emura, *VDI-Ber. 358*:273 (1980).
9. G. Takahashi, *Gann 69*:437 (1978).
10. G. Takahashi, K. Kinoshita, K. Hashimoto, and K. Yasuhira, *Cancer Res. 39*:1914 (1979).
11. J. Jacob, A. Schmoldt, and G. Grimmer, *Z. Physiol. Chem. 362*:1021 (1981).
12. J. Jacob, A. Schmoldt, and G. Grimmer, *Carcinogenesis 2*:395 (1981).
13. J. Jacob, G. Grimmer, and A. Schmoldt, *Cancer Lett. 14*:175 (1981).
14. A. Schmoldt, J. Jacob, and G. Grimmer, *Cancer Lett. 13*:249 (1981).
15. J. Jacob, G. Grimmer, G. Raab, and A. Schmoldt, *Xenobiotica 12*:45 (1982).

15

Polycyclic Aromatic Hydrocarbons in River and Lake Water, Biota, and Sediments

JOACHIM BORNEFF AND HELGA KUNTE / Institute of Hygiene, Mainz, Federal Republic of Germany

I. Introduction

In recent years, there have been many studies of polycyclic aromatic hydrocarbons (PAH) in river and lake water, as well as sediments and biota. Several review papers summarize the extensive research in this area [1-4]. In the early 1950s some papers were published from which it could be inferred that PAH would be present in surface waters receiving sewage effluents. Wedgewood and Cooper [5,6] detected PAH in industrial sewage effluents and Mallet and Heros [7] found BaP in urine from inhabitants of Paris. At this time methods were also developed which rendered possible the detection and determination of very small amounts of these substances. Borneff and Fischer [8] extracted activated carbon which had been used in a water treatment plant for purification of bank filtered river water and identified seven PAH at a total concentration of approximately 250 μg/kg (uncorrected value). An extract of fresh carbon contained only a small amount of fluoranthene and a trace of benzo[b]fluoranthene. The same compounds were found in the sludge from backwashing of sand filters from a waterworks using lake water as a source for drinking water and in the phytoplankton of the same lake [9,10]. These results were obtained by extraction of the material with benzene, chromatography on alumina columns followed by paper chromatography and registration of absorption and fluorescence spectra of the substances eluted from the paper.

The above results were an indirect proof of the occurrence of PAH in surface water. Later, direct extraction from water was performed [11] demonstrating the presence of PAH in the aqueous environment.

II. Methods of Analysis

Methods for PAH analysis in water have been greatly improved in the last 15 years. Whereas in 1964 up to 100 liters had to be analyzed, a sample size of 2 liters is now sufficient even for drinking water. Liquid-liquid extraction has proved most effective and is generally used. Cyclohexane is an appropriate solvent. Adsorption on activated carbon cannot be recommended because recovery is never complete and rather dependent on the type and even the batch of the carbon and is influenced by other organic matter present. Adsorption on polyurethane foam has been used [12,13], but for efficient extraction the water has to be heated to 60°C and even then the recovery rate is less than using liquid-liquid extraction.

Separation and detection may be achieved by thin-layer chromatography (TLC) and fluorometry, by gas chromatography (GC) with flame-ionization detection (FID), or high-performance liquid chromatography (HPLC) with detection by fluorescence or UV-absorption. In the Federal Republic of Germany a technique using two-dimensional TLC with fluorescence detection [14] has been developed as a standard method for the determination of six specified PAH (see Table 1) in water [15] and has also been included in the IARC Publications Series [16]. A slightly modified version is being prepared by the International Organization for Standardization (ISO) as a recommended method.

In the USSR and other eastern countries low-temperature fluorescence (quasi-linear fluorescence) is used in combination with TLC [17,18]. Advantages of the TLC/fluorescence methods are a high specificity and sensitivity and a minimum of interferences. A disadvantage is that only compounds with a fairly strong fluorescence are detected.

With GC/FID all PAH are detected at the same level. But because the FID will detect all organic substances, a careful precleaning is necessary and highly efficient columns must be used. Moreover, sensitivity is not sufficient for the low concentrations encountered in water. According to Grimmer and Naujack [19], in routine analysis using high-efficiency packed columns the amount of a single substance in the sample would have to be 100 to 1000 ng [concentrations of, e.g., benzo[a]pyrene (BaP) in surface waters are usually below 100 ng/liter and may be less than 1 ng/liter in filtered river water!].

HPLC has been developing rapidly in the last few years and a large number of papers have been published demonstrating the applicability of this method

Table 1. PAH Concentrations in Selected River Waters

Source	Concentration (ng/liter) Total of 6 PAH[a]	BaP	Ref.
Federal Republic of Germany:			
River Gersprenz (small stream)	93-96	9.6	11
River Schussen (small stream)	533	10	44
River Stockacher Aach (small stream) 4 locations	777-1530	4-43	44
River Rhine at km 30	9-40		24
River Rhine at km 853	132-5680		24
River Main (at Schweinfurt)			
26 samples	197-560	12-68	24
during flood (4 samples)	1090-1740	58-157	24
River Danube (near Ulm)			
28 samples	58-1371	0.4-347	24
Netherlands:			
River Rhine (Lek)	400-910	50-90	45
River Rhine (Waal)	140-2240	<10-580	27,28, 29,30
United Kingdom:			
River Thames	380-1000[b]	100-260	32
River Severn (18 locations)	15-160[c]	1.5-20	31
River Stour (3 locations)	266-1430[c]	48-132	31
River Trent (9 locations) 1975	213-2596[c]	35-504	31
Tributaries to River Trent (11)	26-3789[c]	5-435	31
River Trent (1976/78)	40-1000	–	33
River Gt. Ouse	5-500	–	33
River Ancholme	5-70	–	33
France:			
River Seine (downstream from Paris, 10 locations)	89-233	–	34
Canal at Lille	294-404[d]	35-300	35
Canal at Marquette	215-273[d]	50-55	35
Canal at Violaines	259[d]	32	

(continued)

Table 1 (continued)

Source	Concentration (ng/liter)		Ref.
	Total of 6 PAH[a]	BaP	
United States:			
River Monongahela	600-663	42-77	43
River Ohio	58	5.6	43
River Delaware	352	41	43
Poland:			
River Warta	124	2	37
River Wisla	165	30	37
River Warta (7 locations)	–	1-20	38
River Wrercica (3 locations)	–	1-18	38
River Olawa	–	2-350	36
River Nysa Klodzka	–	3-482	36
River Odra	–	18-823	36
USSR:			
River Plyussa, site of discharge of effluent	–	12000	39
3,5 km down stream	–	1000	39
River Moskwa, above Moscow	–	0.3-0.5	42
A river above large town	–	1	42
below large town	–	40	42

[a]Six PAH: fluoranthene (Flu), benzo[b]fluoranthene (BbF), benzo[k]-fluoranthene (BkF), benzo[a]pyrene (BaP), benzo[ghi]perylene (BghiPer), and indeno[1,2,3-cd]pyrene (IP).

[b]Six PAH as above, but perylene instead of BbF.

[c]Five compounds only; BbF not determined.

[d]Five compounds: Flu, BaP, BkF, perylene, benzo[b]fluorene.

for the determination of PAH in water. At present there is no generally accepted method, but work on this subject has started within the ISO.

Details of methodology are described in other chapters of this book as well as some recent publications [20,21].

III. Solubility of PAH in Water

Solubilities of PAH in water are generally very low but cover a wide range (0.5 μg/liter for perylene to more than 1000 μg/liter for phenanthrene in distilled water of 25 to 27°C [4]. Two-ring aromatics have still higher solubilities. As might be expected, solubility tends to decrease as the number of rings increases. Alkyl groups influence solubility to a considerable extent as well as the configuration of the compound (linear, angular, or pericondensed system). Temperature also has a great effect on PAH solubility [4,22]. A rise in water temperature from 6° to 26°C may increase solubility three times.

Other organic matter present in the water may play a part in solubilizing PAH. Such compounds include bile acids, alkali salts of fatty acids, butyric acid and lactic acid, purines, organic solvents, detergents [4,23]. The latter compounds are only effective if their concentration is at least 20 to 30 mg/liter [23], which is not usually encountered in surface water.

IV. PAH in River Water

PAH concentrations in river water vary widely depending on the extent and kind of pollution. Concentrations found in rivers in different parts of the world are given in Table 1. Values in the first column are not strictly comparable because the number and type of compounds analyzed are not identical. Nevertheless the order of magnitude may well be compared. Values for BaP are given in the second column to allow a comparison with studies in which only this compound was determined.

Borneff [24] investigated PAH in several German rivers in different seasons and during periods of several years. Monthly samples were taken from the River Danube near Ulm from 1964 to 1966; the concentrations of the total of the six PAH (see Table 1) ranged from 58 to 1370 ng/liter and the mean values for these 3 years were 107, 391, and 300 ng/liter, respectively. Concentrations in the River Main at Schweinfurt, where samples were collected every 2 weeks from March 1970 to February 1971, ranged from 197 to 560 ng/liter, with one exception when the river was flooding. On this occasion concentration rose up to 1740 ng/liter. In a longitudinal study of the River Rhine, samples were collected at 12 points in 1969/70 and again in 1976/77. As shown in Table 2, for some of the sampling points reductions in PAH concentrations up to a factor of 20 are noted. A similar study was conducted by the Arbeitsgemeinschaft Rheinwasserwerke (ARW) [25,26]. Continuous sampling was used and analysis carried out on monthly samples. Annual averages ranged from 190 to 670 ng/liter at the five points investigated, the minimum value measured was 40 ng/liter and the maximum value 1270 ng/liter. Unfortunately, due to the difference in sample type and sampling points a direct comparison of the two studies is not possible. PAH concentrations of the Rhine water at Basel were followed up from 1973 to 1979 [24].

Table 2. Influence of Sanitary Measures on PAH Concentration of the River Rhine

Location	River (km)	Total of 6 PAH (ng/liter)	
		1969/70	1976/77
At Basel	170	274	45-258
At Bingen	527	800-3610	42-160
At Cologne	689	980-3465	87-302
At Emmerich	853	1020-5680	132-370

In this case samples were taken every workday and combined to monthly samples. Annual averages are shown in Fig. 1 together with minimal and maximal values. It is evident that the concentration has decreased considerably, to about one-quarter of the values measured in the first two years.

In the Netherlands, PAH concentrations have been determined in the River Waal (a branch of the River Rhine) since 1976 [27-30]. Monthly averages varied between 70 and 2240 ng/liter, but annual averages were 440, 340, 500, and 190 ng/liter, respectively, for the years from 1976 to 1979—the same order of magnitude as was found in the longitudinal study of 1974/75 in the German part of the river.

Several rivers have been investigated in the United Kingdom [31-33]. As in the German rivers, PAH concentrations varied considerably. The samples from the River Severn were taken at 18 locations within 133 river km and PAH concentrations ranged from 15 to 160 ng/liter. Of several tributaries investigated, the River Stour had the highest concentrations up to 1430 ng/liter. A survey of the Trent River covering a stretch corresponding to 237 river km showed much higher concentrations (up to 2596 ng/liter); in one of the 11 tributaries feeding that river (which were investigated at the same time) a maximum concentration of 3790 ng/liter was found. Later analyses of the Trent in 1976-1978 seem to point to a decrease of PAH pollution [33].

Of French rivers, only data on the Seine River downstream from Paris were available to us. The results of four samplings at 10 locations revealed PAH concentrations from 89 to 233 ng/liter [34]. In a canal in the Lille region, slightly higher values were detected [35].

A number of analyses have been carried out on rivers in Poland, but in most cases BaP only was determined. The concentration of this compound ranged from 1 to >800 ng/liter [36-38], which indicates PAH pollution in a similar order of magnitude as in other European rivers.

In the USSR, again BaP only was determined, showing a wide range of concentrations covering five orders of magnitude [39-42]. Extremely high BaP levels were found in river water receiving sewage effluents from industries with high PAH emissions, e.g., shale oil and coke by-product industries and oil refineries.

Data are scarce for non-European rivers. In the United States, Basu and Saxena [43] determined PAH concentrations in the Ohio, Delaware, and Monongahela Rivers, ranging from 58 ng/liter in the Ohio to 663 ng/liter in the Monongahela.

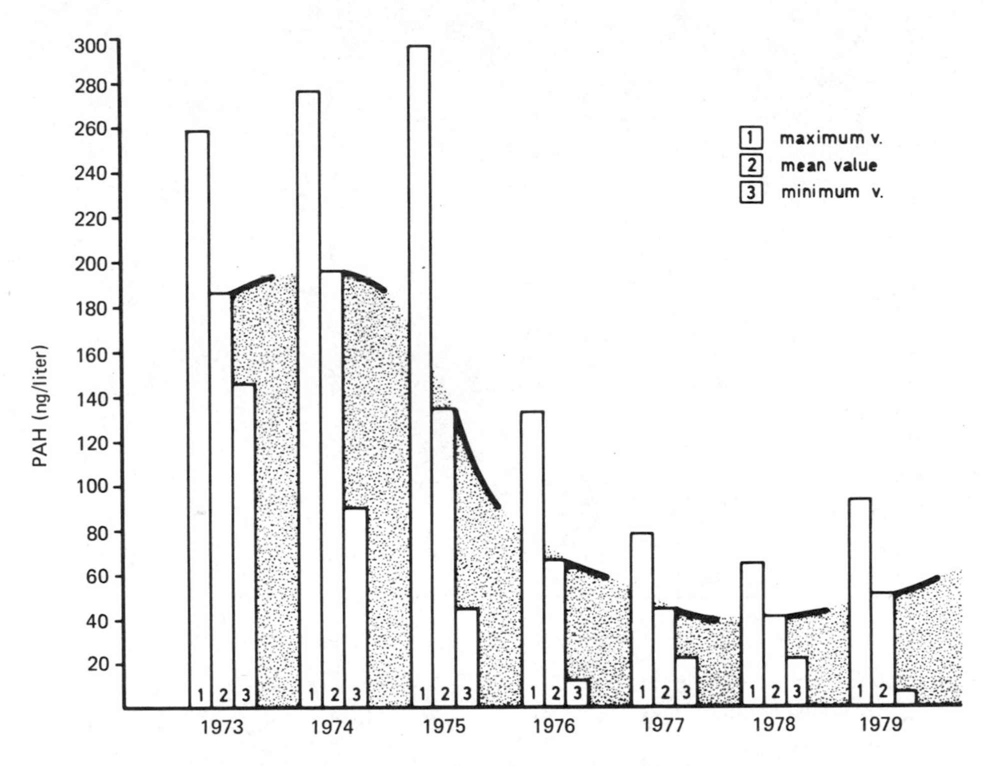

Figure 1. PAH concentrations in the River Rhine at Basel. Annual maximum, mean, and minimum values from 1973 to 1979 for the total of six PAH.

In most cases the concentrations cited refer to water including suspended matter. In the experiments of Lewis [31] in the United Kingdom, the water was filtered through glas fiber filters and the filtrate as well as suspended particles were analyzed separately. The ratio of suspended particles/water was extremely variable and ranged from 1.9 to 50.5 (in one exceptional case, 161) in the River Trent and from 0.6 to 7.9 (with two exceptional values of 13.7 and 77.3) in the River Severn. In the same way, in Germany, water samples from the River Rhine were analyzed [26] and the ratios found ranged from 0.7 to 13.0. Generally high ratios were associated with high pollution levels but there was no clear correlation.

V. PAH in Lake Water

Only few data on natural lakes are available. Perhaps the one most intensively investigated is Lake Zürich [24,46]. Samples were taken at weekly intervals from the lake surface (1970-1976) and from a depth of 30 m (1970-1979) and analyzed for the six PAH. Mean concentrations were calculated for each year ranging from 5 to 34 ng/liter for the 30-m depth and from 11 to 66 ng/liter for the lake surface. During this time a similar decrease in

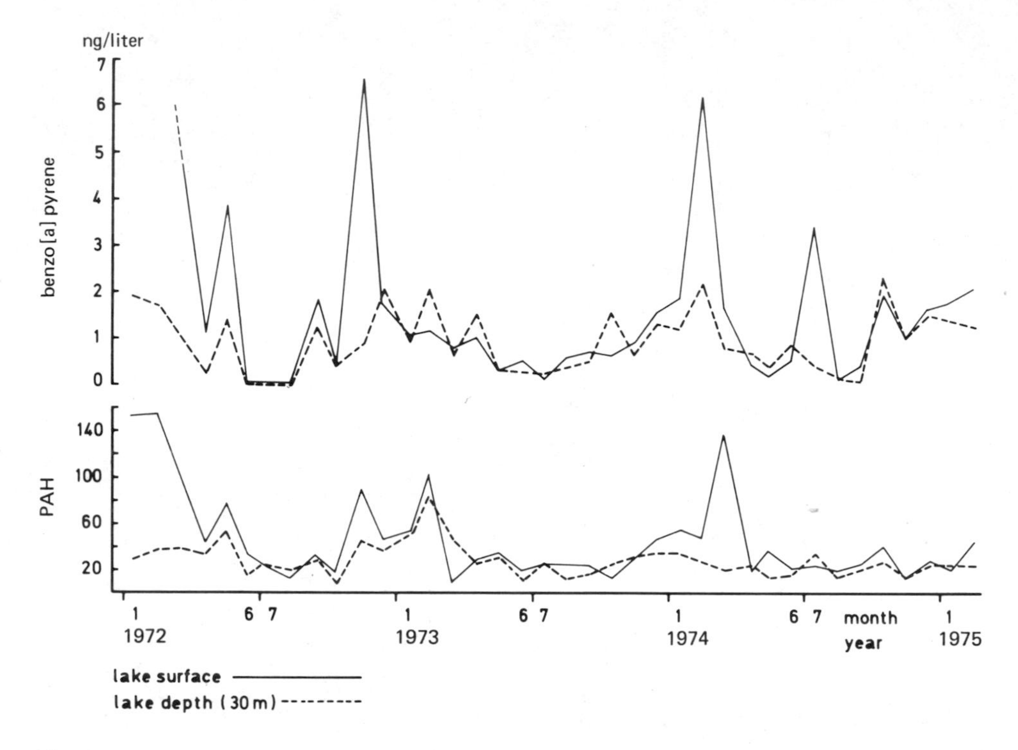

Figure 2. PAH concentrations in the water of Lake Zürich. Monthly samples from 1972 to 1975 for the total of six PAH and BaP.

Table 3. PAH Concentrations in Lake Waters

Source	Concentration (ng/liter) Total of 6 PAH	BaP	Ref.
Lake Constance (1964)	39	1.3	11
Lake Zürich, surface (mean values 1970-1976)	11-66	0.5-5.4	24,46
30-m depth (mean values, 1970-1979)	5-34	0.4-1.5	24,46
Lake in Austria	6.6	0.3	49
Lake Zegrzynski (Poland)	47	5.0	37
Lake Erie (U.S.)	4.7	0.3	43
Lakes in Estonian SSR		0.3-2	50,51
19 Artificial lakes (Fed. Rep. of Germany)	12-2400[a]	—	52

[a]Total of eight compounds: the six PAH specified in Table 1 plus BaA and perylene.

concentration—about 80%—was observed as in the River Rhine at Basel. Further, it was noticed that highest concentrations always occurred in the first months of the year (Fig. 2). This is considered to be due to the snow melting in the Alps. It has been shown that snow may contain considerable amounts of PAH [47,48]. Concentrations found in other lakes are of the same order of magnitude (see Table 3 and references cited therein).

A comprehensive study has been conducted on artificial lakes in the Federal Republic of Germany [52]. PAH concentrations of 19 lakes were followed up for a period of 3 years. With one exception (2400 ng/liter) values for the total of 8 PAH ranged from 12 to 880 ng/liter for single determinations and from 56.5 to 153.2 ng/liter averaged over the period of investigation for the individual lakes. PAH concentrations in the inlets showed some dependence on the area they flow through (agricultural use, industrial area, population density); these differences, however, disappeared in the lakes themselves.

VI. PAH in Sediments

The investigations of Lewis [31] and the ARW [26] showed that a considerable portion of the PAH in water bodies may be associated with suspended particles. These eventually sink to the bottom to form a sediment. Sediments consequently contain PAH, usually in a concentration much higher than the overlying water. Some selected values are given in Table 4. In highly industrialized areas, PAH concentrations are in the ppm range (up to 130 mg/kg have been found), whereas in areas with low population density and far away from industrial emissions they may be smaller by two to three orders of magnitude.

Table 4. Concentrations of PAH in Sediments from Rivers and Lakes

Source	Concentration (mg/kg)	Ref.
River Severn (U.K.)	1.6-23.5[a]	53
River Taff (U.K.)	6.9-15.8[a]	53
River Rhonda Fach (U.K.)	10.3-25.7[a]	53
River Usk (U.K.)	130[b]	54
River Charles (Mass., U.S.)	87[c]	55
River Amazon flood plain	0.017[c]	55
River Amazon	N.D.-0.544[c]	55
Mono Lake (Calif., U.S.)	0.16-0.40[c]	55

[a]GC determination on packed column. Some PAH not separated. Sum includes at least 12 compounds.

[b]GC/MS determination on packed column, with computer analysis; 10 compounds.

[c]GC/MS determination on packed column. Concentrations given here include at least 12 compounds. In the original concentrations are given for groups of isomers. N.D. = not detected.

In some cases cores from deeper layers of sediment were investigated (Table 5). Grimmer and Böhnke [56] studied PAH concentrations at different levels of cores taken 100 m from the north and south shores of the Grosser Plöner See (northern Germany). The southern shore is woodland without settlements, whereas the northern shore has been continuously settled and is crossed by a highway and a main railway line. At the southern shore PAH concentrations in the sections of the core showed no significant variation; at the northern shore an increase was recorded since about 1930, with a maximum in the 1940s. Values have been decreasing since, but in the 1969 layer they are still three times higher than in the older sections before 1920 where they amounted to less than 10 mg/kg for the sum of 28 compounds. A similar investigation was conducted in Lake Constance [57,58]. Again the highest values were not found in the top layer but in sections corresponding to some 15 to 30 years earlier and were much higher near the mouth of a polluted river than in the middle of the lake, which is in agreement with the findings in the Grosser Plöner See. Concentrations in the layers sedimented before 1900 were well below 1 mg/kg for the total of 24 compounds.

From the ratios of individual PAH in the sediments and the relative abundance of certain compounds, the authors concluded that the main source of the elevated concentrations in the last 50 years is coal combustion, other sources such as automotive exhausts and petroleum products being of minor importance.

Wakeham and co-workers [59,60] investigated several cores from three Swiss lakes and from Lake Washington (in Washington State). One core from the Greifensee extended back to 15,000 B.C., and one from Lake Washington to 3000 B.C. Besides the "total aromatic fraction," the sum of more than 33 most abundant PAH (the number of homologs included is not always specified) was determined by GC and GC/MS. In the Swiss lakes increasing PAH concentrations were measured in the core sections later than 1920 to 1930; in Lake Washington the 1920 section already showed a distinct elevation. There were only minor variations in the PAH concentrations before 1900, ranging from 0.6 to 1.0 mg/kg even as far back as 3000 B.C. in Lake Washington and 10,000 B.C. in the Greifensee. But in lower sections, the PAH concentration as determined by GC was less than 0.1 mg/kg. One might argue as to whether the PAH in the sediment above the 10,000 B.C. mark, which corresponded to ~3 m, could have migrated from the more recent and heavily polluted layers. This possibility has also been discussed by Mallet [61]. He determined the BaP concentration in several cores (up to 10 m in length) from the River Seine sediments about 40 km downstream from Paris and in a zone up to 300 m away from the river. BaP concentrations in the order of 10 mg/kg were found in some of the river sediments in the upper 2 to 3 m; in the layer from 8 to 10 m, 0.1 mg/kg or less were detected. Similar concentrations were found in the cores of the adjacent area. A possible migration of PAH in sand or soil has been demonstrated by Borneff and Knerr [62]. A sand filter 48 cm in diameter and 50 cm deep was rinsed with water for several days and then was topped with a layer of sand mixed with dibenz[a,h]anthracene (DBA) and subsequently sprinkled with water at intervals up to a total of 626 mm "rain." Ten individual layers of the filter were then extracted and the amount of DBA determined. Appreciable amounts had penetrated to 35 cm depth, and even in the bottom layer traces could be detected. A similar experiment

Table 5. Concentrations of PAH in Sediment Cores from Rivers and Lakes

Source	Depth or year	PAH concentration (mg/kg)	Ref.
Seine River Core I (1965)	5.9 m	3.0[a]	61
	7.15 m	0.029[a]	61
	9-10 m	0.04[a]	61
Core II	0.8 m	8.5[a]	61
	3 m	6.5[a]	61
	10 m	0.05[a]	61
Core III	2 m	>10[a]	61
	9 m	0.1[a]	61
Seine River (1969)	50 m	0.019[a]	63
Ur-Federsee	70 m	0.03[b]	64
	140 m	0.01[b]	64
Greifensee	1900	0.7[c]	59
	1975	6[c]	59
Lake Zürich	1900	0.6[c]	59
	1945	15[c]	59
Lake Lucerne	1920	1[c]	59
	1975	5[c]	59
Lake Washington	1860	1[c]	59
	1920	2.5[c]	59
	1975	6[c]	59
Grosser Plöner See:			
100 m from north shore	1918	8.99[d]	56
	1945	38.27[d]	56
	1969	27.12[d]	56
100 m from south shore	1919	8.24[d]	56
	1943	10.42[d]	56
	1968	7.74[d]	56
	1972	5.70[d]	56
Lake Constance:			
near mouth of polluted river	0-2 cm	4.3/3.8[e]	57,58
1900 = 35 cm	15-20 cm	17.3/20.4[e]	57,58
	30-35 cm	11.6/15,5[e]	57,58
deepest part of the lake	0-1 cm	2.4[e]	57,58
	3-4 cm	6.0[e]	57,58
1900 = 10 cm	25-30 cm	0.06[e]	57,58
	60-68 cm	0.35[e]	57,58
near south shore	0-2 cm	1.6[e]	57,58
	2-4 cm	3.6[e]	57,58
1900 = 6 cm	20-25 cm	0.04[e]	57,58
	30-35 cm	0.18[e]	57,58

[a]BaP.
[b]Total of 8 PAH (six PAH specified in Table 1 + BaA and BjF).
[c]Approximate values, taken from a graph, for total of 24 PAH.
[d]Total of 28 PAH.
[e]Total of 24 PAH.

in natural soil under field conditions showed that the substance had penetrated up to 25 cm in 12 months. So there is a possibility that substances in a dated core do not always represent the actual situation at the time of sedimentation.

On the other hand, Mallet [63] detected BaP in a concentration of 0.02 mg/kg in a core of 50 m depth underneath the Seine and Oise Rivers. He discusses the possibility of a synthesis by anaerobic bacteria from organic detritus containing fatty acids, waxes, and other organic matter. Borneff and Kunte [64] isolated and determined nine PAH in a core from the Ur-Federsee, an area of a lake which is silting up gradually. Concentrations of the PAH amounted to 0.03 mg/kg at 70 m depth and 0.01 mg/kg at 140 m (100,000 years old). The BaP concentration was 0.0002 mg/kg. In Japan, Ishiwatari and Hanya [65] investigated organic matter in a 200-m core from Lake Biwa, near Kyoto. PAH were determined by mass chromatography and the results expressed as a value equivalent to perylene. Concentrations ranged from 0.5 to 1.6 mg/kg throughout the core, and only at 26 m depth a much lower value of 0.004 mg/kg was found. Seven PAH were identified, perylene being the most abundant.

VII. PAH in Biota

Much information is available for PAH in marine organisms, but it is scarce for freshwater biota. In nearly all cases only BaP has been determined (Table 6). Borneff and Fischer [9] found PAH in sludge from filters of a water treatment plant that, to a large extent, consisted of phytoplankton. Subsequently, they collected phytoplankton of Lake Constance with a plankton net, avoiding the use of combustion engines. They identified seven PAH (Flu, BaA, BbF, BjF, BkF, BaP, BghiPer) in a total concentration of 600 μg/kg dry weight (BaP, 2 μg/kg) [10]. In algae (*Spirulina*) grown in Lake Texcoco (Mexico) BaP concentrations ranged from 2.6 to 3.8 μg/kg [66]. Ilnitzky [42] also reports on BaP concentration of plankton, algae, higher water plants, mussels, and fish from rivers and reservoirs in the USSR. The highest concentration, 475 μg/kg, was found in plankton of a river below a large city; multicellular algae collected at the same location had a BaP concentration of 75 μg/kg, whereas the concentration in algae of the same river above the city was lower by a factor 10 to 100. Several species of higher water plants in reservoirs showed BaP values from 0.1 to 5.0 μg/kg. In freshwater mussels (*Unio pictorum*) 0.03-1.2 μg/kg (fresh weight) were detected in specimens from the Moscow River and 10 to 12 μg/kg in those from a reservoir. In other species of the same reservoir (*Dreissenia polymorpha* and *Viviparus viviparus*) BaP concentrations were 3.36 and 1.68 μg/kg, respectively. Five species of fish (sterlet, bream, *Pelecus cultratus*, *Aspius aspius*, and pike perch) from a river were investigated above and below a sewage discharge. At the first location BaP levels differed little, whereas at the polluted place BaP concentrations were considerably higher, depending on the species (those with high lipid content accumulated more BaP).

In laboratory experiments, accumulation and release of [^{14}C]naphthalene by the freshwater alga *Chlamydomonas angulosa* was studied by Soto and coworkers [71]. In a saturated solution of naphthalene the cells accumulated this hydrocarbon and became nonmotile until cell division began. If they were resuspended in clean medium they began to release naphthalene and

Table 6. BaP Concentrations in Freshwater Biota

Organism	Location	Concentration of BaP (μg/kg)	Ref.
Phytoplankton	Lake Constance	2	9
Plankton	River below city (USSR)		
Algae:		475	42
Algae spec.	River above city (USSR)	0.6-6.0	42
Algae spec.	River below city (USSR)	75	42
Cladophora	Reservoir (USSR)	0.16-1.7	42
Spirulina	Lake Texcoco (Mexico)	2.6-3.8	66
Scenedesmus	Open tank (Bangkok): median value	1.39	67
	Open tank (Dortmund): median value	39.5	67
Mollusks:			
Unio pictorum	Moscow River	0.03-1.3	42
Unio pictorum	Reservoir	10.0-12.0	42
Dreissena polymorpha	Reservoir	3.36	42
Viviparus viviparus	Reservoir	1.68	42
Cipangopaludina	Ponds (Hamilton, Ont., Canada)	2.5	68
Fish:			
Lake trout	Lake Maskinonge (Ont., Canada)	<1.0	69
5 species	River prior to sewage discharge (USSR)		
body		1.06-7.89	42
gills		6.1-9.1	42
4 species	River after sewage discharge (USSR)		
body		2.4-66.8	42
gills (Sterlet only)		62-93	42
predatory and nonpredatory fish	Water bodies in Estonia	0.11-5.96	70

regained motility. Studies on the freshwater crustacean *Daphnia pulex* were conducted by Herbes and Risi [72]. If incubated with [^{14}C]anthracene the animals rapidly accumulated the compound until an equilibrium with the water was reached, which was about 760 times the radiocarbon concentration in water. On return to fresh water, release was rapid for the first 30-35% of total ^{14}C concentration and slowed down subsequently. Eventually levels asymptotically approached 8% of the initial level. Water-soluble metabolites were detected in the medium. The bioaccumulation potential of seven PAH in *Daphnia pulex* was studied by Southworth and co-workers [73]. In all cases an equilibrium concentration was reached within 24 hr. The equilibrium concentration factor was closely related to the chemical structure and increased with the number of aromatic rings: naphthalene is accumulated about 100-fold, and BaA about 10,000-fold.

VIII. Origin of PAH in Water

Several sources of PAH in water must be considered, including: (a) sewage effluents; (b) airborne particulates—either settling on the water, and on the land (from where they may be carried into surface waters by rain), or washed out from the air by rain and snow; (c) runoff from land, particularly roads; (d) biosynthesis.

In the following subsections, each of these is treated separately.

A. Sewage and Effluents

It has been well established that sewage generally contains large amounts of PAH. Borneff and Kunte [74] investigated several sewage treatment plants and found PAH concentrations up to 210 μg/liter in raw sewage; these concentrations were reduced to less than 1 μg/liter with mechanical and biological treatment (Table 7). If, however, treatment is insufficient or lacking, large amounts of PAH may reach surface waters. Sewage from domestic sources may contain high levels of PAH, but in general the larger the industrial contribution to the sewage, the higher the PAH concentrations. PAH in industrial effluents were studied by a number of investigators [5,6,39,74-80]. The concentrations vary widely, dependent on the type of industry and the degree of treatment (Table 7). Gasworks, coke by-product plants, aluminum smelters, iron foundries, the shale oil industry, and oil refineries are some of the industries known to emit large amounts of PAH. A BaP concentration of more than 10 mg/liter was found in untreated wastewater of a shale oil industrial plant and 0.3 mg/liter in the treated effluent [39]. Much lower concentrations were reported by other investigators [78]. In gasworks effluents up to 1.0 mg/liter BaP was detected before treatment and 0.02 mg/liter after filtration through coke beds [75].

B. Atmospheric Fallout

PAH formed in combustion and other high-temperature processes are, to a large extent, primarily released to the atmosphere and are assumed to be almost completely associated with particulate matter. Depending on the size of the particles they are deposited on land and water surfaces in a few days or weeks or are washed down by rain and snow. Rainwater, therefore, may contain considerable amounts of PAH. A mean concentration of 0.545 μg/liter and a maximum of 1.270 μg/liter (the total for six PAH) were detected in rainwater in the Netherlands [81] and from 1 to 688 μg/liter BaP in snow in the Alps at 2239 m, the highest values being found near a road [47]. In the Lithuanian SSR, the average monthly fallout of BaP calculated from concentrations in rain and snow was 0.02-0.15 $\mu g/m^2$ for the summer and 0.15-0.5 $\mu g/m^2$ for the winter. BaP was concentrated in the snow cover and redistributed in the spring by surface waters [82].

Table 7. PAH Concentrations in Raw and Treated Sewage of Domestic and Industrial Origin

Source	Kind of treatment	Concentration (μg/liter)		Ref.
		Inlet	Effluent	
Town + industry (textile, food)	Mech., biol., chem.	4.4[a]	0.83[a]	74
Town + industry (sugar)	Mech., biol.	211; 9.1[a]	0.72; 0.78[a]	74
Town + little industry	Mech., biol.	12.4; 2.7[a]	0.48; 0.52[a]	74
Small village	Mech.			
dry weather		0.77[a]	3.9[a]	44
heavy rainfall		78[a]	62[a]	44
Rural area, domestic	No	–	0.44-5.6	44
Small town, mainly domestic	No	–	12.5	44
Metalworking industry				
hardening	Sodium hypochlorite	1.09[a]	1.76[a]	74
galvanization	Neutralization	0.26[a]	0.24[a]	74
Shale oil industry		10-900[b]	312[b]	39
Shale oil industry	Dephenolization	5-20[b]	2[b]	78
Gasworks	Filtration through coke bed	340-1000[b]	20[b]	75
Gasworks	Dephenolization	–	130-290[b]	76
Coke by-product plant	Biochemical	44-64[b]	12-16[b]	77

[a]Total of 10 PAH: Flu, Pyr, BaA, BbF, BjF, BkF, BaP, BghiPer, Ip, Per.
[b]BaP.

C. Runoff from Land

Considerable amounts of PAH may also be swept into sewers and surface waters from roads, particularly from those with tar surfaces. Dust collected from tarred roads contained up to 750 mg/kg PAH (10 compounds) or 17 mg/kg BaP [44]. Asphalt surfaces contain significantly lower PAH concentrations.

D. Biosynthesis

There has been much controversy about the question of biosynthesis of PAH by microorganisms or plants. Mallet and co-workers [83] incubated bacteria together with lipids from marine plankton and found BaP in the extracts from cultures of anaerobic bacteria (*Clostridium*) but not in those of aerobic strains, and they considered these results to be evidence of biosynthesis. Knorr and Schenk [84] grew several species of bacteria (*Escherichia coli*, *Proteus vulgaris*, *Pseudomonas fluorescens*, and others) on highly purified nutrient agar and glycerin-grape sugar agar and detected BaP and other PAH in the harvested bacterial cells which they also attributed to biosynthesis. Evidence for a bacterial synthesis of PAH was also presented by Niaussat [85] and Ehrhardt et al. [86]. The synthesis of PAH by anaerobic bacteria was questioned by Hase and Hites [87], who investigated the extracts of bacteria and blanks (consisting of all chemicals used to grow the bacteria) with a GC/MS technique. The relative abundance of the alkyl homologs of various PAH was about the same in both extracts. The authors concluded that the PAH present in the media in very low concentrations are accumulated by the bacteria by a factor of approximately 500. An accumulation of PAH by bacteria has also been reported by other authors [88,89].

The possibility of PAH synthesis by algae was investigated by Borneff and co-workers [90]. The alga *Chlorella vulgaris* was cultured in the presence of [^{14}C]acetate, which served as the sole organic carbon source. Seven PAH were isolated from labeled and nonlabeled algae: those from the labeled algae contained radioactive counts at least 10 times higher than the PAH from nonlabeled algae cultured under the same conditions. From these results a biosynthesis of the PAH must be concluded. This conclusion has been questioned on the grounds that PAH measured may not have been completely separated from other organic compounds [4]. It is, however, difficult to imagine that a very similar background level would be present in all seven PAH isolated. Payer and co-workers [67] grew algae (*Scenedesmus acutus*) in open tanks in both Dortmund (Germany), a heavily polluted area, and Bangkok (Thailand), where air pollution is considerably lower. From the statistical distribution patterns they conclude that with the exception of Flu, BghiPer, and BkF in the samples from Bangkok, endogenous formation is not likely to occur to a significant extent. PAH concentrations in the Bangkok algae were higher by a factor of 2 than in the labeled algae mentioned above and were still higher by a factor of 5 to 20 in the Dortmund algae. These results do not exclude the possibility that at least part of the PAH may have been synthesized by the algae.

Many investigations have demonstrated the presence of PAH in higher plants [e.g., 91-93], mainly in those destined for human consumption, and in leaves from different trees. Gräf and Diehl [91] found a significant increase of BaP in seedlings of rye, wheat, and lentils germinated in a hydroponic medium free of PAH in comparison to the concentration in the seeds, and they ascribe this increase to biosynthesis. The cultures were covered with plastic sheets to exclude contamination by airborne PAH.

Wettig and co-workers [94] also cultured seeds of several vegetables and of corn using different nutritive solutions. Seeds germinated in phosphate (NaH_2PO_4) solution alone did not show an increase in BaP concentration; those germinated in phosphate-acetate solution showed a significant increase. Grimmer and Düvel [95] conducted similar experiments in special chambers where the air was carefully filtered and found no detectable levels of PAH in the plants grown there, indicating that all PAH found in the plants originate from airborne contamination.

IX. Degradation of PAH

The PAH present in surface water are subject to degradation through different processes, the most important ones being oxidation and biotransformation. It is well documented that PAH are degraded by photooxidation [96-99]. The effects of several factors on photodegradation were studied by Suess [100]. From the results of these experiments the author concluded that photooxidation in natural waters will depend mainly on temperature, dissolved oxygen, and solar radiation. Chemical oxidation by chlorine and ozone has been investigated in connection with the purification of drinking water and is discussed later (Sec. X).

Biotransformation of PAH by aquatic organisms has been reviewed by Neff [4]. The vast majority of investigations is concerned with marine organisms, and only few were done on freshwater biota. Several species of bacteria isolated from water or sediment have been found capable of oxidizing PAH [101-103]. In most cases either the bacteria were isolated from heavily polluted bodies of water containing high concentrations of PAH or they were adapted by preincubation with PAH. Whereas two- and three-ring compounds are readily utilized by several strains of bacteria and in some cases are completely degraded, higher-molecular-weight compounds are metabolized only by a few organisms and only partially. Main metabolites are dihydrodiols [102]. Oxidation of PAH by activated sludge was found to be negligible [104].

The capability of higher aquatic organisms in the freshwater ecosystem to oxidize PAH has been demonstrated, e.g., for fungi [105] and for fish (trout). The enzyme system found in the fish appeared to be very similar to the mammalian system, and activity was higher than in rat liver when measured per milligram of microsomal protein [106,107].

X. Evaluation of PAH in the Aquatic Environment

Several PAH, e.g., BaP, are known to be carcinogenic, as established in animal experiments. In vivo BaP is transformed to the 7,8-epoxide by oxidative processes in the cells; the next reaction is the addition of water, followed by further enzymatic oxidation, which includes further oxygen incorporation between C-9 and C-10. Thus two diol epoxides are formed; one of these reacts with DNA, thereby producing the essential carcinogenic factor.

Primarily, PAH become effective at the locus of application; that is, a skin tumor is caused by subcutaneous injection. In some cases systemic reactions do occur: descriptions have been given of subpleural adenomas, hyperplasias of the forestomach of mice, ulcerations of the glandular part of the stomach, and other manifestations, but usually no malign tumors are seen. For tumor initiation only a dose of some micrograms is necessary of these highly active compounds.

Though the experimental application of BaP to the skin of human volunteers did not cause any tumors, there is no doubt as to the effect in general. This is proved by occupational skin cancer of tar workers and by the fact that lung cancer of nonsmokers occurs some three to four times more frequently in cities than in the country. In both cases—lung cancer in cities and skin cancer of tar workers—a cocarcinogenic effect of other noxious compounds must be assumed. There are no indications that PAH are a causal factor for tumor promotion in the human digestive tract and its organs; nevertheless, the BaP content of smoked meat and drinking water was limited by law in the Federal Republic of Germany.

In the literature a growth-retarding or lethal effect of PAH on plants has been discussed, but PAH in surface water and river and lake sediments are not likely to cause such harm, since the necessary levels have been reached only experimentally. On the other hand, a hazard cannot be excluded a priori. Theoretically a contamination of humans is possible:

1. Via water
   a. Drinking water
   b. Bath water
2. Via sediments
   a. Following resolubilization and transmission with drinking water
   b. Ingestion of particles as a result of insufficient drinking water purification
   c. Use of dredged sediments as fertilizer and subsequent transmission into food plants
   d. Ingestion via fish and seafood used for human nutrition
3. Via biota used as animal feed (e.g., algal protein)

In order to estimate the risk from the aforementioned sources, quantitative aspects have also to be considered. Basically it must be pointed out that zero standards are unrealistic, which means that nutrition with foodstuffs free of PAH is not possible at present. Furthermore, a setting of standards to satisfy the most rigorous toxicological demands cannot be accomplished at this time. The epidemiological prerequisites are lacking that would permit us to draw a line between harmful and nonharmful concentrations. But a technical standard for PAH concentrations should be laid down on this principle: they must be kept as low as possible.

Surface water is used for drinking water on a large scale (Germany, 31%; Norway, 95%). There are no objections to this source of supply as long as national and international standards (for example, WHO and European Community guidelines) are taken into account.

At present PAH (six substances) are limited to 200 ng/liter in the countries of the European Community. In light of the results of routine analyses in Germany, a reduction of this standard to 100 ng/liter is recommended. Since wholesome drinking water obtained from groundwater normally contains only 1-10 ng/liter, this range should be aimed at also in purifying surface water for use as drinking water.

Possible technical ways are sand and activated carbon filtration, coagulation, and sedimentation; if necessary, ozonization or treatment with chlorine dioxide may be used in addition. The efficiency factor of combined processing ranges from 90 to 99%. Furthermore, the contamination of lake and river water with PAH should not exceed 100 ng/liter as the mean value per year. The data we reported show that the actual concentrations are often higher. In these cases there is probably no measurable increase of carcinogenic risk. However, we have to take into consideration that in addition to drinking water PAH are ingested also by other foodstuffs in doses of about 200 μg/year. The intake via drinking water amounts to 1-10 μg/year; nevertheless, a rather low level is recommended, since the consumer usually has no choice about his drinking water supply. Bathing in surface waters is sometimes judged as being dangerous with respect to the PAH content. For the risk assessment one should consider that the occupational exposure of tar workers (hard coal tar contains up to 1% BaP) leads to the occurrence of skin and lung cancer only after several years or even decades. A short contact, during bathing, with water containing 10^8 times less PAH than tar should pose no carcinogenic risk on humans.

According to the results on hand, river and lake sediments are contaminated with PAH in the order of magnitude of 10-20 mg PAH/kg and sewage sludge up to 15 mg/kg. The problem of solubilization of PAH in sediments has not been sufficiently investigated. Also PAH analyses from cores obtained in Lake Constance do not give a clear answer. Old layers do contain less PAH than younger ones; the youngest, however, are not the most contaminated. Perhaps there is a transmission of PAH from the sediment in rivers into the water stored in reservoirs. However, the necessary analyses are lacking on this point.

When river water is used for drinking water, the purification process should guarantee the elimination of suspended matter. Generally insufficient treatment leads to the persistence of PAH. If the PAH level is regulated by law, as it is in the Federal Republic of Germany, no risk need be feared since the standard of 200 ng/liter includes PAH in solubilized as well as in suspended state. The degree of transmission of PAH through the food chain caused by using dredged sediments for cultivation of land is difficult to assess. Concentration in such sediments is about 10 mg/kg (max. 130 mg/kg); in comparison, sewage sludge-garbage compost contains some 10-20 mg/kg; and animal manure some 1-10 mg/kg [108]. Experiments with fertilization [108-110] outdoors and in laboratory tests led to the conclusion that aboveground plant parts do not show any enrichment of PAH; only radish and carrot roots contained minor amounts. In these cases, probably diffusion and not active resorption is taking place.

The daily intake of PAH is increased by 1% as a consequence of the application of highly (PAH) contaminated fertilizers. River sediments are contaminated at about the same levels as garbage compost; thus the application of such sediments as fertilizer appears to pose no significant hazard to human health, provided there is no high heavy-metal load.

Fish finding their feed in river sediments might be exposed to a raised carcinogenic risk. This might produce a danger to humans eating fish and seafood if PAH concentrations in the musculature are extremely high. There are no objections to feeding algal protein to animals.

If in the near future algal protein is used for human nutrition directly the risk of PAH intake will not be higher than consuming usual vegetables, which show a "normal level" of 0.1 mg/kg. In areas with heavy air pollution (e.g., Dortmund, Germany) cultivated algae contained 0.7 mg/kg, which corresponds to concentrations in vegetables grown in comparable regions.

XI. Recommendations

Judging from the results of the diverse investigations considered here, it must be emphasized that PAH do affect mankind via air, water, and foodstuffs; a carcinogenic risk must be generally considered likely. No epidemiological proof has yet been given concerning the digestive tract; however, as a precaution all possibilities for the reduction of PAH concentrations in the ecosystem should be investigated:

1. Drinking water ought to contain only 1-10 ng of the six PAH per liter; the upper limit should be 100 ng/liter; since surface waters used for drinking water supply are in some cases contaminated with 100-1000 ng/liter, conventional purification procedures reducing them by about 90% should be required. Regulatory control measures concerning PAH in drinking water are necessary.
2. In case of increased PAH concentrations in drinking water, the purification process must be optimized and, moreover, the quality of the raw water should be improved. Sewage purification plants should be constructed.
3. Drinking water from reservoirs fed by river water ought to be carefully analyzed because of possible (but not verified) resolubilization of PAH in sediments.
4. Because there may be a high contamination of river sediments with PAH, dredged material should be analyzed before it is used as fertilizer. A limitation is not necessary as regards fertilization of grain crops. For fertilization of carrots, radishes, asparagus, and the like, river sediments should not contain more than 10 mg/kg for the total six PAH.

References

1. J. B. Andelman and M. J. Suess, *Bull. World Health Org. 43*:479 (1970).
2. J. B. Andelman and J. E. Snodgras, *CRC Critical Rev. Environ. Control 4*:69 (1974).
3. R. M. Harrison, R. Perry, and R. A. Wellings, *Water Res. 9*:331 (1975).
4. J. M. Neff, *Polycyclic Aromatic Hydrocarbons in the Aquatic Environment*, Applied Science Publ., London, 1979.
5. P. Wedgewood and R. L. Cooper, *Analyst 78*:170 (1953).
6. P. Wedgewood and R. L. Cooper, *Analyst 79*:163 (1954).
7. L. Mallet and M. Héros, *C.R. Acad. Sci., Paris 250*:943 (1960).
8. J. Borneff and R. Fischer, *Arch. Hyg. Bakteriol. 146*:1 (1962).
9. J. Borneff and R. Fischer, *Arch. Hyg. Bakteriol. 146*:183 (1962).
10. J. Borneff and R. Fischer, *Arch. Hyg. Bakteriol. 146*:334 (1962).
11. J. Borneff and H. Kunte, *Arch. Hyg. Bakteriol. 148*:585 (1964).
12. J. Saxena, J. Kozuchowski, and D. K. Basu, *Environ. Sci. Technol. 11*:682 (1977).
13. D. K. Basu and J. Saxena, *Environ. Sci. Technol. 12*:791 (1978).
14. H. Kunte and J. Borneff, *Z. Wasser Abwasser Forsch. 9*:35 (1976).
15. DIN Deutsches Institut fur Normung, *Deutsche Einheitsverfahren zur Wasser-, Abwasser- und Schlammuntersuchung*, No. 38409, Pt. 13 Weinheim (1981).
16. J. Borneff and H. Kunte, Analysis of polycyclic aromatic hydrocarbons in water using thin-layer chromatography and spectrofluorometry, in *Environmental Carcinogens: Selected Methods of Analysis*, Vol. 3: *Analysis of Polycyclic Aromatic Hydrocarbons in Environmental Samples*, H. Egan, M. Castegnaro, P. Bogovski, H. Kunte, E. A. Walker, and W. Davis (Eds.), International Agency for Research on Cancer, Lyons, France, 1979.

17. A. Ya. Khesina, Determination of benzo[a]pyrene in extracts by spectroluminescence, in *Environmental Carcinogens: Selected Methods of Analysis*, Vol. 3: *Analysis of Polycyclic Aromatic Hydrocarbons in Environmental Samples*, H. Egan, M. Castegnaro, P. Bogovski, H. Kunte, E. A. Walker, and W. Davis (Eds.), International Agency for Research on Cancer, Lyons, France, 1979.
18. N. Ya. Yanisheva, and I. S. Kireeva, Determination of benzo[a]pyrene in air using quasi-linear luminescence, in *Environmental Carcinogens: Selected Methods of Analysis*, Vol. 3: *Analysis of Polycyclic Aromatic Hydrocarbons in Environmental Samples*, H. Egan, M. Castegnaro, P. Bogovski, H. Kunte, E. A. Walker, and W. Davis (Eds.), International Agency for Research on Cancer, Lyons, France, 1979.
19. G. Grimmer and K.-W. Naujack, *Vom Wasser 53*:1 (1979).
20. J. Jacob and G. Grimmer, Extraction and enrichment of polycyclic aromatic hydrocarbons (PAH) from environmental matter, in *Environmental Carcinogens: Selected Methods of Analysis*, Vol. 3: *Polycyclic Aromatic Hydrocarbons in Environmental Samples*, H. Egan, M. Castegnaro, P. Bogovski, H. Kunte, E. A. Walker, and W. Davis (Eds.), International Agency for Research on Cancer, Lyons, France, 1979.
21. H. Kunte, Separation, detection and identification of polycyclic aromatic hydrocarbons, in *Environmental Carcinogens: Selected Methods of Analysis*, Vol. 3: *Analysis of Polycyclic Aromatic Hydrocarbons in Environmental Samples*, H. Egan, M. Castegnaro, P. Bogovski, H. Kunte, E. A. Walker, and W. Davis (Eds.), International Agency for Research on Cancer, Lyons, France, 1979.
22. W. E. May, S. P. Wasik, and D. H. Freeman, *Anal. Chem. 50*:997 (1978).
23. J. Borneff und R. Knerr, *Arch. Hyg. Bakteriol. 143*:390 (1959).
24. J. Borneff, unpublished data (1969-1979).
25. Arbeitsgemeinschaft Rheinwasserwerke e.V. (ARW), Rept. No. 31 (1974).
26. Arbeitsgemeinschaft Rheinwasserwerke e.V. (ARW), Rept. No. 32 (1975).
27. Rijncommissie Waterleidingbedrijven (RIWA), Annual Rept. for 1976, Pt. A: *Der Rhein* (1976).
28. Rijncommissie Waterleidingbedrijven (RIWA), Annual Rept. for 1977, Pt. A: *Der Rhein* (1977).
29. Rijncommissie Waterleidingbedrijven (RIWA), Annual Rept. for 1978, Pt. A: *Der Rhein* (1978).
30. Rijncommissie Waterleidingbedrijven (RIWA), Annual Rept. for 1979, Pt. A: *Der Rhein* (1979).
31. W. M. Lewis, *Water Treat. Exam. 24*:243 (1975).
32. M. A. Acheson, R. M. Harrison, R. Perry, and R. A. Wellings, *Water Res. 10*:207 (1976).
33. B. T. Croll, personal communication to M. J. Suess (1980).
34. B. Festy, personal communication to M. J. Suess (1980).
35. J. Vasseur, J. Dequidt, and F. Erb, *Bull. Soc. Pharm. Lille 31*:157 (1975).
36. M. Karlowska-Jasek and E. Rzewuska, *Environ. Protection Eng. 3*:145 (1977).
37. J. Dojlido, personal communication to M. J. Suess (1980).
38. H. Manczak, personal communication to M. J. Suess (1980).
39. P. P. Dikun and A. I. Machinenko, *Gig. Sanit. 28*:10 (1963).

40. L. N. Samoilovich and Y. R. Redkin, *Gig. Sanit.* *33*:6 (1968); see *Chem. Abstr.* *70*:22789 m (1969).
41. A. P. Ilnitski, L. G. Rozhnova, and T. V. Drosdova, *Gig. Sanit.* *36*: 316 (1971); see *Water Pollut. Abstr.* *46*:No. 260 (1973).
42. A. P. Ilnitski, J. L. Lembik, L. G. Solenova, and L. M. Shabad, *Cancer Detection Prevention* *2*:471 (1979).
43. D. K. Basu and J. Saxena, *Environ. Sci. Technol.* *12*:795 (1978).
44. J. Borneff and H. Kunte, *Arch. Hyg. Bakteriol.* *149*:226 (1965).
45. Rijncommissie Waterleidingbedrijven (RIWA), Annual Rept. for 1975, Pt. A: *Der Rhein* (1975).
46. J. Borneff, *Gas- Wasser- Abwasser* *55*:467 (1975).
47. H.-J. Elster, M. Knorr, H. Lehn, R. Mühleisen, and W. J. Müller, *Bodensee-Projekt der Deutschen Forschungsgemeinschaft*, 2nd Rept., July 1, 1963 through Dec. 31, 1966, Steiner Verlag, Wiesbaden, Germany, 1968.
48. R. Herrmann, *CATENA* *5*:165 (1978).
49. H. Woidich, W. Pfannhauser, G. Blaicher, and K. Tiefenbacher, *Mitt. GDCH-Fachgr. Lebensmittelchem. gerichtl. Chem.* *30*:141 (1976).
50. I. Veldre, M. Rahu, A. P. Ilnitski, and L. G. Lokhova, *Vodn. Resur.* *1977*:147; see *Chem. Abstr.* *90*:127156u (1979).
51. I. Veldre, A. Itra, and L. Paalme, *Org. Veshchestvo Biog. Elem. Vnutr. Vodakh, Tezisy Dokl. Vses. Simp.*, 3rd, p. 19 (1978); see *Chem. Abstr.* *91*:62360e (1979).
52. DVGW-Untersuchungsprogramm: Biozidgehalt in 19 deutschen Talsperren, *DVGW-Scheiftenreihe Wasser No. 5* (1974).
53. E. D. John, M. Cooke, and G. Nickless, *Bull. Environ. Contam. Toxicol.* *22*:653 (1979).
54. G. Eglinton, B. R. T. Simoneit, and J. A. Zoro, *Proc. Roy. Soc., Ser. B* *189*:415 (1975).
55. R. E. LaFlamme and R. A. Hites, *Geochim. Cosmochim. Acta* *42*:289 (1978).
56. G. Grimmer and H. Böhnke, *Cancer Lett.* *1*:75 (1975).
57. G. Müller, G. Grimmer, and H. Böhnke, *Naturwissenschaften* *64*:427 (1977).
58. G. Grimmer and H. Böhnke, *Z. Naturforsch.* *32c*:703 (1977).
59. S. G. Wakeham, C. Schaffner, and W. Giger, *Geochim. Cosmochim. Acta* *44*:403 (1980).
60. S. G. Wakeham, G. Schaffner, and W. Giger, *Geochim. Cosmochim. Acta* *44*:415 (1980).
61. L. Mallet, *Bull. Acad. Nat. Med. (Paris)* *149*:656 (1965).
62. J. Borneff and R. Knerr, *Arch. Hyg. Bakteriol.* *144*:81 (1960).
63. L. Mallet, *C. R. Soc. Biol., Paris* *163*:319 (1969).
64. J. Borneff and H. Kunte, *Oberrhein. Geol. Abhandl.* *16*:94 (1967).
65. R. Ishiwatari and T. Hanya, *Proc. Japan Acad.* *51*:436 (1975).
66. G. Bories and J. Tulliez, *Ann. Nutr. Aliment.* *29*:573 (1975).
67. H. D. Payer, C. J. Soeder, H. Kunte, P. Karuwanna, R. Nonhof, and W. Gräf, *Naturwissenschaften* *62*:536 (1975).
68. L. Kalas, A. Mudroch, and F. I. Onuska, *Environ. Sci. Res.* *16*:567 (1980).
69. R. J. Pancirov and R. A. Brown, *Environ. Sci. Technol.* *11*:989 (1977).
70. P. A. Bogovski, I. Veldre, A. Itra, and L. Paalme, *Gig. Sanit.*, p. 111 (1978); see *Chem. Abstr.* *89*:18127u (1978).

71. C. Soto, J. A. Hellebust, and T. C. Hutchinson, *Can. J. Bot. 53*:118 (1975).
72. S. E. Herbes and G. F. Risi, *Bull. Environ. Contam. Toxicol. 19*:147 (1978).
73. G. R. Southworth, J. J. Beauchamp, and K. Schmieder, *Water Res. 12*: 973 (1978).
74. J. Borneff and H. Kunte, *Arch. Hyg. Bakteriol. 151*:202 (1967).
75. P. Wedgewood and R. L. Cooper, *Analyst 81*:42 (1956).
76. S. N. Cherkinskii, P. P. Dikun, and G. P. Yakoleva, *Gig. Sanit. 24*: 11 (1959); see *Chem. Abstr. 54*:7942f (1960).
77. Z. P. Fedorenko, *Gig. Sanit. 29*:17 (1964); see *Chem. Abstr. 61*:4055h (1964).
78. I. Veldre, L. Lahe, and I. Arrol, *Eesti NSV Teaduste Akad. Toimetised, Biol. Seer. 14*:268 (1965); see *Chem. Abstr. 63*:16021h (1965).
79. K. P. Ershova, *Gig. Sanit. 33*:102 (1968); see *Chem. Abstr. 68*:107745c (1968).
80. L. K. Lupanova, P. E. Shkodich, and Zh. L. Lembik, *Vop. Profil. Zagryaznenia Vnesh. Sredy, Chastnosti Vodoemov, Kantserogen. Veshchesto*, p. 28 (1972); see *Chem. Abstr. 79*:9479b (1973).
81. B. C. J. Zoeteman, G. J. Piet, C. F. H. Morra, and F. E. de Grunt, *Drinking Water Quality and Public Health, Papers and Proc., Water Res. Coll., High Wycombe, Bucks., England, Nov. 4-6, 1975*, p. 260 (1976).
82. A. Milukaite and K. Sopanskas, *Rast. Khim. Kantserogeny, Proc. 1st Symposium, 1976*, E. I. Slepyan (Ed.), Leningrad, 1979, p. 194; see *Chem. Abstr. 92*:134290c (1980).
83. L. Mallet, L. Zanghi, and J. Brisou, *C.R. Acad. Sci., Paris, Ser. D 264*:1534 (1967).
84. M. Knorr and D. Schenk, *Arch. Hyg. Bakteriol. 152*:282 (1968).
85. P. Niaussat, *Rev. Intern. Oceanogr. Med. 17*:87 (1970).
86. J. P. Ehrhardt, P. Niaussat, J. Trichet, M. Héros, and N'Guyen Trung Luong, *Rev. Int. Oceanogr. Med. 48*:97 (1977).
87. A. Hase and R. A. Hites, *Geochim. Cosmochim. Acta 40*:1141 (1976).
88. L. Mallet and M. Héros, *C. R. Acad. Sci., Paris 253*:287 (1961).
89. H. Lorbacher, H.-D. Püls, and H. W. Schlipköter, *Zbl. Bakteriol. Hyg. I. Abt. Orig. B 155*:168 (1971).
90. J. Borneff, F. Selenka, H. Kunte, and A. Maximos, *Environ. Res. 2*: 22 (1968).
91. W. Gräf and H. Diehl, *Arch. Hyg. Bakteriol. 150*:49 (1967).
92. G. Grimmer and A. Hildebrandt, *Deut. Lebensmittelrundsch. 61*:237 (1965).
93. L. M. Shabad and Y. L. Cohan, *Arch. Geschwulstforsch. 40*:237 (1972).
94. K. Wettig, A. Ya. Chesina, G. Gelbert, and L. M. Shabad, *Arch. Geschwulstforsch. 46*:634 (1976).
95. G. Grimmer and D. Düvel, *Z. Naturforsch. 25b*:1171 (1970).
96. J. Borneff and R. Knerr, *Arch. Hyg. Bakteriol. 143*:405 (1959).
97. C. B. Allsopp and B. Szigeti, *Cancer Res. 6*:14 (1946).
98. M. Kuratsune and T. Hirohata, *Nat. Cancer Inst. Monogr. 9*:117 (1962).
99. J. Jager and B. Kassowitzova, *Čs. Hyg. 13*:288 (1968); see *Excerpta Med. Publ. Health 15*:No. 981 (1969).
100. M. J. Suess, *Zbl. Bakteriol. Hyg. I. Abt. Orig. B 155*:541 (1972).
101. F. D. Sisler and C. E. ZoBell, *Science 106*:521 (1947).

102. D. T. Gibson, V. Mahadevan, D. M. Jerina, H. Yagi, and H. J. C. Yeh, *Science 189*:295 (1975).
103. J. D. Walker, R. R. Colwell, and L. Petrakis, *Appl. Microbiol. 30*: 1036 (1975).
104. G. W. Malaney, P. A. Lutin, J. J. Cibulka, and L. H. Hickerson, *J. Water Pollut. Control Fed. 39*:2020 (1967).
105. C. E. Cerniglia and D. T. Gibson, *Arch. Biochem. Biophys. 186*:121 (1978).
106. M. G. Pedersen, W. K. Hershberger, and M. R. Juchau, *Bull. Environ. Contam. Toxicol. 12*:481 (1974).
107. J. T. Ahokas, O. Pelkonen, and N. T. Karki, *Biochem. Biophys. Res. Commun. 63*:635 (1975).
108. J. Borneff, G. Farkasdi, H. Glathe, and H. Kunte, *Zbl. Bakteriol. Hyg. I. Abt. Orig. B 157*:151 (1973).
109. H. Müller, *Z. Pflanzenernaehr. Dueng. Bodenk.*, p. 685 (1976).
110. G. Peiwast, Der Einfluss steigender Gaben von anaerob und aerob bereitetem Müllklärschlammkompost zu verschiedenen Gemüsearten auf den Gehalt an polyzyclischen aromatischen Kohlenwasserstoffen in Pflanze und Boden. Dissertation, Universität Giessen, Institut für Pflanzenbau und Pflanzenzüchtung (1976).

16

Polycyclic Aromatic Hydrocarbons in Work Atmospheres

RICHARD B. GAMMAGE / Oak Ridge National Laboratory, Oak Ridge, Tennessee

I. Introduction

The energy crisis and the nation's commitment to producing synthetic fossil fuels in the coming decades could mean the emergence of new industries employing many thousands of workers who would be potentially exposed to fresh sources of polycyclic aromatic hydrocarbons (PAH). This seemingly inevitable event is the greatest driving force for writing a chapter of this kind. Accordingly, the greatest emphasis will be placed on PAH exposures that are, or are likely to be, encountered as a result of a proliferating synfuel industry. Energy sources of PAH in fossil fuels have recently received an excellent review by Guerin [1].

Occupational health control in environments containing PAH is a major objective of PAH measurements. The critical parameter is understanding the relationship of PAH exposures to human cancer. This needs to be coupled to an improving ability to monitor these exposures. Thus the characterization of workplace PAH will be considered from the viewpoint of mutagenicity and carcinogenicity as well as individual PAH composition and its measurement.

During the development of this chapter, the reader should be able to gain a better feeling for the types of spectra of PAH met within different workplace environments and the use of indicator compounds for characterizing these spectra. Attention will also be given to pathways of worker exposure that are important, methods of collecting, monitoring, and analyzing the PAH therein, and the difficulties presently encountered in doing this. It is intended that the reader appreciate the balances and relationships between potential human carcinogenicity, PAH composition, and the technological sources of these PAH.

No paper has been written previously that addresses the specific topic of PAH exposure in the workplace. Two recent papers, however, have dealt with the more generalized topic of multimedia human exposure to PAH from nonoccupational and occupational sources [2,3].

II. PAH in Synfuels

A. Composition

Heavy oils and tars contain a multitude of PAH. Over 300 compounds have been identified in coal tar, and it is estimated that as many as 10,000 compounds may exist [4]. Much of the complexity of PAH in coal-derived materials is conferred by multialkylated PAH [5].

Several PAH compounds are known or suspected carcinogens, or have the capability of synergistic interaction with a carcinogen. Table 1 is a listing of the bioactivities of several PAH [2,6-12].

It is generally known that coal tar is highly polyaromatic and that coal-derived oils are more aromatic than their petroleum-derived counterparts. Absolute quantitation and direct intercomparison, however, are very difficult; reported quantities are highly dependent on analytical procedures, and few of these procedures can be applied without modification to both petroleum and coal oils or to oils of widely differing physical properties [13]. Published comparisons give ranges of total PAH that are 5-15% in coal-derived crudes as compared to 1-2% in petroleums [13,14]. The polyaromatic content of shale oil appears to fall between that of coal-derived crudes and petroleum.

A major difference between coal liquids and petroleum is the greater nitrogen content of the former. Even greater nitrogen content (1.5-2.0%) has been reported for shale oil by Morandi and Poulson [15]. These data suggest the following relative ranking by nitrogen content: shale oil > coal-derived crude > petroleum crude. These differences are important in light of the recent awareness that nitrogenous, basic compounds of PAH are highly mutagenic [16].

Table 1. Biological Activities of Some PAH

Compound	Carcinogenic potential[a]	Bioactivity[b]	Ref.
2-Methylnaphthalene	0	TP	10
Fluoranthene	0	CC	9
2-Methylfluoranthene	+	C,TI	9,12
3-Methylfluoranthene	?	TI	9,12
Pyrene	0	CC	8,9
Benz[a]anthracene	+	TI	9
Chrysene	+	TI	9
Benzo[c]phenanthrene	+++	C	6
3-Methyl Chrysene	+	TI	11
5-Methyl Chrysene	+++	C,TI	11
7,12-Dimethylbenz[a]anthracene	++++	C,TI	2
Benzo[b]fluoranthene	++	C,TI	9
Benzo[j]fluoranthene	++	C,TI	9
Benzo[a]pyrene	+++	C,TI	9
Dibenz[a,h]anthracene	+++	C,TI	9
Indeno[1,2,3-cd]pyrene	+	TI	9
Benzo[ghi]perylene	0	CC	8,9
Picene	+	TI	7

[a]Key: ? = uncertain; 0 = inactive; + to ++++ = active.
[b]CC = cocarcinogenic with BaP; TP = tumor promoter; TI = tumor initiator; C = complete carcinogen.

Benzo[a]pyrene (BaP) is the classically studied five-ring PAH carcinogen whose metabolism to an ultimate carcinogen is understood fairly well. It is ubiquitous to virtually all PAH-bearing samples. For these reasons, BaP is traditionally determined in complex PAH-bearing materials as an indicator of the overall PAH content, or as an index for the potential health risk.

Concentrations of BaP in natural petroleum crudes, retorted shale oils, and coal-derived liquids are quite variable. The degree of variability can be judged from the sets of data [1,17-21] presented in Table 2.

These analyses indicate that both the overall PAH and the BaP contents of natural and synthetic crude oils are in the following order: coal-derived oils >> shale oils ⩾ petroleum. Tars and pitches that are the residuals of crude oils contain much larger concentrations of BaP.

From the PAH and BaP contents of crude oils one would rank the potential carcinogenic risk in the same order: coal-derived crude oils > shale oils >

Table 2. BaP Content of Natural and Synthetic Crude Oils and Tars

Materials	BaP (~ppm)
Crude oils [1]:	
Petroleum	1
Mixture of crudes from Libya, Venezuela, Persian Gulf, and Arabia	
Shale-derived	3
Crude oil	
Hydrotreated	
Coal-derived	3
Catalytic hydrogenation	
Pyrolysis	
Crude oils and tar [17]:	
Gasifier tar	153
Coal-derived fuel oil (Repository No 1701)[a]	87
Shale oil (Repository No 4101)[a]	24
Petroleum crude (Repository No. 5103)[a]	2.6
Shale-derived diesel fuel (Repository No. 4610)[a]	0.045
Natural and synthetic crudes [18]:	
Petroleum crude A (Repository No. CRM3)[a]	3.5
Coal-derived oil A (Repository No. CRM1)[a]	152
Coal-derived vacuum still overhead oil (Repository No. 1310)[a]	456
Crude shale oil A (Repository No. CRM2)[a]	13
Crude shale oil (Repository No. 4601)[a]	11
Tars and pitches:	
Petroleum pitch [19]	2,000
Coal tar [20]	3,000
Coal tar pitch [21]	10,000

[a]USEPA/USDOE designation of synfuel samples from the Sample Repository located at Oak Ridge National Laboratory [22].

petroleum crudes. Differences in mutagenic and carcinogenic effects, however, are often much more dramatic than might be anticipated from BaP and homocyclic PAH contents alone. A missing element in the jigsaw puzzle has recently been identified. Primary amines of PAH have been identified as a major contributor to the mutagenic activity of coal-derived crude oils from direct liquefaction processes [16,22,23]. More detailed qualitation and quantitation of these amines is taking place [24], and their potentially key role in presenting an occupational health hazard is being investigated vigorously.

B. Comparative Chemical and Biological Characteristics

Present-day comparative testing of PAH-bearing products involves a battery of tests that include complex chemical fractionation, with mutagenicity and carcinogenicity testing of the fractionated products. As a result, a firmer picture is emerging of the ranking of PAH-type mutagens and carcinogens to which workers can be potentially exposed. The ability to quantify mutagenic activity is important; at a qualitative level, the correlation between mutagenicity and carcinogenicity is high for the PAH as a class [25].

A number of comparative tests have been conducted with petroleum crude oils and shale- and coal-derived petroleum substitutes [23,26]. To separate, identify, and biologically test these active compounds and groups of compounds within such chemically complex starting materials has been an enormous undertaking. The chemists and biologists have successfully coordinated their research efforts toward providing meaningful health effects data.

Recently presented comparative mutagenicity data [23] from Pacific Northwest Laboratory (PNL) are reproduced in Table 3. These data show that the high-boiling products of coal liquefaction are generally of the highest toxicity.

The heavy distillates and process solvent streams exhibit substantial mutagenic activity. The lower-boiling, light *solvent refined coal* (SRC) products show no significant activity. By comparison, the raw shale oils exhibit limited activity, and the two crude petroleum oils do not show activity in the Ames system [27].

Other biological assays, including mammalian cell culture and skin painting carcinogenicity studies, are being undertaken by the PNL group [23]. Agreement among the different assays for synfuels and pure PAH has been fairly good at a rough quantitative level, as can be judged from the comparative data of Table 4. There are some important differences. For example, the mutagenic activity of 2-aminoanthracene is very high, whereas its tumorigenic activity is only moderate. The reverse is true for SRC heavy distillate: while the mutagenicity is only moderate, relative to standard control compounds, the tumorigenicity is high.

Oak Ridge National Laboratory biologists and analytical chemists have teamed together to correlate chemical class fractions with mutagenic activity [16,26]. The mutagenicities of the fractionated oils listed in Table 5 vary over several orders of magnitude. The authors claim that a newly modified fractionation procedure removes formerly observed artifacts [16] and produces <10 rev*/mg for each of the petroleum crude oils [26]. It is reasonable,

*rev = revertants.

Table 3. Comparison of the Mutagenicity of Solvent-Refined Coal Materials, Shale Oils, and Crude Petroleums in *Salmonella typhimurium* TA98 with S-9 Activation

Materials	Revertants/μg of material
SRC-I:	
Process solvent	12.3 ± 1.9
Wash solvent	<0.01
Light oil	<0.01
SRC-II:	
Heavy distillate	40.0 ± 23
Middle distillate	<0.01
Light distillate	<0.01
Shale oil:	
Paraho-16	0.60 ± 0.19
Paraho-504	0.59 ± 0.13
Livermore L01	0.65 ± 0.22
Crude petroleum:	
Prudhoe Bay	<0.01
Wilmington	<0.01
Pure carcinogens:	
Benzo[a]pyrene	114 ± 5
2-Aminoanthracene	5430 ± 394

Source: Ref. 23.

Table 4. Comparison of Mutagenic and Carcinogenic Activity for Several Crude Fossil-Derived Materials

Material	Ames assay	Mammalian cell culture	Skin tumorigenesis
Light distillate	-	-	-
Heavy distillate	++	++	++++
Shale oil	+	+	++
Crude petroleum	-	slight	+
Benzo[a]pyrene	++	++	+++
2-Aminoanthracene	++++	+++	++

[a]Key: - = no activity; + to ++++ = moderate to high activity.
Source: Ref. 23.

Table 5. Contribution to Mutagenicity of Oils by Chemical Class

Compound	Total (rev/mg)	Mutagenic activity of oil: Percentage of total — Neutrals	Acids	Bases	Other
Petroleum crude oil:					
Composite No. 5107[a]	0	0	0	0	0
Wilmington No. 5301[a]	5	100	0	0	0
Recluse No. 5305[a]	5	100	0	0	–
LMS No. 5101[a]	4	100	0	0	0
Shale-derived petroleum substitute:					
LETC No. 4101[a]	180	54	2	42	2
Paraho No. 4601[a]	390	31	0	69	0
Coal-derived petroleum substitute:					
Synthoil No. 1202[a]	4200	10	2	80	8
SRC-II No. 1701[a]	1100	65	0	35	0
H-Coal Atmospheric Still Bottom No. 1309[a]	970	53	1	46	0
H-Coal Vacuum Still Overhead No. 1310[a]	2400	58	0	42	0
COED HDT No. 1106[a]	530	89	0	11	0

[a]USEPA/USDOE Sample Repository designation.
Source: Refs. 16 and 26.

therefore, to rank the absolute mutagenicities in the following order: petroleum < shale < coal-derived substitutes. The ORNL investigators caution, however, that additional samples of many differing origins and histories must be tested before a firm generalization can be made. Within the milieu of PAH-compounds composing these products are some whose toxicity and biological effects are extremely potent. The causative agents were shown to include polycyclic aromatic primary amines that were isolated in the acetone eluates of the ether-soluble base fractions [16]. When compared to the mutagenic activity of BaP, as in Table 6, these base fractions of shale oil- or coal-derived petroleum substitutes had activities similar to or exceeding that of pure BaP.

More than 50 aromatic nitrogen compounds have been biologically characterized by the Ames test [27]. The following structure-activity relationships were noted [28]: (1) The trends in mutagenic activities of multiring aza-arenes are the same as their aromatic hydrocarbon analogs. (2) The mutagenic activity of multiring aza-arenes is of the same order as their hydrocarbon counterparts, with the activity depending on the number of rings and not so much on the isomeric structure. (3) Nitro and amino compounds are highly mutagenic, with even two-ring compounds having a biological activity similar to BaP. Their mutagenic potential increases with increasing number of rings to a maximum at four or five rings. (4) the secondary and tertiary methyl-substituted amines are as active as their primary amino homologs, at least in the cases of aminofluorene and aminopyrene.

Table 6. Relative Mutagenic Activity in Comparison to BaP

Sample	Relative mutagenic activity
BaP	1.0
Acetone subfraction of the ether-soluble base fraction:	
Petroleum	
Composite No. 5107[a]	0.02
Wilmington No. 5301[a]	0.02
Shale-derived	
LETC No. 4101[a]	0.50
Paraho No. 4601[a]	0.80
Coal-derived	
SRC-II No. 1701[a]	1.35
Synthoil No. 1202[a]	7.20
Polynuclear aromatic primary amines:	
2-Naphthylamine	1.4
2-Aminoanthracene	4.0
3-Aminoperylene	4.0
3-Aminopyrene	52.0

[a]USEPA/USDOE Sample Repository designation.
Source: Ref. 16.

Table 7. Comparison of the Carcinogenicity of Petroleum and Petroleum Substitutes and Their BaP Contents

Material and repository number	Cancer Index [29,30][a]	BaP (ppm)	Ref.
Petroleum			
Composite No. 5107	0	~1[b]	
Shale-derived			
No. 4101	30	24	17
Coal-derived			
Coed Heavy Oil No. 1115	19	18[b]	
Synthoil centrifuged No. 1204	42	85[b]	
Atmospheric Still Bottom No. 1309	13	145	18
Vacuum Separator Overhead No. 1314	22	296	18
Vacuum Still Overhead No. 1310	67	456	18

[a]Cumulative percent incidence ÷ average latency in days × 100.
[b]BaP concentration provided by H. Kubota, ORNL.

The suggestion is made once again that the increased biological activity of coal- and shale-derived substitutes may be caused in large measure by polycyclic aromatic primary amines [16].

C. Dermal Toxicity

Reports of potent skin-tumor-producing propensities of fossil-fuel-derived materials introduce a concern in evaluating the potential carcinogenic hazard to the plant worker. The data shown in Table 7 are from Holland et al. [29] and Holland [30] and result from thrice-weekly applications of 50% (w/v) solutions in 50 μl aliquots to the skin of mice over a period of 22 weeks. The skin carcinogenicity is assessed by the Iball index [31]. Concentrations of BaP are included for the purpose of judging the effectiveness of this classical PAH as an indicator of skin carcinogenicity. The correlation between BaP concentration and the cancer index is only moderate.

Holland [30] observed that the coal-derived heavy distillate (No. 1310) evoked persistent, invasive skin tumors after six weeks. This is the shortest skin tumor induction time ever noted in mice, shorter even than pure BaP or any other potent skin carcinogen. The warning was made that such a material deserves respect, with exposure to the worker being avoided if at all possible. Other measurements of skin tumor incidence in mice after painting with the same and similar fossil fuel materials have been conducted at the Battelle Memorial Institute [32]. The latencies and tumor incidences gave cancer indices very similar to the values reported by Holland.

III. Industrial Health Effects

There is a long list of case histories and epidemiological studies linking morbidity and exposure to coal tar [33] and shale oil products [34]. It is natural to question the relationship between these experiences and those that are or will be met within the petroleum industry and newer technologies that produce synfuels from coal and oil shale. Concerns about the carcinogenic impact associated with the production and introduction of synthetic fuel into the marketplace could become the Achilles' heel in the commercialization of these needed technologies.

A. Petroleum Crude Oils

There is no conclusive evidence that the extraction and transportation of petroleum crude oils are accompanied by a major PAH-related cancer risk [1]. For example, thrice-weekly topical application to mouse skin of a composite of six natural petroleum crudes produced a skin tumor incidence of zero [35]. Under the same conditions, two coal-derived liquids and a shale oil each produced a significant incidence of skin tumors.

B. Petroleum Processing

In oil refineries, crude oils are distilled and then further upgraded by purification, catalytic cracking, hydrotreating, and other processes. It is in the higher-boiling-point distillates and solid residues that the PAH are generally found to concentrate. Petroleum pitch is heavily enriched in PAH. For one pitch [1], 2.5 weight percent was higher-boiling PAH with BaP comprising 0.2% of the total weight. The petroleum pitch volatiles of this material were even more concentrated in PAH.

Although these petroleum oils and pitches can be quite highly carcinogenic to the skin of mice, the impact on human workers in recent years appears to be fairly benign. Eckardt [36] reports that in 15 years of operational experience with catalytic crackers, the workers developed no skin cancers. The medical departments of the operating affiliates made special skin examination of the workers who were potentially exposed during this time period. It was also reported that "tar erythema," indigenous to the coal tar industry, is not seen in the petroleum industry.

Cancers of the type that used to be observed among wax pressmen in the petroleum industry (scrotal cancers) are no longer occurring. This is the result of carrying out modern refining operations in closed units and the elimination of industrial hygiene practices that used to leave much to be desired [37].

C. Shale Oil

Oil shale processing is not a new industry. Knowledge of the carcinogenic potential of shale oil, both in its production and end use, is not new either. It was in 1876 that cancer of the scrotum was first described in Scottish shale oil workers. Between 1920 and 1943, there were over 1000 verified cases of skin cancer in the British mule spinning industry, which then used shale oil to lubricate spindles [38]. Better experience has occurred in the shale oil industry in Estonia, USSR. Because of good hygienic practices, Bogovsky and Jons [39] report no increase in the incidence of cancer.

Shale oils are generally more carcinogenic than any of the petroleum oils with matching physical properties. The elevated carcinogenicity of shale oil when painted on mouse skin, vis-à-vis petroleum oils, is well demonstrated [40]. The BaP content (1-4 ppm) by itself was considered to be too low to account for the high carcinogenic activity. It may turn out that the carcinogenicity of shale oils is associated with the high nitrogen content [15] and the mutagenicity that is concentrated in the nitrogen-base PAH [16].

D. Coal-Derived Products

1. *Coal Tar*

The history of work-related coal-tar cancers begins in 1775 [41]. A compilation of references establishing the epidemiological basis for carcinogenic concern in older traditional industries is contained in Table VII of Ref. 1. For example, persons at risk work in tar distillation factories, gasworks, coke ovens, steel and aluminum plants, or they are pitch product workers, street pavers, roofers, creosoters, or fishermen handling tar-treated nets. Homo- and hetero-PAH are implicated in ailments and morbidities that range from pitch warts to cancer of the skin, scrotum, lung, respiratory system, bladder, kidney, stomach, and pancreas [33].

2. *Liquefied Coal*

There have been few coal liquefaction plant operations where good medical records have been kept and follow-up studies of the workers conducted. The best documented instance of worker exposure to synfuels from direct coal liquefaction pertains to a plant operated at Institute, West Virginia, from 1952-1959 [42]. Several hundred workers were employed in this 300-ton/day, giant pilot plant. The product streams boiling at temperatures higher than 260°C were highly carcinogenic, the degree of carcinogenicity increasing and the length of the median latency period decreasing as boiling points rose. Middle oil, light oil stream residues, pasting oil, and pitch products were carcinogenic to animals. The heavy oil stream boiling above 320°C was both potently carcinogenic and abundant in PAH such as phenanthrene and pyrene.

A comprehensive industrial hygiene program [43] revealed that airborne oil fumes and fallout of droplets of oil caused contamination of the skin. A "tracer" compound, BaP, was selected to monitor the prevalence of PAH in airborne contamination. Values for BaP concentration ranged widely from $<10^{-2}$ to 20 $\mu g/m^3$ of air. Airborne contamination was by no means the entire problem. It was noted [43] that containment of oils within the equipment was extremely difficult. Direct contamination of skin and clothing occurred quite often.

The comprehensive study at Institute is especially important for several reasons: the presence of possible cancer-causing agents was suspected before the unit was started; the products were tested for carcinogenic action as soon as they were available; and with the suspicion confirmed, workers were thoroughly warned, urged to avoid contact, and provided with changes of clothing and information about decontamination procedures. In spite of these best efforts by a team of toxicologists, industrial hygienists, and physicians, cases of precancerous and cancerous skin lesions did occur.

Out of 359 employees who were exposed to oils boiling at high temperatures, 10 were diagnosed initially as having a cutaneous precancerous lesion [44]. Only 5 of the 10 were confirmed as having had skin cancer. The incidence of skin cancer was, therefore, 1.4%. In a recent follow-up study [45] of this subgroup of workers, there was the more comforting finding of no increase of chronic disease and death, and no systemic cancers from the prior exposure to high-boiling oils that contained PAH. The relatively small number of workers should, however, make one wary of too early vindication.

An example of drawing erroneously safe conclusions from the use of too small a cohort was strikingly demonstrated in the case of the strong carcinogen furylfuramide [46]. Furylfuramide, an antibacterial food additive used until recently in Japan, was originally tested on rats and mice. No carcinogenic activity was evident. When it was subsequently tested and found to be a potent mutagen, more thorough animal tests revealed furylfuramide's true carcinogenic character.

A report for a modern direct coal liquefaction facility that is as thoroughly revealing and comprehensive as the one for the Institute plant is yet to appear. Harris et al. [47] give the opinion that the significance of the Institute experience relative to today's direct coal liquefaction processes is questionable. First, these authors point out that there are changes in current process technology. Secondly, hygiene, sanitation, work practices, and controls were not implemented as rigorously as they are now in modern pilot plants. The industrial hygiene practiced at the SRC liquefaction plant is an appropriate example [48]. Many procedures of this new health program, however, are improved designs based on the Institute experiences. It is encouraging that, to date, the SRC employees have not revealed any of the severe problems experienced by the Institute plant population [47]. Less severe skin problems have been encountered. These are discussed in a later section on skin contamination.

Another significant present-day industrial hygiene experience has been reported by the Exxon Research and Engineering Company for their Donor Solvent pilot plant [49]. Workers have been exposed for approximately 11 years. No cases of systemic cancer were reported. A small number of workers (3 out of 190 employees, or 1.5%) developed basal cell carcinomas of the lip, ear, and nose. Once again it is difficult to judge the significance of these cases in light of the small number of individuals involved and a lack of reference data.

Worker contact with high-boiling coal conversion liquid products does seem to be implicated in an increased incidence of cutaneous cancer and other skin abnormalities, albeit the risk is small: 1.5% of the workers at each of two different liquefaction plants [44,49] developed skin cancer. It is doubtful that a special significance is attachable to this particular number. But the emergence of skin abnormalities is not too surprising in light of the previously discussed mutagenicity and carcinogenicity testing. The materials that produced skin cancer at the Institute facility are gone forever. Were they available today and subject to chemical fractionation, toxicity, mutagenicity, and carcinogenicity testing, one doubts that the results would be much different from those listed in Tables 3, 5, and 7.

The worry will remain that as the population of synfuel workers grows in size, a nonzero risk for contracting systemic cancer will become apparent. A close eye will need to be kept on synfuel workers as the industry grows in size.

IV. Standards

Regulatory standards for PAH-containing materials are minimal in number. This situation arises not so much from the lack of a perceived need as from the difficulties of dealing with large numbers of PAH in complex mixtures. The difficulties are compounded by PAH being able to exert strong synergistic effects on one another.

There is only one PAH compound, naphthalene, with a regulated threshold limit value - time weighed average (TLV-TWA) for a normal 8-hr workday [50]. Individual PAH compounds that are recognized human carcinogens, such as β-naphthylamine, are listed by the American Conference of Governmental Industrial Hygienists (ACGIH) [50] without an assigned TLV. Particulate polycyclic aromatic hydrocarbons (PPAH), however, are listed as a group having carcinogenic or cocarcinogenic potential with an assigned TLV [50].

It is inferred that the PPAH is airborne. No standard exists--or has been considered—for PAH on surfaces, either inanimate or living surfaces.

A. Benzene Solubles

In 1968 the ACGIH published a TLV of 0.2 mg/m^3 for the benzene-soluble portion of coal tar pitch volatiles (CPTVs) in air. At this concentration of CPTVs, workers are believed not to be at increased risk to systemic cancers [51]. This standard was promulgated under the Occupational Safety and Health Act of 1970. In 1972, the Parma Standard Method 1013 was accepted as the method of analysis for determining compliance with the standard.

A broadening in the terminology occurred in 1976 when the ACGIH changed the term CPTV to particulate PAH (PPAH). This increases the scope of the TLV to include other than coal tar products. The benzene solubles standard is now used as a method of controlling worker exposure to PAH in a large number of different workplace atmospheres [52] such as aluminum plants, coke plants, iron and steel plants, ferroalloy plants, asphalt production and handling, secondary lead smelters, carbon black plants, oil refineries, production and use of creosote, gasworks, rubber plants, coal gasification and liquefaction plants, and shale oil production.

Although the benzene solubles method is still the only officially accepted method of measuring PAH in the workplace atmosphere, there is widespread criticism of its efficacy. As its name implies, it is nonspecific, analyzing for all soluble compounds. One aspect of this problem was brought to light in a study sponsored by the Aluminum Company of Canada [53]. Very often the workplace air contains traces of machinery lubricating oils, hydraulic oils, or other aliphatic hydrocarbons. Collection of these extraneous benzene solubles, not originally intended to be included in the measurement, may make the spurious difference between reporting compliance or noncompliance with the standard.

Other criticism has been voiced. Schulte et al. [54] found that the measurement does not correlate well with the analytical determination of PAH by gas or liquid chromatography or of BaP by thin-layer chromatography. These authors pointed out the need for a method specific to certain PAH compounds or classes of compounds and a reevaluation of the Federal standard once a good understanding is obtained of which components of coke oven emissions are hazardous and at what concentrations they affect the worker adversely.

Another problem is the inaccuracy that can occur in the gravimetric determination. The estimated weight of benzene-soluble material can be erroneous due to (1) disintegration of the filter, (2) loss of particulates during handling and extraction, or (3) a change in the water content of the filter between measurements of the tare and final weight after extraction [55]. Another question has been raised about the use of the silver membrane in areas where H_2S is present; formation of AgS might clog the pores making the filter ineffective [56].

An item of great concern is breakthrough and subsequent loss of low-boiling PAH. With the addition of a backup support pad of cellulose behind the glass fiber-silver membrane filters, significant quantities of the more volatile PAH have been found [55,57]. Losses of three-ring PAH that sublime from the CTPV filter system were reported by Seim et al. [55]. Broader questions about weight losses of PAH by evaporation and sublimation have been recently dealt with by Lao et al. [58].

Yet more difficulties for the benzene solubles standard came to light in an internal NIOSH* memorandum that reported on the percentage of BaP in the respirable particulate fraction of coke oven emissions [59]. Three hundred samples were statistically analyzed to determine correlations between three variables: particulate loading, mass of benzene solubles, and BaP concentration. The degree of correlation for each pair of these variables was not particularly good. The data strongly suggest that the composition of benzene solubles and particulates vary considerably. Such variation is contrary to the assumption that the composition of that substance is relatively constant, an assumption made when a substance is used to define a standard.

For all these objections, the NIOSH-recommended benzene (or cyclohexane) standard is still the only one that requires enforcement in the workplace. Before this present standard can be improved upon, a practical workplace procedure has to be evolved that is capable of determining the PAH compounds that are toxicologically proven etiological agents.

B. Cyclohexane-Soluble Fraction of Total Particulate Matter

Because of the more recent awareness of the carcinogenic nature of benzene, less toxic cyclohexane has been recommended as a substitute for benzene in the extraction of PAH from particulate matter [60]. The permissible exposure limit recommended by NIOSH is 0.1 mg/m^3 for the cyclohexane-extractable fraction of the sample, determined as a time-weighted average concentration for up to a 10-hr workshift in a 40-hr workweek [60]. This value is the lowest concentration that can be reliably detected by the recommended method of environmental monitoring. It is predicated by portable personal sampling pumps having maximum pumping rates of about 1.6 liters/min, so that approximately 750 liters of air can be sampled during an 8-hr workshift, and because 0.075 mg of cyclohexane-extractable material is required for a reliable analysis (equivalent to 0.1 mg in 10^3 liters, or 1 m^3 of air).

NIOSH states that this permissible exposure limit is the lowest concentration that can be reliably detected by the recommended method of environmental monitoring. It is conceded that there is no safe concentration established for PAH or any other carcinogen. The proposed TLV of 0.1 mg of cyclohexane solubles per cubic meter of air is governed by practical expediency. Compliance with this TLV should substantially reduce—but not necessarily eliminate—the risk of cancer.

The following is the recommended procedure for a determination of cyclohexane solubles (see Appendix II of Ref. 60). Full-shift (8-hr) samples should be collected with a personal sampling pump at a flow rate of at least 1.6 liters/min. Particulates should be collected on 0.8-μm pore size, silver membrane filters (37-mm diameter), preceded by Gelman glass fiber A-E filters encased in three-piece plastic (polystyrene) field monitoring cassettes. Extraction in cyclohexane at room temperature is conducted ultrasonically [61]. After the cyclohexane solution is filtered, the extract is evaporated to dryness. The total extract is determined by weighing a dried aliquot of the extract to the nearest 10 μg.

---

*National Institute for Occupational Safety and Health.

C. Benzo[a]pyrene

In 1974, OSHA* established a Standard Advisory Committee on Coke Oven Emissions. The following year, the committee recommended a limit of 0.2 μg BaP per cubic meter of air [62]. It has been estimated that each nanogram per cubic meter of exposure to airborne BaP may cause 0.4 extra lung cancer cases per year per 10^5 persons [63].

The recommended BaP standard, however, has not been promulgated. The use of a BaP level as an indicator of ambient exposure to carcinogenic PAH in general is a questionable practice on the following grounds [64]: (a) the PAH composition and BaP level can be highly variable depending on their sources, and consequently no constant and significant correlation can be expected between the levels of BaP and that of most other PAH present; (b) BaP levels cannot account entirely for the carcinogenic and mutagenic activity of organic extracts of suspended particulate matter.

The situation existing for indicator compounds of PAH has been discussed lately by Gammage and Bjørseth [65] and is one facet of the discussion in this chapter on PAH profiles.

D. Total PAH

A new administrative threshold value for occupational exposure to total particulate PAH has recently been introduced in Norway [66] and is 40 $\mu g/m^3$. Two-ring PAH are excluded. Phenanthrene is the lowest-boiling PAH included in the definition of total PAH.

V. PAH Measurements in the Workplace

The eventual goal behind measurements of PAH in workplace environments is the introduction of regulatory standards to control worker exposure to specific PAH compounds, or groups of these PAH compounds, that provide measures of health risk. A preceding step will likely be the formulation of guidelines to aid the industrial hygienist, and perhaps a list of priority PAH pollutants similar to the list issued by the Environmental Protection Agency (EPA) for PAH in water [67]. Specificity of PAH compounds raises the possibility of measuring compounds that are indicators of larger groups or classes of PAH.

Much of the data basc is yet to be established upon which regulatory decisions can be made. A start has been made [66], and much of the credit for this goes to Bjørseth and his co-workers, who have reported upon airborne PAH emissions in a number of different industrial atmospheres in Norway.

A. Pathways for Exposure

Pathways for exposure of workers to PAH are the inhalation of particulates or vapors, ingestion, and contamination of the skin. The relative importance of each exposure route varies according to the particular technology being

---

*The Occupational Safety and Health Administration.

utilized or the particular type of work in which the worker is engaged. For example, tarry fumes and their inhalation may be the principal concern at a battery of coking ovens. Conversely, in a high-pressure coal liquefaction plant, skin contact with heavy distillates during maintenance work might be of paramount importance. As for the direct ingestion of PAH, virtually nothing is known about this subject. The present author is inclined to believe, perhaps naively, that good plant hygiene which excludes polluted air and prevents tracking of oily material into eating and drinking areas should negate this particular health concern. The exception might be where a severe accident occurs. Also, the inhalation and swallowing of coarse airborne particulate matter—a more indirect route of ingestion—should not be forgotten. The inhaled larger particles of a nonrespirable size will be trapped in the mucous fluids of the upper and middle respiratory system. After movement to the throat they can be swallowed.

Ideally, PAH measurements should already have been made that are applicable to each of these special situations. This has not been the case. The PAH in fugitive emissions have been measured almost exclusively for airborne samples of respirable-sized particulate matter. Other types of fugitive emission involving other exposure pathways have to be contended with. The production, handling, transportation, and end use of synfuels will be accompanied by potentials for skin contamination, or even ingestion, and the inhalation of the vapors of various products. In the following subsections we will consider each of these hazards in the workplace, viz., aerosols, vapors, and skin contamination.

B. Methods of Measuring Airborne Samples

It is noteworthy that the combination of silver membrane and glass fiber filters recommended by NIOSH [60] for sampling particulates of coal tar products is blind to the problem of loss of some PAH compounds by evaporation. This is because losses due to evaporation occur as air is being drawn through already deposited particulate matter in the particulate sampling device. Several earlier analyses of PAH in particulate matter are deficient for this reason [68].

Jackson and Cupps were two of the first experimenters to point out the problem of the sublimation of PAH during sampling [69]. In the conventional sampling of aerosols, Cautreels and Van Cauwenberghe [70] were able to show that large quantities of lower polyaromatics (up to benzofluorenes) are lost due to this evaporation process. Even BaP is not immune to sublimation. During high-volume air sampling, loss of 60% of the BaP in the particulates has been reported by DeWiest and Rondia [71]. A more comprehensive treatment of the problems of the volatility of PAH and the inefficiency of collection has been made recently by Lao and Thomas [72].

Evaporative problems can be overcome by catching volatile PAH in a backup system. A trapping solvent such as cold ethanol in absorption bottles [73] has been a favorite technique in Norway. In the United States, solid sorbents are favored. These are of two basic types. The first is thermally stable resins, such as Tenax GC [74] and the Chromosorb series [75], which do not decompose at temperatures below 300°C. The second type consists of the resins that do not dissolve in common organic solvents. The XAD resins and chromatographic packings such as Poropaks N and Q are examples of these chemically stable polymers. In a recent review article, Flotard [76] favors a device for sampling workplace air at coal-conversion facilities that consists of a Teflon filter followed by a bed of Amberlite XAD-2 resin. It is also reported in this article that NIOSH is experimenting with an XAD resin as a backup for the 37-mm glass fiber and silver membrane filter used in the NIOSH low-volume personal sampler [60]. There is today a general awareness of the need in both high- and low-volume samplers for nearly quantitative capture of both PAH vapors and PAH particulate matter.

It would be misleading to impart the impression that there are only evaporative problems to be overcome. Extensive degradation of PAH can occur on filter surfaces. Lee et al. [77] reported that extensive oxidation of filter-collected PAH can occur on most filter types, even in the dark. The extent of degradation is heavily dependent on the filter type. The PTFE (Teflon) membrane filter type minimizes, without eliminating, the effect. These same authors question whether spurious effects in filter sampling can ever be truly eliminated.

In spite of these difficulties, several good methods for sample collection and analysis are in use at this time. Each sampling device has the capability of nearly quantitative capture of both vapors and particulate matter while preserving the integrity of the sample to the highest degree possible.

A responsibility for evaluating the potential occupational health hazards for new coal conversion technologies was given to NIOSH. The initial work on this project was conducted by the Bendix Corporation. A silver membrane filter collection system was forthcoming [78], with a Tenax adsorber as backup for collecting all PAH that passed through the sampling train in the vapor state or that sublimed during the collection. The selection of Tenax resulted from recommendations made by Jones et al. [79].

More recent NIOSH-sponsored development of collection systems for PAH has been conducted by Enviro Control, Inc. Both personal and area devices have been developed and successfully tested in the workplace [75]. Chromosorb 102 was chosen as the preferred adsorbent. This choice of resin was made on the advice of J. O. Jackson, who has had extensive experience with this adsorbent medium at the SRC pilot plant at Tacoma, Washington [69].

A schematic drawing of the personal monitoring device for PAH is shown in Fig. 1. It operates at flow rates of between 0.5 and 1.0 liter/min. Another Enviro Control device for area monitoring is a high-volume sampler operating at ~9 liters/min. Both types of device use silver membrane filters in conjunction with Chromosorb 102 backup. The PAH from each trapping medium is extracted separately. The Chromosorb 102 is Soxhlet extracted with methylene chloride (Me_2Cl_2) for 22 hr. The silver membrane filter is subjected to multiple ultrasonic extractions in cyclohexane, each of 15-min duration. The latter procedure is a modification of the NIOSH-validated method P & CAM 217. Extracts from the Chromosorb 102 and the silver membrane are each concentrated to 1 ml and combined prior to analysis for PAH.

Another sampler for PAH in workplace air has been recently developed at the Argonne National Laboratory (ANL) [76]. A schematic drawing is reproduced in Fig. 2. A principal requirement is the need for explosion-proof equipment; there often exists the potential for explosive atmospheres in the environs of a plant that produces synfuels. The ANL sampler is built with

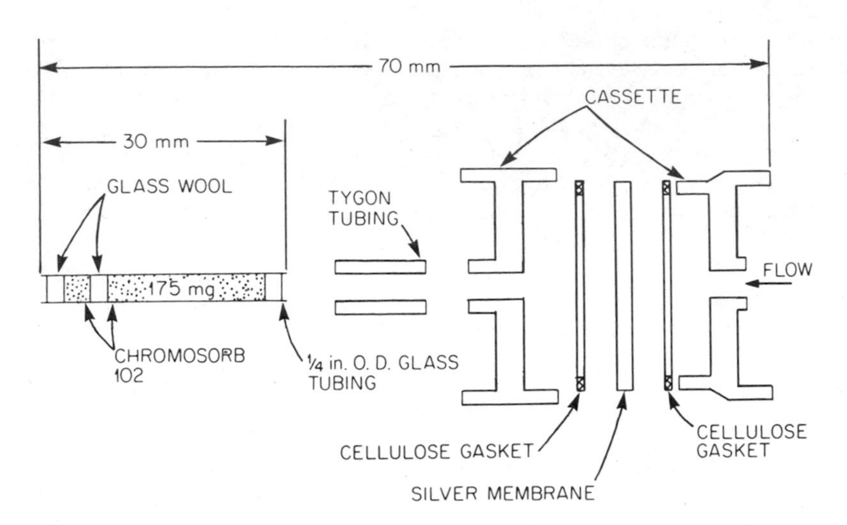

Figure 1. Personal monitoring device for PAH. (From Ref. 75.)

an Amatek explosion-proof vacuum motor. A maximum flow rate of 600 liters/min can be sustained through the resin bed.

The sampling train consists in this instance of a Teflon filter, followed by a bed of Amberlite XAD-2 resin. The more expensive Tenax adsorbent is listed as an alternative choice of resin. Soxhlet extraction with Me_2Cl_2 is recommended for removal of the adsorbed organics. The sample is rinsed repeatedly. The time of extraction, however, is kept as short as possible, consistent with near complete recovery of the PAH. Long extraction times can produce interferences from breakdown products of the XAD-2 resin. The results of workplace air sampling with the ANL sampler have recently been reported [80].

A nearly complete and up-to-date review of the methods for collecting PAH from air samples and of the methods for identifying and quantifying the PAH is contained in the paper by Flotard [76]. The pros and cons of using activated carbon and a variety of polymeric resins and chromatographic adsorbents are also discussed. The preferred system of the Norwegian group is, however, not mentioned; Bjørseth et al. employ Acropore filters (AN-800), with two subsequent absorption bottles cooled in dry ice and containing ethanol [73]. The ethanol absorption technique is used only in area-sampling devices. A backup system is missing from their personal particulate sampler.

Clearly, different techniques and trapping agents for sample collection are favored by different investigators. A need exists for intercomparison testing, with eventual selection and standardization of the best method or methods.

As mentioned already, even the filter medium is not immune to causing problems during the period of collection of airborne particulates. The picture emerges of extensive dark oxidation of PAH collected on most filter types. The PTFE (Teflon) membrane filter appears to be the most satisfactory and minimizes the degradation problem. Thus oxidation is another important parameter for further investigation, intercomparison, and standardization.

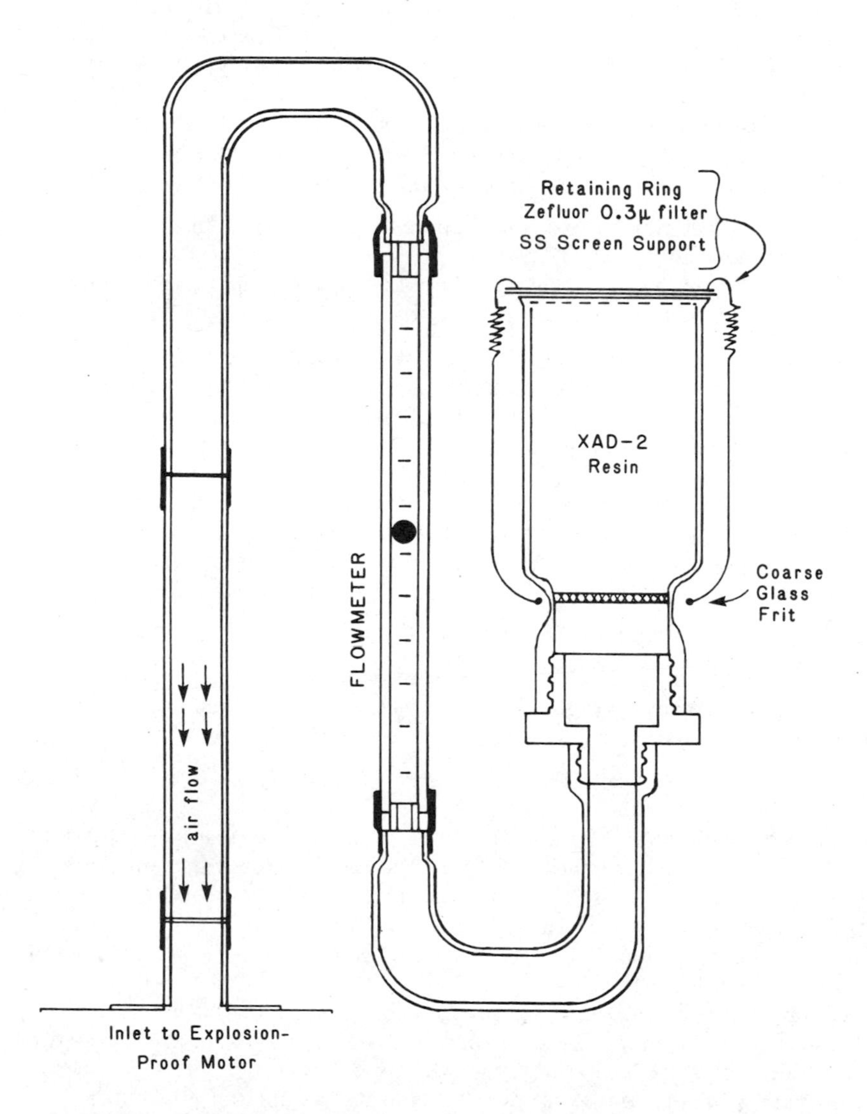

Figure 2. Modified air sampler with explosion-proof motor; cross-section view. (From Ref. 76.)

Several chapters of this book are devoted to the topic of analytical procedures for the determination and analysis of PAH. Readers are referred there to the types of analysis that are available, together with the individual preferences of the authors. Flotard's article [76] also details the spectroscopic and chromatographic methods for identifying and quantifying the PAH recovered from air samples. For example, Burchill et al. claim that despite the rapid progress of high-performance liquid chromatography (HPLC) in recent years there seems to be a consensus that gas chromatography (GC) is still the method of choice for a detailed characterization of PAH, and in particular GC with glass capillary columns [81]. Whatever one's personal preference might be, the requisites for detailed analyses are invariant [52]:

1. High separation efficiency in order to separate positional isomers of PAH that have different biological effects
2. High sensitivity in order to detect and identify minor compounds in the sample that may be of biological significance

As PAH measurements in the workplace evolve through the stages of detailed characterization to routine industrial hygiene, less costly and less sophisticated analytical techniques for PAH will be called for. Suitable on-the-job tests will probably include screening by sensitized fluorescence for total PAH [82]. The method estimates the overall magnitude of PAH with a detection limit of 1 to 10 pg. Another sensitive and simple analytical technique capable of giving compound specific information is phosphorescence at room temperature from samples of PAH spotted on filter paper [83].

C. Profiles and Concentrations of Workplace Aerosol Samples

Examples are presented of PAH profiles for air samples collected in the work environments of plants that engage in coking, reduction of aluminum, manufacture of iron or ferroalloy, and liquefaction or gasification of coal.

1. Coke Ovens

The few measured profiles of PAH emitted by a coke oven operating at about 1200°C show relatively small variation [73]. The nature and constancy of the profile is revealed in Fig. 3; the changes in the relative proportions of the PAH are not too large for the two seasons of spring and autumn. The profile is also similar to the PAH in a tar sample from the electrostatic precipitator of a low-Btu coal gasifier [84].

For these particular examples of tar produced by the devolatilization of coal, the relatively constant profile raises the possibility of simplified measurement with an indicator PAH compound. In this instance (Fig. 3), measurement of the concentration of any one major PAH compound would permit the concentration of any other PAH compound to be estimated within a factor of 2 or 3. However, considerably more data are needed to gauge the degree of constancy of the PAH profile over extended periods of operation of individual plants. The indicator concept for PAH can then be argued with more conviction. The use of indicator compounds under these and other circumstances was discussed recently in considerable detail [65].

An appreciation of PAH exposures to workers at coke ovens is important because of the excess of lung cancers that have been observed in epidemio-

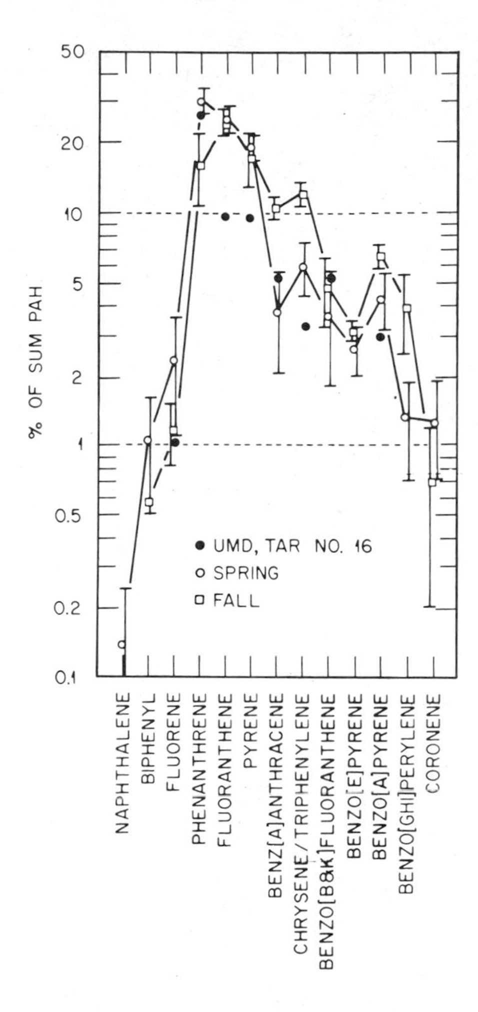

Figure 3. Coke oven (PAH) profiles in spring and fall sampling and of a low-Btu coal gasifier tar. (From Ref. 73,84.)

logical studies. Topside coke oven workers employed 5 or more years were at highest risk. They had about a tenfold increase in lung cancer risk [85]. Typical values for occupational exposure at coke plants are summarized in Table 8 [52]. As expected, occupational exposure to PAH (and the lung cancer risk) is largest at the battery top where the larry cars are operated. Exposures for larry car operators are typically in the range of several hundred micrograms per cubic meter.

These results clearly demonstrate that exposure to PAH varies strongly with the type of job. Likewise the exposure to BaP varies considerably with worker location: 0.5-43 $\mu g/m^3$ for personal sampling and 14-134 $\mu g/m^3$ for stationary sampling. With respect to BaP concentrations, the workers were all being exposed in excess of the recommended level of 0.2 $\mu g/m^3$ [62].

Table 8. Exposures of Coke Oven Workers to PAH and BaP

	Total PAH ($\mu g/m^3$)		Benzo[a]pyrene ($\mu g/m^3$)	
	Range	Arithmetic mean value	Range	Arithmetic mean value
Stationary sampling:				
Battery top[a]	270-2,423	1,000	14-69	37
Personal sampling:				
Larry car operator[b]	168-1,044	370	12-43	22
Coke car operator[c]	4.8-73	37	0.5-5.8	3.1
Push car operator[d]	9-62	33	0.9-4.4	2.9

[a]Average of 25 samples.
[b]Average of 7 samples.
[c]Average of 3 samples.
[d]Average of 4 samples.
Source: Ref. 97.

Other examples of worker exposures in excess of this "safe" level of BaP have been reported. For instance, Jackson et al. [86] reported the BaP concentration at a U.S. coke oven battery top to be in the range of 1.2-15.9 $\mu g/m^3$.

Size fractionation of the airborne particulate matter has been conducted together with the concentration of PAH associated with each size fraction [73]. The bulk of the PAH is in those size fractions that are of a respirable size: 98 and 72% of the particulate PAH is attached to particles below 7 and 3 μm, respectively. This finding explains the epidemiologically proven dangers of the tarry fumes that can be inhaled inside coke oven plants.

2. Aluminum Plants

The types of PAH profile measured in a Norwegian aluminum plant [87] are shown in Fig. 4. There are both vertical pin Soderberg and closed, prebaked anode cells in the primary aluminum smelter. Other associated plant operations with potential for PAH exposure are in the anode prebaking and anode baking plants. A postulate advanced by Bjørseth et al. [87] is that the PAH profile is a characteristic of the process involved, i.e., the basic composition of the organic material, redox conditions, and reaction temperature. The PAH profiles of the prebaked and anode baking plants are relatively parallel. Since these processes use essentially the same carbonaceous material heated to approximately the same temperature, this observation supports the above authors' idea that the profile is characteristic of the PAH source.

The Soderberg plant operation employs a different process from the other two. The PAH profile is accordingly different. In this case there is a significantly higher fraction belonging to the high-boiling PAH.

The profile found for the aluminum vertical pin smelter is quite similar to the PAH profile for the Norwegian coke plant. Bjørseth attributes this likeness to the similarly high operating temperatures of 1000° and 1200°C for the aluminum and coke plants, respectively.

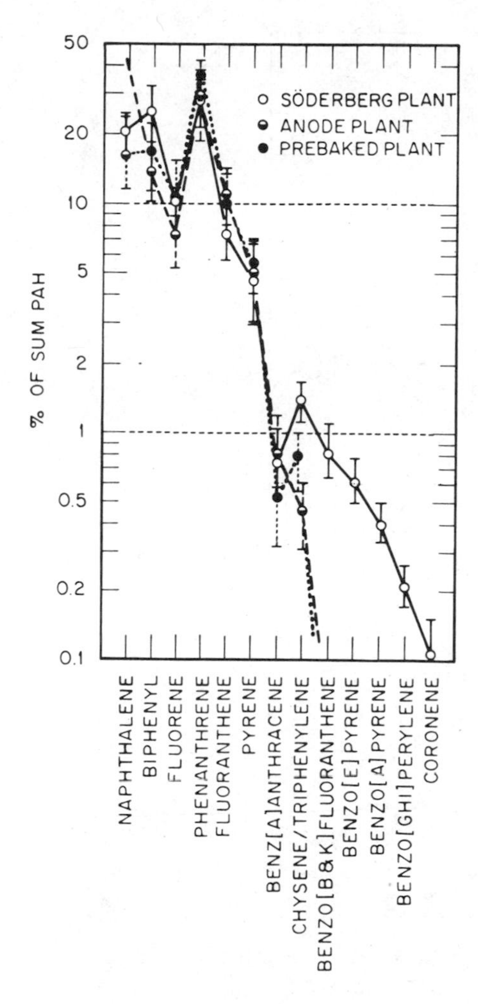

Figure 4. Profiles of polycyclic aromatic hydrocarbons in the anode baking, prebaked, and Söderberg plants. (From Ref. 87.)

In the work atmospheres of the aluminum plants, more than 30 PAH compounds were identified in the total PAH. They varied in size from naphthalene to coronene and dibenzopyrene. The values for total airborne PAH as a function of job type and location are shown schematically in Fig. 5. Some jobs present particularly high risks of exposure to PAH. The highest exposure occurs for the pin pullers at the vertical pin Soderberg plant and for workers in the paste plant. The vertical line at 40 $\mu g/m^3$ indicates the present Norwegian administrative threshold value for occupational exposure to total PAH [66].

Uneven distributions of PAH in the atmospheres of aluminum plant potrooms give rise to significant variations in occupational exposure, even for workers doing the same jobs. As seen in the data of Table 9, there can be large variations in exposure from day to day for the same person doing the same

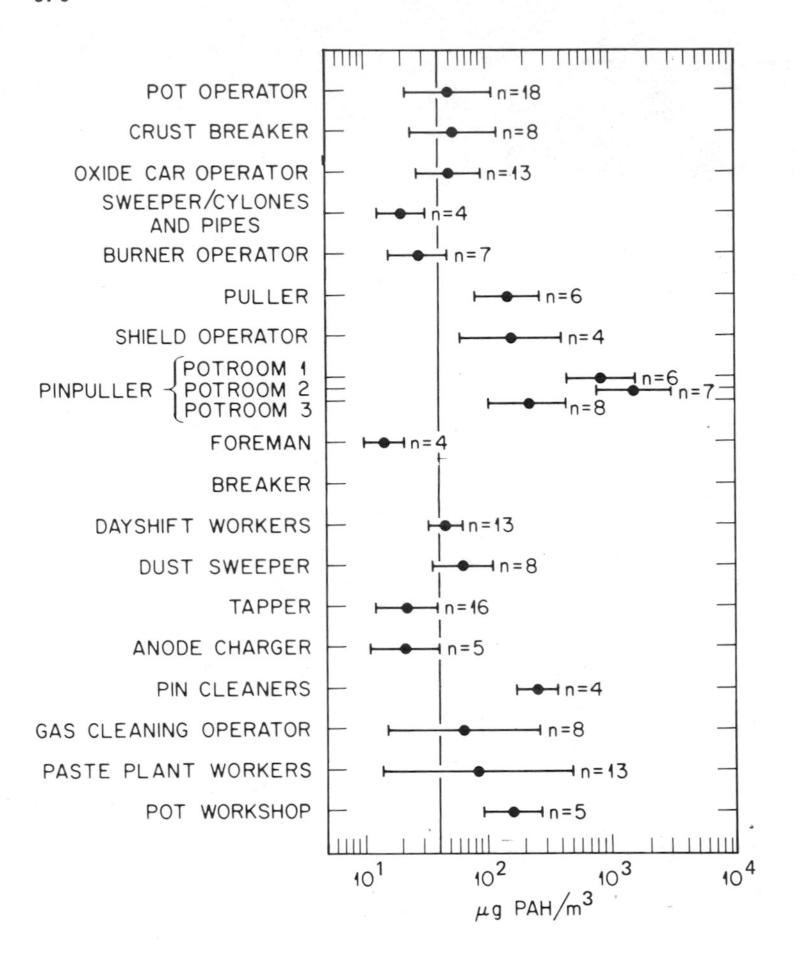

Figure 5. Total PAH in an aluminum plant according to job type, n being the number of workers monitored. (From Ref. 66.)

job. This variability is a factor in 80 in the case of one of the gas-cleaning operators. These individuals may wear respirators on the occasional days with high exposure. Other occupations, such as sweepers and pin pullers, have a generally high level of exposure. With respect to personal exposure to BaP, quantitative measurement in the anode plant (0.8-28 $\mu g/m^3$) and Söderberg potroom (3.4-116 $\mu g/m^3$) are similar to those measured in the Norwegian coke plants (0.5-43 $\mu g/m^3$). Equivalent measurements at a U.S. coke plant [86] were also in a similar range of values (1.2-16 $\mu g/m^3$).

3. Iron- and Steelworks, Foundries, Ferroalloy Plants

Other high-temperature industrial sources of PAH have been examined in Scandinavia. These include iron and steelworks and ferroalloy plants in Norway [66,88], and foundries in Finland [89]. The PAH profiles for typical

Table 9. Variations in Occupational Exposure ($\mu g/m^3$) in an Aluminum Plant

Job type	Day 1	Day 2	Day 3	Day 4	Day 5
Gas cleaning operator	39.0	12.0	79.9	234.0	
Gas cleaning operator	47.8	11.3	854.0		
Foreman	17.1	17.0	18.7		
Dayshift worker		93.9	54.9	29.9	42.8
Dayshift worker		51.2	53.6	31.9	47.0
Dayshift worker		54.0	28.3	37.0	61.6
Sweeper		31.9	93.6	28.6	147.0
Sweeper		86.8	81.5	45.3	64.2
Tapper	12.7	31.6	16.7		
Tapper	11.3	20.1		34.3	
Pot operator	31.9	82.6	65.2		
Burner operator			26.1	16.2	25.8
Puller		141.0	71.1	316.0	
Crust breaker	25.8	49.2	13.8		
Oxide car operator	32.5	63.2	39.2		
Ocidc car operator	16.6	54.6	21.8		
Sweeper/cyclones and pipes		19.4		12.8	17.3

Source: Ref. 66.

airborne samples taken from each of these work environments are given in Fig. 6. The spectrum for PAH in the iron and steel works roughly parallels those for the coking and aluminum plants. Notable differences exist in the cases of foundries and ferroalloy plants. The PAH profile for foundry atmospheres is exceptional for the absence of BaA. The PAH spectrum for the ferroalloy plants is dominated by four compounds—phenanthrene, anthracene, fluoranthene, and pyrene. The balance, due to higher-boiling PAH compounds, is less than 3%. The lack of high-boiling PAH is attributable to electrodes of anthracite and pitch-anthracene oil mixtures that are processed at the low temperature of 120-190°C.

Quantitatively the occupational exposures to PAH are lower than those generally met with in the coke and aluminum plants. Only the tappers and stampers in iron and steel works have occupational exposures that sometimes exceed 40 $\mu g/m^3$.

Occupational exposure to total PAH in these industries is in the following order: coke oven ≃ aluminum plant > iron and steel works > ferroalloy plants. Quantitative data are not available for foundry workers. Bjørseth was able to draw several conclusions from these studies [52]:

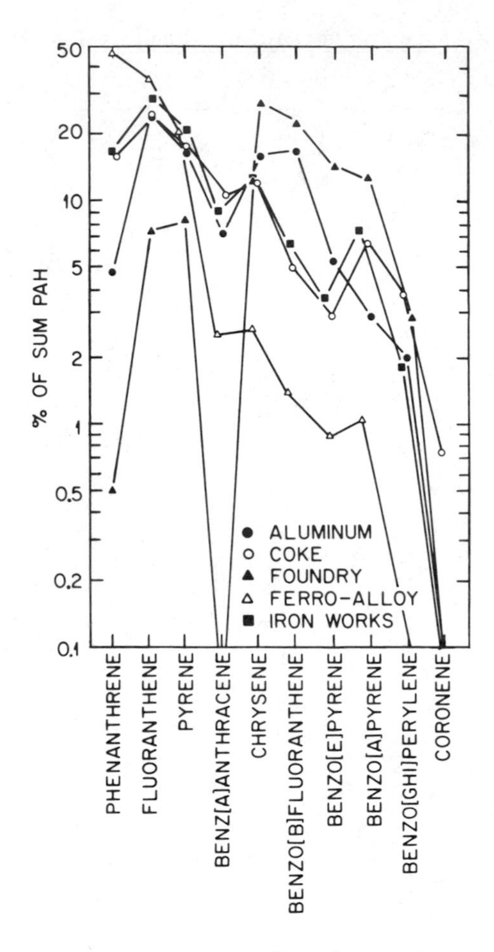

Figure 6. PAH profiles for various work atmospheres. (From Ref. 52.)

1. There exist analytical methods for PAH with sufficient precision and accuracy to allow industries to monitor their work atmospheres.
2. Each source at the plants seems to have a characteristic and fairly constant PAH profile. This opens the possibility for selecting and measuring a proxy compound (such as BaP or another major PAH compound) or compounds that are indicative of the total amount of PAH or other individual PAH.
3. Accurate and fairly frequent registrations of occupational exposures to PAH are needed to take fuller advantage of future epidemiological studies.

4. *Coal Liquefaction*

Comprehensive surveys of airborne PAH have recently been conducted [90] at three pilot liquefaction facilities: SRC-I and -II and Consol Synthetic Fuel (CSF) process plants. Both area and personal samples were collected using the silver membrane Chromosorb 102 backup devices described previously [75]. The PAH that were collected and analyzed quantitatively ranged in size from naphthalene and quinoline to the benzopyrenes and dibenzanthracene. The analytical method was not sensitive enough for analyzing trace quantities of larger multiring PAH.

The most extensive data were taken at the SRC-II plant at Fort Lewis, Washington. A few examples will be selected that point out some general trends. The data shown in Table 10 give the makeup of the total PAH, by individual compounds, collected as area samples in different plant locations. If these data are plotted as PAH profiles, one is struck by the great variability in the relative proportions of PAH. The profiles are quantitatively dominated by the low-boiling two- and three-ring PAH. Naphthalene, its alkyl derivatives, and phenanthrene/anthracene are abundant. It is also noteworthy that quinoline is abundant in some samples; an area sample collected in the vicinity of the solvent recovery equipment contained a very high 43% by weight of quinoline. The presence of significant amounts of quinoline (and the related heterocyclic PAH, acridine and carbazole) is a matter for some speculation. In a liquefaction plant of this type, different products are being produced or processed in different areas of the plant. It is to be expected, therefore, that fugitive emissions from different plant areas should have substantially different PAH profiles.

The results presented in Table 11 are for personal sampling in the coal preparation area. Even here there is great variability in the relative proportions of PAH as well as the quantitative exposures to total PAH.

Both the area and personal samples reveal considerable variations in exposure levels and PAH profiles. This feature does not bode well for the easy characterization of long-term exposures of individual workers to PAH. Nor does it seem likely that the characterization and monitoring procedures can be easily simplified by the use of proxy or indicator compounds. This situation contrasts markedly with the one discussed for the coking and aluminum plants where less complicated and variable sources of PAH are encountered.

At the SRC-II plant, large variations in personal exposures occur both for different operators working in the same zone and operators working at different jobs. The ranges of exposure are listed in Table 12. It should be stressed, however, that the bulk of the exposure is due to relatively innocuous two- and three-ring PAH. Quinoline is an exception, it being a well-recognized carcinogen [91].

Personal exposures are lower at the SRC-I plant and much lower at the CSF plant. The exposures for personnel at each plant are listed in Table 12. Exposures are $<1\ \mu g/m^3$ for CSF workers. As at the SRC-II plant, naphthalene and methylnaphthalenes are by far the most abundant PAH. Nitrogen heterocyclics account for only a few percent of the total PAH, reflecting the normal nitrogen content of coal, which is generally below 2% [92].

More recently, results have been published for PAH concentrations measured during surveys at five coal liquefaction pilot plants [93]. Sampling was conducted when plant test-run conditions were as near to test specifications as possible.

Significant differences were found in the average total PAH concentrations measured at the five plants. The differences for area samples were not correlated to plant size or liquefaction process. Other factors, including specific operating conditions at the time of each survey, were felt to be more influential.

Table 10. Individual PAH by Percentage in Area Air Samples Taken at the Solvent Refined Pilot Plant (SRC-II Process), Fort Lewis, Wash.[a]

Compound	Coal preparation area			Mineral separation	Dissolver preheater	Solvent recovery	Product solidification
	006	007	010	041	040	025	026
Naphthalene	10.1	1.1	10.3	4.0	28.3	11.8	13.3
Quinoline	24.3	2.2	1.0	6.5	34.2	43.1	30.0
2-Methylnaphthalene	14.4	2.6	9.1	15.6	10.7	11.3	10.0
1-Methylnaphthalene	1.8	0.7	1.2	0.5	1.6	7.2	3.3
Acenaphthalene	1.1	ND	ND	ND	ND	0.1	ND
Acenaphthene	ND	1.5	2.0	2.2	0.3	1.3	3.3
Fluorene	7.8	14.6	3.2	3.8	3.2	3.9	6.7
Phenanthrene, anthracene	24.3	71.5	52.6	19.3	16.2	17.3	30.0
Acridine	3.2	ND	ND	ND	ND	ND	ND
Carbazole	6.7	0.7	9.1	12.9	1.1	0.2	0.7
Fluoranthene	3.0	1.8	5.3	10.3	1.1	0.6	1.2
Pyrene	2.7	0.7	3.2	11.1	1.1	0.7	1.7
Benzo[a]fluorene	0.5	0.4	0.8	4.7	0.4	1.3	ND
Benzo[b]fluorene	ND	2.2	0.8	6.1	ND	1.5	ND
Benz[a]anthracene	0.04	ND	0.8	1.7	1.6	ND	ND
Chrysene, triphenylene	0.02	ND	0.8	0.9	ND	ND	ND
Total PAH ($\mu g/m^3$)	56.3	27.4	50.6	224	18.7	92.9	6.0

[a] ND = compound not detected.
Source: Ref. 75,90.

Table 11. Individual PAH by Percentage in Personal Air Samples Taken in the Coal Preparation Area of the Solvent Refined Coal Pilot Plant (SRC-II)[a]

Compound	Coal preparation worker		
	Welder	Operator	Operator
Naphthalene	59.9	0.5	18.4
Quinoline	3.9	18.9	ND
2-Methylnaphthalene	10.9	49.5	49.0
1-Methylnaphthalene	1.8	5.4	2.1
Acenaphthalene	3.1	13.7	ND
Acenaphthene	ND	0.4	1.3
Fluorene	7.0	3.2	10.4
Phenanthrene, anthracene	12.4	ND	15.8
Acridine	0.05	0.2	ND
Carbazole	ND	0.5	ND
Fluoranthene	0.7	6.8	ND
Pyrene	1.6	ND	0.5
Benzo[a]fluorene	0.3	0.5	ND
Benzo[b]fluorene	ND	ND	ND
Benz[a]anthracene	ND	ND	0.5
Chrysene, triphenylene	ND	ND	ND
Total PAH ($\mu g/m^3$)	127.0	19.0	38.6

[a]ND = compound not detected.
Source: Ref. 75,90.

Three different worker groups, operators, maintenance workers, and laboratory technicians were monitored at each plant. As with the area samples, there were significant differences in the mean PAH exposures for all five plants. Exposure levels of each worker group, however, were dependent on the type of activity being performed and the time spent in a process area. Maintenance activities that involved breaking into process equipment or the handling of contaminated equipment resulted in higher exposures to airborne PAH.

The following findings may prove to be of considerable consequence in the assessment of health risk. In the airborne samples collected at all five liquefaction plants, naphthalene and its methyl derivatives comprised the major constituent PAH. The same air samples showed very low levels of 4- and 5-ring PAH. The suggestion was made that inhalation exposure to PAH may not present as great a hazard as hitherto had been supposed. Presumably

Table 12. Total PAH by Personal Sampling in Coal Liquefaction Pilot Plants

Facility	Plant operation	Job type	Range of exposure ($\mu g/m^3$)
SRC-II	Coal preparation	Operators, welders, mechanics	3.5-127
	Mineral separations	Operators	14-71
	Solvent recovery	Operators, technicians	13-115
	Product solidification	Operators	18-264
SRC-I	All types	Operators	0.05-21
		Maintenance workers	4.5-29
CSF	Solvent extraction	Operators	~0.2
	Solid separation	Millwrights	0.05-0.4
	and carbonization	Operators	0.1-0.2
	Solvent recovery and fractionation	Operators	0.05-0.3
	Hydrogenation and utilities	Operators	0.02-0.3
	Laboratory	Chemists	0.01-0.2
		Technicians	0.02-0.3
		Shift supervisors	0.08-0.2

Source: Ref. 75,90.

the sources of the airborne PAH are the low-ring and higher volatility PAH compounds evaporating from process materials lying on exposed contaminated surfaces.

Wipe samples were collected from contaminated equipment and tools. In contrast to the airborne samples, 4-, 5-, and sometimes higher-ring PAH could be measured in the surface contamination. These surface contaminations are not easily quantifiable. Nevertheless, dermal contact with process liquids that contain high-ring PAH is a source of chronic exposure in these liquefaction plants, and deemed to be a significant health hazard. Dermal rather than respiratory exposure could, therefore, be considered to be the greater health concern.

Coal Gasification Plants

Little data are available concerning the types and quantities of PAH in airborne samples at coal gasification plants. In the United States, commercial coal gasifiers do not exist today; there are only a number of experimental and demonstration units. Nevertheless, some useful comparisons can be made between the nature of process stream tars produced by gasifiers and airborne emissions from coking and aluminum-reducing operations.

As shown in Fig. 3, the tar from the electrostatic precipitator of a low-Btu coal gasifier is quite similar in composition to the PAH profile for tarry fumes associated with coke oven emissions. Another type of gasifier that uses waste sources, such as residential garbage or agricultural wastes, also produces tarry fumes with a PAH profile similar to a coke oven sample [94]. These PAH profiles can be compared in Fig. 7. The coking process employs reaction conditions and temperatures similar to those used in this low-Btu gasifier. It is, therefore, not too surprising that the PAH profiles are quite similar. One might suppose, therefore, that the potential PAH-associated health risks in low-pressure coal gasification operations are probably on a par with those associated with coking and aluminum reduction.

A recent report of an industrial hygiene assessment of three coal gasification plants [95] points to a possible fault in such a line of reasoning, at least with respect to airborne PAH exposures. In collected air samples, the overwhelmingly dominant PAH was naphthalene and its methyl derivatives. There were almost no 4- and 5-ring PAH present. The three gasifiers were of the fixed bed, entrained bed, and fluidized bed types for producing low- or medium-Btu gas. Another significant finding was that measured levels of PAH were relatively low compared to coal liquefaction plants and around the cracking and coking units in petroleum refineries. The conclusion was drawn that inhalation exposure to carcinogenic PAH is not likely to be a significant hazard in the coal gasification plants.

In contrast. higher-ring PAH characteristic of process liquids were found in wipe samples from contaminated equipment, tools, and protective clothing. Dermal contact was judged to be a significant source of chronic exposure to PAH, especially at gasification plants producing large quantities of tar.

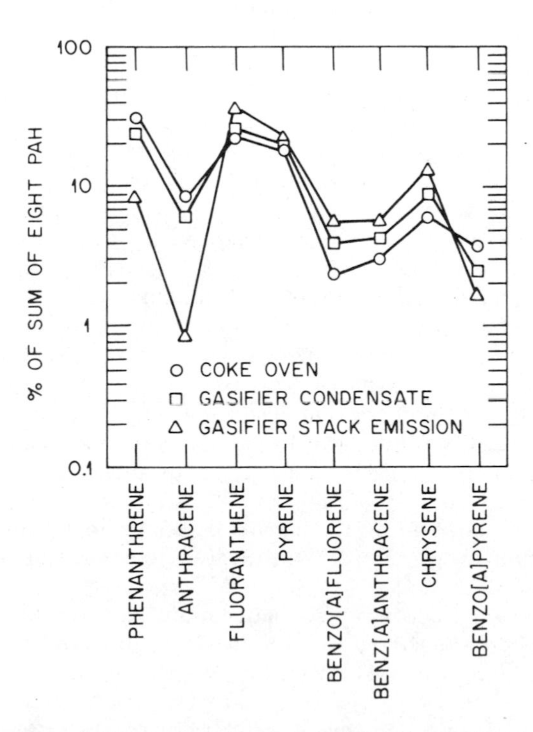

Figure 7. A comparison of PAH profiles of gasifier samples with the profile obtained from a sample collected at the top of a Norwegian coke oven battery. (From Ref. 94.)

The following scenario might explain the circumstances of relatively small and benign inhalation, but potentially dangerous dermal exposure to PAH. During gasification of coal, the necessity of containing the product carbon monoxide also requires containment of concommitant vapors of coal tar volatiles. It is after rather than during the primary gasification that fugitive emissions of tar are more likely to take place. Opportunities arise during maintenance and repair, and from sources such as dripping valves and taps. Airborne PAH may arise largely from evaporation, and condensation onto dust particles, of volatile naphthalenic PAH whose source is tar on exposed surfaces. On the other hand, the low-volatility, high-ring PAH will remain behind in surface contaminant. As a result, the latter's carcinogenic potential will tend to be preserved. This scenario bears a close resemblance to the one advanced in the previous section for explaining the character of fugitive emissions of PAH in coal liquefaction plants.

D. Mutagenicity of Airborne Samples

It would be hard to reject the notion that a detailed knowledge of the concentrations of compounds of PAH is of utmost importance for estimating the potential hazard of PAH. There is, however, reason to question whether reliance solely on chemical analysis is sufficient or even the best method for determining the health effects of PAH-bearing airborne samples.

A few recent studies of workplace samples have sought to emphasize measurements of mutagenic activity as a means of categorizing biological significance. Gasifier particulate emissions have been analyzed [93] for both PAH and their mutagenic activity via the Ames test [27]. Eight abundant PAH (phenanthrene through BaP) that were quantified accounted for only a few percent of the mutagenic activity. Synergistic effects and additional unidentified mutagenic compounds were suggested as the reasons for the noncorrespondence between PAH concentrations and biological activity.

A similar result has been published for particulate air samples collected from steel foundry atmospheres [96]. This study was prompted by preliminary data indicating no direct correlation between lung cancer risk in various parts of the foundry and the levels of benzene-soluble material, "marker" PAH, total or respirable particulate matter, various fumes, or silica. The bacterial mutagenicity assay [27] was used as the primary tool to examine the possibility of unidentified carcinogens being present. Extracts from particulate matter were divided into aqueous, acidic, basic, and neutral fractions. On occasions when high levels of total mutagens were present, the basic and aqueous fractions had more biological activity than the neutral fraction. The authors concluded that foundry air particulates contain mutagens belonging to several chemical classes. Day-to-day variations in foundry activities led to occasions when PAH represent a relatively small proportion of the total mutagenic material present.

Instead of direct examination of workplace samples, mutagenic activity in the urine of exposed workers has been sought [97]. The study thus far has been restricted to a small number of smoking and nonsmoking coke oven workers.

The future will probably see a shifting of emphasis away from solely chemical analysis of particulate air samples to combined analyses where the biological activity is accorded better balance.

E. Vapors of PAH

Consideration of the health risk posed by the inhalation of vapors of PAH has received little or no attention. One reason for this neglect is that carcinogenic potential is generally appreciable only for high-boiling PAH that have correspondingly low vapor pressures at ambient temperatures [98]. Equilibrium vapor pressures of pure PAH decrease rapidly as the aromatic rings increase in number. Data on equilibrium vapor pressures and concentrations for a few important PAH of varying size are presented in Table 13 [98,99].

In the workplace these values should be regarded as upper limits. Vapor molecules of PAH have a proclivity to adsorb strongly on particulate matter. The equilibrium vapor pressure in the adsorbed state may be considerably less than for pure PAH [100]. Another reason for equilibrium vapor pressures not being reached is a kinetic one, i.e., slow rates of vaporization.

For each of these reasons, the health risk from the direct inhalation of PAH vapors is usually taken for granted as insignificant compared to inhalation of the same PAH in the form of particulate matter. Some of these PAH vapors, however, could exert promotional or cocarcinogenic responses in concert with other vapors or with inhaled particulates containing PAH. Perhaps more attention needs to be given to the inhalation of vapors of two-, three-, and even four-ring PAH.

1. *Low-Boiling PAH*

Naphthalene is the only PAH listed in Table 13 with a high enough equilibrium vapor concentration to produce atmospheric concentrations of greater than 1 ppm at ambient temperatures. Naphthalene is also the only PAH with an OSHA concentration limit, which is 10 ppm (8-hr time-weighted average) [50]. With increasing size of PAH, the maximum atmospheric concentrations fall rapidly; for BaA and BaP the values are sub-ppb.

To monitor PAH vapors in the workplace requires highly sensitive instruments. A derivative ultraviolet-absorption spectrometer (DUVAS) has been

Table 13. Equilibrium Vapor Pressures and Concentrations of Pure PAH in Air

PAH	Equilibrium vapor pressure (torr)	Equilibrium vapor concentration ($\mu g/m^3$)	Concentration by volume in air
Naphthalene [98]	6.7×10^{-2} (295.5K)	4.6×10^5	88 ppm
Anthracene [98]	2.9×10^{-4} (295K)	2.8×10^3	0.38 ppm
Pyrene [99]	6.8×10^{-6}	74 (298K)	9 ppb
BaA [99]	8.6×10^{-8}	1.05 (298K)	1.1×10^{-1} ppb
BaP [99]	6.3×10^{-9}	8.5×10^{-2} (298K)	8×10^{-3} ppb

Source: Refs. 98 and 99. Adapted with permission from C. Pupp, R. C. Lao, J. J. Murray, and R. F. Pottie, *Atmos. Environ.* 8:915-929. Copyright 1974, Pergamon Press, Ltd.

Table 14. Approximate Detection Limits for Some Aromatic Vapors Using DUVAS

Compound	TLV[a] (ppm)	Approximate detection limit (ppm)[b]
Benzene	10 (1)	0.06
Toluene	200	0.01
Cresol	5	0.06
Phenol	5	0.06
Naphthalene	10	0.001
Methylnaphthalenes	—	0.001
Indan	—	0.07

[a]Threshold limit value. Proposed limit in parentheses.
[b]Total path length of 4 m.
Source: Ref. 102.

developed for the task of monitoring monocyclic and bicyclic aromatic hydrocarbons [101]. A list of lower limits of detection for these compounds is shown in Table 14.

The unit can be used as a portable survey instrument [102] or as a remotely operating area monitor that records and stores the results of real-time analysis over the course of a day. The latter mode of operation is the one most appropriate for recording human exposures to PAH that are in the category of high risk but not immediately hazardous [103].

The spectra of vapors issuing from oils and tars have been measured with the DUVAS. A variety of coal conversion and shale oil products have been examined. For a given product, a characteristic signature is obtained that is determined by the relative proportions of those volatile aromatic compounds that give intense second-derivative peaks. The spectra, in the wavelength range of 200-300 nm, are gratifyingly simple for the purposes of compound identification and quantitation. Two examples are given in Fig. 8. For a low-Btu gasifier tar, the spectral "fingerprints" of benzene and phenol are dominant. These are two compounds entrained in the tar. A heavy distillate oil produced by a coal-liquefaction plant gives a different signature; the most prominent peaks are caused by naphthalene and indan.

The DUVAS has been evaluated in the field. A few results of area surveys for naphthalene and methylnaphthalenes [102,104] are reported in Table 15. Even though most of these surveys were conducted in the dirtiest locations of the plants, the vapor concentrations of these two compounds are generally quite low and orders of magnitude lower that the TLV for naphthalene (10 ppm) [50].

2. Higher-Boiling PAH

In PAH-containing work atmospheres, the DUVAS has been unable to detect the vapors of higher-boiling, multiring PAH. A passive device capable of making integrated exposures is being developed for such a purpose. The

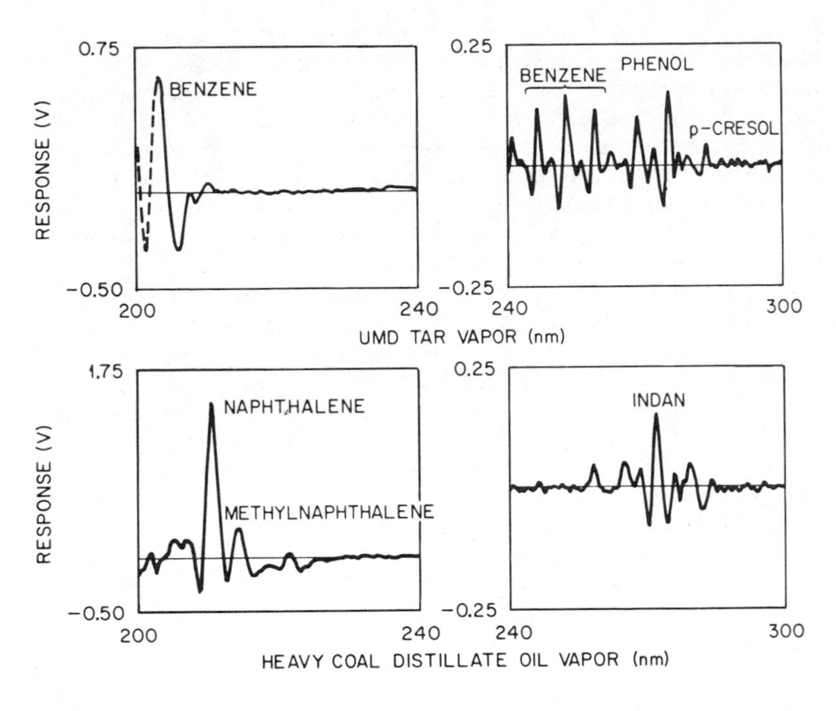

Figure 8. Spectral "fingerprints" of aromatic compounds in oil and tar vapors as measured by the DUVAS.

Table 15. Concentration of Naphthalene and Methylnaphthalene Vapors in Synfuel Plants

Plant type and sampling locality	Naphthalene (ppb)	Methylnaphthalene (ppb)
Low-Btu gasifier:		
Tar storage tank and manifolds	5-31	6-25
Coal liquefaction by catalytic conversion:		
Leaking pumps and compressors	10-65	12-77
Small scale hydrocarbonization of coal:		
Open condensate vessel	20	35

Source: Refs. 102 and 104.

detecting element is filter paper treated with a heavy atom salt [105] for the subsequent induction of phosphorescence at room temperature (RTP) during readout [106]. The filter paper acts as the collecting element during exposure and as the substrate from which the RTP of PAH can be induced after placement in a phosphorescence spectrometer. Selective PAH vapors such as pyrene or fluoranthene can be measured in units of concentration × hr. The device is calibrated by exposure to the vapors of pure PAH compounds. At equilibrium vapor concentrations, the intensity of the RTP increases linearly with time of exposure for up to 8 hr [106].

Small, lightweight, and unobtrusive, the monitoring device can easily be worn clipped to a pocket. It is suitable, therefore, for use as a personal monitor. A number of small RTP devices have been tested as area monitors at a coal liquefaction pilot plant. The devices were exposed for a day at various plant locations. Where area contamination with heavy distillate was evident, positive readings of RTP were forthcoming. An example is shown in Fig. 9. The corresponding time-weighted average concentration of pyrene is about 70% of the equilibrium vapor concentration at 30°C [98]. The devices have been worn recently by two sampling crew members for periods of several hours [108]. The exposure to pyrene vapor was determined to be in the range of concentration 4-8 ppb. Trace amounts of carbazole were also detected.

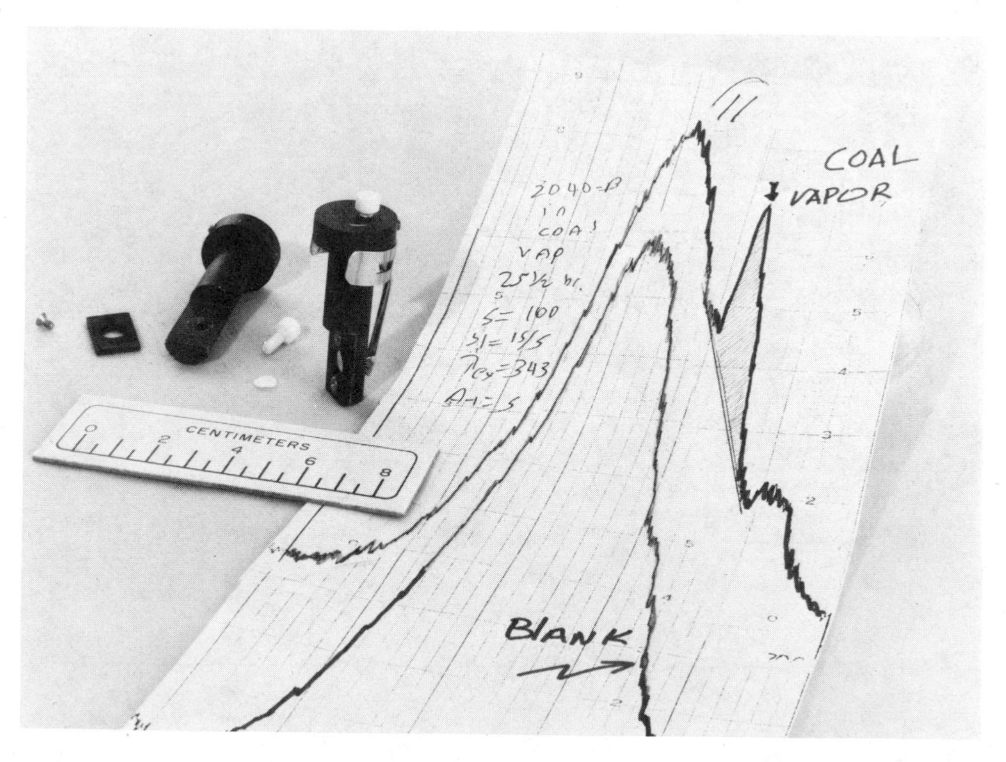

Figure 9. Detection by RTP of pyrene in vapors of liquefied coal collected by a passive filter monitor at a coal conversion pilot plant.

The filter paper is usually covered with a honeycomb to obtain diffusion controlled sampling of PAH vapors. There is also the potential for collecting and analyzing liquid droplets of respirable size on a separate open portion of the filter paper element. It would then be a relatively simple matter to separate the contributions of vapor and aerosol in the combined exposure.

A much more sophisticated and expensive means of measuring PAH vapors is available in the form of an atmospheric pressure, chemical ionization mass spectrometer [99]. The spectrometer can perform essentially instantaneous analyses using air at ambient temperatures. Current detection limits are in the low parts per trillion (ppt) range for most PAH. This should allow measurements to be made on most PAH with boiling points below that of BaP. The mobile spectrometry system will permit surveys and assessments of PAH vapors to be conducted in workplace atmospheres.

F. Skin Contamination

It is the considered opinion of some that the most probable route of human exposure to high-boiling carcinogens produced by synfuel technologies is through the skin after leakage and contact with the skin [29,93,95,109]. Respiratory exposure by the airborne route is viewed as a lesser hazard. In synfuel plants this is due largely to the open-air environment of most plant areas and the tight sealing of equipment used in high-pressure operations [109].

1. Industrial Experience

Experience at the high-pressure, high-temperature hydrogenation plant at Institute, West Virginia [42] provided convincing evidence linking skin contamination to skin lesions. The same affected but small group of workers have not developed any systematic cancer [45]. Nevertheless, these medical experiences suggest that, for these types of synfuel plant, skin contact may indeed be the exposure pathway for PAH that is of major concern. Discussions about PAH in the workplace would be remiss without adequately addressing this problem.

The best-documented industrial hygiene experiences on skin contamination in a modern-day facility are available in reports from SRC pilot plants. The emphasis of the personal hygiene programs is avoidance of contact with coal tar to the extent that is practical [110]. To date this approach has been successful since there have been no cancer cases that are occupationally connected at the SRC pilot plant [111]. The only problems reported at the SRC pilot plant are mild transient dermatitis from skin contact with coal-derived materials. Table 16 summarizes the medical observations that were reported in 1979 [111]. About 25 cases of transient erythema and multiple cases of mild foliculitis (mechanics' acne) have been observed. Each of these cases responded well to temporary suspension of exposure. The duration of exposure and the length of follow-up of individually exposed workers has, however, been too short to draw any conclusions concerning skin cancer.

Anyone who has experience with such plants knows, nevertheless, that "the oily fractions produced by the hydrogenation of coal are extremely difficult to confine within the equipment. Maintenance and repair operations are particularly difficult to perform without rather widespread contamination of

Table 16. Solvent Refined Coal Pilot Plant Medical Observations

Description	Number of incidents	Number related to SRC work
Eye irritation	50-60	Most
Erythema	25	25
Foliculitis (mechanics' acne)	Multiple	Most
Skin cancer	1	0

Source: Ref. 111.

nearby equipment and even walkways, floors, stairs, and railings. Direct contamination of employee's skin and clothing is a continuous problem" [43]. Constant vigilance is required in the teaching and practice of cleanliness if workers are to be adequately safeguarded. In older traditional industries that produce or handle coal tar, the medical evidence indicates incomplete safeguarding. Among pitch workers [112] and tar distillers [113] there are reports of skin and scrotal cancer, warts, and photosensitization. There is also well-documented evidence that liquids derived from shale oil have caused skin cancer in workers of the Scottish shale oil industry [34].

In addition to skin contamination by direct contact with contaminated surfaces there is also the potential for surface adsorption of vapors and fallout droplets from airborne oil fumes [43].

With the emergence of a large synfuels industry, much larger numbers of workers will have to contend with these problems. Satellite industries involved with the transportation and utilization of synfuels would likely find that skin contamination of their workers was the principal occupational risk.

2. *Nature of Interaction*

It is not only at the site of skin contamination that damage by PAH can occur. Aromatic hydrocarbons are highly lipophilic and readily penetrate through the skin and into cells. It is not surprising, therefore, that repeated topical applications of PAH in large doses to the skin of mice and rabbits causes systemic effects. Percutaneous absorption causes pathological changes in the blood, spleen, lymph nodes, and bone marrow [114]. These systemic effects are additional to the expression of malignantly transformed epidermal cells at the site of application of the PAH.

Following topical application to the skin, penetration is extremely rapid. Within 5 min fluorescence can be detected in the sebaceous glands of hair follicles [29]. A schematic representation of a vertical section of mouse skin with fluorescing contaminant is reproduced in Fig. 10. A photomicrograph of a corresponding vertical frozen section of mouse skin treated with fluorescing coal-derived oil is shown in Fig. 11 [29]. Fluorescent material is seen penetrating the antrum of the follicle and extending into the sebaceous duct and gland. The predilection of fluorescing compounds for sebaceous lipids is very striking. For this particular coal-derived oil (~200 $\mu g/cm^2$) that was rich in PAH, en-

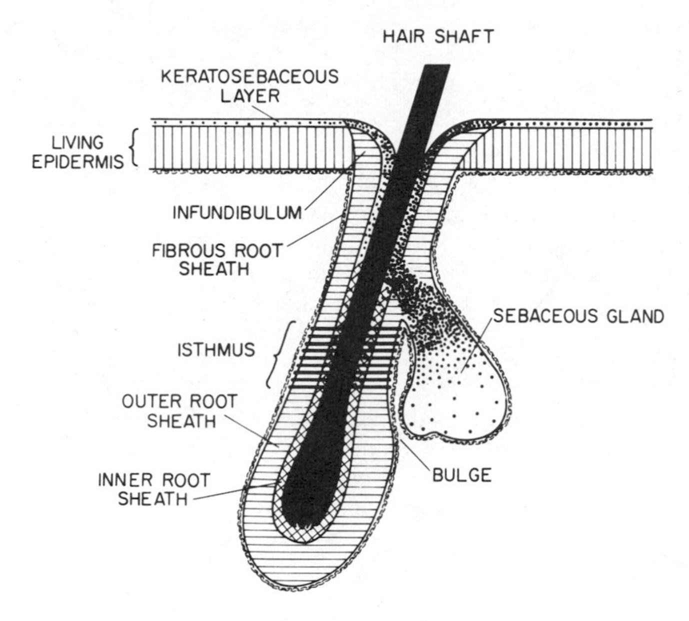

Figure 10. Schematic representation of the distribution and concentration (strippling) of fluorescence in mouse skin as seen in a frozen section. (From Ref. 29.)

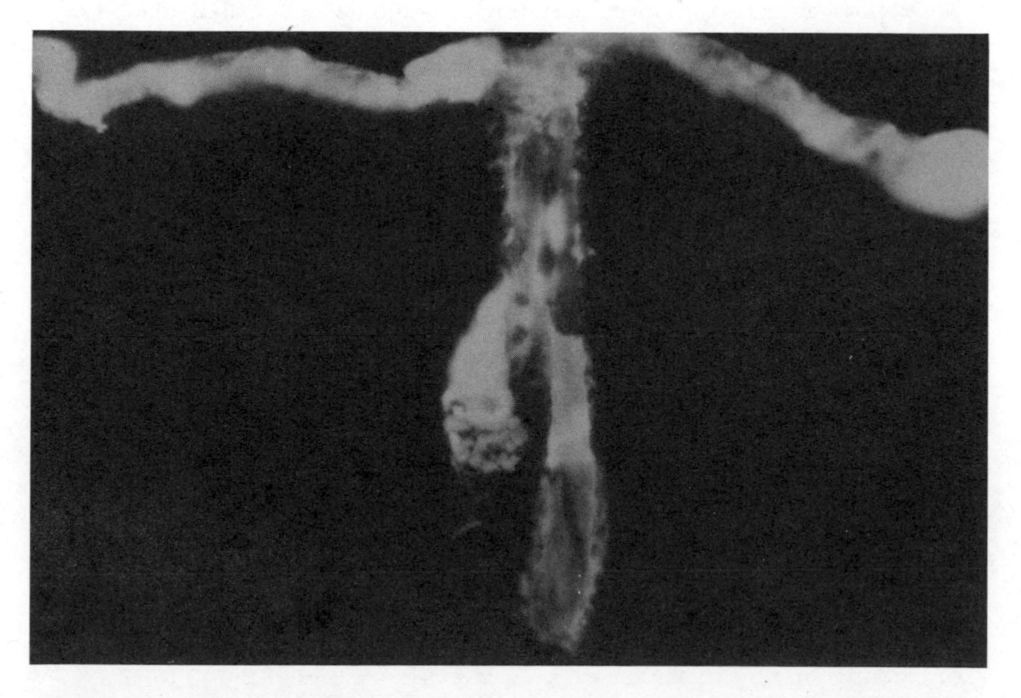

Figure 11. Photomicrograph of a frozen section of fluorescing mouse skin contaminated with synfuel oil. (From Ref. 29.)

hanced fluorescence was still evident 14 days after application. Similar human experiences for persistence of fluorescence for up to 8 days after exposure have been reported [115]. In this latter case, fluorescence was recorded from only the outer keratosebaceous layer of the human's skin.

3. *Monitoring Skin Contamination*

It used to be the case that skin examinations were made freely using 254 nm ultraviolet (UV) light [43], taking care to avoid illuminating the eyes. Fluorescence is most easily observable at this excitation wavelength. The clean skin shows a slight bluish fluorescence against which the fluorescing oils are contrasted sharply. The fluorescence of concentrated oils generally has a golden hue that shades to yellow and white with dilution or solvation by natural oils of the skin.

More recently, a gradual change in philosophy has occurred. Now skin examination with UV light is used in only restricted circumstances.

Threshold limit values (TLVs) exist for occupational exposure to ultraviolet radiation to protect workers against erythemal effects. For radiations of 254 and 365 nm, the TLVs are 6.0 and 1000 mJ/cm^2, respectively. This difference was instrumental in popularizing the use of near-UV light (365 nm) for the detection of skin contamination. Until as late as May 1975, the regular use of this longwave UV light was being recommended in an Oak Ridge National Laboratory booklet to detect PAH on the exposed portions of the body [116]. Its use was advocated before and after work and after showering [117].

Unfortunately PAH-containing oils are phototoxified by near-UV light (320-400 nm) [118] to produce acute effects. Shorter wavelength UV light (290-320 nm) does not produce short-term phototoxicity. Similar photosensitizing effects of coal tar were reported by Crow et al. [119] for radiation of 330-440 nm. These authors also noted a "smarting reaction" in patients being treated for psoriasis. An immediate burning sensation is experienced when tar-treated sites are irradiated with UVA (315-400 nm) but not when the irradiation is with UVB (280-315 nm).

Because of the recognition of phototoxic and synergistic effects, there is now a reluctance to advocate the routine use of near-UV light for detecting oils and tars on the skin. Heightened awareness of phototoxic and other synergistic effects is the underlying reason. The aforementioned ORNL booklet [116] has been amended and now states that UV light will be used routinely to detect PAH contamination only on tools and equipment. Examination of skin with UV light is conducted at the discretion of a plant physician.

The health information pamphlet published by the Gulf Oil Corporation [110] likewise neglects to recommend skin examinations by near-UV light at its SRC plant. Should contamination occur, instructions are to wash the affected area immediately. An interesting chronological trend can be detected in the periodic industrial hygiene reports issued for the SRC plant. The report for the period ending June 1977 [120] states that skin examinations with a (near) UV light are periodically given by the plant nurse to determine how well the employees are cleaning themselves. The tone changes in the June 1978 report [48] where the statement is made that quarterly after-shower skin examinations by UV light reveal that workers frequently fail to remove all oil and tar residues from the face, hair, forearms, and hand areas. The statement is also made that complete removal of process solvent by soap and water is impossible; a fluorescence residue remains for approximately 2 days. By December 1979 [115], UV light is no longer advocated for routine examination of skin. The same report indicates the near demise of UV light by saying that its use is unnecessary for detecting coal tar on inanimate objects; tools are clean enough if they appear uncontaminated by the unaided eye.

This fall from favor of the black light unfortunately leaves us without any established means for the routine monitoring of skin contamination.

a. UV Light:

(1) Portable UV lamps These devices are commercially available and inexpensive. Detection is quick, easy, and obtained in real time. The near-UV-irradiated and subsequently fluorescing skin is viewed directly by the unaided eye in a darkened room or viewing cabinet. The information obtained is qualitative and the limit of detection suffers from variable keenness of eyesight between different observers.

The hand-held black-light lamp that is used for skin monitoring produces power in the near-UV region roughly equivalent to that of sunlight in the same wavelength region. This assumes that the lamp is being held a few inches away from the skin. Therein lies much of the problem. The radiant flux is too high. This leads us to a new developmental instrument that produces much lower radiant fluxes—a lightpipe luminoscope.

(2) Lightpipe luminoscope During the last decade, the realization has emerged that significant cocarcinogenic or promotional effects are possible between UV light and skin contaminants. Not only is there enhancement of the effects of proven epidermal tumor initiators such as 7,12-dimethylbenz[a]anthracene (DMBA) [121], but normally harmless compounds can also be imbued with carcinogenic properties. For example, several n-paraffins produce skin tumors in mice in conjunction with light of wavelength >350 nm [122]. Another seemingly innocuous compound with photocarcinogenic properties is anthracene. Enhanced covalent binding of phototoxified anthracene to DNA in tissue culture can be detected at 365 nm for exposures as low as 700 ergs/cm^2 [123].

The TLV for occupational exposure to near-UV light (320-400 nm) is set at 10^7 ergs/cm^2 [50] over exposures times of up to 1000 sec. This is the equivalent of exposure to the near-UV components of sunlight at sea level where the radiant flux is about 1 mW/cm^2 or 10^4 ergs/(cm^2)(sec) [124]. It is implicitly stated, however, that this TLV does not apply to photosensitized individuals or workers who are concomitantly exposed to photosensitizing chemicals. Dermatitis and melanosis due to photosensitization has long been recognized in industries where tar and pitch are handled [125].

While acknowledging that no level of intensity of near-UV light is absolutely safe, workers who are contaminated have to live in a world that contains UV light. It seems reasonable to assert that if the detection of skin contamination with UV light is to be used, then the intensity should be as low as possible and well below the intensities found in sunlight. It can then be argued that the UV light detection and the subsequent removal of contamination will present the lesser of two evils.

With these considerations in mind for operation of the luminoscope, the intensity of the excitation at 360 nm, with a 50-nm bandwidth, was set at 10^2 ergs/cm^2 sec [126]. For a maximum 10-sec survey of a particular area of skin, the radiance produced by the luminoscope is then only 10^3 ergs/cm^2.

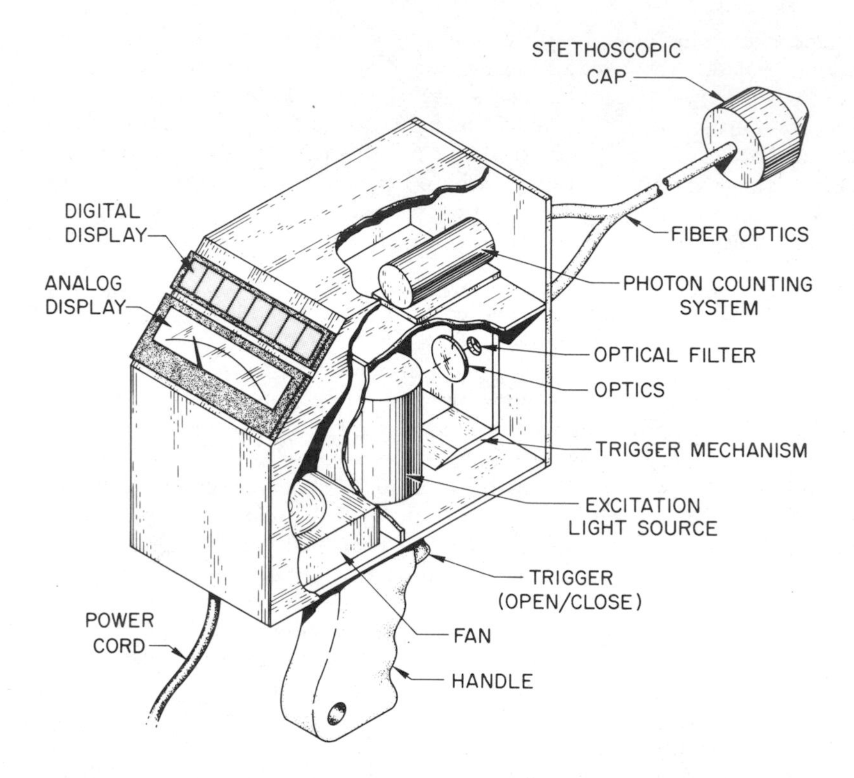

Figure 12. Layout of the main components of the lightpipe luminoscope. (From Ref. 126.)

This value is only 1/100 of the radiant flux of sunlight (350-400 nm) at sea level [124] and 10^{-4} of the TLV. This selection is judged to be a safe compromise between minimizing phototoxic effects and maintaining adequate sensitivity for the detection of contamination.

A schematic diagram of a prototype hand-held "luminoscope" is shown in Fig. 12. Further details concerning this instrument can be obtained from recent papers [126,127].

A newer version of the luminoscope incorporates several improvements [128]. There is a background-nulling circuit that allows one to record the net fluorescence of contamination with greater convenience and sensitivity; the background signal from an area of clean skin is measured, stored, and then subtracted from the gross signal. This second luminoscope also uses quartz instead of plastic fiber optic lightguides. This allows excitation and emission with less acutely phototoxifying UVB (313 nm) to be studied. The output is also used to drive a small speaker that provides an audio signal whose frequency increases with the net fluorescence intensity. Scanning an object like the arm can now be accomplished quickly without glancing back and forth at a front-panel meter.

The luminoscopes have been used as field monitors for detecting skin contamination in lieu of hand-held black lamps. Tests have been made successfully at several coal-conversion facilities [126]. The sensitivity is good enough for detecting residual contamination even after affected areas of skin have been washed several times.

For the time being, however, the luminoscope will find its greatest use in broadening our understanding of the course of skin contamination events and evaluating the effectiveness of various remedial actions.

Limits of detection and ranges of linear response have been studied by spotting diluted oils and tars on filter paper. Two examples are shown in Figs. 13 and 14. The lower limits of detection are about 50 ng/cm^2 and 2 nl/cm^2 for the tar and recycle solvent, respectively. On a unit mass basis, the limit of detection is several hundredfold less for the tar than for the recycle solvent; the tar contains a higher proportion of highly fluorescing compounds.

Linearity of response extends to several hundred nanograms per square centimeter in the case of the tar (Fig. 13). Below a few hundred nanograms per square centimeter, such a tar is not directly visible to the unaided eye. The potential exists, therefore, for quantifying invisible traces of contamination still residing in the skin after washing. Saturation effects that produce the nonlinearity are caused by quenching, energy transfer, and self-absorption.

Similar exposure-response behavior has been obtained after painting hairless mice with 2 μl amounts of a tumorigenic coal-derived liquid diluted in acetone (Fig. 15) [129]. The properties of this material, No. 1314, are listed in Table 7. Also for a related group of these materials, the fluorescence

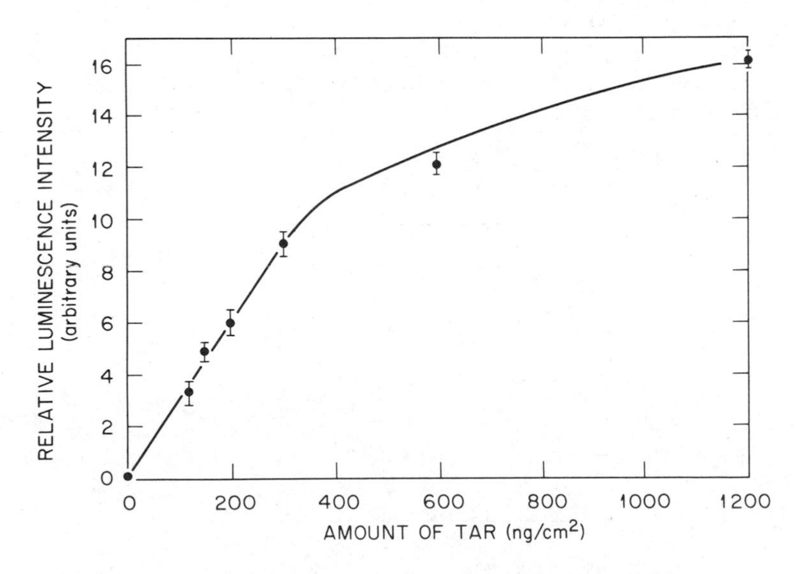

Figure 13. Fluorescence response of UMD coal tar dissolved in solvent and spotted on filter paper. (From Ref. 126,127.)

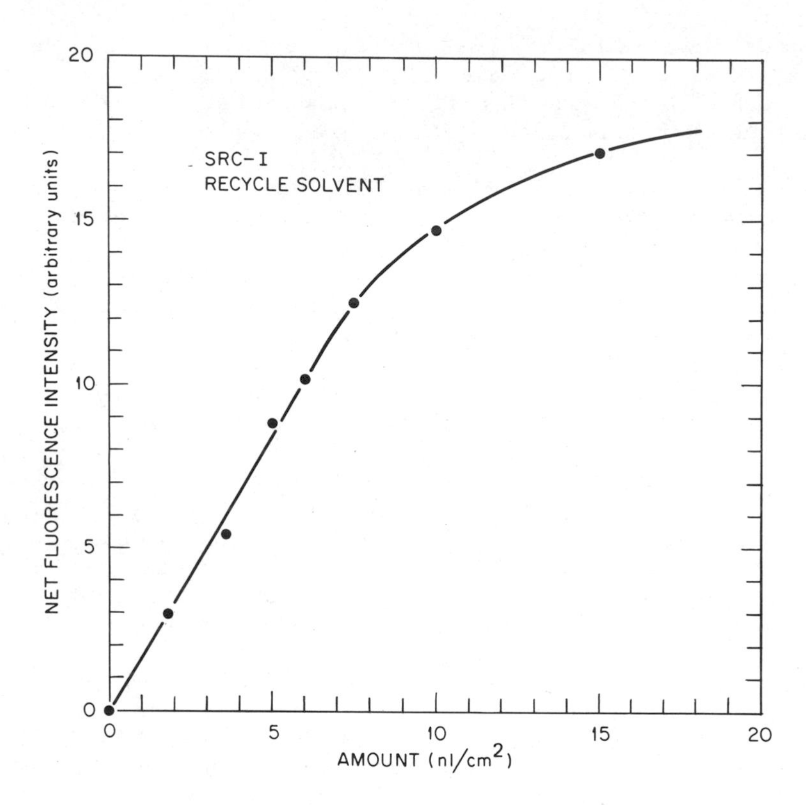

Figure 14. Fluorescence response of diluted SRC-I recycle solvent spotted on filter paper. (From T. Vo-Dinh and R. Gammage, in *Chemical Hazards in in the Workplace: Measurement and Control*, G. Choudhary (Ed.), ACS Symposium Ser. No. 149, American Chemical Society, Washington, D.C., 1981, p. 269. Reprinted by permission.)

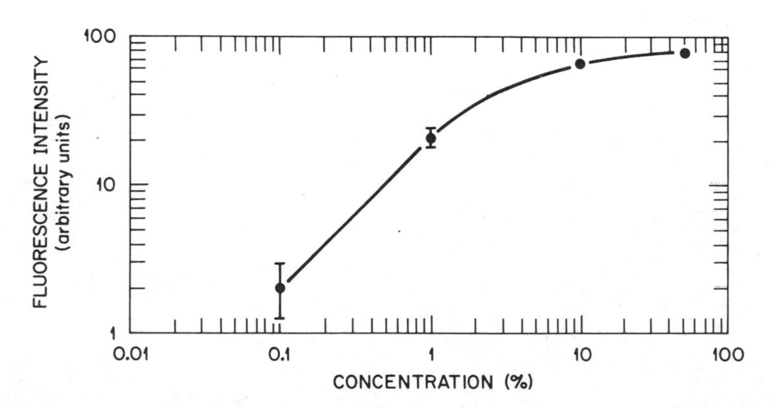

Figure 15. Exposure vs. response for tumorigenic coal-derived liquid No. 1314 painted on mouse skin.

Table 17. Fluorescence vs. Biological Activity

Material	Fluorescence [129] (arbitrary units)	BaP (μg/g) [18]	Tumorigenicity (%) [130]	
			Male	Female
1312 (atmospheric overheads)	0.9	1.5	0.0	0.0
1313 (atmospheric bottoms)	2.7	43.0	4.0	24.0
1314 (vacuum overheads)	6.3	296.0	100.0	100.0

intensity from the mouse skin correlates in a general way with both the BaP content and the tumorigenicity [18,130,131]. These data are compared in Table 17.

Temporal measurements of the painted mouse skin (Fig. 16) have also been made. With the more dilute solutions, a clearance mechanism operates to produce a monotonically decreasing fluorescence intensity. For exposures to higher concentrations of oil, clearance is preceded by increases in fluorescence intensity. This is attributed to a lessening of quenching as spreading and solvation occur. The magnitude of the changes of fluorescence intensity with time are such that exposures can be measured quantitatively with reasonable reliability (10-20%) if made within the first 2 hr. The luminoscope is, therefore, capable of quantifying exposures if the identity and fluorescent properties of the contaminant are known.

(3) Fluorescence spill spotter Surface contamination of inanimate objects can lead to transfer of PAH to the skin by direct contact. This aspect of monitoring general types of surface is, therefore, an appropriate one for consideration in this section. It is important, for example, to detect surface contamination being tracked into changing areas, lunchrooms, control rooms, and other "clean" areas. Another need is to locate and define the boundaries of spilled PAH. A means is also needed to measure the effectiveness of clean-up.

A hand-held, fluorescence spill spotter with telephoto lens and photomultiplier tube has been developed to meet these needs [131]. The manner of operation is demonstrated in Fig. 17. The illuminating beam, and hence the fluorescence, is modulated at 1 kHz. The characteristic fluorescence of PAH is separated from other background signals in the blue-green region of the spectrum. This requires electronic filtering, demodulation, and low-pass filtering. Fluorescence can be detected that is only 3% as intense as the background illumination in the optical wavelength band of interest. Hence the spill spotter can be operated outdoors in direct sunlight or indoors in strong background illumination. This instrument has greater versatility than the conventional hand-held black lamp.

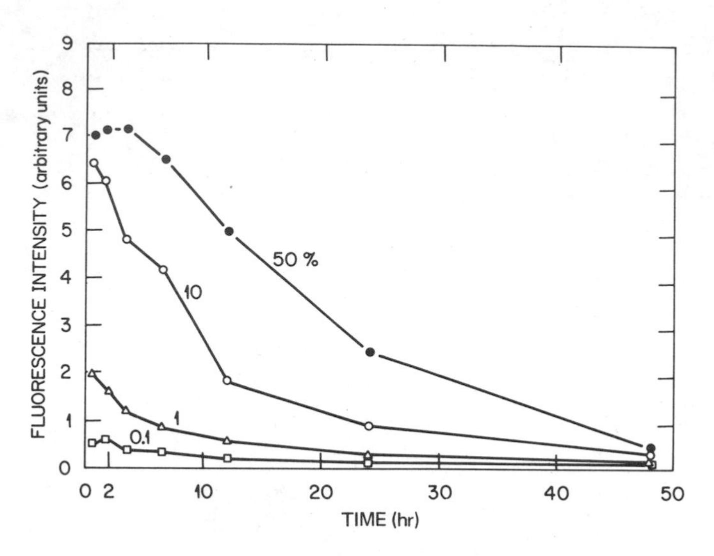

Figure 16 Temporal measurements of fluorescence from coal-derived liquid No. 1314 painted on mouse skin at different concentrations (weight/volume acetone).

Figure 17. Fluorescence spill spotter for general surface contamination.

Under field-test conditions, the buildup and spread of contamination coal-conversion facilities has been measured [132]. The detection was often made of fluorescent PAH in amounts too small to be seen with the unaided eye. The collimation of the beam through the telephoto lens is such that remote monitoring of work-area surfaces can be conducted at distances of up to 3 m. These attributes make the spill spotter an important component of industrial hygiene hardware for monitoring PAH.

b. Infrared Spectrometry In principle, the infrared (IR) spectrometer with a commercially available multiple internal reflectance attachment [133] can be used to examine oil or tar on the skin. The measurement could be conducted directly by placing the affected area of skin against a slit in the reflectance attachment. Alternately a wipe specimen could be placed on a flat plate of ZnSe to achieve better spectral resolution. Although cosmetics, bactericides, antiperspirants, and natural secretions on the skin have been examined in this manner, no data have been published for oils and tars.

An intrinsic drawback of IR spectroscopy is the lack of unique spectral features for PAH. Nevertheless, it would seem worthwhile to examine the potentials of IR detection in more detail. Two good reasons are the dearth of currently available techniques for measuring skin contamination and the benign nature of IR light when used to irradiate either clean or PAH-contaminated skin.

c. Skin Wash New methods for measuring skin contamination have been developed by the Pittsburg and Midway Mining Company, who operate the SRC plant. Instead of measuring fluorescence in vivo, as with the luminoscope, the skin is washed with a solvent (cyclohexane) that is subsequently analyzed by measuring the total fluorescence of the cyclohexane solution irradiated at 295 nm [115]. An open-cup, cyclohexane spray device [115] is pressed against

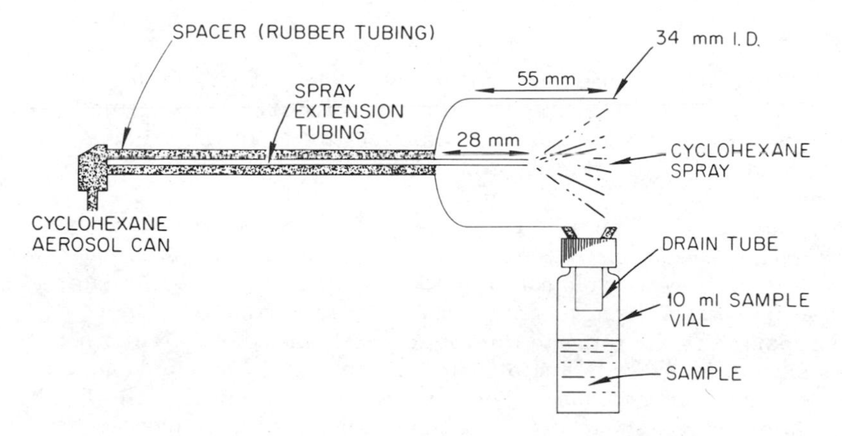

Figure 18. Skin-wash apparatus for oil and tar contamination. (From Ref. 115,134.)

Table 18. Preliminary Results of Cyclohexane Skin Wash Samples Collected from SRC Pilot Plant Workers: Sept. 25, 1979[a]

Skin wash sample collected from:	Fluorescence expressed as "heavy distillate" equivalent in skin wash sample (μg) Before shift	After shift	After shower
#1 Maintenance Worker	ND	ND	ND
#2 Maintenance Worker	ND	24	ND
#3 Maintenance Worker	ND	ND	ND
#4 Maintenance Worker	ND	44	ND
010 Operator #1	—	29	ND
010 Operator #2	—	ND	ND
020 Operator	—	54	ND
095 Operator	—	ND	ND
Office Worker #1 (Control)	ND	—	—
Office Worker #2 (Control)	ND	—	—
Office Worker #3 (Control)	ND	—	—
Office Worker #4 (Control)	ND	—	—
Office Worker #5 (Control)	ND	—	—

[a]Sample matrix, cyclohexane skin wash; analytical method, fluorescence; ND, not detected (detection limit = 1.82 μg/sample); sample volume, 12 ml; skin area, 8.55 cm^2.
Source: Ref. 115.

the skin and operated in the manner shown in Fig. 18. It was found that wash samples from office workers and "after-shower workers" contained natural skin oils and dead skin cells that did not fluorescence. On the other hand, before-shower extracts from process-area workers showed the positive results listed in Table 18.

There are radical differences between the skin wash and in vivo fluorescence methods that need to be borne in mind. A device such as the luminoscope measures residual, nonremovable contamination, a quantity that is related to skin dose. A skin wash measures only removable contamination, before it has had an opportunity for intimate interaction with the skin. One is not measuring a skin dose and it is a moot point whether this removable material should even be called an exposure. Potential exposure would seem to be the most appropriate terminology to use. On the other hand, the fluorescence intensity of the PAH extracted into the cyclohexane solution can be measured accurately. One's ability to measure contamination directly in the skin via induction of fluorescence, and interpret the results, is limited by penetration of PAH along hair follicles and into the living skin. In such locations the PAH becomes more difficult to detect and quantify.

Perhaps the most disturbing feature of the skin wash approach is the likelihood for enhanced penetration [134]. Cyclohexane might promote the penetration of PAH at the site of the measurement. This event could exacerbate the original insult to the skin.

There are limiting facets to all methods currently being examined for skin monitoring. Further development and testing is obviously in order. The need will remain for reproducible, simple, and safe means for measuring contamination by PAH.

References

1. M. R. Guerin, in *Polycyclic Hydrocarbons and Cancer*, H. V. Gelboin and P. O. P. Ts'o (Eds.), Vol. 1, Academic Press, New York, 1978, pp. 3-42.
2. K. Bridbord and J. G. French in *Carcinogenesis*, Vol. 3: *Polynuclear Aromatic Hydrocarbons*, P. W. Jones and R. I. Freudenthal (Eds.), Raven Press, New York, 1978, pp. 451-463.
3. J. Santodonato, D. Basu, and P. H. Howard, in *Polynuclear Aromatic Hydrocarbons: Chemistry and Biological Effects*, A. Bjørseth and A. J. Dennis (Eds.), Battelle Press, Columbus, Ohio, 1980, pp. 435-454.
4. J. P. Weideman, *Roofing Spec.* 7:10-15 (1974).
5. W. H. Griest, B. A. Tomkins, J. L. Epler, and T. K. Rao, in *Polynuclear Aromatic Hydrocarbons*, P. W. Jones and P. Leber (Eds.), 2nd ed., Ann Arbor Science Publs., Ann Arbor, Mich., 1979, pp. 395-400.
6. *Particulate Polycyclic Matter*, National Academy of Sciences, Washington, D.C., 1972, pp. 6-12.
7. D. W. Jones and R. S. Matthews, *Progr. Med. Chem. 10*:159 (1974).
8. B. L. Van Duuren, C. Katz, and B. M. Goldschmidt, *J. Nat. Cancer Inst. 51*:703 (1973).
9. D. Hoffmann, I. Schmeltz, S. S. Hecht, and E. L. Wynder, in *Modifying the Risk for the Smoker*, U.S. Dept. of Health and Human Services Publ. No. (NIH)76-1221, Vol. 1, pp. 125-145 (1976).
10. I. D. Schmeltz, *Trace Substances in Environmental Health—VIII*, D. D. Hemphill (Ed.), University of Missouri, Columbia, Mo., 1974, pp. 281-295.
11. S. S. Hecht, W. E. Bondinell, and D. Hoffman, *J. Nat. Cancer Inst. 53*:1121 (1974).
12. D. Hoffman, G. Rathkamp, S. Nesnow, and E. L. Wynder, *J. Nat. Cancer Inst. 49*:1165 (1972).
13. M. R. Guerin, J. L. Epler, W. H. Griest, B. R. Clark, and T. K. Rao, in *Carcinogenesis*, Vol. 3: *Polynuclear Aromatic Hydrocarbons*, P. W. Jones and R. I. Freudenthal (Eds.), Raven Press, New York, 1978, pp. 21-33.
14. W. H. Griest, M. R. Guerin, B. R. Clark, C.-h. Ho, I. B. Rubin, and A. R. Jones, in *Proc. Symposium on Assessing the Industrial Hygiene Monitoring Needs for the Coal Conversion and Oil Shale Industries*, O. White, Jr. (Ed.), BNL-51002 (1979), pp. 61-78.

15. J. R. Morandi and R. E. Poulson, *ACS Preprints, Div. Petrol. Chem. 20*(2):162-174 (1975).
16. M. R. Guerin, C.-h. Ho, T. K. Rao, B. R. Clark, and J. L. Epler, *Environ. Res. 23*:42-53 (1980).
17. B. A. Tomkins, H. Kubota, W. H. Griest, J. E. Caton, B. R. Clark, and M. R. Guerin, *Anal. Chem. 52*:1331-1334 (1980).
18. B. A. Tomkins, R. R. Reagan, J. E. Caton, and W. H. Griest, *Anal. Chem. 53*:1213-1217 (1981).
19. R. A. Grienke and I. C. Lewis, *Anal. Chem. 47*:2151-2155 (1975).
20. W. Lijinsky, I. Domsky, G. Mason, H. Y. Ramahi, and T. Safavi, *Anal. Chem. 35*:952-956 (1963).
21. L. Wallcave, H. Garcia, R. Feldman, W. Lijinski, and P. Shubik, *Toxicol. Appl. Pharmacol. 18*:41-49 (1971).
22. D. L. Coffin, M. R. Guerin, and W. H. Griest, in *Proc. Symposium on Potential Health and Environmental Effects of Synthetic Fossil Fuels Technologies*, CONF-780903, NTIS, Springfield, Va. (1979), p. 153.
23. W. D. Felix, D. D. Mahlum, W. C. Weimer, P. A. Pelroy, and B. W. Wilson, in *Symposium Proc.: Environmental Aspects of Fuel Conversion Technology, V* (Sept. 1980, St. Louis, Mo.), EPA 600/9-81-006, pp. 134-158 (1981).
24. B. A. Tomkins and C.-h. Ho, *Anal. Chem.* (in press).
25. J. McCann, E. Choi, E. Yamansaka, and B. N. Ames, *Proc. Nat. Acad. Sci. (U.S.) 72*:5135-5139 (1979).
26. M. R. Guerin, I. B. Rubin, T. K. Rao, B. R. Clark, and J. L. Epler, *Fuel 60*:282-288 (1981).
27. B. N. Ames, J. McCann, and E. Yamasaki, *Mutation Res. 31*:374-390 (1975).
28. C.-h. Ho, B. R. Clark, M. R. Guerin, B. D. Barkenbus, T. K. Rao, and J. L. Epler, *Mutation Res. 85*:335-345 (1981).
29. J. M. Holland, M. S. Whitaker, and J. W. Wesley, *Amer. Ind. Hyg. Assoc. J. 40*:496-503 (1979).
30. J. M. Holland, Preliminary report of the chronic dermal toxicity of process coal liquids. Internal Report, Oak Ridge National Laboratory, Oak Ridge, Tenn., 1980.
31. J. Iball, *Amer. J. Cancer 35*:188-90 (1939).
32. Appendix to biomedical studies on solvent refined coal (SRC-II) liquefaction materials: A status report. Battelle Memorial Institute, Pacific Northwest Laboratory, Richland, Wash., PNL-3189, Appendix (1979).
33. NIOSH criteria for a recommended standard: Occupational exposure to coal tar products. U.S. Dept. of Health and Human Services (NIOSH) Publ. No. 78-107 (1977).
34. A. Scott, *Eighth Scientific Report of the Imperial Cancer Fund*, Taylor & Frances, London, 1923, pp. 85-142.
35. J. M. Holland, M. S. Whitaker, and J. W. Wesley, in *Proc. Symposium on Potential Health and Environmental Effects of Synthetic Fossil Fuel Technologies*, CONF-780903 (1979), pp. 137-142.
36. R. E. Eckardt, *Arch. Environ. Health 1*:232-233 (1960).
37. R. E. Eckardt, *Intern. J. Cancer 2*:656-661 (1967).
38. B. Commoner, *Hospital Practice 10*:138-141 (1975).

39. P. Bogovsky and H. J. Jons, Toxicological and carcinogenic studies of oil shale dust and shale oil, presented at *Workshop on Health Effects of Coal and Oil Shale Mining, Conversion, and Utilization*, Dept. of Environmental Health, Kettering Laboratory, University of Cincinnati, Jan. 27-29, 1975.
40. W. Barkley, D. Warshawsky, R. R. Suskind, and E. Bingham, in *Proc. Symposium on Potential Health and Environmental Effects of Synthetic Fossil Fuel Technologies*, CONF-780903 (1979), pp. 157-162.
41. P. Pott, in *Chirurgical Observations*, Hawes, Clark, and Collings, London, 1775, p. 63
42. R. J. Sexton, C. S. Weil, N. O. Condra, N. H. Ketcham, R. W. Norton, and R. E. Eckardt, *Arch. Environ. Health 1*:181-233 (1960).
43. N. H. Ketcham and R. W. Norton, *Arch. Environ. Health 1*:194-207 (1960).
44. R. J. Sexton, *Arch. Environ. Health 1*:208-231 (1960).
45. A. Palmer, *J. Occup. Med. 21*:41-45 (1979).
46. B. N. Ames, *Science 204*:587-593 (1979).
47. L. R. Harris, J. A. Gideon, S. Berardinelli, L. R. Reed, R. D. Dobbin, J. M. Evans, D. R. Telesca, and R. K. Tanita, *Amer. Ind. Hyg. J. 4*: A50-A60 (1980).
48. *Solvent Refined Coal (SRC) Process: Health Programs*, Interim Rept. No. 28, Vol. 3, Pt. 4: Industrial hygiene, chemical, and toxicological programs. U.S. Dept. of Energy, FE/496-T19, April (1979).
49. R. F. Bauman, Coal Conversion Hygiene Program, Exxon Research and Engineering Company, Baytown Research and Development Division: Advisory Workshop on Carcinogenic Effects of Coal Conversion, Asilomar, Calif. (1978).
50. *TLVs: Threshold Limit Values for Chemical Substances and Physical Agents in the Workroom Environment with Intended Changes for 1980*. American Conf. of Governmental Industrial Hygienists. ISBN: 0-936712-29-5, P. O. Box 2937, Cincinnati, Ohio, 1980.
51. S. Mazumdar, C. Redmond, W. Sollecito, and N. Sussman, *J. Air Pollut. Control Assoc. 25*: 382-389 (1975).
52. A. Bjørseth, in *Luftverunreinigung durch Polycyclische Aromatische Kohlenwasserstoffe: Erfassung und Bewertung*. VDI-Ber. No. 358 (1980), pp. 81-93.
53. W. E. MacEachen, H. Boden, and C. Larivière, in *Proc. 107th AIME Annual Meeting, Denver* (Feb. 26-Mar. 2, 1978), pp. 509-517.
54. K. A. Schulte, D. J. Larsen, R. W. Hornung, and J. V. Crable, *Amer. Ind. Hyg. Assoc. J. 36*: 131 (1965).
55. H. J. Seim, W. W. Hanneman, L. R. Barsotti, and T. J. Walker, *Amer. Ind. Hyg. Assoc. J. 35*:718-723 (1974).
56. J. Campbell and W. Porter, in *Proc. Symposium on Assessing the Industrial Hygiene Monitoring Needs of the Coal Conversion and Oil Shale Industries*, O. White, Jr. (Ed.), BNL-51002 (1979), pp. 146-151.
57. J. O. Jackson and J. A. Cupps, in *Carcinogenesis*, Vol. 3: *Polynuclear Aromatic Hydrocarbons*, P. W. Jones and R. I. Freudenthal (Eds.), Raven Press, New York, 1978, pp. 183-191.
58. R. C. Lao and R. S. Thomas, in *Polynuclear Aromatic Hydrocarbons: Chemistry and Biological Effects*, A. Bjørseth and A. J. Dennis (Eds.), Battelle Press, Columbus, Ohio, 1980, pp. 829-839.

59. J. R. Burg and A. W. Teass, A statistical analysis of 300 personal samples of respirable particulate coke-oven emissions. Memorandum to the Deputy Director, NIOSH, Dec. 18, 1975.
60. NIOSH Criteria for a Recommended Standard: Occupational Exposure to Coal Tar Products. U.S. Dept. of Health and Human Services (NIOSH) Publ. No. 78-107 (Sept. 1977).
61. C. Golden and E. Sawicki, *Intern. J. Environ. Anal. Chem.* *4*:9-23 (1975).
62. *Federal Register 41*:46742-87 (Oct. 2, 1976).
63. M. C. Pike, R. J. Gordon, B. E. Henderson, H. R. Menck, and J. Soohoo, in *Persons at High Risk of Cancer: An Approach to Cancer Etiology and Control*, J. F. Franmeui (Ed.), Academic Press, New York, 1975.
64. *Polynuclear Aromatic Hydrocarbons: A Background Report Including Available Ontario Data.* Ontario Ministry of the Environment, ARB-TDA-Rept. No. 58-79 (Sept. 1979).
65. R. B. Gammage and A. Bjørseth, in *Polynuclear Aromatic Hydrocarbons: Chemistry and Biological Effects*, A. Bjørseth and A. J. Dennis (Eds.), Battelle Press, Columbus, Ohio, 1980, pp. 565-578.
66. A. Bjørseth, O. Bjørseth, and P. E. Fjeldstad, Characterization of PAH in work atmospheres. Rept. No. 6, Royal Norwegian Council for Scientific and Industrial Research, Oslo, Norway (1978) [in Norwegian].
67. L. H. Keith and W. A. Telliard, *Environ. Sci. Technol. 13*:416-423 (1979).
68. A. Bjørseth and G. Lunde, *Amer. Ind. Hyg. Assoc. J. 38*:224-228 (1977).
69. J. O. Jackson and J. A. Cupps, in *Carcinogenesis*, Vol. 3: *Polynuclear Aromatic Hydrocarbons*, P. W. Jones and R. I. Freudenthal (Eds.), Raven Press, New York, 1978, pp. 188-191.
70. W. Cautreels and K. Van Cauwenberghe, *Atmos. Environ. 12*:1131-1141 (1978).
71. F. DeWiest and D. Rondia, *Atmos. Environ. 10*:487-489 (1976).
72. R. C. Lao and R. S. Thomas, in *Polynuclear Aromatic Hydrocarbons: Chemistry and Biological Effects*, A. Bjørseth and A. J. Dennis (Eds.), Battelle Press, Columbus, Ohio, 1980, pp. 829-840.
73. A. Bjørseth, O. Bjørseth, and P. E. Fjeldstad, *Scand. J. Work Environ. Health 4*:212-23 (1978).
74. P. W. Strup, R. D. Giammar, T. B. Stanford, and P. W. Jones, in *Carcinogenesis*, Vol. 1: *Polynuclear Aromatic Hydrocarbons: Chemistry, Metabolism and Carcinogenesis*, R. I. Freudenthal and P. W. Jones (Eds.), Raven Press, New York, 1976, pp. 241-251.
75. *Sampling and Analytical Procedures for the Industrial Hygiene Characterization of Coal Gasification and Liquefaction Pilot Plants.* Submitted to NIOSH by Enviro Control, Inc., under contracts 210-78-0101 and 0040 (May 1980).
76. R. D. Flotard, Sampling and analysis of trace-organic constituents in ambient and workplace air at coal-conversion facilities, Argonne National Laboratory, ANL/PAG-3 (July 1980).
77. F.S.-C. Lee, W. R. Pierson, and J. Ezike, in *Polynuclear Aromatic Hydrocarbons: Chemistry and Biological Effects*, A. Bjørseth and A. J. Dennis (Eds.), Battelle Press, Columbus, Ohio, 1980, pp. 543-563.
78. R. D. Philips, in *Proc. Second ORNL Workshop on Exposure to Polynuclear Aromatic Hydrocarbons in Coal Conversion Processes*, CONF-770361 (1977), pp. 49-58.

79. P. W. Jones, R. D. Giammar, P. E. Strup, and T. B. Stanford, *Environ. Sci. Technol. 10*:806-810 (1976).
80. R. D. Flotard, J. R. Stetter, and V. C. Stamoudis, Workplace air sampling at coal-conversion facilities. Presented at the *20th Hanford Life Sciences Symposium, Coal Conversion and the Environment: Chemical, Biomedical, and Ecological Considerations*, Richland, Wash. (Oct. 19-23, 1980).
81. P. Burchill, A. A. Herod, and R. G. James, in *Carcinogenesis*, Vol. 3: *Polynuclear Aromatic Hydrocarbons*, P. W. Jones and R. I. Freudenthal (Eds.), Raven Press, New York, 1978, pp. 35-45.
82. E. M. Smith and P. L. Levins, in *Polynuclear Aromatic Hydrocarbons: Chemistry and Biological Effects*, A. Bjørseth and A. J. Dennis (Eds.), Battelle Press, 1980, pp. 973-982.
83. T. Vo-Dinh, R. B. Gammage, and P. R. Martinez, *Anal. Chim. Acta 118*: 313-323 (1980).
84. W. H. Griest, internal Oak Ridge National Laboratory memorandum, June 11, 1980.
85. J. W. Lloyd, *J. Occup. Med. 13*:53-68 (1971).
86. J. O. Jackson, P. O. Warner, and T. F. Mooney, *Amer. Ind. Hyg. Assoc. J. 35*:276-281 (1974).
87. A. Bjørseth, O. Bjørseth, and P. E. Fjeldstad, *Scand. J. Work Environ. Health 4*:212-223 (1978).
88. A. Bjørseth, O. Bjørseth, and P. E. Fjeldstad, Characterization of PAH in work atmospheres. Rept. No. 5, Royal Norwegian Council for Scientific and Industrial Research, Oslo, Norway (1978) [in Norwegian].
89. R. W. Schirmberg, P. Pfäffli, and A. Tossavainen, *Staub-Reinhalt Luft 38*:273-276 (1978).
90. R. Tanita, Enviro Control, Inc., personal communication on PAH, Aug. 26, 1980.
91. M. Dong, I. Schmeltz, E. LaVoie, and D. Hoffmann, in *Carcinogenesis*, Vol. 3: *Polynuclear Aromatic Hydrocarbons*, P. W. Jones and R. I. Freudenthal (Eds.), Raven Press, New York, 1978, pp. 97-108.
92. L. Horton and R. B. Randall, *Fuel 26*:127-132 (1947).
93. D. A. Cubit and R. K. Tanita, Industrial Hygiene Assessment of Coal Liquefaction Plants, Final Summary Report by Enviro Control Division, Dynamac Corporation, for the National Institute for Occupational Safety and Health, March 1982.
94. O. M. Bjørseth, C. P. Flessel, N. M. Monto, T. R. Parker, and P. K. Ouchida, in *Safe Handling of Chemical Carcinogens, Mutagens, Teratogens, and Highly Toxic Substances*, D. D. Walters (Ed.), Vol. 2, Ann Arbor Science Pubs., Ann Arbor, Mich., 1980, Chap. 33.
95. D. A. Cubit and R. K. Tanita, Industrial Hygiene Assessment of Coal Gasification Plants, Final Summary Report by Enviro Control Division, Dynamac Corporation, for the National Institute for Occupational Safety and Health, March 1982.
96. C. Kaiser, A. Kerr, D. R. McCalla, J. N. Lockington, and E. S. Gibson, in *Polynuclear Aromatic Hydrocarbons: Chemistry and Biological Effects*, A. Bjørseth and A. J. Dennis (Eds.), Battelle Press, Columbus, Ohio, 1980, pp. 579-588.
97. A. Bjørseth, in *Polynuclear Aromatic Hydrocarbons* (P. W. Jones and P. Leber, eds.), Ann Arbor Science Publishers, Ann Arbor, Michigan, pp. 371-381 (1979).

98. J. J. Murray and R. F. Pottie, *Can. J. Chem.* *52*:557-563 (1974).
99. D. A. Lane, T. Sakuma, and E. S. K. Quan, in *Polynuclear Aromatic Hydrocarbons: Chemistry and Biological Effects*, A. Bjørseth and A. J. Dennis (Eds.), Battelle Press, Columbus, Ohio, 1980, pp. 199-214.
100. C. Pupp, R. C. Lao, J. J. Murray, and R. F. Pottie, *Atmos. Environ.* *8*:915-925 (1974).
101. A. R. Hawthorne and J. H. Thorngate, *Appl. Spectrosc.* *33*:301-305 (1979).
102. A. R. Hawthorne, *Amer. Ind. Hyg. Assoc. J.* *41*:915-921 (1980).
103. *Industrial Hygiene Monitoring Needs for the Coal Conversion and Oil Shale Industries: Study Group Report.* U.S. Dept. of Energy, Office of Health and Environmental Research, DOE/EV-0058 (Nov. 1979).
104. A. R. Hawthorne, C. E. Metcalfe, and R. B. Gammage, to be published.
105. T. Vo-Dinh and J. R. Hooyman, *Anal. Chem.* *51*:1915-1921 (1979).
106. T. Vo-Dinh, R. B. Gammage, and P. R. Martinez, *Anal. Chim. Acta* *118*: 313-323 (1980).
107. T. Vo-Dinh, G. Miller, and R. B. Gammage, to be published.
108. K. E. Cowser, Life Sciences Synthetic Fuels Semiannual Progress Report for the Period Ending December 31, 1981, ORNL Report TM-8229, May 1980.
109. J. M. Evans, in *Proc. Second ORNL Workshop on Exposure to PAH in Coal Conversion Processes*, CONF-770361 (1977), pp. 23-29.
110. *Coal-Derived Materials: A Health Information Booklet.* Gulf Oil Corp., 1977.
111. C. R. Moxley and D. K. Schmalzer, in *Proc. 4th Symposium on Environmental Aspects of Coal Conversion Technology, Hollywood, Fla., April 1979*, EPA-600/7-79-217 (Sept. 1979), pp. 357-382.
112. G. A. Hodgson and H. J. Whiteley, *Brit. J. Ind. Med.* *27*:160-166 (1970).
113. J. Rosamanith, *Prakto Lekar* *5*:270-272 (1953).
114. P. Shubik and G. Della Porta, *Arch. Pathol.* *64*:691-703 (1957).
115. *Solvent Refined Coal (SRC) Process: Health Programs*, Interim Rept. No. 32: Industrial hygiene, clinical, and toxicological programs. U.S. Dept. of Energy, FE/496-T23 (Aug. 1980).
116. *Health Information for Employees Involved in the Coal Technology Program.* Oak Ridge National Laboratory Health Div. Booklet (May 1975).
117. N. E. Bolton, in *Proc. Second ORNL Workshop on Exposure to PAH in Coal Conversion Processes*, CONF-770361 (1977), pp. 13-18.
118. L. Tanenbaum, J. A. Parrish, M. A. Parthak, R. R. Anderson, and T. B. Fitzpatrick, *Arch. Dermatol.* *111*:467-470 (1975).
119. K. D. Crow, E. Alexander, W. H. L. Buck, B. E. Johnson, I. A. Magnus, and A. D. Porter, *Brit. J. Dermatol.* *73*:220-232 (1961).
120. *Solvent Refined Coal (SRC) Process: Health Programs*, Interim Rept. No. 24: Industrial hygiene, clinical, and toxicological programs. U.S. Dept. of Energy, FE/496-T15 (Jan. 1978).
121. J. H. Epstein and W. L. Epstein, *J. Invest. Dermatol.* *39*:455-460 (1965).
122. E. Bingham and P. J. Nord, *J. Nat. Cancer Inst.* *58*:1099-1101 (1977).
123. G. M. Blackburn and P. E. Taussig, *Biochem. J.* *149*:289-291 (1975).
124. D. M. Gates, *Amer. Scientist* *51*:327-348 (1963).
125. H. R. Foerster and L. Schwartz, *Arch. Dermatol.* *39*:55-68 (1939).
126. T. Vo-Dinh and R. B. Gammage, *Amer. Ind. Hyg. J.* *42*:112-120 (1981).

127. R. B. Gammage and T. Vo-Dinh, *Nucl. Instr. Methods 175*:236-238 (1980).
128. T. Vo-Dinh, to be published.
129. T. Vo-Dinh and D. D. Schuresko, to be published.
130. M. J. Holland, to be published.
131. D. D. Schuresko, *Anal. Chem. 52*:371-373 (1980).
132. D. D. Schuresko, in *Polynuclear Aromatic Hydrocarbons: Chemical Analysis and Biological Fate*, M. Cooke and A. J. Dennis (Eds.), Battelle Press, Columbus, Ohio, 1981, pp. 551-560.
133. *Wilkes Infrared Spectroscopic Accessories*. Foxboro Co. (Feb. 1981).
134. R. R. Keenan and S. B. Cole, *Amer. Ind. Hyg. J. 43*:473-476 (1982).

Appendix: List of Dicyclic and Polycyclic Aromatic Hydrocarbons, Their Structure, Molecular Weight, Melting and Boiling Points

Structure	IUPAC nomenclature (synonyms)	Molecular weight	Melting point (°C)	Boiling point (°C)[760]
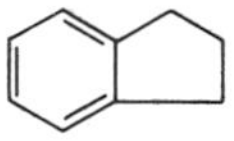	Indan Hydrindene 2,3-Dihydroindene	118.18	-51	178
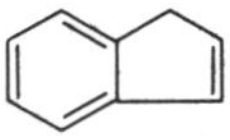	Indene Indonaphthene	116.16	-2	183
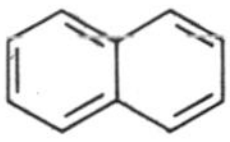	Naphthalene Tar Camphor White Tar Moth Flakes	128.19	81	218
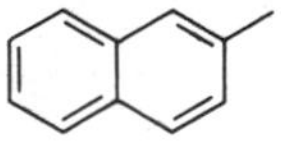	2-Methylnaphthalene β-Methylnaphthalene	142.20	35	241
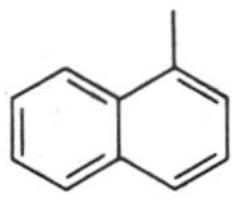	1-Methylnaphthalene α-Methylnaphthalene	142.20	-22	245
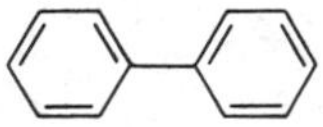	Biphenyl Diphenyl Phenylbenzene Bibenzene	154.21	71	255
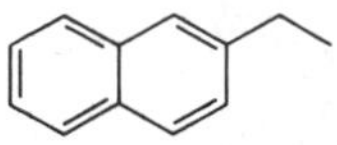	2-Ethylnaphthalene β-Ethylnaphthalene	156.23	-7	258

(continued)

Structure	IUPAC nomenclature (synonyms)	Molecular weight	Melting point (°C)	Boiling point $(°C)^{760}$
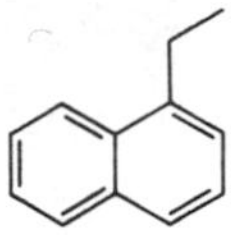	1-Ethylnaphthalene	156.23	-14	259
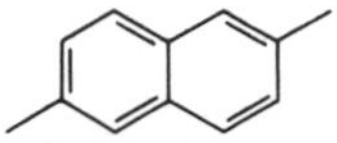	2,6-Dimethylnaphthalene	156.23	110	262
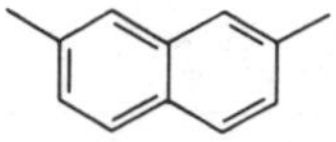	2,7-Dimethylnaphthalene	156.23	97	262
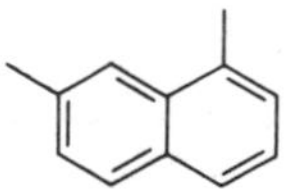	1,7-Dimethylnaphthalene	156.23		263
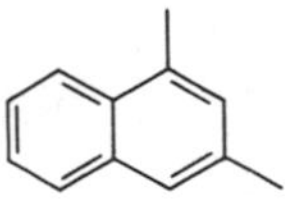	1,3-Dimethylnaphthalene	156.23		265
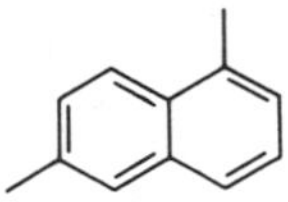	1,6-Dimethylnaphthalene	156.23		266
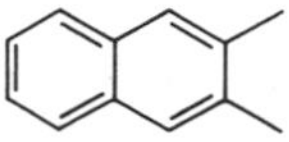	2,3-Dimethylnaphthalene Guaiene	156.23	105	268
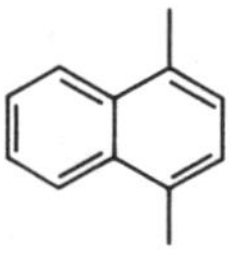	1,4-Dimethylnaphthalene α-Dimethylnaphthalene	156.23	8	268
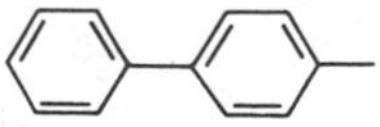	4-Methylbiphenyl	168.24	50	268

Structure	IUPAC nomenclature (synonyms)	Molecular weight	Melting point (°C)	Boiling point $(°C)^{760}$
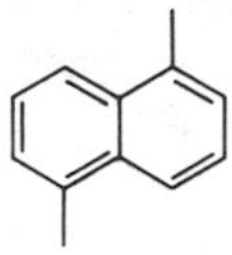	1,5-Dimethylnaphthalene	156.23	80	269
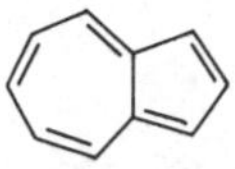	Azulene	128.19	100	270 d
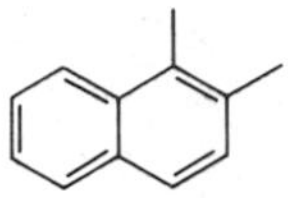	1,2-Dimethylnaphthalene	156.23	-4	271
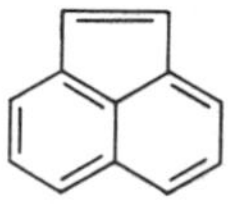	Acenaphthylene	152.21	93	~270 par d
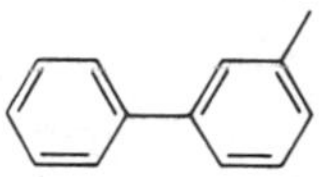	3-Methylbiphenyl	168.24	5	273
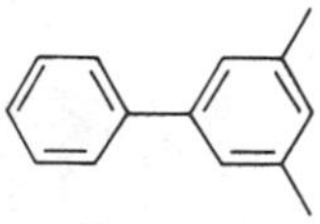	3,5-Dimethylbiphenyl	182.27		275
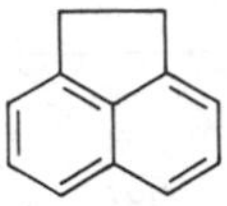	Acenaphthene Naphthyleneethylene	154.21	96	279
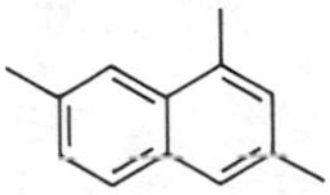	1,3,7-Trimethylnaphthalene	170.25	14	280
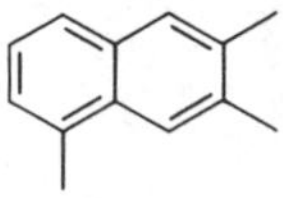	2,3,5-Trimethylnaphthalene	170.25	25	285

(continued)

Structure	IUPAC nomenclature (synonyms)	Molecular weight	Melting point (°C)	Boiling point (°C)760
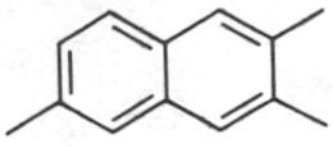	2,3,6-Trimethylnaphthalene	170.25	101	286
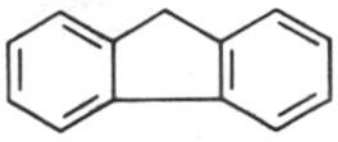	Fluorene 2,3-Benzindene Diphenylenemethane	166.23	117	294
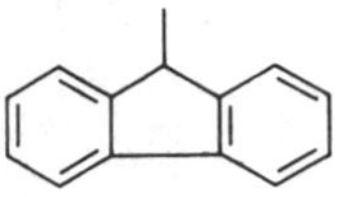	9-Methylfluorene	180.25	47	
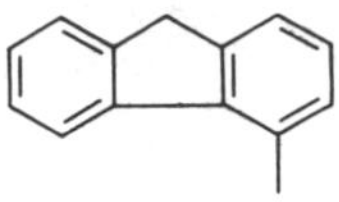	4-Methylfluorene	180.25		
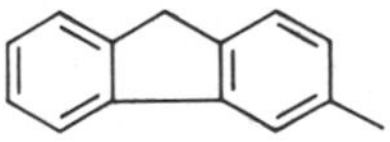	3-Methylfluorene	180.25	85	316
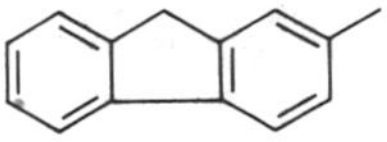	2-Methylfluorene	180.25	104	318
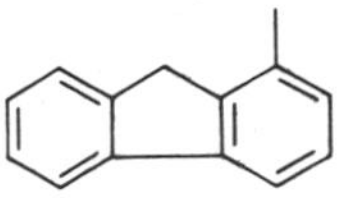	1-Methylfluorene	180.25		~318
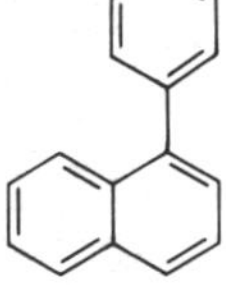	1-Phenylnaphthalene α-Phenylnaphthalene	204.28	~45	334
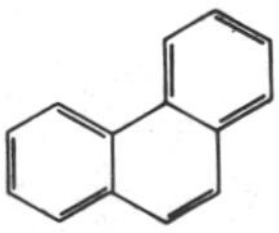	Phenanthrene o-Diphenyleneethylene	178.24	101	338
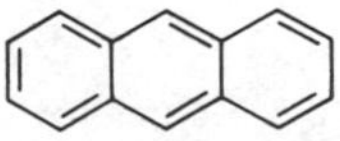	Anthracene	178.24	216	340

Structure	IUPAC nomenclature (synonyms)	Molecular weight	Melting point (°C)	Boiling point $(°C)^{760}$
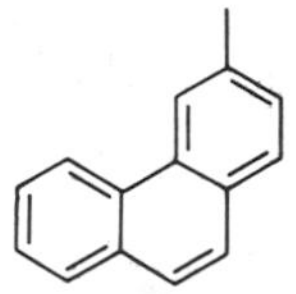	3-Methylphenanthrene	192.26	65	352
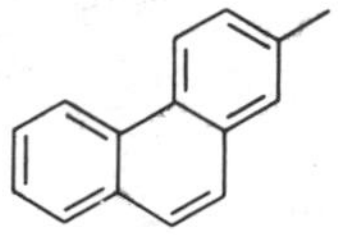	2-Methylphenanthrene	192.26		355
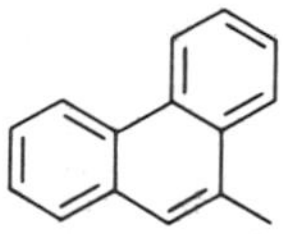	9-Methylphenanthrene	192.26	92	355
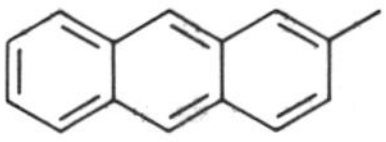	2-Methylanthracene	192.26	209	359 sub
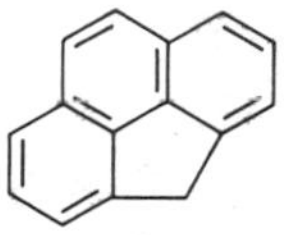	4,5-Methylenephenanthrene 4H-Cyclopenteno[def]phenanthrene 4H-Cyclopenta[def]phenanthrene 4,5-Phenanthrylenemethane	190.24	116	359
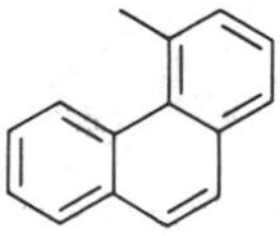	4-Methylphenanthrene	192.26		
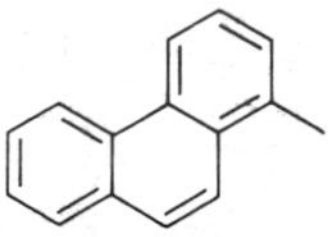	1-Methylphenanthrene	192.26	123	359
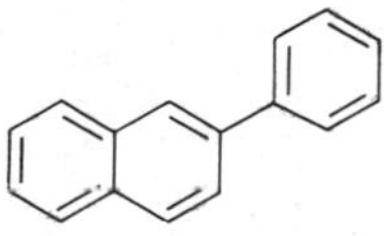	2-Phenylnaphthalene β-Phenylnaphthalene	204.28	104	360
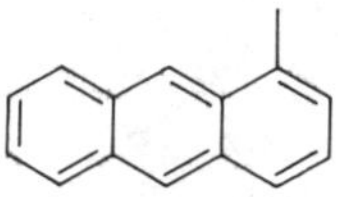	1-Methylanthracene	192.26	86	363

(continued)

Structure	IUPAC nomenclature (synonyms)	Molecular weight	Melting point (°C)	Boiling point (°C)^{760}
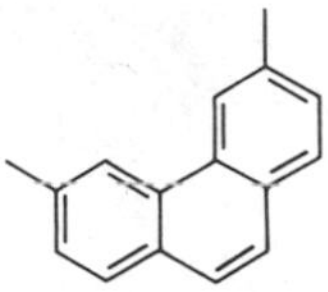	3,6-Dimethylphenanthrene	206.29		363
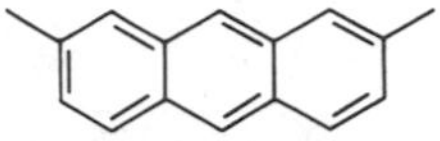	2,7-Dimethylanthracene	206.29	241	~370
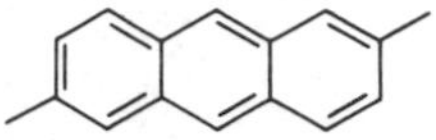	2,6-Dimethylanthracene	206.29	250	~370
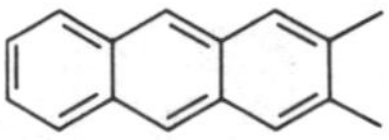	2,3-Dimethylanthracene	206.29	252	
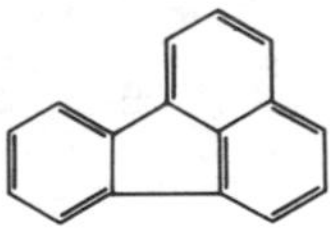	Fluoranthene Idryl 1,2-Benzacenaphthene Benzo[jk]fluorene Benz[a]acenaphthylene	202.26	111	383
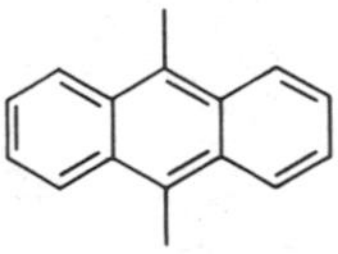	9,10-Dimethylanthracene	206.29	183	
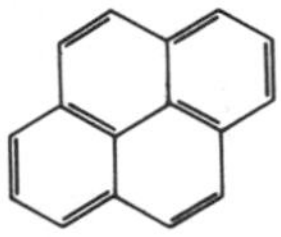	Pyrene Benzo[def]phenanthrene	202.26	156	393
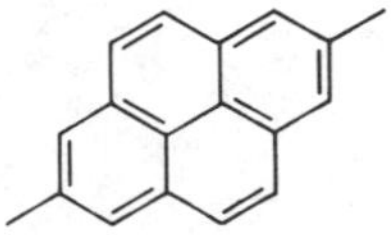	2,7-Dimethylpyrene	230.32		396
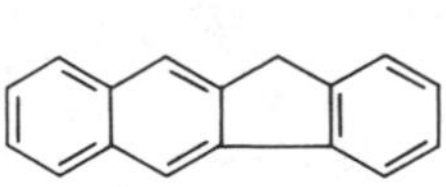	Benzo[b]fluorene 11H-Benzo[b]fluorene 2,3-Benzofluorene Isonaphthofluorene	216.29	209	402
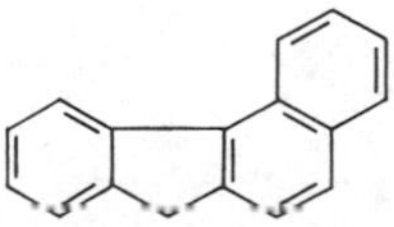	Benzo[c]fluorene 7H-Benzo[c]fluorene 3,4-Benzofluorene	216.29		406

Structure	IUPAC nomenclature (synonyms)	Molecular weight	Melting point (°C)	Boiling point $(°C)^{760}$
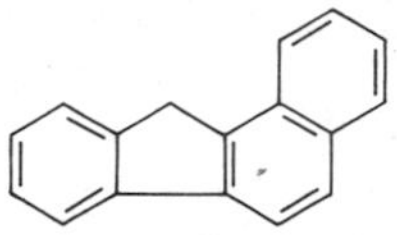	Benzo[a]fluorene 11H-Benzo[a]fluorene 1,2-Benzofluorene Chrysofluorene	216.29	190	407
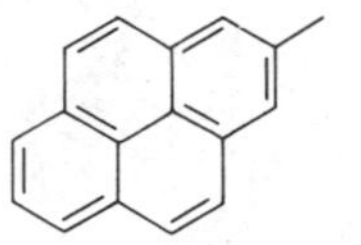	2-Methylpyrene 4-Methylpyren	216.29		410
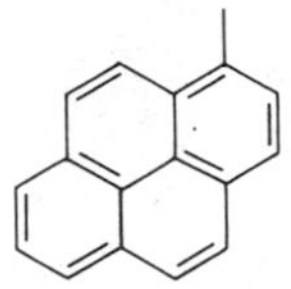	1-Methylpyrene 3-Methylpyren	216.29		410
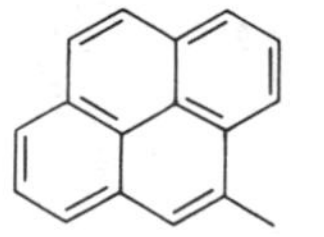	4-Methylpyrene 1-Methylpyren	216.29		410
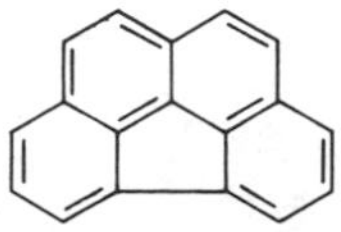	Benzo[ghi]fluoranthene	226.28		432
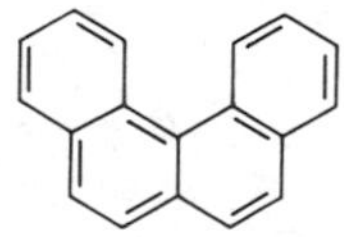	Benzo[c]phenanthrene 3,4-Benzophenanthrene	228.30	68	
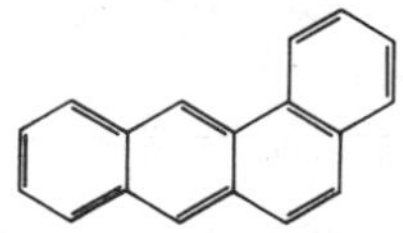	Benz[a]anthracene 1,2-Benzanthracene Tetraphene 2,3-Benzophenanthrene Naphthanthracene	228.30	162	435 sub
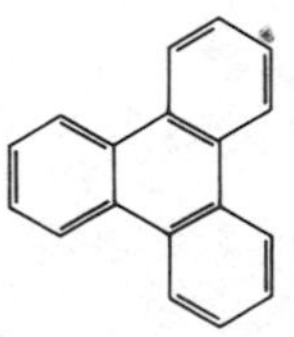	Triphenylene 9,10-Benzophenanthrene Isochrysene	228.30	199	439
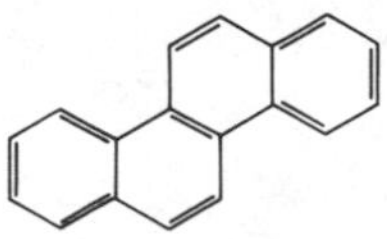	Chrysene 1,2-Benzophenanthrene Benzo[a]phenanthrene	228.30	256	441

(continued)

Structure	IUPAC nomenclature (synonyms)	Molecular weight	Melting point (°C)	Boiling point (°C)[760]
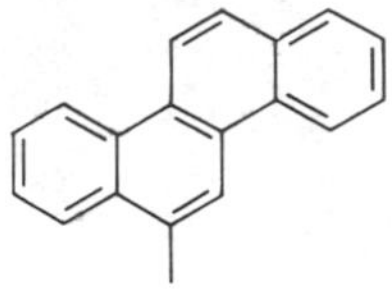	6-Methylchrysene	242.32		
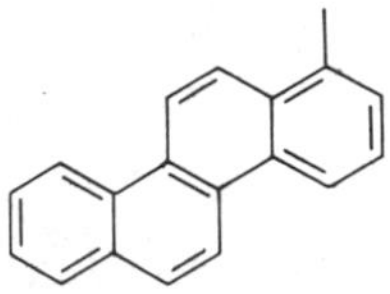	1-Methylchrysene	242.32	257	
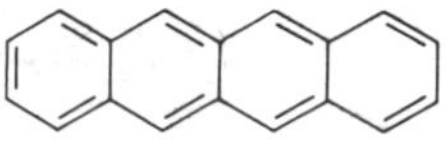	Naphthacene Benz[b]anthracene 2,3-Benzanthracene Tetracene	228.30	257	450 sub
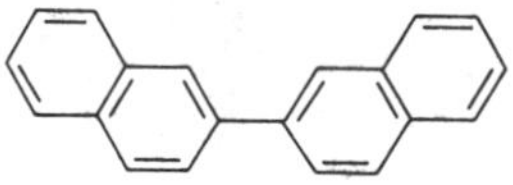	2,2'-Dinaphthyl 2,2'-Binaphthyl β,β'-Binaphthyl β,β-Dinaphthyl	254.34	188	452[753] sub
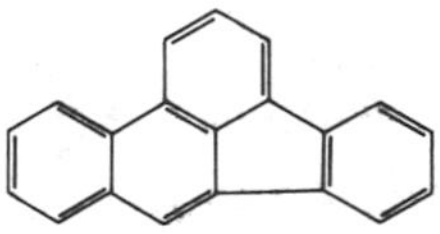	Benzo[b]fluoranthene 2,3-Benzofluoranthene 3,4-Benzofluoranthene Benz[e]acephenanthrylene	252.32	168	481
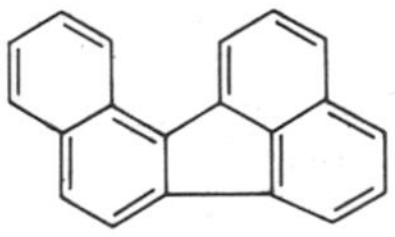	Benzo[j]fluoranthene 7,8-Benzofluoranthene 10,11-Benzofluoranthene	252.32	166	~480
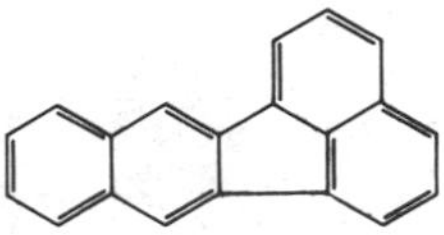	Benzo[k]fluoranthene 8,9-Benzofluoranthene 11,12-Benzofluoranthene	252.32	217	481
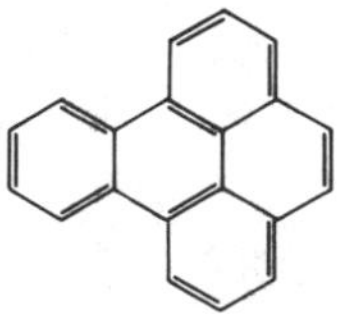	Benzo[e]pyrene 4,5-Benzpyrene 1,2-Benzopyrene	252.32	179	493
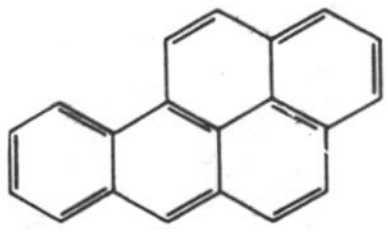	Benzo[a]pyrene 1,2-Benzpyrene 3,4-Benzopyrene Benzo[def]chrysene	252.32	177	496

Structure	IUPAC nomenclature (synonyms)	Molecular weight	Melting point (°C)	Boiling poing (°C)^{760}
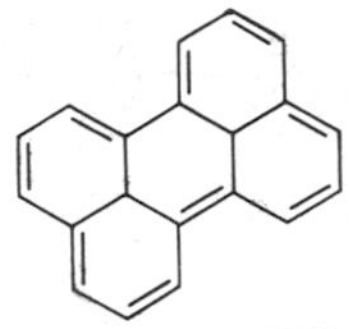	Perylene peri-Dinaphthalene	252.32	278	
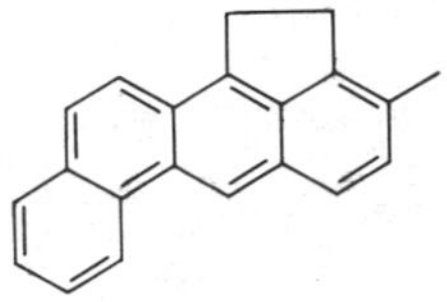	3-Methylcholanthrene 20-Methylcholanthrene	268.38	180	
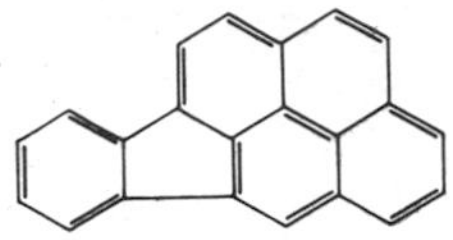	Indeno[1,2,3-cd]pyrene o-Phenylenepyrene	276.34		
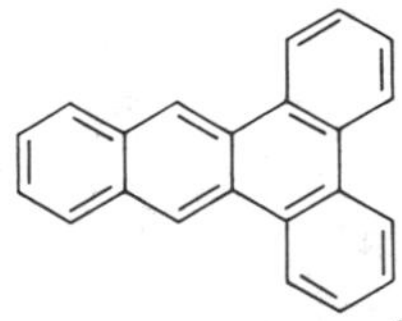	Dibenz[a,c]anthracene 1,2:3,4-Dibenzanthracene Naphtho-2',3',:9,10-phenanthrene	278.36	205	
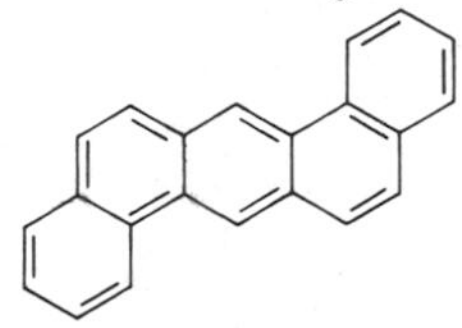	Dibenz[a,h]anthracene 1,2:5,6-Dibenzanthracene	278.36	270	
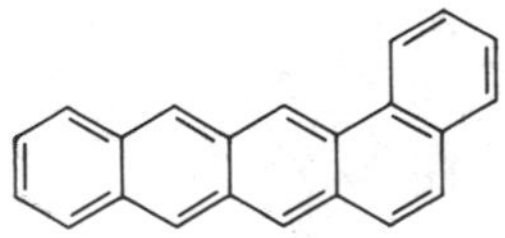	Dibenz[a,i]anthracene 1,2:6,7-Dibenzanthracene 1,2-Benzonaphthacene Isopentaphene	278.36	264	
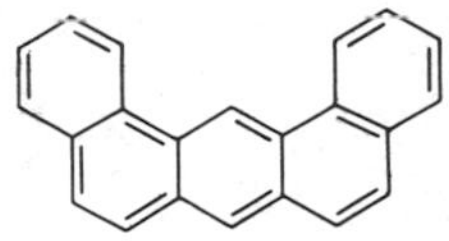	Dibenz[a,j]anthracene 1,2:7,8-Dibenzanthracene $\alpha.\alpha'$-Dibenzanthracene Dinaphthanthracene	278.36	198	
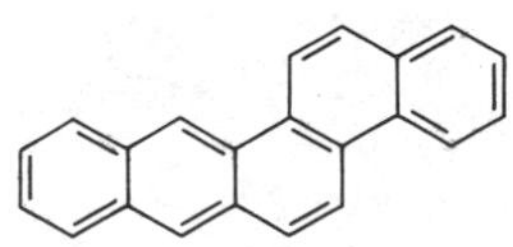	Benzo[b]chrysene 1,2:6,7-Dibenzophenanthrene 3,4-Benzotetraphene Naphtho-2',1':1,2-anthracene	278.36	294	

(continued)

Structure	IUPAC nomenclature (synonyms)	Molecular weight	Melting point (°C)	Boiling point $(°C)^{760}$
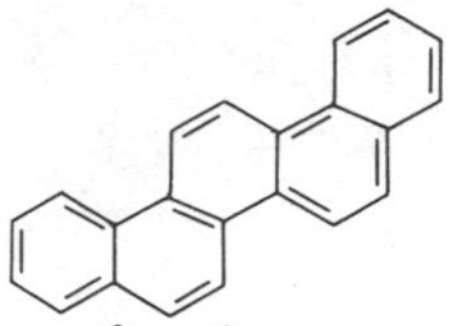	Picene Dibenzo[a,i]phenanthrene 3,4-Benzochrysene 1,2:7,8-Dibenzophenanthrene	278.36	368	519
	Benzo[ghi]perylene 1,12-Benzoperylene	276.34	278	
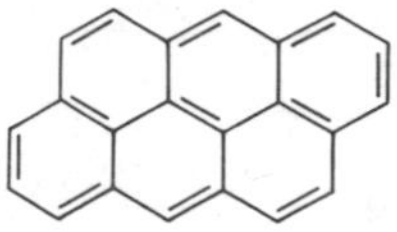	Anthanthrene Dibenzo[def,mno]chrysene	276.34		
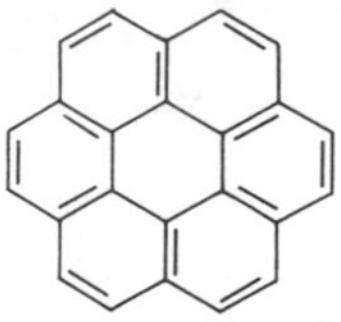	Coronene Hexabenzobenzene	300.36	439 cor	525 ?
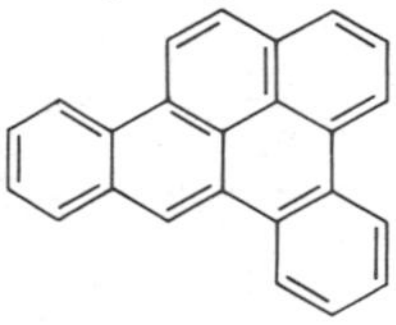	Dibenzo[a,e]pyrene	302.38	234	

*Key: d = decomposes; par d = partly decomposes; sub = sublimes.

Index

B

C

D

Q

R